Lecture Notes in Mathematics

Edited by A. Dold and B. Eckmann

743

Romanian-Finnish Seminar on Complex Analysis

Proceedings, Bucharest, Romania,
June 27–July 2, 1976

Edited by
Cabiria Andreian Cazacu, Aurel Cornea,
Martin Jurchescu and Ion Suciu

Springer-Verlag
Berlin Heidelberg New York 1979

Editors

Cabiria Andreian Cazacu
Aurel Cornea
Martin Jurchescu
Ion Suciu

Department of Mathematics
University of Bucharest
str. Academiei 14
Bucuresti 1/Romania

AMS Subject Classifications (1970): 20Hxx, 30-xx, 31-xx, 32-xx, 34-xx, 35-xx, 40-xx, 41-xx, 46-xx, 47-xx

ISBN 3-540-09550-0 Springer-Verlag Berlin Heidelberg New York
ISBN 0-387-09550-0 Springer-Verlag New York Heidelberg Berlin

Printed in Germany

Printing and binding: Beltz Offsetdruck, Hemsbach/Bergstr.
2141/3140-543210

Lecture Notes in Mathematics

Edited by A. Dold and B. Eckmann

743

Romanian-Finnish Seminar on Complex Analysis

Proceedings, Bucharest 1976

Edited by
Cabiria Andreian Cazacu, Aurel Cornea,
Martin Jurchescu and Ion Suciu

Springer-Verlag
Berlin Heidelberg New York

Lecture Notes in Mathematics

For information about Vols. 1–521, please contact your bookseller or Springer-Verlag.

Vol. 522: C. O. Bloom and N. D. Kazarinoff, Short Wave Radiation Problems in Inhomogeneous Media: Asymptotic Solutions. V. 104 pages. 1976.

Vol. 523: S. A. Albeverio and R. J. Høegh-Krohn, Mathematical Theory of Feynman Path Integrals. IV, 139 pages. 1976.

Vol. 524: Séminaire Pierre Lelong (Analyse) Année 1974/75. Edité par P. Lelong. V, 222 pages. 1976.

Vol. 525: Structural Stability, the Theory of Catastrophes, and Applications in the Sciences. Proceedings 1975. Edited by P. Hilton. VI, 408 pages. 1976.

Vol. 526: Probability in Banach Spaces. Proceedings 1975. Edited by A. Beck. VI, 290 pages. 1976.

Vol. 527: M. Denker, Ch. Grillenberger, and K. Sigmund, Ergodic Theory on Compact Spaces. IV, 360 pages. 1976.

Vol. 528: J. E. Humphreys, Ordinary and Modular Representations of Chevalley Groups. III, 127 pages. 1976.

Vol. 529: J. Grandell, Doubly Stochastic Poisson Processes. X, 234 pages. 1976.

Vol. 530: S. S. Gelbart, Weil's Representation and the Spectrum of the Metaplectic Group. VII, 140 pages. 1976.

Vol. 531: Y.-C. Wong, The Topology of Uniform Convergence on Order-Bounded Sets. VI, 163 pages. 1976.

Vol. 532: Théorie Ergodique. Proceedings 1973/1974. Edité par J.-P. Conze and M. S. Keane. VIII, 227 pages. 1976.

Vol. 533: F. R. Cohen, T. J. Lada, and J. P. May, The Homology of Iterated Loop Spaces. IX, 490 pages. 1976.

Vol. 534: C. Preston, Random Fields. V, 200 pages. 1976.

Vol. 535: Singularités d'Applications Differentiables. Plans-sur-Bex. 1975. Edité par O. Burlet et F. Ronga. V, 253 pages. 1976.

Vol. 536: W. M. Schmidt, Equations over Finite Fields. An Elementary Approach. IX, 267 pages. 1976.

Vol. 537: Set Theory and Hierarchy Theory. Bierutowice, Poland 1975. A Memorial Tribute to Andrzej Mostowski. Edited by W. Marek, M. Srebrny and A. Zarach. XIII, 345 pages. 1976.

Vol. 538: G. Fischer, Complex Analytic Geometry. VII, 201 pages. 1976.

Vol. 539: A. Badrikian, J. F. C. Kingman et J. Kuelbs, Ecole d'Eté de Probabilités de Saint Flour V-1975. Edité par P.-L. Hennequin. IX, 314 pages. 1976.

Vol. 540: Categorical Topology, Proceedings 1975. Edited by E. Binz and H. Herrlich. XV, 719 pages. 1976.

Vol. 541: Measure Theory, Oberwolfach 1975. Proceedings. Edited by A. Bellow and D. Kölzow. XIV, 430 pages. 1976.

Vol. 542: D. A. Edwards and H. M. Hastings, Čech and Steenrod Homotopy Theories with Applications to Geometric Topology. VII, 296 pages. 1976.

Vol. 543: Nonlinear Operators and the Calculus of Variations, Bruxelles 1975. Edited by J. P. Gossez, E. J. Lami Dozo, J. Mawhin, and L. Waelbroeck, VII, 237 pages. 1976.

Vol. 544: Robert P. Langlands, On the Functional Equations Satisfied by Eisenstein Series. VII, 337 pages. 1976.

Vol. 545: Noncommutative Ring Theory. Kent State 1975. Edited by J. H. Cozzens and F. L. Sandomierski. V, 212 pages. 1976.

Vol. 546: K. Mahler, Lectures on Transcendental Numbers. Edited and Completed by B. Diviš and W. J. Le Veque. XXI, 254 pages. 1976.

Vol. 547: A. Mukherjea and N. A. Tserpes, Measures on Topological Semigroups: Convolution Products and Random Walks. V, 197 pages. 1976.

Vol. 548: D. A. Hejhal, The Selberg Trace Formula for PSL (2, $\mathbb{R}$). Volume I. VI, 516 pages. 1976.

Vol. 549: Brauer Groups, Evanston 1975. Proceedings. Edited by D. Zelinsky. V, 187 pages. 1976.

Vol. 550: Proceedings of the Third Japan – USSR Symposium on Probability Theory. Edited by G. Maruyama and J. V. Prokhorov. VI, 722 pages. 1976.

Vol. 551: Algebraic K-Theory, Evanston 1976. Proceedings. Edited by M. R. Stein. XI, 409 pages. 1976.

Vol. 552: C. G. Gibson, K. Wirthmüller, A. A. du Plessis and E. J. N. Looijenga. Topological Stability of Smooth Mappings. V, 155 pages. 1976.

Vol. 553: M. Petrich, Categories of Algebraic Systems. Vector and Projective Spaces, Semigroups, Rings and Lattices. VIII, 217 pages. 1976.

Vol. 554: J. D. H. Smith, Mal'cev Varieties. VIII, 158 pages. 1976.

Vol. 555: M. Ishida, The Genus Fields of Algebraic Number Fields. VII, 116 pages. 1976.

Vol. 556: Approximation Theory. Bonn 1976. Proceedings. Edited by R. Schaback and K. Scherer. VII, 466 pages. 1976.

Vol. 557: W. Iberkleid and T. Petrie, Smooth S^1 Manifolds. III, 163 pages. 1976.

Vol. 558: B. Weisfeiler, On Construction and Identification of Graphs. XIV, 237 pages. 1976.

Vol. 559: J.-P. Caubet, Le Mouvement Brownien Relativiste. IX, 212 pages. 1976.

Vol. 560: Combinatorial Mathematics, IV, Proceedings 1975. Edited by L. R. A. Casse and W. D. Wallis. VII, 249 pages. 1976.

Vol. 561: Function Theoretic Methods for Partial Differential Equations. Darmstadt 1976. Proceedings. Edited by V. E. Meister, N. Weck and W. L. Wendland. XVIII, 520 pages. 1976.

Vol. 562: R. W. Goodman, Nilpotent Lie Groups: Structure and Applications to Analysis. X, 210 pages. 1976.

Vol. 563: Séminaire de Théorie du Potentiel. Paris, No. 2. Proceedings 1975–1976. Edited by F. Hirsch and G. Mokobodzki. VI, 292 pages. 1976.

Vol. 564: Ordinary and Partial Differential Equations, Dundee 1976. Proceedings. Edited by W. N. Everitt and B. D. Sleeman. XVIII, 551 pages. 1976.

Vol. 565: Turbulence and Navier Stokes Equations. Proceedings 1975. Edited by R. Temam. IX, 194 pages. 1976.

Vol. 566: Empirical Distributions and Processes. Oberwolfach 1976. Proceedings. Edited by P. Gaenssler and P. Révész. VII, 146 pages. 1976.

Vol. 567: Séminaire Bourbaki vol. 1975/76. Exposés 471–488. IV, 303 pages. 1977.

Vol. 568: R. E. Gaines and J. L. Mawhin, Coincidence Degree, and Nonlinear Differential Equations. V, 262 pages. 1977.

Vol. 569: Cohomologie Etale SGA 4½. Séminaire de Géométrie Algébrique du Bois-Marie. Edité par P. Deligne. V, 312 pages. 1977.

Vol. 570: Differential Geometrical Methods in Mathematical Physics, Bonn 1975. Proceedings. Edited by K. Bleuler and A. Reetz. VIII, 576 pages. 1977.

Vol. 571: Constructive Theory of Functions of Several Variables, Oberwolfach 1976. Proceedings. Edited by W. Schempp and K. Zeller. VI. 290 pages. 1977

Vol. 572: Sparse Matrix Techniques, Copenhagen 1976. Edited by V. A. Barker. V, 184 pages. 1977.

Vol. 573: Group Theory, Canberra 1975. Proceedings. Edited by R. A. Bryce, J. Cossey and M. F. Newman. VII, 146 pages. 1977.

Vol. 574: J. Moldestad, Computations in Higher Types. IV, 203 pages. 1977.

Vol. 575: K-Theory and Operator Algebras, Athens, Georgia 1975. Edited by B. B. Morrel and I. M. Singer. VI, 191 pages. 1977.

Vol. 576: V. S. Varadarajan, Harmonic Analysis on Real Reductive Groups. VI, 521 pages. 1977.

Vol. 577: J. P. May, E_∞ Ring Spaces and E_∞ Ring Spectra. IV, 268 pages. 1977.

Vol. 578: Séminaire Pierre Lelong (Analyse) Année 1975/76. Edité par P. Lelong. VI, 327 pages. 1977.

Vol. 579: Combinatoire et Représentation du Groupe Symétrique, Strasbourg 1976. Proceedings 1976. Edité par D. Foata. IV, 339 pages. 1977.

continuation on page 719

Math. Sep.

PREFACE

During an International meeting in Helsinki 1957, Professor Stoilow, the wellknown founder of the romanian school of topology and function theory, first proposed to establish a cooperation between the romanian and the finnish mathematicians. It was Professor Cabiria Andreian-Cazacu who later took up this idea again. On her initiative, three romanian-finnish symposia have been held, in Braşov in 1969, in Jyväskylä, organized by Professor I.S.Louhivaara, in 1973 and in Bucharest in 1976.

The program of these meetings has been in the first place devoted to questions of complex Analysis and the theory of quasiconformal mappings. Later the subjects have been enlarged to other problems of Analysis. The meetings have been attended by many mathematicians from other countries as well. This was especially the case during the last symposium held in Bucharest.

The main part of the lectures of the Bucharest meeting is collected in the present volume. The organizing committee of the meeting expresses its gratitude to the Springer-Verlag for publishing the proceedings of the lectures in the "Lecture Notes .

Helsinki, February 1978

Rolf Nevanlinna

C O N T E N T S

I Section

Quasiconformal and Quasiregular mappings.

Teichmüller spaces and Kleinian groups.

II Section

Function theory of one complex variable

III Section

Several complex variables

IV Section

Potential theory

V Section

Function theoretical methods in functional analysis (Operators and differential operators)

LECTURES NOT APPEARING IN THIS PROCEEDINGS

I Section

T.Sorvali,

On Teichmüller spaces of tori, Ann.Acad.Sci.Fenn. Series A.I.Vol.I, 1975, 7-11.

P.Tukia,

The space of quasisymmetric mappings, Math.Scandinavica vol.4o, 1977, p.127-142.

II Section

P.Mocanu,

Hardy spaces for some classes of analytic functions.

U.Pirl,

Normalformen für endlich-vielfach zusammenhängende in einer Kreisringfläche eingelagerte Gebiete., Math.Nachr. Band 76, 1977, 181-194.

P.Russev,

On the representation of analytic functions by means of series in orthogonal polynomials.

III Section

C.Hatvany,

Deformations on complex structures.

J.Siciak,

Maximal analytic extensions of Riemannian domains over topological vector spaces; joint work with K. Rusek. Infin. Dim. Holom. Appl. Proc. Int. Symp. Univ. Estadual de Campinos/Brasil, 1975, 347-377. North Holland 1977.

M.Skwarczynski,

A new notion of boundary in the theory of several complex variables, Bull.Acad.Polonaise Sc., Série sc.math.astr.phys. Vol.XXIV, No.5, 1976, 327-33o.

IV Section

J.Bliedtner,

Characterization of resolutive, semi-regular and regular sets, joint work with W.Hansen. To appear in Math.Zeitschrift under the title "Cones and hyperharmonic functions".

W.Hansen,

A simplicial characterization of elliptic harmonic spaces, Joint work with J.Bliedtner.To appear in Math. Annalen.

I.Laine,

Full-harmonic structures on harmonic spaces, Math. Ann. 235, 1978, 267-290; Math. Z. 160, 1978, 1-26.

V Section

I.Ciorănescu,

La construction de la solution fondamentale pour les équations de convolution dans les espaces de hyperfonctions,

I.Ciorănescu and L.Zsidó

Analytic function method as a theory of ultradistributions.

C.Foiaş,

Quasi-similarity

St.Frunză,

Some applications of complex-analysis in the spectral theory for several operators.

P.Kopp,

A non normal boundary value problem for elliptic 1-st order system in the plane(see thesis Techn.Hochschule,Darmstadt,1977)

Gh.Marinescu,

On the finite element method.

M.N,Roşculeţ,

Algèbres avec radical associées aux équations aux dérivées partielles itérées.

I.A.Rus,

Sur une équation fonctionnelle.

N.Teodorescu,

Dérivées spatiales et opérateurs différentiels linéaires généralisés.

D.Voiculescu,

Closure of unitary and similarity orbits of Hilbert space operators.

THE ORGANIZING COMMITTEE

R.Nevanlinna - Finnish Academy

O.Lehto - University of Helsinki

I.Louhivara-University of Jyväskylä

I.Laine - University of Joensuu

G.Ciucu- University of Bucharest

G.Vranceanu-Academy of Romania Section of mathematics

N.Teodorescu-Mathematical Society of the S.R. of Romania

Cabiria Andreian Cazacu- Central Institute of Mathematics

WORKING COMMITTEE

I Section - Quasiconformal and Quasiregular mappings, Teichmüller spaces and Kleinian groups.

Cabiria Andreian Cazacu,P.Caraman,D.Ivaşcu

II Section - Function theory of one complex variable

P.Mocanu

III Section - Several complex variables

M.Jurchescu,C.Bănică,G.Gussi,O.Stănăşilă

IV Section - Potential theory

A.Cornea, Gh. Bucur

V Section - Functional theoretical methods in Functional Analysis (Operators and Differential Operators)

C.Foiaş, D.Pascali, I.Suciu

Secretariat : Simona Pascu,Virginia Zamă
Marina Marineş ,Carmen Vasilescu

The editorial committee expresses its sincere thanks to Simona Pascu for the contribution given in preparing this volume.

SEMINAR PARTICIPANTS

AUSTRIA
Stegbuchner H.-Univ.Salzburg-Dept.Math.

BULGARIA
Dimiev S.-Inst.of Math.Mech.-Academy of Sc.Sofia
Russev P.-Inst.of Math.Mech.-Academy of Sc.Sofia

CANADA
Gauthier P.M.-Univ.Montreal-Dept.Math.
Zaidman S.- Univ.Montreal -Dept.Math.

CZECHOSLOVAKIA
Lukeš J.-Matem.-fysikálni fakulta KU-Prague
Netuka I.-Matem.-fysikálni fakulta KU-Prague

FINLAND
Kiikka Maire -Univ.of Helsinki-Dept.Math.
Laine I.-Univ.of Joensuu-Dept.Math.
Laine (Mrs).
Lehtinen M.-Univ.of Helsinki-Dept.Math.
Lehto O.-Univ.of Helsinki -Dept.Math.
Mattila P.-Univ.of Helsinki-Dept.Math.
Näätänen Marjatta-Univ.of Helsinki-Dept.Math.
Nevanlinna R.-Univ.of Helsinki-Dept.Math.
Rickman S.-Univ.of Helsinki-Dept.Math.
Sointu Marjatta-Univ.of Helsinki-Dept.Math.
Sorvali T.-Univ.of Joensuu-Dept.Math.
Tukia P.- Univ.of Helsinki-Dept.Math.
Vuorinen M.-Univ.of Helsinki-Dept.Math.

FRANCE
Dolbeault Pierre- Univ. Paris VI- Math.
Dolbeault Simone- Univ. de Poitiers- Math.
Lelong-Ferrand - Jacqueline -Univ. Paris VI- Math.

D.R.GERMANY
Kühnau R.-Martin Luther Univ.Halle-Wittenberg,Sekt.Math.
Pirl U.-Humboldt Univ.-Berlin,Sekt.Math.
Schulze B.W.-Akad.der Wiss.der DDR ZI Math.-Mech.Berlin.

F.R.GERMANY
Begehr H.-Freie Univ.Berlin -Inst.Math.I
Bliedtner J.-Univ.Bielefeld-Fak. Math.
Gackstatter F.-Lehrstuhl II für Math.der RWTH Aachen
Hansen W.-Univ.Bielefeld-Fak.Math.
Kopp P.-Technische Hochschule Darmstadt-Math.
Schirmeier Ursula-Univ.Erlangen-Nürnberg,Math.Inst.

GREECE
Mallios A.-Univ.of Athens-Math.Institute

ITALY
Gay Laura-Ist.Mat.G.Castelnuovo-Univ.Roma
Guaraldo Fr.-Ist.di Mat.Appl.Roma
Macri Patrizia-Ist.di Mat.Appl.-Roma
Succi Fr.-Ist.Mat.G.Castelnuovo-Roma

JAPAN

Kuroda T.- Tôhoku Univ.-Math.Inst.Sendai

POLAND

Ławrynowicz J.-Inst.Math.PAN Lodz
Skwarcinski M-Warsaw Univ .-Dept.Math.
Wojciechowska Maria-Univ.Lodz-Dept.Mat

SWITZERLAND

Holmann H.-Fribourg Univ.Inst.of Math.
Pfluger A.-ETH Zürich-Math.Inst.

U.S.A.

Accola R.-Brown Univ.Providence R.I.
Arsove M.-Univ.of Washington-Dept.Math.
Bers L.-Columbia Univ.New York-Dept.Math,
Frank Evelyn-Univ.of Illinois-Dept.Math.Chicago
Loeb P.-Univ.of Illinois-Dept.Math.Urbana
Miller S.S.-State Univ.Collete at Brockport.New York.
Renggli H.-State Univ.Ohio-Dept.Math.,Kent
Rodin B.-Univ.of North California,Dept.Math.-San Diego.
Schober G.-Univ.of North Carolina at Chapel Hill, Dept.Math.

YUGOSLAVIA

Perović M.Tehnicki fak.Univ.Titograd

ROMANIA

Andreian Cazacu ,Cabiria -Univ.Bucureşti-Fac.Mat.
Apostol C.-INCREST -Bucureşti
Arsene Gr.-INCREST-Bucureşti
Bantea R.-Univ.Bucureşti-Fac.Mat.
Bănică C.-INCREST-Bucureşti
Bărcănescu S.-IPA-Bucureşti
Borcea C.-Univ.Bucureşti-Fac.Mat.
Borcea Viorica-Univ.Iaşi-Fac.Mat.
Brînzănescu V.-Instit.Politehnic Bucureşti
Bucur Gh.-INCREST-Bucureşti
Burcea N.-Univ.Bucureşti-Fac.Mat.
Caraman P.-Univ.Iaşi-Fac.Mat.
Călugăreanu Gh.-Acad.Română-Fil.Cluj-Napoca
Ceauşescu Zoia-INCREST-Bucureşti
Chiţescu I.Univ.Bucureşti-Fac.Mat.
Ciorănescu Ioana-INCREST-Bucureşti
Ciucu G.-Univ.Bucureşti-Fac.Mat
Colojoară I.-Univ.Bucureşti-Fac.Mat.
Cornea A.-INCREST-Bucureşti
Dincă Gh.-Univ.Bucureşti-Fac.Mat.
Fekete O.-Univ.Cluj-Napoca-Fac.Mat.
Flondor P.-Instit.Politehnic-Bucureşti
Foiaş C.-Univ.Bucureşti-Fac.Mat.
Frunză Monica-Univ.Iaşi-Fac.Mat.
Frunză St.-Univ.Iaşi-Fac.Mat.
Georgescu Adelina-IMFCA-Bucureşti
Gheorghiţă St.I.-Univ.Bucureşti
Ghişa D.-Univ.Timişoara-Fac.Mat.

Godini Gliceria-INCREST -Bucureşti
Grigore Gh.-Univ.Bucureşti-Fac-Mat.
Grosu Corina-Univ.Bucureşti-Fac.Mat.
Gussi Gh.-INCREST-Bucureşti
Hatvany Csada-Instit.Politehnic Timişoara
Ichim I.-Univ.Bucureşti-Fac.Mat.
Iordănscu R.-IFTAR-Măgurele
Ivaşcu D.-ISPE-Bucureşti
Jurchescu M.-Univ.Bucureşti-Fac.Mat.
Licea Gabriela-Univ.Bucureşti-Fac-Mat.
Liess Otto-INCREST-Bucureşti
Marcus S.-Univ.Bucureşti-Fac.Mat.
Marinaş Marina-Instit.Construcţii-Bucureşti
Marinescu Gh.-Univ.Bucureşti-Fac.Mat.
Maruşciac I.-Univ.Cluj-Napoca-Fac-Mat.
Mocanu P.-Univ.Cluj-Napoca-Fac.Mat.
Morozan T.-INCREST-Bucureşti
Mustaţă P.-Univ.GalaţiMatematică
Onicescu O.-Univ.Bucureşti-Fac.Mat.
Pascali D.-Inst.Matem.-Bucureşti
Pascu F.-Univ.Bucureşti-Fac.Mat.
Pascu N.-Univ.Cluj-Napoca-Fac.Mat.
Păltineanu G.-Acad.Militară-Cat.Matematică
Petrescu Ruxandra-Univ.Bucureşti-Fac.Mat.
Popa E.-Univ.Iaşi-Fac.Mat.
Popa N.-INCREST-Bucureşti
Popovici I.-Instit.Matem.Bucureşti
Radu Nicolae.-Univ.Bucureşti-Fac.Mat.
Roşculeţ M.N.-Instit.Politehnic-Bucureşti
Roşu Radu-ISPE-Bucureşti
Sălăgean G-Univ.Cluj-Napoca-Fac.Mat.
Sburlan S.F.-Instit.Inv.Super.Constanţa
Silvestru Monica-IPCD-Bucureşti
Stănăşilă O.-Instit.Politehnic-Bucureşti
Stoia Manuela-Instit.Matem.Bucureşti
Stoica L.-INCREST-Bucureşti
Suciu I.-INCREST-Bucureşti
Şabac M.-Univ.Bucureşti-Fac-Mat.
Ţeleman S.-INCREST-Bucureşti
Teodorescu N.-Univ.Bucureşti-Fac.Mat.
Teodorescu P.P.-Univ.Bucureşti.Fac-Mat.
Tomescu D.V.-Univ.Bucureşti-Fac.Mat.
Valuşescu I.-INCREST-Bucureşti
Vasilescu Carmen-Instit.Agronomic Bucureşti
Verona A.-INCREST-Bucureşti
Voiculescu D.-INCREST-Bucureşti
Vrănceanu Gh.-Univ.Bucureşti-Fac-Mat.
Zamfirescu A.-Univ.Bucureşti-Fac.Mat.
Zsidó L.-INCREST-Bucureşti

On certain subloci of Teichmüller space

Robert D. M. Accola*
Brown University
Providence, R. I. 02906

For $p \geq 0$ let W_p stand for a Riemann surface of genus p, and let T_p stand for the Teichmüller space of marked Riemann surfaces of genus p. If p is greater than one the complex dimension of T_p is $3p - 3$. This paper will consider some problems that arise in studying subloci of Teichmüller space that correspond to Riemann surfaces having special properties. Three types of subloci will be considered: 1) G - loci, or loci of surfaces admitting certain non-trivial automorphism groups; 2) θ - loci, or loci of surfaces whose theta functions have certain unusual vanishing properties; and 3) C - loci, or surfaces where equality holds in Castelnuovo's inequality. For G - loci we will consider defining equations in terms of theta functions and for θ - loci and C - loci we determine some of the dimensions of these subloci.

The G - loci are well-known subloci of Teichmüller space. For $p \geq 3$ G - loci are the fixed points of the Teichmüller modular group. The dimension of a particular G-locus is easy to derive from the Riemann - Hurwitz formula. If G is a group of automorphisms on W_p where q is the genus of the space of orbits and s is the number of ramified orbits then the dimension of the corresponding G-locus in T_p is $3q - 3 + s$. [3] Moreover, G-loci are known to be closed submanifolds of T_p. Let $(q\text{-}H)_p$ denote those Riemann surfaces in T_p which admit an involution whose orbit space has genus q. Thus

* Research supported by the National Science Foundation

$(0\text{-}H)_p$ is the locus of hyperelliptic surfaces in T_p and $(1\text{-}H)_p$ is the locus of elliptic-hyperelliptic surfaces. We will use the simpler notation H_p for $(0\text{-}H)_p$. The codimension of $(q\text{-}H)_p$ in T_p is $p + q - 2$. A type of problem which naturally presents itself is that of finding $p + q - 2$ equations which define $(q\text{-}H)_p$.

In order to discuss this and related questions we now consider the theta functions. [6] For W_p let $(\pi iE, B)$ be a $p \times 2p$ period matrix where E and B are $p \times p$ matrices, E being the identity. For u in C^p let $\theta[\eta](u;B)$ be the first order theta function defined on $C^p \times T_p$ with half-integer theta-characteristic $[\eta]$. Since such a function is odd or even as a function of u we shall define the theta-characteristic $[\eta]$ to be odd or even according to the parity of $\theta[\eta](u;B)$. The order of vanishing of this function at $u = 0$ thus has the parity of $[\eta]$. By Riemann's solution to the Jacobi inversion problem there is a one to one correspondence between complete half-canonical linear series g^r_{p-1} ($2g^r_{p-1} = K$, the canonical series) and theta-characteristics $[\eta]$ where $\theta[\eta](u;B)$ vanishes to order $r+1$ at $u = 0$. Now consider for each $[\eta]$ $\theta[\eta](u;B)$ and its partial derivatives with respect to the u variables evaluated at $u = 0$ as analytic functions on T_p. It is known that for the general Riemann surface in T_p $\theta[\eta](0;B)$ will vanish only for the obvious $[\eta]$ which are odd, and only there to order one. If $[\eta]$ is even then for $\theta[\eta](0;B)$ to be zero to order two it is only required that $\theta[\eta](0;B) = 0$. If $[\eta]$ is odd, for $\theta[\eta](u;B)$ to vanish to order three at $u = 0$ it is required that the p equations $(\partial/\partial u_j)\theta[\eta](0;B) = 0$ be satisfied. And so on.

The two classical characterizations of G-loci that I know are as follows. $W_3 \in H_3$ if and only if $\theta[\eta](0;B) = 0$ for one even characteristic $[\eta]$ (Riemann). [8] $W_4 \in H_4$ if and only if $\theta[\eta_i](0;B) = 0$ for two even characteristics (Weber). [9] These two examples are highly suggestive and lead to a "p-2 conjecture" for the defining equations of H_p in terms of $\theta[\eta](0;B) = 0$ for $p - 2$ even characteristics $[\eta]$. The conjecture is most probably false in general but a modified form of it is true for genus 5. [1,2] Also $(1-H)_5$ is characterized by four such vanishing properties. [1,2] leading to a "p-1 conjecture" for elliptic-hyperelliptic surfaces, again a conjecture which is most probably false. In cases of genus three and five several groups which are elementary abelian groups of order 2^n $(n = 1,2,3,4)$ have been characterized by the correct number of vanishing properties of $\theta[\eta_i](0,B)$ except that for $p = 3$ we must allow $[\eta_i]$ to be one-quarter integer theta-characteristics. [2] Most probably the most interesting caseshere are the Humbert surfaces, surfaces of genus 5 which admit a group of order sixteen generated by four commuting elliptic-hyperelliptic involutions. This G-locus has dimension two and is characterized by ten equations $\theta[\eta_i](0;B) = 0$, $i = 1,2,\ldots,10$, where there are further conditions on the $[\eta_i]$.

In order to discribe other results in this direction, define θ^r_p to be the locus in T_p of those surfaces where $\theta[\eta](u;B)$ vanishes to order $r + 1$ at $u = 0$ for some $[\eta]$. Thus Riemann's characterization says $H_3 = \theta^1_3$ and Weber's characterization says that H_4 is a set of self intersections of θ^1_4. Using this terminology we can state Marten's characterizations of H_p. [7] If p is odd then $H_p = \Theta^{(p-1)/2}{}_p$, and if p is even and $p \geq 8$ then $H_p = \Theta^{(p-2)/2}{}_p$. Thus we have defining equations for H_p, but for $p \geq 5$ the number of defining equations exceeds the codimension of H_p in T_p. Further results of a similar type include characterizations of elliptic-hyperelliptic surfaces for $p \neq 4$ and other $(q - H)_p$ where p is about 6q or larger. [2]

That one is able to show that some G-loci are Θ-loci, that is, that one can characterize the existence of certain (abelian) automorphism groups in terms of vanishing properties of the theta function is most probably not a general phenomenom. We now shift our attention to the Θ-loci and ask whether it is possible to find the dimension of other Θ-loci which are not G-loci. Fortunately, there is some information here, and the procedures lead to interesting problems in the classical theory of algebraic curves.

Let us first consider $\Theta^2{}_p$, that is, the loci of surfaces W_p where $\Theta[\eta](u;B)$ vanishes to order three at $u = 0$ for some odd characteristic $[\eta]$. By Riemann's solution to the Jacobi inversion problem this means that W_p admits a half-canonical $g^2{}_{p-1}$. For $p = 5$ this characterizes H_p, of codimension three. If we assume for $p = 6$ that $g^2{}_5$ is simple, this characterizes those W_6 which have plane models which are non-singular plane quintics. This family of plane curves has dimension 20. Subtracting 8 for the family of plane collineations and knowing that such a W_6 admits a unique simple $g^2{}_5$, we see that the family of Riemann surfaces of genus 6 admitting a simple half-canonical $g^2{}_5$ has dimension 12 (= dim T_6 - 3). For $p = 7$ a simple half-canonical $g^2{}_6$ leads to a plane sextic with three collinear nodes. The family of such curves has dimension 23 so the family of Riemann surfaces has dimension 15 ($= \dim T_7 - 3$).

To discuss the appropriate conjecture let us denote by Θ^{r*}_p those surfaces in Θ^r_p where the corresponding g^r_{p-1} is simple. Notice that Θ^{r*}_p need not be closed in T_p. The conjecture is this: $\operatorname{codim} \Theta^{2*}_p = 3$. This number is considerably less than p the number of equations defining Θ^2_p. One arrives at the conjecture by making Kraus' observation that such a g^2_{p-1} leads to a plane curve of degree p-1 where there is necessarily an adjoint of degree p - 6. [5] Since any Riemann surface can admit only a finite number of such g^2_{p-1}'s a simple modification of the usual counting of dimensions gives 3p - 6. However, there is a real problem in constructing such a curve for general p; that is, it does not seem to be known whether or not Θ^{2*}_p is empty for arbitrary p. The conjecture has been confirmed for $p \leq 10$. Perhaps the conjecture is unreasonable since it is known that Θ^{3*}_{10} has at least two components of different dimension.

Now we turn to Θ^r_p where r is large. A theorem of Castelnuovo, which we shall discuss in more detail later, states that if g^r_{p-1} is half-canonical and simple then $p \geq 3r$. Thus if $p < 3r$, a half-canonical g^r_{p-1} must be composite, and it is not difficult to show that the covering $W_p \to W_q$ which gives rise to g^r_{p-1} being composite is two-sheeted; that is, W_p admits an involution. One concludes that if $p < 3r$ then every surface in Θ^r_p must admit an involution. This is the main technique used in characterizing $(q\text{-}H)_p$ in terms of vanishing properties of the theta function.

It turns out that if $p = 3r$ $(r \geq 3)$ then $\dim \Theta^{r*}_p = 5r + 3$

except for components of Θ^{5*}_{15} corresponding to non-singular plane septics which have dimension 27.

First we give examples of plane curves of genus $p = 3r$ where the simple g^r_{3r-1} is fairly obvious. (It is known that such a g^r_{3r-1} must be unique on W_{3r}.) Assure r is even. Then a plane curve of degree $r + 3$, C_{r+3}, with an ordinary $(r-1)$-fold singularity at say Q is an example. g^r_{3r-1} is cut out by (rational) curves $f_{r/2}$ of degree $r/2$ with an $((r-2)/2)$-fold singularity at Q. To obtain examples for r odd, one adds a triple point P to the other singularity of C_{r+3} and requires the $f_{r/2}$ to pass through P. That the dimension of Riemann surfaces corresponding to such curves is $5r + 3$ is again the classical counting arguement and the fact that W_{3r} has at most one simple g^r_{3r-1}.

Notice in the above example that the lines through Q cut out a g^1_4 on C_{r+3}. If a W_{3r} admits a simple g^r_{3r-1} <u>and</u> a g^1_4 then in general one can expect models as above. For in this case g^1_4 must impose two linear conditions on g^r_{3r-1}; that is, $g^r_{3r-1} - g^1_4 = g^{r-2}_{3r-5}$. Continuing to subtract g^1_4 from g^{r-2}_{3r-5}, g^{r-4}_{3r-9}, etc., and assuming that g^1_4 imposes two conditions on each $g^{r-2l}_{3r-1-4l}$ we obtain (if r is even) the g^2_{r+3} $(l = (r-2)/2)$. g^1_4 still imposes two conditions on g^2_{r+3} which accounts for the $(r-1)$-fold singularity Q. For r odd we subtract g^1_4 $(r-3)/2$ times and then subtract an arbitrary point x. The triple point P is the other three points in the divisor of g^1_4 determined by x.

That the general surface in Θ^{r*}_{3r} admits such a model now follows from the following theorem. Suppose W_{3r} admits a simple g^r_{3r-1} (which is necessarily half-canonical) where $r \geq 2$ and $r \neq 5$; then W_{3r} admits a g^1_4 without fixed points.

An extension of Castelnuovo's method leads to the following results. Suppose W_p admits two simple half-canonical g^r_{p-1}'s; then $p \geq 3r + 2$. Suppose W_p admits four simple half-canonical g^r_{p-1}'s whose sum is bicanonical; then $p \geq 3r + 3$. Again we can ask for the dimension in T_p of those surfaces where we have equality in the above inequalities. One first shows that W_p admits a g^1_4 and that this leads to certain plane models. The following results are true. The locus of surfaces in T_{3r+2} admitting two simple half-canonical g^r_{3r+1}'s has dimemsion $3r + 5$. The locus of surfaces in T_{3r+3} admitting four simple half-canonical g^r_{3r+2}'s whose sum is bicanonical has dimension $3r + 6$.

As a final type of sublocus of Teichmüller space let us generalize those examples just preceding the last paragraph. First we shall need a precise statement of Castelnuovo's general theorem. [4] Suppose W_p admits a simple g^r_n. then

$$p \leq (n-r+\varepsilon)(n-1-\varepsilon)/2(r-1)$$

where $0 \leq \varepsilon \leq r-2$ and $n-r+\varepsilon \equiv 0 \pmod{r-1}$. Another way of writing this is as follows. If $n = (r-1)m + q$, $q = 2,3,\ldots,r$ then $\varepsilon = r - q$ and (1) $p \leq m((m-1)(r-1) + 2q - 2)/2$.

Suppose we have equality in Castelnuovo's theorem; that is, a surface W_p where we have equality in (1) and W_p admits a simple $g^r_{m(r-1)+q}$. Let us call the locus of such surfaces in T_p a C-locus. The problem is to determine the dimension of this C-locus. As in the half-canonical case such a g^r_n $(n = m(r-1) + q)$ must be unique on W_p.

From the proof of Castelnuovo's theorem, equality in (1), and the Riemann-Roch theorem it follows that

$$K - (m-1)g^r_n = g^{q-2}_{(m+1)(q-2)}$$

where K is the canonical series. Thus if $q = 2$ and $m = 3$ we are in the half-canonical case and $p = 3r$. If $q = 2$ and m is arbitrary, $m \geq 3$, then g^r_n is $(1/(m-1))$-canonical. If $q = 3$ then W_p admits a g^1_{m+1} which can be shown to be without fixed points. As in the half-canonical case the main problem is to show that such a surface admits a g^1_{m+1}. If $q \geq 4$ one has the divisor $g^{q-2}_{(m+1)(q-2)}$ to work with, and in fact one can show for fixed m and q and large r that $g^{q-2}_{(m+1)(q-2)} = (q-2)g^1_{m+1}$. Then one shows easily that g^1_{m+1} imposes two linear conditions on g^r_n. As in the half-canonical case, one subtracts g^1_{m+1} from g^r_n a number of times to obtain a plane model. If r is even one obtains a plane curve of degree $n - (r-2)(m+1)/2$ with a single singularity Q of multiplicity $n - r(m+1)/2$. g^r_n is cut out by (rational) curves of degree $r/2$ with a singularity of order $(r-2)/2$ at Q. Again by adding a singularity P of multiplicity m and requiring the curves of degree $r/2$ to pass through P we get models for the case r odd. We can then compute by the classical counting arguements the dimension of the corresponding C-locus which turns out to be:

$$(m+2)(m-1)(r-1)/2 \; + \; q(m+2) \; + \; (m-5)$$

References

[1] Accola, R. D. M., Some loci of Teichmüller space for genus five defined by vanishing theta nulls. Contributions to Analysis Academic Press, 1974 pp 11 - 18.

[2] Accola, R. D. M., Riemann surfaces, theta functions, and abelian automorphism groups Lecture notes in Mathematics 483 (1975) Springer - Verlag

[3] Baily, W. L. On the automorphism group of a generic curve of genus > 2. Journal of Mathematics of Kyoto University Vol 1 (1961/2) pp 101 - 108. Correction p 325.

[4] Castelnuovo, G., Sui multipli du une serie lineare di gruppi di punti, etc. Rendiconti del circolo Matematico di Palermo Vol 7 (1893) pp 89 - 110.

[5] Kraus, L., Note über ausgewohnliche special Gruppen auf algebraischen Kurven Mathematische Annalen Vol 15 (1880) p 310

[6] Krazer, A., Lehrbuch der Thetafunktionen Chelsea

[7] Martens, H. H., On the varieties of special divisors on a curve, II Journal fur die reine und angewandt Mathematik Vol 233 (1968) pp 89 - 100

[8] Riemann, B., Gesammelte Mathematische Werke Dover

[9] Weber, H., Über gewisse in der Theorie der Abel'schen Funktionen auftretende Ausnahmfalle Mathematische Annalen Vol 13 (1878) pp 35 - 48

Limit theorems and estimates for extremal rings of high dimension

Glen D. Anderson

1. Introduction

1.1. Summary of results. In the first part of this paper we study how the modulus of certain space rings depends upon the dimension of the space and the dimension of the bounded component of the complement as the dimension of the space increases.

For $n \geq 2$ and $1 \leq p \leq n$, the ring $R_{n,p}(a)$ may be described as the open unit ball in n-space minus a closed concentric p-dimensional ball of radius a. In §2 we show that

$$\lim_{n\to\infty} \operatorname{mod} R_{n,n-1}(a) = \log \frac{1}{a} ,$$

thus proving a conjecture of [2, p. 20]. This result gains its significance partly from the fact that $\operatorname{mod} R_{n,n}(a)$ is known to be $\log 1/a$ for each n [10].

In §3 we show that if p/n tends to 0 as n tends to ∞, then

$$\lim_{n\to\infty} \frac{p}{n} \operatorname{mod} R_{n,p}(a) = 1 .$$

This result extends Theorem 5 of [2, §6], which states that

$$\lim_{n\to\infty} \frac{1}{n} \operatorname{mod} R_{n,1}(a) = 1 .$$

In the latter part of the paper we obtain improved upper bounds for the constant λ_n related to the Grötzsch ring in n-space. The estimates for general n are derived in §4 as explicit double integrals, and in §5 close approximations are found for these for $n = 3$ and 4. For $n = 3$ we then have $9.19\ldots \leq \lambda_3 \leq 9.90\ldots$; the best previously known upper bound for λ_3 was $12.4\ldots$ [10]. For $n = 4$ we have $21.6\ldots \leq \lambda_4 \leq 26.0\ldots$, compared with the previously known upper bound $35.5\ldots$ for λ_4 ([6], [11]). In conclusion we state without proof the corresponding results for $\operatorname{mod} R_{n,n-1}$.

1.2. Definitions and notation. For the most part we follow the notation of [2]. By a ring R is meant a domain in euclidean n-space R^n whose complement consists of two components C_0 and C_1, where C_0 is bounded. We let $B_0 = \partial C_0$ and $B_1 = \partial C_1$ be the boundary components of R. The conformal capacity (cf. [13]) of R is

$$\operatorname{cap} R = \inf_{\varphi} \int_R |\nabla \varphi|^n \, d\omega ,$$

where ∇ denotes the gradient, and where the infimum is taken over all real-valued C^1 functions φ in R with boundary values 0 on B_0 and 1 on B_1. Then the modulus of the ring R is defined by

$$\text{(1)} \qquad \operatorname{mod} R = \left(\frac{\sigma_{n-1}}{\operatorname{cap} R} \right)^{\frac{1}{n-1}} ,$$

where for each positive integer p we let σ_p denote the p-dimensional measure of the unit sphere $S^p = \left\{ (x_1, \ldots, x_{p+1}) : \sum_{j=1}^{p+1} x_j^2 \right\}$.

Then

$$\text{(2)} \qquad \sigma_p = 2\, \pi^{(p+1)/2}\, \Gamma((p+1)/2)^{-1}$$

(cf. [11], [15]), where Γ denotes the classical Gamma function; moreover, the relation

$$\text{(3)} \qquad \int_0^{\pi/2} \cos^p u \, du = \frac{\sigma_{p+1}}{2\, \sigma_p}$$

holds for each positive integer p.

For integers n and p, $n \geq 2$, $1 \leq p \leq n$, and for $0 < a < 1$, we define the ring $R_{n,p}(a)$ to be the set $B^n(1) \setminus \overline{B^p(a)}$, where

$$B^n(1) = \left\{ (x_1, \ldots, x_n) : \sum_{j=1}^{n} x_j^2 < 1 \right\}, \quad \overline{B^p(a)} = \left\{ (x_1, \ldots, x_p) : \sum_{j=1}^{p} x_j^2 \leq a^2 \right\},$$

and where $\setminus$ denotes set-theoretic difference. When we regard a subset of R^p as a subset of R^n, $p < n$, we usually identify R^p with the subspace $\{(x_1,\ldots,x_p,0,\ldots,0)\}$ of R^n.

If Γ is a family of arcs in a ring R, the <u>modulus</u> of Γ is defined as

$$M(\Gamma) = \inf_{\rho} \int_R \rho^n \, d\omega ,$$

where the infimum is taken over all non-negative Borel measurable functions ρ in R for which $\int_\gamma \rho \, ds \geq 1$ for each $\gamma \in \Gamma$.

If Σ is a family of compact sets in a ring R, the <u>modulus</u> of Σ is defined by

$$M(\Sigma) = \inf_{\rho} \int_R \rho^n \, d\omega ,$$

where the infimum is taken over all ρ which are non-negative and Borel measurable in R and for which $\int_E \rho^{n-1} \, d\sigma \geq 1$ for each $E \in \Sigma$.

A compact set E is said to <u>separate the boundary components</u> of a ring R if $E \subset R$ and if B_0 and B_1 lie in different components of $C(E)$. It can be shown [17] that if Γ is the family of all arcs in a ring R joining B_0 and B_1 and if Σ is the family of all compact sets in R separating B_0 and B_1 then

$$M(\Gamma) = M(\Sigma)^{1-n} = \operatorname{cap} R .$$

From this and (1) it then follows that

$$(4) \qquad \operatorname{mod} R = \left(\frac{\sigma_{n-1}}{M(\Gamma)}\right)^{\frac{1}{n-1}} = \sigma_{n-1}^{\frac{1}{n-1}} M(\Sigma) = \left(\frac{\sigma_{n-1}}{\operatorname{cap} R}\right)^{\frac{1}{n-1}} .$$

2. Limit of mod $R_{n,n-1}$ as n tends to ∞

In Theorems 1 and 6 of [2] we proved that

$$\log \frac{1}{a} \leq \text{mod } R_{n,n-1}(a) \leq \text{mod } R_{2,1}(a)$$

for each $n \geq 2$, where $R_{n,n-1}(a)$ is the ring $B^n(1) \setminus \overline{B^{n-1}(a)}$. Since these bounds are independent of n, it was natural to conjecture that $\lim_{n\to\infty} \text{mod } R_{n,n-1}(a) = \log 1/a$. In this section that conjecture is proved.

Theorem 1. *For fixed* a, $0 < a < 1$,

$$\lim_{n\to\infty} \text{mod } R_{n,n-1}(a) = \log \frac{1}{a} .$$

Proof. Since $\log 1/a \leq \text{mod } R_{n,n-1}(a)$ it is obviously sufficient to prove that

$$\limsup_{n\to\infty} \text{mod } R_{n,n-1}(a) \leq \log \frac{1}{a} . \tag{5}$$

To this end we let f_k denote the one-to-one plane conformal mapping

$$z = f_k(w) = k^{\frac{1}{2}} \text{sn}(2K'w/\pi, k)$$

which carries the rectangle $T_k = \{w = u+iv: 0 \leq u \leq \alpha_k, 0 < v < \pi/4\}$ onto the set $Q_k = \{z = x+iy: |z| < 1, x \geq 0, y \geq 0\} \setminus \{z = x: 0 \leq x \leq a\}$ (cf. [4], [5]). Here sn denotes Jacobi's elliptic sine function, $k = a^2$, and $\alpha_k = \pi K/2K'$, where K and K' are the elliptic integrals

$$K = K(k) = \int_0^1 [(1-t^2)(1-k^2t^2)]^{-\frac{1}{2}} dt ,$$

$$K' = K(k') , \quad k' = (1-k^2)^{\frac{1}{2}} .$$

For each fixed u, $0 \leq u \leq \alpha_k$, let $\gamma' = \gamma'(u;)$ be the vertical segment $\{w = u+iv: 0 < v < \pi/4\}$ joining the horizontal sides of T_k and let $\gamma = f_k(\gamma')$. Then γ is an arc joining the segment $\{z: 0 \le x \le a,\ y = 0\}$ to the quarter circle $\{z: |z| = 1,\ x \ge 0,\ y \ge 0\}$. If we reflect Q_k in the x-axis and then rotate (cf. [2, §1.2]) the resulting plane set about the y- (that is, x_2-) axis in R^n we obtain the ring $R_{n,n-1}(a)$, and the collection of all such arcs γ generates a family of arcs $\Gamma_a \subset \Gamma$, where Γ is the family of all arcs joining the components of $\partial R_{n,n-1}(a)$ in $R_{n,n-1}(a)$. Hence $M(\Gamma) \ge M(\Gamma_a)$, and by the methods employed in [2, §4] we may show that

$$M(\Gamma) \ge M(\Gamma_a) \ge 2\,\sigma_{n-2} \int_0^{\alpha_k} \left(\int_0^{\pi/4} \left(\frac{|f_k'|}{x} \right)^{\frac{n-2}{n-1}} dv \right)^{1-n} du .$$

Using (4), we thus obtain

$$\operatorname{mod} R_{n,n-1}(a) \le \left(\frac{\sigma_{n-1}}{2\,\sigma_{n-2}} \right)^{\frac{1}{n-1}} \left[\int_0^{\alpha_k} \left(\int_0^{\pi/4} \left(\frac{|f_k'|}{x} \right)^{\frac{n-2}{n-1}} dv \right)^{1-n} du \right]^{\frac{1}{1-n}} .$$

The author is indebted to F. W. Gehring for suggesting the use of the arc family Γ_a in the study of such rings.

Next, an application of Hölder's inequality to the inner integral in this upper bound, together with use of the limit

$$\lim_{n\to\infty} \left(\frac{\sigma_{n-1}}{2\,\sigma_{n-2}} \right)^{\frac{1}{n-1}} = 1$$

[2, p. 19], gives

$$\text{(6)} \quad \limsup_{n\to\infty} \operatorname{mod} R_{n,n-1}(a) \le \limsup_{n\to\infty} \left[\int_0^{\alpha_k} \left(\int_0^{\pi/4} \frac{|f_k'|}{x}\, dv \right)^{2-n} du \right]^{\frac{1}{1-n}} = \limsup_{n\to\infty} \|h_k\|_{n-2}^{\frac{2-n}{n-1}} ,$$

where $\|\cdot\|_{n-2}$ denotes the $L_{n-2}[0, \alpha_k]$ norm and

$$\text{(7)} \quad h_k(u) = \left(\int_0^{\pi/4} \frac{|f_k'|}{x}\, dv \right)^{-1} , \quad 0 \le u \le \alpha_k .$$

We claim that h_k is bounded on $[0, \alpha_k]$. To see this, we use [4, pp. 40, 41] or [5, ##125.01, 121.00] to write

$$(8) \qquad \left(\frac{|f_k'|}{x}\right)^2 = \left(\frac{2K'}{\pi}\right)^2 \frac{(1 - s^2 d_1^{\,2})(d_1^{\,2} - k^2 s^2)}{s^2 d_1^{\,2}} ,$$

where

$$(9) \qquad \begin{aligned} s &= \mathrm{sn}(2K'u/\pi, k), & c &= \mathrm{cn}(2K'u/\pi, k), & d &= \mathrm{dn}(2K'u/\pi, k), \\ s_1 &= \mathrm{sn}(2K'v/\pi, k'), & c_1 &= \mathrm{cn}(2K'v/\pi, k'), & d_1 &= \mathrm{dn}(2K'v/\pi, k') . \end{aligned}$$

For later reference we also record that

$$(10) \qquad \left(\frac{|f_k'|}{y}\right)^2 = \left(\frac{2K'}{\pi}\right)^2 \frac{(1 - s^2 d_1^{\,2})(d_1^{\,2} - k^2 s^2)}{c^2 d^2 s_1^{\,2} c_1^{\,2}} .$$

By using the identities in [4, p. 9] or [5, # 121.00] we may rewrite (8) and (10) in the form

$$(11) \qquad \left(\frac{|f_k'|}{x}\right)^2 = \left(\frac{2K'}{\pi}\right)^2 \left[\frac{1 + k^2 s^4}{s^2} - \frac{k^2 + d_1^{\,4}}{d_1^{\,2}}\right]$$

and

$$(12) \qquad \left(\frac{|f_k'|}{y}\right)^2 = \left(\frac{2K'}{\pi}\right)^2 \left[\frac{d_1^{\,2}}{s_1^{\,2} c_1^{\,2}} + \frac{k'^4 s^2}{c^2 d^2}\right] .$$

Differentiating (11) partially with respect to s^2 and using the fact that $0 \leq s^2 \leq 1$, we can easily see that $|f_k'|/x$ is monotone decreasing as a function of s^2, $0 \leq s^2 \leq 1$. Since $s^2 = \mathrm{sn}^2(2K'u/\pi, k)$ is monotone increasing as a function of u for $0 \leq u \leq \alpha_k$, it follows that in $[0, \alpha_k]$, $|f_k'|/x$ has its minimum value when $u = \alpha_k$. Thus, applying the special values $\mathrm{sn}\, K = 1$, $\mathrm{cn}\, K = 0$, $\mathrm{dn}\, K = k'$ ([4, p. 9], [5, #122.02]) and the identities $d_1^{\,2} = 1 - k'^2 s_1^{\,2}$ and $d_1^{\,2} - k^2 = k'^2 c_1^{\,2}$

([4 , p. 9], [5,#121.00]) we obtain for (8) the lower bound

$$(13) \qquad \frac{|f_k'|}{x} \geq \frac{2K'}{\pi} k'^2 \frac{s_1 c_1}{d_1} \, , \quad 0 \leq u \leq \alpha_k \, ,$$

with equality when $u = \alpha_k$. For later reference we remark that an analogous procedure shows that

$$(14) \qquad \frac{|f_k'|}{y} \geq \frac{2K'}{\pi} \frac{d_1}{s_1 c_1} \, , \quad 0 \leq u \leq \alpha_k \, ,$$

with equality when $u = 0$.

Finally, if we use the differentiation formula for dn ([4, p. 9], [5,#731.03]) and the special values $\mathrm{dn}(K'/2, k') = \sqrt{k}$ and $\mathrm{dn}\, 0 = 1$ ([4, pp. 14, 9], [5,## 122.10, 122.01]) together with the lower bound (13) we obtain

$$\int_0^{\pi/4} \frac{|f_k'|}{x} \, dv \geq - \int_0^{\pi/4} \frac{d}{dv} \log \mathrm{dn}(2K'v/\pi, k') \, dv = \log \frac{1}{\sqrt{k}} \, ,$$

with equality when $u = \alpha_k$. It therefore follows that the function h_k defined in (7) above is bounded and that

$$(15) \qquad \|h_k\|_\infty = \sup_{0 \leq u \leq \alpha_k} \left(\int_0^{\pi/4} \frac{|f_k'|}{x} \, dv \right)^{-1} = (\log k^{-\frac{1}{2}})^{-1} \, .$$

Then using (6), (15), the relation $k = a^2$, and the fact that $\lim_{n \to \infty} \|h_k\|_{n-2} = \|h_k\|_\infty$, we arrive at (5) . This concludes the proof of Theorem 1.

3. Limit for $\frac{p}{n}$ mod $R_{n,p}(a)$ as n tends to ∞

In Theorem 5 of [2,§6] we showed that for fixed a, $0 < a < 1$, $(1/n)$ mod $R_{n,1}(a)$ tends to 1 as n approaches ∞. In this section we consider the rings $R_{n,p}(a)$, $p = p(n)$, as n tends to ∞ with a fixed, $0 < a < 1$. If $p = p(n)$ depends upon n in such a way that the ratio p/n tends to 0 as n approaches ∞ we shall write $p = o(n)$. Our goal in this section will be the following result.

Theorem 2. For fixed a, $0 < a < 1$, and $p = o(n)$,

$$\lim_{n\to\infty} \frac{p}{n} \bmod R_{n,p}(a) = 1 .$$

Proof. First, we shall use the plane conformal mapping $z = f_k(w)$ introduced in § 2 ; f_k carries the rectangle $T_k = \{w = u+iv: 0 \leq u \leq \alpha_k, 0 < v < \pi/4\}$ onto the set $Q_k = \{z = x + iy: |z| < 1, x \geq 0, y \geq 0\} \setminus \{z = x: 0 \leq x \leq a\}$, where $k = a^2$. Let Σ denote the family of all compact sets in $R_{n,p}(a)$ separating the boundary components of $R_{n,p}(a)$, and let Σ_a denote the subfamily of surfaces generated by images $\gamma = f_k(\gamma')$ of horizontal segments $\gamma' = \gamma'(v;) = \{w = u+iv: 0 \leq u \leq \alpha_k\}$ under the rotation that produces $R_{n,p}(a)$ from Q_k (cf. [2, pp. 12, 13] for a description of this geometric process). Then $\Sigma \supset \Sigma_a$, and by the methods employed in [2, §§3, 4] we may show that

$$M(\Sigma) \geq M(\Sigma_a) \geq (\sigma_{n-p-1}\sigma_{p-1})^{\frac{1}{1-n}} \int_0^{\pi/4} \left(\int_0^{\alpha_k} \left(\frac{x}{|f_k'|}\right)^{p-1} \left(\frac{y}{|f_k'|}\right)^{n-p-1} du \right)^{\frac{1}{1-n}} dv .$$

Employing (4), we then have

$$\text{(16)} \qquad \bmod R_{n,p}(a) \geq \left(\frac{\sigma_{n-1}}{\sigma_{p-1}\sigma_{n-p-1}}\right)^{\frac{1}{n-1}} \int_0^{\pi/4} \left(\int_0^{\alpha_k} \left(\frac{x}{|f_k'|}\right)^{p-1} \left(\frac{y}{|f_k'|}\right)^{n-p-1} du \right)^{\frac{1}{1-n}} dv .$$

We wish to show that the first factor on the right side of (16) tends to 1 . For this, if we use (2) and Binet's first expression for the logarithm of the Gamma function in terms of an infinite integral [16, pp. 248-249] we see, after some simplifications, that

$$\log\left(\frac{\sigma_{n-1}}{\sigma_{p-1}\sigma_{n-p-1}}\right)^{\frac{1}{n-1}} = \frac{1}{2}\left[\frac{n-p}{n-1}\log\left(1-\frac{p}{n}\right) - \frac{1}{n-1}\log(n-p)\right] + \varepsilon(n) ,$$

where $\varepsilon(n)$ tends to 0 as n tends to ∞ with $p = o(n)$. We conclude that

$$(17) \qquad \lim_{\substack{n\to\infty \\ p=o(n)}} \left(\frac{\sigma_{n-1}}{\sigma_{p-1}\sigma_{n-p-1}}\right)^{\frac{1}{n-1}} = 1 .$$

Next, from (16), (13), and (14) we have

$$\operatorname{mod} R_{n,p}(a) \geq \left(\frac{\sigma_{n-1}}{\sigma_{p-1}\sigma_{n-p-1}}\right)^{\frac{1}{n-1}} \left(\frac{2K'}{\pi}\right)^{\frac{n-2}{n-1}} (k')^{\frac{2p-2}{n-1}} \alpha_k^{\frac{1}{1-n}} \int_0^{\pi/4} \left(\frac{d_1}{s_1 c_1}\right)^{\frac{n-2p}{n-1}} dv .$$

In light of (17) this says that

$$(18) \qquad \liminf_{\substack{n\to\infty \\ p=o(n)}} \frac{p}{n} \operatorname{mod} R_{n,p}(a) \geq \frac{2K'}{\pi} \liminf_{\substack{n\to\infty \\ p=o(n)}} \frac{p}{n} \int_0^{\pi/4} \left(\frac{d_1}{s_1 c_1}\right)^{\frac{n-2p}{n-1}} dv .$$

To determine the behavior of the integral in (18) as a function of n , we begin by using Landen's transformation ([4, p. 72], [5, p. 39]) to write

$$\frac{s_1 c_1}{d_1} = \operatorname{sn}[(1+k)2K'v/\pi, (1-k)/(1+k)]/(1+k) .$$

Making the change of variable $t = (1+k)2K'v/\pi$, we then have

$$(19)\int_0^{\pi/4}\left(\frac{d_1}{s_1c_1}\right)^{\frac{n-2p}{n-1}}dv = (1+k)^{\frac{1-2p}{n-1}}\frac{\pi}{2K'}\int_0^{K^*}\operatorname{sn}(t,\ k^*)^{\frac{2p-n}{n-1}}\,dt\quad ,$$

where $k^* = (1-k)/(1+k)$ and $K^* = K(k^*)$. Thus the integral in (18) is asymptotically equal to

$$\frac{\pi}{2K'}\int_0^{K(k^*)}\operatorname{sn}(t,k^*)^{\frac{2p-n}{n-1}}\,dt$$

as n tends to ∞ with $p = o(n)$.

Now we can employ the argument used on the middle of page 19 of [2] . It is easy to see that $0 \leq \operatorname{sn}(t,k^*) \leq t$ for $0 \leq t \leq K^*$ and that $(\operatorname{sn} t)/t$ has limit 1 as t tends to 0 . Choose $\varepsilon > 0$. Then there exists δ , $0 < \delta < K^*$, such that $1 < t/\operatorname{sn} t < 1+\varepsilon$ for $0 < t < \delta$. Direct computation thus shows that

$$1 \leq \liminf_{\substack{n\to\infty\\ p=o(n)}} \frac{p}{n}\int_0^{\delta}(\operatorname{sn} t)^{\frac{2p-n}{n-1}}dt \leq \limsup_{\substack{n\to\infty\\ p=o(n)}} \frac{p}{n}\int_0^{\delta}(\operatorname{sn} t)^{\frac{2p-n}{n-1}}dt \leq 1+\varepsilon\ ,$$

while

$$\lim_{\substack{n\to\infty\\ p=o(n)}} \frac{p}{n}\int_{\delta}^{K^*}\operatorname{sn}(t,k^*)^{\frac{2p-n}{n-1}}dt = 0$$

since the integrand is bounded on $(\delta,\ K^*)$. Therefore

$$\lim_{\substack{n\to\infty\\ p=o(n)}} \frac{p}{n}\int_0^{K^*}\operatorname{sn}(t,k^*)^{\frac{2p-n}{n-1}}dt = 1\ .$$

Combining this limit with (18) and (19) , we arrive at the desired lower limit

$$\liminf_{\substack{n\to\infty\\ p=o(n)}} \frac{p}{n}\operatorname{mod} R_{n,p}(a) \geq 1\quad .$$

Next, we prove that $\limsup (p/n) \bmod R_{n,p}(a) \leq 1$ as n approaches ∞ with $p = o(n)$. For this we appeal to the upper bound

$$(20)\quad \bmod R_{n,p}(a) \leq \left(\frac{2^{p-1}\,\sigma_{n-1}}{\sigma_{p-1}\sigma_{n-p-1}}\right)^{\frac{1}{n-1}} \left(\int_1^b (r^2-1)^{p-1} r^{-p} dr\right)^{\frac{1}{1-n}} \int_0^{\pi/2} (\sin t)^{\frac{1+p-n}{n-1}} dt.$$

derived in [2, §4], where $b = (1+a)/(1-a)$, $0 < a < 1$.

By an argument almost identical to the one used to establish the second limit in (28) of [2] it is easy to see that

$$\lim_{\substack{n\to\infty\\ p=o(n)}} \frac{p}{n}\int_0^{\pi/2} (\sin t)^{\frac{1+p-n}{n-1}} dt = 1 ,$$

while it follows from (17) that the first factor on the right side of (20) has limit 1. Hence

$$\limsup_{\substack{n\to\infty\\ p=o(n)}} \frac{p}{n} \bmod R_{n,p}(a) \leq \limsup_{\substack{n\to\infty\\ p=o(n)}} \left(\int_1^b (r^2-1)^{p-1}\, r^{-p}\, dr\right)^{\frac{1}{1-n}} .$$

By making the change of variable $\tau = \log r$ we may write this last expression as

$$(21)\qquad \left(\int_1^b (r^2-1)^{p-1}\, r^{-p}\, dr\right)^{\frac{1}{1-n}} = \| 2 \sinh \tau \|_{p-1}^{\frac{p-1}{1-n}} ,$$

where $\|\cdot\|_p$ denotes the $L_p[0, \log b]$ norm. If for some sequence n_j the sequence $p(n_j)$ tends to ∞, then $\|2 \sinh \tau\|_{p-1}$ tends to $\|2 \sinh \tau\|_\infty = b - 1/b$, and the expression in (21) tends to 1. If a sequence $p(n_j)$ is bounded, it is clear that the limit is also 1. The upper limit is now proved.

4. Improved bounds for λ_n , $n \geq 3$

Let $R_{G,n}(a)$ denote the n-dimensional Grötzsch ring, that is, the ring whose complementary components are

$$C_0 = \{(x_1,\dots,x_n): 0 \leq x_1 \leq a,\ x_j = 0,\ 2 \leq j \leq n\}$$

and

$$C_1 = \{(x_1,\dots,x_n): \sum_{j=1}^{n} x_j^2 \geq 1\} .$$

In [10] Gehring proved that $\operatorname{mod} R_{G,3}(a) + \log a$ is monotone decreasing in the interval $0 < a < 1$ and he obtained bounds $4 \leq \lambda_3 \leq 12.4\ldots$, where $\log \lambda_3 = \lim_{a\to 0}(\operatorname{mod} R_{G,3}(a) + \log a)$. Using analogous methods in higher dimensions, Caraman [7] and Ikoma [11] have shown that the limit

$$\log \lambda_n = \lim_{a\to 0}(\operatorname{mod} R_{G,n}(a) + \log a)$$

exists for each $n \geq 3$.

At the same time these authors provided bounds for λ_n , although the lower bound $\lambda_n \geq 4$ was very imprecise. It is now known [1], in fact, that $\lim_{n\to\infty} \lambda_n^{1/n} = e$. The lower bound for λ_3 was later improved by Gehring [9] to $\lambda_3 \geq 8$. In [2] a set of lower bounds for λ_n was obtained which specialize, in the cases $n = 3$ and 4 , to $\lambda_3 \geq 9.1942\ldots$ and $\lambda_4 \geq 21.685\ldots$.

The previously known upper bounds for λ_n are apparently also imprecise. For $n = 3$ and 4 these are, respectively, $\lambda_3 \leq 12.4\ldots$ [10] and $\lambda_4 \leq 35.5\ldots$ ([6], [11]) . In this section we provide a set of improved upper bounds for λ_n . In the next section we show that when $n = 3$ and 4 , these bounds reduce to $\lambda_3 \leq 9.9002...$ and $\lambda_4 \leq 26.046...$.

Theorem 3. For each $n \geq 3$,

$$\text{(22)} \qquad \log \frac{\lambda_n}{4} \leq \frac{2\,\sigma_{n-2}}{\sigma_{n-1}} \int_0^{\pi/2} \cos^{n-2} u \int_0^{\infty} [(1+\cos^2 u \operatorname{csch}^2 v)^{\frac{n-2}{2n-2}} - 1]\,dv\,du$$

and

$$\text{(23)} \qquad \log \frac{\lambda_n}{4} \geq \int_0^{\infty} [\left(\frac{2\,\sigma_{n-2}}{\sigma_{n-1}} \int_0^{\pi/2} (\sec^2 u + \operatorname{csch}^2 v)^{\frac{2-n}{2}}\,du \right)^{\frac{1}{1-n}} - 1]\,dv \quad .$$

Proof. The lower bound (23) was obtained in [2] . For the upper bound (22) , fix $n \geq 3$. Following Gehring [10] (cf. [6], [11]) we introduce the ring $R_E = R_E(b)$ whose complementary components are the segment

$$C_0 = \{(x_1,\ldots,x_n) : |x_1| \leq 1,\ x_j = 0,\ 2 \leq j \leq n\}$$

and

$$C_1 = \{(x_1,\ldots,x_n) : \frac{x_1^2}{\cosh^2 b} + \frac{1}{\sinh^2 b} \sum_{j=2}^{n} x_j^2 \geq 1\} ,$$

the complement of an open "prolate" ellipsoid in R^n . We remark that the parameter b above corresponds to $\sinh^{-1} a$ in the work of Gehring cited. Next, for $\sinh b > 4$, let R' and R'' be the rings bounded by the segment C_0 and by the spheres

$$S' = \{(x_1,\ldots,x_n) : (x_1+1)^2 + \sum_{j=2}^{n} x_j^2 = (\sinh b - 2)^2\}$$

and

$$S'' = \{(x_1,\ldots,x_n) : (x_1+1)^2 + \sum_{j=2}^{n} x_j^2 = (\sinh b + 2)^2\},$$

respectively. Then R' separates the boundary components of R_E, while R_E separates those of R''. Hence

$$\text{mod } R' \leq \text{mod } R_E \leq \text{mod } R'' ;$$

moreover, it is easy to see (cf. [6], [11]) that

$$\log \frac{\lambda_n}{2} = \lim_{b\to\infty} (\text{mod } R_E(b) - \log \sinh b) \quad . \tag{24}$$

Next, we let $\varphi = \varphi_b$ be the plane conformal mapping $z = \varphi(w) = \sin w$ of the rectangle $Q_b = \{w = u+iv: |u| \leq \pi/2 , 0 < v < b\}$ onto the plane set $D_b = \{z=x+iy: x^2/\cosh^2 b + y^2/\sinh^2 b < 1, y \geq 0\} \setminus \{z=x: |x| \leq 1\}$, with boundary correspondence $\varphi(0) = 0$, $\varphi(\pm\pi/2) = \pm 1$, $\varphi(\pm\pi/2+ib) = \pm \cosh b$, $\varphi(ib) = i \sinh b$.

For each fixed u, $|u| \leq \pi/2$, let $\gamma' = \gamma'(u;)$ be the vertical segment $\{w = u+iv: 0 < v < b\}$ joining the horizontal sides of Q_b, and let γ denote the hyperbolic arc $\varphi(\gamma')$ joining the segment $\{z: |x| \leq 1 , y = 0\}$ to the half ellipse $\{z: x^2/\cosh^2 b + y^2/\sinh^2 b = 1, y \geq 0\}$. If D_b is rotated about the x- (that is, x_1-) axis in R^n (cf. [2, §1.2]), we obtain $R_E(b)$, and the arcs γ generate a family of arcs $\Gamma_b \subset \Gamma$, where Γ denotes the family of all arcs joining the boundary components of R_E. Then $M(\Gamma) \geq M(\Gamma_b)$, and using the techniques of [2, §§3, 4] we may show that

$$M(\Gamma) \geq M(\Gamma_b) \geq 2\, \sigma_{n-2} \left(\int_0^{\pi/2} \int_0^b \left(\frac{|\varphi'|}{y} \right)^{\frac{n-2}{n-1}} dv \right)^{1-n} du \quad .$$

Using (4) we are then led to

$$\operatorname{mod} R_E(b) \leq \left(\frac{\sigma_{n-1}}{2\,\sigma_{n-2}}\right)^{\frac{1}{n-1}} \left[\int_0^{\pi/2} \left(\int_0^b \left(\frac{|\varphi'|}{y}\right)^{\frac{n-2}{n-1}} dv\right)^{1-n} du\right]^{\frac{1}{1-n}} .$$

Now

$$(25) \qquad \frac{|\varphi'|}{y} = (\sec^2 u + \operatorname{csch}^2 v)^{\frac{1}{2}} , \quad \frac{|\varphi'|}{x} = (\csc^2 u - \operatorname{sech}^2 v)^{\frac{1}{2}} ,$$

where the second formula is listed for later reference. Thus

$$\operatorname{mod} R_E(b) \leq \left(\frac{\sigma_{n-1}}{2\,\sigma_{n-2}}\right)^{\frac{1}{n-1}} \|F_b\|^{-1} ,$$

where

$$(26) \qquad F_b(u) = \left(\int_0^b (\sec^2 u + \operatorname{csch}^2 v)^{\frac{n-2}{2n-2}} dv\right)^{-1} \in L_{n-1}[0, \tfrac{\pi}{2}]$$

and where $\|\cdot\|$ denotes the $L_{n-1}[0, \pi/2]$ norm. Since $\lim_{b\to\infty} (b - \log \sinh b) = \log 2$ we now have

$$\lim_{b\to\infty} (\operatorname{mod} R_E(b) - \log \sinh b) \leq \log 2 + \limsup_{b\to\infty} \left[\left(\frac{\sigma_{n-1}}{2\,\sigma_{n-2}}\right)^{\frac{1}{n-1}} \|F_b\|^{-1} - b\right] .$$

Comparing this with (24), we see that

$$(27) \qquad \log \frac{\lambda_n}{4} \leq \limsup_{b\to\infty} \left[\left(\frac{\sigma_{n-1}}{2\,\sigma_{n-2}}\right)^{\frac{1}{n-1}} \|F_b\|^{-1} - b\right] .$$

We now seek to estimate the upper limit in (27). First, for fixed u,

$$(28) \qquad \frac{1}{b}\int_0^b (\sec^2 u + \operatorname{csch}^2 v)^{\frac{n-2}{2n-2}} dv \geq \lim_{b\to\infty} \frac{1}{b}\int_0^b (\sec^2 u + \operatorname{csch}^2 v)^{\frac{n-2}{2n-2}} dv = (\sec u)^{\frac{n-2}{n-1}} ,$$

since the expression on the left is the average of a decreasing function over $[0, b]$. The value of the limit is easily obtained by l'Hospital's Rule. Recalling the definition (26) of F_b , appealing to the Monotone Convergence Theorem, and using (3) , we see that this gives

$$(29) \qquad b\,\|F_b\| \leq \lim_{b\to\infty} b\,\|F_b\| = \left(\frac{\sigma_{n-1}}{2\,\sigma_{n-2}}\right)^{\frac{1}{n-1}} .$$

Thus (27) and (29) yield

$$(30) \qquad \log\frac{\lambda_n}{4} \leq \limsup_{b\to\infty} b\left[1 - b\left(\frac{2\,\sigma_{n-2}}{\sigma_{n-1}}\right)^{\frac{1}{n-1}} \|F_b\|\right] .$$

Next, using the identity $1 - r = (1 - r^{n-1}) / \sum_{j=0}^{n-2} r^j$ with $r = b(2\sigma_{n-2}/\sigma_{n-1})^{1/(n-1)}\|F_b\|$, and taking into account the limit in (29) we may rewrite (30) as

$$(31) \qquad \log\frac{\lambda_n}{4} \leq \frac{1}{n-1} \limsup_{b\to\infty} b\left[1 - b^{n-1}\,\frac{2\,\sigma_{n-2}}{\sigma_{n-1}}\,\|F_b\|^{n-1}\right] .$$

Now we use (3) to write the bracketed expression in (31) as an integral over $[0, \pi/2]$, then factor the integrand, and use the bound in (28) , arriving at

$$(32) \quad \log\frac{\lambda_n}{4} \leq \frac{2\,\sigma_{n-2}}{\sigma_{n-1}} \limsup_{b\to\infty} b\int_0^{\pi/2} \cos^{n-2}u\left[1-\left(\frac{1}{b}\int_0^b (1+\cos^2 u\,\mathrm{csch}^2 v)^{\frac{n-2}{2n-2}}dv\right)^{-1}\right]du.$$

Factoring $\left(\frac{1}{b}\int_0^b (1 + \cos^2 u\ \mathrm{csch}^2 v)^{\frac{n-2}{2n-2}}dv\right)^{-1}$

from the integrand in (32) and employing the bound (28), we are able to replace (32) by (22). This concludes the proof of the theorem.

Remark. It is easy to show that this result gives an improvement over the previous upper bound for λ_n. For replacing $\cos u$ by 1 in the inner integrand of (22) and taking (3) into account gives

$$\log \frac{\lambda_n}{4} \leq \int_0^\infty [(\coth v)^{\frac{n-2}{n-1}} - 1]\, dv \quad .$$

Now, making the change of variable $t = (\coth v)^{1/(n-1)}$, we get

$$\log \frac{\lambda_n}{4} \leq (n-1) \int_1^\infty \frac{t^{2n-4} - t^{n-2}}{t^{2n-2} - 1}\, dt \quad ,$$

which is the upper bound previously known ([6], [11]).

5. Numerical bounds for λ_3 and λ_4

For purposes of numerical evaluation of the upper bound (22) we make the change of variable $t = (1 + \cos^2 u \operatorname{csch}^2 v)^{(2-n)/(2n-2)}$ in the inner integral. Then using Fubini's Theorem to change the order of integration, and making the change of variable $r = \sin u$ in the new inner integral, we are able to rewrite (22) in the following equivalent form:

$$(33) \qquad \log \frac{\lambda_n}{4} \leq \frac{n-1}{n-2} \frac{2\,\sigma_{n-2}}{\sigma_{n-1}} \int_0^1 \int_0^1 \frac{(1-t)\,t^{\frac{n-1}{n-2}}(1-r^2)^{\frac{n-2}{2}}}{t^2\left(1-t^{\frac{2n-2}{n-2}}\right)\left(1-r^2 t^{\frac{2n-2}{n-2}}\right)^{\frac{1}{2}}}\, dr\, dt \ .$$

When n is even, the inner integral in (33) is elementary. If n is odd it is an elliptic integral.

<u>Theorem 4</u>. $9.1942\ldots \leq \lambda_3 \leq 9.9002\ldots$.

<u>Proof</u>. The lower bound was obtained in [2], and follows from (23) above. For the upper bound, we substitute $n = 3$ in (33), getting

$$\log \frac{\lambda_3}{4} \leq 2\int_0^1 \frac{1-t}{1-t^4} \int_0^1 \left(\frac{1-r^2}{1-r^2t^4}\right)^{\frac{1}{2}} dr\, dt \ .$$

Now it is easy to check that

$$\int_0^1 \left(\frac{1-r^2}{1-r^2t^4}\right)^{\frac{1}{2}} dr = t^{-4}[E(t^2) - (1-t^4)K(t^2)] \ ,$$

where E and K are complete elliptic integrals of the second and first kinds, respectively. Thus

(34) $$\log \frac{\lambda_3}{4} \leq I = \int_0^1 \frac{1 - k^{\frac{1}{2}}}{k^{5/2} k'^2} (E - k'^2 K)\, dk ,$$

where we have substituted $k = t^2$, $k'^2 = 1 - t^4$ and written $E - k'^2 K$ for $E(k) - k'^2 K(k)$. Now integrating by parts in (34) , integrating the factor $(E - k'^2 K)/kk'^2$ by means of [4, p. 21] or [5,#710.00] , we have

(35) $$I = \left[\frac{1 - k^{\frac{1}{2}}}{k^{3/2}} K - \frac{1}{2}\int k^{-5/2}(-3 + 2k^{\frac{1}{2}}) K\, dk\right]_0^1 .$$

Employing the reduction formula [12, (4a)]

(36) $$k^{n-1}(nk'^2 K - E) = (n-1)^2 \int k^{n-2} K\, dk - n^2 \int k^n K\, dk$$

with $n = -\frac{1}{2}$ and $n = 0$, we may replace (35) by

$$I = \left[\frac{1 - k^{\frac{1}{2}}}{k^{3/2}} K - \frac{k'^2}{3k^{3/2}} K - \frac{2}{3k^{3/2}} E + \frac{E}{k}\right]_0^1 + \frac{1}{6}\int_0^1 k^{-\frac{1}{2}} K\, dk .$$

The expression in brackets may be written as

(37) $$[3(1 - k^{\frac{1}{2}}) - k'^2](K - E)/(3k^{3/2}) + k^{\frac{1}{2}} E/3 .$$

Referring to the hypergeometric form of K and E we see that $\lim_{k\to 0}(K - E)/k^2 = \pi/4$ ([5,##900.00, 900.07], [16, p. 521]), so that the first term of (37) has limit 0 as k tends to 0 . The second term also has limit 0 because E is bounded. Moreover, from the behavior of $K(k)$ near $k = 1$ ([5,#112.01], [16, p. 521]) and the fact that $E(1) = 1$ ([4, p. 20], [5,#111.05]) we get

$$(38) \qquad I = \frac{1}{3} + \frac{1}{6}\int_0^1 k^{-\frac{1}{2}} K\,dk \quad ,$$

where I is defined by (34). The value of the integral in (38) is tabulated in [12] to 10 decimal places. Using this we have $I = 0.9062654847\ldots$, and combining this value with (34) we arrive at the upper bound in Theorem 4.

We now turn our attention to λ_4.

Theorem 5. $21.685\ldots \leq \lambda_4 \leq 26.046\ldots$.

Proof. The lower bound was obtained in [2] and follows from (23) above. For the upper bound, we use (33) with $n = 4$, which reduces to

$$\log \frac{\lambda_4}{4} \leq \frac{6}{\pi}\int_0^1 \frac{1-t}{t^2(1-t^3)}\int_0^1 \frac{1-r^2}{(t^{-3}-r^2)^{\frac{1}{2}}}\,dr\,dt \quad .$$

The inner integral is elementary; evaluating it and substituting gives

$$(39) \quad \log \frac{\lambda_4}{4} \leq I = \frac{3}{\pi}\int_0^1 \frac{1-t}{t^5(1-t^3)}\left[(2t^3-1)\arcsin t^{3/2} + t^{3/2}(1-t^3)^{\frac{1}{2}}\right]dt \quad .$$

The following method of approximating the upper bound in (39) is due to J. S. Frame in private consultation. First, making the change of variable $t^{3/2} = \sin\theta$ and then using the identity

$$(40) \quad 1 - \frac{\theta\cos 2\theta}{\sin\theta\cos\theta} = (1-\theta\cot\theta) + \theta\tan\theta\sin^2\theta + \theta\cot\theta\sin^2\theta$$

splits I into the sum of three terms $I = I_1 + I_2 + I_3$

corresponding to the three terms of (40) and having exact values

$$I_1 = (1/4\pi)B(1/6,\ 1/2) - 1/8 \quad ,$$
$$I_2 = (1/12)(9 \log 3 - \pi\sqrt{3}) + \log 2 - G(1/6) \quad ,$$
$$I_3 = -3/2 + (3/2\pi)B(1/6,\ 1/2) - \log 2 \quad ,$$

respectively. Here B represents the classical Beta function, and the function G appearing in I_2 is defined by

$$(41) \qquad G(\beta) = \frac{2}{\pi}\int_0^{\pi/2} \cos^{2\beta}\theta \,\frac{\theta}{\sin\theta}\, d\theta \quad , \ \beta > -\frac{1}{2} \ .$$

Integration by parts can be employed in the evaluation of I_1 and I_3, while the substitution $\alpha = \pi/2 - \theta$ splits I_2 into two parts, one of which can be reduced to the integral of a rational function by a second change of variable $t^3 = \cos\alpha$; the other part is the sum of $-G(1/6)$ and an integral whose value is $\log 2$ (cf. [14,#704, p. 467]) .

Finally, we estimate $G(1/6)$. For this, if we multiply the integrand of (41) by $\cos^2\theta + \sin^2\theta$, $G(\beta)$ is split into two parts, one of which is $G(\beta+1)$ and the other reducible to a Beta function by an integration by parts. Thus we are led to the recursion relation

$$(42) \qquad G(\beta) = G(\beta+1) + (\pi(2\beta+1))^{-1}B(\beta+1,\ \tfrac{1}{2}) \ .$$

It is easy to obtain the values $G(\frac{1}{2}) = \log 2$ and $G(0) = 1.1662436167\ldots$ (cf. [14,##704, 652, pp. 467, 464]) . After this we use (42) to determine G at each integer and half integer in [0, 4] , apply Newton's interpolation formula to obtain the value of the function $G(\beta) - (\log 2)(\beta+1/2)^{-2/3}$ to seven places at $\beta = 13/6$, and finally get $G(13/6) = 0.3738644\ldots$. Then applying (42) twice and using the difference equation for the Gamma function gives

$$G(1/6) = G(13/6) + ((3/16 + 21/400)/\pi)\, B(1/6,\ 1/2) \quad .$$

Using the value $B(1/6,\ 1/2) = 7.285913\ldots$ (cf. [8]), collecting the above results and inserting in (39) gives $\log \lambda_4/4 \leq I = 1.873604\ldots$. This implies that $\lambda_4 \leq 26.046\ldots$, which is the upper bound in the theorem.

Remark. In conclusion we state without proof that by using methods analogous to those employed in Theorem 3 above we may establish the following result.

Theorem 6. *As a function of* a, $\operatorname{mod} R_{n,n-1}(a) + \log a$ *is monotone decreasing in the interval* $0 < a < 1$, *and*

$$\lim_{a\to 0} (\operatorname{mod} R_{n,n-1}(a) + \log a) = A_n ,$$

where

$$A_n \leq \log 2 + \frac{2\,\sigma_{n-2}}{\sigma_{n-1}} \int_0^{\pi/2} \sin^{n-2}u \int_0^{\infty} [(1 - \sin^2 u \operatorname{sech}^2 v)^{\frac{n-2}{2n-2}} - 1]\, dv\, du$$

and

$$A_n \geq \log 2 + \int_0^{\infty} [\left(\frac{2\,\sigma_{n-2}}{\sigma_{n-1}} \int_0^{\pi/2} (\csc^2 u - \operatorname{sech}^2 v)^{\frac{2-n}{2}} du\right)^{\frac{1}{1-n}} - 1]\, dv \quad .$$

In particular,

$$A_3 \leq \log 2 - \int_0^1 \frac{k(1-k^2)^{\frac{1}{2}}}{k'^2} E(k)\, dk = 0.4535\ldots ,$$

$$A_3 \geq \log 2 - \int_0^{\infty} \left[\frac{t}{\sinh t} - \left(\frac{2t}{\sinh 2t}\right)^{\frac{1}{2}}\right] \frac{dt}{t} = 0.4129\ldots ,$$

$$A_4 \leq \log 2 - 3\int_0^1 \frac{t^2(1-t)}{1-t^3} (1 - \frac{t^3}{2})\, dt = 0.3455\ldots ,$$

$$A_4 \geq \log 2 - \int_0^1 \frac{1 - (t(1+t)/2)^{\frac{1}{3}}}{1 - t^2}\, dt = 0.2886\ldots \quad .$$

The bounds for A_3 stated above may be compared with the estimates $0.254... \leq A_3 \leq 0.693...$ obtained earlier [3] by other methods.

The principal differences between the proof of the upper bound of Theorem 6 and that of the upper bound in Theorem 3 are that here all rings in the proof are centered at the origin, one rotates about the y-axis instead of the x-axis and uses the second formula in (25) instead of the first, and one uses the Bounded Convergence Theorem at the point where the Monotone Convergence Theorem was invoked previously. The proof of the lower bound involves a family of confocal ellipsoidal surfaces but is otherwise similar to the proof of Theorem 5 of [2,§5].

In the upper bound for A_3 the function E is a complete elliptic integral of the second kind; the bound may be approximated to as many as ten places by means of [12] . The integral in the upper bound for A_4 is elementary, while the lower bound for A_4 may be evaluated exactly by a method similar to that used for (24) of [2] .

The numerical estimate of the lower bound for A_3 was computed by J. S. Frame by the following techniques. The integral over $[0, 2]$ may be computed to six-place accuracy by Simpson's Rule. Over $[2, \infty)$ the first term may be integrated exactly. For the integral over $[2, \infty)$ of the second term, first make the change of variable $t = u^2$ and expand the integrand in powers of $\exp(-u^2)$. For six-place accuracy the integrals of all but the first term are negligible, and the integral of the first term may be expanded in a continued fraction to provide an estimate of its value, correct to six places.

Michigan State University
East Lansing, Michigan USA 48824

References

[1] Anderson, G. D., Dependence on dimension of a constant related to the Grötzsch ring. Proc. Amer. Math. Soc. 56 (1976).

[2] ______________, Extremal rings in n-space for fixed and varying n . Ann. Acad. Sci. Fenn. A. I. 575 (1974), 1-21.

[3] ______________, Symmetrization and extremal rings in space, Ann. Acad. Sci. Fenn. A I 438 (1969), 1-24.

[4] Bowman, F., Introduction to Elliptic Functions with Applications. Dover Publications, New York, 1961.

[5] Byrd, P. F., and M. D. Friedman, Handbook of Elliptic Integrals for Engineers and Physicist. Springer-Verlag, Berlin, 1954.

[6] Caraman, P., n-Dimensional Quasiconformal (QCf) Mappings. Abacus Press, Tunbridge Wells, Kent, England, 1974.

[7] __________, On the equivalence of the definitions of the n-dimensional quasiconformal homeomorphisms (QCfH) . Rev. Roumaine Math. Pures Appl. 12 (1967), 889-943.

[8] Frame, J. S., An approximation to the quotient of Gamma functions. Amer. Math. Monthly 41 (1949), 529-535.

[9] Gehring, F. W., Inequalities for condensers, hyperbolic capacity, and extremal lengths. Mich. Math. J. 18 (1971), 1-20.

[10] ______________, Symmetrization of rings in space. Trans. Amer. Math. Soc. 101 (1961), 499-519.

[11] Ikoma, K., An estimate for the modulus of the Grötzsch ring in n-space. Bull. Yamagata Univ. Natur. Sci. 6 (1967), 395-400.

[12] Kaplan, E. L., Multiple elliptic integrals. J. Math. Phys. 29 (1950), 69-75.

[13] Loewner, C., On the conformal capacity in space. J. Math. Mech. 8 (1959), 411-414.

[14] Selby, S. M., editor, Standard Mathematical Tables, 21st ed. Chemical Rubber Co., Cleveland, 1973.

[15] Väisälä, J., Lectures on n-Dimensional Quasiconformal Mappings. Springer-Verlag, Berlin, 1971.

[16] Whittaker, E. T., and G. N. Watson, Modern Analysis. Cambridge, 1958.

[17] Ziemer, W. P. Extremal length and conformal capacity. Trans. Amer. Math. Soc. 126 (1967), 460-473.

Michigan State University
Department of Mathematics
East Lansing, Michigan 48824
USA

Some problems in quasiconformality

Cabiria Andreian Cazacu

§1.

The first part of this paper contains some problems related to the theory of qc(quasiconformal)mappings.

In 1969 at the first Romanian-Finnish Seminar in Braşov I presented a paper [3] concerning the influence of the orientation of the characteristic ellipses on the properties of the qc mappings in the plane, a subject which preoccupied me for a long time. What about this research direction in the n-dimensional case?

Let $f: G \longrightarrow G^*$ be a sense-preserving homeomorphism of a domain G in $\mathbb{R}^n$ onto the domain G^* in $\mathbb{R}^n$. As usual a regularity point or an A-point of f is a point x in G at which f is differentiable with non-vanishing Jacobian $J_n(x) > o$. If $A(f,x)$ denotes the affine transformation associated to f at x and $B^n(x)$ the n-dimensional unit ball with the center at x, then the image of $B^n(x)$ under $A(f,x)$ will be an n-dimensional ellipsoid $E^n(x^*)$ centered at $x^*=f(x)$. This ellipsoid, as well as the ellipsoid $\mathcal{E}^n(x)$ mapped by f onto the unit ball $B^n(x^*)$, will be called characteristic.

In the plane case, where balls and ellipsoids are reduced to cercles and ellipses respectively, we used instead of the classical ratio $p = \frac{a}{b} \geqslant 1$ of the semi-axes a and b of the ellipse $\mathcal{E}^2(x)$ the function

$$d = \frac{\cos^2\alpha}{p} + p\sin^2\alpha .$$

Here α represents the angle of the great axis of $\mathcal{E}^2(x)$ with a certain direction $\tau(x)$ through x given by a field of directions defined in G and related to the discussed problem. Thus d expresses more adequately than p the action of the mapping f on the direction τ through the point x. Working with d one obtains a more precise solution of different geometric problems,since d can remain bounded or can increase slowly, even when p has not this property [3], [5].

It is easy to generalize the function d to the n-dimensional case by starting from its interpretation connected with the extremal length or the module of curve families, since d intervenes as a weight in a transformation formula of the module of a curve family in G under f([3], p.68, formula (4)).Thus we shall have in the n-dimensional case for each dimension q= =1,2,...,n-1 a function d ([4], p.97), the directions being replaced by q-planes for $q \geq 2$.

Convention 1. For the sake of simplicity we shall generally write q-surface or q-plane, for q=1,2,...,n-1 and understand that for q=1 we deal with a curve or a straight line, a direction.

If x is a regular point of the homeomorphism $f:G \longrightarrow G^*$, $\Pi = \Pi_q$ a q-plane through x and $B^q_\Pi(x) = B^n(x) \cap \Pi$, the affine transformation A(f,x) maps $B^n(x)$ onto the n-ellipsoid $E^n(x^*)$ with the semi-axes $a_1 \geq \dots \geq a_n$ and $B^q_\Pi(x)$ onto the q-ellipsoid $E^q_{\Pi^*}(x^*)$ with the semi-axes $v_1 \geq \dots \geq v_q$, $\Pi^* = A(f,x)(\Pi)$. Let $J_q(x)$ designate the Jacobian of $A(f,x)|_\Pi : \Pi \to \Pi^*$. Then for every $m > 0$ we define

$$(1) \qquad d_{f,\Pi,m}(x) = \frac{J_q(x)^{\frac{m}{q}}}{J_n(x)} = \frac{(v_1 \dots v_q)^{\frac{m}{q}}}{a_1 \dots a_n},$$

the most important case being m=n.

Convention 2. In what follows we work with the Lebesgue measure and integral, and the classical module.We convene to consider for each q-surface family Σ the subfamily Σ_0 of all the q-surfaces of Σ for which the Lebesgue measure and integral are defined, and we put for the module of Σ

$$M(\Sigma) = \begin{cases} M(\Sigma_o) & \text{if } \Sigma_o \neq \emptyset \quad \text{and} \\ 0 & \text{if } \Sigma_o = \emptyset . \end{cases}$$

Similarly for the module of order m in Fuglede's sense and with a weight π in Ohtsuka's sense, which will be designated by $M_{\pi, m}(\Sigma)$ ([4] ,p.88). M corresponds to $\pi = 1$ and m=n. By an index zero we always denote the regular subfamily mentioned. Further $d\sigma_q$ will designate the element under the corresponding integral sign.

Suppose that f is differentiable with non-vanishing Jacobian n-a.e. in G and that a field of q-planes $\Pi(x)$ is given n-a.e. in G such that $d = d_{f,\Pi,m}$ be defined n-a.e. in G finite, positive and measurable. Let us suppose, for instance, that Σ is a sufficiently regular family of q-surfaces such that through n-a.e. point x of G passes a unique q-surface $S \in \Sigma$ with a well defined tangent q-plane Π at x. In this case we shall also write $d_{f,S,m}$ instead of $d_{f,\Pi,m}$, and whenever one of the indices f, S, Π or m is well known (especially when m=n) we shall drop it.

Let further π^* be a function defined n-a.e. in G^* finite, positive and measurable, and $\pi = \pi^* \circ f$.

Under sufficient regularity conditions on f and on the family of q-surfaces Σ in G ([4] , Theorems 2 and 3, Consequence 1) we established the following transformation formula for the module of the families Σ and $\Sigma^* = f(\Sigma)$:

$$(2) \qquad M_{\pi^*, m}(\Sigma^*) = M_{\pi d^{-1}, m}(\Sigma),$$

where π^* and πd^{-1} are Ohtsuka's weights and m is Fuglede's order.

This formula is an example of the importance of the oriented q-dimensional dilatation d. Evidently, at each regularity point x of f one can introduce the outer and inner q-dilatations of f at x with respect to m

$$(3)\quad H_{O,q,m}(f'(x)) = \frac{(a_1\cdots a_q)^{\frac{m}{q}}}{a_1\cdots a_n} \quad \text{and} \qquad H_{I,q,m}(f'(x)) = \frac{a_1\cdots a_n}{(a_{n-q+1}\cdots a_n)^{\frac{m}{q}}},$$

where f'(x) denotes the derivative of f at x, and write

$$(4)\quad H^{-1}_{I,q,m}(f'(x)) \leq d_{f,\Pi,m}(x) \leq H_{O,q,m}(f'(x)).$$

The dilatations (3) have been considered by several authors under various notations (even in my papers where I used Agard's notation [4], then a notation inspired by [15]).

After the monotony principle of the module with respect to the weight [4] one can replace d in (2) by means of (4) and one obtains instead of (2) two inequalities. But working with the dilatations (3) one loses exactly the advantage of considering d which must precise the results since it expresses the dilatation on some q-plane, which is specific for the discussed problem.

Further if f and Σ are sufficiently regular ([4], §2) and if d satisfies n-a.e. in G the inequality

$$(5)\quad d \leq Q \quad \text{or} \quad (5')\quad d \geq \frac{1}{Q}$$

for a positive constant Q, one obtains from (2) as well as from [4] (Theorem 4, Consequence 3) the inequality of Grötzsch's type.

$$(6)\quad M_{\Pi,m}(\Sigma) \leq Q M_{\Pi^*,m}(\Sigma^*) \quad \text{or} \quad (6')\, M_{\Pi^*,m}(\Sigma^*) \leq Q\, M_{\Pi,m}(\Sigma)$$

respectively.

In particular, if f is a K-qc mapping after Väisälä's definitions ([15], 13.1, 34.6) at each regularity point of f (i.e. n-a.e. in G), for each q-plane Π and for m=n the inequalities (5) and (5') hold with Q=K and, for each sufficiently regular family Σ as above, the inequalities (6) and (6') are also true with Q=K, [4].

Let us mention that Grötzsch's inequalities for sufficiently regular q-surface families (q > 1) have been proved by several authors, beginning with the case of smooth surfaces and C^1 - mappings with non-vanishing Jacobian [14], and then for different classes of admissible surfaces ([7] and [8]), especially qc surfaces ([11], [13], [2]).

These considerations point out some problems:

I. One could study the homeomorphisms which satisfy, besides regularity conditions as ACL_n, or ACL and differentiability n-a.e., or conditions as in [4], the inequalities (5)-(5') or one of them n-a.e. in G but only with respect to a certain field of q-planes.

Starting this time from the geometric definition of the qcty instead of the analytic one, we could also look for the consequences of the inequalities (6)-(6'), supposing they are verified only for certain sufficiently regular families Σ which correspond, for instance, to a field of q-planes.

In this way one could generalize to the n-dimensional case the classes of homeomorphisms $\mathcal{G}$ or $\mathcal{G}'$ and $\mathcal{G}^*$ [5].

II. Further one could examine the case when the inequalities mentioned above are fulfilled for several q-plane fields or even for all of them. Does this last case imply the qcty for m=n?

The answer for the inequalities (5)-(5') can be sometimes simple. For instance, if f is ACL in G, differentiable with non-vanishing Jacobian n-a.e. in G, and satisfies (5) with m=n for all q-planes n-a.e. in G, then we have $H_{O,q} \leq Q$ n-a.e. in G, therefore $H \leq Q^{\max\left(\frac{q}{n-q},1\right)}$ n-a.e. in G [2], and, since H_O (and H_I) $\leq H^{n-1}$ in every A-point of f, this homeomorphism is

$Q^{\max\left(\frac{q(n-1)}{n-q},\, n-1\right)}$ - qc after Väisälä's analytic definition.

On the contrary, the characterization of the qcty by means of Grötzsch's inequalities (6)-(6') for q=2,...,n-1 arised difficulties as soon as one passed beyond the frame of the diffeomorphisms, when the problem was solved affirmatively. These difficulties are due to the irregular behaviour of a qc mapping even on some smooth q-surfaces so that the classical integral of Lebesgue does not permit to define satisfactory the module of an arbitrary q-surface family. Several mathematicians dealt with this problem ([7] and [8] ,[13] ,[2] and others). Agard [2] especially made a profound analysis of the different aspects of it. He developed the idea of using the module defined by means of the Hausdorff measure or of the measure derived from the Lebesgue area but also the idea of considering only particular families of q-surfaces, such as the test-families. Choosing a class of "admissible" families for which (6) and (6') hold if f is qc and which permit the calculation of the module of cylinders and of rings, P.Caraman solved also affirmatively the problem in the case q=n-1, [8].

III. In connection with the above problem, the analysis of the different methods to demonstrate analytic properties of the mappings, for instance the differentiability a.e., points out another question : how to obtain for the dimension q=2,...,n-2 a concept as adequate as that of the ring. Namely the boundary of a ring A in $\mathbb{R}^n$ has two components B_0 and B_1. One can consider the family of curves $\Gamma = \Delta(A, B_0, B_1)$ which join B_0 and B_1 in A as well as the family Σ of (n-1)-surfaces which separate B_0 and B_1 in A, and define the module of the ring by means of each of these families .For this module one obtains estimations which are fundamental in the proofs ([15],[9]). The geometrical intuition seems to show that the problem is how to replace the verbs "join" and "separate " for q=2,..,n-2. Much simpler is the case of the cylinder, where one has again families joining or separating the (n-1)- bases and defining the module.This case is related to that of an n-segment which can be regarded as a direct product of a q-and an (n-q)-segment and to which we associate in a natural way the module of a q-surface family as well as the module of an (n-q)-surface family.This fact will be considered in §2,where we deal with affine images which we call n-parallelotopes.

§2.

In this second part of our paper we give some results relative to the problems I-III in §1. Our aim is to generalize (in a double sense from $\mathbb{R}^2$ to $\mathbb{R}^n$ and from modules of curve families to modules of q-surface families) our paper [6] .More precisely we establish connections between Grötzsch's and Rengel's inequalities and starting from this we deduce results concerning the verification of Grötzsch's inequality for one q-plane field.By this way one goes deeper in the definition of the qcty and one simplifies it requiring the modular inequality which enters in this definition to be satisfied only for certain families of q-surfaces. Such aspects of the theory were studied in the 2-dimensional case [1o] , [6] as well as in the n-dimensional one [17] .

1. Modules of a topological cube.

In what follows the main role will be played by the topological cubes, i.e.topological images of the unit n-dimensional cube in $\mathbb{R}^n$, which can be also conceived as a natural extension of the quadrilaterals and for which we want to define q-modules.

The unit n-cube and in general every n-segment or rectangular n-parallelotope I_n may be regarded for each $q=1,\dots,n-1$ in $\binom{n}{q}$ ways as the direct product of a q-segment I_q by an (n-q)-segment I_{n-q} orthogonal to I_q. Let us choose one of these possibilities and consider the corresponding I_q as a "base ", I_{n-q} playing the role of a "height ". We call I_{n-q} associated face with respect to the base I_q and in order to avoid any confusion we shall denote it, if necessary, $I_{n-q,\perp}$.

Let $\Sigma_{n-q} = \Sigma(I_n,I_q)$ be the family of all the (n-q)-segments in I_n parallel to I_{n-q}. They "join" in I_n the 2^{n-q} q-faces paralles to I_q and are in the same time "orthogonal sections " in I_n with respect to I_q. Further let $\mathcal{S}_{n-q} = \mathcal{S}(I_n,I_q)$ be the family of all (n-q)-surfaces "joining" in I_n the q-faces parallel to I_q and having in I_n the same position as the (n-q)-segments of $\Sigma(I_n,I_q)$. With this notation $\Sigma_q = \Sigma(I_n,I_{n-q})$ will consist of q-segments parallel to I_q and "separating" I_q from the other q-faces parallel to it in the sense that every (n-q)-surface which "joins " in I_n all these faces (i.e. belongs to $\mathcal{S}(I_n,I_q)$) must intersect every q-segment in $\Sigma(I_n,I_{n-q})$. The same remark for $\mathcal{S}_q = \mathcal{S}(I_n,I_{n-q})$.

Convention 3. We suppose, by a suitable choice of coordinates, that $I_n = \{x=(x_1,\dots,x_n) \in \mathbb{R}^n \mid 0 \le x_i \le \ell_i,\ i=1,\dots,n\}$. Writing $\xi = (x_1,\dots,x_q)$ and $\eta = (x_{q+1},\dots,x_n)$ we identify $I_q =$

$= \{\xi \in \mathbb{R}^q \mid 0 \le x_i \le l_i,\ i=1,\dots,q\}$ with the face $\{x \in \mathbb{R}^n \mid x= =(\xi,0),\ \xi \in I_q\}$ of I_n and $I_{n-q} = \{\eta \in \mathbb{R}^{n-q} \mid 0 \le x_i \le l_i,\ i=q+1,\dots,n\}$ with the face $\{x \in \mathbb{R}^n \mid x=(0,\eta),\ \eta \in I_{n-q}\}$. Sometimes it will be useful to have a notation for both the families Σ_q and Σ_{n-q} in I_n and we put $\Sigma = \{S_\xi\}_{\xi \in I_q}$ with $S_\xi = \{x \in I_n \mid x= =(\xi,\eta),\ \eta \in I_{n-q}\}$ for Σ_{n-q} and $\Gamma = \{C_\eta\}_{\eta \in I_{n-q}}$ with $C_\eta = =\{x \in I_n \mid x=(\xi,\eta),\ \xi \in I_q\}$ for Σ_q.

By means of the family $\Sigma(I_n,I_q)$ we shall define the <u>module of I_n with respect to the base I_q</u>, denoting it by $M(I_n,I_q)$ and briefly by M_{n-q} or even only by M. Namely we put

$$(1.1)\quad M=M\left[\Sigma(I_n,I_q)\right] \text{ hence } M = \frac{v_n}{h_{n-q}^{n/(n-q)}} = \frac{a_q}{h_{n-q}^{q/(n-q)}},$$

where v_n is the volume of I_n, a_q the q-area of I_q and h_{n-q} the (n-q)-area of I_{n-q}, $v_n = a_q h_{n-q}$.

With the Convention 2, §1, on the module of surface families we also have

$$(1.1')\qquad M = M\left[\mathcal{S}(I_n,I_q)\right].$$

Formula (1.1) was used by several authors. It also holds, [2], if one considers instead of $I_n = I_q \times I_{n-q}$ a direct product $D_n = D_q \times D_{n-q} \subset \mathbb{R}^n$, where $D_q \subset \mathbb{R}^q$ and $D_{n-q} \subset \mathbb{R}^{n-q}$ are domains of q-area a_q and (n-q)-area h_{n-q} respectively, $\Sigma(D_n,D_q) = =\left\{\{\xi\} \times D_{n-q}\right\}_{\xi \in D_q}$, or more general configurations fulfilled by the test-families of Agard. In what follows formula (1.1) will be applied for the case of an n-parallelotope $P_n = P_q \times P_{n-q}$, where P_q is a q-parallelotope lying in a q-plane Π_q and P_{n-q} an (n-q)-

parallelotope lying in the (n-q)-plane Π_{n-q} orthogonal to Π_q. We shall call such an n-parallelotope quasirectangular.

Now let us take a topological n-cube $\mathcal{P}_n = \varphi(I_n)$, where φ is a homeomorphism $I_n \to \mathcal{P}_n$, which defines also the vertices and the faces of $\mathcal{P}_n$. We associate to $\mathcal{P}_n$ for each "base" $\mathcal{P}_q = \varphi(I_q)$ the family $\mathcal{S}(\mathcal{P}_n, \mathcal{P}_q) = \varphi[\mathcal{S}(I_n, I_q)]$ and the module

$$(1.2) \quad M(\mathcal{P}_n, \mathcal{P}_q) = M\left[\mathcal{S}(\mathcal{P}_n, \mathcal{P}_q)\right].$$

Again $\mathcal{P}_{n-q} = \varphi(I_{n-q})$ will be called an associated face of $\mathcal{P}_n$ with respect to the base $\mathcal{P}_q$.

However this definition, based on Convention 2, presents obviously difficulties as soon as we want to obtain for a K-qc mapping $f: \mathcal{P}_n \to \mathcal{P}_n^*$ inequalities of Grötzsch's type.

Let us remark that there are two kinds of Grötzsch's inequalities: one relative to the modules of q-surface families as (6) and (6') and another one relative to the modules of certain configurations as rings, cylinders. With the definition (1.2) it is not possible to obtain inequalities for $M(\mathcal{P}_n, \mathcal{P}_q)$ and $M(\mathcal{P}_n^*, \mathcal{P}_q^*)$, since the relation between $f\left[\mathcal{S}(\mathcal{P}_n, \mathcal{P}_q)_o\right]$ and $\mathcal{S}_o^* = \mathcal{S}(\mathcal{P}_n^*, \mathcal{P}_q^*)_o$ remains unknown.

If $\mathcal{P}_n$ has a regular extremal subfamily $\mathcal{E}(\mathcal{P}_n, \mathcal{P}_q)$ of $\mathcal{S}(\mathcal{P}_n, \mathcal{P}_q)$ such that

$$(1.3) \quad M(\mathcal{P}_n, \mathcal{P}_q) = M\left[\mathcal{E}(\mathcal{P}_n, \mathcal{P}_q)\right],$$

then we can write the inequality (6) for $\mathcal{E}(\mathcal{P}_n, \mathcal{P}_q)$ and $f\left[\mathcal{E}(\mathcal{P}_n, \mathcal{P}_q)\right]$ and, since

$$M(\mathcal{P}_n^*, \mathcal{P}_q^*) \geq M\left[f(\mathcal{E}(\mathcal{P}_n, \mathcal{P}_q))\right],$$

we deduce

(1.4) $M(\mathcal{P}_n, \mathcal{P}_q) \leq K\ M(\mathcal{P}_n^*, \mathcal{P}_q^*)$.

Similarly, if $\mathcal{P}_n^*$ has a regular extremal family for its module with respect to $\mathcal{P}_q^*$ one obtains the other inequality

(1.4') $M(\mathcal{P}_n^*, \mathcal{P}_q^*) \leq K\ M(\mathcal{P}_n, \mathcal{P}_q)$.

An important case when (1.4) applies is that of a quasiconformal n-cube $\mathcal{P}_n = \varphi(I_n)$, where φ is a Q-qc mapping. Then we have

(1.5) $M(I_n, I_q) \leq Q\ M(\mathcal{P}_n, \mathcal{P}_q)$

and, evidently, by writing $\Sigma(\mathcal{P}_n, \mathcal{P}_q) = \varphi[\Sigma(I_n, I_q)]$,

(1.6) $Q^{-1}M(I_n, I_q) \leq M[\Sigma(\mathcal{P}_n, \mathcal{P}_q)] \leq Q\ M(I_n, I_q)$.

2. Rengel's inequalities.

One knows that if P_2 is a parallelogram with the sides of length a and A, the corresponding heights h and H respectively, and with the module $M(P_2)$ with respect to the base of length a, Rengel's inequalities have the form

$$\frac{H}{A} \leq M(P_2) \leq \frac{a}{h}.$$

The generalization to the n-dimensional case, especially of the inequality in the right side was made by different authors, beginning for instance with Zimmermann [18]. Väisälä wrote it for cylinders in the fundamental paper [16] and for the general case in his book [15]. Gehring gave both Rengel's inequalities for rings in [9]. In the case of the module M of a family of q-surfaces with an area $\geq a_q$ and which are contained in a domain with the volume v_n the right side inequality was used, for instance, by C.Stan (in a communication in 1974) under the form

$$M \leq \frac{v_n}{a_q^{n/q}}.$$

In the following we shall consider an n-parallelotope P_n, choose a base P_q and designate by O_{n-q} the orthogonal section in P_n with respect to P_q, which we define by intersecting with an (n-q)-plane orthogonal to P_q the infinite n-parallelotope obtained from P_n when we replace the faces parallel to P_q by the q-planes containing them. Let v_n denote the volume of P_n, a_q the q-area of P_q and h_{n-q} the (n-q)-area of O_{n-q}, $v_n=a_q h_{n-q}$. Then the right side of Rengel's inequality for the module $M=M(P_n,P_q)$ will be

$$(2.1) \qquad M \leq \frac{v_n}{h_{n-q}^{n/(n-q)}} = \frac{a_q}{h_{n-q}^{q/(n-q)}} .$$

By a standard method of module theory we prove also the other Rengel's inequality

$$(2.2) \qquad \frac{H_q}{A_{n-q}^{q/(n-q)}} = \frac{v_n}{A_{n-q}^{n/(n-q)}} \leq M ,$$

where, if P_n is obtained from I_n by the affine transformation T and $P_q = T(I_q)$, then A_{n-q} is the (n-q)-area of $P_{n-q}=T(I_{n-q})$ and H_q the q-area of the orthogonal section O_q in P_n with respect to P_{n-q}.

Indeed, for each function ρ, which is admissible for $M[\mathcal{S}(P_n, P_q)]$, and each $\xi \in I_q$, one has with the notation in Convention 3

$$\int_{T(S_\xi)} \rho^{n-q}\, d\sigma_{n-q} \geq 1$$

and by Hölder's inequality

$$\int_{P_n} \rho^n d\sigma_n = \int_{O_q} \left(\int_{T(S_\xi)} \rho^n d\sigma_{n-q} \right) d\sigma_q \geq \frac{H_q}{A_{n-q}^{q/(n-q)}} .$$

The equality occurs in (2.1) or (2.2) if P_n is quasi-rectangular.

3. <u>Connections with the oriented q-dilatation d.</u>

In order to investigate the behaviour of a homeomorphism f at a regular point x_0 we shall consider the associated affine transformation $T=A(f,x_0)$.

Let Π_q be a q-plane through x_0 and Π_{n-q} the (n-q)-plane through x_0 orthogonal to Π_q.

Further we shall denote by Ω_n a rectangular n-parallelotope obtained from I_n by a translation $0 \mapsto x_0$ and a rotation such that I_q becomes the base Ω_q of Ω_n situated in Π_q and I_{n-q} becomes Ω_{n-q} in Π_{n-q}, and write v_n, a_q and h_{n-q} for the volume of Ω_n, the area of Ω_q and that of Ω_{n-q} respectively.

Applying T we obtain $\Omega_j^*=T(\Omega_j)$, j=n,q and n-q. Let us denote by v_n^*, a_q^*, A_{n-q}^*, h_{n-q}^*, H_q^* the volume of Ω_n^*, the area of Ω_q^*, of Ω_{n-q}^*, and the areas of the orthogonal sections in Ω_n^* with respect to Ω_q^* and Ω_{n-q}^* respectively.

Then the Jacobians at the point x_0 of the functions f, $f|_{\Pi_q}$ and $f|_{\Pi_{n-q}}$ are

$$(3.1)\qquad J_n=\frac{v_n^*}{v_n}, \qquad J_q=\frac{a_q^*}{a_q} \quad\text{and}\quad J_{n-q}=\frac{A_{n-q}^*}{A_{n-q}},$$

so that the q- and the (n-q)-dilatation of f at x_0 are given by

$$(3.2)\qquad d_{f,\Pi_q}=\frac{J_q^{n/q}}{J_n} \quad\text{and}\quad d_{f,\Pi_{n-q}}=\frac{J_{n-q}^{n/(n-q)}}{J_n}$$

respectively.

If $M=M(\Omega_n, \Omega_q)$ and $M^*=M(\Omega_n^*, \Omega_q^*)$, then Rengel's in -

equalities (2.1) and (2.2) together with (1.1) imply

$$(3.3) \qquad d_{f,\Pi_{n-q}}^{-1} \leq \frac{M^*}{M} \leq d_{f,\Pi_q}^{q/(n-q)} .$$

Let us convene to write $h_{n-q} \to o$ when the length of each side of I_{n-q} tends to zero, and analogously $H_q \to o$, (which is for our problem equivalent to $h_{n-q} \to \infty$ together with the length of each side of I_{n-q}). Then

$$(3.4) \qquad \lim_{h_{n-q} \to o} \frac{M^*}{M} = d_{f,\Pi_q}^{q/(n-q)}$$

and

$$(3.5) \qquad \lim_{H_q \to o} \frac{M^*}{M} = d_{f,\Pi_{n-q}}^{-1} .$$

In order to prove (3.4) ((3.5) respectively) it is sufficient to suppose h_{n-q} (H_q) small enough and consider the quasi - rectangular parallelotope $\underset{\sim}{\Omega}_n$ ($\widetilde{\Omega}_n$) described by the orthogonal sections in Ω^*_n with respect to Ω^*_q(Ω^*_{n-q}) which intersect all the q-((n-q)-) faces of Ω^*_n parallel to Ω^*_q (Ω^*_{n-q}). The area $\underset{\sim}{b}_q$ ($\widetilde{b}_{n-q}$) of the parallelotope $\underset{\sim}{\Omega}_q = \underset{\sim}{\Omega}_n \cap \Omega^*_q$ ($\widetilde{\Omega}_{n-q} = \widetilde{\Omega}_n \cap \Omega^*_{n-q}$) tends to a^*_q (A^*_{n-q}) when $h_{n-q}(H_q) \to o$. The module of $\underset{\sim}{\Omega}_n$ with respect to $\underset{\sim}{\Omega}_q$, $\underset{\sim}{M} = \dfrac{\underset{\sim}{b}_q}{h^{*\,q/(n-q)}_{n-q}}$, verifies the inequality $\underset{\sim}{M} \leq M^*$.(Let $\widetilde{\Omega}_q$ be one of the q-faces of $\widetilde{\Omega}_n$, which is an orthogonal section in Ω^*_n with respect to Ω^*_{n-q}. The module of $\widetilde{\Omega}_n$ with respect to $\widetilde{\Omega}_q$, $\widetilde{M} = \dfrac{H^*_q}{\widetilde{b}^{\,q/(n-q)}_{n-q}}$, verifies the inequality $M^* \leq \widetilde{M}$). From here the assertion follows by passing to the limit.

Remark 1 . Let us now consider Ω_n with the base Ω_{n-q} in the (n-q)-plane Π_{n-q} and put $M_q = M(\Omega_n, \Omega_{n-q})$ and $M_{n-q} = M(\Omega_n, \Omega_q)$. Then $M_q^q \; M_{n-q}^{n-q} = 1$ and Rengel's inequalities imply for $M_q^* = M(\Omega_n^*, \Omega_{n-q}^*)$ and $M_{n-q}^* = M(\Omega_n^*, \Omega_q^*)$

$$(3.6) \qquad \Delta^{-n} \leqslant (M_{n-q}^*)^{n-q} (M_q^*)^q \leqslant \Delta^n$$

with $\Delta = (d_{f,\Pi_q}^{q} \; d_{f,\Pi_{n-q}}^{n-q})^{1/n} = \dfrac{J_q \; J_{n-q}}{J_n}$.

Remark 2. All the modular relations and inequalities given above for the usual module in $\mathbb{R}^n (m=n)$, can be written for the module of Fuglede's order $m > 0$, which we shall denote here(to avoid confusion and since $\pi = 1$) by $M_{1,m}$.

Thus instead of (1.1),(2.1) and (2.2), and (3.3) we have

$$M_{1,m}(I_n, I_q) = \frac{v_n}{h_{n-q}^{m/(n-q)}} = \frac{a_q}{h_{n-q}^{\mu/(n-q)}}, \quad \mu = m-n+q ,$$

$$\frac{v_n}{A_{n-q}^{m/(n-q)}} \leqslant M_{1,m}(P_n, P_q) \leqslant \frac{v_n}{h_{n-q}^{m/(n-q)}},$$

and for Ω_n and Ω_n^* by putting $m' = qm/\mu$

$$d^{-1}_{f,\Pi_{n-q},m} \leqslant \frac{M_{1,m}^*}{M_{1,m}} \leqslant \left(d_{f,\Pi_q,m'}\right)^{\mu/(n-q)}.$$

Analogous relations to (3.4) and (3.5) hold too and,since $(M_{1,m}(\Omega_n, \Omega_q))^{n-q} (M_{1,m'}(\Omega_n, \Omega_{n-q}))^{\mu} = 1$, the inequalities (3.6) can be directly extended.

The results in the next nr.4 can be also formulated for every order m.

4. Connections between Grötzsch's and Rengel's inequalities.

Let us consider again a homeomorphism $f:G \longrightarrow G^*$, a point $x_o \in G$, a q-plane Π_q through x_o and an arbitrary n-parallelotope Ω_n obtained (as in nr.3) from a segment $I_n = I_q \times I_{n-q}$ by a translation and a rotation such that $x_o \mapsto o$ and Ω_n should have the base Ω_q in Π_q.

We shall say that the Grötzsch inequality is verified by f at x_o for the q-plane Π_q, if for every Ω_n the modules $M = M(\Omega_n, \Omega_q)$ and $M^* = M(\Omega^*_n, \Omega^*_q)$, $\Omega^*_n = f(\Omega_n)$ and $\Omega^*_q = f(\Omega_q)$, satisfy the inequality

$$(4.1) \quad M \leq QM^* \quad \text{or} \quad (4.1') \quad M^* \leq QM,$$

where Q is a positive constant.

Lemma 1. Let x_o be an A-point for f and suppose that f verifies at x_o one or both of Grötzsch's inequalities with respect to a q-plane Π_q. Then so does the associated affine mapping $A(f,x_o)$, which we briefly denote by T.

The proof of this Lemma is a direct extension of that of Lemma 1 in [6], which was based on a device of A.Pfluger [12]. Namely, we use the sequences of mappings $\zeta_\nu : x \mapsto x_o + \frac{x-x_o}{\nu}$ and $\mathcal{T}_\nu : x^* \mapsto x^*_o + \nu(x^* - x^*_o)$ ($\nu = 1,2,\ldots$), and the sequence of topological n-cubes $\tilde{\Omega}_{n\nu} = \mathcal{T}_\nu \circ f \circ \zeta_\nu(\Omega_n)$. Let us denote by $\tilde{\Omega}_{q\nu} = \mathcal{T}_\nu \circ f \circ \zeta_\nu(\Omega_q)$, by $\tilde{\Omega}_n = T(\Omega_n)$ and by $\tilde{\Omega}_q = T(\Omega_q)$. Since according to a generalization of Mori's Lemma (formulated below)

$$(4.2) \quad \lim_{\nu \to \infty} M(\tilde{\Omega}_{n\nu}, \tilde{\Omega}_{q\nu}) = M(\tilde{\Omega}_n, \tilde{\Omega}_q),$$

the inequality (4.1) (or(4.1')) implies

$$M(\tilde{\Omega}_n, \tilde{\Omega}_q) \geq Q^{-1} M(\Omega_n, \Omega_q) \quad (\text{or} \leq Q\, M(\Omega_n, \Omega_q) \text{respectively}).$$

Lemma 2. (A generalization of Mori's lemma). Let $\tilde{\Omega}_n$ be an n-parallelotope with the base $\tilde{\Omega}_q$ and the associated face $\tilde{\Omega}_{n-q}$, and $\tilde{\Omega}_{n\nu}$ a sequence of topological n-cubes, which converges to $\tilde{\Omega}_n$ in the sense of Fréchet, i.e. for each ν there exists a homeomorphism $\chi_\nu : \partial\tilde{\Omega}_n \to \partial\tilde{\Omega}_{n\nu}$ which preserves the structure of topological cube and such that if $\chi_\nu(x) = x^\nu$ then $\max_{x \in \partial\tilde{\Omega}_n} |x - x^\nu| \to 0$ when $\nu \to \infty$. Then (4.2) holds for $\tilde{\Omega}_{q\nu} = \chi_\nu(\tilde{\Omega}_q)$.

The proof of this special case of Mori's Lemma is easy to be done directly by means of Ahlfors' module inequality so that we squetch it. For every small enough $\varepsilon > 0$ let us construct two n-parallelotopes Ω_n^i and Ω_n^σ with the same center as $\tilde{\Omega}_n$, with each face parallel to the corresponding face of $\tilde{\Omega}_n$ and at the distance ε from it, and such that the q-((n-q)-) faces of Ω_n^i (Ω_n^σ) parallel to $\tilde{\Omega}_q$ ($\tilde{\Omega}_{n-q}$) be exterior to $\tilde{\Omega}_n$ and the (n-q)-faces of $\tilde{\Omega}_n$ parallel to $\tilde{\Omega}_{n-q}$ (as well as the q-faces of $\tilde{\Omega}_n$ parallel to $\tilde{\Omega}_q$) be exterior to Ω_n^i (Ω_n^σ).

Therefore $\mathcal{S}(\Omega_n^i, \Omega_q^i) \succ \mathcal{S}(\tilde{\Omega}_n, \tilde{\Omega}_q) \succ \mathcal{S}(\Omega_n^\sigma, \Omega_q^\sigma)$,
where $\succ$ is the minoration sign. However if ν is great enough we can replace in these relations $\mathcal{S}(\tilde{\Omega}_n, \tilde{\Omega}_q)$ by $\mathcal{S}(\tilde{\Omega}_{n\nu}, \tilde{\Omega}_{q\nu})$. Since

$$\lim_{\varepsilon \to 0} M(\Omega_n^i, \Omega_q^i) = \lim_{\varepsilon \to 0} M(\Omega_n^\sigma, \Omega_q^\sigma) = M(\tilde{\Omega}_n, \tilde{\Omega}_q),$$

one obtains (4.2).

According to Lemma 1 we can work with T instead of f and combine Rengel and Grötzsch inequalities. Thus from (3.3)-(3.5) we deduce the main result of §2 :

Theorem 1. Suppose that f verifies at the regular point x_o one of the Grötzsch inequalities with respect to Π_q. Then at the point x_o

1°. Grötzsch's inequality (4.1) implies

(4.3) $d_{f,\Pi_{n-q}} \leq Q$ i.e. (4.3') $J^n_{n-q} \leq Q^{n-q} J^{n-q}_n$

and

(4.4) $d_{f,\Pi_q} \geq Q^{-\frac{n-q}{q}}$ i.e. (4.4') $J^n_q \geq Q^{-(n-q)} J^q_n$,

2°. Grötzsch's inequality (4.1') implies

(4.5) $d_{f,\Pi_q} \leq Q^{\frac{n-q}{q}}$ i. e. (4.5') $J^n_q \leq Q^{n-q} J^q_n$

and

(4.6) $d_{f,\Pi_{n-q}} \geq Q^{-1}$ i.e. (4.6') $J^n_{n-q} \geq Q^{-(n-q)} J^{n-q}_n$

Evidently, if f is a non-degenerate affine mapping, then (4.1) is equivalent to (4.3) and even when (4.1) is verified only for a single Ω_n with the base Ω_q in Π_q it implies (4.4); similarly for (4.1').

Remark 3. Theorem 1 constitutes a refinement of Theorem 15.2 in [15] as well as of Theorem 2 in [2] . It yields an upper bound (4.3) for $d_{f,\Pi_{n-q}}$ starting from Grötzsch's inequality (4.1) verified at the A-point x_o only on a single q-plane Π_q through x_o. If one supposes (4.1) verified for all the q-planes through x_o, one deduces at the point x_o the inequality (4.3) for all this q-planes, hence $H_{0,n-q} \leq Q$ and further $H \leq Q^{\max(\frac{n-q}{q},1)}$.

Theorem 1 can be completed by the following Lemma 3, which shows that if f is differentiable at the point x_o then (4.1) implies (4.3') and (4.1') implies (4.5').

Lemma 3. Let x_o be a differentiability point of f where the Jacobian $J_n=o$. If f verifies Grötzsch's inequality (4.1') or (4.1) for the q-plane Π_q through x_o then one has at this point $J_q = o$ for Π_q or $J_{n-q} = o$ for Π_{n-q} respectively.

Both assertions follow as in Lemma 2 from [6] by considering again $\tilde{\Omega}_{n\nu}$ and $\tilde{\Omega}_n$ in Lemma 1. This time however $\tilde{\Omega}_n$ will be degenerate. Suppose that in the case (4.1') we had $J_q \neq o$ (similarly, in the case (4.1), $J_{n-q} \neq o$) and further that the lengths of the sides of Ω_{n-q} (Ω_q) are chosen so small that it is possible to construct as in Lemma 2 for each sufficiently small $\varepsilon > o$ an n-parallelotope Ω_n^i (Ω_n^σ) for which

$$\mathcal{S}(\Omega_n^i, \Omega_q^i) > \mathcal{S}(\tilde{\Omega}_{n\nu}, \tilde{\Omega}_{q\nu}) \quad \left(\mathcal{S}(\tilde{\Omega}_{n\nu}, \tilde{\Omega}_{q\nu}) > \mathcal{S}(\Omega_n^\sigma, \Omega_q^\sigma)\right)$$

if ν is great enough and

$$\lim_{\varepsilon \to o} M(\Omega_n^i, \Omega_q^i) = \infty \qquad (\lim_{\varepsilon \to o} M(\Omega_n^\sigma, \Omega_q^\sigma) = o).$$

It follows

$$\lim_{\nu \to \infty} M(\tilde{\Omega}_{n\nu}, \tilde{\Omega}_{q\nu}) = \infty \quad (= o).$$

But the inequality (4.1) (or(4.1')) implies

$$M(\tilde{\Omega}_{n\nu}, \tilde{\Omega}_{q\nu}) \leq Q\, M(\Omega_n, \Omega_q) \quad (\geq Q^{-1} M(\Omega_n, \Omega_q))$$

and leads thus to a contradiction.

Remark 4. If f is a Q-qc mapping and x_o and A-point then, according to (1.5), f verifies (4.1) for each Π_q, hence (4.3) and (4.4) hold for each Π_{n-q} and Π_q respectively.

Remark 5. An upper bound for d_{f,Π_q} can be deduced from two Grötzsch's inequalities. Namely, with the notations in Remark 1, the inequality (4.1') $M^*_{n-q} \leq QM_{n-q}$ for Ω_n with the base in Π_q implies (4.5) $d_{f,\Pi_q} \leq Q^{(n-q)/q}$, while the inequality (4.1) $M_q \leq Q\, M^*_q$ for Ω_n with the base in Π_{n-q} implies (4.3) $d_{f,\Pi_q} \leq Q$.

Theorem 1 permits to obtain from Grötzsch's inequalities informations about the qcty coefficient. In another paper :"On the Grötzsch and Rengel inequalities" we shall develop this idea.

References

1. Agard, S. : Angles and quasiconformal mappings in space, J.Analyse Math. 22(1969), 177-2oo.

2. " : Quasiconformal mappings and the moduli of p-dimensional surface families, Proceedings of the Romanian-Finnisch Seminar on Teichmüller spaces and quasiconformal mappings, Braşov , Romania 1969, Publishing House of the Academy of RSR (1971), 9-48.

3. Andreian Cazacu, C. : Influence of the orientation of the characteristic ellipses on the properties of the quasiconformal mappings, Proceedings of the Romanian-Finnish Seminar quoted above, 65-85.

4. " : Some formulae on the extremal length in n-dimensional case, Proceedings of the Romanian-Finnish Seminar quoted above, 87-1o2.

5. " : A generalization of the quasiconformality, Topics in Analysis, Colloquium on Mathematical Analysis, Jyväskylä 197o, Lecture Notes in Mathematics 419,

Springer-Verlag,Berlin-Heidelberg-New York (1974) 4-17.

6. " Sur les inégalités de Rengel et la définition géométrique des représentations quasi-conformes, Rev.Roumaine Math.Pures Appl., 9(1964),141-155.

7. Caraman,P.: On the equivalence of the definitions of the n-dimensional quasiconformal homeomorphisms(OCfH), Rev.Roumaine Math.Pures Appl., 12(1967),889-943.

8. " : About the characterization of the quasiconformality (QCfH) by means of the moduli of q-dimensional surface families, Rev.Roumaine Math.Pures Appl., 16(1971), 1329-1348.

9. Gehring,F.W.: Rings and quasiconformal mappings in space, Trans.Amer.Math.Soc., lo3(1962), 353-393.

lo. Gehring,F.W.,and Väisälä,J.:On the geometric definition for quasiconformal mappings, Comment.Math.Helv.,36 (1961),19-32.

11. Копылов, А.П.: Поведение пространственного квазиконформного отображения на плоских сечениях области определения, Д.А.Н. С.С.С.Р., 167(1966), 743-746.

12. Pfluger, A: Über die Äquivalenz der geometrischen und der analytischen Definition quasikonformer Abbildungen,Comment.Math.Helv.33(1959),23-33.

13. Reimann, H.M.: Über das Verhalten von Flächen unter quasikonformen Abbildungen im Raum, Ann.Acad.Sci. Fenn., A I 47o (197o), 1-27.

14. Шабат, Б.В.: Метод модулей в пространстве, Д.А.Н. С.С.С.Р., 130 (1960), 1210-1213.

15. Väisalä,J.: Lectures on n-Dimensional Quasiconformal Mappings, Lecture Notes in Math., 229, Springer-Verlag, Berlin-Heidelberg-New York, 1971.

16. " : On quasiconformal mappings in space, Ann.Acad. Sci.Fenn., A I, 298(1961), 1-36.

17. " : Two new characterizations for quasi-conformality, Ann.Acad.Sci.Fenn., A I 362 (1965),1-12 .

18. Zimmermann E.: Quasikonforme schlichte Abbildung im 3-dimensionalen Raum, Wiss.Z.Martin-Luther-Univ. Halle Wittenberg, Math.Naturw.Reihe 5(1955), 109-116.

University of Bucharest
Faculty of Mathematics

AN APPLICATION OF QUASICONFORMAL MAPPINGS TO TOPOLOGY*

Lipman Bers

The aim of this talk is to explain, with a minimum of technical details, a function-theoretical proof of a beautiful topological theorem by Thurston [7]. The full proof will be found in my paper [4]; I refer to [2] for a statement and a demonstration of Teichmüller's theorem which is the main tool in the proof, and to Matelsky's paper [5] and to [3] for needed information from the theory of Riemann surfaces.

The proof is based on a variational problem, a rather natural extension of the Grötzsch-Teichmüller problem, of which one should have (but did not) thought many years ago.

Let S be an oriented surface of type (p,m), i.e., obtained from a closed surface of genus p by removing m disjoint continua. If σ is a conformal structure on S (an orientation-preserving homeomorphism of S onto a Riemann surface $\sigma(S) = S_\sigma$) and $g : S \to S$ is an orientation preserving homeomorphism, the deviation of $\sigma \circ g \circ \sigma^{-1}$ from conformality is measured by the dilatation

$$K_\sigma(g) = K(\sigma \circ g \circ \sigma^{-1}),$$

with $K_\sigma(g) = 1$ if and only if $\sigma \circ g \circ \sigma^{-1}$ is conformal. We consider the variational problem of minimizing $K_\sigma(g)$ by varying σ over all conformal structures on S, and by varying g over the isotopy class of a <u>given</u> self-mapping $f : S \to S$.

---

* Work partially supported by the National Science Foundation.

We assume that

$$(1) \qquad 2p - 2 + m > 0;$$

this excludes the cases (p,m) = (0,0), (0,1), (0,2), (1,0) which are special and not very interesting.

We also assume that f is <u>essentially</u> <u>non-periodic</u>, i.e., not isotopic to a mapping f_o with $f_o^n = \mathrm{id}$ for some $n > 0$. Indeed, if f is isotopic to a periodic mapping, and only in this case, our variational problem can be solved quite simply: there is a conformal structure σ and a conformal map $f_o : S_\sigma \to S_\sigma$, isotopic to f. The proof of this statement is elementary.

Finally, we agree to consider only such conformal structures σ on S for which $\sigma(S)$ has no ideal boundary curves and therefore m punctures. This condition is vacuous if S is closed (that is, if $m = 0$) and involves no loss of generality if $m > 0$. For one can show, using properties of the Nielsen extension of a Riemann surface [3], that the above condition on σ must be fulfilled for a σ solving the variational problem.

After these preliminaries, let Σ be the set of all conformal structures on S, F the isotopy class of f, and set

$$(2) \qquad \hat{K} = \inf K_\sigma(g), \quad \sigma \in \Sigma, \ g \in F.$$

There is a sequence $\{\sigma_j, g_j\} \subset \Sigma \otimes F$ with

$$(3) \qquad \lim_{j \to \infty} K_{\sigma_j}(g_j) = \hat{K}.$$

Let (σ_j) be the conformal equivalence class of $\sigma_j(S)$. If there is a limit

(4) $$\lim_{j\to\infty} (\sigma_j) = (\sigma_0)$$

(in the topology of the space of moduli of Riemann surfaces of genus p with m punctures), then one can show, using known compactness properties of quasiconformal mappings, that there is an F_0 isotopic to f with $K_{\sigma_0}(f_0) = \hat{K}$. Hence the existence of a solution of our variational problem is assured provided we can conclude that (3) implies that the sequence $\{(\sigma_j)\}$ contains a convergent subsequence.

For every j, let ε_j denote the length of the shortest simple closed geodesic on $\sigma_j(S)$, in the Poincaré metric; this number depends only on (σ_j). A compactness theorem proved by Mumford [6] for closed Riemann surfaces, but valid in all cases (cf. [2,5]), shows that if $\{(\sigma_j)\}$ contains no convergent subsequence, then

(5) $$\lim_{j\to\infty} \varepsilon_j = 0.$$

Let us see what (5) implies.

On a Riemann surface of type (p,m) one can draw not more than $3p - 3 + m$ dijoint simple closed geodesics. On the other hand, there is a number $\eta > 0$, depending only on (p,m) such that any two distinct simple closed geodesics of length at most η are disjoint; this follows from the "collar lemma" (cf. [5] and the references given there). Finally (cf. Wolpert [8]), if C is a simple closed geodesic of length ℓ on any Riemann surface with a Poincaré metric, and g is a quasiconformal self-mapping of the surface, the $g(C)$ can be deformed into a geodesic of length not exceeding $K(g)\ell$.

Assuming (5), let j be a fixed integer so large that

(6) $$\varepsilon_j K_{\sigma_j}(g_j)^{3p - 3 + m} < \eta$$

Let C be a simple closed geodesic on S_{σ_j} of length ε_j, and let C_n be the unique geodesic in the free homotopy class of $g_j^n(C_o)$. Since the length of C_n is at most $\varepsilon_j K_\sigma(g_j)^n$, the curves C_0, C_1, ..., $C_{3p - 3 + m}$ cannot be distinct. We conclude from (6) that there is a number r, $0 < r < 3p - 3 + m$, such that no two of the curves C_o, $g_j(C_o)$, ..., $g_j^{r - 1}(C_o)$ are freely homotopic, but $g_j^r(C_o)$ is freely homotopic to C_o.

Thurston calls a self-mapping F of S reducible if there are $r > 0$ disjoint Jordan curves Γ_1, ..., Γ_r on S such that (i) no Γ_i can be deformed into a point of S, a boundary continuum of S, or a Γ_k with $i \neq k$, and (ii) there is a mapping f_1 isotopic to f such that $f_1(C_1 \cup \ldots \cup C_r) = C_1 \cup \ldots \cup C_r$. Conditions (ii) is equivalent to the requirement: $f(C_i)$ is freely homotopic to $C_{\pi(i)}$, $i = 1, \ldots, r$ where π is a permutation of $1, \ldots, r$. The preceding considerations contain the proof of

Theorem A. _If_ f _is irreducible_, _the variational problem has a solution_.

Recall now that the space of moduli (conformal equivalence classes) of Riemann surfaces of genus p with m punctures can be compactified by adjoining conformal equivalence classes of so-called Riemann surfaces with nodes. This suggests that, when f is reducible, the variational problem has a generalized solution involving a mapping σ of S onto a Riemann surface with $r > 0$ nodes, such that r Jordan curves $\Gamma_1, \ldots, \Gamma_r$ on S are taken by σ onto the nodes of $\sigma(S)$, and the restriction of σ to the complement of $\Gamma_1 \cup \ldots \cup \Gamma_r$ is a homeomorphism. This is indeed so, and the generalized solution can be used to establish

Theorem B. _If_ f _is essentially non-periodic and reducible, the variational problem has no solution._

The actual construction of the generalized solution, however, and the proof of Theorem B, use not the compactification of the moduli space but a characterization of the solution of the variational problem for an irreducible f.

This solution leads to Riemann surface X and a self-mapping $g : X \to X$ with the property: the dilatation of g cannot be decreased by simultaneously (i) changing the conformal structure of X and (ii) deforming g within its isotopy class. We call such a g an _absolutely extremal mapping_.

Of course, the dilatation of g must also be a minimum if only the change (ii) is permitted; an absolutely extremal mapping must be _extremal_ in the sense of Teichmüller. Extremal self-mappings of a Riemann surface of type (p,m), $2p - 2 + m > 0$, are completely characterized by Teichmüller's theorem. If such a mapping g is not conformal, then there are on S two uniquely determined quadratic differentials Φ and Ψ with the following properties:

(α) Φ and Ψ are holomorphic; locally $\Phi = \varphi(z)dz^2$, $\Psi = \psi(z)dz^2$ where z is a local parameter on X, and φ and ψ are holomorphic functions.

(β) Φ and Ψ are integrable and normalized:

$$\iint_S |\Phi| = \iint_S |\Psi| = 1. \tag{6}$$

This requires that the singularities of Φ and Ψ at the punctures of S be at worst simple poles.

(γ) The mapping g takes the horizontal and vertical trajectories of Φ onto those of Ψ. (The horizontal trajectories of Φ are the curves $\Phi > 0$, the vertical trajectories are the curves $\Phi < 0$. Those of Ψ are defined similarly.) This condition implies that the order of Φ at a point $P \in S$ equals that of Ψ at $g(P)$.

(δ) If $\Phi \neq 0$ at $P \in S$ and $z = x + iy$ is a local parameter such that

$$z = 0 \text{ at } P,\ \Phi = dz^2 \text{ near } P, \tag{7}$$

then there exists a local parameters such that

$$\zeta = 0 \text{ at } g(P),\ \Psi = d\zeta^2 \text{ near } g(P), \tag{8}$$

and near P the mapping g can be written as

$$\xi = K^{1/2}x,\ \eta = K^{-1/2}y \text{ where } K = K(g). \tag{9}$$

We call Φ and Ψ the initial and terminal quadratic differentials of g, respectively.

We must now characterize these extremal self-mappings of a Riemann surface which are absolutely extremal. The problem can be attacked rather directly and this attack, which we shall not describe in detail, leads to a satisfying answer.

Theorem C. A self-mapping of a Riemann surface is absolutely extremal if and only if it is either conformal or an extremal mapping whose initial and terminal quadratic differentials conicide.

We obtain Thurston's theorem if we observe that the horizontal trajectories of an integrable meromorphic quadratic differential Φ form what Thurston calls a measured foliation. It is an actual foliation, except at the zeroes and poles of Φ, where there are singularities of a specified nature (an "r + 2 pronged singularity" at a point where the order of Φ is $r \neq 0$), and there is a consistent way to measure the distance between the leaves (namely, by using the Riemannian metric $ds = |\Phi|^{1/2}$). The vertical trajectories of Φ form, of course, another measured foliation, transversal to the first. If Φ is the initial quadratic differential of an absolutely extremal mapping g, then g permutes the leaves of each of these two foliations, multiplying the distance between the leaves by $K^{1/2}$ and by $K^{-1/2}$, respectively.

Such a mapping Thurston calls a pseudo-Anosov diffeomorphism. Theorem A, B, C contain the following statement.

Theorem D. (Thurston) *An essentially non-periodic self-mapping of a surface is either reducible or isotopic to a pseudo-Anosov diffeomorphism, but not both.*

Department of Mathematics
Columbia University
New York, New York

REFERENCES

[1] L. Bers. Quasiconformal mappings and Teichmuller's theorem; Analytic Functions, Princeton University Press, Princeton (1960), 89-119.

[2] ______. A remark on Mumford's compactness theorem, Israel J. of Math., 12 (1972), 400-407.

[3] ______. Nielsen extensions of Riemann surfaces, Ann. Acad. Sci. Fenn., 2 (1976), 197-22.

[4] ______. An extremal problem for quasiconformal mappings and a theorem by Thurston, to appear.

[5] J. P. Matelski. A compactness theorem for Fuchsian groups of the second kind, Duke Math. J., 43 (1976), 829-840.

[6] D. Mumford. A remark on Mahler's compactness theorem, Proc. Amer. Math. Soc., 28 (1971), 289-294.

[7] W.P. Thurston. On the geometry and dynamics of diffeomorphisms of surfaces, I, to appear.

[8] S. Wolpert. The length spectrum of a compact Riemann surface, I, to appear.

Estimate of exceptional sets for quasiconformal mappings in space

by Petru Caraman

Introduction

Let $f:B\rightleftarrows D^*$ be a qcf (quasiconformal mapping) of the unit ball B onto the domain D^* (B,D^* contained in the n-space R^n) and let for instance E_0 be the set of points of the unit sphere S corresponding to the boundary points of D^* inaccessible by rectifiable arcs. The problem is to find the best estimate of E_0 as well as of other exceptional sets of this kind.

First of all, we shall introduce several concepts, important for our lecture.

From the different definitions of the quasiconformality, let us choose first Väisälä's geometric definition.

Let Γ be an arc family and $F(\Gamma)$ the class of admissible functions ρ satisfying the following conditions: $\rho(x)\geqq 0$ is Borel measurable in R^n and $\int_\gamma \rho ds \geqq 1$ $\forall\gamma\in\Gamma$ ($\forall$ means "for each"). Then, the modulus of Γ is given as

$$M(\Gamma) = \inf_{\rho\in F(\Gamma)} \int_{R^n} \rho^n d\tau ,$$

where $d\tau$ is the volume element (corresponding to the n-dimensional Lebesgue measure).

A homeomorphism $f:D\rightleftarrows D^*$ ($D,D^*\subset R^n$) is said to be a K-qc ($1\leqq K<\infty$) according to Väisälä's definition if

$$\frac{M(\Gamma)}{K} \leqq M(\Gamma^*) \leqq KM(\Gamma) ,$$

where Γ is an arbitrary arc family contained in D and $\Gamma^*=f(\Gamma)$.

A homeomorphism $f:D\rightleftarrows D^*$ is said to be locally qc in D if, $\forall x\in D$, there exists a neighbourhood V_x of x in which f is qc.

Now, let us define, following Zorič [49], the boundary elements (a generalization of Carathéodory's prime ends).

A sequence $\{U_m\}$ of domains $U_m \subset D$ is said to be regular if $\forall$ m: a) $\bar{U}_{m+1} \subset U_m$; b) $\bigcap_{m=1}^{\infty} \bar{U}_m \subset \partial D$; c) $\sigma_m = \partial U_m \cap D$ (i.e. the relative boundary of U_m in D) is connected and d) there is at most an accessible boundary point of D, which is accessible boundary point of each domain of the sequence $\{U_m\}$. Two sequences of domains are called equivalent if every term of each of them contains all the terms of the other one beginning with a sufficiently large index.

A boundary element of a domain D is the pair $(F, \{U_m\})$ consisting of a regular sequence $\{U_m\}$ and a continuum $F = \bigcap_{m=1}^{\infty} \bar{U}_m$. The boundary elements $(F, \{U_m\})$, $(F', \{U'_m\})$ are considered as identical if the 2 regular sequences $\{U_m\}$, $\{U'_m\}$ defining them are equivalent. In this way, any of the equivalent sequences determines uniquely a boundary element.

Proposition 1. $\forall$ K-qc $f: B \rightleftarrows D^*$, it is possible to establish a one-to-one correspondence between the points of S and the boundary elements $(F^*, \{U^*_m\})$ of D^*, so that to each boundary element $(F^*, \{U^*_m\})$, there correspond on S the point determined by the sequence $\{U_m\} = f^{-1}(\{U^*_m\})$ (Zorič [49]).

A surface $\sigma \subset D$ homeomorphic to R^{n-1} is called a cross-cut of D if $D - \sigma = D' \cup D''$, $D' \cap D'' = \emptyset$ and $\partial_D D' = \partial_D D'' = \sigma$, where $\partial_D E = \partial E \cap D$. A sequence $\{\sigma_m\}$ of cross-cuts is said to be a chain of cross-cuts of D if $\forall$ m=2,3,... the cross-cuts σ_{m-1} and σ_{m+1} are contained in different components of $D - \sigma_m$. One says that a chain of cross-cuts is determining if $d_D(\sigma_m, \sigma_{m+1}) > 0$ (m=1, 2,...) and $d(\sigma_m) \to 0$ as $m \to \infty$, where the relative distance $d_D(E_1, E_2)$ is the infimum of the diameters of the arcs $\gamma \subset D$ joining E_1 to E_2. Each determining chain of cross-cuts, in a natural way, generates a regular sequence of subdomains of D determining a boundary element.

We say that a point $\xi^* \in \partial D^*$ is principal (Hauptpunkt) if there exists a determining chain of cross-cuts converging to ξ^*. The other points of ∂D^* are named subsidiary (Nebenpunkte). Each boundary element contains at least a principal point.

The preceding definitions allow us to introduce the different exceptional sets we intended to estimate.

The exceptional set E_γ. Let γ_ξ be an endcut of B from ξ (i.e. an arc $\gamma_\xi \subset B$ with an endpoint $\xi \in S$). A value $\xi^* \in R^n$ is said to be an asymptotic value of f at ξ if $\lim_{x\to\xi} f(x)=\xi^*$ for $x \to \xi$ along γ_ξ. Then, the exceptional set $E_\gamma = \{\xi \in S; \nexists \lim_{x\to\xi} f(x)=\xi^* \neq \infty$ along any $\gamma_\xi\}$. ($\nexists$ means "it does not exist".)

The exceptional set E_σ. Let us recall that a sequence $\{x_m\}$ of points of B is said to converge in a cone to a point $\xi \in S$ if x_m converge to ξ and there exists a constant a, $1 \leq a < \infty$, such that

$$|x_m - \xi| \leq a d(x_m, S)$$

for all m. ξ is said to be a point at which f has an angular boundary value $\xi^* \in R^n$ if ξ^* is the only finite boundary value corresponding to all the sequences $\{x_m\}$ converging in a cone to ξ, i.e. $\lim_{x\to\xi} f(x) \neq \infty$ exists if $x \to \xi$ in an arbitrary way in a cone. Then

$E_\sigma = \{\xi \in S; \nexists \lim_{x\to\xi} f(x)=\xi^* \neq \infty$ for $x \to \xi$ in a cone$\}$.

The exceptional set $E_0 = \{\xi \in S; f(\gamma_\xi)$ is non-rectifiable $\forall \gamma_\xi\}$.

The exceptional set $E_1 = \{\xi \in S; f(r_\xi)$ is non-rectifiable for almost every radius r_ξ of $B(\xi, \frac{1}{2})$ that lie in B$\}$. Here "almost every radius" means that if E_ξ is the union of all the radii r_ξ of $B(\xi, \frac{1}{2})$ that lie in B and such that $f(r_\xi)$ is rectifiable, then $mE_\xi = 0$, where "m" is the n-dimensional Lebesgue measure and $B(x_0, r) = \{x \in R^n; |x - x_0| < r\}$.

The exceptional set $E_2 = S - \{\xi \in S; f(r_\xi)$ is rectifiable for almost every radius r_ξ of $B(\xi, \frac{1}{2})$ that lie in B$\}$.

The exceptional set $E_3 = \{\xi \in S; f(\gamma_\xi)$ is non-rectifiable for at least an endcut $\gamma_\xi\}$.

The exceptional set $E_r = \{\xi \in S; f(\overline{O\xi})$ is non-rectifiable$\}$.

The exceptional set $E_{a^*} = \{\xi \in S; a^* \in \Pi(\xi)\}$, where $\Pi(\xi)$ is the set of the principal points of the boundary element corresponding to ξ. In particular E_∞ is E_{a^*} for $a^* = \infty$.

Now, we establish some inclusion relations between the preceding exceptional sets.

Let $C_c(f,\xi)$ be the cluster set of all points ξ^* (including possibly the point at infinity) for which there exists a sequence $\{x_m\}$ converging in a cone to ξ and such that $\lim_{m\to\infty} f(x_m)=\xi^*$. If $\xi \notin E_c$, then $C_c(f,\xi)=\xi^* \neq \infty$ ($\notin$ means "does not belong").

Let $C_\gamma(f,\xi)$ be the cluster set of all points ξ^* for which there exists a sequence $\{x_m\}$ of points $x_m \in \gamma$ converging to ξ and such that $\lim_{m\to\infty} f(x_m)=\xi^*$. In the particular case in which γ is the radius of B with an endpoint at ξ, the corresponding cluster set is denoted by $C_\rho(f,\xi)$.

Proposition 2. $f: B \rightleftharpoons D^*$ K-qc => $C_c(f,\xi)=\Pi(\xi)$ (=> means "implies").

(For the proof, see our paper [10], theorem 2.)

Proposition 3. $E_\gamma = E_c$ ([10], corollary 3 of proposition 6).

Proposition 4. $E_\gamma \subset E_0$ and the inclusion is strict (corollary 4 of proposition 6 in our paper [10]).

Lemma 1. $E_\infty \subset E_c = E_\gamma \subset E_0 \subset E_1 \subset E_2 \subset E_3$.

In order to prove the first inclusion, suppose $\xi \in E_\infty$. Then $\infty \in \Pi(\xi)$ and, from proposition 2, we deduce that $\infty \in C_c(f,\xi)$, hence $\xi \in E_c$, since $C_c(f,\xi)$ does not consists of a single finite point, i.e. ξ is not a point at which f had an angular boundary value.

The inclusions $E_\gamma = E_c$ and $E_\gamma \subset E_0$ are given by the preceding 2 propositions, while the other 3 inclusions follow directly from the definitions of the corresponding exceptional sets.

Remark. If $\xi \in E_r$, then there are 3 possibilities: I. $C_\rho(f,\xi) \neq \{\xi^*\}$ (i.e. the corresponding cluster set contains at least 2 points), II. $C_\rho(f,\xi)=\{\infty\}$ or III. $C_\rho(f,\xi)=\{\xi^*\} \neq \infty$. Let us denote the corresponding subsets of E_r by E^I, E^{II} and E^{III}, respectively.

Proposition 5. If $f: B \rightleftharpoons D^*$ is K-qc and $\gamma_1 \subset B$ is an endcut of B from $\xi \in S$ which is not tangent to S, then $C_{\gamma_1}(f,\xi)=\bigcap_\gamma C_\gamma(f,\xi)=C_c(f,\xi)$, where the intersection is taken over all endcuts γ of B from ξ ([10], corollary 10 of proposition 6).

Corollary 1. $f: B \rightleftharpoons D^*$ K-qc => $C_\rho(f,\xi)=C_c(f,\xi)$.

Corollary 2. $E_r^I \cup E_r^{II} = E_c = E_\gamma$.

<u>Remark</u>. In the case $n=2$, M. Tsuji [38] established that if almost all the radii of $B(\xi,\frac{1}{2})$ that lie in B have rectifiable images by a qc $f:B \rightleftharpoons D^{*}$, then even all the radii have rectifiable images, hence $E_2=E_3=E_r$ for $n=2$, but this remains an open question for $n>2$.

Now, let us introduce the Φ-capacity, the main tool used in this communication.

Let $\mu \geq 0$ be a measure in R^n. The <u>support</u> S_μ of μ is a closed set $F \subset R^n$ such that, $\forall x \in F$ and $\forall$ neighbourhood V_x of x, $\mu(V_x)>0$.

The Φ-<u>potential</u> of a measure μ is given as

$$u^{\mu}_{\Phi}(x) = \int_{R^n} \Phi(|x-y|)\, d\mu(y) ,$$

where the kernel $\Phi(r)$ is supposed to be strictly decreasing, continuous and to satisfy the condition $\lim_{r\to 0}\Phi(r)=\infty$. And by means of the energy integral

$$I_{\Phi}(\mu) = \int_{R^n} u^{\mu}_{\Phi}(x)\, d\mu(x) ,$$

we define the Φ-<u>capacity</u>

$$C_{\Phi}(E) = [\inf_{\mu} I_{\Phi}(\mu)]^{-1} ,$$

where the infimum is taken over all the measures $\mu \geq 0$ with total mass 1 and the support S_μ E. The diameter $d(E)$ is supposed to be less than r_0, where $\Phi(r_0)=0$. However, for an arbitrary Borel set E, $C_{\Phi}(E)=0$, iff (i.e. if and only if) $C_{\Phi}(E \cap B_r)=0$ $\forall$ ball $B_r=B(x,r)$ $(0<r<r_0)$. If $\Phi(r)=\frac{1}{r^{\alpha}}$ $(\alpha>0)$, the corresponding capacity is called the α-<u>capacity</u> C_α (in particular C_1 is the <u>Newtonian capacity</u>) and if $\Phi(r)=\log\frac{1}{r}$, then, we have the <u>logarithmic capacity</u> C_0.

We shall use also the fact (see for instance B. Fuglede [15], remark 2 of theorem 2.4, p. 160) that, given Φ as above and a compact set F with $d(F) \geq r_0$, then there exists a unique measure $\tau_{\Phi} \geq 0$ having the total mass 1 and $S_{\tau_\Phi} \subset F$, such that the infimum of the energy

$$C_\Phi(F)^{-1} = \inf_\mu I_\Phi(\mu) = \inf_\mu \int_{R^n}\int_{R^n} \Phi(|x-y|)d\mu(x)d\mu(y)$$

is attained for $\mu = \tau_\Phi$, where μ ranges over the class of all $\mu \geq 0$, with total mass 1 and $S_\mu \subset F$. τ_Φ is called the capacitary distribution with total mass 1 and kernel Φ of F. If the kernel is $\frac{1}{r^\alpha}$ or $\log\frac{1}{r}$, then the corresponding capacitary distributions will be denoted by τ_α and τ_0, respectively.

Now, let us recall that the conformal capacity of a bounded set E is

$$\mathrm{cap}E = \inf_u \int_{R^n} |\nabla u|^n d\tau ,$$

where $\nabla u = (\frac{\partial u}{\partial x_1}, \ldots, \frac{\partial u}{\partial x_n})$ is the gradient of u and the infimum is taken over all functions $u \in C^1$ (i.e. continuously differentiable), with a compact support S_u contained in a fixed ball and which are equal to 1 on E.

We remind that the support of a function u is the closure of the set $E\{x; u(x) \neq 0\}$ and is denoted by S_u.

Now, we shall remind a few results about Bessel potential of order k of Φ, i.e. about the convolution

$$(g_k * \varphi)(x) = \int g_k(x-y)\varphi(y)d\tau_y,$$

where the kernel g_k of order k is defined as

$$g_k(x) \equiv g_k^{(n)}(x) = \frac{K_{\frac{n-k}{2}}(|x|)\,|x|^{\frac{k-n}{2}}}{2^{\frac{n+k-2}{2}}\pi^{\frac{n}{2}}\Gamma(\frac{k}{2})},$$

where K_m is called the Bessel function of the third kind and $\Gamma(z)$ is Euler integral of the second kind. (For their definition, see for instance the mathematical dictionary of J.Naas and H.L.Schmid [28].) The kernel g_k is a positive strictly decreasing function of $r=|x|$, continuous outside the origin, $g_k \in L^1(R^n)$ and as $x \to 0$,

$$g_k(x) = \frac{\Gamma(\frac{n-k}{2})\,|x|^{k-n}}{2^k\pi^{\frac{n}{2}}\Gamma(\frac{k}{2})} + o(|x|^{k-n}) , \qquad 0 < k < n ,$$

$$g_n(x) = \frac{\log\frac{1}{|x|}}{2^{n-1}\pi^{\frac{n}{2}}\Gamma(\frac{n}{2})} + O(1),$$

while, as $x \to \infty$,

$$g_k(x) \sim \frac{|x|^{\frac{k-n-1}{2}} e^{-|x|}}{2^{\frac{n+k-1}{2}}\pi^{\frac{n-1}{2}}\Gamma(\frac{k}{2})} .$$

The (k,p)-<u>capacity</u> of a set $E\subset R^n$ is

$$\text{cap}_{(k,p)}E = \inf_{\varphi}\|\varphi\|_{L_p} \equiv \inf_{\varphi}\left[\int|\varphi(x)|^p d\tau\right]^{\frac{1}{p}} ,$$

where the infimum is taken over all $\varphi\geqq 0, \varphi\in L^p$ $(p>1)$, such that $\forall x\in E$,

$$(g_k * \varphi)(x) \geqq 1 .$$

Let L^p_k be the class of all functions u of the form $u=(g_k * v)$, where v is L^p-integrable (i.e. $\int|v|^p d\tau<\infty$). A mapping f is said to belong to L^p_k if all its components $x^{*i}\in L^p_k$.

Next, let $\mathfrak{M}^+(E)$ be the cone of all Radon measures $\mu\geqq 0$ carried by E, i.e. with $\mu(R^n-E)=0$, and $L^+_1(E)$ its subspace composed of all measures μ with

$$\|\mu\|_1 = \text{the total variation of } \mu < \infty .$$

For $\mu\in L^+_1(E)$, we have the convolution

$$\Phi * \mu(x) = \int\Phi(x-y)\,d\mu(y) .$$

For simplicity sake, we use $\Phi(x)=\Phi(|x|)$ (as above for the kernel g_k). Let $\mathcal{L}_1$ be the σ-algebra of all sets which are measurable $\forall\mu\in\mathfrak{M}^+$. For

$E \in \mathcal{L}_1$, we define

$$c_\Phi E = \sup_\mu \|\mu\|_1 ,$$

the supremum being taken over all $\mu \in L_1^+(E)$ so that

$$\Phi * \mu(x) \leq 1 \quad \forall\, x \in R^n,$$

<u>Proposition 6.</u> $\forall$ <u>compact</u> $F \subset R^n$,

$$\frac{1}{\lambda} C_\Phi F \leqq c_\Phi F \leqq \lambda C_\Phi F .$$

(For the proof, see our paper [11], lemma 13.)

The notion of capacity is connected with the <u>generalized transfinite diameter</u>, which historically was considered before capacities (see for instance G.Polya and G.Szego [30]). We shall give a more general definition according to L.Carleson([12], p.37).

Let F be a compact set, then

$$D_m^\Phi(F) = \inf_{x_p \in F} \frac{2}{m(m-1)} \sum_{1 \leqq p < q < m} \Phi(x_p - x_q)$$

is called <u>generalized diameter</u> and $D^\Phi(F) = \lim_{m\to\infty} D_m^\Phi(F)$ the <u>generalized transfinite diameter</u>.

<u>Proposition 7. If</u> $c_\Phi F > 0$ <u>and</u> F <u>is compact, then</u> $\lim_{m\to\infty} D_m^\Phi(F) = D^\Phi(F)$ <u>exists and</u>

$$[D_\Phi(F)]^{-1} = c_\Phi F .$$

<u>If</u> $c_\Phi F = 0$, <u>then</u> $D_m^\Phi(F) \to \infty$ <u>as</u> $m \to \infty$ (L.Carleson [12], theorem 6, p.37).

<u>Corollary. If</u> $C_\Phi F > 0$ <u>and</u> F <u>is compact, then</u> $\lim_{m\to\infty} D_m^\Phi(F) = D^\Phi(F)$ <u>exists and</u>

$$\frac{1}{\lambda} C_\Phi F \leqq [D^\Phi(F)]^{-1} \leqq \lambda C_\Phi F .$$

<u>If</u> $C_\Phi F=0$, <u>then</u> $D_m^\Phi(F) \to \infty$ <u>as</u> $m \to \infty$.

Finally, let us give the definition of the <u>Hausdorff</u> h-<u>measure</u> $H_h(E)$ of a set E with respect to the <u>measure function</u> h. Let h be a continuous nondecreasing function in some interval $(0,r_1)$, $r_1>0$, and assume that $h(r) \to 0$ as $r \to 0$. Let, for $0<\varepsilon<r_1$, $\{E'_m\}$ be a countable covering of E by sets E'_m having a diameter $d(E'_m)\leq\varepsilon$, then the Hausdorff h-measure of E is the non-negative number

$$H_h(E) = \lim_{\varepsilon\to 0} \inf_{\{E'_m\}} \Sigma_m h[d(E'_m)] .$$

Now, using the concepts and the notations introduced above, we shall be able to speak in a more explicite way about the results obtained in the present paper.

After some historical accounts (§1), we establish (in §2) that all the exceptional sets considered above may be denumerable (at least for n=2). Next, since $\operatorname{cap} E_0=0$ (as it was proved by us in [8]) and on account of some relations between conformal capacity, Hausdorff h-measures, Φ-capacities and Bessel capacities (established in our recent paper [11]), we deduce that $H_h(E_0)=C_\Phi E_0=0$, where

$$(1)\quad h(r) = \begin{cases} \prod_{k=1}^{m-1}(\log_k\frac{1}{r})^{1-n}(\log_m\frac{1}{r})^{-\alpha} & \text{for } r\leq r_m, \\ \prod_{k=1}^{m-1}(\log_k\frac{1}{r_m})^{1-n}(\log_m\frac{1}{r_m})^{-\alpha} & \text{for } r>r_m \end{cases} \qquad (m = 1,2,\ldots)$$

$\forall\alpha>n-1$ and

$$(2)\quad \Phi(r) = \begin{cases} \log\frac{1}{r}\prod_{k=1}^{m-1}(\log_k\frac{1}{r})^{n-2}(\log_k\frac{1}{r})^{\beta} & \text{for } r\leq r_m, \\ \log\frac{1}{r_m}\prod_{k=1}^{m-1}(\log\frac{1}{r_m})^{n-2}(\log\frac{1}{r_m})^{\beta} & \text{for } r>r_m \end{cases} \qquad (m = 1,2,\ldots)$$

$\forall\beta>n-2$, respectively, with $\log_k\frac{1}{r}=\underbrace{\log\ldots\log}_{k}\frac{1}{r}$ and $\log_m\frac{1}{r_m}>1$. Hence, and on account of lemma 1, $H_h(E_\infty)=H_h(E_c)=H_h(E_\gamma)=0$ $\forall\alpha>n-1$ and $C_\Phi E_\infty=C_\Phi E_c=C_\Phi E_\gamma=0$ $\forall\beta>n-2$. By another argument, we establish that $H_h(E_\infty)=H_h(E_c)=H_h(E_\gamma)=H_h(E_0)=0$ $\forall$ measure function h satisfying the condition

$$(3)\qquad \int_0^R [h(r)]^{\frac{1}{n-1}}\frac{dr}{r} < \infty .$$

Next, we establish that $C_\Phi E_c=0$ $\forall$ kernel Φ satisfying the condition

$$(4) \qquad \int_0^R \Phi(r)^{\frac{1}{n-1}} \frac{dr}{r} < \infty$$

and, in particular, $\forall$ Φ of the form

$$\Phi(r) = \begin{cases} \prod_{k=1}^{m-1} (\log_k \frac{1}{r})^{n-1} (\log_m \frac{1}{r})^{\beta} & \text{for } r \leq r_m, \\ \prod_{k=1}^{m-1} (\log_k \frac{1}{r_m})^{n-1} (\log_m \frac{1}{r_m})^{\beta} & \text{for } r > r_m \end{cases} \quad (m = 1,2,\ldots) \ \forall \beta > n-1 .$$

This implies, on account of the preceding lemma also $C_\Phi E_1=0$ for the same Φ. The results concerning the last 2 exceptional sets are even weaker, i.e. $H^{n-1}(E_r)=H^{n-1}(E_0)=0$.

In the last part of the paper, we propose the following problem: Are the estimates $H_h(E_0)=C_\Phi E_\sigma=0$, with h and Φ given by (1) $\forall \alpha > n-1$ and (2) $\forall \beta > n-2$, respectively, best possible? We established in [8] that $\operatorname{cap} E_0=0$ and in [11] that, $\forall$ compact $F \subset R^n$, $\operatorname{cap} F=0 \Rightarrow H_h(F)=0$, this implication being best possible (and a similar result for the Φ-capacity). But, in order that also the estimate $H_h(E_0)=0$ be best possible, it would be necessary to prove that, given a compact set $F \subset S$, there exists a qc of B for which F is an exceptional set of the type E_0. We have been able to prove only the existence of a locally qc for which F is of the type E_0.

Some open questions are mentioned (§4).

§1. Historical accounts.

$n = 2$

The French mathematician P. Fatou [13] is the first (in 1906) to consider the problem of the estimate of exceptional sets of one of the types from above, that is why this kind of results is called "theorems of the type of Fatou's theorem". He established that $m_1 E_r=0$, where m_1 is the linear Lebesgue measure and E_r is taken with respect to the class of bounded analytic functions. F. and M. Riesz [34] (from

Hungary) obtained (in 1916) also that $m_1E_{a^*}=0$. The Swedish mathematician A. Beurling [4] (1941) showed that for a meromorphic function of the unit disc onto a Riemann surface of finite area $C_0E_r=0$. That is why, one uses the term of "generalization of Beurling's theorem" for a result aserting that some of the exceptional sets are of logarithmic capacity zero, as does for instance M. Tsuji [39] (1950), which proved that $C_0E_\sigma=0$ for the analytic functions and that $C_0E_{a^*}=0$ for the meromorphic functions under some additional conditions. Using his method, J.A. Lohwater [23,24] (1955) established that $C_0E_c=0$ for qc and in the same conditions and the same year, A.J. Jenkins [20] obtained that $C_0E_\gamma=C_0E_{a^*}=0$. An year later, the same results have been published by A. Mori [27].

n > 2

For n>2, there are only a few results. The first to be mentioned was communicated by M. Reade [34] in 1957 and asserts that for a differentiable qc in R^3, $C_1E_1=0$ and $H^2E_r=0$. Later (1963), D. Storvick [36] shows that if $f\in C^1$ is a 3-dimensional qc, D^* is simply connected and with connected complement, then $H^2E_\sigma=0$. He gives, in the same paper, a proof for the same estimate for qc in R^3 (but without any other restrictive condition) belonging to F. Gehring. At the end of the paper, D. Storvick mentions F. Gehring's conjecture that the evaluation $C_0E_\sigma=0$ (which may be deduced from Lohwater's proof of $C_0E_\sigma=0$) still holds for n>2. An important step further was done by V.A. Zorič [43] (1967), who got $C_\alpha E_\sigma=0$ $\forall\alpha>0$, and this result was extended for quasiregular mappings in 1972 by V.M. Mikljukov [25]. Finally, we communicated at the II Romanian-Finnish Seminar and the Conference of Complex analysis (Juväskylä - Finland, 13-18.VIII.1973) that $C_\alpha E_\sigma=0$. I wish to point out that this result gives something more than Zorič's result $C_\alpha E_\sigma=0$ ($\alpha>0$) since, according to the propositions 3 and 4 the inclusion $E_c\subset E_0$ is strict. Then, we obtained in [8] that $\text{cap}E_\sigma=0$. Next, at the "Conference on analytic functions" (Krakow - Poland, 4-11.IX.1974) we announced (and then published in [9]) that $C_\Phi E_c=C_\Phi E_\gamma=C_\Phi E_0=C_\Phi E_2=0$ for $\Phi(r)=(\log\frac{1}{r})^\beta$ ($\beta>n-1$)

or for $\Phi(r)$ satisfying the condition (4) under the additional condition that $mD^* < \infty$ and also that if

$$\int_B |J|(\log\frac{1}{r-1})^{\beta} d\tau < \infty ,$$

then $C_0E_\sigma = C_0E_\gamma = C_0E_\sigma = C_0E_2 = 0$. Finally, in [11], following a suggestion of L.I.Hedberg, we established that a closed set F with $\mathrm{cap}F=0$ is of Hausdorff h-measure and of Φ-capacity zero, where h and Φ are given by (1) $\forall\ \alpha>n-1$ and by (2) $\forall\ \beta>n-2$, respectively. We proved also that the implication $\mathrm{cap}F=0 \Rightarrow H_h(F)=0$ with h given by (1) and $\forall\ \alpha>n-1$ is best possible in the sense that there exist compact sets F' with $\mathrm{cap}F'=0$, but with $0<H_h(F')<\infty$ and h given by (1) $\forall\ \alpha\leq n-1$. We succeeded to show also that there exist compact sets F'' with $\mathrm{cap}F''=0$, but with $C_\Phi F''>0$ and Φ given by (2) $\forall\ \beta<n-2$.

And now, we wish to mention the following general result published by P.P.Belinskiĭ in his joint paper with M.A.Lavrent'ev [3]:

Proposition 8. There exists a constant $k=k(n)$ such that, $\forall K<k$, each K-qc of $B\subset R^n$ can be extended to a homeomorphism of the closed ball $\bar{B}$ (P.P.Belinskiĭ and M.A.Lavrent'ev [3]).

From this result we deduce that, for $n\geq 3$ and $K-1$ sufficiently small, the exceptional sets E_∞, E_γ and E_σ come to at most one point (the point at the infinity) and if D^* is bounded, then $E_\infty=E_\gamma=E_\sigma=\emptyset$. We conclude this historical account by

Proposition 9. There exists a constant $k_2=k_2(n,x_1,x_2,x_3)$ such that every K-qc $f:B\to R^n$ with $f(x_i)=x_i$ $(i=1,2,3)$ is bounded in B (P.P.Belinskiĭ and M.A.Lavrent'ev [3]).

§3. Estimate of the exceptional sets.

Theorem 1. All the exceptional sets considered above may be non-denumerable for $n=2$.

We prove this for the set E_{a^*} (in particular for E_∞), hence lemma 1 and corollary 2 of proposition 5 imply that the result still holds for $E_\gamma, E_\sigma, E_0, E_1, E_2, E_3$ and E_r.

We establish the fact that E_{a^*} may be non-denumerable by means of an example given to me by D. Gaier at the "Conference of Analytic Functions" held in Krakow (4-11.IX.1974).

Let a^* be a point of the unit circumference S_2^* and let T_{a^*} be the tangent to S_2^* at a^*. Let us divide the angle of measure π, with vertex at a^* and on the same side of T_{a^*} as the unit disc B_2^*, in 2 equal parts by means of a segment of length 1, then, each of these 2 angles of measure $\frac{\pi}{2}$ is divided again in 2 equal parts by a segment of length $\frac{1}{2}$ and so on. Next, let us consider a sequence of circumferemces of center a^* and radii $1, \frac{1}{2}, \frac{1}{2^2}, \ldots$. If we take out of B_2^* all the above segments, we get a simply connected domain D^*, which (by Riemann theorem of conformal mappings) may be mapped conformally onto the unit disc. The number of boundary elements containing a^* is equal to the number of the non-equivalent regular sequences of subdomains of D^* generated by the circular cross-cuts corresponding to the sequence of circumferences from above.

Let us establish now the following one-to-one correspondence: The 2 subdomains obtained by means of the unit circumference with center a^* are associated to the numbers 0 and 0.1; next, the 4 domains generated by the circumference of radius $\frac{1}{2}$ correspond to the numbers 0, 0.01, 0.1, and 0.11 and so on. But the infinite sequences of this kind represent all the real numbers of $[0,1]$ written in the basis 2, hence we conclude that the set of the considered boundary elements is non-denumerable and, since the corresponding sequences of circular arcs are determining chains of cross-cuts, it follows that a^* is a principal point for all these boundary elements, so that the points of the unit circumference corresponding to them belong to E_{a^*}, whence E_{a^*} is non-denumerable.

Finally, transforming D^* by an inversion with center a^*, we obtain another simply connected domain with the property that E_∞ is non-denumerable, as desired.

Corollary. $k=k(2)=1$ in proposition 8.

Proposition 10. $f: B \rightleftharpoons D^*$ K-qc $\Rightarrow$ $\mathrm{cap}E_\sigma=0$ ([8]).

Corollary. $f: B \rightleftharpoons D^*$ K-qc $\Rightarrow$ $\mathrm{cap}E_\infty=\mathrm{cap}E_\gamma=\mathrm{cap}E_\sigma=0$.

This corollary is a direct consequence of the preceding lemma and proposition.

Proposition 11. CapE=0 => $H_h(E)=0$ for all measure functions h satisfying condition (3) (R.D.Adams [1], V.P.Havin and V.G.Maz'ja [18,19] or H.Wallin [41]).

Corollary 1. $H_h(E_\infty)=H_h(E_\gamma)=H_h(E_c)=H_h(E_\sigma)=0$, where h satisfies condition (3).

Corollary 2. $H_h(E_\infty)=H_h(E_\gamma)=H_h(E_c)=H_h(E_\sigma)=0$, where h is given by (1) $\forall\alpha>n-1$.

This result may be deduced from proposition 10, its corollary and

Proposition 12. F compact and capF=0 => $H_h(F)=0$ with h given by (1) $\forall\alpha>n-1$ (our paper [11], corollary of theorem 4).

And this result is best possible, as it follows from

Proposition 13. There exist Cantor sets $F\subset R^n$ of conformal capacity zero, but with $0<H_h(F)<\infty$ and h given by (1) $\forall\alpha\leq n-1$ (our paper [11], corollary of theorem 5).

Now, let us consider

Proposition 14. $H_h(E)<\infty$ with $h(r)=[\Phi(r)]^{-1}$ => $C_\Phi E=0$ (S.Kametani [21,22], lemma 3, or S.J.Taylor [37], theorem 1).

From proposition 14 and corollary 1 of proposition 11, we deduce

Corollary 1. $f:B\rightleftarrows D^*$ K-qc => $C_\Phi E_\infty=C_\Phi E_\gamma=C_\Phi E_c=C_\Phi E_\sigma=0$, where Φ satisfies condition (4).

Corollary 2. $C_\Phi E_\infty=C_\Phi E_\gamma=C_\Phi E_c=C_\Phi E_\sigma=0$, where

$$(5)\qquad \Phi(r) = \begin{cases} \prod_{k=1}^{m-1}(\log_k\frac{1}{r})^{n-1}(\log_m\frac{1}{r})^{\beta} & \text{for } r\leq r_m, \\ \prod_{k=1}^{m-1}(\log_k\frac{1}{r_m})^{n-1}(\log_m\frac{1}{r_m})^{\beta} & \text{for } r>r_m \end{cases} \qquad (m = 1,2,\dots)$$

$\forall\ \beta>n-1$ and $\log_m\frac{1}{r_m}>1$.

This is a direct consequence of the preceding corollary since the kernel Φ given by (5) satisfies condition (4). However, in order to obtain a stronger result, we use

Proposition 15. F compact and capF=0 => $C_\Phi E=0$, where Φ is given by (2) $\forall\ \beta>n-2$, with equality for n=2 (when F is of logarithmic capacity zero), and there exist Cantor sets $F'\subset R^n$ such that capF'=0, while

$C_\Phi(F')>0$ with Φ given by (2) $\forall\ \beta<n-2$ (our paper [11], corollary 2 of lemma 13).

Hence, on account of proposition 10 and its corollary, we deduce the

Corollary. $C_\Phi(E_\infty)=C_\Phi(E_\gamma)=C_\Phi(E_c)=C_\Phi(E_0)=0$, where Φ is given by (2) $\forall\beta>n-2$, with equality for $n=2$ (when $E_\infty, E_\gamma, E_c, E_0$ are of logarithmic capacity zero).

Remark. This result, which is stronger than corollary 2 of proposition 14, cannot be deduced from its corollary 1, since the corresponding kernel Φ does not satisfy condition (4). But, it is possible to prove even something more (in connection with our estimates), i.e. the existence of Cantor sets $F\subset S$ satisfying the conditions of the preceding proposition. In order to do this, let us remind first some preliminary results.

A set $E\subset R^n$ is said to have a positive lower spherical h-density at a point x if

$$\varliminf_{r\to 0}\frac{H_h[B(x,r)\cap E]}{h(2r)} > 0 .$$

Proposition 16. Let h be a measure function such that

$$\int_0^R [h(r)]^{\frac{1}{n-1}}\frac{dr}{r} = \infty$$

and let E be a Borel set satisfying $0<H_h(E)<\infty$ such that E had positive lower spherical h-density at every point $x\in E$. Then $\operatorname{cap}E=0$ (H. Wallin [41]).

Proposition 17. Let h be a measure function satisfying

$$(6)\qquad h(2r)\leq Q h(r)$$

let us say for $0\leq r\leq\frac{1}{2}$ and where $0<Q<\infty$. If $E\subset R^n$ is a Cantor set, then

$$\frac{1}{Q}\varliminf_{m\to\infty}2^{nm}h(l_m)\leq H_h(E)\leq Q\varliminf_{m\to\infty}2^{nm}h(l_m) .$$

(For the proof, see D.R.Adams and N.G.Meyers [2], proposition 5.1.)

Proposition 18. The Borel set $E\subset R^n$, satisfying $0<H_h(E)<\infty$, has positive lower spherical h-density at every point $x\in E$ if, for $0<r<r_1$ (r_1 sufficiently small), (6) holds, where the costant $Q\in[1,2^n)$ and E is a set of Cantor type $E=\bigcap_{q=1}^{\infty}E_q$, where E_q, obtained in the qth step, consists of 2^{nq} n-dimensional intervals with edges of length l_q, $2l_{q+1}<l_q$, obtained in the usual symmetrical way, and l_q is chosen such that

$$(7)\qquad c_1 < 2^{nq}h(l_q) < c_2,\qquad c_1, c_2 \text{ constants.}$$

(For the proof, see H.Wallin [44].)

Since the projection is a qc and the sets of conformal capacity zero are invariant with respect to the qc, we have

Proposition 19. The sets of conformal capacity zero are invariant with respect to the projections (F.Gehring and J.Väisälä [17]).

Proposition 20. If F is compact and $H_h(F)>0$ for a measure function h such that

$$\int_0^R \Phi(r)\,dh(r) < \infty ,$$

then, $C_\Phi F>0$ (S.J.Taylor [37], theorems 1, and 2).

Lemma 2. There exist compact sets $F\subset S$ with cap $F=0$, but such that $C_\Phi F>0$, where Φ is given by (2) $\forall\beta<n-2$.

Let F' be an (n-1)-dimensional set of the Cantor type, $F'=\bigcap_{q=1}^{\infty}F'_q$, where F', obtained in the mth step, consists of $2^{q(n-1)}$ (n-1)-dimensional intervals with edges of length l_q, $2l_{q+1}<l_q$, obtained in the usual symmetrical way, and l_q is chosen such that

$$(8)\qquad 0 < c_1 < 2^{q(n-1)}h(l_q) < c_2 \quad (c_1, c_2 \text{ constants}) .$$

Let us establish first that h given by (1) satisfies (6), i.e. that

$$\prod_{k=1}^{m-1}(\log_k\frac{1}{2r})^{1-n}(\log_m\frac{1}{2r})^{-\alpha} \leqq Q\prod_{k=1}^{m-1}(\log_k\frac{1}{r})^{1-n}(\log_m\frac{1}{r})^{-\alpha} ,$$

with $0<Q<2^{n-1}$. Let us take Q of the form $Q=Q_m^{\alpha}\prod_{k=1}^{m-1}Q_k^{n-1}$. It is enough to prove that

$$\log_k\frac{1}{r} \leqq Q_k\log_k\frac{1}{2r} \quad (k=1,\ldots,m) .$$

We use an induction argument. It is easy to verify that this inequality holds for $k=1$ with $Q_1=1+\varepsilon$ ($\varepsilon>0$ as small as one pleases), and suppose it is true for $k-1$. Then

$$\log_k\frac{1}{r}=\log(\log_{k-1}\frac{1}{r})\leqq\log(Q_{k-1}\log_{k-1}\frac{1}{2r})\leqq\log[(\log_{k-1}\frac{1}{2r})^{Q_k}] = Q_k\log_k\frac{1}{2r}$$

since, for r sufficiently small and $Q_k>1$,

$$1 \leqq Q_{k-1} \leqq (\log_{k-1}\frac{1}{2r})^{Q_k-1}.$$

For $Q_k=1+\varepsilon$ (with $\varepsilon>0$ sufficiently small), $1<Q<2^{n-1}$ and we are in the hypotheses of proposition 18 asserting that F' is of positive lower spherical h-density. Next, according to (8), from proposition 17, it follows that $0<H_h(F')<\infty$ and, since

$$\int_0^{r_m} h(r)^{\frac{-1}{n-1}}\frac{dr}{r} = \int_0^{r_m}\prod_{k=1}^{m-1}(\log_k\frac{1}{r})^{-1}(\log_m\frac{1}{r})^{\frac{-\alpha}{n-1}}\frac{dr}{r} = \infty \quad \forall\ \alpha \leqq n-1$$

and F' is of positive lower spherical h-density, proposition 16 implies $\operatorname{cap}F'=0$.

On the other hand, since

$$\int_0^{r_m}\Phi(r)\,dh(r)=\int_0^{r_m}(\log\frac{1}{r})^{n-1}\prod_{k=2}^{m-1}(\log_k\frac{1}{r})^{n-2}(\log_m\frac{1}{r})^{\beta}d[\prod_{k=1}^{m-1}(\log_k\frac{1}{r})^{1-n}(\log_m\frac{1}{r})^{-\alpha}]$$

$$=(n-1)\sum_{p=1}^{m-1}\int_{\log\frac{1}{r_m}}^{\infty}\prod_{k=1}^{p-1}(\log_k u)^{-2}\prod_{k=p}^{m-2}(\log_k u)^{-1}(\log_{m-1}u)^{\beta-\alpha}d(\log u) +$$

$$\alpha \int_{\log\frac{1}{r_m}}^{\infty} \prod_{k=1}^{m-2}(\log_k u)^{-2}(\log_{m-1}u)^{\beta-\alpha-1}d(\log u) <$$

$$(m-1)(n-1) \int_{\log\frac{1}{r_m}}^{\infty} \prod_{k=1}^{m-2}(\log_k u)^{-1}(\log_{m-1}u)^{\beta-\alpha}d(\log u) +$$

$$\alpha \int_{\log\frac{1}{r_m}}^{\infty} \prod_{k=1}^{m-2}(\log_k u)^{-1}(\log_{m-1}u)^{\beta-\alpha-1}d(\log u) = \frac{(m-1)(n-1)}{\beta-\alpha+1}v^{\beta-\alpha+1}\Big|_{\log_m\frac{1}{r_m}}^{\infty} +$$

$$\frac{\alpha}{\beta-\alpha}v^{\beta-\alpha}\Big|_{\log\frac{1}{r_m}}^{\infty} < \infty \quad \forall\ \beta<\alpha-1 \leq n-2 ,$$

then, the preceding proposition implies $C_\Phi F'>0$ $\forall \beta\leq n-2$.

Now, suppose the diameter of F' is sufficiently small and that the unit sphere is tangent to the plane Π containing F' at a point $x'_0\varepsilon F'$, and let $C_{x'_0}$ be a cap less than a semi-sphere and with center at x'_0. If $F\subset C_{x'_0}$ is the set with the projection F' on Π, then, on account of proposition 19, cap$F=0$.

On the other hand, if $x'_1, x'_2\varepsilon F$ and x_1, x_2 are the corresponding points of F, then, clearly, if the diameter $d(F')$ is sufficiently small, $\frac{1}{2}|x_1-x_2|\leq|x'_1-x'_2|\leq|x_1-x_2|$. Hence, and taking into account the definition of the transfinite diameter (from the introduction), we deduce that

$$D_i^\Phi(F') = \inf_{x'_p\varepsilon F'} \frac{2}{i(i-1)}\sum_{p<q}(\log\frac{1}{|x'_p-x'_q|})\prod_{k=1}^{m-1}(\log_k\frac{1}{|x'_p-x'_q|})^{n-2}(\log_m\frac{1}{|x'_p-x'_q|})^{\beta} \geq$$

$$\inf_{x_p\varepsilon F} \frac{2}{i(i-1)}\sum_{p<q}(\log\frac{1}{|x_p-x_q|})\prod_{k=1}^{m-1}(\log_k\frac{1}{|x_p-x_q|})^{n-2}(\log_m\frac{1}{|x_p-x_q|})^{\beta} = D_i^\Phi(F) ,$$

hence, by corollary of propositions 6,7 and since $C_\Phi F'>0$,

$$(C_\Phi F)^{-1} \leq \lambda D^\Phi(F) = \lambda \lim_{i\to\infty} D_i^\Phi(F) \leq \lambda \lim_{i\to\infty} D_i^\Phi(F') \leq \lambda^2(C_\Phi F')^{-1} < \infty ,$$

so that $C_\Phi F>0$, as desired.

Now, we intend to estimate $E_{a^{\infty}}$.

<u>Theorem 2.</u> $C_\Phi E_{a^{\infty}}=0$, <u>where</u> φ <u>is given by (2)</u> $\forall\beta>n-2$, **with** <u>equality</u>

for $n=2$ (when E_{a^*} is of logarithmic capacity zero).

The particular case $C_\Phi E_\infty=0$, with Φ satisfying (4) or given by (2) $\forall\beta>n-2$, was established by corollary 1 of proposition 14 and corollary of proposition 15, respectively.

If $a^*\neq\infty$, then, by means of an inversion $x^{**}=j(x^*)$ with center a^*, D^* is mapped onto a domain D^{**} and $j(a^*)=\infty\in\partial D^{**}$. The composite mapping $j\circ f: B\rightleftarrows D^{**}$ is K-qc. But an inversion transforms regular sequences again into regular sequences and equivalent sequences again into equivalent sequences, so that it establishes a one-to-one correspondence between the boundary elements of the corresponding domains and, in particular, between the boundary elements of D^* containing a^* as a principal point and the boundary elements of D^{**} containing ∞ as a principal point. Hence, $E_{a^*}\subset S$ considered with respect to f coincides with $E_\infty\subset S$ considered with respect to $j\circ f$, so that the conclusion of the theorem follows directly from corollary of proposition 15.

Arguing as above, we have also

Theorem 3. $C_\Phi E_{a^*}=0$, where Φ satisfies condition (4).

Next, since we have not been able to prove that $\operatorname{cap}E_2=0$, or at least $\operatorname{cap}E_1=0$, we have not an analogous of the corollary of proposition 15 for these exceptional sets, but only corollary 1 of proposition 14 still holds. We shall prove this result first in the more particular case in which D^* is bounded and then, by means of corollary 1 of proposition 10, we shall get rid of this restrictive condition.

We recall first another characterization of the K-qc (F. Gehring [16]).

A homeomorphism $f: D\rightleftarrows D^*$ is said to be K-qc ($1\leq K<\infty$) according to Gehring's metric definition if the linear local dilatation $\delta_L(x)$ is bounded in D and $\delta_L(x)\leq K$ a.e. in D, where

$$L(x,r)=\max_{|\Delta x|=r}|f(x+\Delta x)-f(x)|,\quad l(x,r)=\min_{|\Delta x|=r}|f(x+\Delta x)-f(x)|,$$

$$\delta_L(x)=\overline{\lim_{r\to 0}}\frac{L(x,r)}{l(x,r)}.$$

A property is said to hold a.e. (almost everywhere) in a set E if the subset of E where this property does not hold is of n-dimensional Lebesgue measure zero.

Proposition 20. A K-qc according to Väisälä's geometric definition is K-qc also according to Gehring's metric definition (with the same K).

For the proof, see our paper [6], p.147, or our monograph [7], formula (8), p.134.

Proposition 21. If f is K-qc (according to Gehring's metric definition) and differentiable at a point x_0, then the maximal dilatation

$$\Lambda_f(x_0) \equiv \overline{\lim_{x\to x_0}} \frac{|f(x)-f(x_0)|}{|x-x_0|} \leq K^{\frac{n-1}{n}} \sqrt[n]{|J(x_0)|} \leq K \sqrt[n]{|J(x_0)|} .$$

For the proof, see our book [7] (corollary of theorem 6, p.136).

Proposition 22. If f is K-qc (according to Väisälä's geometric definition) in D, then it is differentiable with the Jacobian $J(x) \neq 0$ a.e. in D.

The proof follows from lemma 6.3 and theorem 6.10 of Väisälä's paper [40] and from the equivalence of all Väisälä's definitions of the K-qc.

Proposition 23. The potential $u_\Phi^{\tau_\Phi}(x) \leq MC_\Phi(F)^{-1}$ everywhere, for $0<M<\infty$ and F a closed set (T. Ugaheri [39]).

Lemma 3. Let $A=\{x; r_1<|x|<r_2\}$ be a spherical ring, $E\subset S(r_2)$ and Γ_E^A the family of radial segments $\gamma_e=\{x; x=re, r_1<r<r_2\}$ joining the boundary components of A, parallel to the unit vectors e such that the endpoints $r_2e \in E$. Then $M(\Gamma_E^A)=0 \iff H^{n-1}(E)=0$.

We prove first the implication

$$(9) \qquad M(\Gamma_E^A) = 0 \Rightarrow H^{n-1}(E) = 0 .$$

$\forall \xi = r_2 e \in S(r_2)$ and $\forall \rho \in F(\Gamma_E^A)$, on account of Hölder inequality,

$$(10) \qquad \chi_E(\xi) \leq (\int_{\gamma_e} \rho ds)^n \leq (\log\frac{r_2}{r_1})^{n-1}\int_{r_1}^{r_2} \rho^n r^{n-1} dr ,$$

where χ_E is the characteristic function of E given as

$$\chi_E(x) = \begin{cases} 1 \text{ if } x \in E , \\ 0 \text{ otherwise} . \end{cases}$$

Now, integrating over $S(r_2)$ the inequality (9) and taking into account Fubini theorem, we get

$$H^{n-1}(E) = \int_{S(r_2)} \chi_E(\xi) d\sigma \leq (\log\frac{r_2}{r_1})^{n-1} \int_A \rho^n d\tau ,$$

where $d\sigma$ is the surface element of $S(r_2)$. Finally, taking the infimum over all $\rho \in F(\Gamma_E^A)$, we are allowed to conclude that

$$H^{n-1}(E) \leq (\log\frac{r_2}{r_1})^{n-1} M(\Gamma_E^A) ,$$

hence, we deduce, in particular, the implication (9).

Now, in order to establish the opposite implication

$$(11) \qquad H^{n-1}(E) = 0 \quad \Rightarrow \quad M(\Gamma_E^A) = 0 ,$$

we observe that

$$\rho_0(x) = \begin{cases} \frac{1}{r_2 - r_1} \text{ if } x \in \gamma_e \in \Gamma_E^A , \\ 0 \quad \text{otherwise} \end{cases}$$

is admissible for $F(\Gamma_E^A)$. Hence,

$$M(\Gamma_E^A) \leq \int \rho^n d\tau = \frac{1}{(r_2 - r_1)^{n-1}} H^{n-1}(E) ,$$

yielding also the implication (11), as desired.

Proposition 24. Almost all the arcs are rectifiable (J. Väisälä [40]).

Remark. On account of the preceding lemma, it is possible to give to E_1 and E_2 the following new definitions, equivalent to those given in the introduction:

$E_1=\{\xi\in S; f(r_\xi)$ is non-rectifiable for almost every radius r_ξ of $B(\xi,\varepsilon)(\forall\varepsilon>0)$ that lie in $B\}$;

$E_2=S-\{\xi\in S; f(r_\xi)$ is rectifiable for almost every radius r_ξ of $B(\xi,\varepsilon)$ $(\forall\varepsilon>0)$ that lie in $B\}$.

Clearly, these new definitions imply the corresponding ones (given in the introduction). Now, let us show also the opposite implications. We shall establish them by reductio ad absurdum.

Let $\xi\in E_1$ (according to the definition in the introduction), but suppose, to prove it is false, that there exists an $\varepsilon_1>0$ such that $f(r_\xi)$ is not non-rectifiable for almost every radius r_ξ of $B(\xi,\varepsilon_1)$ that lie in B. Hence, it follows the existence of a family Γ_1 of radial segments $\gamma^e_\xi=\{x; x=\xi+re, \varepsilon_1<r<\frac{1}{2}\}$ (of the spherical ring $A=\{x; \varepsilon_1<|x-\xi|<\frac{1}{2}\}$) that lie in B such that $f(\gamma^e_\xi)$ is non-rectifiable and $m(\bigcup_e\gamma^e_\xi)>0$. Then, evident, if we denote by $E'\subset S(\frac{1}{2})$ the set of the endpoints belonging to $S(\frac{1}{2})$ of these segments, we have $H^{n-1}(E')>0$. On the other hand, since the arcs of the family $\Gamma_1^*=f(\Gamma_1)$ are non-rectifiable, on account of the preceding proposition, we deduce that $M(\Gamma_1^*)=0$, whence, the K-quasiconformality of f implies $M(\Gamma_1)=0$, so that, on account of the preceding lemma, we obtain that $H^{n-1}(E')=0$. This contradiction shows that also the opposite implication is true for E_1.

Now, we shall do the same for E_2. Let $\xi\in E_2$ (according to the definition from the introduction) and assume that there exists an $\varepsilon_2>0$ so that $f(r_\xi)$ is rectifiable for almost every radius r_ξ of $B(\xi,\varepsilon_2)$ that lie in B. We deduce the existence of a family Γ_2 of radial segments $\gamma^e_\xi=\{x; x=\xi+re, \varepsilon_2<r<\frac{1}{2}\}$ (of the spherical ring $A=\{x; \varepsilon_2<|x-\xi|<\frac{1}{2}\}$) that lie in B such that $f(\gamma^e_\xi)$ are non-rectifiable and $m(\bigcup_e\gamma^e_\xi)>0$. Then, clearly, if we denote by E'' the set of the endpoints belonging to $S(\frac{1}{2})$ of these segments, we have $H^{n-1}(E'')>0$. On the other hand, since the arcs of the family $\Gamma_2^*=f(\Gamma_2)$ are non-rectifiable, then arguing as above (in the case of E_1), we are allowed to conclude that $M(\Gamma_2)=0$, so that, on account of the preding lemma, $H^{n-1}(E'')=0$, and the obtained contradiction establishes the desired implication.

Theorem 4. If $f:B\rightleftharpoons D^*$ is K-qc, then $C_\Phi E_2=0$, where Φ satisfies condition (4).

Suppose, for the moment $mD^*<\infty$.

Let $\xi\in E_2$, let $B_\xi=B(\xi,R)$ be a ball of radius $R<\frac{1}{4}$ and tangent to S at ξ and let γ_e be a segment joining ξ to a point of $S(\xi,R)$, having the direction e and the property that $f(\gamma_e)$ is non-rectifiable. According to the preceding remark, $m(\bigcup_e \gamma_e)>0$, where the union is taken over all such segments. We assume for simplicity sake that $\xi=(1,0,\ldots 0)$. Then

$$(12)\qquad \infty=\int_{\gamma_e^*} ds^* = \int_{\gamma_e}\frac{ds^*}{ds}ds \leq \int_{\gamma_e}\Lambda_f(x)ds = \int_0^{\rho_e}\Lambda_f(x)d\rho\ ,$$

where $\gamma_e^*=f(\gamma_e)$, ρ_e is the length of the segment γ_e and $\rho=|x-\xi|$. If $(\rho,\vartheta_1,\ldots,\vartheta_{n-1})$ are the polar coordinates of a point $x\in B_\xi$ and the origin of this system of coordinates is supposed to be at ξ, then integrating (12) with respect to $\vartheta_1,\ldots,\vartheta_{n-1}$ $(0\leqq\vartheta_1\leqq\frac{\pi}{2}, 0\leqq\vartheta_2\leqq\pi,\ldots$ $0\leqq\vartheta_{n-1}\leqq 2\pi)$ and taking into account Fubini theorem and propositions 20, 21, 22, we obtain $\forall\ \xi\in E_2$,

$$(13)\qquad \infty=\int_{S_1}d\sigma\int_0^{\rho_e}\Lambda_f(x)d\rho = \int_{B_\xi}\frac{\Lambda_f(x)}{\rho^{n-1}}d\tau \leq K\int_{B_\xi}\frac{\sqrt[n]{|J(x)|}}{\rho^{n-1}}d\tau\ ,$$

where S_1 is the corresponding semi-sphere and $d\sigma$ is the spherical element. We precise that the unit vector e indicating the direction of the vector γ_e depends on $\vartheta_1,\ldots,\vartheta_{n-1}$, i.e. $e=e(\vartheta_1,\ldots,\vartheta_{n-1})$.

We prove that $C\ E_2=0$, with Φ satisfying (4), by reductio ad absurdum. Thus, suppose, to prove it is false, that $C_\Phi E_2>0$. Let r_0 be such that $\Phi(r_0)=0$ and let $\Delta\subset B$ be a convex domain with $d(\Delta)<r_0$, $\Delta\supset B_\xi$ and $d(S\cap\bar\Delta)>\frac{r_0}{2}$. Of course S may be covered by a finite number of sets $S\cap\bar\Delta$. Let $E_2^\Delta=E_2\cap\bar\Delta$. Clearly, there is at least a Δ so that $C_\Phi E_2>0$. We put

$$I = \int |J(x)|\, u_\Phi^{\tau_\Phi}(x)d\tau\ .$$

From proposition 23 and taking into account that $\int_B |J(x)|d\tau=mD^*<\infty$, we deduce that

$$(14)\qquad I = \int |J(x)|\, u_{\Phi}^{\tau_\Phi}(x)\,d\tau \leq M(C_\Phi E_2)^{-1}\int |J(x)|\,d\tau < \infty .$$

On the other hand,

$$\int_{B_\xi} \frac{\sqrt[n]{|J|}}{\rho^{n-1}}d\tau = \int_{B_\xi} [|J|\Phi(\rho)]^{\frac{1}{n}} \frac{d\tau}{\Phi(\rho)^{\frac{1}{n}}\rho^{n-1}} \leq \Big(\int_{B_\xi} |J|\Phi d\tau\Big)^{\frac{1}{n}}\Big(\int_{B_\xi} \frac{d\tau}{\Phi^{\frac{1}{n-1}}\rho^{n}}\Big)^{\frac{n-1}{n}} \leq$$

$$\Big(\int_{B_\xi} |J|\Phi d\tau\Big)^{\frac{1}{n}}\Big(\int_0^R \Phi^{\frac{-1}{n-1}} \frac{d\rho}{\rho}\Big)^{\frac{n-1}{n}}(n\omega_n)^{\frac{n-1}{n}} = N\Big(\int_{B_\xi} |J|\Phi d\tau\Big)^{\frac{1}{n}} ,$$

where $N=(n\omega_n)^{\frac{n-1}{n}}\Big(\int_0^R \Phi^{\frac{-1}{n-1}} \frac{d\rho}{\rho}\Big)^{\frac{n-1}{n}}<\infty$ according to (4) and ω_n is the volum

of the unit ball.Hence and by (13),we obtain

$$\infty = \int_{B_\xi} \frac{\sqrt[n]{|J|}}{\rho^{n-1}}d\tau \leq N\Big(\int_{B_\xi} |J|\Phi d\tau\Big)^{\frac{1}{n}} \leq N\int_{B_\xi} |J|\Phi d\tau \leq N\int_\Delta |J|\Phi d\tau .$$

But $\rho=\rho_{x\xi}=|x-\xi|$,so that,integrating with respect to the capacitary distribution τ_Φ over E_2^Δ and taking into account Fubuni theorem,we get

$$\infty=\int_{E_2^\Delta} d\tau(\xi)\int_\Delta |J(x)|\Phi(|\xi-x|)d\tau_x=\int_\Delta |J(x)|\Big[\int_{E_2^\Delta} \Phi(|\xi-x|)d\tau(\xi)\Big]d\tau_x \leq$$

$$\int_\Delta |J(x)|\,u_\Phi^{\tau_\Phi}(x)d\tau_x \leq I ,$$

where we denoted $d\tau$ by $d\tau_x$ in order to point out that x is the variable of integration.But this contradicts (14) and then the hypothesis $C_\Phi E_2>0$ was false,allowing us to conclude that $C_\Phi E_2=0$.

Now,let us show how to get rid of the restrictive condition $mD^*<\infty$.First,let us consider the particular case in which CD^* (the complement of D^*) contains an interior point x_0^*.Let $X=j(x^*)$ be an inversion with center x_0^* and radius r_0^* such that $B(x_0^*,r_0^*)\subset CD^*$.The mapping $\varphi=j\circ f$ is K-qc and maps the ball B onto a bounded domain.But an inversion establishes a one-to-one correspondence between the boundary elements of the corresponding domains (see also the proof of theorem 2).By this inversion,rectifiable arcs are mapped into rectifiable arcs and non-rectifiable into non-rectifiable,except possibly those which have an endpoint at ∞.If we consider the corresponding boundary elements, they all contain ∞ and then, the corresponding points of S belong to E_∞. Indeed, if $\infty\in\partial D^*$ is transformed by the inversion j into a point ξ_0^{**} accessible from $\varphi(B)$ by a rectifiable arc γ^{**}, this means that $\lim_{x\to\xi_0}\varphi(x)=\xi_0^{**}$ if $x\to\xi_0$ along the arc $\gamma=\varphi^{-1}(\gamma^{**})$, hence, we deduce that $\lim_{x\to\xi_0} f(x)=\lim_{x\to\xi_0} j^{-1}\circ\varphi(x)$ exists and is equal to ∞, so that, on account of propositions 5 and 2, it follows that ∞ is a principal point (even the only principal point) of the corresponding boundary element, implying that $\xi_0\in E_\infty$ (where the exceptional set E_∞ is considered with respect to the K-qc f) and, since by corollary 1 of proposition 14, $C_\Phi E_\infty=0$ with Φ satisfying condition (4), we deduce (taking into account also the subadditivity of the Φ-capacity) that the conclusion of the first part of the proof (i.e. $C_\Phi E_2=0$) still holds also in the more general case in which CD^* contains at least an inner point.

And now, we consider the general case. Let us cut the ball B by a plan Π_1, which does not contain its center O and let us denote by B_1 the major spherical segment obtained in this way. Next, let us denote by B_2 the major spherical segment symmetric of B_1 with respect to the plane through O and parallel to Π_1. The ball B may be considered as the union $B=B_1\cup B_2$, while $B_1\cap B_2$ is a spherical segment of 2 bases. Then, let us select from the finite system of domains Δ (defined in the first part of the proof) only those contained in B_1 or in B_2. If B_1 and its symmetric B_2 are sufficiently large and if the radius R of the different balls B_ξ as well as the corresponding Δ are sufficiently small, then all the domains Δ of the initial system (with the property that the corresponding sets $S\cap\bar{\Delta}$ cover S) will be admissible, i.e. will be contained at least in one of the segments B_1, B_2. But then, arguing as above, we obtain the K-qc $\varphi_k=j_k\circ f$, where j_k is an inversion with the center at an inner point of the complement of $f(B_k)$ $(k=1,2)$. Since $m\varphi_k(B_k)<\infty$, we are in the conditions of the first part of the proof, so that we are allowed to conclude that the corresponding sets $E_2^k\subset S^k=\partial B_k\cap S$ $(k=1,2)$ have the Φ-capacity zero, hence

$$C_\Phi E_2 \leq C_\Phi E_2^1 + C_\Phi E_2^2 = 0 ,$$

as desired.

Remark. The argument in the case of a domain D^* with $mD^* = \infty$ was suggested to me by F. Gehring.

The preceding theorem and lemma 1 yield

Corollary 1. $f: B \rightleftharpoons D^*$ K-qc $\Rightarrow C_\Phi E_1 = 0$, where Φ satisfies (4).

Since Φ given by (5) verifies (4) $\forall \beta > n-1$, from the preceding corollary and theorem, we deduce

Corollary 2. $f: B \rightleftharpoons D^*$ K-qc $\Rightarrow C_\Phi E_1 = C_\Phi E_2 = 0$, where Φ is given by (5) $\forall \beta > n-1$.

Theorem 5. $f: B \rightleftharpoons D^*$ K-qc $\Rightarrow H^{n-1}(E_3) = 0$.

According to the definition of the exceptional set $E_3 \subset S$, $\forall \xi \in E_3$, there corresponds a segment γ_ξ with non-rectifiable image. Let us denote by Γ the family of all of them and by Γ_m its subfamily with the property that each segment $\gamma_\xi \in \Gamma_m$ has a subsegment γ_ξ^m joining ξ and $S(r_m)$, where $r_m \to 1$ as $m \to \infty$. The preceding proposition implies $M(\Gamma^*) = M(\Gamma_m^*) = 0$ $(m=1,2,\ldots)$. But since f is K-qc (according to Väisälä's geometric definition), we have also $M(\Gamma) = M(\Gamma_m) = 0$ $(m=1,2,\ldots)$.

Next, let E_m' be the set of the endpoints belonging to S of the segments of Γ_m and let ϑ_ξ be the angle between γ_ξ and $O\xi$. Clearly, $\sqrt{1-r_m^2} \leq \cos\vartheta_\xi \leq 1$. For $\cos\vartheta_\xi = \sqrt{1-r_m^2}$, the segment γ_ξ is tangent to the sphere $S(r_m)$ and for a larger ϑ_ξ, $\gamma_\xi \overline{\in} \Gamma_m$ since it does not meet $S(r_m)$ any more. Now, on account of Hölder inequality, $\forall \rho \in F(\Gamma_m)$,

$$\chi_{E_m'}(\xi) \leq \Big(\int_{\gamma_\xi} \rho ds\Big)^n \leq \Big(\int_{r_m^2}^1 \rho \frac{ds}{dr} dr\Big)^n = \Big(\int_{r_m^2}^1 \rho \frac{dr}{\cos\vartheta_\xi}\Big)^n \leq \frac{1}{(1-r_m^2)^{\frac{n}{2}}} \Big(\int_{r_m^2}^1 \rho dr\Big)^n \leq$$

$$\frac{1}{(1-r_m^2)^{\frac{n}{2}}} \Big(\log\frac{1}{r_m^2}\Big)^{n-1} \int_{r_m^2}^1 \rho^n r^{n-1} dr .$$

Then, integrating over S and taking into account Fubini theorem, we obtain

$$H^{n-1}(E_m') = \int_S \chi_{E_m'}(\xi) d\sigma \leq \frac{2^{n-1}}{(1-r_m^2)^{\frac{n}{2}}} \Big(\log\frac{1}{r_m}\Big)^{n-1} \int_{A_m} \rho^n d\tau ,$$

where $A_m = \{x; r_m^2 < |x| < 1\}$. Finally, taking the infimum over all $\rho \in F(\Gamma_m)$, it follows

$$H^{n-1}(E'_m) = \frac{2^{n-1}}{(1-r_m^2)^{\frac{n}{2}}}\left(\log\frac{1}{r_m}\right)^{n-1}M(\Gamma_m) = 0 \qquad (m = 1,2,\ldots)$$

and, since $E_3 = \bigcup_m E'_m$, the subadditivity of (n-1)-dimensional Hausdorff measure allows us to conclude that $H^{n-1}(E_3)=0$, as desired.

Corollary. $f:B \rightleftharpoons D^*$ K-qc $\Rightarrow H^{n-1}(E_r)=0$.

§3. Is the estimate $C_\Phi E_\sigma = 0$, with Φ given by (2), best possible?

In this paragraph, we show that given an arbitrary closed set $F \subset S$ with cap$F=0$, there exists a locally qc $f:B \rightleftharpoons D^*$ for which F is an exceptional set of the type E_a for f.

In order to establish this result, we have to remind some preliminary results and definitions.

A mapping $\varphi: D \to R^n$ is called a C-isometry if $\forall\, x,y \in D$, the double inequality

$$\frac{1}{C}|x-y| \leq |\varphi(x)-\varphi(y)| \leq C|x-y|$$

holds. If we do not specify the constant C, we call it a quasy-isometry.

Proposition 25. A C-isometry $\varphi: D \to R^n$ has a.e. in D a differential and a Jacobian J satisfying the double inequality $\frac{1}{C^n} \leq |J(x)| \leq C^n$ (Ju. G. Rešetnjak [33]).

A non-void open set $\sigma_q \subset R^n$ is said to be a q-dimensional Lipschitz surface in R^n, if each point $\xi \in \sigma_q$ has a neighbourhood $U_\xi \subset R^n$ such that there exists a quasi-isometry $\varphi: U_\xi \to R^n$ so that $\varphi(U_\xi)$ is a cube $Q = \{x \in R^n;\ |x^i| < 1,\ i=1,\ldots,n\}$, $\varphi(\xi)=0$ and $\varphi(U_\xi \cap \sigma_q) = Q \cap \{x \in R^n;\ x^{q+1} = \ldots = x^n = 0\}$. According to this definition, a connected 1-dimensional Lipschitz surface means an arc, either closed and rectifiable, or open and locally rectifiable, while in the case q=n-1, we have the definition of a Lipschitz surface and then a mapping φ with the above properties is named a **rectifying isometry**. A neighbourhood U of a point $\xi_0 \in \sigma$ ($=\sigma_{n-1}$) admitting a rectifying isometry is termed canonical.

Let σ be a Lipschitz surface in R^n. Suppose that, to each $\xi \in \sigma$, we associate a Jordan arc γ_ξ with an endpoint at ξ. Then, we say that $\{\gamma_\xi\}$ is a **Lipschitz field of directions on** σ if $\forall\ \xi_0 \in \sigma$, there **exists such a rectifying quasi-isometry** $\varphi: U \to R^n$ of a neighbourhood U of ξ_0, that $\forall \eta \in U \cap \sigma$, the image $\varphi(\gamma_\eta \cap U)$ is the segment $\{x; 0 \leq x^n \leq 1, x^i = \text{const.} (i=1,\ldots,n-1)\}$.

Proposition 26. Let σ be an arbitrary Lipschitz surface along which is given a Lipschitz field of directions $\{\gamma_\xi\}$, u an arbitrary function of the class L^p_k, with $kp \leq n$, and $E'_0 \subset \sigma$ the set of points ξ with the property that the limit of u(x) for $x \to \xi$ along γ_ξ does not exist or is infinite. Then $\text{cap}_{(k,p)} E'_0 = 0$ (Ju.G.Rešetnjak [33], theorem 2).

The result is sharp in the following sense:

Proposition 27. For any set E'_0 contained in a Lipschitz surface $\sigma \subset R^n$ and such that $\text{cap}_{(k,p)} E'_0 = 0$, there is a function $u \in L^p_k$ with the property that $\forall\ \xi \in E'_0$, $u(x) \to \infty$ as $x \to \xi$ along γ_ξ (Ju.G.Rešetnjak [33], p. 419).

Corollary 1. For any compact set F contained in a Lipschitz surface $\sigma \subset R^n$ and such that $\text{cap}_{(k,p)} F = 0$, there is a mapping $f \in L^p_k$ with the property that $\forall\ \xi \in F$, $f(x) \to \infty$ as $x \to \xi$ along γ_ξ (from above).

Corollary 2. For any compact set $F \subset S$ such that $\text{cap}_{(k,p)} F = 0$, there is a mapping $f \in L^p_k$ with the property that $\forall \xi \in F$, $f(x) \to \infty$ as $x \to \xi$ along the corresponding radius of B.

Proposition 28. For every compact set $F \subset R^n$, the (k,p)-capacity of F is equal to the infimum of

$$\|u\|^p_{L_p} = \int_{R^n} |u(x)|^p d\tau,$$

taken over all functions $u \in C^\infty(R^n) \cap L^p_k$ such that $u(x) \geq 1$ for $x \in F$ (Ju.G.Rešetnjak [32], lemma 6.4).

We recall that $u \in C^\infty(R^n)$ means that u is infinitely times continuously differentiable in R^n.

And now, we shall give a new definition for the qc, which is equivalent to the other (see our monograph [7], chap.4).

A homeomorphism $f: D \rightleftarrows D^*$ is K-qc in Kreines' sense in D if its components $x^{*i} \in W_1^n(R^n)$ and

$$|\nabla f|^2 \leq nK |J(x)|^{\frac{2}{n}},$$

where

$$|\nabla f|^2 = \sum_{i,k=1}^{n} (x_k^{*i})^2,$$

holds a.e. in D.

We remind that a function u is said to belong to $W_1^n(R^n)$ if u has first order Sobolev derivatives L^n-integrable on every compact set. We say that a function u_i is the Sobolev first order derivative (with respect to the variable x^i) of a real function u, integrable in every bounded subdomain $\Omega \subset D$, if

$$\int_\Omega [u(x)\frac{\partial v(x)}{\partial x^i} + v(x)u_i(x)]d\tau = 0$$

$\forall v \in C^1$ with compact support in D. If a first order partial derivative $\frac{\partial u(x)}{\partial x^i}$ exists at a point x_0, then the corresponding Sobolev derivative $u_i(x_0) = \frac{\partial u(x_0)}{\partial x^i}$.

And now, we shall give a generalization of the modulus of an arc family for q-dimensional surfaces.

A q-dimensional surface $\sigma_q \subset R^n$ is a connected and q-dimensionally locally euclidean set. We recall that a set σ_q is said to be q-dimensionally locally euclidean if $\forall \xi \in \sigma_q$, there exists a neighbourhood $U_\xi \subset \sigma_q$ of ξ homeomorphic to a domain of R^q. Hence, a q-dimensional surface is a continuous image of a domain of R^q. In the case q=1, we have the curves. (For q=n-1, we suppress the index q and the epitet "q-dimensional" from the terminology and the notations). Next, let Σ_q be a family of q-dimensional surfaces $\sigma_q \subset R^n (1 \leq q \leq n-1)$ (q integer) and

$F(\Sigma_q)$ the class of functions $\rho(x)\geqq 0$, Borel measurable and such that $\int_{\sigma_q}\rho^q d\sigma_q \geqq 1\ \forall\sigma_q\in\Sigma_q$. Clearly, $F(\Sigma_q)\neq\emptyset$ since it contains at least the function $\rho(x)\equiv\infty$. The p-modulus of the family Σ_q is

$$M_p(\Sigma_q) = \inf_{\rho\in F(\Sigma_q)} \int_{R^n}\rho^p(x)d\tau .$$

In the case p=n, we drop the index p from the terminology and notations.

A family Σ_q is called p-exceptional if $M_p(\Sigma_q)=0$. A property is said to hold for p-almost every surface of the family Σ_q if the subfamily of the surfaces for which the property does not hold is p-exceptional.

Proposition 29. cap$E=0$ <=> $\mathrm{cap}_{(1,n)}E=0$ (Ju.G.Rešetnjak [39] theorems 6.1 and 6.2, or H.Wallin [41], corollary 2.1).

Proposition 30. $\mathrm{cap}_{(q,p)}E=0$ <=> the family of all q-dimensional Lipschitz surfaces intersecting E is p-exceptional (Ju.G.Rešetnjak [39], theorem 5.8).

Proposition 31. Let $p\geqq 1$ and $qp\leqq n$. In order that the family $\Sigma_q(E)$ of all q-dimensional Lipschitz surfaces which intersect a given set $E\subset R^n$ be p-exceptional, it is necessary and when p>1 also sufficient that there exist a function $\varphi\in L^p$, $\varphi\geqq 0$ whose potential of order q

$$u^{\varphi}_{n-q}(x) = \int\frac{\varphi(y)d\tau_y}{|x-y|^{n-q}}$$

equals $\infty\ \forall\ x\in E$, without being identically infinite.

Remark. For the proof see B.Fuglede [14], theorem 6. He observes also that "a necessary and sufficient condition in order that the potential u^{φ}_{n-q} of order q from above be not identically infinite is $\int\frac{\varphi(x)d\tau}{(1+|x|)^{n-q}}<\infty$, or equivalently $\int_{CB(R)}\frac{\varphi(x-y)d\tau_y}{(1+|y|)^{n-q}}<\infty$ for some, and hence $\forall$ pair (x,R), where $0<R<\infty$ and $x\in R^n$".

From these 2 propositions, we deduce

Corollary 1. $\mathrm{Cap}_{(q,p)}E=0$ $(p>1)$ <=> there exists a function $\varphi\in L^p$, $\varphi\geq 0$, whose potential $u^{\varphi}_{n-q}(x)=\infty\ \forall x\in E$ and $u^{\varphi}_{n-q}(x)\not\equiv\infty$.

Hence and from proposition 29, it follows

Corollary 2. $\mathrm{Cap}E=0$ <=> there exists a function $\varphi\in L^n$, $\varphi\geq 0$, whose potential $u^{\varphi}_{n-1}(x)=\infty\ \forall x\in E$ and $u^{\varphi}_{n-1}(x)\not\equiv\infty$.

(For other proofs, see Y. Mizuta [26], S.P. Preobraženskiĭ [30] and H. Wallin [41].)

Lemma 4. $\mathrm{Cap}E=0$ where $E\subset\overline{B(R_0)}$ <=> there exists a real function $\varphi\in L^n$, $\varphi\geq\varepsilon>0$ in a ball $B(R)$ containing $\overline{E}$, with compact support and whose potential $u^{\varphi}_{n-1}(x)=\infty\ \forall x\in E$ and $u^{\varphi}_{n-1}(x)\not\equiv\infty$.

According to the preceding corollary, there exists a function $v\in L^n$, $v\geq 0$, whose potential $u^{v}_{n-1}(x)=\infty\ \forall x\in E$ and $u^{v}_{n-1}(x)\not\equiv\infty$. Let us prove now that v can be chosen also so that $S_v\subset\overline{B(R)}$, where $B(R)$ is supposed to contain $\overline{E}$. Set $d=d[S(R),E]$ and assume $x\in E$, $x_0\in CE$ and $u^{v}_{n-1}(x_0)<\infty$, which is possible since $u^{v}_{n-1}(x)\not\equiv\infty$. There are 2 possibilities: I. $|x-x_0|\leq d$ and II. $|x-x_0|\geq d$. In the first case, we have

$$\infty>u^{v}_{n-1}(x_0)=\int\frac{v(y)\,d\tau_y}{|x_0-y|^{n-1}}\geq\int\frac{v(y)\,d\tau_y}{(|x_0-x|+|x-y|)^{n-1}}\geq\int_{|x-y|\geq d}\frac{v(y)\,d\tau_y}{(|x-x_0|+|x-y|)^{n-1}}\geq$$

$$\int_{|x-y|\geq d}\frac{v(y)\,d\tau_y}{(d+|x-y|)^{n-1}}\geq\frac{1}{2^{n-1}}\int_{|x-y|\geq d}\frac{v(y)\,d\tau_y}{|x-y|^{n-1}},$$

while, in the second case,

$$\infty>\int\frac{v(y)\,d\tau_y}{|x_0-y|^{n-1}}\geq\int_{|x-y|\geq d}\frac{v(y)\,d\tau_y}{(|x-x_0|+|x-y|)^{n-1}}=\frac{1}{|x-x_0|^{n-1}}\int_{|x-y|\geq d}\frac{v(y)\,d\tau_y}{\left(1+\frac{|x-y|}{|x-x_0|}\right)^{n-1}}\geq$$

$$\frac{1}{|x-x_0|^{n-1}}\int_{|x-y|\geq d}\frac{v(y)\,d\tau_y}{\left(1+\frac{|x-y|}{d}\right)^{n-1}}=\left(\frac{d}{|x-x_0|}\right)^{n-1}\int_{|x-y|\geq d}\frac{v(y)\,d\tau_y}{(d+|x-y|)^{n-1}}\geq$$

$$\left(\frac{d}{2|x-x_0|}\right)^{n-1}\int_{|x-y|\geq d}\frac{v(y)\,d\tau_y}{|x-y|^{n-1}}.$$

Hence, in both cases,

$$(15)\qquad \int_{|x-y|\geq d} \frac{v(y)\,d\tau_y}{|x-y|^{n-1}} < \infty$$

and, since $d[S(R),E]=d$, it follows that $|x-y|\geq d$ $\forall x\in E$ and $y\in CB(R)$, so that

$$\int_{CB(R)} \frac{v(y)\,d\tau_y}{|x-y|^{n-1}} \leq \int_{|x-y|\geq d} \frac{v(y)\,d\tau_y}{|x-y|^{n-1}} < \infty$$

$\forall x\in E$. But, by hypothesis, $u^{v}_{n-1}(x)=\int_{R^n}\frac{v(y)\,d\tau_y}{|x-y|^{n-1}}=\infty$ $\forall x\in E$, so that the preceding inequality yields

$$\int_{B(R)} \frac{v(y)\,d\tau_y}{|x-y|^{n-1}} = \infty$$

$\forall x\in E$, allowing us to conclude that the restriction $\psi(x)=v(x)\,|_{B(R)}=v(x)\chi_{B(R)}(x)$ (where χ_E is again the characteristic function of E) satisfies all the conditions of the preceding corollary, since

$$\int|\psi(x)|^n d\tau = \int_{B(R)}|\psi(x)|^n d\tau = \int_{B(R)}|v(x)|^n d\tau \leq \int|v(x)|^n d\tau < \infty,$$

i.e. $\psi\in L^n$ and

$$u^{\psi}_{n-1}(x) = \int\frac{\psi(y)\,d\tau_y}{|x-y|^{n-1}} = \int_{B(R)}\frac{\psi(y)\,d\tau_y}{|x-y|^{n-1}} = \int_{B(R)}\frac{v(y)\,d\tau_y}{|x-y|^{n-1}} = \infty,$$

without being identically infinite. Finally, let us consider the function

$$\varphi(x) = \begin{cases}\psi(x)+\varepsilon & \text{for } x\in B(R),\\ 0 & \text{otherwise.}\end{cases}$$

Then, using Minkowski's inequality and taking into account also the preceding ones,

$$\left(\int|\varphi|^n d\tau\right)^{\frac{1}{n}}=\left[\int_{B(R)}|\varphi|^n d\tau\right]^{\frac{1}{n}}=\left[\int_{B(R)}|\psi+\varepsilon|^n d\tau\right]^{\frac{1}{n}}\leq\left[\int_{B(R)}|\psi|^n d\tau\right]^{\frac{1}{n}}+\left[\int_{B(R)}\varepsilon^n d\tau\right]^{\frac{1}{n}} =$$

$$\left[\int_{B(R)}|\psi|^n d\tau\right]^{\frac{1}{n}}+\varepsilon\omega_n^{\frac{1}{n}}R < \infty,$$

i.e. $\varphi\in L^n$ and

$$u^{\varphi}_{n-1}(x) = \int \frac{\varphi(y)\,d\tau_y}{|x-y|^{n-1}} = \int_{B(R)} \frac{[\psi(y)+\varepsilon]\,d\tau_y}{|x-y|^{n-1}} \geq \int_{B(R)} \frac{\psi(y)\,d\tau_y}{|x-y|^{n-1}} = u^{\psi}_{n-1}(x) = \infty$$

$\forall x\varepsilon E$ with $u^{\varphi}_{n-1}(x) \not\equiv \infty$.

Remark. In this proof, as in the rest of the paragraph, we use, for simplicity sake, "$\int$" instead of "$\int_{R^n}$".

We recall that a sequence $\{f_m\}$ of functions is said to converge in the mean of order p to a function f if $\int |f_m - f|^p d\tau \to 0$ as $m \to \infty$.

Proposition 32. $\forall \varphi \varepsilon L^n$, there exists a sequence $\{\varphi_k\}$ of functions $\varphi_k \varepsilon C^{\infty}$ convergent in the mean of order p to φ (S.L. Sobolev [35], p.18).

Remark. The functions φ_k (k=1,2,...) are obtained by means of the average functions

$$(16) \qquad \varphi_h(x) = \frac{1}{Ah^n}\int \omega(x-y;h)\,\varphi(y)\,d\tau_y \ ,$$

where

$$\omega(x;h) = \begin{cases} e^{\frac{r^2}{r^2-h^2}} & \text{for } r < h \ , \\ 0 & \text{for } r \geq h \ , \end{cases}$$

$r = |x|$, $A = \int e^{\frac{r^2}{r^2-1}} d\tau$ and $\int \omega(x;h)\,d\tau = Ah^n$.

A property is said to hold (k,p)-quasi-everywhere in R^n if it holds everywhere except in a set of (k,p)-capacity zero.

Proposition 33. Let $\{\varphi_\nu\}$ be a sequence of functions $\varphi_\nu \varepsilon L^p$ such that $\int |\varphi_\nu - \varphi|^p d\tau \to 0$ as $\nu \to \infty$, then there exists a subsequence $\{\varphi_{\nu_m}\}$ so that $g_k * |\varphi_{\nu_m} - \varphi|(x) \to 0$ as $m \to \infty$ (k,p)-quasi-everywhere (Ju.G.Rešetnjak [31], theorem 1.10).

Corollary. Let $\{\varphi_\nu\}$ be a sequence of functions $\varphi_\nu \varepsilon L^n$ such that $\int |\varphi_\nu - \varphi|^n d\tau \to 0$ as $\nu \to \infty$, then there exists a subsequence $\{\varphi_{\nu_m}\}$ so that (1,n)-quasi-everywhere in R^n, $\int \frac{|\varphi_{\nu_m}(y) - \varphi(y)|}{|x-y|^{n-1}} d\tau \to 0$ as $m \to \infty$.

Lemma 5. CapE=0 and E⊂S => there exists a locally qc $f: B \rightleftarrows D^*$ such that

$$(17)\quad \int \frac{\sqrt[n]{|J(x)|}\,d\tau_x}{|\xi - x|^{n-1}} = \infty$$

$\forall\ \xi\varepsilon F$.

First, according to the preceding lemma, there exists a real function $\varphi\varepsilon L^n$, with compact support, $\varphi(x)\geq\varepsilon>0$ in a ball $B(R)\supset S$, such that $u^{\varphi}_{n-1}(x)=\ \forall x\,\varepsilon F$ and $u^{\varphi}_{n-1}(x)\not\equiv\infty$.

Now, let us consider the average function (16). According to Sobolev's proof of proposition 32, it follows that $\varphi_h\,\varepsilon C^{\infty}$ and $\int|\varphi_h-\varphi|^n d\tau \to 0$ as $h \to 0$. Next, since $\varphi>\varepsilon$ in $B(R)$, we deduce that

$$\varphi_h(x)=\frac{1}{Ah^n}\int\omega(x-y;h)\,\varphi(y)\,d\tau_y=\frac{1}{Ah^n}\int_{B(R)}\omega(x-y;h)\,\varphi(y)\,d\tau_y\geq\frac{\varepsilon}{Ah^n}\int_{B(R)}\omega(x-y;h)\,d\tau_y$$

Hence, if $R>h+1$, $\forall x\varepsilon\overline{B}$,

$$\varphi_h(x)\ \geq\frac{\varepsilon}{Ah^n}\int_{B(R)}\omega(x-y;h)\,d\tau_y=\frac{\varepsilon}{Ah^n}\int\omega(x-y;h)\,d\tau_y=\ \varepsilon>0\,.$$

Clearly, u^{φ}_{n-1} is inferior semi-continuous, so that $u^{\varphi}_{n-1}(x)\to\infty$ as $x\to x_0\ \forall x_0\varepsilon F$. Then, evident, $u^{\varphi_h}_{n-1}$ is continuous. Hence, and from the corollary of proposition 33, we deduce that $\lim u^{\varphi_h}_{n-1}(x_0)=\infty\ \forall x_0\varepsilon F$ for a certain sequence $\{h_m\}$.

Now, let us denote $\varphi^n_{h\,1}(x)=\frac{\partial x^{*1}_1(x)}{\partial x^1}$, i.e. let us consider $\varphi^n_{h\,1}$ as being the first order derivative with respect to the variable x^1 of the component x^{*1}_1 of a mapping $f_1(x)=x^{*i}_1(x)e_i$, where $e_i(i=1,\dots,n)$ are the coordinate unit vectors. Then set

$$(18)\qquad x^{*1}_1(x)=\int_{-1}^{x^1}\varphi^n_{h\,1}(x)\,dx^1\,,$$

hence

$$\frac{\partial x^{*1}_1(x)}{\partial x^k}=\frac{\partial}{\partial x^k}\int_{-1}^{x^1}\varphi^n_{h\,1}(x)\,dx^1=\int_{-1}^{x^1}\frac{\partial}{\partial x^k}\varphi^n_{h\,1}(x)\,dx^1\qquad(k=2,\dots,n)$$

are continuous in $\overline{B}$. But then, there exists a constant $M_1<\infty$ such that $|\nabla x^{*1}_1|<M_1$ in $B(1-\frac{1}{N_1})$, where $N_1>2$ and $\lim_{m\to\infty}N_m=\infty$.

Since $|\nabla x^{*1}_1|<M_1$ in B the mapping $x^*=f_1(x)$ given by

$$\begin{cases} x_1^{*1} = x_1^{*1}(x) , \\ x_1^{*k} = x^k \quad (k = 2,\dots,n) , \end{cases}$$

where x_1^{*1} is the function defined by (18), is a homeomorphism in B. The corresponding Jacobian $J_1(x) > \varepsilon^n$ in B and

$$\frac{|\nabla f_1(x)|^2}{n|J_1(x)|^{\frac{2}{n}}} < \frac{M_1+n-1}{n\varepsilon^2} = K_{N_1} < \infty$$

in $B(1-\frac{1}{N_1})$. Hence, and taking into account the last definition of the K-qc from above, we are allowed to conclude that f_1 is K_{N_1}-qc in $B(1-\frac{1}{N_1})$.

Next, by means of φ_{h_1} and φ_{h_2}, let us define $\psi_{h_2} \varepsilon C^1$ by a sewing process as follows

$$\psi_{h_2}(x) = \begin{cases} \varphi_{h_1}(x) & \text{for} \quad x \varepsilon \overline{B(1-\frac{1}{N_1})} , \\ \varphi_{h_2}(x) & \text{for} \quad x \varepsilon \{x; 1-\frac{1}{N}+\varepsilon_2 \leq |x| \leq 1\} \quad (\varepsilon_2 < \frac{1}{N_1}-\frac{1}{N_2}) \end{cases}$$

and $\psi_{h_2}(x) \geq \min[\varphi_{h_1}(x), \varphi_{h_2}(x)]$ in the ring $\{x; 1-\frac{1}{N_1} < |x| < 1-\frac{1}{N_1}+\varepsilon_2\}$ and and such that $\psi_{h_2} \varepsilon C^1$ in B. Denoting, as above,

$$\psi_{h_2}^n(x) = \frac{\partial x_2^{*1}(x)}{\partial x^1}, \quad x_2^{*1}(x) = \int_{-1}^{x^1} \psi_{h_2}^n(x)\,dx^1, \quad \frac{\partial x_2^{*1}(x)}{\partial x^k} = \int_{-1}^{x^1} \frac{\partial \psi_{h_2}^n(x)}{\partial x^k}\,dx^1 \ (k=2,\dots,n)$$

there exists a constant $M_2 < \infty$ such that $|\nabla x_2^{*1}|^2 < M_2^2$ in $B(1-\frac{1}{N_2})$. The mapping $x_2^* = f_2(x)$, where

$$\begin{cases} x_2^{*1} = x_2^{*1}(x) , \\ x_2^{*k} = x^k \quad (k = 2,\dots,n) \end{cases}$$

is clearly a homeomorphism. Its Jacobian $J_2(x) \geq \varepsilon^n$ in B and

$$\frac{|\nabla f_2(x)|^2}{n|J_2(x)|^{\frac{2}{n}}} < \frac{M_2+n-1}{n\varepsilon^2} = K_{N_2} < \infty .$$

Thus f_2 is K_{N_2}-qc in $B(1-\frac{1}{N_2})$.

In general, we define $\psi_{h_m} \varepsilon C^1$ in a similar way:

$$\psi_{h_m}(x) = \begin{cases} \psi_{h_{m-1}}(x) \ \text{for} \ x \varepsilon \overline{B(1-\frac{1}{N_{m-1}})} , \\ \varphi_{h_m}(x) \ \text{for} \ x \varepsilon \{x; 1-\frac{1}{N_{m-1}}+\varepsilon_m < |x| \leq 1\} \quad (\varepsilon_m < \frac{1}{N_{m-1}}-\frac{1}{N_m}) \end{cases}$$

and $\psi_{h_m}(x) \geq \min[\psi_{h_{m-1}}(x), \varphi_{h_m}(x)]$ in the ring $\{x; 1-\frac{1}{N_{m-1}} < |x| < 1-\frac{1}{N_{m-1}}+\varepsilon_m\}$

and such that $\phi_{h_m} \in C^1$ in B. We have again

$$\phi_{h_m}^n(x) = \frac{\partial x_m^{*1}(x)}{\partial x^1}, \quad x_m^{*1}(x) = \int_{-1}^{x^1} \phi_{h_m}^n(x)\,dx^1,$$

$|\nabla x_m^{*1}| < M_m < \infty$ in $B(1-\frac{1}{N_m})$ and the homeomorphism

$$\begin{cases} x_m^{*1} = x_m^{*1}(x) , \\ x_m^{*k} = x^k \quad (k = 2,\dots,n) , \end{cases}$$

with the Jacobian $J_m(x) > \varepsilon^n$ in B and

$$\frac{|\nabla f_m(x)|^2}{n|J_m(x)|^{\frac{2}{n}}} < \frac{M_m+n-1}{n\varepsilon^2} = K_{N_m} < \infty .$$

Finally, the homeomorphism $f(x) = \lim_{m\to\infty} f_m(x)$ is locally qc in B since $\forall x \in B$ and $\forall V_x \subset B$ (i.e. with $\overline{V}_x \subset B$), there is an N_m sufficiently large so that $V_x \subset B(1-\frac{1}{N_m})$ and f be K_N -qc in $B(1-\frac{1}{N_m})$ and then a fortiori in V_x. It is easy to see that $\lim_{m\to\infty} u_{\phi_{h_m}}(x_0) = \infty \; \forall x_0 \in E$ and that $J(x) = \lim_{m\to\infty} J_m(x) = \lim_{m\to\infty} \phi_{h_m}^n(x)$ in B.

Theorem 6. If the compact set $F \subset S$ has the conformal capacity zero, then there exists a locally qc f in B such that $\forall \xi \in F, m(\underset{s}{\cup} l_s) > 0$, where $l_s \subset B$ is the segment with an endpoint at ξ and corresponding to the direction s, for which $f(l_s)$ is a non-rectifiable arc.

According to the proof of the preceding lemma, there esists a mapping f with the properties that $J(x) = \lim_{m\to\infty} \phi_{h_m}^n(x)$ in B and (17) holds $\forall \xi \in F$, hence, $\forall \xi \in F$,

$$\infty = \int_{B_\xi} \frac{\sqrt[n]{|J(x)|}\,d\tau_x}{|\xi - x|^{n-1}} = \int_{S_1} d\sigma \int_0^{s} \sqrt[n]{|J(x)|}\,d\rho ,$$

where $B_\xi = B(x_0, R)$ is a ball tangent to S at ξ, $R < \frac{1}{4}$, ρ_s is the length of the segment joining ξ to the point of the sphere $S_\xi = \partial B_\xi$ corresponding to the direction s, and S_1 is the semi-sphere corresponding to the directions of ρ_s. Hence,

$$(19) \qquad \int_0^{s} \sqrt[n]{|J(x)|}\,d\rho = \infty$$

for a set E of directions s corresponding to a set of points of S_1 of strictly positive (n-1)-dimensional Hausdorff measure, i.e. $m(\bigcup_{s\in E} l_s)>0$.

Finally, since $\forall s\in E$,

$$\left|\frac{\partial f(x)}{\partial s}\right| = \left|\frac{\partial f(x)}{\partial x^i}\cos(s,x^i)\right| \geq \left|\frac{\partial f(x)}{\partial x^1}\right| |\cos(s,x^1)| = \left|\frac{\partial x^{*1}(x)}{\partial x^1}\right| |\cos(s,x^1)| =$$

$$|J(x)|\,|\cos(s,x^1)| \quad ,$$

where

$$\left|\frac{\partial f(x)}{\partial s}\right| = \lim_{|\Delta x|_s\to 0}\left|\frac{\Delta f(x)}{\Delta x}\right|,$$

$\Delta f(x)=f(x+\Delta x)-f(x)$ and $|\Delta x|_s \to 0$ means $\Delta x \to 0$ in the direction s, then (19) yields, $\forall s\in E$,

$$\infty=\int_0^s \sqrt[n]{|J(x)|}\,d\rho \leq \int_0^{\rho_s} \frac{\left|\frac{\partial f(x)}{\partial s}\right| d\rho}{|\cos(s,x^1)|} = \frac{1}{|\cos(s,x^1)|}\int_0^{\rho_s}\left|\frac{\partial f(x)}{\partial s}\right| d\rho =$$

$$\frac{1}{|\cos(s,x^1)|}\int_0^{\rho_s}\frac{ds^*}{d\rho}d\rho = \frac{1}{|\cos(s,x^1)|}\int_{\gamma^*} ds^* \quad ,$$

where $\gamma^*=f(l_s)$.

§4. Open questions.

1. The main open question connected with this paper is if our estimate of E_0 by means of the h-measure is best possible. We established that $\operatorname{cap}E_0=0$ and that $\operatorname{cap}F=0 \Rightarrow H_h(F)=0$ with h given by (1) $\forall\alpha>n-1$ (this last result being best possible). We proved also that given a compact set $F\subset S$ with $\operatorname{cap}F=0$, there is a locally qc for which F is of the type E_0, but, in order that our estimate be best possible, it would be necessary to prove the existence $\forall$ compact $F\subset S$ of a K-qc (not only a locally qc) for which F be of the type E_0.

2. A similar question holds also for the other estimate by means of the Φ-capacity. Only that, in this case, there is something more to show, i.e. we obtained that $\operatorname{cap}F=0 \Rightarrow C_\Phi F=0$ with Φ given by (2)

$\forall\beta>n-2$ (with possibility of equality if $n=2$) and then proved the existence of Cantor sets F' with cap$F'=0$, but $C_\Phi F'>0$ $\forall\beta<n-2$. It remains also to settle the case $\beta=n-1$ when $n>2$.

3. It may be of interest to consider also the exceptional set $E'_0=\{\xi\in S; f(\gamma_\xi)$ non-rectifiable $\forall$ linear segment γ_ξ of B from $\xi\}$. Clearly, $E_0\subset E'_0$, but we ask if we have not even $E_0=E'_0$ (i.e. if the non-rectifiability of the image of all segments of B with an endpoint at ξ does not imply the non-rectifiability of the image of all the endcuts of B from ξ).

4. M.Tsuji [38] established in the case n=2 that if <u>almost all</u> the radii of $B(\xi,\frac{1}{2})$ that lie in B have rectifiable images by a qc $f: B\rightleftarrows D^*$, then even <u>all</u> the radii have rectifiable images, hence $E_2=E_3=E_r$. Is this true also for $n>2$?

5. Another question is if cap$E_r=0$, cap$E_3=0$, cap$E_2=0$, or at least cap$E_1=0$?

6. Let $f: B\rightleftarrows D^*$ be a qc. A point $M\in R^n$ is said to be a <u>natural value of</u> f <u>at</u> $x_0\in\bar{B}$ if

$$\lim_{r\to 0}\frac{1}{r^n}\int_{B(x_0,r)}|f(x)-M|\,d\tau = 0 .$$

If f is continuous at x_0, then $f(x_0)$ is a natural value of f at x_0, so that the mapping f from above has a natural value everywhere in B. M.V.Mikljukov [25] established that the exceptional set of the points of S, which do not admit a natural value is of α-capacity zero $\forall\alpha>0$. However, it remains an open question if also the Φ-capacity is zero with Φ given by (2) $\forall\beta>n-2$.

7. What about the exceptional sets from above in Hilbert spaces?

R E F E R E N C E S

1. Adams David, Traces of potentials. Indiana Univ. Math. J. 22(1973) 907-919.

2. — and Meyers Norman, Bessel potentials. Inclusion relations among classes of exceptional sets. Indiana Univ. Math. J. 22(1973) 873-905.

3. Белинский П.П. и Лаврентьев М.А., Некоторые проблемы геометрической теории функций. Труды Мат. Унив. Стеклова 128(1972) 34-40.

4. Beurling Arne, Ensembles exceptionels. Acta Math. 72(1941) 1-13.

5. Caraman Petru, A new definition of the n-dimensional quasiconformal mappings. Nagoya Math. J. 26(1966) 145-165.

6. — On the equivalence of the definitions of the n-dimensional quasiconformal homeomorphisms (QCfH). Rev. Roumaine Math. Pure Appl. 12(1967) 889-943.

7. — n-dimensional quasiconformal mappings. Edit. Acad. București (Romania) and Abacus Press, Tunbridge Wells, Kent (England) 1974, 553p.

8. — Quasiconformality and boundary correspondence. Conf. on Constructive function theory. Cluj 6-12.IX.1973; Mathematica. Revue Anal. Numér. Théorie Approximation (Cluj) 5(1976) 117-126.

9. — About a conjecture of F.W. Gehring on the boundary correspondence. Conf. on Analytic Functions. Krakow 4-11.IX.1974; Ann. Polon. Math. 35(1976) 21-33.

10. — Exceptional sets for boundary correspondence of quasiconformal mappings. Proc. Inst. of Math. Iași 1976, 117-123.

11. — Relations between capacities, Hausdorff h-measures and p-modules. Mathematica. Revue Anal. Numér. Théorie

Approximation (Cluj)(in print).

12. Carleson Lenart, Selected problems on exceptional sets. Van Nostrand Math. Studies Nr. 43. Univ. of Upsala 1957, Princeton - New Jersey - Toronto - London - Melbourne 151 p.

13. Fatou Pierre, Séries trigonométriques et séries de Taylor. Acta Math. 30(1906)335-400.

14. Fuglede Bent, Extrémal length and functional completion. Acta Math. 98(1957)171-219.

15. - On the thory of potentials in locally compact spaces. Acta Math. 103(1960)139-215.

16. Gehring Frederick, The Carathéodory convergence theorem for quasiconformal mappings. Ann. Acad. Sci. Fenn. Ser. A I 336/11(1963)1-21.

17. - and Väisälä Jussi, The coefficients of quasiconformality of domains in space. Acta Math. 114(1965)1-70.

18. Хавин В.П. и Мазья В.Г., Нелинейный аналог ньютоновского потенцала и метрические свойства (p,1)-емкости. Докл. Акад. Наук СССР 194 (1970)770-773.

19. — Нелинейная теория потенциала. Успехи Мат. Наук 27(168)(1972)67-133.

20. Jenkins J.A., On quasiconformal mappings. J. Rat. Mech. Anal. 5(1956) 343-352.

21. Kametani S., On some properties of Hausdorff's measure and the concept of capacity in generalized potentials. Proc. Imp. Acad. Japan. 18(1942)157-179.

22. - A note on a metric property of capacity. Mat. Sci. Rep. Ochanomizu Univ. 4(1953)51-54.

23. Lohwater A.J., Beurling theorem for quasiconformal mappings. Bull. Amer. Math. Soc. 61(1955)223, abstr. 414t.

24. - The boundary behaviour of a quasiconformal mapping. J. Rat. Mech. Anal. 5(1956)335-342.

25. Миклюков В.М., Об одном граничном свойстве n-мерных отображений с ограниченным искажением. Мат. Заметки 11(1972)159-174.

26. Mizuta Y., Integral representation of Beppo-Levi functions of higher order. Hiroshima Math. J. 4(1974)375-396.

27. Mori Akira, On quasiconformality and pseudo-analyticity. Trans. Amer. Math. Soc. 84(1957)56-77.

28. Naas J. and Schmid H.D., Mathematisches Wörterbuch. I. Akedemie Verlag Berlin und B.C. Teubner Leipzig 1967.

29. Polya G. und Szegö G., Über transfiniten Durchmesser (Kapazitätskonstanten) von ebenen und räumlichen Funktionen. J. reine angew. Math. 165(1931)4-49.

30. Преображенский С.П., О множествах расходимости интегралов типа потенциала с плотностями из L^p. Зап. Научн. Сем. Мат. Инст. Акад. Наук СССР 22(1971)196-198.

31. Reade Maxwell, On quasiconformal mappings in three spaces. Bull. Amer. Math. Soc. 63(1957)193, abstr. 371t.

32. Решетняк Ю.Г., О понятии емкости в теории функций с обобщенными производными. Сибирск. Мат. Ж. 10(1969)1109-1138.

33. — О граничном поведение функций с обобщенными производными. Сибирск. Мат. Ж. 13(1972)411-419.

34. Riesz Frédérick und Marcel, Über die Randwerte analytischer Funktionen. C.R. 4-ème Congrès Scandinave. Stockholm 1916.

35. Соболев С.Л., Некоторые применения функционального анализа к математической физике. Ленинград 1950, 255 с.

36. Storvick David, The boundary correspondence of a quasiconformal mapping in space. Math. Research Center U.S. Army. The Univ. of Wisconsin NRC Technical Summary Report 426(1963)1-8.

37. Taylor S.J., On the connexion between Hausdorff measures and generalized capacity. Proc. Cambridge Philos. Soc. 57(1961) 524-531.

38. Tsuji Masatsugu, Beurling's theorem on exceptional sets. Tôhoku Math. J. 2(1950)113-125.

39. Ugaheri Tadashi, On general potential and capacity. Japan. J. Math. 20 (1950)37-43.

40. Väisälä Jussi, On quasiconformal mappings in space. Ann. Acad. Sci. Fenn. Ser. I A 298(1961)1-36.

41. Wallin Hans, Metrical characterization of conformal capacity zero. J.Math.Anal.Appl.58(1977)298-311.

42. Зорич В.А., Граничные свойства одного класса отображений в пространстве. Докл.Акад.Наук СССР 153(1963)23-26.

43. - Об угловых граничных значениях квазиконформных отображений шара. Докл.Акад.Наук СССР 177(1967)771-773.

Institute of Mathematics
Iaşi

THE BEHAVIOUR OF SOME METRICS ON RIEMANN SURFACES IN RESPECT WITH QUASICONFORMAL MAPPINGS

by

Dorin Ghişa

The behaviour of hyperbolic and Harnack metrics on Riemann surfaces in respect with quasiconformal mappings may be described by some simple inequalities which are verified by some associated metrics. The manner of association is general and it will be done in the lemma 3. Before that we prove a property of some classes of numerical sequences.

Let $a \in (0,1)$ be a given number. We define step by step the following sequence :

$$a_0 = a \ , \ a_{n+1} = \left(\frac{1-\sqrt{1-a_n^2}}{a_n}\right)^2 \quad , \ n = 0,1,2,\ldots$$

Lemma 1 : The sequences constructed by means of two numbers $a,b \in (0,1)$ and of the number ab are related by :

$$(ab)_n > a_n b_n \qquad (1)$$

Proof : We proceed by induction. For $n = 1$ this inequality is equivalent with

$$(1-\sqrt{1-a^2})(1-\sqrt{1-b^2}) + a^2b^2 < 0 \ ,$$

or after some calculations, with

$$\sqrt{1-a^2} + \sqrt{1-b^2} > \sqrt{1-a^2b^2} + \sqrt{(1-a^2)(1-b^2)} \ .$$

By denoting $1-a^2 = a'$ and $1-b^2 = b'$, this inequality is written

$$\sqrt{a'} + \sqrt{b'} > \sqrt{a'+b'-a'b'} + \sqrt{a'b'} \ ,$$

which after some more calculation becomes

$$a' + b' - a'b' < 1$$

which is equivalent with $ab > o$.

Taking into account that the function $\left(\frac{1-\sqrt{1-x^2}}{x}\right)^2$ is increasing and having in view the previous result, the induction hypothesis $(ab)_n > a_n b_n$ permites us to write

$$(ab)_{n+1} = \left(\frac{1-\sqrt{1-(ab)_n^2}}{(ab)}\right)^2 > \left(\frac{1-\sqrt{1-a_n^2 b_n^2}}{a_n b_n}\right)^2 >$$

$$> \left(\frac{1-\sqrt{1-a_n^2}}{a_n}\right)^2 \cdot \left(\frac{1-\sqrt{1-b_n^2}}{b_n}\right)^2 = a_{n+1} b_{n+1} ,$$

and the desired inequality is completely proved.

Now let $\mu(r)$ the modulus of the extremal domain $B(r)$ of Grötzsch, that is the double connected domain obtained by taking out from the unity disk $E = \{ z ; |z| < 1\}$ the segment $[o,r]$, $o < r < 1$. It is known ([4], pp.64) that for every $a \in (o,1)$,

$$2^{-n}\log \frac{1+\sqrt{1-a_n^2}}{a_n} < \mu(a) < 2^{-n}\log \frac{4}{a_n} \qquad (2)$$

where (a_n) is the anterior defined sequence.

<u>Lemma 2</u> : If $\mu(r)$ is the modulus of the extremal domain $B(r)$ of Grötzsch, then for every $a,b \in (o,1)$ we have

$$\mu(ab) < \mu(a) + \mu(b) \qquad (3)$$

<u>Proof</u> : Let (a_n) and (b_n) the sequences constructed by means of a and b . Writing the inequalities (2) for a and b , we obtain particularly that $a_n \to o$ and $b_n \to o$ as $n \to \infty$. Then there is a n_o so that for $n \geq n_o$ we have

$$(1+\sqrt{1-a_n^2})(1+\sqrt{1-b_n^2}) > 2 \qquad (4)$$

On the ground of the inequalities (1),(2),(4), for $n \geq n_o$ we may write :

$$\mu(ab) < 2^{-n}\log\frac{4}{(ab)_n} < 2^{-n}\log\frac{4}{a_n b_n} < 2^{-n}\log\frac{(1+\sqrt{1-a_n^2})^2(1+\sqrt{1-b_n^2})^2}{a_n b_n} =$$

$$= 2^{-n}\log\frac{(1+\sqrt{1-a_n^2})^2}{a_n} + 2^{-n}\log\frac{(1+\sqrt{1-b_n^2})^2}{b_n} < \mu(a) + \mu(b).$$

The relation (3) may be completed for $a = 1$ or $b = 1$, in which case it becomes equality if we put $\mu(1) = o$.

Lemma 3 : Let (X,d) be a metric space. Then $d_\mu : X \times X \to R$ defined by $d_\mu(x,y) = \mu(e^{-d(x,y)})$ is also a metric on X.

Proof : We have $\mu(t) = o$ iff $t = o$, hence $d_\mu(x,y) = o$ iff $e^{-d(x,y)} = 1$, that is $d(x,y) = o$. In other words $d_\mu(x,y) = o$ iff $x = y$.

The equality $d_\mu(x,y) = d_\mu(y,x)$ is obvious.

Finally, from (2) and the fact that μ is decreasing it results : $d_\mu(x,y) = \mu(e^{-d(x,y)}) \leq \mu(e^{-d(x,z)-d(y,z)}) = \mu(e^{-d(x,z)}e^{-d(y,z)}) \leq \mu(e^{-d(x,z)}) + \mu(e^{-d(y,z)}) = d_\mu(x,z) + d_\mu(y,z)$ and lemma is completely proved.

Let now W be a hyperbolic Riemann surface and $\widetilde{W}$ his universal covering surface. Then $\widetilde{W}$ is conformally equivalent with the unity disk E. It is known that the element of arc $ds = \frac{|dz|}{1-|z|^2}$ is invariant in respect with the conformal mappings of E on E. This invariance permits us to define a distance on W in the following way. Let $\pi : \widetilde{W} \to W$ the canonical projection and $\gamma : E \to \widetilde{W}$ be a conformal mapping. By hyperbolic element of arc $\widetilde{ds}$ on $\widetilde{W}$ it is meant γ-image of the anterior defined element ds on E. From the invariance of ds at a conformal mapping of E on E it results that ds does not depend on γ. By hyperbolic element $d\sigma$ on W it is meant π-image of $\widetilde{ds}$. The hyperbolic distance between two points $p,q \in W$ is $\delta(p,q)=$

$= \inf_C \int_C d\sigma$, where C runs into the family of continuous differentiable arcs on W which connect p and q . If we choose Ψ so that one of the points $\tilde{p}$ over p pass by Ψ^{-1} in zero and the corresponding $\tilde{q}$ in z , then

$$\delta(p,q) = \frac{1}{2} \log \frac{1+|z|}{1-|z|} \tag{5}$$

Theorem 1 : Let W and W' be two hyperbolic Riemann surfaces and f a K-quasiconformal mapping of W on W' . Then for every $p,q \in W$, we have :

$$\frac{1}{2K} \delta_\mu(p,q) \le \delta_\mu(p',q') \le 2K\, \delta_\mu(p,q) \tag{6}$$

where p',q' are the images of p,q by f .

Proof : Let $\Psi : E \to W$ and $\Psi' : E \to W'$ be so selected that a point $\tilde{p}$ over p pass by Ψ^{-1} in zero and a point $\tilde{p}'$ over p' pass by Ψ'^{-1} in zero. Let z and z' be the images of the corresponding $\tilde{q}$ and $\tilde{q}'$. Then

$$\begin{aligned} \delta(p,q) &= \delta = \frac{1}{2} \log \frac{1+|z|}{1-|z|} \\ \delta(p',q') &= \delta' = \frac{1}{2} \log \frac{1+|z'|}{1-|z'|} \end{aligned} \tag{7}$$

To f correspond a K-quasiconformal mapping w of E on E with w(o) = o . According to (5),[1] , we have :

$$\varphi_{1/K}\left(\frac{1-|z|}{1+|z|}\right) \le \frac{1-|z'|}{1+|z'|} \le \varphi_K\left(\frac{1-|z|}{1+|z|}\right) \text{ , or}$$

$$\varphi_{1/K}(e^{-2\delta}) \le e^{-2\delta'} \le \varphi_K(e^{-2\delta}) \tag{8}$$

where $\varphi_K(t) = \mu^{-1}\left(\frac{\mu(t)}{K}\right)$.

The last inequality (8) may be written

$$\mu(e^{-2\delta'}) \geq \frac{\mu(e^{-2\delta})}{K}$$

and from (3) and the continuity and monotony of μ , it follows that

$$2\mu(e^{-\delta'}) \geqslant \mu(e^{-2\delta'}) \geqslant \frac{\mu(e^{-2\delta})}{K} \geqslant \frac{\mu(e^{-\delta})}{K}, \text{ or}$$

$$\frac{\mu(e^{-\delta})}{2K} \leq \mu(e^{-\delta'}), \text{ hence}$$

$$\frac{1}{2K}\,\delta_\mu(p,q) \leq \delta_\mu(p',q')$$

which is exactly the first inequality (6) .

In an analogous way, from $\varphi_{1/K}(e^{-2\delta}) \leq e^{-2\delta'}$, we obtain

$$2K\mu(e^{-\delta}) \geqslant K\mu(e^{-2\delta}) \geqslant \mu(e^{-2\delta'}) \geqslant \mu(e^{-\delta'}), \text{ or}$$

$$\delta_\mu(p',q') \leq 2K\,\delta_\mu(p,q)$$

Hence the theorem is completely proved.

Now let Ω be an arbitrary set and let $(0,\infty)^\Omega$ be the family of positive real functions on Ω . It is said that $\mathcal{S} \subset (0,\infty)^\Omega$ verifies the Harnack inequality iff for every x and y from Ω there is a positive constant $a = a(x,y)$ so that $\frac{1}{a} \leq \frac{h(x)}{h(y)} \leq a$ for every $h \in \mathcal{S}$.

By denoting $D(x,y) = \inf\left\{a \,;\, \frac{1}{a} \leq \frac{h(x)}{h(y)} \leq a,\ h \in \mathcal{S}\right\}$ and $d(x,y) = \log D(x,y)$ it is found that d is a semimetric on Ω . If $\mathcal{S}$ separate the points of Ω , that is for every $x,y \in \Omega$, $x \neq y$ there exists $h \in \mathcal{S}$ with $h(x) \neq h(y)$, then d is a metric on Ω , so called the Harnack metric.

If W is a hyperbolic Riemann surface and $H^+(W)$ the family of positive harmonic functions on W , then $H^+(W)$ verifies the Harnack inequality, hence it defines a Harnack metric d on W.

According to the theorem 4.2, [3] , if W is a simple connected Riemann surface then $d = 2\delta$.

In a similar way with the proof of the theorem 1 we can prove:

<u>Theorem 2</u> : Let W and W' two simple connected Riemann surfaces and f a K-quasiconformal mapping of W on W' . Then for every $p,q \in W$ we have :

$$\frac{1}{K} d_{\mu}(p,q) \leq d_{\mu}(p',q') \leq K d_{\mu}(p,q) \qquad (9)$$

where p',q' are the images of p,q by f .

Remarks : 1^{0}. The theorem 1 is proved in [3] for the particular case of a simple connected domain W .

2^{0}. Taking $K = 1$ in the theorem 2 , we obtain the conformal invariance of Harnack metric for simple connected riemann surface.

BIBLIOGRAPHY

1. D.Ghişa : Remarks on Hersch-Pfluger Theorem. Math.Z. 136 (1974), p.291-293

2. J.Hersch : Contribution à la thèorie des functions pseudo-analitiques. Comment.Math.Helv. 3o (1956),p.1-19

3. J.Köhn : Die Harnacksche Metric in der Theorie der harmonischen Functionen. Math.Zeitschr. 91 (1966), p. 5o-64

4. O.Lehto und K.I.Virtanen : Quasiconforme Abbildungen. Berlin-Heidelberg-New York ; Springer 1965.

University of Timişoara
Faculty of Mathematics

ON KLEIN - MASKIT COMBINATION THEOREMS

by D. Ivaşcu

In this paper we intend to discuss an extension of the well-known combination theorems that F. Klein and B. Maskit have proved for Kleinian groups. This theorems give sufficient conditions that the group G generated by the union of two discontinuous groups G_1, G_2 to be discontinuous and allow to construct a fundamental domain for G by means of two corresponding fundamental domains of the groups G_1, G_2. Such theorems can be very useful in a classification theory of the discontinuous groups acting on a fixed space. They also can be used to construct groups with some special properties.

First of all it is necessary to remind some definitions.

<u>Definition 1</u> Let X be a locally compact, connected topological space and G a group of homeomorphisms of X. We say that G acts discontinuously on the open set $U \subset X$ if for every compact set $K \subset U$ the set $G_K = \{T \mid T \in G, \ TK \cap K \neq \emptyset\}$ is finite. By $\Omega(G)$ we denote the maximal open set on which G acts discontinuously. The group G will be called discontinuous if $\Omega(G) \neq \emptyset$.

<u>Definition 2</u> Let G be a discontinuous group acting on X and D an open subset of $\Omega(G)$. WE say that D is a fundamental domain for G if the following two conditions are satisfied.

i) $TD \cap D = \emptyset$ for all $T \in G - I$

ii) $\bigcup_{T \in G} T\bar{D} = \Omega(G)$ ($\bar{D}$ being the closure of D in $\Omega(G)$)

<u>Definition 3</u> The one point compactification of R^n will be denoted by $M(n)$ and will be called the Moebius space of dimension n.

<u>Definition 4</u> The group of the transformations of $M(n)$ generated by reflexions in the spheres and hyperplanes of R^n will be denoted by $GM(n)$ and will be called the Moebius group of dimension n.

<u>Definition 5</u> A sequence (G,D,H,Δ,S) in which G is a dis - continuous subgroup of $GM(n)$, H is a subgroup of G, D (Δ) is a fundamental domain for G (H) and S is a compact hypersurface of $M(n)$, will be called conglomerate if the following conditions are satisfied

a) $D \subset \Delta$

b) S is invariant under H

c) there exists a neighbourhood V of S such that $\Delta \cap V \subset D$.

<u>Definition 6</u> The sequence (G,D,H,Δ,S,B) will be called a big conglomerate if (G,D,H,Δ,S) is a conglomerate, B is one of the two conexe components of $M(n) - S$, invariant under H and the following conditions are satisfied:

a) $\Delta \cap (B \cup S) = D \cap (B \cup S)$

b) $D \cap (B \cup S) \neq D$

c) $T(S) \cap S = \emptyset$ if $T \in G - H$.

The last two definitions were introduced by B. Maskit in the case of the the Kleinian groups.

Now, if we remark that the group $GM(n)$ have the same geometrical nature as the group all conformal mappings of the compact complex plane onto itself, we can restate some results, concerning the Kleinian groups, for the discontinuous subgroups of $GM(n)$.

<u>Lemma 1</u> If $T \in GM(n)$ and $T(\infty) \neq \infty$ then there exists a n-1 dimensional sphere S_T centered in the point $O_T = T^{-1}(\infty)$ such that the restriction of T on S_T is an isometric mapping. If B_T is the ball for which $\operatorname{Fr} B_T = S_T$, then $T(B_T) = \operatorname{Ext} B_T$.

<u>Lemma 2</u> If $T \in GM(n)$, $O_T \neq \infty$ and r_T is the radius of the ball B_T then there exists a linear isometric mapping of R^n denoted by A_T such that $T(x) = O_{T^{-1}} + r_T^2 A_T(x - O_T)/|x - O_T|^2$.

<u>Proposition 1</u> If G is a subgroup of $GM(n)$ such that $G_\infty = \{T \mid T \in G,\ O_T = \infty\} = I$, then the following two conditions are equivalent:

i) there exists $R > 0$ such that $T(\complement B_R) \cap \complement B_R = \emptyset$

ii) $D(G) = \operatorname{Int} (\bigcap_{T \in G-I} \operatorname{Ext} B_T) \neq \emptyset$

<u>Corollary 1</u> If the subgroup G of the group $GM(n)$ satisfies one of the two equivalent conditions of the preceding proposition, then the set $\{r_T\}_{T \in G-I}$ has no point of accumulation different from 0.

Further all the subgroups of $GM(n)$ taken into account will satisfy the conditions of the proposition 1.

<u>Proposition 2</u> Let G be a discontinuous subgroup of $GM(n)$. If $\Lambda(G)$ is the set of limit points of G then $\Lambda(G) = \complement\Omega(G) = \{O_T\}'_{T \in G-I}$.

As a consequence of the proposition 2 we can state the following proposition.

<u>Proposition 3</u> Let G be a discontinuous subgroup of $GM(n)$. If $\lambda \in \Lambda(G)$ then there exists $\lambda' \in \Lambda(G)$ such that $\lambda \in (Gx)'$ for all $x \in M(n) - \{\lambda, \lambda'\}$.

<u>Corollary 2</u> If G is a discontinuous subgroup of $GM(n)$ and A is a subset of $M(n)$ with more than two points then $\Lambda(G) \subset \overline{G(A)}$. If A is closed and G invariant then $\Lambda(G) \subset A$.

If we consider in R^n the usual metric given by the formula $ds^2(x) = \sum_{i=1}^{n} dx_i^2$, then it is easy to see, that under a transformation $T \in GM(n)$ for which $0_T \neq \infty$,it changes as follows $ds_T^2(x) = r_T^4/|x - 0_T|^4 ds^2(x)$. Denoting by $\lambda_T(x)$ the coefficient $r_T^2/|x-0_T|^2$ then we have the following relation: $\lambda_{T_1 \circ T_2} = (\lambda_{T_1} \circ T_2)\lambda_{T_2}$.

Lemma 3 If G is a discontinuous subgroup of GM(n) and $x \in \Omega(G)$ then the set $\{\lambda_T(x)\}_{T \in G-I}$ has no point of accumulation different from 0 .

Corollary 3 If G is a discontinuous subgroup of GM(n) then for every $x \in \Omega(G)$ there exists a transformation $T_x \in G$ such that $\lambda_{T_x}(x) \geqslant \lambda_T(x)$ for all $T \in G - I$.

Because it is easy to see that $\overline{D(G)} = \bigcap_{T \in G-I} \overline{\mathrm{Ext}\, B_T}$ and $\overline{\mathrm{Ext}\, B_T} = \{x \mid \lambda_T(x) \leq 1\}$ we obtain the following proposition:

Proposition 4 If G is a discontinuous subgroup of GM(n) then D(G) is a fundamental domain for G.

Theorem 1 If G is a discontinuous subgroup of GM(n) then the series $\sum_{T \in G-I} r_T^{2n}$ is convergent.

Proof Let x^* be a point of D(G) and r a positive real number such that $B_r(x^*) \subset D(G)$ and $d(B_r(x^*), \{0_T\}_{T \in G-I}) \geqslant d > 0$. As we have seen, we can find $R > 0$ such that $TB_r(x^*) \subset B_R$ for all $T \in G - I$. Because $TB_r(x^*) \cap \tilde{T}B_r(x^*) = \emptyset$ if $T \neq \tilde{T}$ ($T, \tilde{T} \in G$) we can write: $\sum_{T \in G-I} \int_{TB_r(x^*)} dv \leq V(B_R)$ (dv being the volume element in R^n and $V(B_R)$ the volume of the ball B_R)

Since $\int_{TB_r(x^*)} dv = \int_{B_r(x^*)} (r_T^{2n}/|x - 0_T|^{2n} dv$ it follows that

$$\sum_{T \in G-I} \int_{TB_r(x^*)} dv = \int_{B_r(x^*)} \sum_{T \in G-I} r_T^{2n}/|x - 0_T|^{2n} dv \leqslant V(B_R)$$

Because $|x - O_T| \leqslant |x| + |O_T| \leq |x^*| + r + R$ it results that

$$\Big(\sum_{T \in G-I} r_T^{2n}\Big)/(|x^*| + r + R)^{2n} \int_{B_r(x^*)} dv \leqslant \int_{B_r(x^*)} \sum r_T^{2n}/|x - O_T|^{2n} dv \leqslant V(B_R)$$

and the theorem is proved.

Now we prove the following extension of a theorem of Koebe.

<u>Theorem 2</u> Let G be a discontinuous subgroup of $GM(n)$, H a subgroup of G and K a compact subset of R^n. If the following two condition are satisfied

i) there exists a complet system of representatives S for G/H ($G = \bigcup_{T \in S} TH$, $TH \cap \tilde{T}H = \phi$ if $T \neq \tilde{T}$ and $T, \tilde{T} \in S$) such that $d(K, \{O_T\}_{T \in S}) > 0$.

ii) K is invariant under H.

then the series $\sum_{T \in S} d^n(TK)$ is convergent. (Here $d(TK)$ is the diameter of the set TK).

<u>Proof</u> As a consequence of the condition (ii) there exists $d > 0$ such that $d(K, O_T) \geqslant d$ for all $T \in S$. Therefore $d(TK) \leqslant r_T^2/ d^2(K, O_T) \leqslant \leqslant r_T^2/ d^2$. Since the series $\sum_{T \in G-I} r_T^{2n}$ is convergent the theorem is proved.

<u>Corollary 3</u> If G_1 is an infinite subset of discontinuous subgroup G and K is a compact set that satisfies the condition $d(K, \{O_T\}_{T \in G-I}) > 0$, then the set $\{d(TK)\}_{T \in G}$ has no accumulation point different from 0. As a consequence if $x_j \in K$, $T_j \in G_1$ ($T_j \neq T_k$ if $j \neq k$) and $T_j(x_j) \longrightarrow x$ then for every $\varepsilon > 0$ there exists N_ε such that $T_j(K) \subset B_\varepsilon(x)$ if $j \geqslant N_\varepsilon$.

The Klein combination theorem can be extended to the case of the discontinuous subgroups of $GM(n)$ as follows.

Theorem 3 Let $\{G_j\}_{j\in J}$ ($J\subset N$) be a family of discontinuous subgroups of GM(n). If the fundamental domains $D(G_j)$ satisfies the following conditions

a) $\mathrm{Ext}\, D(G_i) \subset D(G_j)$ if $i \neq j$

b) $D = \mathrm{Int}\,(\bigcap_{j\in J} D(G_j)) \neq \emptyset$

then it can be proved that:

a) The group $G = G(\bigcup_{j\in J} G_j)$ generated by the union of the groups G_j is a discontinuous group isomorphic with the free product $*\prod_{j\in J} G_j$.

b) D is a fundamental domain for G.

c) If the set J is finite and $m_n(\Lambda(G_j)) = 0$ for all $j\in J$ then $m_n(\Lambda(G)) = 0$. (Here $m_n(A)$ denotes the n - dimensional Hausdorff measure of the set A)

Proof The proof of the first two statements is the same as in the case of the Kleinian groups. The last statement is a consequence of one of the Maskit combination theorems that can also be extended to the case of the discontinuous subgroups of GM(n).

Concerning the first statement of the theorem we remark that it remains true in a much more general situation. Namely it can be proved the following theorem:

Theorem 3′ Let $\{G_j\}_{j\in J}$ be a family of discontinuous groups acting on a locally compact, connected space X. If there exists a family of open sets $\{D_j\}_{j\in J}$ $(D_j \subset X)$ such that

a) $\mathrm{Ext}\, D_i \subset D_j$ if $i \neq j$

b) $D = \mathrm{Int}\,(\bigcap_{j\in J} D_j) \neq \emptyset$

c) $(G_j - I)(D_j) \cap D_j = \emptyset$ for all $j\in J$

then the group $G = G(\bigcup_{j \in J} G_j)$ is a discontinuous group isomorphic with the free product $\ast\prod_{j \in J} G_j$.

The Maskit first combination theorem, extended to the case of the discontinuous subgroups of $GM(n)$ can be stated as follows.

<u>Theorem 4</u> Let $(G_i, D_i, H, \Delta, S, B_i)$ with $i \in Z_2$ be two big conglomerates. If $B_0 \cap B_1 = \emptyset$ then

1) $G = G(G_0 \cup G_1)$ is a discontinuous group isomorphic with the free product $G_0 \ast_H G_1$ with amalgamated subgroup H.

2) $D = D_0 \cap D_1$ is a fundamental domain for G.

3) (G, D, H, Δ, S) is a conglomerate.

4) if $m_n(\Lambda(G_0) \cup \Lambda(G_1)) = 0$ then $m_n(\Lambda(G)) = 0$

<u>Proof</u> First of all we prove that $(G_i - H)(\overline{B}_i) \subset B_{i+1} - D$ and $(G_i - H)(D) \subset B_{i+1} - D$. In order to verify the first relation we proceed as follows:

$$(G_i - H)H(D_i \cap B_i) \cap H(D_i \cap B_i) = H^{-1}(G_i - H)H(D_i \cap B_i) =$$
$$=(G_i - H)(D_i \cap B_i) \cap (D_i \cap B_i) = \emptyset .$$

The last equality is due to the fact that $(G_i - H)(D_i) \cap D_i = \emptyset$.

If we remark that $H(D_i \cap B_i)$ and $(G_i - H)H(D_i \cap B_i)$ are open sets and $H(D_i \cap B_i)$ is a dens subset of B_i then we obtain the relation $(G_i - H)H(D_i \cap B_i) \cap B_i = \emptyset$. From this relation it follows that $(G_i - H)(\overline{B}_i) \subset B_{i+1}$. By analogy from the relation $(G_i - H)H(D_i \cap B_i) \cap D_i = \emptyset$ it results that $(G_i - H)(\overline{B}_i) \cap D_i = \emptyset$. Consequently $(G_i - H)(\overline{B}_i) \subset B_{i+1} - D$. Now we remark that

$$(G_i - H)(D_i \cap B_{i+1}) \cap H(D_i \cap B_i) = (G_i - H)H(D_i \cap B_{i+1}) \cap H(D_i \cap B_i) =$$

The proof follows closely the proof given by B. Maskit for the case of the Kleinian groups.

$= H^{-1}(G_i - H)H(D_i \cap B_{i+1}) \cap (D_i \cap B_i) = (G_i - H)(D_i \cap B_{i+1}) \cap (D_i \cap B_i) = \emptyset$.
Because $H(D_i \cap B_i)$ is a dense subset of B_i and $(G_i - H)(D_i \cap B_{i+1}) \cap D_i = \emptyset$ it follows that $(G_i - H)(D_i \cap B_{i+1}) \subset B_{i+1} - D_i$. Hence $(G_i - H)(D_i \cap B_{i+1}) \subset B_{i+1} - D$. As a consequence of the preceding relatioms we obtain $(G_i - H)(D) = (G_i - H)(D \cap \overline{B}_i) \cup (G_i - H)(D \cap B_{i+1}) \subset B_{i+1} - D$. Further we use the following notations: $\hat{G} = G_0 *_H G_1$, $\hat{G}_0 = H$, $\hat{G}_k^+ = \{T_{k-1} * \cdots\cdots * T_0 \in \hat{G} \mid T_j \in G_{[j]} - H\}$ $\hat{G}^+ = \bigcup_{k \geq 1} G_k^+$

$\hat{G}_k^- = \{T_k * \cdots\cdots * T_1 \in \hat{G} \mid T_j \in G_{[j]} - H\}$ $\hat{G}^- = \bigcup_{k \geq 1} G_k^-$

$h : \hat{G} \longrightarrow G$ will be the canonical morphism. Because $h(\hat{G}_k^-) = (G_k - H) \circ \cdots\cdots \circ (G_1 - H)$ we can write $h(\hat{G}_k^-)(D) = (G_{[k]} - H) \circ \cdots \cdots \circ (G_1 - H)(D) \subset (G_{[k]} - H) \circ \cdots \circ (G_0 - H)(B_0 - D) \subset \cdots \subset (G_{[k]} - H)(B_{[k]} - D) \subset B_{[k+1]} - D$. Hence $h(\hat{G}_k^-)(D) \cap D = \emptyset$. Analogously $h(\hat{G}_k^+)(D) \cap D = \emptyset$. Consequently $I \notin h(\hat{G}^+ \cup \hat{G}^-)$ and h is an isomorphism. If we make the following notations: $G_k^+ = h(\hat{G}_k^+)$, $G_k^- = h(\hat{G}_k^-)$, $L_k = G_k^+(B_0) \cup G_k^-(B_1)$ $L = \bigcap_{k \geq 0} L_k$ then we can prove that for every $x \in \Omega(G) - L$ there exists $T \in G$ such that $T(x) \in \overline{D}$ ($\overline{D}$ being the closure of D in $\Omega(G)$) and $\Lambda(G) = L \cup (\bigcup_{k \geq 0} [G_k^+ \Lambda(G_1) \cup G_k^- \Lambda(G_0)])$.

If $x \in M(n) - L \cup G(S)$ then there exists $k \geq 0$ such that x $L_k - L_{k+1}$ Hence $x = T_k \circ \cdots \circ T_1(y)$ with $y \in B_1$ and $T_1 \in G_1 - H$ or $x = T_{k-1} \circ \cdots \circ T_0(y)$ with $y \in B_0$ and $T_0 \in G_0 - H$. If $y \in B_1$ ($y \in B_0$) then as a consequence of the relation $x \notin L_{k+1} \cup G(S)$ it follows that $G_0(y) \subset B_1$ ($G_1(y) \subset B_0$). If $y \in B_1 - \Lambda(G_0)$ ($y \in B_0 - \Lambda(G_1)$) then there exists $z \in \overline{D}_0 \cap B_1$ ($z \in \overline{D}_1 \cap B_0$) and $T \in G_0$ ($T \in G_1$) such that $Tz = y$. If we remark that $(\overline{D}_0 \cap B_1) \cup (\overline{D}_1 \cap B_0) \subset \overline{D}$, we can conclude that for all $x \in M(n) - L \cup G(S) \cup (\bigcup_{k \geq 0} [G_k^+ \Lambda(G_1) \cup G_k^- \Lambda(G_0)])$ there exists $T \in G$ such that $T(x) \in \overline{D}$. (In the preceding relations $\overline{D}$ ($\overline{D}_i$) is the closure of D (D_i) in $\Omega(G)$ ($\Omega(G_i)$)).

Because $\Delta \cap S = D \cap S$ and $T(D) \cap D = \emptyset$ for all $T \in G - I$ it is easy to see that $\Lambda(G)\cap S = \Lambda(H)$, $\Omega(G)\cap S = \Omega(H)\cap S$. Hence for every $x \in G(S) - \bigcup_{k>0}[G_k^+\Lambda(G_1)\cup G_k^-\Lambda(G_0)]$ there exists $T \in G$ such that $T(x) \in \bar{\Delta}\cap S = \bar{D}\cap S$. Consequently $\Omega(G) \supset \bigcup_{T\in G} T\bar{D} \supset$ $\supset M(n) - L\cup(\bigcup_{k>0}[G_k^+\Lambda(G_1)\cup G_k^-\Lambda(G_0)])$. Further we prove that $L \subset \Lambda(G)$ and $m_n(L) = 0$. Clearly we can suppose that $\infty \in D_0\cap D_1$. If V is a neighbourhood of S such that $\Delta\cap V \subset D$ and T is a transformation of $G - H$ that satisfies the condition $O_T \in V$ then we can find $\tilde{T} \in H$ such that $\tilde{T}\circ T^{-1}(\infty) \in \bar{\Delta}$. Because $\infty \in D$, $\Delta \cap V \subset D$ and $\tilde{T}\circ T^{-1} \neq I$ it follows that $\tilde{T}\circ T^{-1}(\infty) \in \bar{\Delta} - V$. Con - sequently we can find a complete system of representatives σ for $(G - H)/H$ such that $O_T \notin V$ if $T \in \sigma$. Now as a consequence of the Koebe theorem the series $\sum_{T\in\sigma} d^n(TS)$ is convergent. Denoting by σ_k^+ (σ_k^-) the set $\sigma\cap G_k^+$ ($\sigma\cap G_k^-$) we can write:

$$\sum_{k>0}\Big(\sum_{T\in\sigma_k^+} d^nT(S) + \sum_{T\in\sigma_k^-} d^nT(S)\Big) < \infty$$

If we remark that $dT(B_0) = dT(S)$ if $T \in G_k^+$ and $dT(B_1) =$ $= dT(S)$ if $T\in G_k^-$ then the preceding relation takes the form:

$$\sum_{k>0}\Big(\sum_{T\in\sigma_k^+} d^nT(B_0) + \sum_{T\in\sigma_k^-} d^nT(B_1)\Big) < \infty$$

As a consequence $\lim_{k\to\infty}\Big(\sum_{T\in\sigma_k^+} d^nT(B_0) + \sum_{T\in\sigma_k^-} d^nT(B_1)\Big) = 0$. The last relation shows that $m_n(L) = 0$. If we remark that $L_k = G_k^+(B_0)\cup G_k^-(B_1)$ $= \sigma_k^+(B_0)\cup\sigma_k^-(B_1)$ then for every $x \in L$ we can suppose that $x = \lim_{k\to\infty} T_k(x_k)$ with $x_k \in B_0$ and $T_k \in \sigma_k^+$ (or $x_k \in B_1$ and $T_k \in \sigma_k^-$). Since $\lim_{k\to\infty} dT_k(B_0) = \lim_{k\to\infty} dT_k(S) = 0$ it follows that $x = \lim_{k\to\infty} T_k(y)$ with $y \in B_0$ (or $y \in B_1$).

Consequently $L \subset \Lambda(G)$. The preceding considerations shows that D is a fundamental domain for G, $\Lambda(G) = L\cup(\bigcup_{k>0}[G_k^+\Lambda(G_1)\cup G_k^-\Lambda(G_0)])$ and $m_n(\Lambda(G)) = 0$. Because clearly (G,D,H,Δ,S) is a conglomerate the theorem is proved.

Finally we remark that the second combination theorem of B. Maskit can also be proved in the case of the discontinuous subgroups of $GM(n)$. It can be stated as follows:

<u>Theorem 5</u> Let $(G,D,H_0,\Delta_0,S_0,B_0)$, $(G,D,H_1,\Delta_1,S_1,B_1)$ be big conglomerates and G_2 be a cyclic group generated by the transformation T_0. If we suppose that

1) $\mathrm{Int}(D - [(D \cap B_0) \cup (D \;\; B_1)]) \neq \emptyset$

2) $[T(B_0 \cup S_0)] \cap (B_1 \cup S_1) = \emptyset$ if $T \in G$

3) $T_0(S_0) = S_1$

4) $T_0 H_0 T_0^{-1} = H_1$

5) $T_0(B_0) \cap B_1 = \emptyset$

then it can be proved that

a) the group $\tilde{G}$ generated by the set $G \cup G_2$ is a discontinuous subgroup of $GM(n)$.

b) the set $D \cap (M(n) - (B_0 \cup B_1 \cup S_1))$ is a fundamental domain for G.

c) if $m_n(\Lambda(G)) = 0$ then $m_n(\Lambda(\tilde{G})) = 0$

d) there exists a fundamental domain $\tilde{D}$ for $\tilde{G}$ such that $(\tilde{G},\tilde{D},H_0,\Delta_0,S_0)$ is a conglomerate.

The proof can be made by analogy with the proof of the theorem 4.

References

1 W. Abikoff, Some Remarks on Kleinian Groups, "Advances in The Theory of Riemann Surfaces" Ann. Math. Studies No. 66, pp. 1-7 New Jersey, 1971

2 W. Abikoff, On The Decomposition and Deformation of Kleinian Groups, Contributions to Analysis, A Collection of Papers Dedicated to Lipman Bers, Academic Press New York 1974.

3 B. Maskit, On Klein's combination theorem, III, "Advances in The Theory of Riemann Surfaces" Ann. Math. Studies No. 66, pp. 297-316 New Jersey, 1971.

4 G. Mostow, Quasiconformal mappings in n-Space and The Rigidity of Hyperbolic Space Forms, Inst. Hautes Études Sci. Publ. Math. 34, 1968, pp. 53-104.

Institute of Mathematics
Bucharest

REMARKS ON A CLASS OF QUASISYMMETRIC MAPPINGS

by D. Ivaşcu

In this paper we want to discuss a problem concerning the extension of a quasisymmetric mapping of the real line R to a quasiconformal transformation of the complex plane. If Q_R denotes the group of quasiconformal transformation of the complex plane that maps R onto itself and H_{qs} denotes the group of quasisymmetric mappings of R then the mapping $r:Q_R \longrightarrow H_{qs}$ $(r(w) = w_{|R})$ is an epimorphism of groups. This fact was proved by Beurling and Ahlfors who constructed an extension mapping $e:H_{qs} \longrightarrow Q_R$ $(r \circ e = 1_{H_{qs}})$. It is easy to see that this mapping is not a group morphism. Thus it remains an open question if the mapping r has a right inverse in the category of groups. Concerning this problem, we consider a subgroup H_{fqs} (that will be indicated further) of the group H_{qs} and define an extension mapping $e':H_{fqs} \longrightarrow Q_R$ which is also a group morphism.

We begin by proving the following theorem :

Theorem 1 If $f:R \longrightarrow R$ is an increasing homeomorphism, then the following conditions are equivalent:

a) there exists a $k \geqslant 1$ such that $\frac{1}{k} \leqslant \frac{f(y) - f(x)}{f(v) - f(u)} \leqslant k$, for all $x,y,u,v \in R$ that satisfy the condition $0 < y - x = v - u$.

b) for each $s \geqslant 1$ there exists a $k(s) \geqslant 1$ such that $\frac{1}{k(s)} \leqslant \frac{f(y) - f(x)}{f(v) - f(u)} \leqslant k(s)$, for all $x,y, u, v \in R$ that satisfy the condition $\frac{1}{s} \leqslant \frac{(y - x)}{(v - u)} \leqslant s$.

c) there exists an $k \geq 1$ such that

$$1/k \leq \frac{[f(x + r\cos\theta) - f(x)]^2 + [f(u + r\sin\theta) - f(u)]^2}{[f(x + r\cos\theta_1) - f(x)]^2 + [f(u + r\sin\theta_1) - f(u)]^2} \leq k$$

<u>Proof</u> To see that (a) and (b) are equivalent it is clearly suficient to prove that (b) is a consequence of (a). If $1/s \leq \frac{y - x}{v - u} \leq s$ then $y \leq x + ([s]+1)(v - u)$ ($[s]$ being the greatest natural number smaller or equal to s). Because f is increasing $f(y) \leq f(x + ([s]+1)(v - u))$. Hence $f(y) - f(x) \leq$
$\leq \sum_{j=0}^{[s]} f(x + (j + 1)(v - u)) - f(x + j(v - u))$. If f satisfies the condition (a) then $1/k \leq \frac{f(x + (j+1)(v-u)) - f(x + j(v-u))}{f(v) - f(u)} \leq k$

Consequently $\frac{f(y) - f(x)}{f(v) - f(u)} \leq \sum_{j=0}^{[s]} \frac{f(x + (j+1)(v-u)) - f(x + j(v-u))}{f(v) - f(u)} \leq$
$\leq ([s]+1)k$. Now changing y by v and x by u we obtain the inequality $\frac{f(v) - f(u)}{f(y) - f(x)} \leq ([s]+1)k$, so the condition (b) is a consequence of (a) if we take $k(s) = ([s]+1)k$.

Further we prove that (a) is a consequence of (c) and (c) a consequence of (b). If in the relation (c) we take $r = y - x = v - u$ $\theta = 0$ and $\theta_1 = \pi/2$ then we obtain $1/k \leq \frac{[f(y) - f(x)]^2}{[f(v) - f(u)]^2} \leq k$.
Because f is increasing the last relationis equivalent with the condition $1/\sqrt{k} \leq \frac{f(y) - f(x)}{f(v) - f(u)} \leq \sqrt{k}$ for all x, y, u, v such that $0 < y - x = v - u$. Hence (a) follows from (c).

In order to prove that (c) follows from (b) we remark that
$[f(x + r\cos\theta) - f(x)]^2 + [f(u + r\sin\theta) - f(u)]^2 \geq$
$\geq \min\{[f(x + r\sqrt{2}/2) - f(x)]^2, [f(x - r\sqrt{2}/2) - f(x)]^2,$
$[f(u + r\sqrt{2}/2) - f(u)]^2, [f(u - r\sqrt{2}/2) - f(u)]^2\}$ This inequality is a direct consequence of the facts that f is an increasing function

and $|\max(\cos\theta, \sin\theta)| \geq \sqrt{2}/2$. Since $[f(x + r\cos\theta) - f(x)]^2 + [f(u + r\sin\theta) - f(u)]^2 \leq 2\max\{[f(x \pm r) - f(x)]^2, [f(u \pm r) - f(u)]^2\}$ we obtain the following inequality:

$$\frac{[f(x + r\cos\theta) - f(x)]^2 + [f(u + r\sin\theta) - f(u)]^2}{[f(x + r\cos\theta_1) - f(x)]^2 + [f(u + r\sin\theta_1) - f(u)]^2} \leq$$

$$\leq \frac{2\max\{[f(x\pm r) - f(x)]^2, [f(u\pm r) - f(u)]^2\}}{\min\{[f(x \pm r\sqrt{2}/2) - f(x)]^2, [f(u \pm r\sqrt{2}/2) - f(u)]^2\}}$$

Because f satisfies the condition (b) it follows that

$$1/k^2(2/\sqrt{2}) \leq \frac{\max\{[f(x\pm r) - f(x)]^2, [f(u\pm r) - f(u)]^2\}}{\min\{[f(x \pm r\sqrt{2}/2)-f(x)]^2, [f(u\pm r\sqrt{2}/2)-f(u)]^2\}} \leq k^2(2/\sqrt{2})$$

Consequently $\dfrac{[f(x+r\cos\theta)-f(x)]^2+[f(u+r\sin\theta)-f(u)]^2}{[f(x+r\cos\theta_1)-f(x)]^2+[f(u+r\sin\theta_1)-f(u)]^2} \leq 2k^2(2/\sqrt{2})$.

Since in the last relation we can interchange x and u it follows that (c) is a consequence of (b).

Definition 1 Let $f:R \longrightarrow R$ be an increasing homeomorphism. We say that f is free quasisymmetric if f satisfies the condition (a) of theorem 1.

Remark 1 Every free quasisymmetric mapping is quasisymmetric.

Remark 2 The inverse of a free quasisymmetric mapping is also free quasisymmetric.

As a direct consequence of the condition (b) of theorem 1 it follows that for any two free quasisymmetric mappings f and g the mapping $f \circ g$ is also free quasisymmetric. Consequently we can state the following proposition:

Proposition 1 The set H_{fqs} of all the free quasisymmetric mappings is a subgroup of the group H_{qs}.

Now we can prove the following theorem:

Theorem 2 The mapping $e': H_{fqs} \longrightarrow Q_{\mathbb{R}}$ defined by the formula $e(f)(z) = f(\mathrm{Re}z) + if(\mathrm{Im}z)$ is a monomorphism of the group H_{fqs}

onto a subgroup of Q_R. This mapping provides a quasiconformal extension to the complex plane for each quasisymmetric mapping f that sarisfies the condition $f(0) = 0$.

Proof Because $e'(f^{-1}) = e'(f)^{-1}$ and $e'(f)$ is continuous it follows that $e'(f)$ is an homeomorphism. If $H_f(z)$ denotes the dilatation of $e'(f)$ in z then

$$H_f^2(z) = \limsup_{r \to 0} \frac{\sup_\theta([f(x+r\cos\theta) - f(x)]^2 + [f(y+r\sin\theta) - f(y)]^2)}{\inf_\theta([f(x+r\cos\theta) - f(x)]^2 + [f(y+r\sin\theta) - f(y)]^2)}$$

Hence, as a consequence of the theorem 1, it follows that $e'(f)$ is quasiconformal. Since $e'(f\circ g) = e'(f)\circ e'(g)$ and $e'(f)|_R = f$ if $f(0) = 0$, the theorem is proved.

Finally we remark that our proof of the theorem 2 needs the inequality (c) only for small values of r. This fact suggests the following definition.

Definition 2 An increasing homeomorphism $f:R \longrightarrow R$ is called (k,ε) free quasisymmetric $(k \geq 1,\ \varepsilon > 0)$ if

$1/k \leq \frac{f(y) - f(x)}{f(v) - f(u)} \leq k$ for all y, x, v, u that satisf the relation $0 < y - x = v - u < \varepsilon$.

Theorem 1 If $f:R \longrightarrow R$ is an increasing homeomorphism then the following conditions are equivalent.

(a) there exists $k \geq 1$ and $\varepsilon > 0$ such that

$$1/k \leq \frac{f(y) - f(x)}{f(v) - f(u)} \leq k \quad \text{if} \quad 0 < y - x = v - u < \varepsilon .$$

b) there exists $\varepsilon > 0$ such that for each $s \geq 1$ there is a $k(s) \geq 1$ for which $1/k(s) \leq \frac{f(y) - f(x)}{f(v) - f(u)} \leq k(s)$ if $1/s \leq \frac{y - x}{v - u} \leq s$ and $0 < y - x < \varepsilon,\ 0 < v - u < \varepsilon$.

c) there exists $k \geq 1$ and $\varepsilon > 0$ such that

$$1/k \leq \frac{[f(x+r\cos\theta) - f(x)]^2 + [f(y+r\sin\theta) - f(y)]^2}{[f(x+r\cos\theta_1) - f(x)]^2 + [f(y+r\sin\theta_1) - f(y)]^2} \leq k$$

for all $0 < r < \varepsilon$ and $\theta, \theta_1 \in R$.

<u>Proof</u> The same proof as in the case of theorem 1.

<u>Lemma1</u> If f is a (k,ε) free quasisymmetric mapping then f, f^{-1} are uniformly continuous.

<u>Proof</u> Because any (k,ε) free quasisymmetric mapping is clearly uniformly continuous it is suficient to prove that the inverse of a (k,ε) free quasisymmetric mapping is also uniformly continuous. If we suppose that f^{-1} is not uniformly continuous, then there exists $\eta > 0$ such that for each $\delta > 0$ we can find y_δ, x_δ satisfying the conditions $0 < y_\delta - x_\delta < \delta$ and $f^{-1}(y_\delta) - f^{-1}(x_\delta) > \eta$ Clearly we can suppose that $\eta < \varepsilon$. If $f^{-1}(y_\delta) - f^{-1}(x_\delta) \geq \varepsilon$ we can choose $\tilde{y}_\delta \in [x_\delta, y_\delta]$ such that $\eta \leq f^{-1}(y_\delta) - f^{-1}(x_\delta) < \varepsilon$. Now let $v, u, s \in R$ satisfy the conditions $0 < v - u < \varepsilon$ $[\eta/(v-u), \varepsilon/(v-u)] \subset [1/s, s]$. Since f is (k,ε) free quasisymmetric we can write $1/k(s) \leq \frac{f(f^{-1}(y_\delta)) - f(f^{-1}(x_\delta))}{f(v) - f(u)} \leq k(s)$ Consequently $y_\delta - x_\delta \geq \frac{f(v) - f(u)}{k(s)}$. The last condition cannot be satisfied because $0 < y_\delta - x_\delta < \delta$ and δ can be chosen arbitrarily. Thus the lemma is proved.

<u>Lemma 2</u> If f is (k,ε) free quasisymmetric then f is (k_1, ε_1) free quasisymmetric.

<u>Proof</u> We prove that f^{-1} satisfies the condition (b′) of the theorem 1′. Since f^{-1} is uniformly continuous there exists $\varepsilon_1 > 0$ such that $f^{-1}(y) - f^{-1}(x) < \varepsilon$ if $0 < y - x < \varepsilon_1$. If we suppose that f^{-1} does not satisfy the condition (b′) of the theorem 1′ we can find $s \geq 1$ and $y_n, x_n, v_n, u_n \in R$ such that $0 < y_n - x_n < \varepsilon_1$,

$$0 < v_n - u_n < \varepsilon_1 \ ; \quad 1/s \leqslant \frac{y_n - x_n}{v_n - u_n} \leqslant s \quad \text{and} \quad \lim_{n\to\infty} \frac{f^{-1}(y_n) - f^{-1}(x_n)}{f^{-1}(v_n) - f^{-1}(u_n)} = \infty .$$

Now we choose the points $y_{n,j}$, with $0 \leqslant j \leqslant m_n+1$, such that

$$y_{n0} = f(x_n), \quad y_{n,m_n+1} = f(y_n), \quad y_{n,j} < y_{n,j+1} \quad \text{and} \quad 1/s \leqslant \frac{y_{n,j+1} - y_{n,j}}{f(v_n) - f(\bar{u}_n)} \leqslant s.$$

Since f is (k,ε) free quasisymmetric it follows that

$$1/k(s) \leqslant \frac{f(y_{nj+1}) - f(y_{nj})}{v_n - u_n} \leqslant k(s). \quad \text{Consequently}$$

$$\sum_{j=0}^{m_n} \frac{f(y_{nj+1}) - f(y_{nj})}{v_n - n} = \frac{y_n - x_n}{v_n - u_n} \geqslant (m_n + 1)/k(s). \text{ Because}$$

obviously, $\lim m_n = \infty$, the last relation contradicts the condition $1/s \leqslant \frac{y_n - x_n}{v_n - u_n} \leqslant s$. This contradiction proves the lemma.

<u>Lemma 3</u> If f, g are (k,ε), (k_1, ε_1) free quasisymmetric mappings, then there exists $k \geqslant 1$, $\varepsilon > 0$ such that $f \circ g$ is a (k,ε) free quasisymmetric mapping.

<u>Proof</u> g being uniformly continuous the lemma follows directly from the condition (b') of the theorem $1'$. Now we can state the following proposition.

<u>Proposition 2</u> The set $\tilde{H}_{fqs}$ of all increasing homeomorphisms f of R that satisfy one of the three equivalent conditions of the theorem $1'$ is a subgroup of the group H_{qs}. Clearly H_{fqs} is a subgroup of $\tilde{H}_{fqs}$.

<u>Theorem $2'$</u> The mapping $\tilde{e}: \tilde{H}_{fqs} \longrightarrow Q_R$ defined by the same formula as the mapping e' from theorem $2'$ is a monomorphism of the group $\tilde{H}_{fqs}$ onto a subgroup of Q_R. This mapping provides a quasi-conformal extension to the complex plane for each $f \in \tilde{H}_{fqs}$ that verifies the condition $f(o) = o$.

<u>Proof</u> The same proof as in the case of theorem 2.

References.

Andreian Cazacu Cabiria, O cvaziconformnîh otobrajeniah, D.A.N. S.S.S.R., 1959, 126, 2, 235-238.

- , Sur les inégalités de Rengel et la définition géométrique des représentations quasi-conformes, Revue Roum. Math. Pures Appl., 9,2,141-155.

Beurling,A. and L.Ahlfors - The boundary correspondence under quasiconformal mappings. Acta Math. 96, 125-142 (1956).

Lehto,O. and K.I.Virtanen, Quasiconformal mappings in the plane, Springer-Verlag, Berlin Heidelberg, New-York (1973).

Martio,O., Boundary values and injectiveness of the solutions of Beltrami equations. Ann.Acad.Sci.Fenn.AI 402 (1967).

Reed,T.J., Quasiconformal mappings with given boundary values, Duke Math.J 33,44-48 (1962).

Institute of Mathematics
Bucharest

S.L.Krushkal'

TO THE PROBLEM OF THE SUPPORTS OF BELTRAMI DIFFERENTIALS FOR KLEINIAN GROUPS.

In the theory of Kleinian groups of the plane there exists the following well-known and so far unsolved problems: whether Beltrami differentials $\mu(z)d\bar{z}/dz \not\equiv 0$, $\mu \in L_\infty(\mathbb{C})$, relative to the given group G, concentrated only on the limit set $\Lambda(G)$ (when its plane measure is positive), can exist, or whether any quasiconformal deformation $f^\mu : \overline{\mathbb{C}} \to \overline{\mathbb{C}}$ of group G(i.e. such one that $f^\mu G (f^\mu)^{-1} \subset SL(2,\mathbb{C})/\{\pm 1\}$) must be conformal on $\Lambda(G)$?

In this paper we shall give an account of some results concerning the mentioned problem. The paper consists of two parts. In the first part the consideration is given for any $\mathbb{R}^n$, $n \geq 2$, and the characteristics of fundamental set in $\Lambda(G)$ are investigated. The second part is connected immediately with the problem mentioned above.

§ I. Fundamental sets in $\Lambda(G)$.

I. In Euclidean space $\mathbb{R}^n$, $n \geq 2$, let us consider unit ball $B^n = \{x \in \mathbb{R}^n : |x| < 1\}$ and group $\mathcal{M}\ddot{o}b\, B^n$ of conformal(Möbius) automorphisms $x \to Ax$ of that ball. Let G be a discrete subgroup $\mathcal{M}\ddot{o}b\, B^n$, or, in other words, Kleinian group. Group G acts discontinuously in B^n and, perhaps, on some set $\Omega(G) \subset \partial B^n = S^{n-1}$ too, open in S^{n-1}.

The limit set $\Lambda = \Lambda(G)$ of the group is the set of orbit accumulation points $Gx = \{Ax\}, x \in B^n$. It is closed, and either coincides with S^{n-1}, or is dense nowhere in S^{n-1}. We shall assume that $\Lambda(G)$ is infinite, and, for the simplicity, we suppose that G does not contain rotations of S^{n-1}.

Let us denote here the spherical (n-1)-dimensional Lebesgue measure on S^{n-1} by μ_{n-1}; Jacobi matrix of mapping A – by $A'(x)$, and the linear dilation at that mapping in x point – by $|A'(x)| = |dAx|/|dx|$. It should be noted that the consideration of spherical metric instead of Euclidean one is unessential, it is more convenient in case when $\Lambda(G) = S^{n-1}$. If $\Omega(G) \neq \emptyset$, then, at $n > 2$, it is easier to consider Möbius automorphisms of $\mathbb{R}^{n-1}$, without continuing them in $\mathbb{R}^n_+ = \{x = (x_1, \ldots, x_n) \in \mathbb{R}^n : x_n > 0\}$, which allows to assume that point $x = \infty$ lies in $\Omega(G)$ and, hence, $\Lambda(G)$ is bounded in $\mathbb{R}^{n-1}$, so we may take (n-1)-dimensional Lebesgue measure m_{n-1} in $\mathbb{R}^{n-1}$ instead of μ_{n-1}.

As G is discontinuous on $B^n \cup \Omega(G)$, so we can always choose in this set the fundamental set of the group, containing one point from every orbit Gx, $x \in B^n \cup \Omega(G)$. For instance, isometric fundamental polyhedron $P_G = \{x \in B^n \cup \Omega(G) : \sup_{A \in G \setminus \{1\}} |A'(x)| < 1\}$, bounded by the parts of isometric spheres $I_A = \{x : |A'(x)| = 1\}$ (circumferences at n = 2) may serve as such a set.

2. The analogous procedure, to some extent, can be performed in $\Lambda(G)$ as well. Indeed, let us remove from Λ all the fixed points of elements G, belonging to it. Set F of such points is dense in $\Lambda(G)$; it is countable if $n \leq 3$, and at $n > 3$ it has the dimensionality not more than $n-3$. The remaining set $\Lambda_0 = \Lambda(G) \setminus F$ does not contain points x, for which equality $A_1 x = A_2 x$ is satisfied at different $A_1, A_2 \in G$, i.e. any two, not

equivalent concerning G points from Λ_0 have non-intersecting orbits. Using the axiom of arbitrary choice we can form set e_Λ, containing one point from every orbit $Gx, x \in \Lambda_0$. However, this set possesses the following characteristics:

Theorem 1. If $\mu_{n-1}\Lambda(G) > 0$ and $\mu_{n-1}(\partial P_G \cap \Lambda(G)) = 0$ then each of the sets e_Λ is not measurable (relative to measure μ_{n-1}).

We shall illustrate this statement with the following example, describing the situation in the simpliest case. Let G be the Fuchsian group without torsion in unit disc $U = \{z \in \mathbb{C} : |z| < 1\}$ with the compact fundamental domain. Let us assume that on $|z| = 1$ there exist measurable fundamental set e_Λ. Then, taking arbitrary bounded measurable on e_Λ function $\psi(z) \not\equiv const$, we shall continue it as far as all $\Lambda(G) = \{z : |z| = 1\}$ by the equality $\psi(Az) = \psi(z), A \in G$. Having built harmonic in U function $u(z)$ with $u|_\Lambda = \psi$, according to it, resulting from uniqueness, we shall obtain that $u(z) \not\equiv const$ and is automorphic concerning G, which is impossible.

Proof of Theorem 1 for the general case is produced in the author's paper [3]. It is based on the properties of polyhedron P_G. And it turns out that any subset $e' \subset e_\Lambda$ of positive exterior measure cannot be measurable.

3. Theorem 1 gives, in particular, the negative solution of one problem of Ahlfors, formulated in the Proceedings of Romanian-Finnish Seminar on Teichmüller Spaces and Quasiconformal Mappings of 1969([5] , p. 505).

Let us note that the problem of the supports of Beltrami differentials, formulated above, presents the particular case of the following general problem: whether there exist at $\mu_{n-1}\Lambda(G) > 0$ non-trivial, measurable on Λ forms $\varphi = \sum \varphi_{j_1,\ldots,j_n}(x_1,\ldots,x_n)\, dx_1^{j_1} \ldots$

$\ldots dx_n^{j_1}$, invariant relative to G, with supports $\operatorname{supp}\varphi \subset \Lambda$ (here $j_1, \ldots, j_n \in \mathbb{Z}$ are fixed numbers, and the sign of sum denotes that there might be several summands of the mentioned type) or not?

§ 2. B-property.

1. Let us now assume that group G acts on Riemann sphere $\overline{\mathbb{C}} = \mathbb{C} \cup \{\infty\}$, i.e. elements $A \in G$ are linear fractional transformations:

$$Az = (az+b)/(cz+d), \quad ad-bc=1.$$

Alongside with it, G acts discontinuously on the open set $\Omega = \Omega(G)$, dense in $\overline{\mathbb{C}}$.

For G-invariant measurable set $\Delta \subset \overline{\mathbb{C}}$ let $M(\Delta, G)$ denote the space of G-differentials $\mu(z)d\bar{z}/dz$, $\mu \in L_\infty(\mathbb{C})$, which are equal to zero outside Δ.

Non-measurability of set e_Λ, resulting from Theorem 1, greatly complicates the solution of the question we are interested in, i.e. the question of the possible support of Beltrami differentials. Let us say that group G possesses property(B), if such G-differentials, concentrated on $\Lambda(G)$ only, do not exist. We shall show some classes of Kleinian groups, for which this property is satisfied.

It should be noted that many-dimensional analogue of this problem(if $\Omega(G) \neq \emptyset$) is solved easily and in the most general case. Namely, any compatible with G quasiconformal homeomorphism $f: \mathbb{R}^n \to \mathbb{R}^n$, conformal in $\Omega(G)$, is conformal everywhere in $\mathbb{R}^n$. This follows from Liouville's theorem, according to which homeomorphism f (as it is conformal) is Möbius on $\Omega(G)$ - the superposition of the finite number of inversions, as for continuity, it will be such everywhere in $\mathbb{R}^n$.

Thus, the plane case is the most difficult one. The results for it, formulated below, may be considerably strengthened.

2. Let us begin with f i n i t e l y g e n e r a t e d groups. Here we have Ahlfors hypothesis, which states that for all such groups it must be $m_2 \Lambda(G)=0$. This statement in the general case has not been proved yet, neither has it been disproved. It has not been stated for a more particular case of the so-called quasiconformal stable Kleinian groups, too. However, it is possible to prove the following result about B-property.

We shall consider the system of generators $A_1,\dots,A_r$ of group G; these elements may be connected by the finite number of relations $P_\ell(A_{j_1},\dots,A_{j_k})=1$. Let $a^0=(a_1^0,\dots,a_m^0)$ be some system of numbers(parameters), defining elements $A_1,\dots,A_r$ (their coefficients) fully, and, hence, the whole group G.

T h e o r e m 2 . If some neighbourhood of point a^0 in $\mathbb{C}^m$ is filled with quasiconformal deformations $f^\mu G (f^\mu)^{-1}$ of group G with $\mu\in M(\Omega,G)$, then G possesses B-property.

For proving it, note that if $M(\Lambda,G)\neq\{0\}$, then for any $\mu\in M(\Lambda,G)$ we would find, using the condition of the theorem, $\mu_0\in M(\Omega,G)$ and such that $f^{\mu_0}|_\Lambda = f^\mu|_\Lambda$, and this involves $f^\mu_{\bar z}=0$ on $\Lambda(G)$, against hypothesis.

The condition imposed in Theorem 2, does not seem essential; it is also possible that B-property must be satisfied for all the Kleinian groups.

At any rate, Theorem 2 deliberately satisfies for (quasiconformally) stable groups. We shall show the following groups as well. Assume that G is the group without torsion and that $z=\infty$ is the ordinary point of the group, i.e. $\infty\in\Omega(G)$ and $\{0,1\}\subset\Lambda(G)$ (this does not restrict generality). As parameters of each element A_j let us take its fixed points z_{1j}, z_{2j} (which coincide if A_j is parabolic) and centre z_{3j} of its isometric circumference $\{z\in\mathbb{C}: |A_j'(z)|=1\}$, and let $a^0=(a_1^0,\dots,a_m^0)$ be the minimum set of points from system $\{z_{1j}, z_{2j}, z_{3j}\}_{j=1}^{r}$, defining these elements $A_1,\dots,A_r$.

Let us consider the functions:

$$\psi_k(z)=a_k^0(a_k^0-1)\sum_{A\in G}\frac{A'^2(z)}{Az(Az-1)(Az-a_k^0)},\quad k=1,\dots,m.$$

Let group G be such that functions $\psi_1,\dots,\psi_m$ are linearly independent in $\Omega(G)$. Then the condition of Theorem 2 is satisfied. For the proof the variation formula is used:

$$f^\mu(z)=z-\frac{z(z-1)}{\pi}\iint_\Omega\frac{\mu(\zeta)\,d\xi\,d\eta}{\zeta(\zeta-1)(\zeta-z)}+O(\|\mu\|_\infty^2),\quad \mu\in M(\Omega,G)$$

(where the estimation of the remainder term is uniform in any disc $|z|\le R<\infty$) and the properties of holomorphic mappings in $\mathbb{C}^m$.

3. As for the case of infinitely generated groups, we shall show here a rather particular criterion, providing the satisfaction of B-property.

Let Δ be the open G-invariant set in $\Omega(G)$. Let us denote as $B_2(\Delta,G)$ the complex Banach space of holomorphic in Δ functions $\varphi(z)$, such ones that $\varphi(Az)A'^2(z)=\varphi(z)$, $A\in G$, and $\|\varphi\|=\sup_\Delta\lambda_\Delta^{-2}(z)|\varphi(z)|<\infty$, where $\lambda_\Delta(z)|dz|$ is Poincaré metric in (components of) Δ.

We shall also require that near $\partial\Delta$ uniform estimation $\lambda_\Delta(z)=O\left(\frac{1}{\rho(z,\partial\Delta)}\right)$ should be fulfilled. Then we have the mapping:

$$K: \mu \longrightarrow \iint\limits_{\Omega\setminus\bar{\Delta}} \frac{\mu(\zeta)\,d\xi\,d\eta}{(\zeta - z)^4}, \quad z \in \Delta,$$

acting from $M(\Omega\setminus\bar{\Delta}, G)$ into $B_2(\Delta, G)$.

T h e o r e m 3 . If for some open set $\Delta \subset \Omega$ with $\partial\Delta \supset \Lambda(G)$ mapping

$$\hat{K}(\psi) = K(\lambda^{-2}_{\Omega\setminus\bar{\Delta}}\,\overline{\psi}): \; B_2(\Omega\setminus\bar{\Delta}, G) \to B_2(\Delta, G)$$

is surjective, then group G possesses the property(B).

The proof of this statement employs the technique of quasiconformal continuation of conformal deformations(cf. [1] , [2] , [4]).

We shall not dwell upon other cases.

REFERENCES

1. L.V.Ahlfors, Lectures on quasiconformal mappings, Van Notstrand, Princeton, 1966.
2. L.Bers, A non-standard integral equation with applications to quasiconformal mappings, Acta Math., v.116(1966), 113-134.
3. S.L.Krushkal', On a property of limit sets of Kleinian groups, Dokl. Akad. Nauk SSSR, Tom 225(1975), No. 3, 500-502.
4. S.L.Krushkal', Quasiconformal mappings and Riemann surfaces, "Nauka", Novosibirsk, 1975.
5. Proceedings of the Romanian-Finnish Seminar on Teichmüller spaces and quasiconformal mappings (Braşov, Romania, 1969), 1971.

Institute of Mathematics
Academy of Science Novosibirsk

GAUSS - THOMSONsches Prinzip minimaler Energie, verallgemeinerte transfinite Durchmesser und quasikonforme Abbildungen

Von Reiner Kühnau in Halle an der Saale

§ 1. Einleitung

Beim GAUSSschen Prinzip minimaler Energie, das als Spezialfall des THOMSONschen Prinzips aufgefasst werden kann (vgl. [26], 2.8), wird eine Charakterisierung elektrostatischer Gleichgewichtsverteilungen von Ladungen auf Leiteroberflächen durch eine Extremaleigenschaft gegeben. Dies führt im Falle ebener Probleme und einheitlicher Dielektrizitätskonstante z.B. zu einer Charakterisierung des konformen Radius' einfach zusammenhängender Gebiete. Durch Diskretisierung gelangt man so bekanntlich zum Begriff des transfiniten Durchmessers [7], [30]. In vorliegender Mitteilung soll aufgezeigt werden, wie das GAUSS - THOMSONsche Prinzip bei Zugrundelegung ortsabhängiger Dielektrizitätskonstante allgemeiner auch nutzbringend eingesetzt werden kann zu Extremalcharakterisierungen gewisser quasikonformer Normalabbildungen, die in [15], [18], [20] in Verallgemeinerung klassischer konformer Normalabbildungen (Parallel -, Kreisbogenschlitz -, Radialschlitzabbildungen u.ä.) auftraten. Das führt wieder zu einfachen a priori - Abschätzungen für Gebietsfunktionale. In einem besonders explizit behandelbaren Spezialfalle erhalten wir insbesondere eine einparametrige Schar von transfiniten Durchmessern, die in dem einen Grenzfalle auf den gewöhnlichen (euklidischen) transfiniten Durchmesser, im anderen Grenzfalle auf den hyperbolischen transfiniten Durchmesser [29], [30], [17] führt.

Eine zentrale Rolle spielt bei diesen Überlegungen die Differentialgleichung

$$(1) \qquad w_{\bar{z}} = \nu \cdot \overline{w_z}$$

mit $z = x + iy$, $w = u + iv$, wobei $\nu = \nu(z)$ reell ist und

$|\nu(z)| \leq \nu_0 < 1$ erfüllt. Die Funktion $\nu(z)$ und die auftretenden Randkomponenten mögen stückweise einschlägige, die Anwendung des GAUSSschen Integralsatzes (zweckmäßig gleich in der der Differentialgleichung (1) angepassten Form in [2], S. 11 oder S. 156) unmittelbar ermöglichende Glattheitsvoraussetzungen erfüllen (z.B. im in [18] oder [20] präzisierten Sinne). Auf gewisse in der sog. modernen Potentialtheorie übliche allgemeinere Auffassungen soll hier der Kürze halber verzichtet werden. Im übrigen läßt sich die unten für Linienbelegungen durchgeführte Überlegung bei entsprechenden Glattheitsvoraussetzungen auch auf Flächenbelegungen übertragen, wie das für $\nu \equiv 0$ z.B. in [4] (S. 774 ff.), [8] durchgeführt wurde.

§ 2. Bezeichnungen und Hilfsfunktionen

Die Funktion $\nu(z)$ sei für alle z erklärt und also stückweise glatt incl. in $z = \infty$ im in [18], [20] präzisierten Sinne. Ferner sei $\nu(z)$ auch stückweise glatt in diesem Sinne in dem endlich vielfach zusammenhängenden Gebiet $\mathcal{G}$ mit dem inneren Punkt $z = \infty$, wobei $\nu(z)$ jeweils gleich einer Konstanten sei in einem beidseitigen Umgebungsstreifen jeder Randkomponente von $\mathcal{G}$. Wir setzen noch

$$p(z) = \frac{1 + \nu(z)}{1 - \nu(z)}.$$

Der Rand $\mathcal{L}$ von $\mathcal{G}$ sei so orientiert, daß $\mathcal{G}$ zur Linken liegt. Die Randkomponenten $\mathcal{L}_k$ ($\sum \mathcal{L}_k = \mathcal{L}$), $k = 1,\dots,n$, $n \geq 1$, seien analytische Jordankurven (ebenso wie die möglichen Kurven, längs deren $\nu(z)$ bzw. $p(z)$ Sprünge besitzt).

Im Anschluß an [18], [20] definieren wir dann einige Hilfsfunktionen.

Mit $\mathfrak{w}(z, \zeta)$ bezeichnen wir diejenige stetige schlichte Abbildung der Vollkugel auf sich, die $\mathfrak{w}(\infty, \zeta) = \infty$, $\mathfrak{w}(\zeta, \zeta) = 0$ und für die $i \cdot \log \mathfrak{w}(z, \zeta)$ in z die Differentialgleichung (1) erfüllt, wobei bei $\log \mathfrak{w}(z, \zeta)$ der Entwicklungstypus ist

(2) $\quad (1 + \nu(\infty))^{-1}\left[\log z - \nu(\infty)\,\overline{\log z}\right] + \varepsilon(z) \qquad$ in $z = \infty$,

(3) $\quad (1 + \nu(\zeta))^{-1}\left[\log (z - \zeta) - \nu(\zeta)\,\overline{\log (z - \zeta)}\right]$

$$+ \text{const} + \varepsilon(z) \qquad \text{in } z = \zeta,$$

letzteres nur, wenn ζ nicht auf einer Sprunglinie von $\nu(z)$ liegt.
$\varepsilon(z)$ bezeichne hier und fürder irgendeine nach 0 strebende Funktion, für die im Falle (2) noch $z^{\alpha} \cdot \varepsilon(z)$, im Falle (3) noch $(z - \zeta)^{-\alpha} \cdot \varepsilon(z)$ nach 0 strebt zu jedem α mit $0 < \alpha < 1$.

Für die "Pseudodistanz"

(4) $\quad [z,\zeta] = |w(z, \zeta)|$

gilt nach [20] (Satz 5 und § 15)

(5) $\quad [z,\zeta] = [\zeta, z]$ und $[z,\zeta] \geqq 0$

mit Gleichheit genau für $z = \zeta$.

Nach dem TEICHMÜLLER - WITTICH - BELINSKIǏschen Verzerrungssatze, angewandt (nach Stürzung) auf die Abbildung $w(z, \zeta)$ nach Affinität in der Logarithmusebene (so daß in $z = \infty$ wie im Spezialfalle $p(\infty) = 1$ Konformität vorliegt nach (2)), ergibt sich in Verbindung mit allgemeinen Verzerrungssätzen für quasikonforme Abbildungen, daß $\log w(z, \zeta)$ beschränkt bleibt, wenn z und ζ in fixierten Umgebungen von Punkten z_0 und ζ_0 ($z_0 \neq \zeta_0$) variieren. Nach dem (lokal gleichmäßige Konvergenz liefernden) Häufungsprinzipe für pseudo - analytische Funktionen, angewandt auf Ausdrücke $i \cdot \log w(z, \zeta_k)$ als Lösungen von (1) mit $\zeta_k \to \zeta_0$, ergibt sich daraus die Stetigkeit von $\log w(z, \zeta)$ und damit auch von $[z,\zeta]$ in z und ζ (d.h. reell - vierdimensional) für $z \neq \zeta$. Nach dem genannten Verzerrungssatze ergibt sich in ähnlicher Schlußweise wie eben ferner, daß in (2) die (auch von ζ abhängende) Größe $\varepsilon(z)$ gleichmäßig nach 0 strebt für alle ζ einer beliebigen festen beschränkten Menge.

Weiter bezeichnen wir mit $E(z)$ diejenige stetige schlichte Abbildung von $\mathcal{G}$, die $\mathcal{L}_1$ in einen zu 0 konzentrischen Kreis und $\mathcal{L}_2, \ldots, \mathcal{L}_n$ (sofern vorhanden) in hierzu konzentrische Kreisbogenschlitze überführt, die $E(\infty) = \infty$ und für die i log E(z) die Differentialgleichung (1) erfüllt, wobei bei log E(z) in $z = \infty$ die Entwicklung (2) vorliegt.

Schließlich sei $j_\theta(z)$ diejenige (nur bis auf eine additive Konstante eindeutig bestimmte) stetige schlichte Abbildung von $\mathcal{G}$, die die $\mathcal{L}_k$ in Strecken des Neigungswinkels θ gegen die positiv reelle Achse überführt, die $j_\theta(\infty) = \infty$ und für die $e^{-i\theta} j_\theta(z)$ die Differentialgleichung (1) erfüllt, wobei der Entwicklungstypus

$$(6) \quad \left(z + e^{2i\theta}\,\nu(\infty)\,\bar{z}\right)\cdot\left(1 + \varepsilon(z)\right) \qquad \text{in } z = \infty$$

vorliegt.

Endlich werde $\mathfrak{j}_\theta(z)$ genau wie $j_\theta(z)$ definiert, wobei lediglich $\mathcal{G}$ durch die Vollebene zu ersetzen ist. Da $j_\theta(z) - \mathfrak{j}_\theta(z)$ in $z = \infty$ z.B. nach [25] eine hebbare Singularität besitzt, können wir noch das Verschwinden dieser Funktion in $z = \infty$ fordern.

§ 3. Das GAUSS - THOMSONsche Prinzip

Dieses Prinzip benötigen wir hier in der folgenden Form, die sich unmittelbar aus [2] (S. 162/163) ergibt.

Es sei Φ^* in $\bar{\mathcal{G}}$ eine Lösung von

$$(7) \qquad \operatorname{div}\,(p \operatorname{grad} \Phi^*) = 0,$$

die nebst der mit p multiplizierten Normalableitung von Φ^* längs der Sprunglinien von p noch stetig ist, mit $\underline{\Phi}^* = \text{const}$ auf $\mathcal{L}_k$ (mit einer von k abhängigen Konstanten),

$$(8) \qquad - \int_{\mathcal{L}_k} p \frac{\partial \Phi^*}{\partial n}\, ds = c_k, \qquad k = 1, \ldots, n$$

(n durchweg Außennormale aller Kurven, außer beim unten definierten $\mathfrak{K}$). Für alle in $\bar{\mathcal{G}}$ stückweise glatten Vektorfunktionen $\mathfrak{q}$ mit gleichen Sprunglinien wie p(z), wobei die Normalkomponente von $p\mathfrak{q}$ dort aber stetig bleibt, mit

$$(9) \qquad \operatorname{div}\,(p\,\mathfrak{q}) = 0,$$

$$(10) \qquad \int_{\mathcal{L}_k} p\, \mathfrak{q}\mathfrak{n} \; ds = c_k, \quad k = 1, \ldots, n,$$

und für die der Grenzwert

$$(11) \qquad \lim_{R \to \infty} \int_{\mathfrak{K}} p\, \Phi^* (\mathfrak{q}\mathfrak{n} + \frac{\partial \Phi^*}{\partial \mathfrak{n}}) \; ds \quad \text{mit } \mathfrak{K} = \{|z| = R\}$$

existiert, gilt dann

$$(12) \qquad \iint_{\mathcal{G}} p \cdot \left[\mathfrak{q}^2 - (\mathrm{grad}\ \Phi^*)^2\right] dx\, dy - \lim_{R \to \infty} 2 \int_{\mathfrak{K}} p\, \Phi^* (\mathfrak{q}\mathfrak{n} + \frac{\partial \Phi^*}{\partial \mathfrak{n}}) \; ds \geqq 0,$$

mit Gleichheit genau für $\mathfrak{q} = -\,\mathrm{grad}\, \Phi^*$. (Das Doppelintegral in (12) existiert als Grenzwert des Integrals über das durch $\mathfrak{K}$ abgegrenzte Teilstück $\mathcal{G}_R$ bei $R \to \infty$; die Konvergenz wird i. allg. zerstört bei Trennung des Integrals in die Differenz zweier Einzelintegrale.)

§ 4. Eine Verallgemeinerung des konformen Radius'

a.) In unserer ersten Anwendung des GAUSS - THOMSONschen Prinzips wählen wir zunächst $\Phi^* = -\log |E(z)|$.

Dann ist in der Tat (8) erfüllt und $\Phi^* = \mathrm{const}$ auf den $\mathcal{L}_k$. Man berechnet

$$(13) \qquad - \int_{\mathcal{L}_k} p \frac{\partial \Phi^*}{\partial \mathfrak{n}} ds = - \int_{\mathcal{L}_k} \frac{\partial}{\partial s} \arg E \; ds = -\left[\arg E\right]_{\mathcal{L}_k} = \begin{cases} 2\pi & \text{für } k = 1, \\ 0 & \text{für } k = 2, \ldots n, \end{cases}$$

wenn mit $[\;]_{\mathcal{L}_k}$ der Zuwachs der betreffenden Funktion längs $\mathcal{L}_k$ bezeichnet wird. Die Stetigkeit der mit p multiplizierten Normalableitung von Φ^* längs der Sprunglinien von p(z) ergibt sich wegen (1) aus der Stetigkeit der Tangentialableitung von arg E(z).

b.) Es gilt auch die Darstellung

$$(14) \qquad \Phi^*(z) = - \int_{\mathcal{L}} \mu^*(\zeta) \log |\varkappa(z, \zeta)| \, ds_\zeta$$

mit

$$\mu^{*} = \frac{p}{2\pi}\frac{\partial}{\partial \mathfrak{n}} \log |E| = -\frac{1}{2\pi}\frac{\partial}{\partial s} \arg E. \tag{15}$$

Denn es ist zunächst bei $z \in \mathfrak{G}$, wenn $\mathfrak{k}$ ein zu z konzentrischer Kreis mit Radius r ist, der wie $\mathfrak{L}_k$ und $-\mathfrak{K}$ mathematisch negativ orientiert ist,

$$2\pi \int_{\mathfrak{L}} \mu^{*}(\zeta) \log |\mathfrak{w}(z, \zeta)| \, ds_{\zeta} = \int_{\mathfrak{L}} p(\zeta) \log |\mathfrak{w}(z, \zeta)| \frac{\partial}{\partial \mathfrak{n}} \log |E(\zeta)| \, ds_{\zeta}$$

$$= \int_{\mathfrak{L}} p(\zeta) \log |\mathfrak{w}(\zeta, z)| \frac{\partial}{\partial \mathfrak{n}} \log |E(\zeta)| \, ds_{\zeta}$$

$$= \int_{\mathfrak{L}+\mathfrak{K}+\mathfrak{k}} p(\zeta) \log |E(\zeta)| \frac{\partial}{\partial \mathfrak{n}} \log |\mathfrak{w}(\zeta, z)| \, ds_{\zeta}$$

$$- \int_{\mathfrak{K}+\mathfrak{k}} p(\zeta) \log |\mathfrak{w}(\zeta, z)| \frac{\partial}{\partial \mathfrak{n}} \log |E(\zeta)| \, ds_{\zeta} .$$

Hier gilt

$$\int_{\mathfrak{L}} p(\zeta) \log |E(\zeta)| \frac{\partial}{\partial \mathfrak{n}} \log |\mathfrak{w}(\zeta, z)| \, ds_{\zeta}$$

$$= \sum_{k} \log |E(\mathfrak{L}_k)| \int_{\mathfrak{L}_k} p(\zeta) \frac{\partial}{\partial \mathfrak{n}} \log |\mathfrak{w}(\zeta, z)| \, ds_{\zeta}$$

$$= -\sum_{k} \log |E(\mathfrak{L}_k)| \int_{\mathfrak{L}_k} d \arg \mathfrak{w}(\zeta, z) = 0.$$

Der Rest lässt sich zusammenfassen zu

$$\int_{\mathfrak{K}+\mathfrak{k}} p(\zeta) \Big[\log |E(\zeta)| \frac{\partial}{\partial \mathfrak{n}} \log |\mathfrak{w}(\zeta, z)| - \log |\mathfrak{w}(\zeta, z)| \frac{\partial}{\partial \mathfrak{n}} \log |E(\zeta)|\Big] ds_{\zeta}$$

$$= -\int_{\mathfrak{K}+\mathfrak{k}} \Big[\log |E(\zeta)| \, d \arg \mathfrak{w}(\zeta, z) - \log |\mathfrak{w}(\zeta, z)| \, d \arg E(\zeta)\Big]$$

$$= \operatorname{Im} \int_{\mathfrak{K}+\mathfrak{k}} \log \mathfrak{w}(\zeta, z) \, d \log E(\zeta) - \Big[\log |E(\zeta)| \arg \mathfrak{w}(\zeta, z)\Big]_{\mathfrak{K}+\mathfrak{k}} .$$

Nun kann man weiter schreiben

$$\left[\log |E(\zeta)| \cdot \arg \mathfrak{w}(\zeta, z)\right]_{\mathfrak{k}} \to -2\pi \log |E(z)| \quad \text{für } r \to 0,$$

$$\int_{\mathfrak{k}} \log \mathfrak{w}(\zeta, z)\, d \log E(\zeta) \to 0 \qquad \text{für } r \to 0 \quad (\text{vgl. } (3)).$$

Schließlich gilt

$$\mathfrak{Im} \int_{\mathfrak{K}} \log \mathfrak{w}(\zeta, z)\, d \log E(\zeta) - \left[\log |E(\zeta)| \arg \mathfrak{w}(\zeta, z)\right]_{\mathfrak{K}}$$

$$= \mathfrak{Im} \int_{\mathfrak{K}} \log \mathfrak{w}\ d(\log E - \log \mathfrak{w}) + \mathfrak{Im} \left[\frac{1}{2}(\log \mathfrak{w})^2\right]_{\mathfrak{K}}$$

$$- \left[(\log |E|) \cdot \arg \mathfrak{w}\right]_{\mathfrak{K}}$$

$$= \mathfrak{Im} \int_{\mathfrak{K}} \log \mathfrak{w}\ d(\log E - \log \mathfrak{w}) + \left[(\log |\mathfrak{w}| - \log |E|) \arg \mathfrak{w}\right]_{\mathfrak{K}} .$$

Hier strebt für $R \to \infty$ wegen (2) der zweite Term nach 0. Dies tut auch der erste Term, da die partiellen Ableitungen von $\log E - \log \mathfrak{w}$ in $z = \infty$ von zweiter Ordnung abklingen. Denn diese Funktion ist in der Umgebung von $z = \infty$ eindeutig und besitzt einen Grenzwert (nämlich 0), also in $z = \infty$ nach z.B. [25] eine hebbare Singularität, womit sie nach Stürzung stetig differenzierbar ist.

Damit ist (14), (15) gezeigt.

c.) Als Vergleichsvektoren $\mathfrak{q}$ nehmen wir speziell diejenigen der Form

(16) $\quad \mathfrak{q} = -\operatorname{grad} \Phi \quad$ mit $\quad \Phi(z) = -\int_{\mathfrak{L}} \mu(\zeta) \log |\mathfrak{w}(z, \zeta)|\, ds_{\zeta}$,

wobei $\mu(\zeta)$ eine auf $\mathfrak{L}$ definierte gleichmäßig hölderstetige Funktion sei mit

(17) $$\int_{\mathfrak{L}_k} \mu\, ds = \begin{cases} 1 & \text{für } k = 1, \\ 0 & \text{für } k = 2, \ldots, n. \end{cases}$$

Da man durch Bildung von (finiten) Näherungssummen des Integrals

$$i \int_{\mathcal{L}} \mu(\zeta) \log \varkappa(z, \zeta)\, ds_\zeta$$

Folgen von Lösungen von (1) bilden kann, die punktweise (d.h. zu festem z) gegen dieses Integral konvergieren und lokal gleichmäßig beschränkt sind (wegen der Stetigkeit von $\log \varkappa(z,\zeta)$ in z und ζ - vgl. nach (5)), ergibt sich nach dem Häufungsprinzip für pseudoanalytische Funktionen, daß dieses Integral (1) erfüllt. Daher erfüllt $\phi(z)$ auch (9) und ist insbesondere stetig und stückweise glatt. Auf $\mathcal{L}$ zeigt $\phi(z)$ wegen unserer dortigen Voraussetzungen über p gleiches Verhalten wie im klassischen Falle $p \equiv 1$.

Wegen (9) ist weiter die Stetigkeit der Normalkomponente von $p\mathfrak{v}$ erfüllt. Ferner gilt (10) mit $c_1 = 2\pi$, $c_2 = \dots = c_n = 0$ wegen

$$\int_{\mathcal{L}_k} p\, \mathfrak{q}_n\, ds = -\int_{\mathcal{L}_k} p \frac{\partial \phi}{\partial n}\, ds = -\int_{\mathcal{L}_k'} p\, \frac{\partial \phi}{\partial n}\, ds = -\int_{\mathcal{L}_k'} \frac{\partial}{\partial s_z}\left(\int_{\mathcal{L}} \mu(\zeta) \arg \varkappa(z,\zeta)\, ds_\zeta\right) ds_z$$

$$= -\left[\int_{\mathcal{L}} \mu(\zeta) \arg \varkappa(z,\zeta)\, ds_\zeta\right]_{\mathcal{L}_k'}$$

$$= -\int_{\mathcal{L}} \mu(\zeta) \left[\arg \varkappa(z,\zeta)\right]_{\mathcal{L}_k'} ds_\zeta = \begin{cases} \int_{\mathcal{L}_1} \mu(\zeta)\, 2\pi\, ds = 2\pi & \text{für } k = 1, \\ 0 & \text{für } k = 2, \dots n \end{cases}$$

($\mathcal{L}_k'$ = Nachbarkurve von $\mathcal{L}_k$ innerhalb $\mathfrak{G}$).

Ferner gilt für $R \to \infty$

$$(18) \qquad \int_{\mathfrak{K}} p\, \phi^*\left(\mathfrak{q}_n + \frac{\partial \phi^*}{\partial n}\right) ds = \int_{\mathfrak{K}} p\, \phi^* \frac{\partial(\phi^* - \phi)}{\partial n}\, ds \longrightarrow 0,$$

da ϕ^* gemäß (2) anwächst und die partiellen Ableitungen von $\phi^* - \phi$ von zweiter Ordnung abklingen. $\phi^* - \phi$ ist nämlich Realteil der Funktion

$$\int_{\mathcal{L}} (\mu - \mu^*) \log |\mathfrak{r}(z, \zeta)| \, ds + i \cdot \int_{\mathcal{L}} (\mu - \mu^*) \arg \mathfrak{r}(z, \zeta) \, ds,$$

die nach Multiplikation mit i die Differentialgleichung (1) erfüllt, wobei offenbar auch der Imaginärteil eindeutig ist bei Umlaufung von $z = \infty$. Dadurch besitzt diese Funktion, da sie höchstens logarithmisch anwächst, in $z = \infty$ z.B. nach [25] eine hebbare Singularität, ist also nach Stürzung stetig differenzierbar. $\Phi^* - \Phi$ strebt dabei nach 0, da in (2) $\mathcal{E}(z)$ gleichmäßig für $\zeta \in \mathcal{L}$ nach 0 strebt und

$$\int_{\mathcal{L}} (\mu(\zeta) - \mu^*(\zeta)) \log |z| \, ds_\zeta = 0.$$

d.) Für diese Vergleichsvektoren der Form (16) wird die linke Seite von (12) nach dem GAUSSschen Integralsatz

$$(18') \qquad \iint_{\mathfrak{G}} p \cdot \left[\operatorname{grad}^2 \Phi - \operatorname{grad}^2 \Phi^*\right] dxdy = - \int_{\mathcal{L}} p \left[\Phi \frac{\partial \Phi}{\partial \mathfrak{n}} - \Phi^* \frac{\partial \Phi^*}{\partial \mathfrak{n}}\right] ds,$$

da

$$\lim_{R \to \infty} \int_{\mathfrak{K}} p \left[\Phi \frac{\partial \Phi}{\partial \mathfrak{n}} - \Phi^* \frac{\partial \Phi^*}{\partial \mathfrak{n}}\right] ds = 0.$$

Letzteres ergibt sich wegen (18) und

$$- \int_{\mathfrak{K}} p (\Phi - \Phi^*) \frac{\partial \Phi}{\partial \mathfrak{n}} ds = \int_{\mathfrak{K}} (\Phi - \Phi^*) \, d\Psi = - \int_{\mathfrak{K}} \Psi \, d(\Phi - \Phi^*) + \left[\Psi(\Phi - \Phi^*)\right]_{\mathfrak{K}} \to 0$$

bei

$$\Psi(z) = - \int_{\mathcal{L}} \mu(\zeta) \arg \mathfrak{r}(z, \zeta) \, ds_\zeta = - \arg E(z) + \text{beschränkte Funktion},$$

da $\Phi - \Phi^*$ nach 0 strebt, die partiellen Ableitungen hiervon von zweiter Ordnung (vgl. oben unter c.)).

Wendet man entsprechend den GAUSSschen Integralsatz auf das Komplement $\mathfrak{G}'$ von $\mathfrak{G}$ an, so kommt ($\mathfrak{n}'$ = Innennormale) mit Φ^* gemäß (14)

(18'') $\iint\limits_{\mathcal{G}'} p\,[\mathrm{grad}^2\phi - \mathrm{grad}^2\phi^*]\,dx\,dy = -\int\limits_{\mathcal{L}} p[\phi\frac{\partial\phi}{\partial n'} - \phi^*\frac{\partial\phi^*}{\partial n'}]ds.$

Dabei ist in $\mathcal{G}'$ durchweg grad $\phi^* \equiv 0$, da auf $\mathcal{L}$ für die (durchweg stetige) Funktion ϕ^* gilt $\phi^* \equiv$ const. Durch Addition von (18') und (18'') ergibt sich nach (12) bei Benutzung von (18)

$$-\int\limits_{\mathcal{L}} p\cdot[\phi(\frac{\partial\phi}{\partial n} + \frac{\partial\phi}{\partial n'}) - \phi^*(\frac{\partial\phi^*}{\partial n} + \frac{\partial\phi^*}{\partial n'})]\,ds$$

$$= \iint\limits_{\mathcal{G}} p\cdot[\mathrm{grad}^2\phi - \mathrm{grad}^2\phi^*]dx\,dy + \iint\limits_{\mathcal{G}'} p\,\mathrm{grad}^2\phi\,dx\,dy \geqq 0.$$

Hierbei kann man noch schreiben

$$\frac{\partial\phi}{\partial n} + \frac{\partial\phi}{\partial n'} = -2\pi\mu/p.$$

Schließlich ist noch

$$\int\limits_{\mathcal{L}} p\,\phi^*(\frac{\partial\phi^*}{\partial n} + \frac{\partial\phi^*}{\partial n'})\,ds = -2\pi\int\limits_{\mathcal{L}}\phi^*\mu^*\,ds$$

$$= \int\limits_{\mathcal{L}} p\log|E|\frac{\partial}{\partial n}\log|E|ds = -\sum_k \log|E(\mathcal{L}_k)|\int\limits_{\mathcal{L}_k} d\arg E$$

$$= 2\pi\log|E(\mathcal{L}_1)|.$$

Die Größe $|E(\mathcal{L}_1)|$ ist im Falle $p \equiv 1$ und $n = 1$ der gewöhnliche konforme Radius bzw. transfinite Durchmesser.

Wir erhalten zusammenfassend den

Satz 1. Für alle auf dem Rand $\mathcal{L}$ erklärten gleichmäßig hölderstetigen Funktionen μ, die (17) erfüllen, gilt

(19) $\int\limits_{\mathcal{L}}\int\limits_{\mathcal{L}} \mu(z)\,\mu(\zeta)\log|\mathfrak{r}(z,\zeta)|\,ds_z\,ds_\zeta \leqq \log|E(\mathcal{L}_1)|,$

mit Gleichheit genau für $\mu \equiv \mu^*$, definiert gemäß (15).

Bemerkung. Diese Ungleichung (19) gilt also unabhängig von der Art der Definition von $\nu(z)$ innerhalb der $\mathcal{L}_k$. Insbesondere wird damit die obere Grenze der Werte der linken Seite von (19) von diesen Werten innerhalb der $\mathcal{L}_k$ nicht beeinflußt, obwohl zunächst die Funktion $\mathfrak{r}(z,\zeta)$ von diesen Werten abhängt.

§ 5. Eine Verallgemeinerung der Spanne eines Gebietes

a.) In der folgenden Anwendung von (12) wählen wir

$$\phi^* = \Re\left(i\, e^{-i\theta}\, j_\theta\right). \tag{20}$$

Dann ist (7) erfüllt und $\phi^* = \text{const}$ auf den $\mathcal{L}_k$, ferner in (8) $c_k = 0$ $(k = 1, \ldots, n)$ wegen

$$-\int_{\mathcal{L}_k} p \frac{\partial \phi^*}{\partial n}\, ds = \int_{\mathcal{L}_k} \frac{\partial}{\partial s} \Im\left(ie^{-i\theta}\, j_\theta\right) ds = \left[\Im\, ie^{-i\theta}\, j_\theta\right]_{\mathcal{L}_k} = 0.$$

b.) Es gilt auch die Darstellung

$$\phi^*(z) = \Re\left(ie^{-i\theta}\, \mathfrak{j}_\theta\right) + \int_{\mathcal{L}} \mu^*(\zeta) \log |\varkappa(z,\zeta)|\, ds_\zeta \tag{21}$$

mit

$$\mu^* = \frac{p}{2\pi} \frac{\partial}{\partial n} \Re\left(ie^{-i\theta}\, j_\theta\right) = -\frac{1}{2\pi} \frac{\partial}{\partial s} \Im\left(i\, e^{-i\theta}\, j_\theta\right). \tag{22}$$

Der Beweis, d.h. die Berechnung des Integrals in (21), ist im wesentlichen wie bei (14), wobei log E zu ersetzen ist durch $ie^{-i\theta}\, j_\theta$. Im Verlaufe der Rechnung tritt hier dann auf

$$\Im \int_{\mathfrak{K}+\mathfrak{k}} \log \varkappa(\zeta, z)\, d\left(ie^{-i\theta}\, j_\theta(\zeta)\right) - \left[\left(\Re\, ie^{-i\theta}\, j_\theta\right) \arg \varkappa(\zeta, z)\right]_{\mathfrak{K}+\mathfrak{k}}.$$

Hier gilt

$$\left[\left(\Re\, ie^{-i\theta}\, j_\theta(\zeta)\right) \cdot \arg \varkappa(\zeta, z)\right]_{\mathfrak{k}} \to -2\pi \Re\, ie^{-i\theta}\, j_\theta(z) \quad \text{für } r \to 0,$$

$$\int_{\mathfrak{k}} \log \varkappa(\zeta, z)\, d\left(ie^{-i\theta}\, j_\theta(\zeta)\right) \to 0 \quad \text{für } r \to 0 \text{ (vgl. (3))}.$$

Durch entsprechende Rechnungen, bei denen j_θ durch $\mathfrak{j}_\theta$ ersetzt wird, erhält man

$$0 = \Im \int_{\mathfrak{K}+\mathfrak{k}} \log \varkappa(\zeta, z)\, d\left(ie^{-i\theta}\, \mathfrak{j}_\theta(\zeta)\right) - \left[\left(\Re\, ie^{-i\theta}\, \mathfrak{j}_\theta\right) \cdot \arg \varkappa(\zeta, z)\right]_{\mathfrak{K}+\mathfrak{k}}.$$

Die Anteile über $\mathfrak{k}$ werden wie eben bzw. wie an der entsprechenden Stelle in § 4 berechnet. Man hat noch

$$\mathfrak{Im}\int_{\mathfrak{C}} \log \mathfrak{w}(\zeta, z)\, d\, ie^{-i\theta}(j_\theta - \mathfrak{j}_\theta) - \left[(\mathfrak{Re}\, ie^{-i\theta}(j_\theta - \mathfrak{j}_\theta))\arg \mathfrak{w}(\zeta, z)\right]_{\mathfrak{C}}$$

$$\to 0 \quad \text{für } R \to \infty ,$$

da $j_\theta - \mathfrak{j}_\theta$ in $z = \infty$ eine hebbare Singularität hat nach z.B. [25], also die partiellen Ableitungen dieser Differenz von zweiter Ordnung abklingen.

Das - alles gesammelt - liefert (21).

c.) Als Vergleichsvektoren $\mathfrak{q}$ nehmen wir jetzt speziell diejenigen der Form

$$(23) \quad \mathfrak{q} = -\operatorname{grad}\phi \text{ mit } \phi(z) = \mathfrak{Re}(ie^{-i\theta}\mathfrak{j}_\theta(z)) + \int_{\mathfrak{L}} \mu(\zeta) \log |\mathfrak{w}(z,\zeta)| ds_\zeta ,$$

wobei μ eine auf $\mathfrak{L}$ definierte gleichmäßig hölderstetige Funktion sei mit

$$(24) \qquad \int_{\mathfrak{L}_k} \mu \, ds = 0 \qquad \text{für } k = 1, \ldots, n.$$

Dann ist (9) usw. erfüllt und (10) mit $c_k = 0$ wegen

$$\int_{\mathfrak{L}_k} p\, \mathfrak{q}\mathfrak{n}\, ds = \int_{\mathfrak{L}'_k} p \frac{\partial \phi}{\partial \mathfrak{n}} ds = \int_{\mathfrak{L}'_k} \frac{\partial}{\partial s_z}\Big[\mathfrak{Im}(ie^{-i\theta}\mathfrak{j}_\theta(z)) + \int_{\mathfrak{L}} \mu(\zeta) \arg \mathfrak{w}(z,\zeta) ds_\zeta\Big] ds_z = 0.$$

Die Forderung der Existenz von (11) ist noch erfüllt wegen der Darstellung

$$\int_{\mathfrak{C}} p\, \phi^*(\mathfrak{q}\mathfrak{n} + \frac{\partial \phi^*}{\partial \mathfrak{n}})\, ds = \int_{\mathfrak{C}} p\, \phi^* \frac{\partial(\phi^* - \phi)}{\partial \mathfrak{n}} ds = -\int_{\mathfrak{C}} \phi^* d(\Psi^* - \Psi),$$

wenn wieder Ψ^* und Ψ die zu ϕ^* und ϕ "konjugierten" Funktionen bezeichnen. Es ist nämlich wegen (6) in Umgebung von $z = \infty$

$$\phi^* = \mathfrak{Re}\left[ie^{-i\theta}(z + e^{2i\theta}\nu(\infty)\bar{z})(1 + \varepsilon(z))\right]$$

$$= \frac{i}{2}(1 - \nu(\infty))\, e^{-i\theta} z - \frac{i}{2}(1 - \nu(\infty))\, e^{i\theta}\, \bar{z} + z \cdot \varepsilon(z),$$

$$\Psi^*_z - \Psi_z = Cz^{-2}(1 + \varepsilon(z)),\ \Psi^*_{\bar z} - \Psi_{\bar z} = \overline{Cz}^{\,-2}(1 + \varepsilon(z))$$

mit einer gewissen Konstanten C, da $\Psi^* - \Psi$ in $z = \infty$ eine hebbare Singularität besitzt, also nach Stürzung stetig differenzierbar ist.

d.) Es wird nun die linke Seite von (12) gebildet als Grenzwert von

$$\iint_{\mathcal{G}_R} p\left[\operatorname{grad}^2\phi - \operatorname{grad}^2\phi^*\right]dx\,dy - 2\int_{\mathfrak{K}} p\,\phi^*\frac{\partial(\phi^*-\phi)}{\partial n}\,ds$$

$$= -\int_{\mathcal{L}+\mathfrak{K}} p\left[\phi\frac{\partial\phi}{\partial n} - \phi^*\frac{\partial\phi^*}{\partial n}\right]ds - 2\int_{\mathfrak{K}} p\phi^*\frac{\partial(\phi^*-\phi)}{\partial n}\,ds$$

$$= -\int_{\mathcal{L}} p\left[\phi\frac{\partial\phi}{\partial n} - \phi^*\frac{\partial\phi^*}{\partial n}\right]ds + \int_{\mathfrak{K}} p\left[\phi^*\frac{\partial\phi}{\partial n} - \phi\frac{\partial\phi^*}{\partial n}\right]ds - \int_{\mathfrak{K}} p(\phi^* - \phi)\frac{\partial(\phi^*-\phi)}{\partial n}ds$$

$$= -\int_{\mathcal{L}} p\left[\phi\left(\frac{\partial\phi}{\partial n} + \frac{\partial\phi}{\partial n'}\right) - \phi^*\left(\frac{\partial\phi^*}{\partial n} + \frac{\partial\phi^*}{\partial n'}\right)\right]ds + \int_{\mathcal{L}} p\left[\phi\frac{\partial\phi}{\partial n'} - \phi^*\frac{\partial\phi^*}{\partial n'}\right]ds$$

$$-\int_{\mathcal{L}} p\left[\phi^*\frac{\partial\phi}{\partial n} - \phi\frac{\partial\phi^*}{\partial n}\right]ds - \int_{\mathfrak{K}} p(\phi^* - \phi)\frac{\partial(\phi^*-\phi)}{\partial n}\,ds$$

$$= -2\pi\int_{\mathcal{L}}(\phi\mu - \phi^*\mu^*)\,ds - \iint_{\mathcal{G}'} p\left[\operatorname{grad}^2\phi - \operatorname{grad}^2\phi^*\right]dx\,dy$$

$$-2\pi\int_{\mathcal{L}}(\phi^*\mu - \phi\mu^*)\,ds - \int_{\mathfrak{K}} p(\phi^* - \phi)\frac{\partial(\phi^*-\phi)}{\partial n}\,ds$$

$$\leqq 2\pi\int_{\mathcal{L}}\phi\,\mu^*\,ds - 2\pi\int_{\mathcal{L}}\phi\mu\,ds - \int_{\mathfrak{K}} p(\phi^* - \phi)\frac{\partial(\phi^*-\phi)}{\partial n}\,ds.$$

Dabei benutzen wir, daß hier wieder (7) für ϕ und ϕ^* erfüllt ist, daß grad $\phi^* \equiv 0$ in $\mathcal{G}'$, ϕ^* = const auf den $\mathcal{L}_k$ und bezeichnen mit $\mathcal{G}'$ wieder das Komplement von $\mathcal{G}$. Da für $R \to \infty$ das zuletzt angeschriebene Integral über $\mathfrak{K}$ nach 0 strebt, entsteht aus (12) also

$$0 \leqq \int_{\mathcal{L}}\phi\mu^*\,ds - \int_{\mathcal{L}}\phi\mu\,ds$$

$$= \Re\left(ie^{-i\theta}\int_{\mathcal{L}}\mu^*\,\mathfrak{z}_\theta\,ds\right) + \iint_{\mathcal{L}\,\mathcal{L}}\mu^*(\zeta)\,\mu(z)\,\log|\kappa(z,\zeta)|\,ds_\zeta\,ds_z$$

$$-\Re\left(ie^{-i\theta}\int_{\mathcal{L}}\mu\,\mathfrak{z}_\theta\,ds\right) - \iint_{\mathcal{L}\,\mathcal{L}}\mu(\zeta)\,\mu(z)\,\log|\kappa(z,\zeta)|\,ds_\zeta\,ds_z.$$

Hier läßt sich das zweite Integral noch umschreiben als

$$\int_{\mathcal{L}} \mu(z)\left(\phi^*(z) - \mathfrak{Re}\, ie^{-i\theta}\, \mathfrak{j}_\theta\right) ds = -\mathfrak{Re}\, ie^{-i\theta} \int_{\mathcal{L}} \mu\, \mathfrak{j}_\theta\, ds.$$

Das Endergebnis ist damit

$$(25)\quad \int_{\mathcal{L}} \mu(\zeta)\,\mu(z) \log|w(z,\zeta)|\, ds_\zeta\, ds_z + 2\,\mathfrak{Re}\left(ie^{-i\theta}\int_{\mathcal{L}} \mu\, \mathfrak{j}_\theta\, ds\right)$$

$$\leqq \mathfrak{Re}\left(ie^{-i\theta}\int_{\mathcal{L}} \mu^*\, \mathfrak{j}_\theta\, ds\right)$$

$$= \frac{1}{2\pi}\int_{\mathcal{L}} p\left(\mathfrak{Re}\, ie^{-i\theta}\, \mathfrak{j}_\theta\right)\frac{\partial}{\partial n}\left(\mathfrak{Re}\, ie^{-i\theta}\, j_\theta\right) ds,$$

mit Gleichheit genau für $\mu \equiv \mu^*$.

Die rechte Seite wird bei Addition eines verschwindenden Termes

$$= \frac{1}{2\pi}\int_{\mathcal{L}} p\Big[(\mathfrak{Re}\, ie^{-i\theta}\, \mathfrak{j}_\theta)\frac{\partial}{\partial n}(\mathfrak{Re}\, ie^{-i\theta}\, j_\theta) - (\mathfrak{Re}\, ie^{-i\theta} j_\theta)\frac{\partial}{\partial n}(\mathfrak{Re}\, ie^{-i\theta}\, \mathfrak{j}_\theta)\Big] ds$$

$$= -\frac{1}{2\pi}\int_{\mathfrak{K}} p\Big[(\mathfrak{Re}\, ie^{-i\theta}\mathfrak{j}_\theta)\frac{\partial}{\partial n}(\mathfrak{Re}\, ie^{-i\theta} j_\theta) - (\mathfrak{Re}\, ie^{-i\theta}\, j_\theta)\frac{\partial}{\partial n}(\mathfrak{Re}\, ie^{-i\theta}\, \mathfrak{j}_\theta)\Big] ds$$

$$= -\frac{1}{2\pi}\,\mathfrak{Im}\int_{\mathfrak{K}} (ie^{-i\theta}\, j_\theta)\, d(ie^{-i\theta}\, \mathfrak{j}_\theta) = \frac{1}{2\pi} K_{j_\theta}$$

mit

$$(26)\qquad K_{j_\theta} = \mathfrak{Im}\, e^{-2i\theta}\int_{\mathfrak{K}} j_\theta\, d\mathfrak{j}_\theta,$$

wobei $\mathfrak{K}$ wieder ein hinreichend großer Kreis (oder auch eine topologisch äquivalente Kurve) ist in mathematisch positiver Orientierung.

Vermöge der Umformung

$$K_{j_\theta} = \mathfrak{Im}\, e^{-2i\theta}\int_{\mathfrak{K}} (j_\theta - \mathfrak{j}_\theta)\, d\mathfrak{j}_\theta = -\mathfrak{Im}\, e^{-2i\theta}\int_{\mathfrak{K}} \mathfrak{j}_\theta\, d(j_\theta - \mathfrak{j}_\theta)$$

$$= -\mathfrak{Im}\, e^{-2i\theta}\int_{\mathfrak{K}} \left(z + e^{2i\theta}\,\nu(\infty)\,\bar{z}\right)\cdot\left(1 + \varepsilon(z)\right) d(j_\theta - \mathfrak{j}_\theta)$$

$$= \mathfrak{Im}\, e^{-2i\theta}\int_{\mathfrak{K}} (j_\theta - \mathfrak{j}_\theta)\left(dz + e^{2i\theta}\,\nu(\infty)\,\overline{dz}\right)$$

$$- \mathfrak{Im}\, e^{-2i\theta}\int_{\mathfrak{K}} \left(z + e^{2i\theta}\,\nu(\infty)\,\bar{z}\right)\varepsilon(z)\, d(j_\theta - \mathfrak{j}_\theta)$$

erkennt man noch die Darstellung

(27) $$K_{j_\theta} = \mathrm{Re} \lim_{R\to\infty} \frac{1}{i} e^{-2i\theta} \int_{\mathcal{L}} (j_\theta - \mathfrak{j}_\theta)\left(dz + e^{2i\theta}\, \nu(\infty)\, \overline{dz}\right).$$

Hier sieht man die Übereinstimmung von K_{j_θ} im Falle $\theta = 0$ und $p(z) \geqq 1$ mit dem in [18] unter (3) eingeführten Funktional. Speziell für den Fall, es ist $\nu(z) \equiv 0$ in einer Umgebung von $z = \infty$, so daß Entwicklungen der Form

(28) $$j_\theta = z + \frac{a_{1,\theta}}{z} + \dots, \qquad \mathfrak{j}_\theta = z + \frac{\mathfrak{a}_{1,\theta}}{z} + \dots$$

vorliegen, folgt

(29) $$K_{j_\theta} = 2\pi\, \mathrm{Re}\, e^{-2i\theta} (a_{1,\theta} - \mathfrak{a}_{1,\theta}),$$

also bis auf eine additive Konstante für $p \geqq 1$ ein in [12] (Satz 1) betrachtetes Funktional, bzw. im Spezialfalle $p \equiv 1$ ein klassisches von H. GRÖTZSCH und R. DE POSSEL studiertes Funktional.

Zusammenfassend haben wir damit den

Satz 2. Für alle auf dem Rande $\mathcal{L}$ erklärten gleichmäßig hölderstetigen Funktionen μ, die (24) erfüllen, gilt

(30) $$\int_{\mathcal{L}}\int_{\mathcal{L}} \mu(z)\,\mu(\zeta)\, \log|\mathfrak{w}(z,\zeta)|\, ds_z\, ds_\zeta + 2\,\mathrm{Re}\left(ie^{-i\theta}\int_{\mathcal{L}} \mu\, \mathfrak{j}_\theta\, ds\right) \leqq \frac{1}{2\pi} K_{j_\theta},$$

mit Gleichheit genau für $\mu \equiv \mu^*$, definiert gemäß (22). K_{j_θ} ist dabei durch (26) oder (27) definiert, bzw. durch (29), falls speziell $\nu(z) \equiv 0$ in Umgebung von $z = \infty$.

Durch Addition der aus (30) für $\theta = 0$ und $\theta = \pi/2$ entstehenden Ungleichungen erhält man bei $\nu \geqq 0$ und $\nu \equiv 0$ in Umgebung von $z = \infty$ die folgende scharfe Abschätzung für die (positiv reelle) "quasikonforme Spanne" $S = a_{1,0} - a_{1,\pi/2}$ von $\mathfrak{G}$ (vgl. [12], [21]):

$$(31) \quad \iint_{\mathcal{L}\mathcal{L}} \mu_1(z)\,\mu_1(\zeta)\,\log|\mathfrak{w}(z,\zeta)|\,ds_z\,ds_\zeta \; + $$

$$+ \iint_{\mathcal{L}\mathcal{L}} \mu_2(z)\,\mu_2(\zeta)\,\log|\mathfrak{w}(z,\zeta)|\,ds_z\,ds_\zeta$$

$$- 2\int_{\mathcal{L}} \mu_1\,\mathfrak{Im}\,\mathfrak{j}_0\,ds + 2\int_{\mathcal{L}} \mu_2\,\mathfrak{Re}\,\mathfrak{j}_{\pi/2}\,ds$$

$$\leqq S - (\alpha_{1,0} - \alpha_{1,\pi/2})$$

für alle (24) erfüllenden μ_1, μ_2.

Wenn sogar durchweg $\nu \equiv 0$ ist für alle z, dann gilt in (30), (31) $\mathfrak{w}(z,\zeta) \equiv z - \zeta$, $\mathfrak{j}_0 \equiv \mathfrak{j}_{\pi/2} \equiv z$, $\alpha_{1,0} = \alpha_{1,\pi/2} = 0$.

Eine andere a priori - Abschätzung dieser Größe S wurde in [21] angegeben.

§6. Zusatzbemerkungen

a.) Es sei das Gebiet $\mathfrak{G}$ in Satz 2 jetzt zentrisch symmetrisch zum inneren Punkt z = 0, wobei sich die Randkomponenten zu Paaren gruppieren (ihre Anzahl n also gerade ist). Dabei gelte $\nu(-z) = \nu(z)$, was auf Grund der zugehörigen Unitätssätze $j_\Theta(-z) = -j_\Theta(z)$, $\mathfrak{j}_\Theta(-z) = -\mathfrak{j}_\Theta(z)$ nach sich zieht bei $j_\Theta(0) = \mathfrak{j}_\Theta(0) = 0$.
Sei $\mathcal{L}'$ eine Auswahl von n/2 Randkomponenten, so daß bei Hinzunahme der jeweils zentrisch symmetrischen ganz $\mathcal{L}$ entsteht. Dann ist die linke Seite von (30) bei auf der Randkomponentenmenge $\mathcal{L}$ ungerade angesetzter Vergleichsfunktion μ (μ^* ist dann nämlich auch ungerade)

$$\iint_{\mathcal{L}\mathcal{L}} \mu(z)\,\mu(\zeta)\,\log|\mathfrak{w}(z,\zeta)|\,ds_z\,ds_\zeta + 2\,\mathfrak{Re}\Big(ie^{-i\Theta}\int_{\mathcal{L}} \mu\,\mathfrak{j}_\Theta\,ds\Big)$$

$$= 2\iint_{\mathcal{L}'\mathcal{L}'} \mu(z)\,\mu(\zeta)\,\log|\mathfrak{w}(z,\zeta)|\,ds_z\,ds_\zeta$$

$$- 2\iint_{\mathcal{L}'\mathcal{L}'} \mu(z)\,\mu(\zeta)\,\log|\mathfrak{w}(z,-\zeta)|\,ds_z\,ds_\zeta$$

$$+ 4\,\mathfrak{Re}\Big(ie^{-i\Theta}\int_{\mathcal{L}'} \mu\,\mathfrak{j}_\Theta\,ds\Big).$$

Somit schreibt sich dann (30) so:

$$(32)\quad \int_{\mathcal{L}'}\int_{\mathcal{L}'} \mu(z)\,\mu(\zeta)\,\log\left|\frac{w(z,\zeta)}{w(z,-\zeta)}\right| ds_z\, ds_\zeta + 2\Re\left(ie^{-i\theta}\int_{\mathcal{L}'} \mu\, j_\theta\, ds\right) \leqq \frac{1}{4\pi} K_{j_\theta}.$$

Ähnlich lassen sich andere symmetrische Konfigurationen (natürlich auch im Zusammenhang mit (19)) behandeln, d.h. Konfigurationen, die bei einer Gruppe direkt oder indirekt konformer Abbildungen in sich übergehen. Hierunter fallen z.B. auch doppelperiodische Anordnungen (entsprechend Translationsgruppen) oder auch einfach zur reellen Achse symmetrische Anordnungen (vgl. hierzu [6]).

b.) Durch die Substitution

$$(33)\qquad z = \sqrt{\mathfrak{z} - \mathfrak{z}_1}$$

erhält man aus (32) auch eine a priori - Abschätzung eines weiteren Funktionals. Der Einfachheit halber sei dies nur für den von H. GRÖTZSCH behandelten klassischen konformen Spezialfall ($\nu = 0$ für alle z) angeschrieben.

Sei dazu $\mathfrak{G}$ ein Gebiet der $\mathfrak{z}$ - Ebene mit den inneren Punkten $\mathfrak{z} = \infty$ und $\mathfrak{z} = \mathfrak{z}_1$ und dem Rand $\mathfrak{r}$ (der obige Glattheitsvoraussetzungen befriedigt). Dann ist bezüglich aller hydrodynamisch normierten schlichten konformen Abbildungen $\mathfrak{w} = \mathfrak{w}(\mathfrak{z})$ von $\mathfrak{G}$ der genaue Wertebereich von $\mathfrak{w}(\mathfrak{z}_1)$ nach H. GRÖTZSCH eine abgeschlossene Kreisscheibe mit Mittelpunkt m und Radius R, wobei der Punkt $m - Re^{2i\theta}$ nur von einer gewissen Parabelschlitzabbildung angenommen wird, die vermöge (33) mit obigem $j_\theta(z)$ zusammenhängt. Das liefert bei

$$(34)\qquad \mu_0(\mathfrak{z}) = \mu(z)\cdot|dz/d\mathfrak{z}|$$

die Ungleichung

$$(35)\quad \int_{\tau}\int_{\tau} \mu_0(\zeta)\,\mu_0(\zeta')\,\log\left|\frac{\sqrt{\zeta-\zeta_1}-\sqrt{\zeta'-\zeta_1}}{\sqrt{\zeta-\zeta_1}+\sqrt{\zeta'-\zeta_1}}\right| ds_{\zeta}\, ds_{\zeta'}$$

$$+\,2\int_{\tau} \mu_0(\zeta)\,\mathfrak{Re}(ie^{-i\theta}\sqrt{\zeta-\zeta_1}\,)\,ds_{\zeta} = \frac{1}{4}R + \frac{1}{4}\mathfrak{Re}\,e^{-2i\theta}(z_1-m)$$

für alle auf τ erklärten gleichmäßig hölderstetigen $\mu_0(\zeta)$, bei denen längs jeder Komponente von τ gilt

$$(36)\qquad \int \mu_0(\zeta)\,ds_{\zeta} = 0.$$

Das Gleichheitszeichen steht in (35) genau für dasjenige $\mu_0(\zeta)$, das gemäß (34) aus $\mu^*(z)$ entsteht. Durch Wahl von $\theta = 0$ und $\theta = \pi/2$ erhält man aus (35) insbesondere eine scharfe Abschätzung des "Verschiebungsradius" R. Im Falle, τ ist eine Kreislinie, ist bekanntlich [7] (Seite 129) R analytisch als Ausdruck in vollständigen elliptischen Integralen anschreibbar. Wenn τ aus zwei Kreislinien besteht, ist die exakte Bestimmung von R erheblich komplizierter (vgl. [10], Seiten 101 - 104).

c.) Die in § 5 durchgeführten Überlegungen lassen sich in naheliegender Weise verallgemeinern für die nichtschlichten Parallelschlitzabbildungen, die in [15] (dort $g_0^{(k)}(z)$ und $g_{\pi/2}^{(k)}(z)$) im Zusammenhang mit verallgemeinerten GRUNSKYschen Koeffizientenbedingungen auftraten, ferner für die Extremalfunktionen des Satzes 1 in [15] (insbesondere für die verallgemeinerten Radial- und Kreisbogenschlitzabbildungen). Auch lässt sich so z.B. durch Übertragung von [1] das extremale Funktional von Satz 2 in [9] charakterisieren.

d.) Ungleichungen (19), (30), (31) und (35) liefern zu beliebig eingesetztem zulässigen μ eine a priori - Abschätzung der zugehörigen rechten Seite. Durch Konstruktion einer Maximalfolge zulässiger μ lässt sich überdies die rechte Seite jeweils mit beliebiger Genauigkeit berechnen.

e.) Durch Einführung von N bzw. 2N "diskreten Massen" erhält man in sinngemäß entsprechender Weise wie beim klassischen transfiniten Durchmesser aus der linken Seite von (19) bzw. (30)

$$(37) \qquad \log \left\{ \prod_{k \neq l} [z_k, z_l] \right\}^{1/N^2}$$

bzw. (für $\Theta = 0$)

$$(38) \quad \log \left\{ \prod_{k \neq l} \left([z_k, z_l] [z_k^*, z_l^*] [z_k, z_l^*]^{-2} \right)^{\lambda^2} \times \right.$$

$$\left. \times \exp \left(2 \lambda \sum_k \operatorname{Im} j_o(z_k^*) - 2 \lambda \sum_k \operatorname{Im} j_o(z_k) \right) \right\}.$$

Dabei seien $z_1, \ldots, z_N, z_1^*, \ldots, z_N^*$ untereinander verschiedene Punkte auf $\mathcal{L}$, wobei $\mathcal{L}$ zur Vereinfachung aus nur einer Randkomponente bestehe (n = 1). λ sei ein ebenfalls mitvariierender reeller Parameter (den man evtl. zunächst "optimal" wählen wird). Zu beachten ist allerdings, daß analog zum klassischen Falle (vgl. z.B. [26], 2.11) durch diese Diskretisierung nicht unmittelbar untere Schranken für die rechte Seite von (19) bzw. (30) geliefert werden.

Man kann auch Verfeinerungen dieser Betrachtung ähnlich wie in [23], [24] durchführen. Man vgl. auch die dort zitierten Arbeiten, insbesondere von CHR. POMMERENKE.

f.) Es werde auf die Möglichkeit hingewiesen, das GAUSSsche Prinzip in Form des obigen Satzes 1 bzw. 2 für Verzerrungsaussagen nutzbar zu machen, ähnlich wie das bei konformen Abbildungen in [3] geschehen ist. Dabei bietet sich z.B. auch an, die geometrische Gestalt des in [11] betrachteten Extremalkontinuums näher durch Satz 1 (bzw. den hiermit entsprechend obiger Bemerkung e.) zusammenhängenden transfiniten Durchmesser) zu charakterisieren (z.B. durch Konvexitätsaussagen), ähnlich wie das bei einem GRÖTZSCHschen Extremalkontinuum im konformen Falle in [27] (S. 100) bzw. bei der hyperbolischen und elliptischen Übertragung in [17] (S. 93/94) geschah.

g.) Die Betrachtungen von §4 bzw. § 5 liefern noch unmittelbar folgendes Nebenresultat.

Satz 3. Die Integralgleichung

(39) $$\int_{\mathcal{L}} \mu(\zeta) \log |\kappa(z, \zeta)| ds_\zeta = C_k \qquad \text{für } z \in \mathcal{L}_k$$

besitzt bei der Nebenbedingung (17) genau die durch (15) definierte Funktion μ^* als (gleichmäßig hölderstetige) Lösung.

Die Integralgleichung

(40) $$\mathfrak{Re}\left(ie^{-i\Theta} \mathfrak{j}_\Theta(z)\right) + \int_{\mathcal{L}} \mu(\zeta) \log |\kappa(z, \zeta)| ds_\zeta = C_k$$

besitzt bei der Nebenbedingung (24) genau die durch (22) definierte Funktion μ^* als Lösung.

Das Konstantensystem C_k ist dabei variabel, d.h. zusammen mit μ zu bestimmen.

Denn ist z.B. (39) für μ erfüllt, dann ist mit dem gemäß (16) bzw. (14) gebildeten Φ bzw. Φ^* die Differenz $\Phi - \Phi^*$ nach § 4.e.) im Gebiet eine Lösung von (7), die in $z = \infty$ verschwindet, auf jeder Randkomponente jeweils konstant ist und eindeutige komplexe Ergänzung hat wegen (17). $\Phi - \Phi^*$ ist also eine Konstante, mithin $\equiv 0$.

Ist μ^* als Lösung von (39) bzw. (40) bestimmt, ergeben sich nach (14) bzw. (21) die zugehörigen Abbildungen E(z) bzw. $j_\Theta(z)$ aus

(41) $$\log E(z) = \int_{\mathcal{L}} \mu^*(\zeta) \log \kappa(z, \zeta) \, ds_\zeta$$

bzw.

(42) $$ie^{-i\Theta} j_\Theta(z) = ie^{-i\Theta} \mathfrak{j}_\Theta(z) + \int_{\mathcal{L}} \mu^*(\zeta) \log \kappa(z, \zeta) \, ds_\zeta \, .$$

Es sei noch bemerkt, daß sich die Integralgleichung (39) bzw. (40) durch die Bedingung des Verschwindens der ersten Variation (bei Variation von μ) der linken Seite von (19) bzw. (30) ergibt.

Im klassischen Falle $\nu \equiv 0$ enthält (39) eine Integralgleichung, die von SYMM zur numerischen Realisierung der RIEMANNschen Abbildungsfunktion benutzt wurde - vgl. GAIER [5] (und dort zitierte Literatur), wo die theoretischen Zusammenhänge geklärt wurden.

h.) Ein letzter allgemeiner Hinweis in Bezug auf Verallgemeinerungsmöglichkeiten der hier betrachteten Art gelte [28].

§ 7. Ein Spezialfall

Im allgemeinen Falle ist eine explizite Darstellung der eine zentrale Rolle spielenden Funktion $w(z,\zeta)$ nicht bekannt. Im Spezialfalle, es ist

$$(43)\qquad \begin{aligned} \nu(z) &\equiv q_1 \qquad (-1 < q_1 < 1) \text{ für } |z| < 1, \\ \nu(z) &\equiv q_2 \qquad (-1 < q_2 < 1) \text{ für } |z| > 1, \end{aligned}$$

erhält man leicht für $|\zeta| > 1$

$$(44)\quad w(z,\zeta) = \begin{cases} \left[(z-\zeta)(\bar z-\bar\zeta)^{-q_2}\left(1-\dfrac{1}{\bar\zeta z}\right)^{(q_2-q_1)/(1-q_1q_2)} \right. \\ \qquad \left. \times\left(1-\dfrac{1}{\zeta\bar z}\right)^{-q_2(q_2-q_1)/(1-q_1q_2)}\right]^{1/(1+q_2)} \\ \qquad\qquad \text{für } |z| \geqq 1, \\ \left[(z-\zeta)(\bar z-\bar\zeta)^{-q_2}\left(1-\dfrac{\bar z}{\bar\zeta}\right)^{(q_2-q_1)/(1-q_1q_2)} \right. \\ \qquad \left. \times\left(1-\dfrac{z}{\zeta}\right)^{-q_2(q_2-q_1)/(1-q_1q_2)}\right]^{1/(1+q_2)} \\ \qquad\qquad \text{für } |z| \leqq 1, \end{cases}$$

dagegen für $|\zeta| < 1$

$$
(45)\quad \mathfrak{w}(z,\zeta) = \begin{cases} \left[(z-\zeta)^{1-q_1}(\bar{z}-\bar{\zeta})^{-q_2(1-q_1)} z^{(q_1-q_2)/(1+q_2)} \right. \\ \left. \times\, \bar{z}^{-q_2(q_1-q_2)/(1+q_2)}\right]^{1/(1-q_1 q_2)} & \text{für } |z| \geqq 1, \\ \left[(z-\zeta)(\bar{z}-\bar{\zeta})^{-q_1}(1-z\bar{\zeta})^{(q_1-q_2)/(1-q_1 q_2)} \right. \\ \left. \times (1-\bar{z}\zeta)^{-q_1(q_1-q_2)/(1-q_1 q_2)}\right]^{1/(1+q_1)} & \text{für } |z| \leqq 1. \end{cases}
$$

Denn diese Ausdrücke besitzen offenbar alle in der Definition verlangten Eigenschaften, und $\mathfrak{w}(z,\zeta)$ ist ja durch diese eindeutig bestimmt. In dem Sonderfalle $q_2 = 0$ (auf den (39) übrigens durch eine Affinität in der Logarithmusebene zurückführbar ist) sind die Ausdrücke (40) bereits in [12], [13] angegeben worden.

Damit wird

$$
(46)\quad [z,\zeta] = \begin{cases} |z-\zeta|^{1/Q_2} \left|1 - \frac{1}{\bar{\zeta} z}\right|^{Q_{12}/Q_2} & \text{für } |z| \geqq 1,\ |\zeta| \geqq 1, \\ |z-\zeta|^{1/Q_2} \left|1 - \frac{z}{\bar{\zeta}}\right|^{Q_{12}/Q_2} & \text{für } |z| \leqq 1,\ |\zeta| \geqq 1, \\ |z-\zeta|^{1/Q_2} \left|1 - \frac{\bar{\zeta}}{z}\right|^{Q_{12}/Q_2} & \text{für } |z| \geqq 1,\ |\zeta| \leqq 1, \\ |z-\zeta|^{1/Q_1} \left|1 - z\bar{\zeta}\right|^{-Q_{12}/Q_1} & \text{für } |z| \leqq 1,\ |\zeta| \leqq 1 \end{cases}
$$

bei

$$
(47)\quad Q_1 = \frac{1+q_1}{1-q_1},\quad Q_2 = \frac{1+q_2}{1-q_2},\quad Q_{12} = \frac{q_2-q_1}{1-q_1 q_2}.
$$

Speziell für $q_1 = 0$ gilt für $|z| < 1$, $|\zeta| < 1$

$$(48) \qquad [z, \zeta] = |z - \zeta| / |1 - z\bar{\zeta}|^{q_2}.$$

Im Grenzfall $q_2 \to 1$ entsteht also die von M.TSUJI [29], [30] (vgl. auch [17]) betrachtete hyperbolische Pseudodistanz. Dadurch erhält man also bei Zugrundelegung von (44) eine einparametrige Schar von transfiniten Durchmessern, die sich - wenn q_2 von 0 nach 1 variiert - zwischen den klassischen gewöhnlichen von FEKETE eingeführten transfiniten Durchmesser und den hyperbolischen transfiniten Durchmesser von TSUJI einspannt.

Entsprechend wie (40), (41) konstruiert man noch im Spezialfalle (39) den analytischen Ausdruck

$$(49) \qquad \mathfrak{z}_\theta(z) = \begin{cases} z + q_2 e^{2i\theta}\bar{z} - Q_{12}e^{2i\theta} z^{-1} - Q_{12}q_2\bar{z}^{-1} & \text{für } |z| \geq 1, \\ z + q_2 e^{2i\theta}\bar{z} - Q_{12}e^{2i\theta}\bar{z} - Q_{12}q_2 z & \text{für } |z| \leq 1. \end{cases}$$

Es sei noch bemerkt, daß man die analytischen Ausdrücke für $w(z, \zeta)$ und $\mathfrak{z}_\theta(z)$ nach [14], [22], [19] allgemeiner für den Fall erhält, in dem $\nu(z)$ eine nur von $|z|$ abhängige Funktion ist, z.B. jeweils in zu 0 konzentrischen Kreisringen konstant vorgegeben wird.

$\mathfrak{z}_\theta(z)$ wurde auch noch in [16] konstruiert für den Fall, es ist $\nu(z) \equiv 0$ außerhalb einer CASSINIschen Linie und $\nu(z) \equiv \text{const} \neq 0$ innerhalb derselben.

S c h r i f t t u m

[1] T.BAGBY, The modulus of a plane condenser. J.Math.Mech.17,315-329 (1967).

[2] S.BERGMAN and M.SCHIFFER, Kernel functions and elliptic differential equations in mathematical physics. New York 1953.

[3] G.FABER, Über Potentialtheorie und konforme Abbildung. Sitzungsber. Bayer.Akad.Wiss.München,math.-nat.Kl.1920, 49 - 64.

[4] P.FRANK und R.von MISES, Die Differential- und Integralgleichungen der Mechanik und Physik, I.Teil. Braunschweig 1930.

[5] D.GAIER, Integralgleichungen erster Art und konforme Abbildung. Math.Z. 147, 113 - 129 (1976).

[6] A.GIROUX, Sur une conjecture de R.M.Robinson. Canadian math.Bull. 13, 281 - 282 (1970).

[7] G.M.GOLUSIN, Geometrische Funktionentheorie. Berlin 1957.

[8] H.GRUNSKY, Die Energie einer Ladungsverteilung beim logarithmischen Potential. Deutsche Math.3, 501 - 504 (1938).

[9] R.KÜHNAU, Über gewisse Extremalprobleme der quasikonformen Abbildung. Wiss.Z.d.Martin-Luther-Univ.Halle-Wittenberg, Math.-Nat. Reihe 13, 35-39 (1964).

[10] -- , Über die analytische Darstellung von Abbildungsfunktionen, insbesondere von Extremalfunktionen der Theorie der konformen Abbildung. J.reine angew.Math. 228, 93 - 132 (1967).

[11] -- , Quasikonforme Abbildungen und Extremalprobleme bei Feldern in inhomogenen Medien. J.reine angew.Math.231, 101 - 113 (1968).

[12] -- , Wertannahmeprobleme bei quasikonformen Abbildungen mit ortsabhängiger Dilatationsbeschränkung. Math.Nachr.40,1-11 (1969).

[13] -- , Einige Extremalprobleme bei differentialgeometrischen und quasikonformen Abbildungen. II. Math.Z.107, 307 - 318 (1968).

[14] -- , Bemerkungen zu den GRUNSKYschen Gebieten. Math.Nachr.44, 285 - 293 (1970).

[15] -- , Verzerrungssätze und Koeffizientenbedingungen vom GRUNSKYschen Typ für quasikonforme Abbildungen. Math.Nachr.48,77-105 (1971).

[16] -- , Eine funktionentheoretische Randwertaufgabe in der Theorie der quasikonformen Abbildungen. Indiana Univ.Math.J.21,1-10(1971)

[17] -- , Geometrie der konformen Abbildung auf der hyperbolischen und der elliptischen Ebene. Berlin 1974.

[18] -- , Zur Methode der Randintegration bei quasikonformen Abbildun-

gen. Ann.Polon.Math. 31, 269 - 289 (1975/76).

[19] -- , Extremalprobleme bei quasikonformen Abbildungen mit kreisringweise konstanter Dilatationsbeschränkung. Math.Nachr.66, 269 - 282 (1975).

[20] -- , Identitäten bei quasikonformen Normalabbildungen und eine hiermit zusammenhängende Kernfunktion. Math.Nachr.73,73-106 (1976)

[21] -- , Die Spanne von Gebieten bei quasikonformen Abbildungen. (In Vorbereitung).

[22] J.O.McLEAVEY, Extremal problems in classes of analytic univalent functions with quasiconformal extensions. Trans.Amer.math.Soc. 195, 327 - 343 (1974).

[23] K.MENKE, Extremalpunkte und konforme Abbildung. Math.Ann.195, 292 - 308 (1972).

[24] -- , Zur Approximation des transfiniten Durchmessers bei bis auf Ecken analytischen geschlossenen Jordankurven. Israel J.Math.17, 136 - 141 (1974).

[25] G.N.POLOŽIĬ, Theorie und Anwendung p-analytischer und (p,q)-analytischer Funktionen. 2., überarb.u.erg.Aufl., Kiew 1973 [Russ.].

[26] G.PÓLYA and G.SZEGŐ, Isoperimetric inequalities in mathematical physics. Princeton 1951.

[27] E.REICH and M.SCHIFFER, Estimates for the transfinite diameter of a continuum. Math.Z.85, 91 - 106 (1964).

[28] L.SARIO and K.NOSHIRO, Value distribution theory. Princeton 1966.

[29] M.TSUJI, Some metrical theorems on Fuchsian groups. Jap.J.Math. 19, 483 - 516 (1947).

[30] -- , Potential theory in modern function theory. Tolyo 1959.

Parametrization and Boundary Correspondence for Teichmüller Mappings in an Annulus

Julian Ławrynowicz
Institute of Mathematics
Polish Academy of Sciences
The Łódź Branch
Łódź, Poland

Introduction

Under a Teichmüller mapping of the closed unit disc $\Delta = \{s \in \mathbb{C}: |s| \leq 1\}$ onto itself we mean any quasiconformal (cf.e.g. [8], p. 17) mapping $f(,t)$ whose complex dilatation μ has almost everywhere the form $t\bar{\varphi}/|\varphi|$, where $0 \leq t < 1$, and φ is a function meromorphic in the interior of Δ whose only singularities may be poles of the first order. In Section 1 we consider only normalized Teichmüller mappings, i.e. mappings satisfying a condition

(1) $f(s_j, t) = s_j$, s_j distinct and $|s_j| = 1$, $j = 1,2,3$, $0 < t \leq T$,

with fixed s_j, $j = 1,2,3$, and suppose that $\varphi = \Phi'^2$, where Φ is holomorphic. Therefore we have almost everywhere in Δ

(2) $f_{\bar{s}}(s,t) = \mu(s,t) f_s(s,t)$, $\mu(s,t) = t\overline{\Phi'(s)}/\Phi'(s)$, $0 \leq t < 1$.

We say that the mapping $f(,t)$, uniquely determined (cf.e.g. [8], p. 204) by the conditions (1) and (2), corresponds to Φ and t.

Investigation of the mappings in question was originated by Teichmüller [13, 14]. They play an important role in variational problems. Simple examples show that normalized Teichmüller mappings of Δ onto itself keep, in general, the boundary points invariant. This interesting result is due to Reich and Strebel [10] and the proof is based on an ad hoc parametrical method leading to a parametrical equation valid at most for $t<\frac{1}{2}$, i.e. for normalized Teichmüller mappings with Q bounded by some constant (depending on Φ) less than 3 (the final result concerning invariance is proved, however, for all Q). The general result, without the restriction $Q\leq Q_0<3$ for the parametrical equation has been obtained in [5] and is reproduced in Section 1 below as a consequence of the general theorems on parametrization in Δ [4]. The result in question (Theorem A and B) shows the power and beauty of the parametrical method.

In analogy to the case of mappings defined in Δ, under a normalized Teichmüller mapping of an annulus $\Delta_r=\{s\in\mathbb{C}: r\leq|s|\leq 1\}$ onto $\Delta_{R(t)}$ we mean any quasiconformal mapping satisfying $f(1,t)=1$ and (2) almost everywhere in Δ_r. We say that the mapping $f(\ ,t)$, uniquely determined (cf. [3], p. 26) by the conditions $f(1,t)=1$ and (2) a.e. in Δ_r, corresponds to Φ and t. In this paper we prove the counterpart of the result mentioned above, for the case of mappings defined in Δ_r (Section 3, Theorem 1 and Section 5, Theorem 2). A particular case, under rather very restrictive continuability conditions for Φ is already known [6]. These theorems are shown to be consequences of the general theorems on parametrization in Δ_r [4] which are reproduced in Section 2 below (Theorems C and D).

Our present procedure applies also to the limit case $r=0$, i.e. to the normalized Teichmüller mappings of Δ onto itself keeping the origin 0 invariant (Section 6, Theorems 3 and 4) and to the general case of normalized Teichmüller mappings of Δ onto itself. This second case will be discussed in detail in [7].

1. The case of the unit disc

In the case of the unit disc we have the following result [2], [5]:

Theorem A. Suppose that Φ is holomorphic in int Δ, Φ' has zeros at s_j, $j=1,2,\dots$ (we do not exclude the case where $\Phi'(s)\neq 0$, $|s|<1$),

(3) $$\iint_\Delta \frac{\overline{\Phi'(z)}}{\Phi'(z)}\frac{dxdy}{s-z}=0 \quad \text{for } |s|=1 \text{ and } s=s_j,\ j=1,2,\dots$$

($x=\mathrm{re}\,z$, $y=\mathrm{im}\,z$), and

(4) $$\left|\frac{\Phi'(w)-\Psi_w(w,t)}{t\Phi'(w)}\right|\leq M,\quad 1<M<+\infty,\ \text{for } |w|<1,$$

with

(5) $$\Psi(w,t)=\Phi(w)+t\left[\overline{\Phi(w)}-\frac{1}{\pi}\Phi'(w)\iint_\Delta\frac{\overline{\Phi'(z)}}{\Phi'(z)}\frac{dxdy}{w-z}\right],\quad 0\leq t<1.$$

Then there exists a unique solution $w=f(s,t)$ of

(6) $$w_t=\frac{1}{\pi}\iint_\Delta\frac{\varphi(z,t)}{w-z}\,dxdy,\ \text{where}\ \varphi(w,t)=\frac{1}{1-t^2}\frac{\overline{\Psi_w(w,t)}}{\Psi_w(w,t)},$$

or $\Delta\times[0;1)$, subject to the initial condition $f(s,0)=s$, which for every t represents the Teichmüller mapping of Δ onto itself normalized by the conditions (1) and corresponds to Φ and t. Moreover, for every t we have

(7) $$\iint_\Delta\frac{\overline{\Psi_z(z,t)}}{\Psi_z(z,t)}\frac{dxdy}{w-z}=0 \quad \text{for } |w|=1 \text{ and } w=s_j,\ j=1,2,\dots,$$

and the mapping $f(\ ,t)$ satisfies the relations

(8) $f(s,t) = s$ for $|s| = 1$ and $s = s_j$, $j = 1,2,\dots$,

as well as

(9) $\Psi(f(s,t),t) = \Phi(s) + t\overline{\Phi(s)}$ for $|s| < 1$.

Theorem A gives a sufficient condition for a quasiconformal mapping of Δ onto itself to keep the boundary points invariant. The inverse problem is even simpler. We have the following result (relations (3) and (7) in this context are already due to Ahlfors [1], and relation (9) to Strebel [12]):

Theorem B. Suppose that Φ is holomorphic in int Δ and $f(\ ,t)$, $0 \leq t < 1$, are Teichmüller mappings of Δ onto itself, normalized by the conditions (1), which correspond to Φ and t. Then (8) with $|s_j| < 1$, $j = 1,2,\dots$ (we do not exclude the case where the set of points s_j is empty), implies (3), (7), (6), and (9), where Ψ is given by (5).

Quasiconformal mappings which keep the boundary points invariant and, more generally, quasiconformal mappings with prescribed boundary values were extensively studies by Reich and Strebel [9-11], and R.S. Hamilton [2].

2. General theorems on parametrization in an annulus

In this section we recall two general theorems on parametrization of quasiconformal mappings in an annulus. Hereafter a function f is said to be of the class $S_Q^{r,R}$ if it maps Δ_r onto Δ_R Q-quasiconformally with $f(1) = 1$; in the degenerate case $r = 0$ we assume, in addition, that $f(0) = 0$. It is obvious (cf.e.g. [8], pp. 40-41) that the class $S_Q^{r,R}$ is nonempty if and only if $r^Q \leq R \leq r^{1/Q}$

$0 \leq r \leq 1$ (we shall always assume that $r < 1$). The following theorems in their general form are proved in [4], where references concerning their earlier versions are given:

Theorem C. Suppose that the functions $f(\ ,t)$, $0 \leq t \leq T$, belong to $S^{r,R(t)}_{Q(t)}$ as functions generated by the complex dilatations $\mu(\ ,t)$ having the partial derivative μ_t almost everywhere in Δ_r, where Q is given by

$$(10)\quad Q(t) = \{1 + \|\mu(\ ,t)\|_\infty\}/\{1 - \|\mu(\ ,t)\|_\infty\},$$

the integral being taken over Δ_r. Then $w = f(s,t)$ is a solution of the differential equation

$$(11)\quad w_t = \frac{w}{\pi} \iint_{\Delta_{R(t)}} \sum_{n=-\infty}^{+\infty} \Big\{ \frac{\varphi(z,t)}{z^2} \Big[\frac{w + R^{2n}(t)z}{w - R^{2n}(t)z} - \frac{1 + R^{2n}(t)z}{1 - R^{2n}(t)z} \Big] - \frac{\overline{\varphi(z,t)}}{\bar{z}^2} \Big[\frac{1 + R^{2n}(t)w\bar{z}}{1 - R^{2n}(t)w\bar{z}} - \frac{1 + R^{2n}(t)\bar{z}}{1 - R^{2n}(t)\bar{z}} \Big] \Big\} dxdy,$$

where $x = \operatorname{re} z$, $y = \operatorname{im} z$, $R^{2n}(t) = [R(t)]^{2n}$, the notation $\ldots + a_{-1} + a_0 + a_1 + \ldots$ is applied for the sake of simplicity instead of $a_0 + (a_1 + a_{-1}) + \ldots$ provided that the last series converges and φ is defined by

$$(12)\quad \varphi(w,t) = \frac{\mu_t(f^{-1}(w,t),t)}{1 - |\mu(f^{-1}(w,t),t)|^2} \exp\{-2i \arg f^{-1}_w(w,t)\}.$$

Moreover $\varrho = R(t)$, $0 \leq t \leq T$, is a differentiable function which satisfies the differential equation

$$(13)\quad \varrho' = (1/2\pi) \iint_{\Delta_\varrho} \varrho\, [\varphi(z,t)/z^2 + \overline{\varphi(z,t)}/\bar{z}^2]\, dxdy.$$

Corollary A. If in Theorem C we additionally assume that, for all $s \in \Delta_r$, $\mu(s,\ )$ is of the class C^1 or, even less, that

(14) $\|\mu_t(\ ,t+\tau) - \mu_t(\ ,t)\|_p \to 0$ as $t \to 0+$ for $p \geq 1$,

the integral being taken over Δ_r, then also $f(s,\ )$ and R are of the class C^1 for all $s \in \Delta_r$.

Theorem D. Suppose that φ is a measurable function defined in $\Delta \times [0;T]$ and such that $\|\varphi(\ ,t)\|_\infty \leq \frac{1}{2}$ for $0 \leq t \leq T$. Then there exists a unique solution $\varrho = R(t)$ of (13) on $[0;T]$ subject to the initial condition $R(0) = r$. Furthermore, there exists also a unique solution $w = f(s,t)$ of (11) on $\Delta_r \times [0;T]$ subject to the initial condition $f(s,0) = s$ which for every t represents a mapping belonging to $S_{Q(t)}^{\tau, R(t)}$, where $Q(t) \leq \exp t$.

3. Application to Teichmüller mappings of an annulus

We are going to formulate now our main result:

Theorem 1. Suppose that Φ is holomorphic in int Δ_r, Φ' has zeros at s_j, $j = 1,2,\dots$ (we do not exclude the case, where $\Phi'(s) \neq 0$, $r < |s| < 1$),

(15) $F(s) = 0$ for $|s| = 1$ and $s = s_j$, $j = 1,2,\dots$,

(16) $$F(s) = \frac{1}{\pi} \iint_{\Delta_r} \frac{\overline{\Phi'(z)}}{\Phi'(z)} \frac{dxdy}{s-z}$$
$$+ \frac{1}{\pi} \iint_{\Delta_r} \sum_{n=1}^{+\infty} r^{4n} \left[\frac{\overline{\Phi'(z)}/\Phi'(z)}{s - r^{2n} z} - \frac{1}{\bar{z}^3} \frac{\Phi'(z)/\overline{\Phi'(z)}}{r^{2n} - s\bar{z}} \right] dxdy$$

($x = \operatorname{re} z$, $y = \operatorname{im} z$). Let further $\Phi(s) = \int (r^{4n}/s)\overline{\Phi'(r^{2n}/\bar{s})}ds$, $r^{2n} < |s| < r^{2n+1}$, be a holomorphic function, and let $\Phi(s) = \Phi(s/r^{2n})$, $r^{2n+1} < |s| < r^{2n}$, where $n = 1,2$. Finally, suppose an estimate

(17) $$\left| \frac{\Phi'(w) - \Psi_w(w,t)}{t\Phi'(w)} \right| \leq M,\ 1 < M < +\infty, \text{ for } r < |w| < 1,$$

with Ψ given by (5). Then the following statements hold.

(i) There exists a unique solution $\varrho = R(t)$ of (13) with φ as in (6) on $[0;1)$ subject to the initial condition $R(0) = r$.

(ii) There exists a unique solution $w = f(s,t)$ of

$$(18)\quad w_t = G(w,t),$$

$$(19)\quad G(w,t) = \frac{1}{\pi}\iint_{\Delta_{R(t)}} \frac{\varphi(z,t)}{w-z}\,dxdy + \frac{1}{\pi}\iint_{\Delta_{R(t)}} \sum_{n=1}^{+\infty} R^{4n}(t)\left[\frac{\varphi(z,t)}{w-R^{2n}(t)z} - \frac{1}{\bar z^3}\frac{\overline{\varphi(z,t)}}{R^{2n}(t)-w\bar z}\right]dxdy,$$

where $R^n(t) = [R(t)]^n$, on $\Delta\times[0;1)$, subject to the initial condition $f(s,0) = s$, which for every t represents the Teichmüller mapping of Δ_r onto $\Delta_{R(t)}$ normalized by the condition $f(1,t) = 1$ and corresponds to Φ and t.

(iii) For every t we have

$$(20)\quad G(w,t) = 0 \quad \text{for} \quad |w| = 1 \quad \text{and} \quad w = s_j,\ j = 1,2,\dots,$$

and the mapping $f(\ ,t)$ satisfies the relations

$$(21)\quad f(s,t) = s \quad \text{for} \quad |s| = 1 \quad \text{and} \quad s = s_j,\ j = 1,2,\dots,$$

as well as

$$(22)\quad \Psi(f(s,t),t) = \Phi(s) + t\overline{\Phi(s)} \quad \text{for} \quad r < |s| < 1.$$

(iv) If we assume, in addition, one of the cases

$$(23)\quad \operatorname{im}\left\{\frac{1}{s}\,\frac{F(s) - tH(s)\overline{F(s)}}{1 - tH(s)\frac{\partial}{\partial s}[F(s)/H(s)]}\right\} = C(t) \quad \text{for} \quad |s| = R(t)$$

with

$$(24)\quad \lim_{\substack{z\to s\\ \Phi'(z)\neq 0}} \frac{\overline{\Phi'(z)}}{\Phi'(z)} = H(s) \quad \text{for} \quad r^{2n} < |s| < r^{2n-2},\ n = 1,2,\dots,$$

(25) $\frac{\partial}{\partial s}\frac{F(s)}{H(s)} = -\frac{s}{\bar{s}}\frac{1+(\bar{s}/s)^2H^2(s)}{1-(\bar{s}/s)^2H^2(s)}, \quad \frac{\overline{F(s)}}{F(s)} = -\frac{\bar{s}}{s}$ for $|s| = r$

with

(26) $\lim\limits_{\substack{z\to s\\ \Phi'(z)\neq 0}} \frac{\overline{\Phi'(z)}}{\Phi'(z)} = H(s),\ F(s)\neq 0$ for $|s| = r$;

and

(27) $F(s) = 0$ for $|s| = r$,

then we obtain correspondingly

(23') $\Theta(s,t) = \vartheta(s) + \int_0^t (1-t^2)^{-2}C(t)dt$ for $|s| = r$,

(25') $R(t) = r$ for $|s| = r$,

and

(27') $f(s,t) = s$ for $|s| = r$,

where $\vartheta(s) = \arg s$, $\Theta(s,t) = \arg f(s,t)$, $\Theta(s,\ )$ being supposed to be continuous for $|s| = r$, $-\pi < \arg s < \pi$.

4. Proof of the main result

For greater clarity the proof of Theorem 1 is divided into seven steps.

Step A. <u>We begin with proving</u> (i). The change of variable $t = T_0\tau$ leads to an analogue of (13), where $\varphi(w,t)$ is replaced by

$$T_0\varphi(w,T_0\tau) = T_0(1-T_0^2\tau^2)^{-1}\overline{\Psi_w(w,T_0\tau)}/\Psi_w(w,T_0\tau).$$

By the generalized Cauchy integral formula (cf. e.g. [8], p. 163) and formula (22), $\Psi(\ ,t)$ is holomorphic in int $\hat{\Delta}$ for any fixed t, so φ is measurable. Next we easily check that, given $\varepsilon > 0$, if we choose $T_0 = \varepsilon(1-\frac{1}{2}\varepsilon)$, then $\| T_0\varphi(\ ,T_0\tau)\|_\infty \leq \frac{1}{2}$ for $0 \leq t \leq 1-\varepsilon$. In other words, if we take T_0, $T_0 > 0$, sufficiently small, we can get t, $t<1$, arbitrarily close to 1. Therefore, by Theorem D, the assertion (i) follows.

Step B. The same theorem implies the existence of a unique solution $w = f(s,t)$ of (11) on $\Delta_r \times [0;1)$ with φ as in (6), subject to the initial condition $f(s,0) = s$, which for every t represents a quasiconformal mapping of Δ_r onto $\Delta_{R(t)}$, normalized by the condition $f(1,t) = 1$. Hence also $\Psi(f(\ ,t),t)$ is unique. We are going to try to guess $\Psi(f(\ ,t),t)$. Relations (15), (16), and (5) naturally suggest that $\Psi(f(\ ,t),t)$ is perhaps given by (22). If so, then $f(\ ,t)$ would be uniquely determined by (22) and $f(1,t) = 1$. Indeed, by holomorphy of $\Psi(\ ,t)$, the complex dilatation $\mu(\ ,t)$ of $f(\ ,t)$ clearly satisfies (2), so the required uniqueness would trivially follow from the theorem on existence and uniqueness of quasiconformal mappings with a preassumed complex dilatation in doubly connected domains (cf. [3], p. 26).

Step C. Now we have to verify that the quasiconformal mappings $w = f(s,t)$ of Δ_r onto $\Delta_{R(t)}$, given by (22) and $f(1,t) = 1$, really satisfy (11) with φ as in (6).

By Theorem C, $w = f(s,t)$ is a solution of the differential equation (11), where φ is defined by (12). Hence, by (2),

$$\varphi(w,t) = t^{-1}(1-t^2)^{-1}\mu(f^{-1}(w,t),t)\exp\{-2i \arg f_w^{-1}(w,t)\}.$$

On the other hand, by (22) and holomorphy of $\Psi(\ ,t)$, we have $\Psi_w(w,t) = \Phi'(f^{-1}(w,t))f_w^{-1}(w,t)$, whence, by (2) again,

$$\mu(f^{-1}(w,t),t)\exp\{-2i \arg f_w^{-1}(w,t)\} = t\,\overline{\Psi_w(w,t)}/\Psi_w(w,t).$$

Consequently we conclude that φ is as given in (6).

Step D. In order to check that also the differential equation (18) is satisfied, we have to study more closely the boundary correspondence under $f(\ ,t)$, so <u>we check the formula</u> (21).

By (22) we have $(d/dt)\Psi(f(s,t),t)=\overline{\Phi(s)}$, $r<|s|<1$, and

$$\Phi(s)=(1-t^2)^{-1}[\Psi(f(s,t),t)-t\overline{\Psi(f(s,t),t)}], \quad r<|s|<1.$$

Hence $w=f(s,t)$ satisfies almost everywhere in Δ_r the differential equation

$$\Psi_w(w,t)w_t+\Psi_t(w,t)=(1-t^2)^{-1}[\overline{\Psi(w,t)}-t\Psi(w,t)]$$

i.e., in view of (5) and (15),

$$(28) \quad \Psi_w(w,t)w_t=\frac{\Phi'(w)}{1-t^2}[F(w)-t\frac{\overline{\Phi'(w)}}{\Phi'(w)}\overline{F(w)}] \quad \text{if } r<|w|<1,\ w\neq s_j,$$

$$\text{and} \quad w_t=0 \quad \text{if} \quad |w|=1 \quad \text{or} \quad w=s_j,\ j=1,2,\ldots$$

Therefore, by (5) again, (17), and the initial condition $f(s,0)=s$, we conclude that (21) certainly holds for $0\leq t<1/M$ since

$$\left|\frac{\Psi_w(w,t)}{\Phi'(w)}\right|\geq 1-t\left|\frac{\Phi'(w)-\Psi_w(w,t)}{t\Phi'(w)}\right|\geq 1-tM$$

(note that the expression estimated by M does not depend on t). On the other hand, by (22) and (5), any function $f(s, \ )$, $r\leq|s|\leq 1$, is real-analytic, so (21) holds in the whole interval $[0;1)$.

Step E. Now we are ready <u>to verify the relation</u> (20). To this end we begin with continuing quasiconformally the mappings $f(\ ,t)$, $0\leq t<1$, into the inner disc Δ_r by the formulae

$$(29) \quad f^*(s,t)=R^{2n}(t)/\overline{f(r^{2n}/\bar{s},t)} \quad \text{for} \quad r^{2n}\leq|s|<r^{2n-1}, \ n=1,2,\ldots,$$

$$(30) \quad f^*(s,t)=R^{2n}(t)f(s/r^{2n},t) \quad \text{for} \quad r^{2n+1}\leq|s|<r^{2n}, \ n=1,2,\ldots$$

(obviously we admit $f^*(s,t)=f(s,t)$ for $r\leq|s|\leq 1$ and $f^*(0,t)=0$).

Then the corresponding complex dilatations are given by

$$\mu^*(s,t) = e^{4i \arg s}\overline{\mu(r^{2n}/\bar{s},t)} \quad \text{for} \quad r^{2n} \leq |s| < r^{2n-1}, \ n=1,2,\ldots,$$

$$\mu^*(s,t) = \mu(s/r^{2n},t) \quad \text{for} \quad r^{2n+1} \leq |s| < r^{2n}, \ n=1,2,\ldots$$

(and, obviously, $\mu^*(s,t) = \mu(s,t)$ for $r \leq |s| \leq 1$). Consequently, the corresponding function φ^*, defined by

$$\varphi^*(w,t) = \frac{\mu_t^*(f^{*-1}(w,t),t)}{1-|\mu^*(f^{*-1}(w,t),t)|^2} \exp\{-2i \arg f^{*-1}_w(w,t)\},$$

satisfies

(31) $$\varphi^*(w,t) = e^{4i \arg w}\overline{\varphi(R^{2n}(t)/\bar{w},t)} \quad \text{for} \quad R^{2n}(t) \leq |w| < R^{2n-1}(t),$$
$$n=1,2,\ldots,$$

(32) $$\varphi^*(w,t) = \varphi(w/R^{2n}(t),t) \quad \text{for} \quad R^{2n+1}(t) \leq |w| < R^{2n}(t), \quad n=1,2,\ldots$$

But, by (2), this yields

(33) $\varphi^*(\ ,t) = \varphi(\ ,t)$ almost everywhere, where φ is as in (6).

Therefore we may replace (11) by

(34) $$w_t = \frac{1}{\pi}\iint_\Delta \frac{\varphi(z,t)}{w-z}\,dxdy + h(w,t) \quad \text{almost everywhere in } \Delta,$$

where $h(\ ,t)$ is holomorphic in int Δ.

Then, by the generalized Cauchy integral formula (cf. e.g. [8], p. 163), the derivative $f_{t\bar{s}}$ exists almost everywhere and, by Step C, we have almost everywhere in Δ

$$f_{t\bar{s}}(s,t) = (1-t^2)^{-1}\overline{\Psi_w(w,t)}/\Psi_w(w,t), \quad w=f(s,t).$$

We utilize now the Green's formula which, by (21), gives

$$\iint_\Delta \frac{\overline{\Psi_z(z,t)}}{\Psi_z(z,t)}\frac{dxdy}{w-z} = -\frac{1}{2}i(1-t^2)\int_{\partial\Delta} f_t(s,t)|_{s=f^{-1}(z,t)}\frac{dz}{w-z}.$$

(the Green's formula under sufficiently weak assumptions is proved e.g. in [8], pp. 156-158). Hence, by (21), we arrive at (3).

Let us observe next that, by (31), (32), and (33), we have

$$\iint_{\Delta} \frac{\varphi(z,t)}{w-z}\,dxdy = \iint_{\Delta_{R(t)}} \frac{\varphi(z,t)}{w-z}\,dxdy$$

$$+\sum_{n=1}^{+\infty}\Big[\iint_{\Delta(2n)} \frac{\varphi(z/R^{2n}(t),t)}{w-z}\,dxdy + \iint_{\Delta(2n-1)} e^{4i\arg z}\,\frac{\overline{\varphi(R^{2n}(t)/\bar z,t)}}{w-z}\,dxdy\Big],$$

where $\Delta(2n-1)$ and $\Delta(2n)$ are given by

$$\Delta(2n-1)=\{w:\ R^{2n}(t)\le |w|\le R^{2n-1}(t)\},\ \Delta(2n)=\{w:\ R^{2n+1}(t)\le |w|\le \le R^{2n}(t)\}.$$

Now the change of variables $z\longmapsto R^{2n}(t)/\bar z$ in the integral over $\Delta(2n-1)$ and $z\longmapsto z/R^{2n}(t)$ in the integral over $\Delta(2n)$, $n=1,2,\dots$, leads to the condition

$$(35)\quad \iint_{\Delta} \frac{\varphi(z,t)}{w-z}\,dxdy = \iint_{\Delta_{R(t)}} \frac{\varphi(z,t)}{w-z}\,dxdy$$

$$+\sum_{n=1}^{+\infty} R^{4n}(t)\iint_{\Delta_{R(t)}}\Big[\frac{\varphi(z,t)}{w-R^{2n}(t)z} - \frac{1}{\bar z^{3}}\,\frac{\overline{\varphi(z,t)}}{R^{2n}(t)-w\bar z}\Big]\,dxdy.$$

It can easily be checked with help of the well known Weierstrass' test that the series of integrands in (35) is uniformly convergent in $\Delta_{R(t)}$. Since all integrals appearing in (35) are continuous in w and $\bar w$ (cf. e.g. [15], p. 27), we can interchange the order of integration and summation in (35). Consequently, by (19), the definition of φ given in (6), and (7), the desired relation (20) follows.

Step F. In order to finish the proof of (ii) and (iii) it still remains to show that the equation (11) may be simplified to the form (18). Owing to Step E we already know that (11) may be replaced, by the continuation (29) and (30) of $w=f(s,t)$, with (34), where $h(\ ,t)$ is holomorphic in int Δ. Since, by (19), (20), and (35), $h(\ ,t)$ vanishes on the unit circle, it must vanish identically, and this completes the proof of (ii) and (iii).

Step G. We proceed to prove (iii). Given t, let $n(t)$ be the nonnegative integer chosen so that $r^{n(t)+1} < R(t) \leq r^{n(t)}$. Hence the first relation in (28) remains valid if $r^{n+1} < |w| < r^n$, $n=1,\dots, n(t)-1$ (of course this set may be empty), and, in the case where $R(t) < r^{n(t)}$, also if $R(t) < |w| < r^{n(t)}$. By (5) this relation may be rewritten as

$$\left\{1 - t\frac{\overline{\Phi'(w)}}{\Phi'(w)}\frac{\partial}{\partial w}\left[\frac{\Phi'(w)}{\overline{\Phi'(w)}}F(w)\right]\right\}w_t = \frac{1}{1-t^2}\left[F(w) - t\frac{\overline{\Phi'(w)}}{\Phi'(w)}\overline{F(w)}\right].$$

If we suppose (24) and t_o denotes the smallest solution of the equation $R(t)=r^2$ in $[0;1)$ (if it has no solutions, we put $t_o=1$), then, by (17), for $0 \leq t < \min(t_o, 1/M)$ we have

$$(36)\quad w_t = \frac{1}{1-t^2}\,\frac{F(w) - tH(w)\overline{F(w)}}{1 - tH(w)\frac{\partial}{\partial w}[F(w)/H(w)]},\qquad \begin{array}{l} w=f(s,t),\ R(t)\leq |w| < 1,\\ w \neq r^{2n},\quad 1 \leq n \leq \tfrac{1}{2}n(t). \end{array}$$

Let us observe now that

$$f_t(s,t) = R(t)e^{i\Theta(s,t)}\{[1/R(t)]R'(t) + i\Theta_t(s,t)\}.$$

Therefore, if we further suppose (23), then, by the initial condition $f(s,0)=s$, (24) and (36) yield (23') for $0 \leq t < \min(t_o, 1/M)$. On the other hand, by (22) and (5), any function $f(s,\ )$, $r \leq |s| \leq 1$, is real-analytic, so (23') holds in the whole interval $[0;1)$. Similarly, in the case of (25) and (26) we arrive at

$$(37)\quad \operatorname{re}\left\{\frac{1}{s}\,\frac{F(s) - tH(s)\overline{F(s)}}{1 - tH(s)\frac{\partial}{\partial s}[F(s)/H(s)]}\right\} = 0 \quad \text{for } |s| = r$$

and, consequently, (36) yields (25'). Finally, in the case of (27) we obtain (23) with $C(t)=0$ and $R(t)$ replaced by r, and (37), whence, by (36), the relation (27') follows. In this way the assertion (iii) is proved.

5. Further results on Teichmüller mappings of an annulus

Theorem 1 implies

Corollary 1. Suppose that Φ is holomorphic in Δ_r, Φ' has zeros at s_j, $j=1,2,\ldots$ (we do not exclude the case where $\Phi'(s)\neq 0$, $r<|s|<1$),

$$(38)\quad \iint_{\Delta_r} \frac{\overline{\Phi'(z)}}{\Phi'(z)} \frac{dxdy}{s-z} = 0 \quad \text{for } |s|=1,\ |s|=r, \text{ and } s=s_j,\ j=1,2,\ldots,$$

and an estimate (17) holds with Ψ given by (5). Then there exists a unique solution $w=f(s,t)$ of

$$(39)\quad w_t = \frac{1}{\pi}\iint_{\Delta_r} \frac{\varphi(z,t)}{w-z}\,dxdy, \text{ where } \varphi(w,t) = \frac{1}{1-t^2}\,\frac{\overline{\Psi_w(w,t)}}{\Psi_w(w,t)},$$

on $\Delta_r\times[0;1)$, subject to the initial condition $f(s,0)=s$, which for every t represents the Teichmüller mapping of Δ_r onto itself normalized by the condition $f(1,t)=1$ and corresponds to Φ and t. Moreover, for every t we have

$$(40)\quad \iint_{\Delta_r} \frac{\overline{\Psi_z(z,t)}}{\Psi_z(z,t)} \frac{dxdy}{w-z} = 0 \quad \text{for } |w|=1,\ |w|=r, \text{ and } w=s_j, j=1,2,\ldots,$$

and the mapping $f(\ ,t)$ satisfies the relations

$$(41)\quad f(s,t)=s \quad \text{for } |s|=1,\ |s|=r, \text{ and } s=s_j,\ j=1,2,\ldots,$$

as well as (22).

Proof. The corollary is an immediate consequence of Theorem 1 in the case (27). In fact, in this case (18) and (19) with φ as in

(6) reduce to the form (39) since any function

$$G(w,t) - \frac{1}{\pi}\iint_{\Delta_{R(t)}} \frac{\varphi(z,t)}{w-z}\,dxdy, \quad R(t) < |w| < 1,$$

being holomorphic in int $\Delta_{R(t)}$ and vanishing or its boundary, must vanish identically.

Actually Corollary 1 is the most direct and natural analogue of Theorem A for an annulus and its direct proof would be much simpler than those of Theorem 1. The way we have chosen is motivated by the fact that the class of mappings covered by Corollary 1 is rather small.

As in the case of Δ the inverse problem is even simpler. We have

Theorem 2. Suppose that Φ is holomorphic in int Δ_r, and $f(\ ,t)$, $0 \leq t < 1$, are Teichmüller mappings of Δ_r onto $\Delta_{R(t)}$, normalized by the condition $f(1,t) = 1$, which correspond to Φ and t. Then (21) with $r < |s_j| < 1$, $j = 1,2,\ldots$ (we do not exclude the case where the set of points s_j is empty), implies (15), (20), (18), and (22), where F, G, and Ψ are given by (16), (19), and (5), respectively. If we assume, in addition, one of the cases (23′) with (24), (25′) with (26), and (27′), then we obtain correspondingly (23), (25), and (27).

Proof. Let us observe that since for $0 \leq t < 1$ there is exactly one quasiconformal mapping of Δ_r onto $\Delta_{R(t)}$ satisfying (2) and $f(1,t) = 1$ (cf. Step B of the preceding proof), where $R(t)$ is uniquely determined as a solution $\varrho = R(t)$ of (13) with φ as in (6) on $[0;1)$ subject to the initial condition $R(0) = r$, the same concerns $\Psi(f(\ ,t),t)$, where Ψ is given by (5). On the other hand, if $\Psi(f(\ ,t),t)$ were given by (22), then, by holomorphy of $\Psi(\ ,t)$ (cf.

the generalized Cauchy integral formula, e.g. in [8], p. 163), f would satisfy (2), the relation $f(1,t)=1$ being fulfilled by (21). Therefore $\Psi(f(\ ,t),t)$ is indeed given by (22). Now we can follow the argument given in Step C of the preceding proof to conclude that the quasiconformal mappings $w=f(s,t)$ of Δ_r onto $\Delta_{R(t)}$ satisfy (11) with φ as in (6). Next the argument of Step E leads to (20) which for $t=0$ reduces to (15). Finally, by the argument given in Step F we conclude that the equation (11) may be simplified to the form given in (18), as desired. The truth of the last statement in the theorem can easily be checked by taking into account the relation (36).

Let us finally notice the following immediate

Corollary 2. Suppose that Φ is holomorphic in int Δ_r, and $f(\ ,t)$, $0\leq t<1$, are Teichmüller mappings of Δ_r onto $\Delta_{R(t)}$, normalized by the condition $f(1,t)=1$, which correspond to Φ and t. Then (41) with $r<|s_j|<1$, $j=1,2,\ldots$ (we do not exclude the case where the set of points s_j is empty), implies (38), (40), (39), and (22), where Ψ is given by (5).

6. The case of a punctured disc

All results of Sections 3-5 have their natural analogues in the case $r=0$, i.e. for mappings discussed in Section 1 with an additional invariant point 0. By the classical theorems of Riemann and Osgood-Carathéodory this can be extended to the case of arbitrary punctured discs.

Theorem 3. Suppose that Φ is holomorphic in int $\Delta\setminus\{0\}$, Φ' has zeros at s_j, $j=1,2,\ldots$ (we do not exclude the case where

$\Phi'(s)\neq 0,\ 0<|s|<1$),

$$(42)\qquad \iint_{\Delta} \frac{\overline{\Phi'(z)}}{\Phi'(z)}\,dxdy = 0 \quad \text{for } |s|=1,\ s=0, \text{ and } s=s_j,\ j=1,2,\ldots,$$

$$(43)\qquad \left|\frac{\Phi'(w)-\Psi_w(w,t)}{t\,\Phi'(w)}\right| \leq M,\quad 1<M<+\infty, \text{ for } 0<|w|<1,$$

with Ψ given by (5). Then there exists a unique solution $w=f(s,t)$ of (6) on $\Delta_r\times[0;1)$, subject to the initial condition $f(s,0)=s$, which for every t represents the Teichmüller mapping normalized by the conditions $f(0,t)=0$, $f(1,t)=1$, and corresponds to Φ and t. Moreover, for every t we have

$$(44)\qquad \iint_{\Delta} \frac{\overline{\Psi_z(z,t)}}{\Psi_z(z,t)}\,\frac{dxdy}{w-z} = 0 \quad \text{for } |w|=1,\ w=0, \text{ and } w=s_j,\ j=1,2,\ldots,$$

and the mapping $f(\ ,t)$ satisfies the relations (8) (whence it is also normalized in the sense of Introduction and Section 1), as well as

$$(45)\qquad \Psi(f(s,t),t)=\Phi(s)+t\overline{\Phi(s)} \quad \text{for } 0<|s|<1.$$

Proof: completely analogous to that of Theorem 1.

Theorem 4. Suppose that Φ is holomorphic in int $\Delta\setminus\{0\}$ and $f(\ ,t)$, $0\leq t<1$, are Teichmüller mappings of Δ onto itself, normalized by the conditions $f(0,t)=0$, $f(1,t)=1$, which correspond to Φ and t. Then (8) with $|s_j|<1$, $j=1,2,\ldots$ (we do not exclude the case where the set of points s_j is empty; in any case, by (8), every mapping $f(\ ,t)$ is also normalized in the sense of Introduction and Section 1), implies (42), (44), (6), and (45), where Ψ is given by (5).

Proof: completely analogous to that of Theorem 2.

References

[1] L.V. Ahlfors, Some remarks on Teichmüller's space of Riemann surfaces, Ann. Math. 74 (1961), 171-191.

[2] R.S. Hamilton, Extremal quasiconformal mappings with preseribed boundary values, Trans. Amer. Math. Soc. 138 (1969), 339-406.

[3] J. Ławrynowicz, On the parametrization of quasiconformal mappings in an annulus, Ann. Univ. Mariae Curie-Skłodowska Sect. A 18 (1964), 23-52.

[4] ——— , On arbitrary homotopies in parametrization theorems for quasiconformal mappings, Bull. Acad. Polon. Sci. Sér. Sci. Math. Astronom. Phys. 20 (1972), 733-737.

[5] ——— , On the parametrization of quasiconformal mappings with invariant boundary points in the unit disc, ibid. 20 (1972), 739-744.

[6] ——— , On the parametrization of quasiconformal mappings with invariant boundary points in an annulus, Comment. Math. Helv. 47 (1972), 213-219.

[7] ——— in cooperation with J. Krzyż, The parametrical method for quasiconformal mappings in the plane, to appear.

[8] O. Lehto und K.I. Virtanen, Quasikonforme Abbildungen (Grundlehren Math. Wissensch. 126), Springer-Verlag, Berlin - Heidelberg - New York 1973.

[9] E. Reich and K. Strebel, On quasiconformal mappings which keep the boundary points fixed, Trans. Amer. Math. Soc. 138 (1969), 211-222.

[10] ——— und ——— , Einige Klassen Teichmüllerscher Abbildungen, die die Randpunkte festhalten, Ann. Acad. Sci. Fenn. Ser. A I 457 (1970), 19 pp.

[11] E. Reich and K. Strebel, Extremal quasiconformal mappings with given boundary values, Bull. Amer. Math. Soc. 79 (1973), 488-490.

[12] K. Strebel, Zur Frage der Eindeutigkeit extremaler quasikonformer Abbildungen des Einheitskreises I - II, Comment. Math. Helv. 36 (1962), 306-329 and 39 (1964), 77-89.

[13] O. Teichmüller, Extremale quasikonforme Abbildungen und quadratische Differentiale, Abh. Preuss. Akad. Wiss. Math.-Naturwiss. Kl. 22 (1940), 197 pp.

[14] ——— , Ein Verschiebungssatz der quasikonformen Abbildung, Deutsche Mathematik 7 (1944), 336-343.

[15] I.N. Vekua, Generalized analytic functions (Internat. Ser. Pure Appl. Math. 25), Pergamon Press, Oxford - London - New York - Paris 1962.

On the boundary value problem for quasiconformal mappings

Olli Lehto
University of Helsinki
Department of Mathematics

In this survey lecture, I consider the boundary value problem for quasiconformal mappings in the plane, with emphasis on more recent developments. The connection with the theory of Riemann surfaces will be discussed and some open problems pointed out.

Mappings of plane domains

1. Solution of the boundary value problem. The now classical boundary value problem is as follows: Let A and A′ be Jordan domains of the extended complex plane, and h a homeomorphism between the boundary curves of A and A′, preserving positive orientation with respect to the domain. When does there exist a quasiconformal mapping of A onto A′ with boundary values h?

The problem makes sense, because every quasiconformal mapping of A onto A′ can be extended to a homeomorphism between the closures of A and A′.

A necessary condition for h can be found in a very straightforward manner. Take four points z_1, z_2, z_3, z_4 on the boundary of A so that A and these points form a quadrilateral $A(z_1,z_2,z_3,z_4)$ Let M denote the conformal module of a quadrilateral. If $f: A \to A'$ is a K-quasiconformal mapping which agrees with h on the boundary, and $w_i = h(z_i)$, i=1, 2, 3, 4, then

$$M(A(z_1,z_2,z_3,z_4))/K \leq M(A'(w_1,w_2,w_3,w_4)) \leq K\, M\,(A(z_1,z_2,z_3,z_4)).$$

If the points z_1, z_2, z_3, z_4 are so chosen that

$$(1) \qquad M(A(z_1,z_2,z_3,z_4)) = 1,$$

the condition simplifies to

(2) $$\frac{1}{K} \leq M(A'(w_1,w_2,w_3,w_4)) \leq K.$$

This becomes much more explicit if A and A' are upper half-planes and $z_4 = w_4 = \infty$. In this case (1) is fulfilled if and only if z_1, z_2, z_3 are equidistant: $z_1 = x-t$, $z_2 = x_1$, $z_3 = x+t$, $t>0$. The module of $M(A'(w_1,w_2,w_3,w_4))$ can be expressed by means of the module $\mu(r)$ of the Grötzsch ring domain bounded by the unit circle and the line segment $0 \leq x \leq r$, $r < 1$. We have

$$M(A'(w_1,w_2,w_3,w_4)) = \frac{2}{\pi}\mu(((w_2-w_1)/(w_3-w_1))^{1/2}).$$

Hence, it follows from (2) that

(3) $$\frac{1}{\lambda(K)} \leq \frac{h(x+t)-h(x)}{h(x)-h(x-t)} \leq \lambda(K),$$

where

$$\lambda(K) = (\mu^{-1}(\tfrac{1}{2}\pi K))^{-2} - 1.$$

The inequality (3) is sharp for every x and t and for every K. If $K \to 1$, then $\lambda(K) \to 1$, and if $K \to \infty$, we have

(4) $$\lambda(K) = \frac{1}{16} e^{\pi K} - \frac{1}{2} + O(e^{-\pi K}).$$

For the details, we refer to [9].

The necessary condition (3) is also sufficient, in the following sense:

T h e o r e m 1. Let h be a strictly increasing continuous function on the real axis, growing from $-\infty$ to $+\infty$ and satisfying for some $\rho \geq 1$ the condition

(5) $$\frac{1}{\rho} \leq \frac{h(x+t)-h(x)}{h(x)-h(x-t)} \leq \rho.$$

Then there is a quasiconformal self-mapping of the upper half-plane with boundary values h, whose maximal dilatation is bounded by a number depending only on ρ.

This is a famous result of Beurling and Ahlfors [3]. They proved it by showing that the mapping f defined by

$$(6)\qquad f(x+iy) = \int_0^1 \tfrac{1}{2}[h(x+ty) + h(x-ty)]dt + i\int_0^1 \tfrac{1}{2}[h(x+ty) - h(x-ty)]dt$$

has the desired properties.

The best estimates known for the maximal dilatation K of a solution are

$$(7)\qquad K \leq \min(\rho^2, 8\rho).$$

The bound ρ^2 was derived by Beurling and Ahlfors; to obtain it the imaginary part of (6) must be multiplied by a suitable positive constant. This bound has the good property that it tends to 1 as $\rho \to 1$. The bound 8ρ, which is due to Reed [10], is of the correct order of magnitude as $\rho \to \infty$. It is of interest to compare this result with (4).

Thus condition (5) characterizes the boundary functions of the quasiconformal self-mappings of the upper half-plane which fix ∞; such functions are called quasisymmetric. Beurling and Ahlfors [3] proved that a function satisfying (5) may be completely singular for every $\rho>1$. This was a surprising result at the time of its discovery, but it turned out later that these badly behaving quasisymmetric functions appear frequently in situations related to lifts of mappings between Riemann surfaces (cf. Section 6).

Having solved the boundary value problem for upper half-planes, we obtain the solution for arbitrary Jordan domains using suitable conformal mappings.

2. Smooth solutions. With the solution of the boundary value problem, new questions arise. We first discuss here the possibility of finding smooth solutions. As before, we consider self-mappings of the upper half-plane. The Beurling-Ahlfors construction (6) shows that, no matter how badly h behaves, there are always solutions which are in class C^1 in the upper half-plane. This

result can be easily improved: Replace in the integrals in (6) the constant $\frac{1}{2}$ by $\kappa(t-\frac{1}{2})$, where the function κ is defined by $\kappa(t) = a\exp((t^2-\frac{1}{4})^{-1})$ for $|t| \leq 1/2$, by $\kappa(t)=0$ otherwise, and the constant a is so chosen that the integral of κ over the real axis equals 1/2. Then f is a C^∞-quasiconformal self-mapping of the upper half-plane with boundary values h, and its maximal dilatation is bounded by a number depending only on ρ. (The details have been worked out by Matti Lehtinen.)

Actually, we can say even more:

T h e o r e m 2. To every quasisymmetric boundary function, there exist real-analytic quasiconformal solutions.

Proof: Given a quasisymmetric h, we associate with it a pair f_1, f_2 of conformal mappings with the following properties: f_1 maps the upper half-plane H_1 and f_2 the lower half-plane H_2 onto complementary domains of a Jordan curve such that $f = f_1 \circ h$ on the real axis. Using the existence theorem for Beltrami differential equations, we find f_1, f_2 as follows. Let $w: H_1 \to H_1$ be a quasiconformal mapping with complex dilatation μ, and W a quasiconformal mapping of the plane with complex dilatation μ in H_1, and 0 in H_2. Then $f_1 = (W|H_1) \circ w^{-1}$, $f_2 = W|H_2$, have the desired properties.

We now construct a solution of the boundary value problem, making use of the following observation. If ψ is a quasiconformal extension to H_1 of the conformal mapping f_2, then $f_1^{-1} \circ \psi$ is a solution of the boundary value problem.

Let $S_f = f'''/f' - \frac{3}{2}(f''/f')^2$ denote the Schwarzian derivative of a conformal mapping f of H_2, and

$$\|S_f\| = 4 \sup_{z \in H_2} y^2 |S_f(z)|, \qquad z=x+iy,$$

its norm. If w_1, w_2 are solutions of the differential equation $w'' + \frac{1}{2} S_f w = 0$, with $w_1 w_2' - w_2 w_1' = 1$, then the function w_1/w_2 has the same Schwarzian as f. Hence $f = g \circ (w_1/w_2)$, where g is a Möbius transformation. In studying quasiconformal extensions of f, we may assume, therefore, that $f = w_1/w_2$.

Ahlfors and Weill [2] proved: If $f = w_1/w_2$ is a conformal mapping of H_2 and $\|S_f\| < 2$, then ψ,

$$\psi(\bar{z}) = \frac{w_1(z) + (\bar{z}-z)w_1'(z)}{w_2(z) + (\bar{z}-z)w_2'(z)}, \tag{8}$$

is a quasiconformal extension of f to H_1. From the expression (8) it is clear that ψ is real-analytic.

In addition to the construction of Ahlfors and Weill, we need the following well-known estimate:

If f is a conformal mapping of H_2 and has a quasiconformal extension to H_1 with complex dilatation μ, then

$$\|S_f\| \leqq 6\|\mu\|\infty. \tag{9}$$

Suppose first that h satisfies (5) for some $\rho<\sqrt{2}$. By (7), we then have a quasiconformal self-mapping w of H_1, with boundary values h and maximal dilatation K<2. Then $f_1 o w$, which is an extension of f_2, also has these properties. Since $K=(1+\|\mu\|\infty)/(1-\|\mu\|\infty)$, it follows that $6\|\mu\|\infty<2$. Considering (9), we may apply the Ahlfors-Weill method and conclude that f_2 has a real-analytic quasiconformal extension ψ. Then $f_1^{-1} o \psi : H_1 \to H_1$ is a real-analytic quasiconformal mapping with boundary values h.

In the general case, we can always write $h=h_n o \ldots o h_1$, where each h_i is quasisymmetric and satisfies (5) for $\rho<\sqrt{2}$. To prove this, we first continue h to a quasiconformal self-mapping w of H_1, then apply the existence theorem for Beltrami equations to conclude that we can write $w=w_n o \ldots o w_1$, where every $w_i : H_1 \to H_1$ has a maximal dilatation as close to 1 as we please, and finally define h_i as the boundary function of w_i. Now, if $\omega_i : H_1 \to H_1$ is a real-analytic quasiconformal mapping with boundary values h_i, then $\omega=\omega_n o \ldots o \omega_1$ is a real-analytic quasiconformal mapping with boundary values h.

3. Extremal solutions. Let h be quasisymmetric and Q_h the class of quasiconformal self-mappings of the upper half-plane with boundary values h. We call $f \in Q_h$ extremal if f has the smallest maximal dilatation among the mappings of Q_h. A standard normal family argument shows that every Q_h contains extremal mappings. A basic and largely open problem is to study for which boundary functions h there is exactly one extremal in Q_h and what properties the extremal mappings possess.

Let Φ be the class of holomorphic functions φ in H_1 such that

$$(10) \qquad \int_{H_1} |\varphi| = 1.$$

Given a mapping $f \in Q_h$ with complex dilatation μ, we introduce the functional l_μ defined by

$$(11) \qquad l_\mu(\varphi) = \int_{H_1} \mu\varphi;$$

its norm is

$$\| l_\mu \| = \sup_{\varphi \in \Phi} \left| \int_{H_1} \mu\varphi \right|.$$

Trivially, $\| l_\mu \| \leq \| \mu \|_\infty$. Equality characterizes extremality:

T h e o r e m 3. A mapping $f \in Q_h$ with complex dilatation μ is extremal if and only if

$$(12) \qquad \| l_\mu \| = \| \mu \|_\infty .$$

It was Kruškal [5] who first realized the importance of the functional (11) for the extremal problem. In the above formulation, the necessity of condition (12) was proved by Hamilton [4] and the sufficiency by Reich and Strebel [11].

Suppose that f is extremal. By Theorem 3, we then have a sequence of functions $\varphi_n \in \Phi$ such that $\int \varphi_n \mu \to \| \mu \|_\infty$; (φ_n) is called a Hamilton sequence for μ. From (10) it follows that Φ is a normal family. Hence, there is a subsequence (φ_{n_i}) such that φ_{n_i}

converges locally uniformly in H_1 towards an analytic limit function φ_o. If φ_o is identically zero, the sequence (φ_{n_i}) is said to be degenerating. If no Hamilton sequence for μ degenerates, then the limit φ_o is uniquely determined by μ and belongs to Φ,

$$(13) \qquad \mu = ||\mu||_\infty \frac{\varphi_o}{|\varphi_o|},$$

and f is the only extremal.

This result has recently been skilfully exploited by Strebel [12]. Consider all ring domains $A \subset H_1$ whose outer boundary component is the real axis, and let W_h be the class of quasiconformal mappings of domains A into H_1, with boundary values h on the real axis. The infimum L of the maximal dilatations of the mappings belonging to W_h is called the dilatation of h. If K is the maximal dilatation of the extremal in Q_h, we have trivially $L \leqq K$. Strebel proved that if $L<K$, then no Hamilton sequence for μ degenerates, and thus obtained the following result:

T h e o r e m 4. If the dilatation of h is strictly less than the extremal maximal dilatation in Q_h, then Q_h has only one extremal mapping and its complex dilatation is of the form (13).

An immediate corollary says that if h has dilatation 1, then the extremal in Q_h is unique. It is either conformal or its complex dilatation is of the form (13).

The case L=1 allows explicit conclusions. Since ∞ is in a special position, it is preferable to state the result for the unit disc rather than for the upper half-plane. So let h now denote a homeomorphism of the unit circle onto itself which can be extended to a quasiconformal self-mapping of the unit disc; its dilatation is defined in the same manner as for quasisymmetric functions. Define a self-mapping η of the real axis by $\eta(x) = \arg h(e^{ix})$. Strebel [12] proved that h has dilatation 1 if and only if

$$\lim_{t \to 0} \frac{\eta(x+t) - \eta(x)}{\eta(x) - \eta(x-t)} = 1,$$

uniformly in x.

Mappings of Riemann surfaces

4. Teichmüller spaces. The boundary value problem for quasiconformal mappings is intimately connected with the theory of Riemann surfaces. Let S and S′ be Riemann surfaces, both admitting H_1 as the universal covering surface. Let G and G′ be the cover transformation groups of H_1 over S and S′. If the limit set of G is the whole real axis, then G is said to be of the first kind. Otherwise G is of the second kind, in which case S is a bordered Riemann surface.

Every homeomorphism $\varphi : S \to S'$ can be lifted to a self-mapping f of H_1: If $\pi : H_1 \to S$, $\pi' : H_1 \to S'$, are the canonical projections, there is a homeomorphism $f: H_1 \to H_1$ such that $\pi' \circ f = \varphi \circ \pi$. If f is a lift of φ, then all lifts of φ are of the form $g \circ f$, where $g \in G'$.

Two homeomorphisms φ and ψ of S onto S′ are said to be homotopic modulo boundary if there is a homotopy between φ and ψ which is constant on the border of S. In the following, we use the word homotopic to mean homotopic modulo boundary. This is justified by the fact that in the important case in which G is of the first kind, the border is empty.

Let us consider all quasiconformal mappings of S onto Riemann surfaces. Two such mappings φ_1, φ_2 are said to be equivalent if $\varphi_2 \circ \varphi_1^{-1}$ is homotopic to a conformal mapping. The equivalence classes are the points of the Teichmüller space T(S) of S.

The points of T(S) can also be defined in terms of complex dilatations. One calls μ a Beltrami differential of S if μ is complex-valued and measurable in local coordinates of S, $\mu d\bar{z}/dz$ is invariant on S, and the function $|\mu|$ is bounded. If we consider μ as a function of $z \in H_1$, then the invariance of $\mu d\bar{z}/dz$ on S is equivalent to $(\mu \circ g)\bar{g}'/g' = \mu$ for every $g \in G$. A Beltrami differential of S with $\|\mu\|_\infty < 1$ is always the complex dilatation of a quasiconformal mapping φ of S, and up to conformal transformations, φ is uniquely determined by μ. If two complex dilatations are called equivalent whenever the corresponding mappings are, then T(S) is also the set of equivalent μ's. The distance between two points p_1, p_2 of T(S) can be defined as $\inf_{\mu_i \in p_i} \|\mu_1 - \mu_2\|_\infty$.

The equivalence of μ_1 and μ_2 can be expressed, non-trivially, in terms of the lifted mappings. Let μ be a Beltrami differential of S with $||\mu||_\infty < 1$, and $f^\mu: H_1 \to H_1$ the unique quasiconformal mapping which has complex dilatation μ and fixes $0,1,\infty$. We then have the following result (see, e.g.[1]):

T h e o r e m 5. The dilatations μ_1 and μ_2 represent the same point of T(S) if and only if $f^{\mu_1} = f^{\mu_2}$ on the real axis.

Consequently, points of T(S) can be represented by quasisymmetric functions. If G is the trivial group, T(S) consists of all normalized quasisymmetric functions. In the general case, f^μ induces an isomorphism θ of G onto a Fuchsian group G', where

(14) $$\theta(g) = f^\mu \circ f \circ (f^\mu)^{-1}, \qquad g \in G.$$

It follows that the boundary function h of f^μ also satisfies the condition

(15) $$h \circ g \circ h^{-1} = \theta(g)$$

for every $g \in G$.

5. Smooth mappings. The results of Sections 2 and 3 can at least partly be carried over to the case where the given boundary function satisfies a relation (15) and the solutions f are required to fulfil the condition $f \circ g \circ f^{-1} = \theta(g)$. As remarked before, this means that the solutions can be projected to quasiconformal mappings of the Riemann surface $S = H_1/G$ onto $S' = H_1/G'$. The counterpart of Theorem 2 is as follows:

T h e o r e m 6. Let S and S' be quasiconformally equivalent Riemann surfaces. Then every homotopy class of quasiconformal mappings contains a real-analytic quasiconformal mapping.

Proof: Let $\varphi: S \to S'$ be a quasiconformal mapping with complex dilatation μ, and h the boundary function of the mapping $f^\mu: H_1 \to H_1$. Considering Theorem 5, we have to show that there is a real-analytic quasiconformal mapping $\omega: H_1 \to H_1$ which has boundary values h and induces the same isomorphism $\theta: G \to G'$ as f^μ.

The proof goes as in Theorem 2. If $||\mu||_\infty < 1/3$, the Schwarzian of the conformal mapping f_2 satisfies the inequality $||S_{f_2}|| < 2$. Thus f_2 has the Ahlfors-Weill extension ψ. From the fact that μ is a Beltrami differential of G, we deduce that $S_{f_2} dz^2$ is invariant under G. Direct computation then shows that the complex dilatation of ψ is a Beltrami differential of G. It follows that $\omega=f^{-1}o\psi$ induces an isomorphism of G onto some Fuchsian group, and since $\omega=f^{\mu}$ on the boundary, this isomorphism must be θ. Therefore, ω has all the desired properties.

If $\|\mu\|_\infty \geqq 1/3$, we write $f^{\mu}=v_1 o u_1$, where u_1, v_1 are quasiconformal self-mappings of H_1 fixing ∞, and the complex dilatation of u_1 is equal to $\mu/3$. Since $\mu/3$ is a Beltrami differential of G, the mapping u_1 induces an isomorphism θ_1: $G \to G_1$, and so v_1 induces the isomorphism $\theta o \theta_1^{-1}$: $G_1 \to G'$. For the complex dilatation ν_1 of v_1 we have

$$\|\nu_1\|_\infty = 2\|\mu\|_\infty (3-\|\mu\|_\infty^2)^{-1} \leqq a\|\mu\|_\infty,$$

where $a<1$. If $\|\nu_1\|_\infty \geqq 1/3$, we set $v_1=v_2 o u_2$, where again u_2, v_2 are quasiconformal self-mappings of H_1 fixing ∞, and the complex dilatation of u_2 is $\nu_1/3$. This process is continued until the L^∞-norm of the complex dilatation ν_n of v_n is $<1/3$; this will be the case after finitely many steps, because $\|\nu_n\|_\infty \leqq a^n \|\mu\|_\infty$. Then $f^{\mu}=v_n o u_n o \ldots o u_1$. Applying the Ahlfors-Weill method, we now construct a real-analytic quasiconformal mapping ω_i, $i=1, \ldots, n+1$, of H_1 onto itself which agrees on the boundary with u_i for $i=1, \ldots, n$ and with v_n for $i=n+1$. Then $\omega=\omega_n o \ldots o \omega_1$ has the boundary values h and induces θ.

6. Extremal mappings. Given two cover transformation groups G and G′ acting on H_1 and an isomorphism θ: $G \to G'$, let h be a quasisymmetric function satisfying (15). We denote by $Q_h(G)$ the class of quasiconformal self-mappings f of H_1, such that f=h on the boundary and $fogof^{-1}=\theta(g)$ for every $g \in G$. Unless $Q_h(G)$ is empty, it is again clear that it contains extremal mappings.

Suppose first that h represents a point of T(S), $S=H_1/G$. Then $Q_h(G)$ is not void, and its elements project to quasiconformal mappings of S onto H_1/G', which are all in the same homotopy class (cf. Section 3). Let $\Phi(G)$ be the class of holomorphic quadratic differentials φ of S with the property $\int_S |\varphi| = 1$. If μ is the complex dilatation of a mapping $f \in Q_h(G)$, we can define as before $l_\mu(\varphi) = \int_S \mu\varphi$. Again, the condition (12) is necessary for f to be extremal in $Q_h(G)$ (Hamilton [4]). In fact, the proof of the necessity part of Theorem 3 carries over easily to arbitrary Riemann surfaces ([8]).

Theorem 4 does not directly apply to this situation, because it requires the quadratic differentials to be integrable over the whole covering surface H_1. On the other hand, if h represents a point of the Teichmüller space of a compact Riemann surface, then by Teichmüller's classical theorem, the class $Q_h(G)$ contains only one extremal. The extremal is either conformal or its complex dilatation is of the form (13). In the latter case, the boundary function h is always completely singular, as proved by Kuusalo [6].

In conclusion, we ask whether the class $Q_h(G)$ is always non-void. If G and G' are of the first kind, the problem is not trivial. Partial solutions are known: $Q_h(G)$ is not empty if G is finitely generated or if h satisfies (5) for some $\rho < \sqrt{2}$ (see [7]). Also, if the larger class Q_h contains a unique extremal f, then this f is in $Q_h(G)$. It may happen, however, that the extremal in Q_h has a smaller maximal dilatation than the extremal in $Q_h(G)$ (Strebel [13]).

References

[1] Ahlfors, L.V.: Lectures on quasiconformal mappings. Van Nostrand, Princeton, N.J., 1966

[2] Ahlfors, L.V. and G. Weill: A uniqueness theorem for Beltrami equations. Proc.Amer.Math.Soc. 13 (1962), 975-978

[3] Beurling, A. and L.V. Ahlfors: The boundary correspondence under quasiconformal mappings. Acta Math. 96 (1956),125-142

[4] Hamilton, R.: Extremal quasiconformal mappings with prescribed boundary values. Trans.Amer.Math.Soc. 138 (1969), 399-406

[5] Kruškal, S.L.: Teichmüller's theorem on extremal quasiconformal mappings. Siberian Math.J. 8 (1967), 231-244

[6] Kuusalo, T.: Boundary mappings of geometric isomorphisms of Fuchsian groups. Ann.Acad.Sci.Fenn. A I 545 (1973)

[7] Lehto, O.: Group isomorphisms induced by quasiconformal mappings. Contributions to Analysis, A Collection of Papers Dedicated to Lipman Bers, Academic Press, New York and London 1974

[8] Lehto, O.: Application of singular integrals to extremal mappings of Riemann surfaces, to appear

[9] Lehto, O. and K.I. Virtanen: Quasiconformal mappings in the plane. Springer-Verlag, Berlin-Heidelberg-New York 1973

[10] Reed, T.J.: Quasiconformal mappings with given boundary values. Duke Math.J. 33 (1966), 459-464

[11] Reich, E. and K. Strebel: Extremal quasiconformal mappings with given boundary values. Contributions to Analysis, A Collection of Papers Dedicated to Lipman Bers, Academic Press, New York and London 1974

[12] Strebel, K.: On the existence of extremal Teichmüller mappings, to appear

[13] Strebel, K.: On lifts of extremal quasiconformal mappings, to appear

REGULARITY OF CONFORMAL MAPPINGS OF RIEMANNIAN MANIFOLDS.

Jaqueline Lelong-Ferrand
University of Paris VI

The analytic definition of quasiregular mappings can be easily extended to riemannian manifolds, and 1-quasiregular mappings are a natural generalization of conformal mappings. In fact this is but an apparent generalization, since 1-quasiregular mappings of C^∞ n-dimensional manifolds, with $n \geq 3$, are C^∞ local diffeomorphisms, and are therefore "conformal" in the most strict meaning of the word. This fact, which can be considered as an extension of Liouville's theorem, will proceed from the two following theorems :

Theorem A. If $M, \bar{M}$ are two C^1 connected n-dimensional riemannian manifolds, with $n \geq 3$, any non constant 1-quasiregular mapping $\phi : M \to \bar{M}$ is a local homeomorphism.

Theorem B. If $M, \bar{M}$ are two C^∞ [resp. C^ω] riemannian manifolds, any 1-quasiregular homeomorphism $\phi : M \to \bar{M}$ is a conformal C^∞ [resp. C^ω] diffeomorphism.

1. Proof of theorem A. Theorem A is a special case of the following one, which is an easy extension of a deep result of O.Martio, S.Rickman, J.Vaïsälä [6] .

Theorem 1. For any integer $n \geq 3$, there exists a constant $K_n > 1$, only depending on n, such that any K-quasiregular non constant mapping $\phi : M \to \bar{M}$ of C^1 connected n-dimensional manifolds, with $K < K_n$, is a local homeomorphism.

Proof. Let be $\phi : M \to \bar{M}$ a K-quasiregular mapping, with $K < K_n$, and let us choose $\varepsilon > 0$ such that $(1+\varepsilon)^{4(n-1)} K < K_n$. Since M is a C^1 riemannian manifold, every point $a \in M$ admits a neighborhood U such that there exists a $(1+\varepsilon)$ - bilipschitzian mapping θ of U onto a domain G in E^n. Moreover, since ϕ is continuous and M is C^1, we can choose U such that $\phi(U)$ be contained in a neighborhood $\bar{U}$ of $\phi(a)$ in $\bar{M}$, satisfying a similar condition (i.e. such there exists a $(1+\varepsilon)$ - bilipschitzian mapping $\bar{\theta}$ of $\bar{U}$ onto a domain $\bar{G}$ in $\bar{E}^n$). Then θ and $\bar{\theta}$ are $(1+\varepsilon)^{2(n-1)}$ quasiconformal and the mapping $\psi = \bar{\theta} \circ \phi \circ \theta^{-1} : G \to \bar{G}$, is K'-quasiregular, with $K' = (1+\varepsilon)^{4(n-1)} K < K_n$; by theorem (4.6) of [6] we infer thich ψ is a local homeomorpnism or a constant ; and it is the same for $\phi_{|U}$. Then every point $a \in M$ admits a neighborhood U in M such that $\phi_{|U}$ is a constant or a local homeomorphism.

Now let be M_o the set of points $a \in M$ admitting a neighborhood U such that $\phi|_U$ be constant. From the preceding result it follows easily that M_o is open and closed in M. Since M has been supposed connected, we have $M_o = M$ or $M_o = \emptyset$; and if $M_o = M$, φ is constant : then theorem 1 is proved.

2. First order regularity of 1-quasiconformal mappings.

In the euclidean case theorem B has been proved independently by Y.G.Resetnyak [

and G.W.Gehring [2] . We have already given in [5] the general proof of this theorem, and we'll only here emphasize the main original points of this proof.

The underlying idea is the following theorem, which is quite elementary when ϕ is a conformal C^3-diffeomorphism :

Theorem 1. Let (M,g) and $(\bar{M},\bar{g})$ two C^∞ riemannian manifolds of the same dimension $n \geq 3$, and $\phi: M \to \bar{M}$ a 1-quasiconformal mapping. Then the function $u = |\phi'|^{\frac{n-2}{2}}$ satisfies the linear elliptic differential equation

$$(1) \qquad \Delta u = \frac{n-2}{4(n-1)} \left[Ru - (\bar{R} \circ \phi) u^{\frac{n+2}{n-2}} \right]$$

where Δ is the Laplacian on M, and R, $\bar{R}$ denote the scalar curvatures of $M,\bar{M}$.

For getting this result in the general case we prove at first that ϕ is locally bilipschizian, afterwards that the distribution Laplacian $\Delta|\phi'|$ is a measure, and at last that $|\phi'|$ satisfies a relation equivalent with (1). We'll essentially use the fact that a riemannian manifold is approximatively euclidean at the neighborhood of every point, and, more precisely, we'll use the following lemma :

2.1. Let (M,g) be a C^2 [resp. C^3] riemannian manifold, and for any $a \in M$, let be $B(a,r)$ the geodesic ball of center a and radius r. Then, if K is a compact subset of M, there exist two numbers $\rho > 0$ and $A > 0$ such that for any $a \in M$ there exists a bijection h_a of $B(a,\rho)$ onto a domain in E^n, such that the restriction of h_a to each ball $B(a,r)$, with $0 < r < \rho$, be $(1+Ar)$-bilipschitzian [resp. $(1+Ar^2)$ - bilipschitzian].

This lemma immediately follows from the fact that the metric tensor g of M is of class C^1 [resp. C^2]. Moreover, in the C^3 case, the mapping h_a given by normal coordinates of center a satisfies the stated condition. Then we can suppose that $h_a(B(a,r))$ is the euclidean ball $\beta_r = \{ x \in E^n \mid |x| < r\}$. This will be used in § 3.

Since C^1-riemannian manifolds are locally diffeomorphic with E^n, we can easily extend a well known property ([2] th 11, corollary) and assert that quasiconformal mappings of C^1 riemannian manifolds are Hölderian (a direct proof is given in [5]).

Now, we can prove

2.2. Let $M,\bar{M}$ be two C^2 riemannian manifolds, and $\phi : M \to \bar{M}$ a 1-quasiconformal mapping. Then ϕ is locally lipschitzian.

Proof. In all what follows, we denote by δ_M , $\delta_{\bar{M}}$ the geodesic distances on $M,\bar{M}$, and by $B,\bar{B}$ the geodesic balls of M, $\bar{M}$.

Then, let K be a compact subset of M, and (ρ,A), $(\bar{\rho},\bar{A})$ the numbers respectisely associated with K, $\phi(K)$ by statement (2.1) relative to C^2-manifolds. Since ϕ is locally hölderian, there exist three numbers $r_o > 0$, $\gamma > 0$ and $\nu \in]0,1]$ such that for any $a \in K$ and $x \in M$ satisfying $\delta_M(a,x) = r < r_o$, we have

$$\delta_{\bar{M}}(\phi(a),\phi(x)) \leq \gamma r^{\nu}$$

Without lost of generality we may suppose $r_o < \inf(\rho,1)$ and such that $\gamma r_o^{\nu} < \bar{\rho}$.

The point $a \in K$ being fixed, let h be a bijection of $B(a,\rho)$ onto a domain of E^n satisfying $h(a) = 0$ and such that the restriction of h to $B(a,r)$, with $r < \rho$, be $(1+Ar)$-bilipschitzian. Similarly, let $\bar{h}$ be a bijection of $\bar{B}(\phi(a),\bar{\rho})$ onto a domain of E^n, satisfying $\bar{h}(\phi(a)) = 0$ and such that the restriction of $\bar{h}$ to $\bar{B}(\phi(a),r)$, with $r < \bar{\rho}$, be $(1+\bar{A}r)$ bilipschitzian.

For any $r \in]0,r_o]$ we have, by construction :

$$\phi(B(a,r)) \subset \bar{B}(\phi(a),\gamma r^{\nu}) \subset \bar{B}(\phi(a),\bar{\rho}) \quad ;$$

and it easily follows that the restriction to $h^{-1}[B(a,r_o)]$ of the mapping $\psi = \bar{h} \circ \phi \circ h^{-1}$ is $K(r)$-quasiconformal, with

$$(2) \qquad K(r) = (1+Ar)^{2(n-1)} (1+\bar{A}\gamma r^{\nu})^{2(n-1)}$$

(Since a k-bilipschitzian mapping is $k^{2(n-1)}$- quasiconformal).

Now, let us denote by β_r the euclidean ball defined by $|x| < r$, and let us choose $R > 0$ such that $\beta_R \subset h(B(a,r_o))$. By lemma 4.1. of [4] we know that there exists an absolute constant C_n, only depending on n, such that, for any $r \in]0, \frac{R}{2}]$:

$$(3) \qquad \int_r^{2r} \frac{\delta^n(t)}{t}\, dt \leq C_n\, K(2r)\, V(2r)$$

where $\delta(t)$ is the diameter of $\psi(\partial\beta_t)$ and $V(t)$ the volume of $\psi(\beta_t)$.

Now, since $\psi(\beta_t)$ is a domain of E^n, the diameter $\delta(t)$ of $\psi(\partial\beta_t) = \partial\psi(\beta_t)$ is equal to the diameter of $\psi(\beta_t)$, and is therefore an increasing function of t.

Then the inequality (3) involves the existence of a number $t \in [r,2r]$ such that

$$\delta(r) \leq \delta(t) \leq \left(\frac{C_n K(2r)V(2r)}{\text{Log } 2}\right)^{1/n} ,$$

and to set the assertion 2.2. we have but to prove that $\frac{V(r)}{r^n}$ is uniformly bounded when r tends to zero (since K(r) is uniformly bounded).

Let us denote by J_ψ the metric jacobian of ψ, and by $S(r)$ the area of $\psi(\partial\beta_r)$. For a.e. $r \in]0,r_o[$, we have :

$$\text{(4)} \qquad \frac{dV}{dr}(r) = \frac{d}{dr}\int_{\beta_r} J_\psi dx = \int_{\partial\beta_r} J_\psi \, d\sigma$$

where $d\sigma$ denotes the element of area on the sphere $\partial\beta_r$. On the other hand, by the isoperimetric inequality, the area $S(r)$ of $\psi(\partial\beta_r)$ is satisfying

$$\text{(5)} \qquad S^n(r) \geq n^n v_n V^{n-1}(r) \quad ,$$

where $v_n = \frac{\omega_{n-1}}{n}$ is the volume of β_1 . At last, by Hölder's inequality, we have

$$S^n(r) \leq \left[\int_{\partial\beta_r} |\psi'|^{n-1} d\sigma\right]^n \leq \left[\int_{\partial\beta_r} |\psi'|^n d\sigma\right]^{n-1} \int_{\partial\beta_r} d\sigma \quad ,$$

and, since ψ is $K(r)$-quasiconformal on β_r :

$$\text{(6)} \qquad S^n(r) \leq K^{n-1}(r)\left[\int_{\partial\beta_r} J_\psi \, d\sigma\right]^{n-1} n v_n r^{n-1}$$

By comparison of (4), (5), (6) we get :

$$\text{(7)} \qquad \frac{V'(r)}{V(r)} \geq \frac{n}{rK(r)}$$

By coming bach to (2) one can easily prove that the positive function

$$\theta : r \longmapsto \frac{n}{r} - \frac{n}{rK(r)}$$

satisfies $\int_o^R \theta(r)dr = B < +\infty$; the function V being absolutely continuous, we have

$$\text{Log}\,\frac{V(R)}{V(r)} \geq n\,\text{Log}\,\frac{R}{r} - B \qquad (\forall\, r \in [0,R])$$

and therefore $\frac{V(r)}{r^n} \leq \frac{V(R)}{R^n} e^B$.

Now $V(R)$ is uniformly bounded when a runs in K, and the constant B is independant from a. Assertion 2.6 follows.

Remark. By any easy extension, one can see that the same result is true if we only suppose that the metric tensors g, $\bar{g}$ are satisfying a Hölder condition.

3. Higher order regularity.

We first prove the following lemma

3.1. Let M, $\bar{M}$ be C^3 riemannian manifolds of dimension n, and $\phi: M \to \bar{M}$ a 1-quasi-conformal mapping. For any compact subset K of M, there exists $\rho > 0$ and $C > 0$ such that, for any $a \in K$, the increasing function

$$r \longmapsto V_a(r) = \int_{B(a,r)} |\phi'|^n dr$$

admits a.e. on $[0,\rho]$ a derivative satisfying

$$(1) \qquad \frac{V_a'(r)}{V_a(r)} \geq \frac{n}{r}(1 - Cr^2)$$

Proof. We already know that ϕ is satisfying a Lipschitz condition on K, whose ratio will be denoted by k. Then let be (ρ, A) and $(\bar{\rho}, \bar{A})$ the numbers respectively associated with K and $\phi(K)$ by statement 2.1. relative to C^3 manifolds. Without loss of generality we many suppose that ρ is small enough to have $\phi(B(a,\rho)) \subset \bar{B}(\phi(a),\bar{\rho})$ for any $a \in K$; and we many also suppose that, for any $a \in K$ and $r \in]0,\rho[$, B(a,r) admits a $(1+Ar^2)$-bilipschitzian mapping onto the euclidean ball β_r (cf remark following 2.1) : this involves that the boundary of B(a,r) is the geodesic sphere S(a,r)

Then $\phi(B(a,r))$ admits a $(1+\bar{A}k^2r^2)$ - bilipschitzian mapping $\bar{h}$ onto a domain of E^n ; and by applying the isoperimetric inequality to $\bar{h}\,[\phi(B(a,r)]$, we easily see that the volume $V_a(r)$ of $\phi[B(a,r)]$ and the area $S_a(r)$ of its boundary $\phi[S(a,r)]$ are satisfying :

$$(2) \qquad n^n v_n V_a^{n-1}(r) \leq (1+\bar{A}k^2r^2)^{2n(n-1)} S_a^n(r)$$

with $v_n = \frac{1}{n}\omega_{n-1}$ (cf § 2). On the other hand, we have, for a.e. $r \in]0,\rho[$:

$$(3) \qquad \frac{d}{dr} V_a(r) = \int_{S(a,r)} |\phi'|^n \, d\tau$$

and, by Hölder's inequality

$$(4) \qquad S_a(r) = \int_{S(a,r)} |\phi'|^{n-1} d\sigma \leq \left[\int_{S(a,r)} |\phi'|^n d\sigma\right]^{\frac{n-1}{n}} \left[\int_{S(a,r)} d\sigma\right]^{\frac{1}{n}}$$

At last, since B(a,r) admits a $(1+Ar^2)$ - bilipschitzian mapping onto the euclidean ball β_r , we have

$$\int_{S(a,r)} d\sigma \leq (1+Ar^2)^{n-1} \; n v_n r^{n-1} .$$

From (2) (3) and (4) we get :

$$\frac{V_a' r}{V_a r} \geq \frac{n}{r(1+Ar^2)(1+\bar{A}k^2r^2)^{2n}}$$

therefore (1) with $C = A + 2n\bar{A}k^2$.

<u>Corollary</u>. For all $a \in K$, $J(a) = \lim \frac{V_a(r)}{v_n r^n}$ exists, and for any $r \in [0,\rho]$:

(5) $$V_a(r) \geq v_n r^n J(a) \exp\left(\frac{-nCr^2}{2}\right).$$

<u>Proof</u>. The inequality (1) involves that the function $r \longmapsto r^{-n} \exp\left(\frac{nCr^2}{2}\right) V_a(r)$ is increasing, which implies the existence of $J(a)$ and inequality (5). Obviously we have $J(a) = |\phi'(a)|^n$ for any $a \in K$ such that $\phi'(a)$ exists. Moreover, since ϕ is lipschitzian, J is bounded ; and, ϕ^{-1} being also conformal and therefore lipschitzian J is bounded away from zero.

Now, from above results, we can easily infer that there exists a constant α such that the distribution $\Delta J + \alpha d\tau$ be positive; in consequence the distribution Laplacian ΔJ is a measure, and J belongs to the Soboleff class $W_2^1(M)$ (for details cf [5])

To achieve the proof, are can argue like Y.G.Resetnjak in the euclidean case [7] by considering harmonic local coordinates y^α on $\bar{M}$. Then the n functions $\phi^\alpha = \phi^* y^\alpha$ are locally extremals of the integral

$$\int_M |\nabla u|^2 \quad |\phi'|^{n-2} \quad d\tau$$

with

$$|\phi'|^2 = \frac{1}{n} \sum_{i=1}^{n} \sum_{j=1}^{n} \sum_{\alpha=1}^{n} \sum_{\beta=1}^{n} g^{ij} (\bar{g}_{\alpha\beta} \circ \phi) \frac{\partial \phi^\alpha}{\partial x^i} \frac{\partial \phi^\beta}{\partial x^j} \quad ;$$

and the smoothness of functions φ^α follows from the properties of solutions of elliptic systems (since $|\varphi'|$ is locally bounded and bounded away from zero).

We can also use the fact that ϕ is an isometry of $(\bar{M},\bar{g})$ onto $(M,|\phi'|^2 g)$, and extend to $(M,|\phi'|^2 g)$, in terms of distributions, the classical computation of the scalar curvature. From this we infer that $J = |\phi'|^n$ satisfies

(6) $$\Delta J = \frac{n+2}{2n} J^{-1} \operatorname{grad}^2 J + \frac{n}{2n-2}\left(RJ - \bar{R}\circ\varphi\, J^{1+\frac{2}{n}}\right)$$

and by induction we prove that J belongs to $H_p = W_2^p$ for any $p \in \mathbb{N}$, and consequently, that J is C^∞. Then the smoothness of φ follows from the theorem of Myers Steenrod relative to isometries (cf [1]).

Equation (6) is equivalent with (1,1) ; and if $M,\bar{M}$ are C^ω, it implies that J and φ are also C^ω.

[1] E.CALABI and P.HARTMAN. On the smoothness of isometries. Duke Math. J.73 (1970) 741-750.

[2] F.W.GEHRING. Rings and quasiconformal mappings in space. Trans. A.M.S. 103 (1962 353-393.

[3] P.HARTMAN. On isometries and a theorem of Liouville. Math. Z. 69 (1968) 202-210

[4] J.LELONG-FERRAND. Transformations conformes et quasiconformes des variétés riemanniennes compactes. Acad. Roy. Belg. Cl. Sci. Mem. 39 n°5 1971

[5] J. LELONG-FERRAND. Geometrical interpretation of scalar curvature and regularity of conformal homeomorphisms. (To appear.

[6] O.MARTIS - S.RICKMAN - J.VÄISÄLA. Topological and metric properties of quasi-regular mappings. Ann. Acad. Sci. Fennical 448 (1969) 1-40

[7] Y.G.RESETNYAK. On conformal mapping of a space. Doklady 130 (1960) n° 1

[8] Y.G.RESETNYAK. Liouville's theorem for conformal mappings with minimal hypothesis of regularity. Sibirsk Math. J. 8 (1967) n°4 835-840.

QUASIREGULAR MAPPINGS AND VALUE DISTRIBUTION

P. Mattila and S. Rickman

1. Introduction. We want to give a description of what has been done in the study of value distribution of quasiregular mappings. As quasiregular mappings generalize plane analytic functions in a natural sense to real n-dimensional space, one expects that to a certain extent the value distribution theory of dimension two has analogs in higher dimensions. Such analogs have really been established recently, and the study has also given some results which seem to be new for the classical theory too.

A systematic study of value distribution of quasiregular mappings was started in [9] and was then continued in [6]. In both these works questions of the relationship between the "nonintegrated" counting function and its over all average are considered. In addition to a pointwise study of the counting function in [9, Section 5] we have established in [6] an equidistribution theory for its averages with respect to integral dimensional Hausdorff measure. The special case of taking averages over $(n-1)$-dimensional spheres was included in [9]. The theory presented in [6] can be regarded as a generalization of Ahlfors's covering theorems.

On the other hand, we are far from getting defect relations for $n \geq 3$ in analog of Nevanlinna's or Ahlfors's theory. We believe that such relations are true in some form, but even an analog for Picard's theorem is not known. The best known result in this direction is the following: If $f: R^n \to R^n$ is quasiregular and omits a set of positive conformal capacity, f is constant [5].

Although defect relations are not known, it is still interesting to ask whether an inverse problem can be solved. It turns out that a restricted problem, which is formally the analog of the inverse problem for defects in the plane, can really be solved in the affirmative. The proof was written in [8] for $n = 3$ but can be carried out also for other dimensions $n \geq 2$. Compared to Drasin's proof of the corresponding solution for meromorphic functions [1] the proof is simpler in the sense that quasiregularity admits flexibility, but on the other hand, complications arise from the geometry in space.

2. Results on the counting function. We shall now go into more detail and first present relationships between the counting function and its over all average. For simplicity we consider a nonconstant quasimeromorphic mapping $f:R^n \to \bar{R}^n$ which is a fundamental case. For definitions of quasiregular and quasimeromorphic mappings for the euclidean n-space R^n and $\bar{R}^n$ see [4,5], and for Riemannian manifolds [6] or [3]. We shall study value distribution with respect to the exhaustion of R^n by balls $B(r) = \{x \in R^n | \ |x| < r\}$. We define the counting function as

$$n(r,y) = \sum_{x \in f^{-1}(y) \cap \bar{B}(r)} i(x,f).$$

Here $i(x,f)$ is the local topological index. The over all average of the counting function, the spherical average, is denoted

$$A(r) = \frac{1}{\lambda_n} \int_{R^n} \frac{n(r,y)}{(1+|y|^2)^n} \, \mathcal{L}^n(y),$$

where $\mathcal{L}^n$ is the Lebesgue measure and λ_n the total spherical measure of R^n.

If f is a nonconstant meromorphic function, it follows trivially from the Ahlfors-Shimizu form of Nevanlinna's first main theorem that

$$\liminf_{r\to\infty} \frac{n(r,a)}{A(r)} \le 1.$$

For quasimeromorphic mappings the left hand side may exceed 1 but it is not known whether it must be finite. However, one can prove the following result ([9, Theorem 5.16]).

Theorem 1. *Let $f:R^n \to \bar{R}^n$ be a nonconstant quasimeromorphic mapping. Then for each $c > 1$ there exists $\theta > 1$ such that*

$$\sup_{a\in\bar{R}^n} \liminf_{r\to\infty} \frac{n(r,a)}{A(\theta r)} \le c$$

It is interesting not only from the point of view of quasimeromorphic mappings but also of meromorphic functions to study the validity of inequalities of the type

$$\limsup_{r\to\infty} \frac{n(r,a)}{A(\theta r)} \leq c$$

where $c, \theta > 1$. That this need not hold for any choices of c and θ follows by an example of meromorphic functions by Toppila [10, Theorem 4] at least for $n = 2$. However, the existence of an asymptotic value assures a positive result as follows ([9, 5.11]).

Theorem 2. Let $f:R^n \to \bar{R}^n$ be a nonconstant quasimeromorphic mapping and let f have an asymptotic value a_0. Then for each $c > 1$ there exists $\theta > 1$ such that whenever $E \subset \bar{R}^n \setminus \{a_0\}$ is a compact set, there exists $r_0 > 0$ such that

$$\sup_{a \in E} n(r,a) \leq cA(\theta r)$$

holds for $r \geq r_0$.

Corollary. Let $f:R^n \to \bar{R}^n$ be a quasimeromorphic mapping which has at least two asymptotic values. Then for each $c > 1$ there exists $\theta > 1$ such that for some $r_0 > 0$

$$\sup_{y \in \bar{R}^n} n(r,y) \leq cA(\theta r)$$

if $r \geq r_0$.

Toppila has again shown by constructing an example of an entire function that one asymptotic value is not enough here. He has also shown that even in the meromorphic case θ cannot be chosen near 1 in Theorem 2. For these results, see [10].

Detailed proofs of Theorems 1 and 2 are presented in [9] and they depend essentially on two facts: (1) maximal path lifting for discrete open mappings ([7]) and (2) inequalities for moduli of path families for quasiregular mappings ([4,3.2] and [11]). Using these we obtain as an effective tool the following lemma which is repeatedly applied in the proofs of Theorems 1 and 2. It relates the spherical averages of the counting function over concentric spheres. We denote by $\nu(r,t)$ the spherical average of $n(r,y)$ over the sphere $S(t) = \{x \in R^n \mid |x| = t\}$.

Lemma (Theorem 4.1 in [9]). *Let* $\theta, c > 1$, *let* $0 < s,t < \infty$, *and let* $f: R^n \to \bar{R}^n$ *be a nonconstant quasimeromorphic mapping. Then for* $r > 0$

$$c\nu(\theta r,t) \geq \nu(r,s) - \frac{K(f)|\log(t/s)|^{n-1}}{(1-1/c)\ (\log\theta)^{n-1}},$$

where $K(f)$ *is the maximal dilatation of* f.

This lemma is a special case of a general equidistribution theory for averages with respect to integral dimensional Hausdorff measure which will be described briefly in the next section.

3. Averages of the counting function. Let us here consider a more general case and assume that M and N are C^∞, connected, orientable Riemannian n-manifolds of which M is noncompact and N is compact, and $f: M \to N$ a nonconstant quasiregular mapping. We are going to study value distribution of f with respect to a fixed, which we shall call, admissible exhaustion function of M. By an *exhaustion function* of M we mean a function $V: [a,b[\to \mathcal{P}(M)$, where $-\infty < a < b \leq \infty$, such that each $V(t) = V_t \subset M$ is open, connected, the closure $\bar{V}_t$ is compact, $\bar{V}_t \subset V_u$ for $t < u$, and

$$M = \bigcup_{t\in[a,b[} V_t.$$

We shall only use exhaustion functions V with $a > 0$ and which are parametrized such that

$$t = a\exp\left(\left(\frac{\omega_{n-1}}{M(\Gamma_{a,t})}\right)^{1/(n-1)}\right). \tag{3.1}$$

Here $\Gamma_{s,t}$ is for $a \leq s < t < b$ the family of paths in $V_t \setminus \bar{V}_s$ joining

∂V_t and $\bar{V}_s$, $M(\Gamma_{s,t})$ is its n-modulus, and ω_{n-1} is the (n-1)-dimensional measure of the unit sphere in R^n. An exhaustion function V satisfying (3.1) is called <u>admissible</u> if there exist constants $a_o \in]a,b[$, $\theta_o > 1$, $\kappa > 0$, and $\mu \geq n-1$ such that

$$(3.2) \qquad (\log \frac{t}{s})^{\mu} \leq \kappa \frac{\omega_{n-1}}{M(\Gamma_{s,t})}$$

holds for $a_o \leq s < t < b$, $t/s \leq \theta_o$. We always have the inverse inequality

$$(\log \frac{t}{s})^{n-1} \geq \frac{\omega_{n-1}}{M(\Gamma_{s,t})} .$$

In the case $M = R^n$, $V_t = B(t)$, there is equality in (3.2) with $\kappa = 1$, $\mu = n-1$.

The counting function of f with respect to the exhaustion V is defined by

$$n(t,y) = \sum_{x \in f^{-1}(y) \cap \bar{V}_t} i(x,f).$$

We shall now introduce the sets over which averages of the counting function of f are considered. In the following k is an integer, $1 \leq k \leq n$, and $\mathcal{H}^k$ is the k-dimensional Hausdorff measure [2,2.10.2].

A set $E \subset N$ is called $(\mathcal{H}^k,k)$-rectifiable if $\mathcal{H}^k(E) < \infty$ and there are sets $E_o \subset N$, $A_i \subset R^k$, $i = 1,2,\ldots$, and Lipschitz maps $g_i : A_i \to N$ such that $\mathcal{H}^k(E_o) = 0$ and

$$E = E_o \cup \bigcup_{i=1}^{\infty} g_i(A_i).$$

For the properties of these sets, see [2].

Let $C \geq 1$. A Borel set $E \subset R^n$ is called <u>strictly</u> $(\mathcal{H}^k,k,C)$-<u>flat</u> if $0 < \mathcal{H}^k(E) < \infty$ and there exists an orthogonal projection p of R^n onto a k-dimensional linear subspace of R^n such that

$$\mathcal{H}^k(B) \leq C\mathcal{H}^k(pB)$$

for all Borel sets $B \subset E$. A Borel set $E \subset N$ is called $(\mathcal{H}^k,k,C)$-

<u>flat</u> if there exist a finite number of strictly $(\mathcal{H}^k,k,C_i)$ - flat sets $E_i \subset R^n$ and L_i-bilipschitzian maps $g_i : U_i \to g_i(U_i) \subset N$, $i = 1, \ldots, m$, such that each U_i is an open ball in R^n containing E_i, $E = \bigcup_{i=1}^m g_i(E_i)$, and

$$C_i L_i^{2(k+n)} \leq C \text{ for } i = 1, \ldots, m.$$

A Borel set $E \subset N$ is called $(\mathcal{H}^k, k)$ - <u>flat</u> if it is $(\mathcal{H}^k,k,C)$ - flat for some $C \geq 1$.

Examples of sets which are both $(\mathcal{H}^k,k)$ - rectifiable and $(\mathcal{H}^k,k)$ - flat are all compact k-dimensional C^1 - submanifolds of N and in the case $k = n$ all Borel subsets of N with positive Lebesque measure. An $(\mathcal{H}^k,k)$ - rectifiable Borel set E with $\mathcal{H}^k(E) > 0$ need not be $(\mathcal{H}^k,k)$- flat, and vice versa. However, the following lemma holds. Its proof, as well as the proofs of the other results of this section, can be found in [6].

<u>Lemma.</u> <u>If</u> $\varepsilon > 0$ <u>and</u> $E \subset N$ <u>is an</u> $(\mathcal{H}^k,k)$ - <u>rectifiable Borel set with</u> $\mathcal{H}^k(E) > 0$, <u>then there exists</u> $F \subset E$ <u>such that</u> $\mathcal{H}^k(E \setminus F) < \varepsilon$ <u>and</u> F <u>is</u> $(\mathcal{H}^k,k,C)$ - <u>flat for all</u> $C > 1$.

If $E \subset N$ is a Borel set with $0 < \mathcal{H}^k(E) < \infty$, we denote by $\nu^k(t,E)$ the average of $n(t,y)$ over E,i.e.

$$\nu^k(t,E) = \mathcal{H}^k(E)^{-1} \int_E n(t,y) d\mathcal{H}^k y.$$

In particular, for $E = N$ we set

$$A(t) = \nu^n(t,N).$$

The proof of the following theorem is based on the aforementioned path lifting theorem in [7], the modulus inequality [11,3.1], and the above lemma.

<u>Theorem 3.</u> <u>Suppose that</u> $c > 1$ <u>and</u> E <u>is an</u> $(\mathcal{H}^k,k)$ - <u>rectifiable and</u> $(\mathcal{H}^k,k)$ - <u>flat subset of</u> N. <u>Then there exists a positive number</u> d <u>such that for all</u> t <u>and</u> $\theta \leq \theta_o$, $a_o \leq t/\theta < t < \theta t < b$, <u>we have</u>

$$c^{-1}A(t/\theta) - d(\log\theta)^{-\mu} \leq \nu^k(t,E) \leq cA(\theta t) + d(\log\theta)^{-\mu}.$$

The left hand side inequality in the above theorem holds also if E is merely an $(\mathcal{H}^k,k)$-rectifiable Borel set with $\mathcal{H}^k(E) > 0$.

With the help of Theorem 3 and a lemma on real functions one obtains the following theorem, which is similar to well-known results in Ahlfors's theory of covering surfaces.

Theorem 4. *Suppose that either* $b = \infty$ *or* $b < \infty$ *and* $\limsup_{t \to b} (b-t)A(t)^{1/\mu} = \infty$. *Then there exists a set* $B \subset [a,b[$ *such that*

$$\int_B \frac{dt}{t} < \infty \quad \text{if} \quad b = \infty,$$

$$\liminf_{t\to b} \frac{\mathcal{L}^1(B \cap [t,b[)}{b-t} = 0 \quad \text{if} \quad b < \infty,$$

and

$$\lim_{\substack{t\to b\\ t\notin B}} \frac{\nu^k(t,E)}{A(t)} = 1$$

whenever E *is an* $(\mathcal{H}^k,k)$-*rectifiable and* $(\mathcal{H}^k,k)$-*flat subset of* N

Using these results on the averages of the counting function, one can prove the following theorems on the pointwise behavior outside an exceptional set.

Theorem 5. *Suppose that* $b = \infty$ *or* $b < \infty$ *and* $\lim_{t\to b} (b-t)A(t)^{1/\mu} = \infty$. *Then there is a sequence* (t_i) *in* $[a,b[$ *such that* $\lim t_i = b$ *and*

$$\lim \frac{n(t_i,y)}{A(t_i)} = 1$$

for almost every $y \in N$.

Theorem 6. Suppose that $b = \infty$. There exists a set $C \subset N$ of conformal capacity zero such that

$$\liminf_{t\to\infty} \frac{n(t,y)}{A(\theta t)} \leq 1 \leq \limsup_{t\to\infty} \frac{n(\theta t,y)}{A(t)}$$

for all $y \in N \setminus C$.

REFERENCES

[1] Drasin, D. : The inverse problem of the Nevanlinna theory. - To appear in Acta Math..

[2] Federer, H.: Geometric Measure Theory. - Springer - Verlag, Berlin - Heidelberg - New York, 1969.

[3] Lelong - Ferrand, J.: Etude d'une classe d'applications liées à des homomorphismes d'algebres de fonctions, et generalisant les quasi conformes. - Duke Math. J. 40 (1973), 163-186.

[4] Martio, O., Rickman, S. and J. Väisälä: Definitions for quasiregular mappings. - Ann Acad. Sci. Fenn. AI 448 (1969), 1-40.

[5] ________Distortion and singularities of quasiregular mappings. - Ann Acad. Sci. Fenn. AI 465 (1970), 1-13.

[6] Mattila, P. and S. Rickman: Averages of the counting function of a quasiregular mapping. - To appear.

[7] Rickman, S.: Path lifting for discrete open mappings. - Duke Math. J. 40 (1973), 187-191.

[8] ________A quasimeromorphic mapping with given deficiencies in dimension three. - Symposia Mathematica XVIII, 535-549, Instituto Nazionale di Alta Mathematica, Roma, Convegno del Marzo 1974, Academic Press 1976.

[9] ________On the value distribution of quasimeromorphic maps. - Ann. Acad. Sci. Fenn. AI 2, to appear.

[10] Toppila, S. : On the counting function for the a-values of a meromorphic function. - Ann. Acad. Sci. Fenn. AI 2, to appear.

[11] Väisälä, J.: Modulus and capacity inequalities for quasiregular mappings. - Ann. Acad. Sci. Fenn. AI 509 (1972), 1-14.

University of Helsinki
Helsinki, Finland

ISOLATED SINGULARITY OF THE MEAN QUASICONFORMAL MAPPINGS

by

Miodrag Perović
Tehnički fakultet
Titograd
Yugoslavia

In our paper [2] it is proved that theorem of Zorič on quasiconformal mappings is also true for the mean quasiconformal mappings, i.e. a locally homeomorphic mapping of the Euclidean space R^n of dimension $n \geq 3$ into itself which is mean quasiconformal is a homeomorphism, and moreover onto the whole space R^n. In fact, in [2] it is precised the asymptotic behaviour at infinity of the mean dilatation which assures that the mentioned theorem holds.

In this paper we want to give the local version of the theorem on homeomorphism. Let us give the precise formulations.

Let $f: D \to R^n$ be a locally homeomorphic mapping of the unbounded domain $D \subset R^n$, let $x_0 \in D$, $B^n(r) = B^n(x_0, r) = \{x \mid |x - x_0| < r\}$ and $D_r = D \cap B^n(r)$. By the mean dilatation of the mapping f in the domain D_r, we mean a quantity $k(r)$ defined by the relation

$$k^{n-1}(r) = \frac{1}{\operatorname{mes} D_r} \int_{D_r} k^{n-1}(x,f)\, dm ,$$

where $\operatorname{mes} D_r$ is the Lebesgue's measure of the domain D_r and $k(x,f)$ coeficient of quasiconformality of the mapping f at $x \in D$. Let $r_0 > 0$. The function

$$K_D(r) = K_D(r, r_0) = \sup_{r_0 \leq t \leq r} k(t) ,$$

will be used as an index of growth of the mean dilatation $k(r)$ when $r \to \infty$.

Using Lemma 2 from [2] and modulus method we can easily obtain the following

Lemma. Let D be a deleted neighbourhood of ∞, and $f: D \to D'$ a homeomorphism of the domain $D \subset R^n$ onto the domain $D' \subset R^n$, so that f and f^{-1} are ACL^n [3]. Then, if

$$\int^{\infty} \frac{dr}{r K_D(r)} = \infty,$$

it follows that f can be continued to a homeomorphism of the domain $D \cup \{\infty\}$.

We now give the basic result.

Theorem. Let f be locally homeomorphic mapping of a deleted neighbourhood of a point ∞, let f be ACL^n and let for every domain where f is homeomorphism f^{-1} be ACL^n in it's image /domain/. Then, if the index of growth $K(r)$ of mean dilatation is such that for every unbounded subdomain U of D

$$\int^{\infty} \frac{dr}{r K_U(r)} = \infty,$$

it follows that the mapping f is homeomorphic in some deleted neighbourhood V of the point ∞, and can be continued to a homeomorphism of the whole neighbourhood $V \cup \{\infty\}$ of this point.

The proof of this statement can be done in the same way as the proof of Theorem 1 in [1] if we use the preceding lemma and the basic lemma from our paper [2].

We remark that the degree of the growth of the quantity $K_D(r)$ in the preceding theorem is the best possible. This can be easily seen from the example given in [3] and the remark about it in [2].

REFERENCES

[1] В.А.Зорич, Изолированная особенность отображений с ограниченным искажением. Матем. сб. 8I (I23), № 4 (I970), 634–636.

[2] М.Перович, О глобальном гомеоморфизме отображений квазиконформных в среднем, ДАН СССР (в печати).

[3] В.А.Зорич, Теорема М.А.Лаврентьева о квазиконформных отображениях пространства, Матем.сб., 74(II6) (I967), 4I9–433.

[4] В.А.Зорич, О допустимом порядке роста коэффициента квазиконформности в теореме М.А.Лаврентьева, ДАН СССР,I8I, № 3 (I968), 530–533.

[5] J. Väisälä, Lectures on n-Dimensional Quasiconformal Mappings, Lect. Not. in Math., Springer-Verlag, 1971.

QUASICONFORMAL MAPPINGS IN NORMED SPACES

by

Giovanni Porru[(+)]

Summary. In this paper the F. Gehring's metric definition for Frèchet-derivable K-quasiconformal mappings in a normed space is investigated. For such mappings, by means of angle distortion, a geometric characterization is given.

1. Let H be a real normed space with norm $\|\cdot\|$ and T a mapping of a domain $\Omega \subset H$ into H. Suppose T Frèchet-derivable with non vanishing derivative through Ω. If $X \in \Omega$ and r is a positive real number such that the close ball $\bar{B}(X,r)$ with center in X and radius r, is contained in Ω, we set

$$L(X,r) = \sup_{\|X'-X\|=r} \|TX'-TX\|$$

$$l(X,r) = \inf_{\|X'-X\|=r} \|TX'-TX\|$$

$$D(X,r) = \begin{cases} 1 \text{ if } L(X,r)=l(X,r)=o \\ L(X,r)/l(X,r) \text{ otherwise} \end{cases}$$

$$\delta(X) = \limsup_{r \to o} D(X,r).$$

Definition. Let K be a real number not less than 1. The mapping $T:\Omega \to H$ is said to be K-quasiconformal if for every $X \in \Omega$ is $\delta(X) \leq K$. T is quasiconformal (abbreviated qc) if it is K-qc for some K (finite).

Since T is derivable, for every $X \in \Omega$ there exists a bounded linear mapping $A:$ $H \to H$ (depending on X) such that

$$TX'-TX = A(X'-X) + \|X'-X\|\cdot\varepsilon(X')$$

with $\lim_{X' \to X} \|\varepsilon(X')\| = o$.

(+) The A. is a member of G.N.A.F.A. (CNR).

If for some $X \in \Omega$ and for every $o < r \leqslant r_o$ is $L(X,r) = l(X,r) = o$, then T is a constant in $\bar{B}(X,r_o)$; therefore, in X, $A = 0$, but this result does contradict the hypotheses. Consequently, we may write

$$\delta(X) = \limsup_{r\to o} \frac{\sup\limits_{\|X'-X\|=r} \|TX'-TX\|}{\inf\limits_{\|X'-X\|=r} \|TX'-TX\|} = \limsup_{r\to o} \frac{\sup\limits_{\|X'-X\|=r} \|A\frac{X'-X}{r} + \varepsilon\|}{\inf\limits_{\|X'-X\|=r} \|A\frac{X'-X}{r} + \varepsilon\|}$$

Let $\delta(X) \leqslant K$. If $\inf\limits_{\|X'-X\|=r} \|A\frac{X'-X}{r}\| = o$, then we have $\sup\limits_{\|X'-X\|=r} \|A\frac{X'-X}{r}\| = o$, and therefore $A = 0$ in contradiction with the hypotheses. Hence we can write

$$\delta(X) = \frac{\sup\limits_{\|X'-X\|=r} \|A\frac{X'-X}{r}\|}{\inf\limits_{\|X'-X\|=r} \|A\frac{X'-X}{r}\|} = \frac{\sup\limits_{\|Z\|=1} \|AZ\|}{\inf\limits_{\|Z\|=1} \|AZ\|}$$

Therefore we may assert that if T is K-qc in Ω, then its derivative A must satisfy the double inequality

$$o < \sup_{\|Z\|=1} \|AZ\| \leqslant K \inf_{\|Z\|=1} \|AZ\| \tag{1}$$

Moreover, if the mapping T is derivable in $\Omega \subset H$ and if its derivative A satisfies (1) $\forall X \in \Omega$, then A is non vanishing and the inequality $\delta(X) \leqslant K$ holds in Ω. Hence the condition (1) is equivalent to the previous definition of quasiconformality.

Let Ω and Ω' be domains in H. If T_1 is a K_1-qc mapping of Ω into Ω' and if T_2 is a K_2-qc mapping of Ω' into H, then $T_2T_1 : \Omega \to H$ is a K_1K_2-qc mapping. In fact, if for $X \in \Omega$, A_1 is the derivative of T_1 and if for T_1X $(\in \Omega')$, A_2 is the derivative of T_2, then A_2A_1 is the derivative of T_2T_1 at the point X. We have

$$\sup_{\|Z\|=1} \|A_2A_1Z\| \geqslant \inf_{\|Z\|=1} \|A_2A_1Z\| = \inf_{Z\neq 0} \frac{\|A_2A_1Z\|}{\|A_1Z\|} \cdot \frac{\|A_1Z\|}{\|Z\|} \geqslant \inf_{Z\neq 0} \frac{\|A_2Z\|}{\|Z\|} \cdot \inf_{Z\neq 0} \frac{\|A_1Z\|}{\|Z\|} > o,$$

that is, if A_1 and A_2 are both not singulars, then also A_2A_1 is not singular. Moreover we obtain

$$\frac{\sup\limits_{\|Z\|=1}\|A_2A_1Z\|}{\inf\limits_{\|Z\|=1}\|A_2A_1Z\|} = \frac{\sup\limits_{Z\neq 0}\frac{\|A_2A_1Z\|}{\|A_1Z\|}\cdot\frac{\|A_1Z\|}{\|Z\|}}{\inf\limits_{Z\neq 0}\frac{\|A_2A_1Z\|}{\|A_1Z\|}\cdot\frac{\|A_1Z\|}{\|Z\|}} \cdot \frac{\sup\limits_{Z\neq 0}\frac{\|A_2Z\|}{\|Z\|}}{\inf\limits_{Z\neq 0}\frac{\|A_2Z\|}{\|Z\|}}\cdot\frac{\sup\limits_{Z\neq 0}\frac{\|A_1Z\|}{\|Z\|}}{\inf\limits_{Z\neq 0}\frac{\|A_1Z\|}{\|Z\|}} \leqslant K_2K_1 .$$

We have shown that quasiconformal mappings of domains of H form a semi-group with respect the composition law. If we set $K_1 = K_2 = 1$, we obtain that 1-quasiconfor_ mal (that is conformal) mappings form a semi-group.

Examples of conformal mappings are the translations ($TX = X + X_0$, X_0 fixed), the homotheties ($TX = rX$, r fixed positive real number), the inversions in spheres ($TX = r^2X/\|X\|^2$, $X\in H-\{0\}$) and the isometries (T linear and such that $\|TX\| = \|X\|$). We note that translations, homotheties and inversions in spheres have the inver_ se conformal, therefore they form a group. The isometries (according to the previ_ ous definition) may be such that $T(\Omega)$ is not a domain of H, as we see by the fol_ lowing example. Let $H = \ell^2$ (Hilbert space of the real sequences $X = (x_i)$ with norm $\|X\| = (\sum_{i=1}^{\infty} x_i)^{\frac{1}{2}} < +\infty$) and

$$T(x_1, x_2, \ldots) = (o, x_1, x_2, \ldots).$$

The latter mapping is obviously an isometry of ℓ^2 into ℓ^2, but $T(\ell^2)$ is not a do_ main of ℓ^2. Neverthless, the isometries T of H onto H have the inverse, and form a group.

2. Suppose H be a real unitary space (linear space with a real scalar product (,)). Let $X_0, Y_0 \in H-\{0\}$. The angle ψ, $o \leqslant \psi \leqslant \pi$, between X_0 and Y_0 is defined by the relation

$$\cos\psi = \frac{(X_0, Y_0)}{\|X_0\|\cdot\|Y_0\|}$$

Let T be a Frèchet-derivable quasiconformal mapping of a domain $\Omega\subset H$ into H. If $Z_0\in\Omega$ we consider the two line segments (with common endpoint Z_0) $Z = Z_0+\lambda X_0$ and

$Z = Z_o + \mu Y_o$, where λ and μ are real non-negative numbers such that $Z_o+\lambda X_o$ and $Z_o+\mu Y_o$ belong to Ω. Obviously such segments are paralleles to the vectors X_o and Y_o respectively. Consider now the two arcs (images of the corresponding segments) $Z = T(Z_o+\lambda X_o)$, $Z = T(Z_o+\mu Y_o)$ respectively. We say that such arcs form, at the point TX_o, the angle $\hat{\psi}$, $0 \leq \hat{\psi} \leq \pi$, if

$$\cos\hat{\psi} = \lim_{\lambda,\mu\to 0} \frac{(T(Z_o+\lambda X_o) - TZ_o\ ,\ T(Z_o+\mu Y_o) - TZ_o)}{\|T(Z_o+\lambda X_o) - TZ_o\| . \|T(Z_o+\mu Y_o) - TZ_o)\|}$$

It is easy to see that such limit exists and, if A is the derivative of T in Z_o, we have

$$\cos\hat{\psi} = \frac{(AX_o\ ,\ AY_o)}{\|AX_o\| . \|AY_o\|}$$

We prove the following

Theorem 1. Let H be a unitary space and T a derivable K-quasiconformal mapping of a domain $\Omega \subset H$ *into H. If* ψ *is the angle between two line segments with common endpoint* $Z \in \Omega$, *and if* $\hat{\psi}$ *is the angle between the respective arcs (with common end point TZ), images of them, we obtain*

$$K^{-1} \operatorname{tang}\frac{\psi}{2} \leq \operatorname{tang}\frac{\hat{\psi}}{2} \leq K \operatorname{tang}\frac{\psi}{2} . \tag{2}$$

Proof. We may write the double inequality (2) as follows

$$K^{-2}\frac{\|X\|\cdot\|Y\| - (X,Y)}{\|X\|\cdot\|Y\| + (X,Y)} \leq \frac{\|AX\|\cdot\|AY\| - (AX,AY)}{\|AX\|\cdot\|AY\| + (AX,AY)} \leq K^2\frac{\|X\|\cdot\|Y\| - (X,Y)}{\|X\|\cdot\|Y\| + (X,Y)} . \tag{3}$$

The inequalities (3) are trivial for $Y = cX$ (c non vanishing real number). For $Y \neq cX$, relations (3) are equivalent to

$$K^{-2}\frac{\|AX\|\cdot\|AY\| + (AX,AY)}{\|X\|\cdot\|Y\| + (X,Y)} \leq \frac{\|AX\|\cdot\|AY\| - (AX,AY)}{\|X\|\cdot\|Y\| - (X,Y)} \leq K^2\frac{\|AX\|\cdot\|AY\| + (AX,AY)}{\|X\|\cdot\|Y\| + (X,Y)} . \tag{4}$$

Now we have

$$(5)\qquad \frac{\|AX\|\cdot\|AY\| + (AX,AY)}{\|X\|\cdot\|Y\| + (X,Y)} \leqslant \frac{\frac{1}{2}\left(\|AX\|^2\cdot\|Y\|/\|X\| + \|AY\|^2\cdot\|X\|/\|Y\|\right) + (AX,AY)}{\|X\|\cdot\|Y\| + (X,Y)} =$$

$$= \frac{\|A(X/\|X\| + Y/\|Y\|)\|^2}{\|(X/\|X\| + Y/\|Y\|)\|^2} \leqslant \sup_{\|Z\|=1} \|AZ\|^2 ;$$

$$(6)\qquad \frac{\|AX\|\cdot\|AY\| + (AX,AY)}{\|X\|\cdot\|Y\| + (X,Y)} \geqslant \frac{\|AX\|\cdot\|AY\| + (AX,AY)}{\frac{1}{2}\left(\|X\|^2\cdot\|AY\|/\|AX\| + \|Y\|^2\cdot\|AX\|/\|AY\|\right) + (X,Y)} =$$

$$= \frac{\|A(X/\|AX\| + Y/\|AY\|)\|^2}{\|(X/\|AX\| + Y/\|AY\|)\|^2} \geqslant \inf_{\|Z\|=1} \|AZ\|^2$$

Substituting in (5) and (6) Y with -Y we obtain respectively:

$$(7)\qquad \frac{\|AX\|\cdot\|AY\| - (AX,AY)}{\|X\|\cdot\|Y\| - (X,Y)} \leqslant \sup_{\|Z\|=1} \|AZ\|^2 ;$$

$$(8)\qquad \frac{\|AX\|\cdot\|AY\| - (AX,AY)}{\|X\|\cdot\|Y\| - (X,Y)} \geqslant \inf_{\|Z\|=1} \|AZ\|^2 .$$

From (5), (6), (7), (8) and (1) follow the inequalities (4), from which we obtain (3) and (2).

We have also the following

Theorem 2. Let H be a unitary space and T a derivable mapping of a domain $\Omega \subset H$ into H. Further, suppose that for every $Z \in \Omega$ the derivative A of T is non-vanishing and satisfies the inequality

$$(9)\qquad \frac{\|X\|\cdot\|Y\| - (X,Y)}{\|X\|\cdot\|Y\| + (X,Y)} \leqslant K^2 \frac{\|AX\|\cdot\|AY\| - (AX,AY)}{\|AX\|\cdot\|AY\| + (AX,AY)}$$

for every X, Y∈H−{0}. Then T is K-quasiconformal.

Proof. Let $Z\in\Omega$ and A the derivative of T in Z. We may write

$$(10)\qquad \frac{\sup_{\|X\|=1}\|AX\|}{\inf_{\|X\|=1}\|AX\|}=\sup_{\|X\|=\|Y\|=1}\frac{\|AX\|}{\|AY\|}\ .$$

First, we note that the supremum on the right hand side of (10) can be obtained also for those X, Y for which $(X,Y)=o$. In fact, let X_o and Y_o be such that $\|X_o\|=\|Y_o\|=1$ and $(X_o,Y_o)\neq o$. We may suppose $X_o\neq\pm Y_o$. Let $Z_1=(X_o+Y_o)/(\|X_o+Y_o\|)$, $Z_2=(X_o-Y_o)/(\|X_o-Y_o\|)$. We have $\|Z_1\|=\|Z_2\|=1$, $(Z_1,Z_2)=o$. Let now

$$(11)\qquad \begin{cases} X = Z_1\cos\alpha + Z_2\sin\alpha \\ Y = Z_1\cos\beta + Z_2\sin\beta\ . \end{cases}$$

Then we have

$$\frac{\|AX\|^2}{\|AY\|^2}=\frac{\cos^2\alpha\|AZ_1\|^2+\sin 2\alpha(AZ_1,AZ_2)+\sin^2\alpha\|AZ_2\|^2}{\cos^2\beta\|AZ_1\|^2+\sin 2\beta(AZ_1,AZ_2)+\sin^2\beta\|AZ_2\|^2}$$

It is easy to see that the maximum and the minimum of the quantity $\|AX\|^2/\|AY\|^2$ are obtained for $\operatorname{tang}2\alpha=2(AZ_1,AZ_2)/(\|AZ_1\|^2-\|AZ_2\|^2)=\operatorname{tang}2\beta$; that is, for $\alpha=\beta\pm\pi/2$ (we may suppose $\alpha\neq\beta$ and $\alpha\neq\beta\pm\pi$). For α and β satisfying $\alpha=\beta+\pi/2$ or $\alpha=\beta-\pi/2$ the corresponding X and Y, computed according to (11), are such that $(X,Y)=o$. Hence, for X_o and Y_o for which $\|X_o\|=\|Y_o\|=1$, $(X_o,Y_o)\neq o$, there exists X_1 and Y_1 such that $\|X_1\|=\|Y_1\|=1$, $(X_1,Y_1)=o$ and

$$\frac{\|AX_o\|}{\|AY_o\|}\leqslant\frac{\|AY_1\|}{\|AY_1\|}$$

Finally, it follows that the supremum of the right hand side of (10) is obtained also for those X, Y for which $(X,Y)=o$. Let now U and V be such that $\|U\|=\|V\|=1$, $(U,V)=o$. We take

$$\begin{cases} X = U/\|AU\| + V/\|AV\| \\ \\ Y = U/\|AU\| + V/\|AV\| \end{cases} .$$

If $X = 0$ or $Y = 0$ then we have

$$\frac{\|AU\|}{\|AV\|} = 1 \leqslant K.$$

If $X, Y \neq 0$ then we have $(AX, AY) = o$. Therefore, for such X and Y , inequality (9) can be rewrite as

$$\frac{\left\|\frac{U}{\|AU\|} + \frac{V}{\|AV\|}\right\| \cdot \left\|\frac{U}{\|AU\|} - \frac{V}{\|AV\|}\right\| - \left(\frac{U}{\|AU\|} + \frac{V}{\|AV\|}, \frac{U}{\|AU\|} - \frac{V}{\|AV\|}\right)}{\left\|\frac{U}{\|AU\|} + \frac{V}{\|AV\|}\right\| \cdot \left\|\frac{U}{\|AU\|} - \frac{V}{\|AV\|}\right\| + \left(\frac{U}{\|AU\|} + \frac{V}{\|AV\|}, \frac{U}{\|AU\|} - \frac{V}{\|AV\|}\right)} \leqslant K^2$$

from which follows

$$\frac{\|AU\|}{\|AV\|} \leqslant K.$$

Since such inequality holds for every U and V with $\|U\| = \|V\| = 1$, $(U,V) = o$, it follows

$$o < \sup_{\|X\|=1} \|AX\| \leqslant K \inf_{\|X\|=1} \|AX\| .$$

The latter result holds for every $Z \in \Omega$; consequentely T is K-quasiconformal.

Remark 1. Theorem 2 does hold also if, instead of condition (9), we take the condition

$$\frac{\|AX\| \cdot \|AY\| - (AX,AY)}{\|AX\| \cdot \|AY\| + (AX,AY)} \leqslant K^2 \frac{\|X\| \cdot \|Y\| - (X,Y)}{\|X\| \cdot \|Y\| + (X,Y)} .$$

Remark 2. Combining theorems 1 and 2 and taking into account Remark 1 we obtain that the left or the right hand side of the double inequality (2), and the condition of non vanishing of the derivative, are equivalent to the double inequality (1).

BIBLIOGRAFY

[1] S. AGARD *Angles and qc mappings in space* J. Analyse Math. 22, 177-200 (1969).

[2] P. CARAMAN *N-dimensional quasiconformal mappings* Edi. Ac. Bucuresti Romania, Abacus Press Tunbridge Wells, Kent, England (1974).

[3] L. COLLATZ *Functional Analysis and Numerical Mathematics* Academic Press New York and London.(1966).

ON SOME CONVERGENCE PROBLEMS

FOR QUASICONFORMAL MAPPINGS (*)

by Carlo SBORDONE (**)

INTRODUCTION

Let Ω be an open set in R^n ($n \geq 2$) and $f = (f^{(1)}, \ldots, f^{(n)}) \in [H^{1,n}_{loc}(\Omega)]^n$ such that for $M \geq 1$:

$$\Big[\sum_{ij} (D_j f^{(i)}(x))^2\Big]^{n/2} \leq M\, n^{n/2} \det\big[D_j f^{(i)}(x)\big] \quad \text{a.e. in } \Omega.$$

We say then that f is M-quasi regular ($f \in Q(M)$) on Ω.

If $J(x,f) = \det\big[D_j f^{(i)}(x)\big] \neq 0$ a.e., set

$$[a_{ij}(x,f)] = J(x,f)\Big(\big[D_j f^{(i)}(x)\big]^{*}\big[D_j f^{(i)}(x)\big]\Big)^{-1}$$

$$[a^{\circ}_{ij}(x,f)] = J(x,f)^{2/n - 1}\,[a_{ij}(x,f)]$$

and consider the functionals

$$F_f : u \longrightarrow \int_\Omega (a^{\circ}_{ij}(x,f)u_{x_i}u_{x_j})^{n/2}$$

$$G_f : u \longrightarrow \int_\Omega a_{ij}(x,f)u_{x_i}u_{x_j} \quad .$$

(*) This paper was prepared while the author was visiting the Institute of Fluid Dynamics Appl. Math., Univ. of Maryland, College Park, MD, may-june 1976.

(**) Istituto Matematico Univ., Via Mezzocannone 8, Napoli (Italy)

It is known that, $\forall$ i = 1,...,n ([2], [3], [9])

$$F_f(f^{(i)}) = \min_{v-f^{(i)} \in H_o^{1,n}(\Omega)} F_f(v)$$

$$G_f(f^{(i)}) = \min_{v-f^{(i)} \in H_o^{1,n}(\Omega)} G_f(v) .$$

In the case n = 2, $G_f = F_f$ is a continuous functional of f, [11], in the sense that if $f_h, f \in Q(M)$ and $f_h \longrightarrow f$ in L^1_{loc}, then:
$F_{f_h} \xrightarrow{G} F_f$ (see § 2).

We extend here this result to n > 2 with some assumptions of boundedness of the sequence $\left(\int_\Omega J(x,f_h)^{1+\varepsilon} + J(x,f_h)^{-(1+\varepsilon)}\right)$ (ε suitable), in the case of G_f, which are less restrictive of those considered in [11].

For the functionals F_f we have only some relative compactness results.

We emphasize that in both cases of F_f and G_f, the condition $f_h \longrightarrow f$ in L^1_{loc} does not imply any kind of convergence of the functions $a_{ij}(x,f_h)$ (resp. $a^\circ_{ij}(x,f_h)$) to $a_{ij}(x,f)$ (resp. $a^\circ_{ij}(x,f)$) (see example in [11]).

§ 2. NOTATIONS

For any Ω open set in R^n:

$L^1_{loc}(\Omega)$ is the space of locally integrable real functions on Ω.

If $m \in L^1_{loc}(\Omega)$ is non negative, $1 \le p < \infty$, $L^p(\Omega,m)$ is the Banach space of real valued measurable functions u on Ω which satisfy

$$\|u\|_{L^p(\Omega,m)} = \left(\int_\Omega |u(x)|^p m(x)\, dx \right)^{1/p} < \infty .$$

We set $L^p(\Omega) = L^p(\Omega,1)$.

By $H_o^{1,p}(\Omega,m)$ (resp. $H^{1,p}(\Omega,m)$) we shall mean the completion of $C_o^1(\Omega)$ (resp. $C^1(\Omega)$) endowed with the norm

$$\|u\|_{H^{1,p}(\Omega,m)} = \|u\|_{L^p(\Omega,m)} + \|Du\|_{L^p(\Omega,m)},$$

where $Du = (D_1u, \dots D_nu)$ and. $D_iu = \frac{\partial u}{\partial x_i}$.

For any Ω open set in R^n $(n \ge 2)$ and $f = (f^{(1)},\dots,f^{(n)})$ such that $f^{(i)} \in H^{1,n}_{loc}(\Omega)$ we set

$$Df(x) = [D_j f^{(i)}(x)]$$

$$J(x,f) = \text{determinant of } Df(x)$$

$$|D(x,f)| = \left(\sum_{ij} (D_j f^{(i)}(x))^2 \right)^{1/2} .$$

§ 3. A variational convergence of functionals.

Let (X,d) be a metric space and $F_h, F : X \to \overline{R} = R \cup \{-\infty, +\infty\}$. The following definition was given in [4].

DEF.3.1 - F is the $\Gamma^-(d)$ limit of F_h ($F = \Gamma^-(d)\lim F_h$) iff:

(i) $u_h \overset{d}{\to} u \implies F(u) \leq \liminf_h F_h(u_h)$

(ii) $\forall u \ \exists w_h \overset{d}{\to} u : F(u) = \lim_h F_h(w_h)$.

In the following we are interested in two classes of functionals on some Sobolev spaces, connected with quasi conformal mappings.

I) For $S \geq s \geq 1$, Ω open set in R^n, let us denote by $\mathcal{F}_{s,S}(\Omega)$ the set of functionals

(3.1) $$F(u) = \int_\Omega (\breve{a}^\circ_{ij}(x)\, u_{x_i} u_{x_j})^{n/2} \qquad u \in H^{1,n}(\Omega),$$

where a°_{ij} satisfy:

(3.2) $$\begin{cases} a^\circ_{ij} = a^\circ_{ji} \in L^\infty(R^n) \\ s|\xi|^2 \leq \sum_{ij} a^\circ_{ij}(x)\xi_i\xi_j \leq S|\xi|^2 \quad x,\xi \in R^n. \end{cases}$$

II) For any $\Lambda \geq 1$, $p > 1$, $m \in L^p_{loc}(R^n)$ satisfying

(3.3) $$\int_\Omega (m^{-p} + m^p) \leq Q(\Omega) < \infty \qquad \forall \Omega \text{ bounded}$$

let us denote by $M_{\Lambda,p}(m)$ the set of $n \times n$ matrices $[a_{ij}(x)]$ such that

(3.4) $$\begin{cases} a_{ij} = a_{ji} \\ m(x)|\xi|^2 \leq \sum_{ij} a_{ij}(x)\xi_i\xi_j \leq \Lambda\, m(x)|\xi|^2, \end{cases}$$

and denote by $\mathcal{G}_{\Lambda,p}(m)$ the set of functionals (Ω bounded)

(3.5) $$G(\Omega, u) = \int_\Omega a_{ij}\, u_{x_i} u_{x_j} \qquad u \in H^{1,2}(\Omega, m)$$

with $a_{ij} \in M_{\Lambda,p}(m)$.

We state now some results concerning the Γ^- convergence of sequences of functionals (3.5).

THEOREM 3.2- Let $G_h \in \mathcal{G}_{\Lambda,p}(m)$, then, there exist a subsequence (G_{h_r}) and $G \in \mathcal{G}_{\Lambda,p}(m)$ such that $\forall\, \Omega$ open in R^n:

$$G(\Omega,u) = \Gamma^-(d_\Omega) \lim_r G_{h_r}(\Omega,u) \qquad \forall u \in H^{1,2}(\Omega,m),$$

where $d_\Omega(u,v) = \|u-v\|_{L^2(\Omega,m)}$.

PROOF : [6]

The following corollary states the implications of previous result on the convergence of solutions of boundary value problems for G_h.

COROLLARY 3.3- Let G_{h_r} and G as in Theorem 3.2, then, for any Ω, if $H_0^{1,2}(\Omega,m) \subseteq V \subset H^{1,2}(\Omega,m)$ and $\phi \in L^2(\Omega,m)$, $\lambda > 0$, we have

$$\lim_r \operatorname{Min}_{v \in V} \left\{ G_{h_r}(v) + \int_\Omega (\lambda v^2 + \phi v) \right\} =$$

$$= \operatorname{Min}_{v \in V} \left\{ G(v) + \int_\Omega (\lambda v^2 + \phi v) \right\}.$$

Moreover, denoting by $u_r(\phi,\lambda)$, $u(\phi,\lambda)$ the corresponding minimizing vectors, one has

$$u_r(\phi,\lambda) \longrightarrow u(\phi,\lambda) \qquad \text{in } L^2(\Omega,m).$$

PROOF : [6]

§4. Γ-convergence of not uniformly elliptic functionals.

For any $p>2n-1$, $\Lambda\geq 1$ and $m(x)\in L^p_{loc}(R^n)$ such that

$$\int_\Omega m^p + m^{-p} \leq Q(\Omega) \qquad \forall\,\Omega \text{ bounded},$$

let $\mathcal{E}_\Lambda(m,p)$ the set of all operators

$$A = -D_i(a_{ij}D_j)$$

such that $\forall\,(x,\xi)\in R^{2n}$:

$$m(x)|\xi|^2 \leq a_{ij}(x)\xi_i\xi_j \leq \Lambda m(x)|\xi|^2 \qquad a_{ij}=a_{ji}\,. \tag{4.1}$$

We have the following generalization of Lemma 2.1 in [5] (where $m(x)$=cost.)

LEMMA 4.1 - For any $w\in H^1(R^n,m)$, $A\in\mathcal{E}_\Lambda(m,p)$, let w_ε be the solution of

$$\begin{cases} \varepsilon Aw_\varepsilon + mw_\varepsilon = wm \\ w_\varepsilon \in H^1(R^n,m). \end{cases} \tag{4.2}$$

Then:

$$\|Dw_\varepsilon - Dw\|_{L^2(R^n,m)} \leq \Lambda\,\|Dw\|_{L^2(R^n,m)} \tag{4.3}$$

$$\|w_\varepsilon - w\|_{L^2(R^n,m)} \leq \Lambda\sqrt{\varepsilon}\,\|Dw\|_{L^2(R^n,m)} \tag{4.4}$$

and, if $Dw\in L^\infty$

$$\|w_\varepsilon - w\|_{L^\infty} \leq c(m,\Lambda)\sqrt{\varepsilon}\,\|Dw\|_{L^\infty} \tag{4.5}$$

If $Aw|_\Omega \in L^2(\Omega,m^{-1})$ for some open set Ω in R^n, then

$$\|Dw - Dw\|_{L^2(\Omega',m)} \leq c(\Lambda,\Omega',\Omega)\sqrt{\varepsilon}\,(\|Dw\|_{L^2(\Omega,m)} + \|Aw\|_{L^2(\Omega,m^{-1})}), \tag{4.6}$$

for any $\Omega'\subset\Omega$ with $\mathrm{dist}(\partial\Omega,\Omega')>0$.

PROOF. With the position $v_\varepsilon = \varepsilon^{-1}(w_\varepsilon - w)$, (4.2) is equivalent to

$$(4.7) \quad \begin{cases} \varepsilon A v_\varepsilon + m v_\varepsilon = A w \\ v_\varepsilon \in H^1(R^n, m) \end{cases}$$

in which, by multiplying v_ε and integrating, we get easily (4.3) and (4.4). Formula (4.5) is a consequence of [7], [12].

If $f = Aw \in L^2(\Omega, m^{-1})$ and $\Omega' \subset \Omega$, let $\tau \in H^{1,\infty}$ be such that $\tau \equiv 1$ in a neighborhood of $\overline{\Omega'}$ and $\tau = 0$ in $R^n - \Omega$, so that $\tau^2 v \in H_0^{1,2}(\Omega, m)$ $\forall\, v \in H^1(R^n, m)$.

By multiplying in (4.7) $\tau^2 v_\varepsilon$ and integrating:

$$\varepsilon \int_\Omega \tau^2 m(x)|Dv_\varepsilon|^2 + \int_\Omega \tau^2 m(x) v_\varepsilon^2 \leq$$

$$\leq \int_\Omega \tau^2 f v_\varepsilon - 2\varepsilon \int_\Omega a_{ij} D_i v_\varepsilon \tau v_\varepsilon D_j \tau \leq$$

$$\leq \int_\Omega \tau^2 f v_\varepsilon + 2\varepsilon\Lambda \sup_{R^n}|D\tau| \left(\int_\Omega m(x)|Dv_\varepsilon|^2\right)^{1/2} \left(\int_\Omega \tau^2 v_\varepsilon^2 m(x)\right)^{1/2} \leq$$

$$\leq \left(\int_\Omega \tau^2 v_\varepsilon^2 m(x)\right)^{1/2} \Big[\left(\int_\Omega \tau^2 f^2 m(x)^{-1}\right)^{1/2} +$$

$$+ 2\varepsilon\Lambda \sup_{R^n}|D\tau| \cdot \left(\int_\Omega m(x)|Dv_\varepsilon|^2\right)^{1/2} \Big].$$

By which, using (4.3)

$$\varepsilon \int_\Omega \tau^2 m(x)|Dv_\varepsilon|^2 + \int_\Omega \tau^2 m(x) v_\varepsilon^2 \leq$$

$$\left(\int_\Omega \tau^2 m(x) v_\varepsilon^2\right)^{1/2} \Big[\left(\int_\Omega \tau^2 f^2 m^{-1}\right)^{1/2} +$$

$$2\Lambda^2 \sup_{R^n}|D\tau| \, \|Dw\|_{L^2(R^n,m)} \Big],$$

and also

$$\int_\Omega \tau^2 m(x) v_\varepsilon^2 \leq c(\tau, \Lambda) \left(\|f\|_{L^2(\Omega, m^{-1})} + \|Dw\|_{L^2(R^n,m)} \right),$$

$$\varepsilon \int_\Omega \tau^2 m(x) |Dv_\varepsilon|^2 \leq c'(\tau,\Lambda) \left(\|f\|_{L^2(\Omega,m^{-1})} + \|Dw\|_{L^2(R^n,m)} \right)^2$$

and the assertion.

The aim of this section is to prove the following

THEOREM 4.2- If $\Lambda, p, m(x)$ are as in II) of § 3, let $G_h, G \in \mathcal{G}_{\Lambda,p}(m)$ be such that (for any Ω bounded)

(4.8) $$G(\Omega,u) = \Gamma^-(d_\Omega) \lim_h G_h(\ ,u) \qquad \forall\, u \in H^{1,2}(\Omega,m).$$

Let us fix Ω bounded and suppose that $u_h, u \in H^1(\Omega,m)$ satisfy (A_h beeing the eulerian of $G_h(\Omega,\ )$):

(4.9) $$u_h \longrightarrow u \quad \text{in } L^2(\Omega,m)$$

(4.10) $$A_h u_h = 0 \quad \text{on } \Omega.$$

Then (A beeing the eulerian of $G(\Omega,\ )$):

(4.11) $$Au = 0,$$

and $\forall\ \phi \in \mathcal{D}(\Omega)$:

(4.12) $$\lim_h \int_\Omega (a_{ij,h} D_i u_h D_j u_h)\phi = \int_\Omega (a_{ij} D_i u D_j u)\phi.$$

We begin with the following

LEMMA 4.3 - Let G_h, G satisfy (4.8), $u_h, u \in H^1(R^n,m)$, $f_h \longrightarrow f$ in $L^2(m^{-1})$ verify

$$A_h u_h + m u_u = f_h$$

$$Au + mu = f.$$

(A_h = eulerian of G_h, A = eulerian of G).

Then, $\forall\ \phi \in \mathcal{D}$

(4.13) $$\lim_h \int (a_{ij,h} D_i u_h D_j u_h)\phi = \int (a_{ij} D_i u D_j u)\phi.$$

PROOF. We can assume $f_h = f$ as in [5].

Let $\varphi_{h,\varepsilon}$, φ_ε be the solutions in $H^1(R^n,m)$ of

$$\varepsilon A_h \varphi_{h,\varepsilon} + m \varphi_{h,\varepsilon} = m\phi$$

$$\varepsilon A \varphi_\varepsilon + m \varphi_\varepsilon = m\phi,$$

then, by (4.5) we deduce, as we are assuming $p > 2n-1$, by [7]

$$\lim_{\varepsilon \to 0} \int (a_{ij,h} D_i u_h D_j u_h)(\phi - \varphi_{h,\varepsilon}) = 0 \quad \text{uniformly in } h,$$

$$\| \varphi_{h,\varepsilon} \|_{L^\infty} \le c \qquad \| u_h \|_{L^\infty} \le c',$$

which imply $\varphi_{h,\varepsilon} \cdot u_h$, $u_h^2 \in H^1(R^n,m)$.

So we have, as in [5] that

$$\int (a_{ij,h} D_i u_h D_j u_h) \phi_{h,\varepsilon} \longrightarrow \int (a_{ij} D_i u D_j u) \phi$$

and also (4.13).

PROOF (of Th; 4.2) As (4.11) follows by [1], let us consider the solutions in $H^1(R^n,m)$ of the problems

$$\varepsilon A_h u_{h,\varepsilon} + m(x) u_{h,\varepsilon} = \tau m(x) u_h$$

$$\varepsilon A u_\varepsilon + m(x) u_\varepsilon = \tau m(x) u,$$

where $\tau \in \mathcal{D}(\Omega)$, $\tau \equiv 1$ in a neighbour hood of $\mathrm{supp}(\phi)$. By (4.6) we deduce, using classical a priori bounds

$$\lim_{\varepsilon \to 0} \| D u_{h,\varepsilon} - D u_h \|_{L^2(\mathrm{supp}(\phi),m)} = 0 \quad \text{uniformly in } h$$

$$\lim_{\varepsilon \to 0} \| D u_\varepsilon - D u \|_{L^2(\mathrm{supp}(\phi),m)} = 0.$$

This, together with the following:

$$\lim_h \int (a_{ij,h} D_i u_{h,\varepsilon} D_j u_{h,\varepsilon}) \phi = \int (a_{ij} D_i u_\varepsilon D_j u_\varepsilon) \phi, \ \forall \varepsilon > 0,$$

which is a consequence of lemma 4.3, gives the assertion.

§ 5. The continuity theorem .

We begin this section with a definition , using notations in § 2.

DEF. 5.1 The mapping $f \in H^{1,n}_{loc}(\Omega)$ (Ω open set in R^n, $n \geq 2$) is K-quasiregular ($K \geq 1$, $f \in Q(K)$) if a.e. in Ω :

$$|D(x,f)|^n \leq K\, n^{n/2}\, J(x,f) .$$

For $f \in Q(K)$ define as in [11] , [9] the matrices

$$a_{ij}(x,f) = J(x,f)\,[Df^*(x)\cdot Df(x)]^{-1}$$

$$a^\circ_{ij}(x,f) = J(x,f)^{2/n}\,[Df^*(x)\cdot Df(x)]^{-1},$$

and the functionals :

$$F_f(u) = \int_\Omega (a^\circ_{ij}(x,f)\,u_{x_i}u_{x_j})^{n/2}\,dx$$

$$G_f(\Omega',u) = \int_{\Omega'} a_{ij}(x,f)\,u_{x_i}u_{x_j}\,dx \qquad \Omega' \subset \Omega .$$

It is easy to check that $F_f \in \mathcal{F}_{s,S}(\Omega)$, and that, extending $a_{ij}(x,f)$ out of Ω as $a_{ij}(x,f) = \delta_{ij}$ $\forall\, x \in R^n - \Omega$, one has $G_f \in \mathcal{G}_{\Lambda,p}(m)$ if we assume

$$\int_{\Omega'} [J(x,f) + J^{-1}(x,f)] \leq Q(\Omega') \qquad \Omega' \text{ bounded } \subseteq \Omega,$$

with

$$m(x) = \begin{cases} 1 & x \in R^n - \Omega \\ J(x,f)^{1/p}\,\lambda_K^{-1} & x \in \Omega, \end{cases}$$

$p = n/(n-2)$, $\Lambda = \lambda_K^2 = (K + \sqrt{K^2-1}\,)^{4(n-1)/n}$, infact we have a.e; in Ω

$$J(x,f)^{(n-2)/n}\,\lambda_K^{-1}\,|\xi|^2 \leq a_{ij}(x,f)\,\xi_i\xi_j \leq \Lambda_K\,J(x,f)^{(n-2)/n}\,|\xi|^2.$$

We begin with the following

THEOREM 5.2 Let $f_h \in Q(K)$ on Ω, then, there exists a subsequence (f_{h_r}) of (f_h) and a function $\psi(x,w)$ such that:

$$s|w| \leq \psi(x,w) \leq S(1+|w|) \qquad \forall x, w \in R^n$$

$$|\psi(x,w') - \psi(x,w)| \leq S|w-w'| \qquad \forall x, w, w' \in R^n$$

such that, $\forall\ \Omega' \subseteq \Omega$

(5.1) $$\int_{\Omega'} \psi^n(x,Du) = \Gamma^-(\sigma_{\Omega'}) \lim_r F_{f_{h_r}}(u) \qquad \forall u \in H^{1,n}(\Omega'),$$

where $\sigma_{\Omega'}(u,v) = \|u-v\|_{L^n(\Omega')}$.

PROOF. As a consequence of the compactness theorem 1 in [10], one has (5.1) $\forall\ u \in C^1$. By using Prop. 1.5 in [4] we have the assertion.

COROLLARY 5.3 Let $F_{f_{h_r}}$ and ψ as in th. 5.2; then $\forall\ \phi \in L^{n'}(\Omega)$, $(1/n+1/n' = 1)$ if V is a subspace such that $H_0^{1,n}(\Omega) \subset V \subset H^{1,n}(\Omega)$, $\lambda > 0$, we have:

$$\lim_r \ \operatorname*{Min}_{v \in V} \left\{ F_{f_{h_r}}(v) + \int_\Omega (\lambda |v|^n + \phi v) \right\} =$$

$$\operatorname*{Min}_{v \in V} \int_\Omega (\psi^n(x,Dv) + \lambda |v|^n + \phi v) .$$

Moreover, denoting by $u_r(\phi,\lambda)$, $u(\phi,\lambda)$ the corresponding minimizing vectors, one has: $u_r(\phi,\lambda) \longrightarrow u(\phi,\lambda)$ in $L^n(\Omega)$.

For functionals G_f we can show more.

We have in fact the following generalization of Spagnolo's continuity result [11] to $n > 2$.

<u>THEOREM 5.4</u> <u>Let $f_h \in Q(K)$ on Ω satisfy:</u>

$$0 < m_o(x) \leq \mathcal{J}^{(n-2)/n}(x, f_h(x)) \leq M \; m_o(x) \qquad \text{a.e. in } \Omega,$$

<u>with</u>

$$\int_{\Omega'} m_o^{p} + m_o^{-p} \leq g(\Omega') \quad \forall \Omega' \text{ bounded, } p > 2n-1.$$

<u>Then,</u> if $f_h \longrightarrow f$ in L^1_{loc}, $\forall \Omega' \subseteq \Omega$

$$G_f(\Omega', u) = \Gamma^-(d_{\Omega'}) \lim_h G_{f_h}(\Omega', u) \qquad \forall u \in H^{1,2}(\Omega', m),$$

<u>where</u> $d_{\Omega'}(u,v) = \| u - v \|_{L^2(\Omega', m)}$.

<u>PROOF.</u> By the compactness th. 3.2 it will be sufficient to prove that if $G(\Omega, u) = \int_{\Omega} a_{ij} u_{x_i} u_{x_j}$ is the Γ^- limit of a subsequence of G_{f_h} (which we still denote by G_{f_h}), then $G = G_f$.

Denoting by Δ_{f_h} the eulerian of G_{f_h} in Ω, we know, [2], that $\Delta_{f_h}(f_h^{(r)}) = 0$ in Ω $\forall r = 1, \dots, n$.

And so, by th. 4.2 $\forall \phi \in \mathcal{D}$,; $r,s = 1,..,n$

$$\int_{\Omega} (a_{ij}(x, f_h) D_i f_h^{(r)} D_j f_h^{(s)}) \phi \longrightarrow \int_{\Omega} (a_{ij}(x) D_i f^{(r)} D_j f^{(s)}) \phi .$$

But it is easy to check that:

$$a_{ij}(x,f_h)D_i f_h^{(r)} D_j f_h^{(s)} = J(x,f_h)\,\delta_{rs} \qquad \text{a.e. in } \Omega$$

and, by a Reshetnyak's result [8]

$$\int_\Omega J(x,f_h)\,\phi \longrightarrow \int_\Omega J(x,f)\,\phi ,$$

so that

$$a_{ij}(x)D_i f^{(r)} D_j f^{(s)} = J(x,f)\,\delta_{rs} \qquad \text{a.e. on } \Omega$$

and also $a_{ij}(x) = a_{ij}(x,f)$ a.e. in Ω, as $J(x,f)$ is a.e. different by zero.

R E F E R E N C E S

[1] BOCCARDO,L.- MARCELLINI,P. Sulla convergenza delle soluzioni di disequazioni variazionali, Ann. di Matem. pura e appl. (1975)

[2] CIMMINO,G.F. Sulla estensione al caso di tre o più dimensioni di un sistema differenziale del tipo di Beltrami, Rend. Acc. Naz. Lincei, 48 (1970)

[3] CIMMINO, G.F. Sugli operatori differenziali lineari del 2° ordine connessi alle rappresentazioni quasiconformi, Rend. sem. Mat. Fis Univ. Milano (1972).

[4] DE GIORGI,E.-FRANZONI,T. Su un tipo di convergenza variazionale, Rend. Acc. Naz. Lincei, 58 (1975)

[5] DE GIORGI,E.-SPAGNOLO,S. Sulla convergenza degli integrali dell'energia per operatori ellittici del 2° ordine, Boll. Un. Mat. Ital. 8 (1973)

[6] MARCELLINI,P.-SBORDONE,C. An approach to the asymptotic behaviour of elliptic-parabolic operators, to appear on Journal de Math. Pures et Appl.

[7] MURTHY,M.K.V.-STAMPACCHIA,G. Boundary value problems for some degenerate-elliptic operators, Ann. di Matem. Pura e Appl (IV) 80 (1968).

[8] RESHETNYAK, YU.G. Stability theorems for mappings with bounded excursion, Sibirskii Math. Z. 9 (1968)

[9] RESHETNYAK.YU.G. Mappings with bounded deformation as extremals of Dirichlet type integrals, Sibirskii Math. Z. 9 (1968).

[10] SBORDONE,C. -SU alcune applicazioni di un tipo di convergenza variaezionale, Ann. Scu. Norm. Sup. Pisa,IV,2 (1975).

[11] SPAGNOLO, S. - Some convergence problems, Symposia Mathematica, XVIII, Acad. Press London New York (1976)

[12] TRUDINGER ,Linear elliptic operators with measurable coefficients, Ann. Scu. Norm. Sup. Pisa, 27 (1973)

S.K.Vodop'ianov, V.M.Gol'dstein and A.P.Kopylov

On boundary values of quasiconformal mappings in space

Introduction

In the plane case a conformal mapping of a simply connected domain on another can be extended to the boundary, which is meant in Carathéodoy's sense. The realized boundary correspondence is bijective and bicontinuous when one introduces a suitable topology on the boundary.*

For the quasiconformal mappings of the half-plane on itself the following characterization of the boundary behaviour due to A. Beurling and L.Ahlfors is well known [1] . Let f be a continuous and monotone mapping of the boundary straight line on itself. In order that f gives the boundary values of a quasiconformal mapping F of the half-plane on itself, which is normalized by the condition $F(\infty) = \infty$, it is necessary and sufficient that there exists a real number $\rho \geq 1$, such that for each pair of numbers x and h, $-\infty < x < \infty$, $h > 0$, the condition

$$\rho^{-1} \leq \frac{f(x+h) - f(x)}{f(x) - f(x-h)} \leq \rho \qquad (\rho\text{-condition})$$

be fulfilled.

These two results give together a complete characterization of the boundary behaviour of the quasiconformal mapping in the plane case.

<u>Theorem</u> . Let G_1 and G_2 be simply connected domains, whose

---

* For instance, let us consider the conformal mapping φ of a domain on the circle. It induces a boundary mapping, which generates the topology on the Carathéodory boundary of the domain.

boundaries consist in more than a single point. Then every quasiconformal mapping $F: G_1 \longrightarrow G_2$ extends homeomorphically to the (Carathéodory) boundary, and in order that the mapping $f: \partial^* G_1 \longrightarrow \partial^* G_2$ between the Carathéodory boundaries $\partial^* G_1$ and $\partial^* G_2$ of the domains G_1 and G_2 gives the boundary values of a quasiconformal mapping of these domains, it is necessary and sufficient, that f be a homeomorphism and that there exist a number $K \geqslant 1$ such that, for each pair of closed sets H_1 and H_2 of boundary elements of the domain G_1 the inequality

$$K^{-1} \lambda(\Gamma^{G_1}_{H_1, H_2}) \leq \lambda(\Gamma^{G_2}_{f(H_1), f(H_2)}) \leq K \lambda(\Gamma^{G_1}_{H_1, H_2}) ,$$

be fulfilled, where $\Gamma^{G}_{H_1, H_2}$ denotes the family of all simple Jordan curves, which join in the domain G the elements of the sets H_1 and H_2, and $\lambda(\Gamma)$ is the extremal 2-length of the curve family Γ [2].

In this paper one tries to construct a boundary values theory, in the sense of the previous theorem for the quasiconformal mappings in the space. The paper consists in two parts.

First part is devoted to the extension problem of the mappings from the boundary of a domain to its interior and refers essentially to the results of A.P.Kopylov.

Second part deals with the extension problem of the quasiconformal and quasi-isometric homeomorphisms to the boundary, from the point of view of the connection between these mappings and the spaces L^1_p. The majority of the results in this second part of the communication have been obtained by S.K. Vodop'ianov and V.M. Gol'dstein.

I. Part

On the extension of the quasiconformal mappings in the space.

§I. Boundary behaviour of the half-space mappings, which are close to the conformal ones. Let $R^n_+ = \{ x = (x_1, \ldots, x_n) \in R^n \mid x_n > 0 \}$ be a half-space of the n-dimensional Euclidean space R^n, $\tau = \tau_{n-1} = \{ x \in R^n \mid x_n = 0 \}$ - its boundary hyperplane and ε a non-negative number.

Definition 1. The mapping $f: \tau \to R^n$ satisfies an ε-condition, if for each point $x_o \in \tau$ and each number $r > 0$ there exists an isometric mapping
$P_{x_o, r, f} : R^n \to R^n$, $P_{x_o,r,f}(0) = 0$, such that

$$\left| \frac{f(x) - f(x_o)}{\max\limits_{|y-x_o|=r} |f(y)-f(x_o)|} - P_{x_o,r,f}\left(\frac{x-x_o}{r}\right) \right| \leq \varepsilon \tag{1}$$

for $x \in \tau$ and $|x-x_o| \leq r$.

From P.P. Belinskii's results on the stability in the Liouville theorem for conformal mappings in the space [3-5] one directly deduces the following characterization of the boundary behaviour of the quasiconformal mappings of the half-space, which are close to the conformal ones.

Theorem 1. For $n \geq 3$ there exist the numbers $\varepsilon_1 = \varepsilon_1(n) > 0$ and $C_1 = C_1(n) > 0$ such that, if $F: R^n_+ \to R^n$, $F(\infty) = \infty$ is an $1+\varepsilon$ - quasiconformal mapping with $0 \leq \varepsilon < \varepsilon_1$, then F extends to a homeomorphism $F_1 : \overline{R^n_+} \to R^n$ of the closed half-space $\overline{R^n_+}$, whose boundary value $f = F_1|_\tau$ satisfies a $C_1 \cdot \varepsilon$ -condition.

To the complete characterization of the boundary values of the mappings, which are considered in this paragraph, leads together with theorem 1 the following theorem.

Theorem 2. [6] For $n \geq 2$ there are numbers $\varepsilon_2 = \varepsilon_2(n) > 0$ and $C_2 = C_2(n) > 0$ such that each mapping $f: \tau_{n-1} \rightarrow R^n$, which satisfies an ε - condition with $0 \leq \varepsilon < \varepsilon_2$, is the restriction to τ_{n-1} of an $1+ C_2 \cdot \varepsilon$ -quasiconformal mapping $F: R^n \rightarrow R^n$.

In the 3-dimensional case, for the mappings, whose coefficient of quasiconformality * is close to 1, theorems 1 and 2 strengthen and complete the results of the paper [7], which describe the behaviour of the 3-dimensional quasiconformal mappings on the plane sections of the definition domains. From these theorems it follows, for instance, that the restriction of these mappings to the inner sections of the half-space R^n_+ by hyperplanes $\{x \in R^n \mid x_n = b > 0\}$ and their boundary values coincide (see [6]).

Theorem 2 is proved by the explicit construction of the quasiconformal extension $F: R^n \longrightarrow R^n$ of the mapping f. Let us outline this extension. It is sufficient to extend f to the half-space The extension is realized in two steps.

I. step. It follows from definition 1 the existence of the mapping $P_f: \tau \times R^1_+ \longrightarrow \{P\}$ of the Cartesian product $\tau \times R^1_+$ of the hyperplane τ and the set R^1_+ of the positive numbers in the set $\{P\}$ of all the sense-preserving isometric mappings $P: R^n \longrightarrow R^n$, $P(0) = 0$, with the properties : for each point $x_0 \in \tau$ and each

* Throughout all the I. part the K-quasiconformality is to be understood as in [6]. We call quasiconformality coefficient of a mapping the smallest number K, for which the mapping is K-quasiconformal

number $r > 0$ the isometry $P_f(x_0,r)$ satisfies the inequality (1) for the mapping f with $x \in \tau$ and $|x - x_0| \leq r$. Let us fix one of these mappings P_f and construct the extension $\Phi : \overline{R^n_+} \rightarrow R^n$ of the mapping f to the half-space $\overline{R^n_+}$, which coincides with f in the points of the hyperplane τ and is equal to

$$f(z_\tau) + \max_{x \in \tau, |x-z_\tau| = z_n} |f(x)-f(z_\tau)| \cdot P_{z_\tau, z_n}(e_n), \qquad (2)$$

for $z = z_1e_1 + \dots + z_{n-1}e_{n-1} + z_ne_n \in R^n_+$, $z_\tau = z_1e_1 + \dots + z_{n-1}e_{n-1}$ and $P_{z_\tau, z_n} = P_f(z_\tau, z_n)$.

For an arbitrary $\varepsilon > 0$ the mapping Φ can be non-quasiconformal (and even discontinuous !), but it is continuous in the points of the plane τ and has a series of properties of the quasiconformal mappings (see [6]).

<u>II. step</u>. The mapping Φ is replaced by a piecewise affine mapping $F: \overline{R^n_+} \longrightarrow R^n$, which can be considered as a sufficiently good approximation of the mapping Φ [6] . The properties of Φ permit to effectuate this approximation such that it leads to the needed extension of the mapping f.

We shall call this extension method the piecewise linear extension method.

§2. <u>Ahlfors' extension problem and the quasiconformal equivalence of the domains to the ball</u>. L.Ahlfors considered in [8] the following problem : given a quasiconformal mapping $f: \tau_{n-1} \rightarrow \tau_{n-1}$ one has to study the possibility of the extension of f to a quasiconformal mapping of the half-space R^n_+. In the same paper he gave a positive solution to the problem in the case n=3. And for n=2 this problem was solved earlier by L. Ahlfors together with A. Beurling in [1] .

In [9] L. Carleson solved Ahlfors' problem in the case $n=4$.

It is natural to try to apply the method of the piecewise linear extension from §1 in order to solve Ahlfors' problem .The first step in this situation leads to the extension Φ ,which in contrast with the case which was considered at the end of the 1. paragraph is a homeomorphism and is uniquely defined by a formula of the type of formula (2):

$$\Phi(z) = f(z_\tau) + \max_{x \in \tau,\ |x-z_\tau| = z_n} | f(x) - f(z_\tau)| \cdot e_n ,$$

with the same notations as in (2).

And in spite of the fact that Φ can be not a quasiconformal mapping, the fact that it has, exactly as the corresponding mapping in §1, some properties of the quasiconformal mappings,permit us to suppose that a conveniently chosen piecewise linear approximation of the mapping Φ happens to be a quasiconformal extension of f to the half-space.

In [10] V.M.Gol'dstein proposed an approximation method for the quasiconformal mapping in the space by piecewise affine (and quasiconformal)ones. Through a more attentive consideration of the question it becomes clear that this method may be successfully applied for all n except for $n = 4$. The question whether it is convenient even in the case $n = 4$ remains till now open. By means of Gol'dstein's method it is possible to construct the needed approximation of the mapping Φ and, thus to prove the following assertion.

Theorem 3. For all $n > 4$, every K-quasiconformal mapping $f\colon \tau_{n-1} \to \tau_{n-1}$ can be extended to an $\alpha(n,K)$-quasiconformal mapping $F\colon \overline{R^n_+} \longrightarrow \overline{R^n_+}$ of the half-space $\overline{R^n_+}$ on itself. The function $\alpha(n,K)$ depends only on n and K.

Theorem 3 completes the solution of the Ahlfors problem. We have to remark that L. Carleson's extension method is based on the piecewise linear approximation of the (n-1)-dimensional mapping f: $\tau_{n-1} \to \tau_{n-1}$, while our method is based on that of the n-dimensional mapping Φ. The absence of the approximation theorem in the case of the 4-dimensional quasiconformal mappings does not permit to apply L. Carleson's method to solve Ahlfors' problem for n=5 and our method for n=4. However the difference between the two methods brings to a successful conclusion.

Let us formulate the fundamental property among the properties of the mapping Φ, which permit to realize the needed approximation of Φ in the demonstration of theorem 3.

<u>Lemma.</u> Let E_1 be a bounded closed set, lying in the hyperplane $x_n = \ell$, $\ell > 0$, further let be $E_2 = E_1 + h\,e_n$, $h > 0$ and $E'_m = \Phi(E_m)$, $m = 1,2$. Then there is a positive function β of three variables such that

$$\frac{\rho(E'_1, E'_2)}{d(E'_1 \cup E'_2)} \geqslant \beta\left(K, \frac{h}{d(E_1 \cup E_2)}, \frac{h}{\ell}\right);$$

here $\rho(E'_1, E'_2)$ means the distance between the sets E'_1 and E'_2, $d(E)$ - the diameter of the set E and K- the quasiconformality coefficient of the mapping f. The function β depends on the dimension of the space.

For close to conformal mappings one obtains theorem 3 as a special case of theorem 2, if one puts $f(\tau) = \tau$. Thus for these mappings the method of the piecewise linear extension from §1 gives another method to solve the extension problem discussed in this paragraph, which is different from L. Ahlfors'one [8, 11].

The Ahlfors **extension problem** is closely related to the question of the quasiconformal equivalence of the domains to the ball. In order to confirm this idea, let us consider some consequences of Ahlfors' theorem in the 3-dimensional case.

Let $C^1(3)$ be the set of all 3-dimensional domains with smooth boundaries and which are homeomorphic to a ball. From Ahlfors' theorem and from the conformal equivalence to a sphere of the boundary of a domain of the class $C^1(3)$ it follows

Theorem 4. Let the domain $\mathcal{D}$ belong to $C^1(3)$. Then every conformal mapping $f: S_{\mathcal{D}} \to S_B$ of the boundary $S_{\mathcal{D}}$ of the domain $\mathcal{D}$ on the sphere S_B which forms the boundary of the 3-dimensional ball B admits an extension to a quasiconformal mapping $F: \bar{\mathcal{D}} \to \bar{B}$, of the closure $\bar{\mathcal{D}}$ of the domain $\mathcal{D}$ on the closed ball $\bar{B}$, whose coefficient of quasiconformality K(F) satisfies the inequality

$$K(F) \leq K(\mathcal{D}) \cdot \alpha_1(3, K(\mathcal{D})); \qquad (3)$$

here $K(\mathcal{D})$ denotes the quasiconformality coefficient of the domain $\mathcal{D}$, which means the smallest possible value of the coefficients of quasiconformality of the mappings of the domain $\mathcal{D}$ onto the ball B, and α_1 is the smallest among the functions α from Ahlfors' theorem.

By means of theorem 2 and by following the scheme of the L. Ahlfors extension [8] one can show, that there exists a finite right derivative $\varkappa$ of the function $\alpha_1(3, K)$ at the point 1, and that $\alpha_1(3, K) \leq K^{\varkappa}$ for every $K \geq 1$. Taking into account this fact we shall consider instead of the inequality (3) the inequality

$$K(F) \leq [K(\mathcal{D})]^{1+\varkappa}.$$

Suppose that $\mathcal{D} \in C^1(3)$. In what follows we shall call a boundary condenser each pair $\{H_1, H_2\}$ of two closed sets H_1 and H_2, lying on the boundary $S_{\mathcal{D}}$ of the domain $\mathcal{D}$ and having no common points. On the set of the boundary condensers of the domain $\mathcal{D}$ let us construct a condenser function, associating to each such a condenser $\{H_1, H_2\}$ the number $\Theta(\{H_1, H_2\}) = \gamma_{\mathcal{D}}(\{H_1, H_2\})/ /\gamma_B(\{\varphi(H_1), \varphi(H_2)\})$; here $\gamma_{\mathcal{D}}(\{H_1, H_2\}) = \lambda_3(\Gamma^{\mathcal{D}}_{H_1, H_2})$ is the extremal 3-length(see [12]) of the family of all simple Jordan curves contained in $\mathcal{D}$ and joining the sets H_1 and H_2, $\varphi: S_{\mathcal{D}} \to S_B$ is a conformal mapping of the boundary $S_{\mathcal{D}}$ of the domain $\mathcal{D}$ on the boundary sphere S_B of the ball B, and $\gamma_B(\{\varphi(H_1), \varphi(H_2)\})$ is calculated with respect to the ball B in the same way as $\gamma_{\mathcal{D}}(\{H_1,H_2\})$ with respect to the domain $\mathcal{D}$. It is easy to verify that $\Theta(\{H_1,H_2\})$ is independent of the considered conformal mapping φ.

Now one obtains from theorem 4 the following assertion.

Theoren 5. If $\mathcal{D} \in C^1(3)$, then for each boundary condenser $\{H_1, H_2\}$ of the domain $\mathcal{D}$ the inequalities hold

$$[K(\mathcal{D})]^{-2(1+\varkappa)} \leq \Theta(\{H_1, H_2\}) \leq [K(\mathcal{D})]^{2(1+\varkappa)}$$

Close to theorem 4 and 5 is

Theorem 6. Suppose that the domain $\mathcal{D} \in C^1(3)$. Then the following assertions are true.

1) For every conformal mapping $\varphi: S_{\mathcal{D}} \to S_{\mathcal{D}}$ of the boundary $S_{\mathcal{D}}$ of the domain $\mathcal{D}$ on itself there exists its extension to a $[K(\mathcal{D})]^{2(1+\varkappa)}$ - quasiconformal mapping $F: \bar{\mathcal{D}} \to \bar{\mathcal{D}}$ of the closure $\bar{\mathcal{D}}$ of the domain $\mathcal{D}$ on itself.

2) If for every conformal mapping $\varphi: S_{\mathcal{D}} \to S_{\mathcal{D}}$ of the boundary of the domain $\mathcal{D}$ on itself there exists its extension to a K-quasiconformal mapping $F: \bar{\mathcal{D}} \to \bar{\mathcal{D}}$ of the closure $\bar{\mathcal{D}}$ of the domain $\mathcal{D}$

on itself, then $K(\mathcal{D}) \leq K$.

For $n > 3$ it is also possible to realize constructions which are to a certain degree analogous to those made by us in the case $n = 3$, by using this time instead of the conformal mappings of the boundaries extremal quasiconformal mappings, i.e. mappings whose quasiconformality coefficient is minimal.

In connection with the content of this paragraph arrises the problem:

Problem A. Is it possible or not to construct a theory of boundary elements of a domain $\mathcal{D} \subset R^n$, which is homeomorphic to the ball, so that the assertion be true: there is a function $\mu : N \times [1,\infty) \to [1,\infty)$, which is defined on the Cartesian product of the set N of the natural numbers and the interval $[1,\infty)$, satisfies the condition $\mu(n,Q) \to \mu(n,1) = 1$ for $Q \to 1$, and has the following properties.

1) If $F: \mathcal{D}_1 \to \mathcal{D}_2$ is a Q-quasiconformal mapping of the domain $\mathcal{D}_1 \subset R^n$ on the domain $\mathcal{D}_2 \subset R^n$, then it generates a correspondence $f : S^*_{\mathcal{D}_1} \to S^*_{\mathcal{D}_2}$ between the sets $S^*_{\mathcal{D}_1}$ and $S^*_{\mathcal{D}_2}$ of the boundary elements of the domain $\mathcal{D}_1$ and $\mathcal{D}_2$, which verifies the inequalities

$$K^{-1}\lambda_n(\Gamma^{\mathcal{D}_1}_{H_1,H_2}) \leq \lambda_n(\Gamma^{\mathcal{D}_2}_{f(H_1),\, f(H_2)}) \leq K\,\lambda_n(\Gamma^{\mathcal{D}_1}_{H_1,\,H_2}) \qquad (4)$$

with $K = \mu(n,Q)$ for each pair H_1 and H_2 of mutually disjoint closed sets of boundary elements. Here λ_n and $\Gamma^{\mathcal{D}}_{H_1,H_2}$ are defined as in the case of the domains in $C^1(3)$.

2) If the mapping $f: S^*_{\mathcal{D}_1} \to S^*_{\mathcal{D}_2}$ is a homeomorphism, which verifies the inequalities (4) with $K = Q \geq 1$, then it is the boundary correspondence induced by a certain $\mu(n,Q)$-quasiconformal mapping $F: \mathcal{D}_1 \to \mathcal{D}_2$ of the domain $\mathcal{D}_1$ on the domain $\mathcal{D}_2$.

We suppose that the boundary elements theory for which we looked is regular in the following sense : the set of boundary elements is endowed with a topology, and in the case of the domains with smooth boundaries the boundary elements reduce to the points of the boundary while the considered topology coincides with the natural topology on the boundary of the domain.

It is asserted in the theorem from introduction, that in the case of the plane problem A is affirmatively solved, and as the corresponding theory of the boundary elements Carathéodory's theory may be taken. We think that for $n > 2$ the theory of boundary elements, which will be presented in the second part of the paper and partially gives the solution of problem A, may also be useful to the complete solution of problem A.

L. Ahlfors' extension problem is an important special case of the problem A. Hence problem A may naturally be considered as a generalization of Ahlfors' problem.

From the results of this paragraph it follows that for $n \geq 3$ the problem to find conditions for the quasiconformal equivalence of the domains with smooth boundaries and the problem to extend the quasiconformal mappings are of the same order. For $n = 3$, for instance, a positive answer to the problem A would lead to an assertion which is - in a certain sense - a reciprocal theorem to theorem 5: Suppose that the domain $\mathcal{D} \in C^1(3)$ and that for some number $K \geq 1$ and for each boundary condenser $\{H_1, H_2\}$ of the domain $\mathcal{D}$ the inequalities $K^{-1} \leq \Theta(\{H_1, H_2\}) \leq K$ are fulfilled, then $K(\mathcal{D}) \leq \leq \mu(3,K)$ holds.

§ 3. Integral means and Ahlfors' extension problem .

Let $f : R^n \longrightarrow R^n$ be a quasiconformal mapping which is normalized by the condition $f(\infty)= \infty$. Put for each number $\delta > 0$

$$f_\delta (x) = \frac{1}{V^n_\delta} \int\limits_{|y| \leq \delta} f(x+y)dV_y \ , \qquad (5)$$

here the integral is taken over the n-dimensional ball of radius δ, and V_δ^n is the volume of this ball. Then the question arrises whether f_δ is also a quasiconformal mapping. In the general case the answer is negative for all $n \geq 2$. However for the close to conformal mappings it holds

Theorem 7. There exists a number $\varepsilon_3 > 0$ and there are functions $\psi_k : \overline{R^1_+} \longrightarrow \overline{R^1_+}$, $\psi_k(\varepsilon) \rightarrow \psi_k(0)=0$ for $\varepsilon \rightarrow 0$, k=1,2, such that, for an arbitrary $\delta > 0$ the δ-mean f_δ ,defined by the equality (5), of each $1+ \varepsilon$ -quasiconformal mapping $f: R^n \longrightarrow R^n$ with $\varepsilon < \varepsilon_3$ is an $1 + \psi_1(\varepsilon)$ - quasiconformal mapping, and the partial derivative·means satisfy the relations

$$D_k f_\delta (x) = \frac{\max\limits_{|z|=1} |f(x+ \delta z)-f(x)|}{\delta} (P(e_k)+ \gamma_k), \quad k=1,2,\dots,n, (6)$$

where $P = P_{x, \delta, f} : R^n \longrightarrow R^n$ is an isometric mapping, $P(0)=0$ and $|\gamma_k| \leq \psi_2 (\varepsilon)$. The number ε_3 and the functions ψ_1 and ψ_2 depend only on n. The functions ψ_1 and ψ_2 are linear and the first of them can be expressed in terms of the second one by means of the relations (6).

The proof of this theorem is based on stability theorems of conformal mappings of the space, due to P.P. Belinskii [3-5] and Iu. G. Rešetnjak [13-14] .

Theorem 7 also admits a generalisation for the case of the mappings of arbitrary domains, and instead of V.A.Steklov's integral means ,one can consider the S.L.Sobolev means [15] , assuring thus their infinite differentiability. In this form for the case of the mappings of the ball theorem 7 is a stronger form of the result in paper [16] .

A. Beurling and L. Ahlfors proposed the following method to extend the mappings $f: \mathcal{T}_1 \to \mathcal{T}_1$, $f(\infty)=\infty$, from the abscisse axis $\mathcal{T}_1$ to the superior half-plane R^2_+ [1] . For each point $(x,y)\in R^2_+$ put $F(x,y) = (u(x,y), v(x,y))$,

$$u(x,y)= \frac{1}{2y}\int_{x-y}^{x+y} f(w)\,dw, \quad v(x,y)= \frac{1}{y}\int_0^y \left[f(x+w)-f(x-w)\right] dw. \qquad (7)$$

One establishes in particular, in [1] , that if f satisfies a ρ-condition, which can be considered as a condition of 1-dimensional quasiconformality, then its extension F to R^2_+ , given by the formulae (7), is quasiconformal (see the introduction of our paper).

For an arbitrarily fixed $y > 0$ the function $u(x,y) = \frac{1}{2y}\int_{x-y}^{x+y} f(w)\,dw$ represents the integral y-mean f_y of type (5) for the mapping f , and it is easy to verify, that if f is a 1-dimensional quasiconformal mapping, so does f_y too. It was remarked before that the last assertion does not subsist for the n-dimensional quasiconformal mapping if $n > 2$. From our opinion there is this circumstance that represents in the several dimensional case the basic obstacle in the attempts made in order to solve the extension problem of a quasiconformal mapping of the hyperplane on itself to a quasiconformal mapping of the half-space by formulae of type (7).

Theorem 1 removes this obstacle for close to conformal mappings, and, thus, arises the hope in the possibility to solve the extension problem for such mappings by means of the Beurling - Ahlfors method. Indeed it is true

Theorem 8. For each natural number $n \geqslant 3$ there exists a number $\varepsilon_4 > 0$ and a function $\Psi_3 : R^1_+ \longrightarrow R^1_+$, $\Psi_3(\varepsilon) \to \Psi_3(0) = 0$ for $\varepsilon \to 0$, such that for every $1+\varepsilon$ -quasiconformal mapping $f : \tau_{n-1} \to \tau_{n-1}$ with $\varepsilon < \varepsilon_4$, and which is normalized by $f(\infty) = \infty$, its extension F to R^n_+ by means of the formulae

$$y_\tau(x) = \frac{1}{V^{n-1}_{x_n}} \int_{|w| \leqslant x_n} f(x_\tau + w) \, d\, V_w \,, \tag{8}$$

$$y_n(x) = \frac{n}{2(n-1)} \cdot \frac{1}{V^{\,n-1}_{x_n}} \int_{|w| \leqslant x_n} | f(x_\tau + w) - f(x_\tau - w) | \, dV_w \,,$$

where $x = x_\tau + x_n e_n = x_1 e_1 + \ldots + x_{n-1} e_{n-1} + x_n e_n \in R^n_+$, $y(x) = y_\tau(x) + y_n(x) e_n = F(x) = F(x_\tau + x_n e_n)$ and the integral is taken over the balls in the hyperplane τ, is $1 + \Psi_3(\varepsilon)$-quasiconformal. The function Ψ_3 is linear. For $n = 2$ the formulae (8) transform into the formulae (7).

§ 4. On the removability of the singularities for close to conformal mappings .

Let $\mathcal{D}$ be a domain in R^n, $n \geqslant 3$, and $B_t \subset \mathcal{D}$ an n-dimensional ball, which is included in $\mathcal{D}$ together with the ball having the same center but an $1 + t$ times greater radius than the radius of B_t (for a certain $t > 0$). It holds

Theorem 9. There exist functions $\omega : R^1_+ \to R^1_+$ and $\nu : \{ (t, \varepsilon) \in R^2 \mid t > 0, \ \varepsilon \geqslant 0 \} \to R^1_+$ such that, for an arbitrary $t \geqslant 0$ every $1+\varepsilon$ - quasiconformal mapping $f : \mathcal{D} \setminus B_t \to R^n$ with $\varepsilon < \omega(t)$ admits an $1 + \nu(t,\varepsilon)$-quasiconformal extension $F: \mathcal{D} \to \hat{R}^n$ in the domain $\mathcal{D}$. Here $\hat{R}^n$ means the space R^n completed by the infinite point, the function $\nu(t,\varepsilon) \to \nu(t,0) = 0$ for $\varepsilon \to 0$ and both functions ω and ν depend only on the dimension of the space.

This theorem characterizes the specific features of the close to conformal case in the problems which concern the removability of the singularities of the quasiconformal mappings in the space and is related to the results in the 1. paragraph.

A special case of theorem 9, which corresponds to the limit value $t = \infty$ was obtained in the paper [6].

§ 5. On the boundary values of quasi-isometric mappings

Finally let us remark, that in connection with the study of the boundary values in a series of cases the class of the quasiconformal mappings proves to be close to the class of the quasi-isometric mappings (see for the definition the II-d part of this communication). For instance, the boundary values of the mappings of the half-space, whose quasi-isometry coefficients are close to 1, can be characterized in a similar way to that of §1, replacing in the definition of the ε-condition the inequality (1) by the following inequality

$$| f(x) - f(x_0) - P_{x_0, r, f}(x-x_0) | \leqslant r\,\varepsilon .$$

The study of the boundary values of the quasi-isometric mappings is related to the specific difficulties, which appear in particular also in the fact that while the theorem in introduction completely characterizes the boundary values of the quasiconformal mappings of the simply connected domains in the plane, a similar characterization in the quasi-isometric case is unknown.

Part II

Functional approach in the boundary values theory.

Let us try to consider the quasiconformal mappings and the quasi-isometries as the class of the mappings, which leave invariant the functional spaces $L^1_n(G)$, $L^1_p(G)$, $G \subset R^n$, $p > 1$. From this point of view one succeeds to prove the possibility of the extension of the quasiconformal mappings on the NED -sets and to try to construct the boundary values theory for quasiconformal homeomorphisms of arbitrary domains in the space.

2.1. Let G be a domain in R^n. The space $L^1_p(G)$, $p > 1$, consists of all the functions $\varphi : G \longrightarrow R$, which are locally integrable in the domain G and have p - integrable generalized first derivatives [15].

The space $L^1_p(G)$ is complete with respect to the semi-norm

$$\| f \|_{1,p} = \| f \|_{L^1_p(G)} = \left(\int_G | \nabla f |^p dx \right)^{1/p}.$$

Definition 1. The homeomorphism φ of the domain $G \subset R^n$ on the domain $G' \subset R^n$ is called a quasiconformal homeomorphism if it induces an isomorphism $\varphi^* : L^1_n(G') \longrightarrow L^1_n(G)$ of semi-normed spaces by the rule $\varphi^* f = f \circ \varphi$ for all $f \in L^1_n(G')$.

If one replaces in the definition the spaces $L^1_n(G')$, $L^1_n(G)$ by $L^1_p(G')$, $L^1_p(G)$ $(p \neq n,\ p > 1)$, we obtain the definition of the quasi-isometric mapping.

Let us now prove the equivalence of the definition with the traditional one. First of all we pass to the definition of the mappings in terms of the (1,p) - capacity.

For each pair of continua $F_o, F_1, F_o \cap F_1 = \emptyset$, which are contained in the domain G , the (1, p) - capacity $C_p(F_o, F_1 ; G)$ is equal to

$$\inf \int_G |\nabla u|^p \, dx ,$$

where the infinimum is taken over all the continuous functions $u \in L^1_p(G)$, $u(x) \geqslant 1$ on F_1, $u(x) \leqslant 0$ on F_o.

A homeomorphism $\varphi: G \longrightarrow G'$, which keeps the (1,p)-capacity of an arbitrary pair of continua $F_o, F_1 \subset G$ quasi-invariant

$$K^{-1} C_p(F_o, F_1 ; G) \leqslant C_p(\varphi(F_o), \varphi(F_1); G') \leqslant K C_p(F_o, F_1; G) \qquad (1)$$

for some constant K, is a quasiconformal mapping if p = n and a quasi -isometry if $p \neq n$, $p > 1$.

The extremal function u^G_{p,F_o,F_1} for the capacity $C_p(F_o, F_1; G)$ of an arbitrary pair $F_o, F_1, F_o \cap F_1 = \emptyset$, belongs to the space $L^1_p(G)$.

From the boundedness of the operators φ^*, $\varphi^{*^{-1}}$ follows

$$C_p(F_o, F_1; G) \leqslant (\| \varphi^* u^{G'}_{p,\varphi(F_o),\varphi(F_1)} \|_{1,p})^p \leqslant \|\varphi^*\|^p C_p(\varphi(F_o), \varphi(F_1); G'),$$

$$C_p(\varphi(F_o), \varphi(F_1); G') \leqslant (\|\varphi^{*^{-1}} u^G_{p,F_o,F_1} \|_{1,p})^p \leqslant \|\varphi^{*^{-1}}\|^p C_p(F_o, F_1 ; G)$$

for an arbitrary pair of continua $F_o, F_1, F_o \cap F_1 = \emptyset$, i.e. the inequalities (1) are true.

If the homeomorphism is quasiconformal (quasi-isometric),

then it is well known that it induces an isomorphism of the spaces L_n^1 (L_p^1, $p>1$).

The conditions imposed to the mapping in the definition can be essentially weakened.

Theorem 1 [17] Let $\varphi: G \to G'$ be a mapping, defined a.e. in G and which induces the isomorphism $\varphi^*: L_p^1(G') \to L_p^1(G)$ ($p \geq n$) by the rule $(\varphi^* f)(x) = f(\varphi(x))$ a.e. for all $f \in L_p^1(G')$. Then the mapping φ is a quasiconformal homeomorphism for $p=n$ and a quasi-isometric homeomorphism for $p>n$.

In this case the domains $\varphi(G)$ and G' are (1,p)-equivalent.

The domains G_1 and G_2 are called (1,p)-equivalent if the restriction mappings

$$\theta_i: L_p^1(G_1 \cup G_2) \to L_p^1(G_i), \quad i=1,2, \quad \theta_i u = u\big|_{G_i},$$

are isometries.

Instead of the condition that the operators θ_i be isometries one can ask only that they be isomorphisms of linear spaces between $L_p^1(G_1 \cup G_2)$ and $L_p^1(G_i)$.

By introducting some natural limitations on the operator φ^* it is possible not to ask a priori the existence of the mapping φ in definition 1 [18].

The non-coincidence of the domains $\varphi(G)$ and G' in theorem 1 comes from the essence of the thing. One sees this from a simple example. The space L_n^1 on the ball $B \subset R^n$ and on the ball punctured in the point $x \in B$ are isomorphic under the inclusion mapping $i: B \setminus \{x\} \to B$.

The quasiconformal and quasi-isometric mappings extend from a domain to a domain which is (1,p)-equivalent with the given one.

Theorem 2 [18]. If the domains G_1 and G_2 are (1,p)-equivalent, then every quasiconformal (for p=n), (quasi-isometric (for $p>1$)) homeomorphism $\varphi : G_1 \to R^n$ uniquely extends to a quasiconformal (quasi-isometric) homeomorphism $\varphi : G_1 \cup G_2 \to R^n$ without any growth of the quasiconformality (quasi-isometry) coefficient.

For $p \geq n$ theorem 2 follows directly from theorem 1.

Let us denote by $\mathrm{Ext}_p\,(G)$ the set of the extremal functions for the (1,p)-capacity of all pairs of continua F_0, $F_1 \subset G$, $F_0 \cap F_1 = \emptyset$. In the general form theorem 2 is a consequence of the density in $L_p^1\,(G)$ of the linear envelope of $\mathrm{Ext}_p(G)$. More precisely, it holds

Theorem 3. There exists a countable set $\mathfrak{A} = \{ v_i \in \mathrm{Ext}_p\,(G) \}$ such that every function $u \in L_p^1\,(G)$ for an arbitrary $\varepsilon > 0$ can be represented as a series $u = c_0 + \sum_{i=1}^{\infty} c_i\, v_i$, which is absolutely convergent in $L_p^1\,(G)$. By this for the semi-norms the inequality holds

$$\| u \|_{L_p^1(G)} \leq \sum_{i=1}^{\infty} \| c_i\, v_i \|_{L_p^1(G)} \leq \| u \|_{L_p^1(G)} + \varepsilon .$$

Theorem 3 gives a stronger form of the corresponding fact in [19]. The proof coincides in the basic features with that from [19].

Theorem 3 permits to describe the (1,p)-equivalent domains in terms of extremal functions from $\mathrm{Ext}_p(G)$.

Definition 2. A closed with respect to $G \subset R^n$ set M is called NC_p- set, if for an arbitrary pair of continua F_0, F_1 in $G \setminus M$

$$C_p(F_0, F_1\, ;\, G \setminus M) = C_p(F_0,\, F_1;\, G\,). \qquad (2)$$

The definition can be reformulated in terms of the moduli of curve families [2o] . For p=n and $G=R^n$ the class NC_n coincides with the class NED [2,21,2o] .

<u>Theorem 4</u> [19] . In order that the domains G_1 and G_2 be (1,p)-equivalent, it is necessary and sufficient, that the set $G_1 \setminus G_2$ be an NC_p - set in G_1 .

The properties of the NC_p - sets are formulated in the paper [19] .

The necessity is evident.

Sufficiency. It follows from (2), that an arbitrary function in $Ext_p(G \setminus M)$ extends to a function in $Ext_p(G)$ without growth of the semi-norm. From here and from theorem 3 we obtain the extension possibility for each function in L^1_p $(G \setminus M)$ to a function in L^1_p (G) without growth of the semi-norm. In order to end the proof it suffices to show that the Lebesgue measure of the set M is equal to zero.

<u>Consequence of the theorems 2,4</u>. Let be given the domain $G \subset R^n$ and the NC_p - set $M \subset G$.

Then every quasiconformal (for p = n) or quasi-isometric (p > 1) homeomorphism $\varphi : G \setminus M \to R^n$ extends in a unique way to a quasiconformal (for p=n) or quasi-isometric (p > 1) homeomorphism $\tilde{\varphi} : G \to R^n$ without growth of the distortion coefficient.

The removability of the NED-sets under quasiconformal mappings was proved in the plane in [22] and in the space in [23] .

It is known that the removable sets under quasiconformal mappings in the plane are NED-sets [22] . In the space case this problem is unsolved and, we think, that it is related to the question whether the set of the coordinate functions of all the quasiconformal mappings is everywhere dense in L_n^1.

The connection which exists between the quasiconformal mappings and the spaces L_n^1 permits us to suppose that the singularities in the behaviour of a quasiconformal mapping near the boundary are connected with the construction of the space L_n^1 for the domain and with the possible boundary singularities of the functions of this class.

Definition of the capacity distance in the domain G with respect to a closed ball $B \subset G$. Capacity distance between the points x_1 and x_2 $G \setminus B$ is by definition equal to

$$\rho_B^G(x_1, x_2) = \inf_{F_0} \sqrt[n]{C_n(F_0, B; G)},$$

where $F_0 \subset G$ is a continuum which joins the points x_1 and x_2.

In the definition of the metric $\rho_B^G(x_1, x_2)$ instead of the ball B it is possible to use an arbitrary closed set which is contained in the domain G.

$\rho_B^G(x_1, x_2)$ is a metric in $G \setminus B$ and the topology introduced in $G \setminus B$ by means of this metric coincides with the Euclidean one.

If one completes the metric space $G \setminus B$ with respect to the metric ρ, then to $G \setminus B$ are added the boundary points, which are equivalence classes of fundamental sequences in the metric space.

Let H be the set of all " boundary points" of $G \setminus B$. Denote by $\widetilde{G}_B = (G \setminus B) \cup H$. As usually the set $\widetilde{G}_B$ will be transformed into a complete metric space. The topology obtained in this way in the completed space is independent of the choice of the closed set B. The space $\widetilde{G}_B$ is linearly connected.

Proposition . Let us consider a quasiconformal mapping $\varphi : G \to G'$. Fix in the domain G a closed ball B. The mapping φ extends to a quasi-isometric mapping of the space $\tilde{G}_B$ on the space $\tilde{G}_{\varphi(B)}$.

The proof follows directly from the definition of the metrics ρ^G_B, $\rho^{G'}_{\varphi(B)}$.

For simply connected domains in the plane the " boundary elements" introduced above coincide with the Carathéodory prime ends. In the space case the coincidence between the " boundary elements " and the Carathéodory prime ends [24, 25] occurs for the domains which are quasiconformally equivalent to the ball.

In the general case this coincidence does not exist. For example, the edge of the outward directed ridge [26] is a " boundary element" in our sense (for a sufficient " slope " of the ridge).

The capacity topology coincides with the Euclidean one on the closure of the domain for a large class of domains.

Theorem 6. If the domain G has the property that every arbitrary function $g \in L^1_n(G)$ is the restriction to G of a function $f \in L^1_n(R^n)$, then

1) the domain G is locally connected in each boundary point and

2) $\tilde{G}_B$ is topologically equivalent to $\bar{G} \setminus B$.

In the case of the coincidence of the topologies every quasiconformal mapping between two domains extends to a homeomorphism between the closures of these domains.

It is possible to give an example of two domains $G, G' \subset R^n$ and a quasiconformal mapping $\varphi : G \to G'$ which does not extend to a correspondence between the Carathéodory prime ends.

QUOTED BIBLIOGRAPHY

1 Beurling A, Ahlfors L.V., The boundary correspondence under quasiconformal mapping, Acta math., 1956, 96, 125-142.

2 Ahlfors L.V., Beurling A., Conformal invariants and function - theoretic null-sets, Acta Math., 195o, 83, 1o1-129.

3 Belinskii P.P., On the continuity of the quasiconformal mappings in space and on Liouville's theorem (Russian), Dokl.Akad.Nauk SSSR, 1962, 147, 1oo3-1oo4.

4 Belinskii P.P., Stability in Liouville's theorem on quasiconformal mappings in space (Russian), in the volume "Nekotorye problemy mat. i meh.," Nauka, Leningrad, 197o, p.88-1o2.

5 Belinskii P.P., On the degree of closeness of the quasiconformal mappings in the space to the conformal ones (Russian), Dokl.Akad. Nauk SSSR, 1971, 2oo, 759-761.

6 Kopylov A.P., On the behaviour on hyperplanes of spatially quasiconformal close to conformal mappings (Russian), Dokl.Akad.Nauk SSSR, 1973, 2o9, 1278-128o.

7 Kopylov A.P., Behaviour of spatially quasiconformal mappings on plane sections of the domain of definition (Russian), Dokl.Akad. Nauk SSSR, 1966, 167, 743-746.

8 Ahlfors L.V., Extension of quasiconformal mappings from two to three dimensions, Proc.Nat.Acad.Sci. USA, 1964,51, 768-771.

9 Carleson L., The Extension Problem for Quasiconformal Mappings, A Collection of Papers Dedicated to Lipman Bers "Contribution to Analysis", Academic Press New York and London, 1974, pp.39-47.

1o Gol'dstein V.M., Approximation of quasiconformal homeomorphisms by simplicial quasiconformal homeomorphisms (Russian), Dokl.Akad. Nauk SSSR, 1973, 213, 23-25.

11 Sedo R.I., Sycev A.V., On the extension of quasiconformal mappings to multidimensional spaces in higher dimension (Russian),Dokl.Akad. Nauk SSSR, 1971, 198, 1278-1279.

12 Vaisälä J., On quasiconformal mappings in space, Ann.Acad.Sci. Fenn., Ser AI, 1961, 298.

13 Resetnjak Ju.G, Stability in Liouville's theorem on conformal mappings in space (Russian), in the volume "Nekotorye problemy mat. i meh.", Novosibirsk, 1961, p.219-223.

14 Resetnjak Ju.G., On the stability in Liouville's theorem on conformal mappings in space (Russian), Dokl.Akad.Nauk SSSR, 1963, 152, 286-287.

15 Sobolev S.L., Some applications of functional analysis to mathematical physics (Russian), Novosibirsk, 1962.

16 Kopylov A.P., On the approximation of spatially quasiconformal, close to conformal mappings, by smooth quasiconformal mappings (Russian), Sibirsk Mat.Ž. 1972, 13, 94-1o6.

17 Vodop'janov C.K., Gol'dstein V.M., Quasiconformal mappings and spaces of functions with first generalized derivatives (Russian) Sibirsk. Mat. Ž., 1976, 17, nr.3, 515-531.

18 Vodop'janov C.K., Gol'dstein V.M., Structural isomorphisms of the spaces W_n^1 and quasiconformal mappings (Russian), Sibirsk. Mat. Ž, 1975, 16, nr.2, 224-246.

19 Vodop'janov C.K., Gol'dstein V.M., Removability of the sets for the spaces W_p^1 of quasiconformal and quasi-isometric mappings (Russian), Dokl.Akad.Nauk. SSSR, 1975, 22o, nr.4, 769-771.

2o Hesse J., A p-extremal length and p-capacity. Arkiv.mat.,1975, 13, 131-144.

21 Väisälä J.,On null-sets for extremal length. Ann.Acad.Sci.Fenn. Ser AI, 1962, 322.

22 Pesin I.N., Metric properties of Q-quasiconformal mappings (Russian), Mat.Sb., 1956, 4o(82), nr.3, 281-294.

23 Aseev V.V., Syčev A.V., On the removable sets of spatially quasiconformal mappings (Russian), Sibirsk. Mat.Ž., 15, nr.6,1974, 1213-1227.

24. Zorič V.A., Boundary correspondence under Q-quasiconformal mappings of the ball (Russian). Dokl.Akad.Nauk SSSR, 1962, 145, Nr.6, 12o9-1212.

25 Zorič V.A., Definition of boundary elements by means of sections (Russian), Dokl. Akad.Nauk SSSR, 1965, 164,Nr.4, 736-739.

26 Gehring F.W., Väisälä Ju., The coefficients of quasiconformality of domains in space, Acta Math., 1965, 114, 1-7o (Russian translation in Matematika, Sb.per., 1966, 1o, Nr.6, 6o-12o).

Institut Matematiki
SO AN SSSR Novosobirsk
SSSR

ASYMPTOTIC VALUES AND ANGULAR LIMITS OF QUASIREGULAR MAPPINGS

by Matti Vuorinen

University of Helsinki

1. Introduction

The study of quasiregular mappings during the last decade has shown that these mappings are natural generalizations of analytic functions of complex plane to n-dimensional euclidean spaces, $n \geq 2$. The basic results on these mappings can be found in [2] and [3].

In this paper we study the behavior of a quasiregular mapping $f : G \to R^n$, $n \geq 2$, close to a boundary point $x_o \in \partial G$ assuming that f omits a set with positive capacity and that ∂G is small at x_o. As local smallness measures of ∂G we employ the following two conditions. The first one is the continuum criterium which was recently introduced by Martio [1]. See also [4]. The second one is the dispersion condition which was introduced by the author in [5]. It was shown in [5] that the continuum criterium implies the dispersion condition but not conversely.

We shall prove following results. If $f : G \to R^n$ is a quasiregular mapping omitting a set with positive capacity, then

(a) f has an angular limit at $x_o \in \partial G$ provided that the continuum criterium holds at x_o and G contains a cone with vertex at x_o.

(b) f has at most one asymptotic value at $x_o \in \partial G$ provided that the dispersion condition holds at x_o.

(c) f has exactly one asymptotic value at $x_o \in \partial G$ provided that the continuum criterium holds at x_o.

Moreover, an example of Toppila shows that the continuum criterium in (a) and (c) cannot be replaced by the dispersion condition and that f need not have any asymptotic values in (b). The above results (a) - (c) seem to be new even for analytic functions of complex plane. Some of the technical details will not be proved here, for details see [5]. The author is grateful to O.Martio for suggesting this subject and to J.Sarvas for useful discussions.

2. Notation and preliminaries

2.1. We denote by R^n, $n \geq 2$, the n-dimensional euclidean space and by $\overline{R}^n$ its compactification with ∞. R^n is endowed with the inner product $(x \mid y) = \sum x_i y_i$ and with the metric given by the norm $|x| = (x|x)^{\frac{1}{2}}$. In $\overline{R}^n$ we use the spherical metric q. All topological operations are performed with respect to $\overline{R}^n$. The ball $\{x \in R^n \mid |x-x_o| < r\}$ is denoted by $B^n(x_o,r)$ and the sphere $\{x \in R^n \mid |x-x_o| = r\}$ by $S^{n-1}(x_o,r)$. We employ the abbreviations $B^n(r) = B^n(0,r)$, $S^{n-1}(r) = S^{n-1}(0,r)$, $B^n = B^n(1)$, and $S^{n-1} = S^{n-1}(1)$.

2.2. A continuum is a non-empty, compact connected set. A path γ is a continuous mapping $\gamma : \Delta \to A$, $A \subset R^n$, where Δ is an interval on the real axis. We denote $\gamma\Delta$ by $|\gamma|$ and we let $\Delta(E,F;G)$ denote the family of all paths $\gamma : [0,1) \to R^n$ such that $\gamma(0) \in E$, $\gamma(0,1) \subset G$ and $\gamma(t) \to F$ as $t \to 1$.

A condenser is a pair (A,C) where $A \subset R^n$ is open and $C \subset A$ is compact. The capacity cap E of a condenser $E = (A,C)$ can be defined using the modulus of a path family as follows: cap $E = M(\Delta(C,\partial A;A)) = M(\Delta(C,\partial A;A\setminus C)) = M(\Delta(C,\partial A;R^n))$, see [6].

The notation $f : G \to R^n$ includes the assumptions that $G \subset R^n$ is a domain and that f is continuous. For quasiregular mappings as well as for other related concepts we employ the same definitions as in [2], [3].

2.3. Definition. Given any proper subdomain G of R^n we let C denote the set $R^n \setminus G$. Let $x_0 \in \partial G \setminus \{\infty\}$. The continuum criterium holds at x_0 if there exists a non-degenerate continuum $K \subset G \cup \{x_0\}$, $x_0 \in K$ such that $M(\Gamma_K) < \infty$ where $\Gamma_K = \Delta(K, \partial G; G)$ (cf. [1], [4]). The dispersion condition holds at x_0 if there exists a sequence (r_j) of positive real numbers $r_1 > r_2 > \ldots$, $\lim r_j = 0$ such that $S^{n-1}(x_0, r_j) \subset G$, $j=1,2,\ldots$ and

$$\lim \operatorname{cap}(G, S^{n-1}(x_0, r_j)) = 0. \tag{2.4}$$

2.5. Notation. The fact that the continuum criterium holds at $x_0 \in \partial G$ is denoted by $M(x_0, C) < \infty$ (recall $C = R^n \setminus G$). The omission of a continuum K from this notation is motivated by an equivalent definition of the continuum criterium which shows that no reference to any particular continuum is needed, cf. [4, 2.20].

2.6. Lemma. ([5]) *Let G be a domain, $0 \in \partial G$ and $M(0, C) < \infty$. Then the dispersion condition holds at 0.*

3. Asymptotic values and angular limits

A point $z \in \overline{R}^n$ is an asymptotic value of $f : G \to R^n$ at $x_0 \in \partial G$ if there is a path $\gamma : [0,1) \to G$ with $\gamma(t) \to x_0$ and $f(\gamma(t)) \to z$ as $t \to 1$.

3.1. Theorem. *Let $f : G \to R^n$ be a quasiregular mapping with $\operatorname{cap} R^n \setminus fG > 0$. If f has asymptotic values α_1, α_2 at $x_0 \in \partial G$ and if the dispersion condition holds at x_0, then $\alpha_1 = \alpha_2$.*

Proof. Suppose that $\alpha_1 \neq \alpha_2$. Performing an auxiliary Möbius transformation we may assume $x_0 = 0$. Choose paths $\gamma_i : [0,1) \to G$ such that $\gamma_i(t) \to x_0$, $f(\gamma_i(t)) \to \alpha_i$, $i = 1,2$, as $t \to 1$. Let (r_j) be a sequence in the definition of the dispersion condition. Choose an integer j_0 such that $|\gamma_i| \cap S^{n-1}(r_j) \neq \emptyset$, $i=1,2$ and $q(fS^{n-1}(r_j)) \geq \frac{1}{2} q(\alpha_1, \alpha_2) > 0$ for every $j \geq j_0$. By [3, 3.11] there exists $\delta > 0$ such that $\operatorname{cap} fE_j \geq \delta$ for every $j \geq j_0$ where $E_j = (G, S^{n-1}(r_j))$. By [2, 7.1] for $j \geq j_0$

$$0 < \delta \leq \operatorname{cap} fE_j \leq K_I(f) \operatorname{cap} E_j$$

which leads to a contradiction, since $\operatorname{cap} E_j \to 0$ as $j \to \infty$ by (2.4). Hence $\alpha_1 = \alpha_2$ and the proof is complete.

3.2. <u>Remark</u>. It follows from 3.8 that the assumption in 3.1 concerning the existence of asymptotic values α_1, α_2 cannot be dropped.

If $e \in S^{n-1}$ and $\alpha \in (0,\pi)$ then $C(e,\alpha)$ denotes the cone $\{y \in R^n \mid (y|e) > |y| \cos\alpha\}$. Next we give two lemmas which follow easily from the results in [5].

3.3. <u>Lemma</u> ([5]). <u>Let</u> $G \subset R^n$ <u>be a domain such that</u> $0 \in \partial G$ <u>and</u> $G \supset C(e,\alpha)$. <u>If</u> $M(0,C) < \infty$ <u>then</u> $M(\Gamma_K) < \infty$ <u>where</u> $K = \overline{B}^n \cap \overline{C(e,\beta)}$, $\beta \in (0,\alpha)$.

3.4. <u>Lemma</u> ([5]). <u>Let</u> $G \subset R^n$ <u>be a domain</u>, $0 \in \partial G$ <u>and let</u> $M(0,C) < \infty$. <u>If</u> K <u>is a continuum in</u> $G \cup \{0\}$, $0 \in K$, <u>such that</u> $M(\Gamma_K) < \infty$, <u>then</u>

$$\lim \operatorname{cap} (G, C_j) = 0$$

<u>where</u> C_j <u>is any continuum in</u> $G \cap K \cap B^n(1/j)$.

We prove next our main result.

3.5. Theorem. Let $G \subset R^n$ be a domain such that $0 \in \partial G$ and $G \supset C(e,\alpha)$. If $f : G \longrightarrow R^n$ is a quasiregular mapping such that $\operatorname{cap} R^n \setminus fG > 0$ then the angular limit

$$\lim_{\substack{x \to 0 \\ x \in C(e,\beta)}} f(x)$$

exists for $\beta \in (0,\alpha)$ whenever $M(0,C) < \infty$.

Proof. Assume that the limit does not exist. Then we may choose sequences (a_j), (b_j) such that $a_j, b_j \in C(e,\beta) \cap B^n(1/j) = K_j$ and $f(a_j) \longrightarrow a'$, $f(b_j) \longrightarrow b' \neq a'$. Choose an integer j_0 such that $q(f(a_j),f(b_j)) \geq \frac{1}{2} q(a',b') > 0$ for every $j \geq j_0$. Let $C_j \subset K_j$ be a continuum joining a_j and b_j, $j \geq j_0$. By [3, 3.11] there exists $\delta > 0$ such that $\operatorname{cap} fE_j \geq \delta$ for every $j \geq j_0$ where $E_j = (G,C_j)$. By the capacity inequality in [2, 7.1]

$$0 < \delta \leq \operatorname{cap} fE_j \leq K_I(f) \operatorname{cap} E_j$$

which yields a contradiction, since $\operatorname{cap} E_j \longrightarrow 0$ as $j \longrightarrow \infty$ by Lemmas 3.3 and 3.4. Thus f has the asserted angular limit.

The proof of Theorem 3.5 has the following consequence.

3.6. Corollary. Let $G \subset R^n$ be a domain, $0 \in \partial G$ and let $M(0,C) < \infty$. Let $f : G \longrightarrow R^n$ be a quasiregular mapping such that $\operatorname{cap} R^n \setminus fG > 0$. If K_1 and K_2 are continua in $G \cup \{0\}$, $0 \in K_i$ such that $M(\Gamma_{K_i}) < \infty$, $i=1,2$, then the limits

$$\lim_{\substack{x \to 0 \\ x \in K_1}} f(x) \quad \text{and} \quad \lim_{\substack{x \to 0 \\ x \in K_2}} f(x)$$

exist and coincide.

3.7. Theorem. Let f be as in Corollary 3.6. Then f has

exactly one asymptotic value α at 0. The value α is attained along every path $\gamma : [0,1) \to G$, $\gamma(t) \to 0$ as $t \to 1$, which satisfies the condition $M(\Gamma_K) < \infty$, $K = |\gamma|$.

Proof. The proof follows from Lemma 2.6, Theorem 3.1 and Corollary 3.6.

3.8. Remark. Toppila has constructed a bounded analytic function $f : G \to R^2$ having no asymptotic value at $0 \in \partial G$. In this example $M(0,R^2 \setminus G) = \infty$ but the dispersion condition holds at 0. For details, see [5]. Thus the assumption in Theorem 3.1 concerning the existence of asymptotic values cannot be dropped. Moreover, the continuum criterium in Theorems 3.5 and 3.7 cannot be replaced by the dispersion condition.

3.9. An open problem. Does f have in fact a limit at 0 in Theorems 3.5 or 3.7?

References

[1] Martio, O.: Equicontinuity theorem with an application to variational integrals. -Duke Math. J.42(1975),569-581.

[2] Martio, O., Rickman, S. and J. Väisälä: Definitions for quasiregular mappings. -Ann.Acad.Sci.Fenn. AI 448(1969),1-40.

[3] - " - Distortion and singularities of quasiregular mappings. -Ibid. AI 465(1970),1-13.

[4] Martio, O. and J. Sarvas: Density conditions in the n-capacity. -Reports of the Department of Mathematics, University of Helsinki, Ser. A No. 4(1975),1-24.

[5] Vuorinen, M.: Exceptional sets and boundary behavior of quasiregular mappings in n-space. -Ann.Acad.Sci.Fenn. A I/ Dissertationes 11 (1976),1-44.

[6] Ziemer, W. P.: Extremal length and p-capacity. -Michigan Math. J. 16(1969),43-51.

On some subclasses of Bazilevič functions

by Otto Fekete
University of Cluj-Napoca
Faculty of Mathematics

1. Let β be real and suppose that $f(z)=z+\sum_2^{\infty} a_n z^n$ is regular in the unit disc D with $f(z)f'(z)\neq 0$ in $0<|z|<1$. If $J(\beta,f(z))=\mathrm{Re}\,(1-\beta)\frac{zf'(z)}{f(z)}+\beta(\frac{zf''(z)}{f'(z)}+1)>0$ for $z\in D$, then f is said to be a β-convex function. We denote the class of these functions by $\mathcal{M}_\beta$ [6]. $\mathcal{M}_0=S^*$ is the class of starlike functions. If $0\leq\beta''\leq\beta'$, $\mathcal{M}_{\beta'}\subset\mathcal{M}_{\beta''}\subset S^*$; $\mathcal{M}_\infty=\{z\}$ [4].

Let $\mathcal{P}$ denote the class of Charathéodory functions and let $B(\alpha,\mathcal{P},\mathcal{M}_0)$ denote the class of Bazilevič functions $f(z)=(\alpha\int_0^z P(\zeta)g^{\alpha}(\zeta)\zeta^{-1}d\zeta)^{\frac{1}{\alpha}}=z+\dots$ for $z\in D$, where $P\in\mathcal{P}$, $g\in\mathcal{M}_0$ and $\alpha>0$.

In this paper we investigate the subclasses of Bazilevič functions for which $g\in\mathcal{M}_\beta$, β real. We determine the Hardy class of the functions from $B(\alpha,\mathcal{P},\mathcal{M}_\beta)$ and their derivatives.

We remember the result of P.J.Eenigenburg and S.S.Miller about the Hardy class of β-convex functions [1].

Theorem A.

(i) If $f\in\mathcal{M}_\beta$, $|\beta|>2$, then $f\in H^{\infty}$.

(ii) If $f\in\mathcal{M}_\beta$, $|\beta|\leq 2$ and $f\neq k_\tau^\beta$ ($k_\tau^\beta(z)=(\frac{1}{\alpha}\int_0^z\zeta^{\frac{1}{\alpha}-1}(1-e^{i\tau}\zeta)^{-\frac{2}{\alpha}}d\zeta)^{\alpha}$) then $f\in H^{p(\beta)}$ where

$$p(\beta)=\begin{cases}\infty & \text{if } \beta=2\\ \frac{1}{2-\beta}+\varepsilon & \text{if } 0\leq\beta<2\\ \frac{1}{2}+\varepsilon & \text{if } -2<\beta<0\\ \infty & \text{if } =-2\end{cases}$$

and $\varepsilon=\varepsilon(f)>0$.

(iii) If $f=k_\tau^\beta$ then $f\in H^{p(\beta)}$ for all

$$p(\beta) < \begin{cases} \infty & \text{if } \beta = 2 \\ \frac{1}{2-\beta} & \text{if } 0 \leq \beta < 2 \\ \frac{1}{2} & \text{if } -2 < \beta < 0 \end{cases}$$

2. Let $f(z) = \left(\alpha \int_0^z P(\zeta) g^{\alpha}(\zeta) \zeta^{-1} d\zeta\right)^{\frac{1}{\alpha}} \in B(\alpha, \mathcal{P}, \mathcal{M}_{\beta})$. We have

$$f'(z) = P(z) g^{\alpha}(z) z^{-1} f^{1-\alpha}(z)$$

We denote

(1) $$F(z) = \left[\frac{f(z)}{z}\right]^{\alpha}$$

F is regular in $D = \{z \mid |z| < 1\}$ and

$$F'(z) = \alpha \frac{P(z) g^{\alpha}(z)}{z^{\alpha+1}} - \frac{F(z)}{z}\alpha$$

If $0 \leq p \leq 1$, $z = re^{i\theta}$, $0 < r < 1$ then we can show, applying Hölder's inequality, that

(2) $$I(r) = \int_0^{2\pi} |F'(z)|^p d\theta \leq \frac{\alpha^p}{r^{p(\alpha+1)}} \left(\int_0^{2\pi} |P(z)|^{pa} d\theta\right)^{\frac{1}{a}} \left(\int_0^{2\pi} |g(z)|^{\alpha p b} d\theta\right)^{\frac{1}{b}} + \frac{\alpha^p}{r^{p(\alpha+1)}} \int_0^{2\pi} |f(z)|^{\alpha p} d\theta$$

$I(r)$ will be bounded as $r \to 1^-$, if

(3) $$\begin{cases} pa < 1 \\ \alpha p < \frac{1}{2} \\ \alpha p b < p(\beta) \end{cases}$$

where $p(\beta)$ is the corresponding value of β from Theorem A. Solving the system (3) for the particular values of $p(\beta)$ and applying the Hardy–Littlewood theorem we obtain the following lemmas.

<u>Lemma 1.</u> If $f \in B(\alpha, \mathcal{P}, \mathcal{M}_{\beta})$, $|\beta| \geq 2$ then

(i) $f \in H^p$ for all $p < \infty$ if $\alpha \leq \frac{1}{2}$

(ii) $f \in H^p$ for all $p < \frac{\alpha}{2\alpha - 1}$ if $\alpha > \frac{1}{2}$

<u>Lemma 2.</u> If $f(z) = \left(\alpha \int_0^z P(\zeta) g^{\alpha}(\zeta) \zeta^{-1} d\zeta\right) \in B(\alpha, \mathcal{P}, \mathcal{M}_{\beta})$ and $g \neq k_{\tau}^{\beta}$, then

(i) $f \in H^{\frac{1}{2-\beta} + \varepsilon}$ if $\alpha < \frac{1}{\beta}$, $0 \leq \beta < 2$

(ii) $f \in H^p$ for all $p < \frac{\alpha}{2\alpha - 1}$ if $\alpha \geq \frac{1}{\beta}$, $0 \leq \beta < 2$

(iii) $f \in H^{\frac{1}{2-\beta} + \varepsilon}$ if $-2 < \beta < 0$

where $\varepsilon = \varepsilon(f) > 0$.

<u>Theorem 1.</u> Let $f(z) = \left(\alpha \int_0^z P(\zeta) g^{\alpha}(\zeta) \zeta^{-1} d\zeta\right)^{\frac{1}{\alpha}} \in B(\alpha, \mathcal{P}, \mathcal{M}_{\beta})$.

(i) If $|\beta| \geq 2$ then $f \in H^p$ for all $p < \infty$.

(ii) If $|\beta|<2$ and $g\neq k_z^\beta$ then $f\in H^{p(\beta)}$ where

$$p(\beta)=\begin{cases}\frac{1}{2-\beta}+\varepsilon & \text{if } 0\leqq\beta<2\\ \frac{1}{2}+\varepsilon & \text{if } -2<\beta<0\end{cases}$$

and $\varepsilon=\varepsilon(f)>0$.

(iii) If $|\beta|<2$ and $g=k_z^\beta$ then $f\in H^{p(\beta)}$ for all

$$p(\beta)<\begin{cases}\frac{1}{2-\beta} & \text{if } 0\leqq\beta<2\\ \frac{1}{2} & \text{if } -2<\beta<0\end{cases}$$

The result is the best one.

Proof. The proof of the theorem is based on induction where the step of induction is a repetition of the proof of Lemma 1 or Lemma 2. E.g. (i) Let $|\beta|\geqq 2$. If $\alpha\leqq\frac{1}{2}$ the desired result is contained in Lemma 1. If $\frac{1}{2}<\alpha<1$, applying (3) and Lemma 2 we obtain $I(r)$ bounded as $r\to 1^-$ if $p<1$ and repeating the proof of Lemma 1, $f\in H^p$ for all $p<\infty$ in this case too. If $\alpha>1$, $I(r)$ is bounded as $r\to 1^-$, for all $p<\frac{1}{2\alpha-1}$, then $f\in H^p$ for all $p<\frac{\alpha}{2\alpha-2}$

The results of Lemma 1 become

(i_2) $f\in H^p$ for all $p<\infty$, if $0<\alpha\leqq 1$

(ii_2) $f\in H^p$ for all $p<\frac{2}{2\alpha-2}$, if $\alpha>1$.

Suppose now that:

(i_n) $f\in H^p$ for all $p<\infty$, if $0<\alpha\leqq\frac{1}{2}n$ and

(ii_n) $f\in H^p$ for all $p<\frac{\alpha}{2\alpha-n}$, if $\alpha>\frac{1}{2}n$.

We can show that

(ii_{n+1}) $f\in H^p$ for all $p<\frac{\alpha}{2\alpha-n-1}$ if $\alpha>\frac{1}{2}(n+1)$;

hence (i) is true for all $\beta\geqq 2$.

The result is the best possibàe because $\mathcal{M}_\beta\subset B(\alpha,\mathcal{P},\mathcal{M}_\beta)$ for all β real and $f(z)=(\alpha\int_0^z\frac{1+\zeta}{1-\zeta}\zeta^{\alpha-1}d\zeta)^{\frac{1}{\alpha}}\in B(\alpha,\mathcal{P},\mathcal{M}_\beta)$ and $f\notin H^\infty$.

Remarks: The function of $B(\alpha,\mathcal{P},\mathcal{M}_\beta)$ belong to the same Hardy class as the functions of $\mathcal{M}_\beta$, except the case $|\beta|\geqq 2$. For $\beta=0$ we obtain the result of S.S.Miller about the Hardy class of Bazilevič functions [3].

Theorem 2. Let $f(z)=(\alpha\int_0^z P(\zeta)g^\alpha(\zeta)\zeta^{-1}d\zeta)^{\frac{1}{\alpha}}\in B(\alpha,\mathcal{P},\mathcal{M}_\beta)$

$0 < \alpha < 1$.

(i) If $|\beta| \geqq 2$ then $f' \in H^p$ for all $p < 1$.

(ii) If $|\beta| < 2$ and $g \neq k_\zeta^\beta$ then $f' \in H^{p(\beta)}$ where

$$p(\beta)=\begin{cases} \frac{1}{3-\beta}+\varepsilon & \text{if } 0 \leqq \beta < 2 \\ \frac{1}{3}+\varepsilon & \text{if } -2<\beta<0 \end{cases}$$

and $\varepsilon = \varepsilon(f) > 0$.

(iii) If $|\beta| < 2$ and $g = k_\zeta^\beta$ then $f' \in H^{p(\beta)}$ for all

$$p(\beta) < \begin{cases} \frac{1}{3-\beta} & \text{if } 0 \leqq \beta < 2 \\ \frac{1}{3} & \text{if } -2<\beta<0. \end{cases}$$

The result is the best one.

Theorem 3. Let $f(z)=(\alpha\int_0^z P(\zeta)g^{\alpha}(\zeta)\zeta^{-1}d\zeta)^{\frac{1}{\alpha}} \in B(\alpha,\mathcal{P},\mathcal{M}_\beta)$, $\alpha \geqq 1$.

(i) If $|\beta| \geqq 2$ then $f' \in H^p$ for all $p < 1$.

(ii) If $|\beta| < 2$ and $g \neq k_\zeta^\beta$ then $f' \in H^{p(\beta)}$ where

$$p(\beta)=\begin{cases} \frac{1}{1+\alpha(2-\beta)}+\varepsilon & \text{if } 0 \leqq \beta < 2 \\ \frac{1}{1+2\alpha}+\varepsilon & \text{if } -2<\beta<0 \end{cases}$$

and $\varepsilon = \varepsilon(f) > 0$.

(iii) If $|\beta| < 2$ and $g = k_\zeta^\beta$ then $f' \in H^{p(\beta)}$ for all

$$p(\beta) < \begin{cases} \frac{1}{1+\alpha(2-\beta)} & \text{if } 0 \leqq \beta < 2 \\ \frac{1}{1+2\alpha} & \text{if } -2<\beta<0. \end{cases}$$

Remarks: For $0 < \alpha < 1$ the derivatives of functions from $B(\alpha,\mathcal{P},\mathcal{M}_\beta)$ belong to the same Hardy class as the β-convex functions. For $\beta = 0$, we obtain the result of S.S.Miller about the Hardy class of the derivatives of Bazilevič functions [3]. There exists a function $f \in B(\alpha,\mathcal{P},\mathcal{M}_\beta)$ so that $f'' \notin N$ (N denotes the Nevanlinna class).[5]

3. Let $\mathcal{F}$ denote a nonempty collection of functions $f(z)=z+a_2z^2+\dots$ each of which is univalent in $D=\{z \mid |z|<1\}$ and let $J(\alpha,f(z)) = \mathrm{Re}\left((1-\alpha)\frac{zf'(z)}{f(z)}+\alpha\left(1+\frac{zf''(z)}{f'(z)}\right)\right)$, $\alpha > 0$. The real number $\mathcal{R}_\alpha(\mathcal{F}) = \sup\{R \mid J(\alpha,f(z))>0, |z|<R, f\in\mathcal{F}\}$ is called the radius of α-convexity of $\mathcal{F}$. S.S.Miller, P.T.Mocanu and M.O.Reade have shown that $\mathcal{R}_\alpha(B(\frac{1}{\alpha},\{1\},S^*))=1$, $\alpha>0$. Inthe following we determine the

radius of α-convexity of $B(\frac{1}{\alpha},\mathcal{P},\{z\})$.

Theorem 4. $\mathcal{R}_{\alpha}(B(\frac{1}{\alpha},\mathcal{P},\{z\})) = -\alpha+\sqrt{\alpha^2+1}$

Proof. Let g be a regular function in D, $\operatorname{Re} g(z)>0$ in D, σ a complex number with $|\sigma|<1$. The function G given by $G(z)=g(\frac{z+\sigma}{1+\bar{\sigma}z})=g(\sigma)+g'(\sigma)(1-|\sigma|^2)z+\dots$ is regular in D and $\operatorname{Re} G(z)>0$. It is known that for such a function we have

$$|g'(\sigma)(1-|\sigma|^2)| \leqq 2|g(\sigma)| \quad \text{or}$$

$$\frac{\sigma g'(\sigma)}{g(\sigma)} \leqq \frac{2|\sigma|}{1-|\sigma|^2}$$

Since $\operatorname{Re} z \geqq -z$, we have

$$\operatorname{Re} \frac{\sigma g'(\sigma)}{g(\sigma)} \geqq -\frac{2|\sigma|}{1-|\sigma|^2} \tag{4}$$

Let f $B(\frac{1}{\alpha},\mathcal{P},\{z\})$. We have

$$\frac{zP'(z)}{P(z)}=\frac{zf''(z)}{f'(z)}+(\frac{1}{\alpha}-1)\frac{zf'(z)}{f(z)}-(\frac{1}{\alpha}-1) \tag{5}$$

From (4) and (5), considering $g(z)=P(z)$, we obtain

$$\operatorname{Re}\left(\alpha(1+\frac{zf''(z)}{f'(z)}+(1-\alpha)\frac{zf'(z)}{f(z)}\right)\geqq 1-\frac{2\alpha|z|}{1-|z|^2} \tag{6}$$

The left-hand member of the inequality (6) is positive for $1-\frac{2\alpha|z|}{1-|z|^2}>0$, i.e. for $|z|<-\alpha+\sqrt{\alpha^2-1}$.

The equality holds for $P(z)=\frac{1+z}{1-z}$ and $z\in R^-$ what completes our proof.

Remark. For $\alpha=1$ we obtain the radius of convexity of the class R established by Mac-Gregor in [2].

Bibliografie

[1] P.J.Eenigenburg, S.S.Miller, The H^p classes for α-convex functions, Proc.Amer.Math.Soc., 38, 558-562, 1973

[2] T.H.Mac-Gregor, Functions whose derivative has a positive real part, Trans.Amer.Math.Soc., 104, 3, 532-537, 1962

[3] S.S.Miller, The Hardy Class of a Bazilevič function and its derivative, Proc.Amer.Math.Soc., 30, 1, 125-132, 1971

[4] S.S.Miller, P.T.Mocanu, M.O.Reade, Bazilevič functions and generalized convexity, Rev.Roum.de Math.Pures et Appl. 19, 2, 213-224, 1974

[5] S.S.Miller, P.T.Mocanu, Alpha-convex functions and derivatives in the Nevanlinna Class, Studia Univ.Babeş-Bolyai, 35-40, 1975

[6] P.T.Mocanu, Une propriété de convexité généralisée dans la théorie de la représentation conform, Mathematica, 11 (34), 1, 127-133, 1969

A Practical Method for the Computation of the Zeros of Complex Polynomials

by

Evelyn Frank

1. Introduction. In [1], [2], the author gave a method for the computation of the zeros of a polynomial with complex coefficients by successive approximations. The method involves computation with only the repetitive processes of multiplication and division. Consequently, it is exceptionally well adapted for use on computers.

Simultaneously, the modulus of each zero is found with an error bound of 10^{-a}, where a is chosen by the user. One can thus read the number of zeros in any given circle. In addition, the zeros in each half of the complex plane can be read from the results. Furthermore, bounds for the moduli of the zeros are found by this method without any previous knowledge of the bounds or any specific formulas for the bounds. By polynomial evaluation, the error is given when one stops the iterative process.

It is the purpose of this study to set up a computer program for the actual computation and to give a number of examples that illustrate the process. These show the great practicability of the method, as the computation of the zeros of a polynomial is the first step in many applied problems.

In §2, a mathematical description of the method is given, and in §3 a description of the computer program. A number of exceptional polynomials are also treated. In §4, tables are provided that show the output of the computed zeros of sample polynomials. The Fortran program can be provided by the author for the benefit of users.

2. Mathematical description of the method. Let the polynomial whose zeros are to be found be

(2.1) $P(z) = (a_o + ib_o)z^n + (a_1 + ib_1)z^{n-1} + \dots + (a_n + ib_n)$.

The transformation

(2.2) $$z = r\cdot\frac{w-1}{w+1}$$

maps $R(w) > 0$ onto $|z| < r$, $R(w) = 0$ onto $|z| = r$, $R(w) < 0$ onto $|z| > r$.

From (2.1), form

(2.3) $$P_r(w) = (w+1)^n \cdot P(r\cdot\frac{w-1}{w+1}),\ r > 0,$$
$$= (p_0+iq_0)w^n + (p_1+iq_1)w^{n-1} + \dots + (p_n+iq_n),$$

and

(2.4) $$Q_r(w) = \frac{P_r(w) \pm \overline{P}_r(-w)}{2}$$
$$= iq_0w^n + p_1w^{n-1} + iq_2w^{n-2} + p_3w^{n-3} + \dots .$$

The positive or negative sign is used in (2.4) according as the degree of $P_r(w)$ is odd or even, respectively, and $\overline{P}_r(w)$ denotes the polynomial whose coefficients are the complex conjugates of those of $P_r(w)$. If the following continued fraction exists, one can write

(2.5) $$\frac{Q_r(w)}{P_r(w)-Q_r(w)} = \frac{1}{c_1w+k_1} + \frac{1}{c_2w+k_2} + \dots + \frac{1}{c_nw+k_n},\ c_p \neq 0, p=1,2,\dots,n,\ \underline{\text{if}}\ q_0 = 0,$$

where $P_r(w)$ and $Q_r(w)$ are given by (2.3) and (2.4), and where the c_p are real and the k_p are pure imaginary or zero. If the following continued fraction expansion exists, one can write

(2.6) $$\frac{Q_r(w)}{P_r(w)-Q_r(w)} = \frac{1}{k_0} + \frac{1}{c_1w+k_1} + \frac{1}{c_2w+k_2} + \dots + \frac{1}{c_nw+k_n},\ k_0 \neq 0,\ c_p \neq 0,$$
$$p=1,2,\dots,n,\ \underline{\text{if}}\ q_0 \neq 0.$$

By a theorem in [1], $P_r(w)$ has j zeros in $R(w) > 0$ and (n-j) zeros in $R(w) < 0$ if j of the coefficients c_p are negative and the remaining (n-j) are positive. By (2.2), the above theorem takes the following form:

Theorem 2.1. _If either the expansion_ (2.5) _or_ (2.6) _exists, the polynomial_ $P(z)$ (2.1) _has_ j _zeros within_ $|z| = r$ _and_ (n-j) _outside_ $|z| = r$ _if_ j _of the coefficients_ c_p _are negative and the remaining_ (n-j) _are positive_.

In order to avoid computation with complex numbers, one forms

(2.7) $$P^*(w) = i^n P_r(-iw) = U(w) + iV(w),$$

where

(2.8)
$$U(w) = p_o w^n - q_1 w^{n-1} - p_2 w^{n-2} + q_3 w^{n-3} + p_4 w^{n-4} - \cdots, \text{ and}$$
$$V(w) = q_o w^n + p_1 w^{n-1} - q_2 w^{n-2} - p_3 w^{n-3} + q_4 w^{n-4} + \cdots.$$

<u>Provided</u> $q_o = 0$, the continued fraction expansion, if it exists, is of the form

(2.9)
$$\frac{V(w)}{U(w)} = \frac{1}{d_1 w + m_1} + \frac{1}{d_2 w + m_2} + \cdots + \frac{1}{d_n w + m_n}$$
$$= \frac{1}{c_1 w + ik_1} - \frac{1}{c_2 w + ik_2} - \cdots - \frac{1}{c_n w + ik_n},$$

where

(2.10) $$c_p = (-1)^{p+1} d_p,\ ik_p = (-1)^{p+1} m_p,\ p=1,2,\ldots,n.$$

<u>If</u> $q_o \neq 0$, the continued fraction, if it exists, is

(2.11)
$$\frac{V(w)}{U(w)} = \frac{1}{m_o} + \frac{1}{d_1 w + m_1} + \frac{1}{d_2 w + m_2} + \cdots + \frac{1}{d_n w + m_n}$$
$$= \frac{1}{ik_o} - \frac{1}{c_1 w + ik_1} - \frac{1}{c_2 w + ik_2} - \cdots - \frac{1}{c_n w + ik_n},$$

where

(2.12) $$c_p = (-1)^p d_p,\ ik_o = m_o,\ ik_p = (-1)^p m_p,\ p=1,2,\ldots,n.$$

Following are explicit formulas for the computation of expansions (2.9) and (2.11).

In order to expand the rational function

$$\frac{f_1}{f_o} = \frac{\alpha_{11} w^{n-1} + \alpha_{12} w^{n-2} + \cdots + \alpha_{1n}}{\alpha_{oo} w^n + \alpha_{o1} w^{n-1} + \cdots + \alpha_{on}}$$

into a continued fraction of the form (2.5), one computes the numbers d_p and m_p of (2.9), and then by (2.10) the c_p and k_p of (2.5). For this, one uses the following $\alpha\beta$-table, which is obtained by the long-division process involved in the euclidean algorithm for the highest common factor of two polynomials f_o and f_1 (cf. [1]):

$$\alpha_{00} = p_0 \qquad \alpha_{01} = -q_1 \qquad \alpha_{02} = -p_2 \qquad \alpha_{03} = q_3 \qquad \cdots$$

$$\alpha_{11} = p_1 \qquad \alpha_{12} = -q_2 \qquad \alpha_{13} = -p_3 \qquad \alpha_{14} = q_4 \qquad \cdots$$

$$\beta_{11} = \frac{\alpha_{11}\alpha_{01}-\alpha_{00}\alpha_{12}}{\alpha_{11}} \quad \beta_{12} = \frac{\alpha_{11}\alpha_{02}-\alpha_{00}\alpha_{13}}{\alpha_{11}} \quad \beta_{13} = \frac{\alpha_{11}\alpha_{03}-\alpha_{00}\alpha_{14}}{\alpha_{11}} \qquad \cdots$$

(2.13)

$$\alpha_{22} = \frac{\alpha_{11}\beta_{12}-\beta_{11}\alpha_{12}}{\alpha_{11}} \quad \alpha_{23} = \frac{\alpha_{11}\beta_{13}-\beta_{11}\alpha_{13}}{\alpha_{11}} \quad \alpha_{24} = \frac{\alpha_{11}\beta_{14}-\beta_{11}\alpha_{14}}{\alpha_{11}} \qquad \cdots$$

$$\beta_{22} = \frac{\alpha_{22}\alpha_{12}-\alpha_{11}\alpha_{23}}{\alpha_{22}} \quad \beta_{23} = \frac{\alpha_{22}\alpha_{13}-\alpha_{11}\alpha_{24}}{\alpha_{22}} \quad \beta_{24} = \frac{\alpha_{22}\alpha_{14}-\alpha_{11}\alpha_{25}}{\alpha_{22}} \qquad \cdots$$

$$\alpha_{33} = \frac{\alpha_{22}\beta_{23}-\beta_{22}\alpha_{23}}{\alpha_{22}} \quad \alpha_{34} = \frac{\alpha_{22}\beta_{24}-\beta_{22}\alpha_{24}}{\alpha_{22}} \quad \alpha_{35} = \frac{\alpha_{22}\beta_{25}-\beta_{22}\alpha_{25}}{\alpha_{22}} \qquad \cdots$$

$$\cdot \quad \cdot \quad \cdot$$

$$d_p = \frac{\alpha_{p-1,p-1}}{\alpha_{p,p}}, \quad m_p = \frac{\beta_{p,p}}{\alpha_{p,p}}, \quad p=1,2,\ldots,n.$$

The expansion (2.5) exists if and only if $\alpha_{00} \neq 0$ and the determinants of odd order (blocked off by lines) in the array

(2.14)

$$\begin{array}{c|cc|cc|cc}
\alpha_{11} & \alpha_{12} & \alpha_{13} & \alpha_{14} & \alpha_{15} & \alpha_{16} & \cdots \\
\cline{1-1}
\alpha_{00} & \alpha_{01} & \alpha_{02} & \alpha_{03} & \alpha_{04} & \alpha_{05} & \cdots \\
0 & \alpha_{11} & \alpha_{12} & \alpha_{13} & \alpha_{14} & \alpha_{15} & \cdots \\
\cline{1-3}
0 & \alpha_{00} & \alpha_{01} & \alpha_{02} & \alpha_{03} & \alpha_{04} & \cdots \\
0 & 0 & \alpha_{11} & \alpha_{12} & \alpha_{13} & \alpha_{14} & \cdots \\
\cline{1-5}
0 & 0 & \alpha_{00} & \alpha_{01} & \alpha_{02} & \alpha_{03} & \cdots \\
\cdot & \cdot & \cdot & \cdot & \cdot & \cdot & \cdots
\end{array},$$

$$(\alpha_{0p} = \alpha_{1p} = 0, \ p > n),$$

are different from zero (cf. [1]).

Similarly,

$$\frac{F_1}{F_0} = \frac{\alpha_{00}w^n + \alpha_{01}w^{n-1} + \cdots + \alpha_{0n}}{\beta_{00}w^n + \beta_{01}w^{n-1} + \cdots + \beta_{0n}}$$

can be expanded into a continued fraction of the form (2,6) by the following $\alpha\beta$-table:

$$\alpha_{00} = q_0 \qquad \alpha_{01} = p_1 \qquad \alpha_{02} = -q_2 \qquad \alpha_{03} = -p_3 \qquad \cdots$$

$$\beta_{00} = p_0 \qquad \beta_{01} = -q_1 \qquad \beta_{02} = -p_2 \qquad \beta_{03} = q_3 \qquad \cdots$$

$$\alpha_{11} = \frac{\alpha_{00}\beta_{01}-\beta_{00}\alpha_{01}}{\alpha_{00}} \quad \alpha_{12} = \frac{\alpha_{00}\beta_{02}-\beta_{00}\alpha_{02}}{\alpha_{00}} \quad \alpha_{13} = \frac{\alpha_{00}\beta_{03}-\beta_{00}\alpha_{03}}{\alpha_{00}} \quad \cdots$$

(2.15)

$$\beta_{11} = \frac{\alpha_{11}\alpha_{01}-\alpha_{00}\alpha_{12}}{\alpha_{11}} \quad \beta_{12} = \frac{\alpha_{11}\alpha_{02}-\alpha_{00}\alpha_{13}}{\alpha_{11}} \quad \beta_{13} = \frac{\alpha_{11}\alpha_{03}-\alpha_{00}\alpha_{14}}{\alpha_{11}} \quad \cdots$$

$$\alpha_{22} = \frac{\alpha_{11}\beta_{12}-\beta_{11}\alpha_{12}}{\alpha_{11}} \quad \alpha_{23} = \frac{\alpha_{11}\beta_{13}-\beta_{11}\alpha_{13}}{\alpha_{11}} \quad \alpha_{24} = \frac{\alpha_{11}\beta_{14}-\beta_{11}\alpha_{14}}{\alpha_{11}} \quad \cdots$$

$$\beta_{22} = \frac{\alpha_{22}\alpha_{12}-\alpha_{11}\alpha_{23}}{\alpha_{22}} \quad \beta_{23} = \frac{\alpha_{22}\alpha_{13}-\alpha_{11}\alpha_{24}}{\alpha_{22}} \quad \beta_{24} = \frac{\alpha_{22}\alpha_{24}-\alpha_{11}\alpha_{25}}{\alpha_{22}} \quad \cdots$$

$$\cdot \quad \cdot \quad \cdot$$

$$d_p = \frac{\alpha_{p-1,p-1}}{\alpha_{p,p}}, \quad m_0 = \frac{\beta_{00}}{\alpha_{00}}, \quad m_p = \frac{\beta_{p,p}}{\alpha_{p,p}}, \quad p=1,2,\ldots,n.$$

The expansion (2.6) exists if and only if $\alpha_{00} \neq 0$ and the determinants of even order (blocked off by lines) in the array

(2.16)

$$\begin{array}{cc|cc|cc|cc}
\alpha_{00} & \alpha_{01} & \alpha_{02} & \alpha_{03} & \alpha_{04} & \alpha_{05} & \alpha_{06} & \cdots \\
\beta_{00} & \beta_{01} & \beta_{02} & \beta_{03} & \beta_{04} & \beta_{05} & \beta_{06} & \cdots \\
\hline
0 & \alpha_{00} & \alpha_{01} & \alpha_{02} & \alpha_{03} & \alpha_{04} & \alpha_{05} & \cdots \\
0 & \beta_{00} & \beta_{01} & \beta_{02} & \beta_{03} & \beta_{04} & \beta_{05} & \cdots \\
\hline
0 & 0 & \alpha_{00} & \alpha_{01} & \alpha_{02} & \alpha_{03} & \alpha_{04} & \cdots \\
0 & 0 & \beta_{00} & \beta_{01} & \beta_{02} & \beta_{03} & \beta_{04} & \cdots \\
\hline
0 & 0 & 0 & \alpha_{00} & \alpha_{01} & \alpha_{02} & \alpha_{03} & \cdots \\
\cdot & \cdot & \cdot & \cdot & \cdot & \cdot & \cdot & \cdots,
\end{array}$$

$$(\alpha_{0p} = \beta_{0p} = 0, \ p > n),$$

are different from zero.

Now, from the mapping

(2.17)
$$z' = r \cdot \frac{w' - 1}{w' + 1},$$

z' is a zero of $P(z)$ of modulus r if w' is a pure imaginary zero of $P(r \cdot (w'-1)/(w'+1))$, where 0 and ∞ are included. The pure imaginary zeros of $P_r(w)$ (2.3) are the pure imaginary zeros of $D_r(w)$ which is the greatest common divisor of $P_r(w)$ and $Q_r(w)$. By the euclidean algorithm, $D_r(w)$ can be obtained, and consequently the zeros of $P(z)$ by (2.17). The problem can be reduced to computation with real numbers if one computes $D_r^*(w)$, the greatest common divisor of $U(w)$ and $V(w)$, and then lets $D_r(w) = D_r^*(iw)$. In particular, if $-\underset{\sim}{r}$ is a zero of $P(z)$, then $w = 0$ is a common zero of $P_r(w)$ and $Q_r(w)$: if $+\underset{\sim}{r}$ is a zero of $P(z)$, then $w = \infty$ is a common zero of $P_r(w)$ and $Q_r(w)$.

The process for the computation of the zeros is as follows: By successively varying the values of r such that the last remainder approaches zero in the euclidean algorithm and such that simultaneously there is a change in the number of positive and negative signs of the c_p in the continued fraction expansions (2.5) or (2.6), one can approximate as closely as desired the value of $D_r(w)$ from the last non-zero remainder. From the zeros w' of this approximation to $D_r(w)$, one can obtain the zeros of $P(z)$ by (2.17) as closely as desired.

If the continued fraction expansion does not exist, one can proceed as follows: If, in the enclidean algorithm for the greatest common divisor of $Q_r(w)$ and $P_r(w) - Q_r(w)$ (tables (2.13) or (2.15)), a zero remainder is obtained before n steps have been carried out, then one finds the zeros of the first non-zero remainder $D_r(w)$.

If, however, one of the c_p, $p \neq n$, is zero and yet the remainder corresponding to this c_p is not zero, one again cannot expand $Q_r(w)/(P_r(w) - Q_r(w))$ into a continued fraction expansion. This case occurs when certain determinants vanish (cf. (2.14) or (2.16)). In this case, if, for r near r_1, the continued fraction expansion shows a change in the number of positive and negative signs of the c_p and for $r = r_1$ one of the c_p, $p \neq n$, vanishes, then $r = r_1$ is the modulus of the desired zero.

3. Description of the program for the computation of the zeros. In a Fortran program for use on an IBM 370 Model 155, a practical procedure, with the use of the algorithm of § 2, is presented for the computation of the zeros of

a polynomial $P(z)$ with real or complex coefficients.

In the main program, first the degree of the polynomial is read in, as well as the error 10^{-a} allowable in the modulus of each zero, where a is a specified constant.

Next the transformation (2.2) is made on $P(z)$. The program then computes bounds for the moduli r of the zeros, as follows. The minimum bound is $r = 0$, and the maximum bound $r = RUP$ is reached when all signs of the c_p (2.10) or (2.12) are negative. After this, in the subroutine RSTEP, one begins with RLOW = 0 and RUP, and computes RMID = (RLOW + RUP)/2. The signs of the c_p in the interval (RLOW, RMID) are computed, and the signs of the c_p in the interval (RMID, RUP). This process of internal halving is continued until the moduli of all the zeros are separated, with one modulus in each sub-interval, and until, in each such sub-interval, the modulus satisfies the requirement $|RUP - RLOW| \leq 10^{-a}$, the specified error limit. Finally, the value of each modulus r is set equal to (RUP + RLOW)/2, where RUP and RLOW refer to the final bounds of each sub-interval. The <u>multiplicity</u> <u>s</u> of the modulus r of each zero of $P(z)$ is the total number s of zeros with modulus r.

After the RSTEP procedure has been completed, each value of r is used in the appropriate $\alpha\beta$-table (2.13) or (2.15). For each such value of r, the α_{ij} in the last row of the table are approximately zero. Then the polynomial $D_r'(iw)$ with coefficients composed of the α_{ij} in the last non-zero row of α_{ij} in the $\alpha\beta$-table is considered. If $D_r'(iw)$ is a first degree polynomial, the usual method is used in the calculation of its zero iw. If certain determinants (2.14) or (2.16) are zero, in which case the corresponding continued fraction (2.9) or (2.11) is degenerate, the degree of $D_r'(iw)$ is higher than first degree. If $D_r'(iw)$ is a quadratic, the quadratic formula is used for the calculation of the zeros iw. After the zeros of $D_r'(iw)$ have been determined, provision is made for the transformation of these values iw back to the zeros z of $P(z)$ by (2.2).

If the degree of $D_r'(iw)$ is greater than or equal to three, the algorithm is repeated for the calculation of the last non-zero remainder. The RSTEP routine is carried through for $D_r'(iw)$. When an r_2 is found such that there is a zero row of α_{ij} in the $\alpha\beta$-table (2.13) or (2.15) for $D_{r_2}'(it)$, where

$$iw = r_2 \cdot \frac{t-1}{t+1}, \tag{3.1}$$

then the zero t is computed for $D_{r_2}'(it)$ as above. Now the corresponding zero value w is computed from (3.1), and then the corresponding value z is computed from (2.2).

If the degree of $D_{r_2}'(it)$ is greater than or equal to three, the program again tries to find the zeros of equal modulus. The result in the $\alpha\beta$-table (2.13) or (2.15) is a line of zeros in an α_{ij} line. Then

$$\alpha_{ii}w^{n-i} + \alpha_{i,i+1}w^{n-i-1} + \dots + \alpha_{in}$$

is a perfect (n-i)-th power, and the (n-i) equal zeros are $t = -(\alpha_{in}/\alpha_{ii})^{1/n-i}$. Substitution of this value in (3.1) gives the corresponding value of w, and then the corresponding value of z is given by (2.2).

If P(z) has g zeros of value r, then P(w) has g zero coefficients starting with the highest power and moving through decreasing powers of w. The program then prints out these g zeros r with multiplicity g. If P(z) has h zeros of value -r, then P(w) has h zero coefficients starting with the constant term and moving through increasing powers of w, and the program prints h of these zeros -r with multiplicity h. Then the remaining non-zero coefficients form a depressed polynomial, and the zeros of this depressed polynomial are computed by the usual RSTEP and $\alpha\beta$-table methods.

Finally, the value of each zero z is substituted back into P(z) for a determination of the accuracy of the computed value of the zero, by subroutine POLY.

Double precision has been used in order to obtain the values of the zeros with optimal accuracy.

University of Illinois
at Chicago Circle

REFERENCES

1. Frank, E. The location of the zeros of polynomials with complex coefficients, Bulletin of the American Mathematical Society 52 (1946) 890-898

2. ________ .On the zeros of polynomials with complex coefficients, Bulletin of the American Mathematical Society 52 (1946) 144-157

N = 7 ERROR = 10**-14

POLYNOMIAL COEFFICIENTS

Z** 7	0.1000000000D+01	0.0
Z** 6	0.0	0.7070000000D+00
Z** 5	-0.6000000000D+01	0.1200000000D+02
Z** 4	0.2650000000D+02	-0.3710000000D+00
Z** 3	-0.8330000000D+01	0.3350000000D+02
Z** 2	0.2500000000D+02	0.3261000000D+02
Z** 1	0.1650000000D+02	0.1140000000D+01
Z** 0	0.3600000000D+01	0.1967000000D+02

MULTIPLICITY	MODULUS	ZERO		P(ZERO)	
1	0.507958135923 4D+00	0.7752185642806D-01	-0.5020077983724D+00	0.4709033163408D-10	-0.1428546170246D-10
1	0.8515345821145D+00	-0.7976885371105D+00	0.2980002387572D+00	0.1145750161413D-12	0.1243449787580D-12
1	0.8939972386977D+00	0.2270308645361D+00	0.8646895681961D+00	0.2834377177408D-09	0.3301536821709D-10
1	0.1436219985621D+01	0.2963320832679D+00	-0.1405316741352D+01	0.6965761301103D-11	-0.1204369937113D-11
1	0.2068045735053D+01	0.9814582117971D+00	0.1820316714411D+01	0.6335637570132D-08	-0.3071249921049D-08
1	0.4086104174592D+01	-0.3940850507543D+01	0.1079789147386D+01	0.1598943200065D-11	-0.3048228336411D-11
1	0.4260905342260D+01	0.3156196025627D+01	-0.2862471132346D+01	0.5220549697538D-04	-0.2035093143959D-04

EXAMPLE 2

N = 6 ERROR = 10** -7

POLYNOMIAL COEFFICIENTS

Z**	6	0.1000000000D+01	0.0
Z**	5	0.2100000000D+02	0.2100000000D+02
Z**	4	0.0	0.3690000000D+03
Z**	3	-0.1736000000D+04	0.1736000000D+04
Z**	2	-0.9225000000D+04	0.0
Z**	1	-0.1312500000D+05	-0.1312500000D+05
Z**	0	0.0	-0.1562500000D+05

W**	6	-0.4320000000D+06	0.0
W**	5	0.0	-0.1080000000D+07
W**	4	0.1116000000D+07	0.0
W**	3	0.0	0.6100000000D+06
W**	2	-0.1860000000D+06	0.0
W**	1	0.0	-0.3000000000D+05
W**	0	0.2000000000D+04	0.0

MULTIPLICITY	MODULUS	ZERO		P(ZERO)	
6	0.50000000000000D+01	-0.4000059331877D+01	-0.2999920889201D+01	0.3637978807092D-11	0.6366462912410D-11
6	0.50000000000000D+01	-0.4000003968761D+01	-0.2999994708311D+01	-0.1273292582482D-10	-0.3637978807092D-11
6	0.50000000000000D+01	-0.3999933157680D+01	-0.3000089121025D+01	-0.1364242052659D-10	-0.1909938873723D-10
6	0.50000000000000D+01	-0.3000061635907D+01	-0.3999953772328D+01	-0.9094947017729D-12	-0.1091393642128D-10
6	0.50000000000000D+01	-0.3000025014207D+01	-0.3999981239223D+01	0.7275957614183D-11	-0.1364242052659D-10
6	0.50000000000000D+01	-0.2999871280528D+01	-0.4000096536368D+01	0.9094947017729D-11	0.1182343112305D-10

EXAMPLE 3

= 8 ERROR = 10**-14

POLYNOMIAL COEFFICIENTS

Z** 8	0.1000000000D+01	0.0
Z** 7	-0.3600000000D+02	0.0
Z** 6	0.5460000000D+03	0.0
Z** 5	-0.4536000000D+04	0.0
Z** 4	0.2244900000D+05	0.0
Z** 3	-0.6728400000D+05	0.0
Z** 2	-0.6728400000D+05	0.0
Z** 1	-0.1095840000D+06	0.0
Z** 0	0.1181240000D+06	0.0

LTIPLICITY	MODULUS	ZERO		P(ZERO)	
1	0.6633236670557D+00	0.6633236670557D+00	0.0	-0.1688022166491D-08	0.0
2	0.1397538120943D+01	-0.8589354454391D+00	-0.1102425825195D+01	-0.1673470251262D-08	-0.4686626198236D-08
2	0.1397538120943D+01	-0.8589354454391D+00	0.1102425825195D+01	-0.1673470251262D-08	0.4686626198236D-08
2	0.7027162162467D+01	0.2067114116854D+01	-0.6716252473330D+01	-0.3512395778671D-06	0.2750493877102D-07
2	0.7027162162467D+01	0.2067114116854D+01	0.6716252473330D+01	-0.3512395778671D-06	-0.2750493877102D-07
2	0.1167813696377D+02	0.9690789533237D+01	-0.6516707847306D+01	-0.9210467396770D-04	-0.1025642450259D-03
2	0.1167813696377D+02	0.9690789533237D+01	0.6516707847306D+01	-0.9210467396770D-04	0.1025642450259D-03
1	0.1353873992363D+02	0.1353873992363D+02	0.0	0.3606692189351D-06	0.0

EXAMPLE 4

N = 4 ERROR = 10**-14

POLYNOMIAL COEFFICIENTS

Z** 4	0.1000000000D+01	0.0
Z** 3	-0.4000000000D+01	0.0
Z** 2	0.6000000000D+01	0.0
Z** 1	-0.4000000000D+01	0.0
Z** 0	0.1000000000D+01	0.0

MULTIPLICITY	MODULUS	ZERO		P(ZERO)	
4	0.1000000000000D+01	0.1000000000000D+01	0.0	0.0	0.0
4	0.1000000000000D+01	0.1000000000000D+01	0.0	0.0	0.0
4	0.1000000000000D+01	0.1000000000000D+01	0.0	0.0	0.0
4	0.1000000000000D+01	0.1000000000000D+01	0.0	0.0	0.0

EXAMPLE 5

N = 6 ERROR = 10**-14

POLYNOMIAL COEFFICIENTS

Z**	6	0.1000000000D+01	0.0
Z**	5	0.0	0.0
Z**	4	0.0	0.0
Z**	3	0.0	0.0
Z**	2	0.0	0.0
Z**	1	0.0	0.0
Z**	0	0.1000000000D+01	0.0

W**	6	-0.2000000000D+01	0.0
W**	5	0.0	0.0
W**	4	-0.3000000000D+02	0.0
W**	3	0.0	0.0
W**	2	-0.3000000000D+02	0.0
W**	1	0.0	0.0
W**	0	-0.20000000C0D+01	0.0

MULTIPLICITY	MODULUS	ZERO		P(ZERO)	
6	0.10000000000000D+01	-0.8660254037844D+00	-0.50000000000000D+00	-0.4440892098501D-15	-0.2564615186884D-13
6	0.10000000000000D+01	-0.8660254037844D+00	0.50000000000000D+00	-0.4440892098501D-15	0.2564615186884D-13
6	0.10000000000000D+01	-0.3415183907562D-04	0.9999999994168D+00	0.2099426638091D-07	-0.2049110330596D-03
6	0.10000000000000D+01	-0.3415183907562D-04	-0.9999999994168D+00	0.2099426638091D-07	0.2049110330596D-03
6	0.10000000000000D+01	0.8660254037844D+00	-0.50000000000000D+00	-0.4440892098501D-15	-C.2498001805407D-14
6	0.10000000000000D+01	0.8660254037844D+00	0.50000000000000D+00	-0.4440892098501D-15	0.2498001805407D-14

Konstruktion vollständiger Minimalflächen von endlicher Gesamtkrümmung – eine Anwendung der klassischen Theorie der Funktionen und Differentiale auf kompakten Riemannschen Flächen

Fritz Gackstatter
Lehrstuhl II für Mathematik
der RWTH Aachen
Aachen, Deutschland

In diesem Vortrag wird über einige Ergebnisse berichtet, die Herr Kunert, Assistent an der TU Berlin, und ich vor kurzem abgeleitet haben. Eine Arbeit mit dem Titel "Konstruktion vollständiger Minimalflächen von endlicher Gesamtkrümmung" wird veröffentlicht. Dort findet man auch die Beweise zu den hier angegebenen Sätzen.

1. Problemstellung und Inhaltsangabe. Wir beschäftigen uns mit den vollständigen orientierbaren Minimalflächen von endlicher Gesamtkrümmung im reellen dreidimensionalen euklidischen Raum. Man hat viele allgemeine Sätze für diese Flächenklasse, aber man kennt nur sehr wenige Beispiele. Zu jeder Geschlechtszahl p mit $p \geqq 1$ hat man einen oder zwei Vertreter. Wir konstruieren hier zu jeder kompakten Riemannschen Fläche eine "zugehörige" Minimalfläche vom genannten Typ und erhalten so eine reell $(6p-6)$-dimensionale Mannigfaltigkeit von Flächen mit p Henkeln, falls $p \geqq 2$ ist, und eine reell zweidimensionale Mannigfaltigkeit für den Fall $p = 1$. Wir benutzen dabei Existenzsätze aus der klassischen Theorie der Funktionen und Differentiale auf kompakten Riemannschen Flächen, insbesondere den Satz von Riemann-Roch. Für die Flächen mit $p = 1$ ziehen wir die Theorie der elliptischen Funktionen heran. Es tritt ein nicht-lineares Gleichungssystem auf, und dieses lösen wir durch sukzessives Linearisieren.

Der Satz von Riemann-Roch, ein Existenzsatz für Funktionen auf kompakten Riemannschen Flächen, ist somit indirekt auch ein Existenzsatz für vollständige Minimalflächen von endlicher Gesamtkrümmung.

2. Eigenschaften unserer Flächen. S sei eine Minimalfläche vom genannten Typ. Zu dieser Fläche S gehört eine endlich-zusammenhängende Riemannsche Fläche R mit grenzpunktartigen Randkomponenten; R ist der Parameterbereich für isotherme Parameter (siehe A. Huber [3, Theorem 13

und 15] und R. Osserman [7, Theorem 9.1]). Wir haben also die Zuordnung

$$S \longrightarrow R = \bar{R} \setminus \{\gamma_0, \gamma_1, \ldots, \gamma_N\}$$

mit einer kompakten Riemannschen Fläche $\bar{R}$. Da es keine geschlossenen Flächen S gibt, ist immer mindestens ein Randpunkt γ vorhanden.

Zu S gehört weiter ein gewöhnliches Differential $f(\zeta)\,d\zeta$ und eine meromorphe Funktion $g(\zeta)$, beide erklärt auf $\bar{R}$ –

$$S \longrightarrow \{f(\zeta)\,d\zeta\,,\ g(\zeta)\}.$$

Die Randpunkte $\gamma_0, \ldots, \gamma_N$ sind höchstens außerwesentliche Singularitäten von $f\,d\zeta$ und g. Die Koordinatendifferentiale $\phi_k(\zeta)\,d\zeta$ der Fläche S berechnen sich nach den Weierstraßschen Formeln

$$\phi_1\,d\zeta = \frac{1}{2}\,f\,(1-g^2)\,d\zeta\,,$$

$$\phi_2\,d\zeta = \frac{i}{2}\,f\,(1+g^2)\,d\zeta\,,$$

$$\phi_3\,d\zeta = f\,g\,d\zeta\,,$$

und die Koordinatenfunktionen x_k erhält man durch Integration,

(1) $$x_k = \operatorname{Re} \int \phi_k\,d\zeta\,.$$

Die Gesamtkrümmung C der Fläche S ist gleich -4π mal Blätterzahl der g-Funktion. Diese Tatsachen findet man zusammengestellt bei Osserman [7, § 8 und 9].

<u>3. Zwei Existenzsätze.</u> Wir konstruieren zu jeder kompakten Riemannschen Fläche $\bar{R}$ eine zugehörige Minimalfläche S,

$$\bar{R} \longrightarrow S\,.$$

Wir müssen dabei nur zu jedem $\bar{R}$ ein gewöhnliches Differential $f\,d\zeta$ und eine meromorphe Funktion $g(\zeta)$ finden derart, daß der Realteil der drei Integrale in (1) bei Integration über Rückkehrschnitte und um Residuenstellen herum periodisch wird. Außerdem müssen $f\,d\zeta$ und g so gewählt sein, daß S vollständig ausfällt. Dies gelingt.

$\bar{R}$ sei eine kompakte Riemannsche Fläche vom Geschlecht $p = 1$, wir denken sie uns parametrisiert durch das Periodenparallelogramm $(1, \tau)$ und wir schreiben $\bar{R}_\tau$. Es gilt der

Satz 1 : Zu jeder Riemannschen Fläche $\overline{R}_\tau$ existiert eine vollständige Minimalfläche S von der Gesamtkrümmung $C \geqq -132\pi$ und mit dem Parameterbereich $R = \overline{R}_\tau \setminus \left\{0, \frac{1}{2}, \frac{\tau}{2}, \frac{1+\tau}{2}\right\}$. Wir haben damit eine zweidimensionale Mannigfaltigkeit von Minimalflächen vom Typ S und mit einem Henkel.

Wir konstruieren mit $f(\zeta) = p'(\zeta)$, p' ist die Ableitung der Weierstraßschen p-Funktion, und mit

$$(2) \qquad g(\zeta) = \frac{z_1}{p'(\zeta)} + \frac{z_3}{p'(\zeta)^3} + \dots + \frac{z_{11}}{p'(\zeta)^{11}} .$$

Für die Flächen vom Geschlecht $p \geqq 2$ haben wir den

Satz 2 : Zu jeder kompakten Riemannschen Fläche $\overline{R}$ vom Geschlecht $p \geqq 2$ existiert eine vollständige Minimalfläche S von der Gesamtkrümmung $C > -336\, p^3 \pi$. Wir haben damit eine $(6p-6)$-dimensionale Mannigfaltigkeit von Minimalflächen vom Typ S und mit p Henkeln.

Bemerkungen zur Konstruktion. $\mathscr{Y}_0$ sei ein beliebiger Punkt auf $\overline{R}$; $\mathscr{Y}_0$ möge der Einfachheit halber kein Weierstraßpunkt sein. Dann existiert eine Funktion $F(\zeta)$ auf $\overline{R}$, die bei $\mathscr{Y}_0$ einen Pol genau der Ordnung $p+1$ besitzt und sonst überall regulär ist. Durch Ableiten von $F(\zeta)$ erhalten wir ein gewöhnliches Differential. Dieses Differential wird unser $f(\zeta)\,d\zeta$. – Zur Konstruktion der g-Funktion. Die Nullstellen von $f\,d\zeta$ bezeichnen wir mit $\mathscr{Y}_1, \dots, \mathscr{Y}_M$ und die Nullstellenordnung bei $\mathscr{Y}_\nu$ sei r_ν. Nach dem Satz von Riemann-Roch existiert eine Funktion $G(\zeta)$, deren Divisor ein Vielfaches von

$$\vartheta = \frac{\prod_{\nu=1}^{M} \mathscr{Y}_\nu^{\,r_\nu}}{\mathscr{Y}_0^{\,4p}}$$

ist. Diese Funktion G übernimmt die Rolle von p' in (2) und wir setzen

$$g(\zeta) = \frac{z_1}{G(\zeta)} + \frac{z_2}{G(\zeta)^2} + \dots + \frac{z_\Lambda}{G(\zeta)^\Lambda} .$$

Durch hinreichend große Wahl von Λ und durch geeignete Wahl der komplexen Größen $z_1, \dots, z_\Lambda$ kann man alle Bedingungen erfüllen. Ein dabei auftretendes nicht-lineares Gleichungssystem lösen wir durch sukzessives Linearisieren.

4. Zwei Probleme. a) Ist es möglich, Minimalflächen S zu konstruieren, wenn man neben $\overline{R}$ auch noch die Randpunkte $\mathfrak{g}_\nu$ vorschreibt? Dies führt zur Frage: Wieweit kann man bei der Konstruktion von Funktionen auf kompakten Riemannschen Flächen die Verzweigungspunkte steuern?

b) Kann man n nicht-lineare Gleichungen, die alle homogen vom Grade 2 sind, nicht-trivial lösen, wenn man nur $n+1$ Unbekannte zur Verfügung hat?

Literatur

[1] Gackstatter, F., Über abelsche Minimalflächen. Erscheint in den Math. Nachr.

[2] Gackstatter, F. und Kunert, R., Konstruktion vollständiger Minimalflächen von endlicher Gesamtkrümmung.

[3] Huber, A., On subharmonic functions and differential geometry in the large. Comment. Math. Helv. 32, 13-72 (1957).

[4] Klotz, T. and Sario, L., Existence of complete minimal surfaces of arbitrary connectivity and genus. Proc. Nat. Acad. Sci. USA 54, 42-44 (1965).

[5] Kunert, R., Über eine Klasse vollständiger Minimalflächen. Diplomarbeit an der TU Berlin, 1974.

[6] Nitsche, J. C. C., Vorlesungen über Minimalflächen. Grundlehren Bd. 199, Springer-Verlag, 1975.

[7] Osserman, R., A survey of minimal surfaces. Princeton N. J.: Van Nostrand, 1969.

[8] Riemann, B., Theorie der Abelschen Funktionen. Journal f. reine und angew. Math. 54, 115-155 (1857) oder Ges. Math. Werke, Kap. VI.

[9] Riemann, B., und K. Hattendorf (Bearbeiter), Über die Fläche vom kleinsten Inhalt bei gegebener Begrenzung. Abh. Königl. Ges. der Wissensch. Göttingen, Math. Kl. 13, 3-52 (1868) oder Ges. Math. Werke, Kap. XVII.

[10] Weyl, H., Die Idee der Riemannschen Fläche. 4. Aufl., Teubner, 1964.

GENERALIZED POLYNOMIALS OF THE BEST L_p-APPROXIMATION SUBJECT TO INTERPOLATORY CONSTRAINTS

by

I. M a r u ş c i a c

1. Introduction

In this paper we describe an algorithm for the best L_p-approximation on a discrete set of the complex plane by generalized polynomials subject to some interpolatory constraints. The algorithm is an adaptation of Newton-Raphson's method. In the real case, when the interpolatory constraints on the class of polynomials are missing, a similar algorithm has been given in 1972 by Kahng, S.W. [2]. The proposed method is based on a special explicit form of the best weighted L_2-approximation given by the author in [4]. This method can be used as well for finding the best L_p- approximation of a continuous function on a rectifiable curve in the complex plane.

2. Preinterpolatory best L_p-approximation

Let f be a given continuous on a compact point set $K \subset C$ function and $\varphi = (\varphi_j)_0^n$ a Chebyshev system on K. We designe by $\mathcal{P}(\varphi)$ the class of generalized polynomials with respect to φ, i.e. the set of linear combinations of the form:

$$(1) \qquad p(a;z) = \sum_{j=0}^{n} a_j \varphi_j(z) , \quad a \in C^{n+1}$$

We consider a finite set $Z_m = \{z_\nu\}_1^m \subset K$ $(m > n-r)$, a system of points $\zeta_i \in K \setminus Z_m$, $i=0,1,\dots,r$ $(0 \leq r \leq n)$ and a system of complex numbers w_i, $i=0,1,\dots,r$. The problem we will deal with is the following: given a system of positive weights $\mu = (\mu_1, \mu_2, \dots, \mu_m)$ $(\mu_\nu > 0, \sum \mu_\nu = 1)$ and $p \in]2,+\infty[$, find a polynomial $p(a^*;.) \in \mathcal{P}(\varphi)$ satisfying the conditions:

(i) $p(a^*;\zeta_i) = w_i$, $i=0,1,\dots,r$,

(ii) $$\sum_{\nu=1}^{m}\mu_\nu|f(z_\nu)-p(a^*;z_\nu)|^p = \inf_{p\in\mathcal{P}_I(\varphi)} \sum_{\nu=1}^{m}\mu_\nu|f(z_\nu)-p(a;z_\nu)|^p ,$$

where $\mathcal{P}_I(\varphi) \subset \mathcal{P}(\varphi)$ is the subset of polynomials satisfying (i).

<u>D e f i n i t i o n</u> . <u>The polynomial</u> $p(a^*;\cdot) \in \mathcal{P}(\varphi)$ <u>satisfying</u> (i)-(ii) <u>is called preinterpolatory best</u> L_p<u>-approximation (or</u> $L_{p,\mu}$<u>-approximation) to the function f on</u> Z_m.

If $w_i = f(\zeta_i)$, $i = 0,1,\dots,r$, and $r = n$, then $p(a^*;.)$ coincide with Lagrange's interpolatory generalized polynomial. If $r = 0$, then $p(a^*;.)$ is the classical best L_p- approximation generalized polynomial to the function f on Z_m.

We complete the system $\zeta_0, \zeta_1,\dots, \zeta_r$ by other arbitrary different points $\zeta_i \in Z_m$, $i=r+1,\dots,n$ and let $w_{r+1},\dots,w_n$ be arbitrary complex numbers. Then

$$L(w;z) = \sum_{i=0}^{n} w_i L_i(z) ,$$

where

$$L_i(z) = \frac{D\begin{pmatrix}\varphi_0,\dots,\varphi_{i-1},\varphi_i,\varphi_{i+1},\dots,\varphi_n\\ \zeta_0,\dots,\zeta_{i-1}, z ,\zeta_{i+1},\dots,\zeta_n\end{pmatrix}}{D\begin{pmatrix}\varphi_0, \varphi_1,\dots, \varphi_n\\ \zeta_0, \zeta_1,\dots, \zeta_n\end{pmatrix}}$$

is the generalized interpolatory polynomial with respect to the system φ. Here

$$D\begin{pmatrix}\varphi_0, \varphi_1,\dots, \varphi_n\\ \zeta_0, \zeta_1,\dots, \zeta_n\end{pmatrix} = \det(\varphi_i(\zeta_j)),\ i,j=0,1,\dots,n.$$

First we consider $p = 2$, i.e. we have to minimize the function

$$F_2(w) = \sum_{\nu=1}^{m}\mu_\nu|f(z_\nu) - L(w;z_\nu)|^2 =$$

$$= \sum_{\nu=1}^{m}\mu_\nu\big(f(z_\nu) - L(w;z_\nu)\big)\big(\bar{f}(z_\nu) - \bar{L}(w;z_\nu)\big)$$

with respect to $w_{r+1},\dots,w_n$.

From the system

$$\frac{\partial F_2}{\partial \overline{w}_i} = -2\sum_{\nu=1}^{m}\mu_\nu\left(f(z_\nu) - L(w;z_\nu)\right)\overline{L}_i(z_\nu) = 0\ ,\quad i=r+1,\dots,n \tag{2}$$

we have

$$\sum_{j=r+1}^{n} w_j \sum_{\nu=1}^{m}\mu_\nu \overline{L}_i(z_\nu)L_j(z_\nu) = \sum_{\nu=1}^{m}\mu_\nu\left(f(z_\nu) - \sum_{j=0}^{r} w_j L_j(z_\nu)\right)\overline{L}_i(z_\nu)\ ,$$

$$i = r+1,\dots,n.$$

If we denote:

$$\begin{aligned} a_{ij} &= \sum_{\nu=1}^{m}\mu_\nu \overline{L}_i(z_\nu)L_j(z_\nu)\ ,\quad i,j=r+1,\dots,n \\ b_i &= \sum_{\nu=1}^{m}\mu_\nu\left(f(z_\nu) - \sum_{j=0}^{r} w_j L_j(z_\nu)\right)\overline{L}_i(z_\nu)\ ,\quad i=r+1,\dots,n \end{aligned} \tag{3}$$

then the solution $w^* = (w^*_{r+1},\dots,w^*_n)$ of the system (2) is the solution of the linear system

$$Aw = b$$

where $A = (a_{ij})$, $b = (b_i)$, $i,j = r+1,\dots,n$.

Let now $p > 2$. Then we have to minimize the function

$$F_p(w) = \sum_{\nu=1}^{m}\mu_\nu \left| f(z_\nu) - L(w;z_\nu) \right|^p .$$

To minimize F_p it is necessary to find the solution of the system:

$$\frac{\partial F_p}{\partial \overline{w}_i} = -\frac{p}{2}\sum_{\nu=1}^{m}\mu_\nu\left(f(z_\nu)-L(w;z_\nu)\right)^{\frac{p}{2}}\left(\overline{f}(z_\nu)-\overline{L}(w;z_\nu)\right)^{\frac{p}{2}-1}\overline{L}_i(z_\nu) = 0 \tag{4}$$

$$\frac{\partial F_p}{\partial w_i} = -\frac{p}{2}\sum_{\nu=1}^{m}\mu_\nu\left(f(z_\nu)-L(w;z_\nu)\right)^{\frac{p}{2}-1}\left(\overline{f}(z_\nu)-\overline{L}(w;z_\nu)\right)^{\frac{p}{2}} L_i(z_\nu) = 0 \tag{4'}$$

$$i = r+1,\dots,n$$

We see that systems (4) and (4') are equivalent. So we will solve the system (4) by Newton-Raphson's method, i.e. the system (4) will be solved iteratively by finding the solution of the system:

$$(5)\qquad \frac{\partial F_p(w^{k-1})}{\partial \bar{w}_i} + \sum_{j=r+1}^{n} \frac{\partial^2 F_p(w^{k-1})}{\partial w_j \partial \bar{w}_i} \Delta w_j^k = 0\ ,\quad i=r+1,\dots,n\ ,$$

where $\Delta w_j^k = w_j^k - w_j^{k-1}$, $j=r+1,\dots,n$.

From (4') we find that

$$(6)\qquad \frac{\partial^2 F_p(w)}{\partial w_j \partial \bar{w}_i} = \frac{p^2}{4} \sum_{\nu=1}^{m} \mu_\nu |f(z_\nu)-L(w;z_\nu)|^{p-2} L_j(z_\nu)\bar{L}_i(z_\nu).$$

If we put $\mu^k = (\mu_1^k, \mu_2^k, \dots, \mu_m^k)$, where

$$(7)\qquad \mu_\nu^k = \mu^k(z_\nu) = \mu_\nu^{k-1} |f(z_\nu)-L(w^{k-1};z_\nu)|^{p-2}\ ,\quad \mu^0 = \mu\ ,$$

where $L(w^{k-1};.) \in \mathcal{P}_I(\varphi)$ is the best $L_{p,\mu^{k-1}}$-approximation to the function f on Z_m, then (4) and (6) can be written under the form:

$$(8)\qquad \begin{aligned} \frac{\partial F_p(w^{k-1})}{\partial \bar{w}_j} &= -\frac{p}{2} \sum_{\nu=1}^{m} \mu_\nu^k \left(f(z_\nu)-L(w^{k-1};z_\nu) \right) \bar{L}_i(z_\nu) \\ \frac{\partial^2 F_p(w^{k-1})}{\partial w_j \partial \bar{w}_i} &= \frac{p^2}{4} \sum_{\nu=1}^{m} \mu^k \bar{L}_i(z_\nu) L_j(z_\nu) \end{aligned}$$

Substituting (8) in (5) we obtain the system

$$\frac{p}{2} \sum_{j=r+1}^{n} \left(\sum_{\nu=1}^{m} \mu_\nu^k \bar{L}_i(z_\nu) L_j(z_\nu) \right) \Delta w_j^k = \sum_{\nu=1}^{m} \mu_\nu^k \left(f(z_\nu)-L(w^{k-1};z_\nu) \right) \bar{L}_i(z\ \ )$$

$$i = r+1,\dots,n\ ,$$

or equivalently

$$(9)\qquad A^k\left(\frac{p}{2} \Delta w^k + w^{k-1}\right) = b^k,$$

where

$$A^k = (a_{ij}^k) = \left(\sum_{\nu=1}^{m} \mu_\nu^k \bar{L}_i(z_\nu) L_j(z_\nu) \right) \ ,i,j= r+1,\dots,n$$

$$b^k = (b_{r+1}^k, \dots, b_n^k)^T,\quad b_i^k = \sum_{\nu=1}^{m} \mu_\nu^k \left(f(z_\nu) - \sum_{j=0}^{r} w_j L_j(z_\nu) \right) \bar{L}_i(z_\nu)$$

Now, if $L(\eta^k;.) \in \mathcal{P}_I(\varphi)$ is the preinterpolatory best weighted L_{2,μ^k}-approximation to f on Z_m, then

$$A^k \eta^k = b^k$$

Therefore, system (9) can be put under the form:

$$A^k\left(\frac{p}{2}\Delta w^k + w^{k-1}\right) = A^k \eta^k \tag{10}$$

If the matrix A^k is nonsingular (as we will show), then from (10) we obtain

$$\Delta w^k = \frac{2}{p}(\eta^k - w^{k-1}),$$

or

$$w^k = \frac{1}{p}((p-2)w^{k-1} + \eta^k). \tag{11}$$

3. Description of the algorithm

From above we have the following algorithm for the preinterpolatory best L_p - approximation to a function f on a finite set Z_m of the complex plane.

Starting from the initial vector $w^o = (w^o_{r+1},\dots,w^o_n) \in C^{n-r}$ and $\mu^o = \mu$

Step 1. Set $\mu^k = (\mu^k_1,\dots,\mu^k_m)$, where

$$\mu^k_\nu = \mu^{k-1}_\nu \,|\, f(z_\nu) - L(w^{k-1};z_\nu) \,|^{p-2} .$$

Step 2. Find the preinterpolatory best L_{2,μ^k} - approximation to f on Z_m: $L(\eta^k;.) \in \mathcal{P}_I(\varphi)$.

Step 3. Set

$$w^k = \frac{1}{p}((p-2)w^{k-1} + \eta^k)$$

and go to Step 1.

4. Convergence of the algorithm

First we will show that A^k is nonsingular. For this we will use one of our results contained in [4] that

$$(12) \qquad \det A^k = \frac{1}{n!} \sum_{\nu_{r+1},\ldots,\nu_n=1}^{m} \mu^k_{\nu_{r+1}} \cdots \mu^k_{\nu_n} \left| D(z_{\nu_{r+1}},\ldots,z_{\nu_n};L) \right|^2$$

where

$$D(x_{r+1},\ldots,x_n;L) = \det(L_i(x_j))_{r+1}^{n} \quad .$$

L e m m a 1. If A^o is nonsingular, then $\det A^k > 0$ for all $k \in N$.

P r o o f. We denote by

$$Z^k = \{ z \in Z_m \mid \mu^k(z) > 0 \} \quad .$$

Then the proof is by induction. Assume that $\det A^k > 0$. Then from (12) it follows that Z^k contains at least n-r points. If $Z^{k+1} = Z^k$ then no polynomial $p \in P_I(\varphi)$ agrees with f on the set $Z^{k+1} = Z^k$ and, therefore, Z^{k+1} contains at least n-r+1 points. Since $\{L_i\}_o^n$ is a Chebyshev system on $Z_m \cup \{\zeta_i\}_o^r$, from (12) it follows that $\det A^{k+1} > 0$.

If $Z^k \setminus Z^{k+1} \neq \emptyset$, then it is seen that $L(\eta^k;.) \in P_I(\varphi)$ is the best L_{2,μ^k}- approximation to f on Z^{k+1} as well as on Z^k, since $\mu^k_\nu = 0$ for $z_\nu \in Z^{k+1} \setminus Z^k$. This shows again that no polynomial $p \in P_I(\varphi)$ agrees with the function f on Z^{k+1}. So Z^{k+1} contains at least n-r+1 points, and from (12) it follows that $\det A^{k+1} > 0$. This completes the proof.

L e m m a 2. If $f \notin P(\varphi)$ then the function

$$F(w) = F(w_{r+1},\ldots,w_n) = \left(\sum_{\nu=1}^{m} \mu_\nu \left| f(z_\nu) - L(w;z_\nu) \right|^p \right)^{1/p}$$

is strictly convex with respect to w for every $p > 1$.

P r o o f. If $w', w'' \in C^{n-r}$ and $t \in]0,1[$, then from Minkowski's inequality we have

$$F((1-t)w'+tw'') = \left(\sum_{\nu=1}^{m} \mu_\nu \left| f(z_\nu) - (1-t)L(w';z_\nu) - tL(w'';z_\nu) \right|^p \right)^{1/p} =$$

$$= \left(\sum_{\nu=1}^{m} \mu_\nu \left| (1-t)(f(z_\nu) - L(w';z_\nu)) + t(f(z_\nu) - L(w'';z_\nu)) \right|^p \right)^{1/p} \leq$$

$$\leq (1-t)\left(\sum_{\nu=1}^{m}\mu_\nu|f(z_\nu)-L(w';z_\nu)|^p\right)^{1/p} + t\left(\sum_{\nu=1}^{m}\mu_\nu|f(z_\nu)-L(w'';z_\nu)|^p\right)^{1/p} =$$

$$= (1-t)F(w') + tF(w''),$$

hence F is convex. It is strictly convex since Minkowski's inequality becomes equality if and only if there exists a constant $\lambda > 0$, such that

$$(1-t)(f(z_\nu)-L(w';z_\nu)) = \lambda t(f(z_\nu)-L(w'';z_\nu)) \ , \quad \nu = 1,2,\dots,m,$$

i.e.

$$f(z_\nu) = \frac{1}{1-t-\lambda t}(L((1-t)w'+\lambda tw'';z_\nu)) \ , \quad \nu = 1,2,\dots,m,$$

which contradicts the assumption that $f \notin \mathcal{P}(\varphi)$.

T h e o r e m . If the initial weight $\mu = \mu^0$ is positive, then the algorithm described at 3 is always convergent and the convergence of the iterations is quadratic.

P r o o f . Since in this case $\det A^0 > 0$, Lemmas 1 and 2 show that the convergence conditions of Newton-Raphson method are satisfied. Therefore the convergence of the algorithm follows from the convergence of Newton-Raphson method.

If $f \in \mathcal{P}(\varphi)$ then there is a polynomial from $\mathcal{P}(\varphi)$ which agrees with f on Z_m and so the approximation problem is trivial.

Remark 1. Since our algorithm is valid for each $p > 2$, when p tends to infinity we obtain the preinterpolatory best uniform approximate polynomial on Z_m to f.

Remark 2. To compute the preinterpolatory best $L_{2,\mu}$-approximation we can use an explicit form of $L(\eta;\cdot)$ given in [5]:

$$L(\eta;z) = \sum_{1}^{m}\Lambda_{\nu_{r+1}\dots\nu_n}L(\zeta_0,\dots,\zeta_r,z_{\nu_{r+1}},\dots,z_{\nu_n};\varphi;f\,|\,z),$$

where

$$\Lambda_{\nu_{r+1}\cdots\nu_n} = \frac{\mu_{\nu_{r+1}}\cdots\mu_{\nu_n}\left| D\begin{pmatrix}\varphi_o,\ldots,\varphi_r,\varphi_{r+1},\ldots,\varphi_n\\ \zeta_o,\ldots,\zeta_r,z_{\nu_{r+1}},\ldots,z_{\nu_n}\end{pmatrix}\right.}{\sum_1^m \mu_{\nu_{r+1}}\cdots\mu_{\nu_n} D\begin{pmatrix}\varphi_o,\ldots,\varphi_r,\varphi_{r+1},\ldots,\varphi_n\\ \zeta_o,\ldots,\zeta_r,z_{\nu_{r+1}},\ldots,z_{\nu_n}\end{pmatrix}}$$

and $L(x_o,x_1,\ldots,x_n;\varphi;f|\,.)$ means the generalized interpolatory polynomial to f on the knots $x_o,x_1,\ldots,x_n$.

<u>Remark 3</u>. This algorithm can be extended to the case when instead of a finite set Z_m we take a rectifiable curve Γ , replasing the operator "summation over a discrete set" by the operator "integration" on Γ .

REFERENCES

1. Andriančik, A, N.; Rusak, V. N., Rešenie odnoi extremal'noi zadači, Vestzi Akad. Nauk, Beloruskai S.S.R., 3, 1973, pp. 25-29.

2. Kahng, S.W.; Best L_p-approximation, Math. Comp., 28, no. 118, 1972, pp. 505-508.

3. Maruşciac, I., Preinterpolatory best L_p-approximation generalized polynomials, Studia Univ. Babeş-Bolyai, Ser. Math.-Mech., 1975, pp. 60-64.

4. Maruşciac, I., Une forme explicite du polynome de meileure approximation d'une fonction dans le domaine complexe, Mathematica (Cluj), 6(29), no. 2, 1964, pp. 257-263.

5. Maruşciac, I., Sur l'approximation préinterpolatoire, Mathematica (Cluj), 13(36), no. 1, 1971, pp. 113-125.

University of Cluj-Napoca
Faculty of Mathematics

Second Order Differential Inequalities in the Complex Plane

Sanford S. Miller*

Department of Mathematics, State University of New York
Brockport, New York 14420, U.S.A.

Petru T. Mocanu

Department of Mathematics, Babes-Bolyai Univeristy
Cluj-Napoca, Romania

ABSTRACT. Let $w(z)$ be regular in the unit disc U and let $h(r, s, t)$ be a complex function defined in a domain of C^3. The authors determine conditions on h such that $|h(w(z), zw'(z), z^2w''(z))| < 1$ implies $|w(z)| < 1$ and such that $\text{Re}\, h(w(z), zw'(z), z^2w''(z)) > 0$ implies $\text{Re}\, w(z) > 0$. Applications of these results to univalent function theory, differential equations and harmonic functions are given.

1. Introduction and the Fundamental Lemma

Let $w(z)$ be regular in the unit disc U, with $w(0) = 0$, and let $h(r, s)$ be a continuous function defined in a domain of C^2. With some simple conditions on h it has been shown [6] that $|h(w(z), zw'(z))| < 1$, for $z \in U$, implies $|w(z)| < 1$ for $z \in U$. In this paper we extend this result to functions $h(r, s, t)$ defined in a domain of C^3 and prove that if $|h(w(z), zw'(w), z^2w''(z))| < 1$ for $z \in U$ then $|w(z)| < 1$ for $z \in U$. This result and applications of it in the theory of differential equations are given in section 2.

In section 3 we determine conditions on $h(r, s, t)$ such that $\text{Re}\, h(w(z), zw'(z), z^2w''(z)) > 0$ implies $\text{Re}\, h(z) > 0$. Applications of this result in the theory of differential equations

* This work was carried out while the first author was a U.S.A. - Romania Exchange Scholar.

are also given. Corresponding results for harmonic functions are given in section 4. Section 5 is concerned with applications in univalent function theory dealing with convex functions, starlike functions and the Schwarzian derivative.

Our basic tool in determining conditions on $h(r, s, t)$ will be Lemma B. Although the following lemma is a special case of Lemma B we need to prove it first in order to prove Lemma B.

<u>LEMMA A</u>. Let $g(z) = g_n z^n + g_{n+1} z^{n+1} + \ldots$ be regular in U with $g(z) \not\equiv 0$ and $n \geq 1$. If $z_o = r_o e^{i\theta_o}$ $(r_o < 1)$ and

$$(1) \qquad |g(z_o)| = \max_{|z| \leq r_o} |g(z)|$$

then

$$\text{(i) } z_o g'(z_o)/g(z_o) = m \quad \text{and}$$

$$\text{(ii) } \operatorname{Re} \frac{z_o g''(z_o)}{g'(z_o)} + 1 \geq m,$$

where $m \geq n \geq 1$.

<u>PROOF</u>

(i) If we let $g(z) = R(r_o, \theta) e^{i\Phi(r_o, \theta)}$ for $z = r_o e^{i\theta}$ then

$$(2) \qquad \frac{zg'(z)}{g(z)} = \frac{\partial \Phi}{\partial \theta} - i \frac{1}{R} \frac{\partial R}{\partial \theta}$$

Since z_o is a maximum point of R we must have $\partial R(z_o)/\partial \theta = 0$, and so we obtain $z_o g'(z_o)/g(z_o) = m$, where m is real. We need to show $m \geq n$. Let $h(z) = g(z_o z)/(g(z_o) z^{n-1})$ for $z \in U$. Then $h(0) = 0$, $h(z)$ is regular in U and by the maximum principle

$$|h(z)| \leq \frac{1}{|g(z_o)| r^{n-1}} \max_{\theta} |g(z_o r e^{i\theta})| \qquad (|z| \leq r < 1) .$$

Hence by (1) we obtain $|h(z)| \leq 1/r^{n-1}$, and by letting r approach 1 we obtain $|h(z)| \leq 1$. Employing the Schwarz Lemma we obtain $|h(z)| \leq |z|$ and $|g(z_o z)/g(z_o)| \leq |z|^n$. In particular, at the point $z = r$, $0 \leq r < 1$ we have

$$(3) \qquad \operatorname{Re} \frac{g(z_0 r)}{g(z_0)} \leq r^n .$$

Since $m = z_0 g'(z_0)/g(z_0)$ we have

$$m = \frac{d}{dr}\left(\frac{g(z_0 r)}{g(z_0)}\right)\Bigg|_{r=1} = \lim_{r \uparrow 1} \frac{g(z_0 r) - g(z_0)}{(r - 1)g(z_0)}$$

$$= \lim_{r \uparrow 1} \left[1 - \frac{g(z_0 r)}{g(z_0)}\right] \cdot \frac{1}{1 - r} .$$

Taking real parts and using (3) we obtain

$$m = \lim_{r \uparrow 1} \left[1 - \operatorname{Re}\left(\frac{g(z_0 r)}{g(z_0)}\right)\right] \cdot \frac{1}{1 - r} \geq \lim_{r \uparrow 1} \frac{1 - r^n}{1 - r} = n.$$

(ii) From (2) a simple calculation yields

$$i\left[\frac{zg'}{g}\left(\frac{zg''}{g'} + 1\right) - \left(\frac{zg'}{g}\right)^2\right] = \frac{\partial^2 \Phi}{\partial \theta^2} - i\left[\frac{1}{R}\frac{\partial^2 R}{\partial \theta^2} - \frac{1}{R^2}\left(\frac{\partial R}{\partial \theta}\right)^2\right].$$

Since z_0 is a maximum point of $R(r_0, \theta)$ we have $\partial R(z_0)/\partial \theta = 0$ and $\partial^2 R(z_0)/\partial \theta^2 \leq 0$, and since $z_0 g'(z_0)/g(z_0) = m \geq 1$ we obtain

$$i\, m\left(\frac{z_0 g''(z_0)}{g'(z_0)} + 1\right) - m^2 = \frac{\partial^2 \Phi(z_0)}{\partial \theta^2} - i\left[\frac{1}{R(z_0)} \frac{\partial^2 R(z_0)}{\partial \theta^2}\right].$$

Taking imaginary parts we obtain

$$\operatorname{Re} m\left(\frac{z_0 g''(z_0)}{g'(z_0)} + 1\right) - m^2 = - \frac{1}{R(z_0)} \frac{\partial^2 R(z_0)}{\partial \theta^2} \geq 0$$

and

$$\operatorname{Re} \frac{z_0 g''(z_0)}{g'(z_0)} + 1 \geq m,$$

which completes the proof of the lemma.

Part (i) of this lemma is stated in a paper by I.S.Jack [2], and the authors believe there is an error in the proof given. The authors wish to thank Professor L.Brickman for his short proof of this part of the lemma.

LEMMA B. Let $\varphi(z)$ be an injective mapping of $\bar{U}$ onto $\bar{\Omega}$, $\varphi(0) = a$, and such that $\varphi(z)$ is regular on $\bar{U}$ except for at most one pole on ∂U. Denote by $\eta(w)$ the argument of the outer normal to $\partial\Omega$ at a finite boundary point $w \in \partial\Omega$. Let $w(z) = a + w_n z^n + w_{n+1} z^{n+1} + \dots$ be regular in U, with $w(z) \not\equiv a$ and $n \geq 1$. Suppose that there exists a point $z_0 = r_0 e^{i\theta} \in U$ such that

$$w_0 = w(z_0) \in \partial\Omega \quad \text{and} \quad w(|z| < r_0) \subset \Omega.$$

If $\zeta_0 = \varphi^{-1}(w_0)$ then

(a) $\arg(z_0 w'(z_0)) = \arg(\zeta_0 \varphi'(\zeta_0)) = \eta(w_0)$,

(b) $|z_0 w'(z_0)| = m|\zeta_0 \varphi'(\zeta_0)| > 0$, and

(c) $\mathrm{Re}\left[\frac{z_0 w''(z_0)}{w'(z_0)} + 1\right] \geq m\,\mathrm{Re}\left[\frac{\zeta_0 \varphi''(\zeta_0)}{\varphi'(\zeta_0)} + 1\right]$,

where $m \geq n \geq 1$.

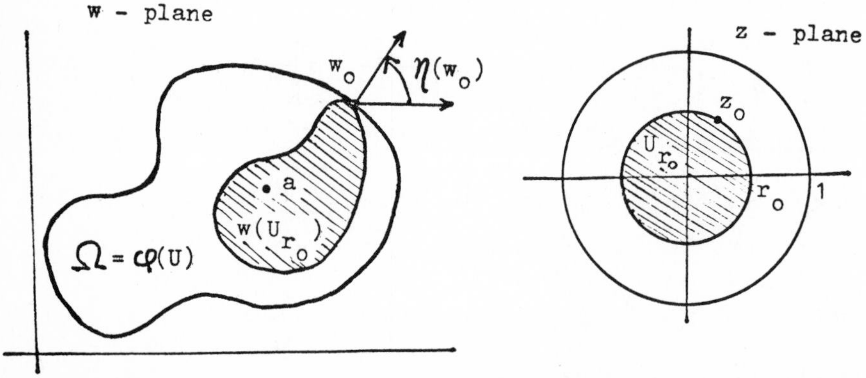

PROOF. Since w_0 is finite and $\varphi(\zeta)$ is univalent at ζ_0 we have $\varphi'(\zeta_0) \neq 0$ and

$$\eta(w_0) = \eta(\varphi(\zeta_0)) = \arg(\zeta_0 \varphi'(\zeta_0)).$$

The function $g(z) = \varphi^{-1}(w(z))$ is regular in $|z| \leq r_0$ and satisfies $|g(z_0)| = 1$, $g(0) = 0$ and $|g(z)| \leq 1$ for $|z| \leq r_0$. A further calculation shows that $g^{(k)}(z_0) = w^{(k)}(z_0) = 0$ for k=

$= 1, 2, \ldots, n-1$. Thus $g(z)$ satisfies the conditions of Lemma A. Since $w(z) = \varphi(g(z))$ we have

(4) $$w'(z) = \varphi'(\zeta)g'(z), \text{ and}$$

$$zw'(z) = \zeta\varphi'(\zeta) \cdot \frac{zg'(z)}{g(z)}$$

By Lemma A we have $z_o g'(z_o)/g(z_o) = m \geq n \geq 1$ and so we obtain $z_o w'(z_o) = m\zeta_o\varphi'(\zeta_o)$. Therefore

$$\arg(z_o w'(z_o)) = \arg(\zeta_o\varphi'(\zeta_o)) = \eta(w_o), \text{ and}$$

$$|z_o w'(z_o)| = m|\zeta_o\varphi'(\zeta_o)| \geq |\varphi'(\zeta_o)| > 0.$$

Differentiating (4) logarithmically we obtain

$$\frac{w''(z)}{w'(z)} = \frac{\varphi''(g(z))g'(z)}{\varphi'(g(z))} + \frac{g''(z)}{g'(z)}, \text{ and}$$

$$\frac{zw''(z)}{w'(z)} + 1 = \frac{\zeta\varphi''(\zeta)}{\varphi'(\zeta)} \cdot \frac{zg'(z)}{g(z)} + \frac{zg''(z)}{g'(z)} + 1 .$$

By using Lemma A we obtain

$$\text{Re}\,\frac{z_o w''(z_o)}{w'(z_o)} + 1 = m\,\text{Re}\,\frac{\zeta_o\varphi''(\zeta_o)}{\varphi'(\zeta_o)} + \text{Re}\,\frac{zg''(z)}{g'(z)} + 1$$

$$\geq m\,\text{Re}\,\frac{\zeta_o\varphi''(\zeta_o)}{\varphi'(\zeta_o)} + m = m\,\text{Re}\left[\frac{\zeta_o\varphi''(\zeta_o)}{\varphi'(\zeta_o)} + 1\right].$$

2. Bounded Functions

THEOREM 1. Let $w(z) = a + w_n z^n + w_{n+1}z^{n+1} + \ldots$ be regular in U with $w(z) \not\equiv 0$ and $n \geq 1$. If $z_o = r_o e^{i\theta}$, $0 < r_o < 1$, and

$$|w(z_o)| = \max_{|z| \leq r_o} |w(z)|$$

then

(i) $z_o w'(z_o)/w(z_o) = m$ and

(ii) $\text{Re}\,\dfrac{z_o w''(z_o)}{w'(z_o)} + 1 \geq m,$

where $$m \geq n\,\frac{|w_o - a|^2}{|w_o|^2 - |a|^2} \geq n\,\frac{|w_o| - |a|}{|w_o| + |a|} .$$

PROOF. If we let $\Omega = \{w \mid |w| \leq |w_o|\}$, where $w_o = w(z_o)$, then $\varphi(\zeta) = w_o(\bar{w}_o\zeta + a)/(w_o + \bar{a}\zeta)$ is a conformal mapping of U onto Ω with $\varphi(0) = a$. Since $\zeta_o = \varphi^{-1}(w_o)$ a simple calculation yields $\zeta_o = (a - w_o)/(\bar{a} - \bar{w}_o)$,

$$\zeta_o\varphi'(\zeta_o) = \frac{w_o|w_o - a|^2}{|w_o|^2 - |a|^2}, \text{ and}$$

$$\operatorname{Re}\frac{\zeta_o\varphi''(\zeta_o)}{\varphi'(\zeta_o)} + 1 = \frac{|w_o - a|^2}{|w_o|^2 - |a|^2} \geq \frac{|w_o| - |a|}{|w_o| + |a|}.$$

Applying Lemma B to these results we obtain (i) and (ii).

We will use this theorem to generate subclasses of bounded functions and also to show that certain second order complex differential equations have bounded solutions. In what follows $J > 0$, n will be a positive integer and a will be a complex number satisfying $|a| < J$. We will also let $\lambda = \lambda(a, n, J) = n(J - |a|)/(J + |a|)$.

THEOREM 2. Let $h(r, s, t) : C^3 \to C$ such that

(i) $h(r, s, t)$ is continuous in a domain $D \subset C^3$,

(ii) $(a, 0, 0) \in D$ and $|h(a, 0, 0)| < J$,

(iii) $|h(Je^{i\theta}, Ke^{i\theta}, L)| \geq J$ when $(Je^{i\theta}, Ke^{i\theta}, L) \in D$, $K \geq J\lambda$ and $\operatorname{Re}[Le^{-i\theta}] \geq K(\lambda - 1)$.

Let $w(z) = a + w_n z^n + w_{n+1}z^{n+1} + \ldots$ be regular in U with $w(z) \not\equiv 0$ and $n \geq 1$. If $(w(z), zw'(z), z^2w''(z)) \in D$ when $z \in U$ and

(5) $\quad |h(w(z), zw'(z), z^2w''(z))| < J$ when $z \in U$

then $|w(z)| < J$ when $z \in U$.

PROOF. $|w(0)| = |a| < J$. Suppose there exists $z_o = r_oe^{i\theta_o} \in U$, $(0 < r_o < 1)$ such that

$$J = |w(z_o)| = \max_{|z| \leq r_o} |w(z)|$$

Then $w(z_o) = Je^{i\theta}$ and since by Theorem 1 $z_o w'(z_o)/w(z_o) = m \geq \lambda$, we have $z_o w'(z_o) = Ke^{i\theta}$ where $K \geq J\lambda$. Also by Theorem 1 we have $\mathrm{Re}[z_o w''(z_o)/w'(z_o)] \geq \lambda - 1$ and this simplifies to $\mathrm{Re}[z_o^2 w''(z_o)/z_o w'(z_o)] = \mathrm{Re}[z_o^2 w''(z_o)/Ke^{i\theta}] \geq \lambda - 1$, or

$$\mathrm{Re}\,\frac{z_o^2 w''(z_o)}{e^{i\theta}} \geq K(\lambda - 1)$$

Therefore at the point $z = z_o$, by (iii) we obtain

$$|h(w(z_o),\ z_o w'(z_o),\ z_o^2 w''(z_o))| \geq J.$$

This contradicts (5) and hence we have $|w(z)| < J$ for $z \in U$.

REMARKS. (1) Condition (5) is not a vacuous concept as $w(z) = a + w_n z^n$ will satisfy this condition for small $|w_n|$.

(2) In the case $a = 0$ and $n = 1$ we have $\lambda(0, 1, J) = 1$ and (iii) simplifies to

(iii') $|h(Je^{i\theta}, Ke^{i\theta}, L)| \geq J$ when $(Je^{i\theta}, Ke^{i\theta}, L) \in D$, $K \geq J$ and $\mathrm{Re}[Le^{-i\theta}] \geq 0$,

a condition much easier to check.

EXAMPLES. (a) Let $h_1(r, s, t) = r + s + t$ with $D = C^3$. Conditions (i) and (ii) are satisfied and we need to show that $|Je^{i\theta} + Ke^{i\theta} + L| \geq J$ or $|J + K + Le^{-i\theta}| \geq J$ when $K \geq J\lambda$ and $\mathrm{Re}[Le^{-i\theta}] \geq J\lambda(\lambda - 1)$. But this follows immediately since

$$K + \mathrm{Re}[Le^{-i\theta}] \geq J\lambda + J\lambda(\lambda - 1) = J\lambda^2 > 0.$$

Hence if $w(z)$ is regular in U, $w(0) = a$, $|a| < J$, and

$$|w(z) + zw'(z) + z^2 w''(z)| < J \quad \text{for } z \in U$$

then $|w(z)| < J$ for $z \in U$.

(b) Let $h_2(r, s, t) = rs(t + r)$, $J \geq 1$ and $a = 0$. Conditions (i) and (ii) are satisfied and we only need to check (iii);

$$|h_2(Je^{i\theta}, Ke^{i\theta}, L)| = JK|Le^{-i\theta} + J| \geq J,$$

when $K \geq J.n$ and $\mathrm{Re}[Le^{-i\theta}] \geq K(n - 1)$. But this follows immediately since $K|Le^{-i\theta} + J| \geq Jn(K(n - 1) + J) \geq Jn(Jn(n - 1) + J) \geq nJ^2(n(n - 1) + 1) \geq J^2 \geq 1$ for $n \geq 1$. Hence if $w(z) = w_n z^n + \ldots$ is regular in U with $w(z) \not\equiv 0$, $n \geq 1$ and

$$|zw(z)w'(z)(z^2w''(z) + w(z))| < J \quad \text{when } z \in U$$

then $|w(z)| < J$ for $z \in U$. This example can be generalized to $h(r, s, t) = r^i s^j (t + r)$ where i and j are positive integers.

In these two examples the results were not dependent on the value n; they held for n = 1, 2, ... This is not always the case as will be seen in Theorem 11 in section 5.

Theorem 2 can be used to show that certain second order differential equations have bounded solutions. For simplicity we will take n = 1. The proof of the following theorem follows immediately from Theorem 2.

THEOREM 3. Let h satisfy the conditions of Theorem 2 with n = 1, and let b(z) be a regular function satisfying $|b(z)| < J$. If the differential equation

$$h(w(z), zw'(z), z^2w''(z)) = b(z) \qquad (w(0) = a)$$

has a solution w(z) regular in U then $|w(z)| < J$.

If we apply this theorem to h_1 we obtain the Euler equation

$$w(z) + zw'(z) + z^2w''(z) = b(z).$$

And if $|b(z)| < J$ then we must also have $|w(z)| < J$. This theorem allows us to obtain bounds on solutions of nonlinear differential equations such as would be obtained from h_2:

$$z^3w''(z)w'(z)w(z) + zw'(z)(w(z))^2 = b(z) \qquad (w(0) = 0)$$

If $|b(z)| < J$ $(J \geq 1)$ and if this equation has a regular solution then $|w(z)| < J$.

3. Functions with Positive Real Part

THEOREM 4. Let $p(z) = a + p_n z^n + p_{n+1} z^{n+1} + \dots$ be regular in U with $p(z) \not\equiv 0$ and $n \geq 1$. If $z_o = r_o e^{i\theta_o}$ $(0 < r_o < 1)$ and $\operatorname{Re} p(z_o) = \min_{|z| \leq r_o} \operatorname{Re} p(z)$ then

$$\text{(i)}\ z_o p'(z_o) \leq - \frac{n|a - p(z_o)|^2}{2 \operatorname{Re}(a - p(z_o))} \leq - \frac{n}{2} \operatorname{Re}(a + p(z_o)),$$

$$\text{(ii)}\ \operatorname{Re} \frac{z_o p''(z_o)}{p'(z_o)} + 1 \geq 0, \text{ and}$$

$$\text{(iii)}\ \operatorname{Re} z_o^2 p''(z_o) + z_o p'(z_o) \leq 0.$$

PROOF. If we let $k = \operatorname{Re} p(z_o)$ and $\Omega = \{ w \mid \operatorname{Re} w \geq k \}$ then

$$\varphi(\zeta) = \frac{a - (2k - \bar{a})\zeta}{1 - \zeta} \tag{6}$$

is a conformal mapping of U onto Ω with $\varphi(0) = a$. Setting $\zeta_o = \varphi^{-1}(p(z_o))$, from (6) we obtain

$$\zeta_o = \frac{p(z_o) - a}{p(z_o) - (2k - \bar{a})},$$

$$\zeta_o \varphi'(\zeta_o) = \frac{-|a - p(z_o)|^2}{2 \operatorname{Re}(a - p(z_o))}, \text{ and} \tag{7}$$

$$\operatorname{Re} \frac{\zeta_o \varphi''(\zeta_o)}{\varphi'(\zeta_o)} + 1 = 0. \tag{8}$$

We now use Lemma B to complete the proof of this theorem. By (a) and (b) of the lemma, and (7) we see that $z_o p'(z_o)$ must be a negative real number and that (i) is satisfied. By applying (8) to (c) we obtain (ii). We obtain (iii) by multiplying (ii) by the negative number $z_o p'(z_o)$.

In the special case when $p(z) = 1 + p_n z^n + p_{n+1} z^{n+1} + \dots$ and $\operatorname{Re} p(z_o) = 0$ then (i) is replaced by the simpler condition

$$\text{(i')}\ z_o p'(z_o) \leq - \frac{n[1 + (\operatorname{Im} p(z_o))^2]}{2} \leq - \frac{n}{2}.$$

We will use this theorem to generate subclasses of functions with positive real part and also to show that certain second order complex differential equations have solutions with positive real part. We need to first describe the class of generating functions.

DEFINITION 1. Let $r = r_1 + r_2 i$, $s = s_1 + s_2 i$ and $t = t_1 + t_2 i$. Let n be a positive integer, a a complex number satisfying Re $a > 0$ and $\Psi_n(a)$ the set of functions $\psi(r, s, t) : C^3 \rightarrow C$ satisfying:

(a) $\psi(r, s, t)$ is continuous in a domain D of C^3,

(b) $(a, 0, 0) \in D$ and Re $\psi(a, 0, 0) > 0$,

(c) Re $\psi(r_2 i, s_1, t_1 + t_2 i) \leq 0$ when

$(r_2 i, s_1, t_1 + t_2 i) \in D$,

$$s_1 \leq -\frac{n|a - r_2 i|^2}{2 \text{ Re } a} \text{ and } s_1 + t_1 \leq 0.$$

We will let $\Psi_n \equiv \Psi_n(1)$.

REMARKS. (1) In the case a = 1 condition (c) simplifies to

(c') Re $\psi(r_2 i, s_1, t_1 + t_2 i) \leq 0$ when

$(r_2 i, s_1, t_1 + t_2 i) \in D$,

$$s_1 \leq -\frac{n}{2}(1 + r_2^2) \text{ and } s_1 + t_1 \leq 0.$$

For most of our applications we will have a = 1 and will need to check (c'). This condition will be satisfied in the shaded area indicated below.

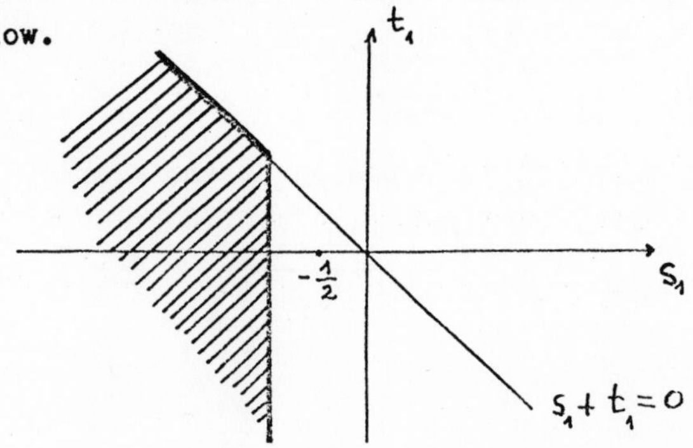

However if Re $\psi(r_2 i, s_1, t_1 + t_2 i) \leq 0$ when $s_1 \leq -1/2$ and $s_1 + t_1 \leq 0$ or when $s_1 \leq 0$, then condition (c) will still hold. These latter conditions are algebraically easier to work with although some generality is lost.

(2) If $\psi(r, s, t) = \lambda(r, s)$, where Re $\lambda(1, 0) > 0$ and Re $\lambda(r_2 i, s_1) \leq 0$ when $s_1 \leq -n(1 + r_2^2)/2$, then (c') will be satisfied.

(3) If $\psi(r, s, t) = \varrho(s, t)$, where Re $\varrho(0, 0) > 0$ and Re $\varrho(s_1, t_1 + t_2 i) \leq 0$ when $s_1 \leq -n/2$ and $s_1 + t_1 \leq 0$, then (c') will be satisfied.

EXAMPLES. It is easy to check that each of the following functions are in Ψ_n for any n = 1, 2, ...

$$\psi_1(r, s, t) = r + 2s + t + (1 - r^2)/2$$

$$\psi_2(r, s, t) = \lambda(r, s) = r + s$$

$$\psi_3(r, s, t) = \varrho(s, t) = 2s + t + \tfrac{1}{2}$$

$$\psi_4(r, s, t) = r + s + t$$

$$\psi_5(r, s, t) = re^s + s + t$$

Note that ψ_2 will satisfy the weaker condition mentioned in Remark (1) whereas ψ_1 requires the stronger original condition. The function $\psi_6(r, s, t) = r + s + (1 - r^2)$ is an example of a function that is in Ψ_2 but is not in Ψ_1.

THEOREM 5. Let $\psi \in \Psi_n(a)$ with corresponding domain D and let $p(z) = a + p_n z^n + p_{n+1} z^{n+1} + \dots$ be regular in U with $p(z) \not\equiv 0$ and $n \geq 1$. If $(p(z), zp'(z), z^2p''(z)) \in D$ when $z \in U$ and

(9) Re $\psi(p(z), zp'(z), z^2p''(z)) > 0$ when $z \in U$,

then Re $p(z) > 0$ for all $z \in U$.

PROOF. Suppose there exists a point $z_0 = r_0 e^{i\theta_0} \in U$, $0 < r_0 < 1$ such that $0 = \text{Re } p(z_0) = \min_{|z| \leq r_0} \text{Re } p(z)$. Applying

Theorem 4 we obtain

$$z_o p'(z_o) \leq - \frac{n|a - p(z_o)|^2}{2 \operatorname{Re}(a - p(z_o))} \quad \text{and}$$

$$\operatorname{Re} z_o^2 p''(z_o) + z_o p'(z_o) \leq 0.$$

Using these results and part (c) of the definition of $\Psi_n(a)$ we must have

$$\operatorname{Re} \psi\left(p(z_o), z_o p'(z_o), z^2 p''(z_o)\right) \leq 0.$$

But this contradicts (9) and so we must have Re $p(z) > 0$ for all $z \in U$.

Note that condition (9) is not a vacuous concept; $p(z) = a + p_n z^n$ will satisfy (9) for small $|p_n|$.

Applying the theorem to ψ_3, ψ_4 and ψ_5 we obtain respectively:

$$\operatorname{Re}\left[2 zp'(z) + z^2 p''(z) + \tfrac{1}{2}\right] > 0 \Rightarrow \operatorname{Re} p(z) > 0,$$

$$\operatorname{Re}\left[p(z) + zp'(z) + z^2 p''(z)\right] > 0 \Rightarrow \operatorname{Re} p(z) > 0,$$

and

$$\operatorname{Re}\left[p(z)e^{zp'(z)} + zp'(z) + z^2 p''(z)\right] > 0 \Rightarrow \operatorname{Re} p(z) > 0.$$

We see that different $\psi \in \Psi_n(a)$ generate, in a sense, functions with positive real part.

This theorem also has an interpretation in terms of differential equations as given in the following theorem. The proof will not be presented as it follows immediately from Theorem 5. For simplicity we take $n = 1$.

<u>THEOREM 6</u>. Let $\psi \in \Psi_1$, and let $q(z)$ be a regular function satisfying Re $q(z) > 0$. If the differential equation

$$\psi(p(z), zp'(z), z^2 p''(z)) = q(z) \quad (p(0) = 1),$$

has a solution $p(z)$ regular in U then Re $p(z) > 0$.

As an example, if we apply this theorem to ψ_4 we obtain the Euler equation

$$p(z) + zp'(z) + z^2p''(z) = q(z) .$$

Hence if $q(z)$ is regular and satisfies $\operatorname{Re} q(z) > 0$ then the regular solution $p(z)$ must satisfy $\operatorname{Re} p(z) > 0$.

4. Harmonic Functions

In this section we use some results of the previous section and the fact that a harmonic function can be represented as the real part of a regular function to obtain some properties of harmonic functions.

THEOREM 7. Let $u(z) = u(x, y)$ be harmonic in U with $u(0) = 1$. If there exists $z_o = r_o e^{i\theta_o} \in U$ such that $u(z_o) = 0$ and $u(z) > 0$ for $|z| < r_o$ then at the point z_o

(a) $x_o u_x + y_o u_y \leq -\frac{1}{2}[1 + v^2(z_o)] \leq -\frac{1}{2}$,

(b) $y_o u_x - x_o u_y = 0$,

(c) $u_x[x_o u_{xx} - y_o v_{xx}] + v_x[y_o u_{xx} + x_o v_{xx}] \geq -(u_x^2 + v_x^2)$,

(d) $(x_o^2 - y_o^2)u_{xx} - 2x_o y_o v_{xx} + x_o u_x - y_o v_x \leq 0$,

where $v(z)$ is the harmonic conjugate of $u(z)$ satisfying $v(0) = 0$.

PROOF. If we let $p(z) = u(z) + iv(z)$ then $p(z)$ will be regular in U, $p(0) = 1$, $\operatorname{Re} p(z_o) = 0$ and $\operatorname{Re} p(z) > 0$ for $|z| < r_o$. By Theorem 4 part (i) $z_o p'(z_o)$ must be real and must satisfy $z_o p'(z_o) \leq -[1 + v^2(z_o)]/2$, that is

$$(x_o + iy_o)(u_x(z_o) + iv_x(z_o)) \leq -[1 + v^2(z_o)]/2.$$

By comparing real and imaginary parts and using $u_y = -v_x$ we obtain (a) and (b). Conditions (c) and (d) follow immediately from parts (ii) and (iii) of Theorem 4.

REMARKS. (1) Conditions (a) and (b) can be written in terms of directional derivatives as

$$\nabla u(z_o) \cdot [x_o, y_o] \leq -\frac{1}{2},$$

$$\nabla u(z_o) \cdot [y_o, -x_o] = 0.$$

From the second result we see that the gradient vector must be parallel to the vector $[x_o, y_o]$ and from the first result we see that it must be in the opposite direction. The bound of -1/2 seems geometrically surprising.

(2) If $x_o \neq 0$ or $y_o \neq 0$ then combining (a) and (b) we obtain respectively

$$\frac{|z_o|^2}{x_o} u_x(z_o) \leq -\frac{1}{2}[1 + v^2(z_o)] \leq -\frac{1}{2} ,$$

$$\frac{|z_o|^2}{y_o} u_y(z_o) \leq -\frac{1}{2}[1 + v^2(z_o)] \leq -\frac{1}{2} .$$

We now use the theorem to generate some positive harm nic functions.

THEOREM 8. Let $g(a, b, c, d, e)$ be a real continuous function defined in a domain D of R^5 and suppose

(i) $(1, 0, 0, 0, 0) \in D$ and $g(1, 0, 0, 0, 0) > 0$,

(ii) $g(0, b, c, d, d) \leq 0$ when $(0, b, c, d, d) \in D$ and $b + c \leq -\frac{1}{2}$.

Let $u(z) = u(x, y)$ be harmonic in U with $u(0) = 1$ and $(u, xu_x, yu_y, xu_y, yu_x) \in D$ when $z \in U$. If

(10) $\quad g(u, xu_x, yu_y, xu_y, yu_x) > 0 \quad$ for $z \in U$

then $u(z) > 0$ for $z \in U$.

PROOF. Since $u(0) = 1$, suppose there exists $z_o \in U$ such that $u(z_o) = 0$ and $u(z) > 0$ for $|z| < |z_o|$. Then by Theorem 7 parts (a) and (b), and from (ii) we would have

$$g(u(z_o), x_o u_x(z_o), y_o u_y(z_o), x_o u_y(z_o), y_o u_x(z_o)) \leq 0.$$

This contradicts (10) and hence we must have $u(z) > 0$.

Note that for any g satisfying (i) and (ii) there are functions $u(z)$ satisfying (10). For example $u = 1 + p_1 x +$

$+ p_2(x^2 - y^2)$ will satisfy (10) for $|p_1|$ and $|p_2|$ sufficiently small.

It is easy to check that the following functions satisfy conditions (i) and (ii) of the theorem:

$$g_1(a, b, c, d, e) = a + b + c + d - e,$$
$$g_2(a, b, c, d, e) = \frac{1}{2} + b + c,$$
$$g_3(a, b, c, d, e) = a^2 + d - e,$$
$$g_4(a, b, c, d, e) = a^2 + b + c + \frac{1}{2}.$$

Hence if $u(z)$ is harmonic in U with $u(0) = 1$ then by Theorem 8 we have respectively:

$$u + (x - y)u_x + (x + y)u_y > 0 \Rightarrow u(z) > 0,$$
$$xu_x + yu_y > -\frac{1}{2} \Rightarrow u(z) > 0,$$
$$u^2 + xu_y - yu_x > 0 \Rightarrow u(z) > 0,$$
$$u^2 + xu_x + yu_y > -\frac{1}{2} \Rightarrow u(z) > 0.$$

5. Applications in Univalent Function Theory

In this section we will demonstrate the usefulness of Theorem 5 by providing some very simple proofs for some well-known classical results dealing with convex and starlike functions. We will then use the theorem to obtain some new results relating the Schwarzian derivative to starlike and convex functions.

Suppose that $f(z) = z + a_2z^2 + \ldots$ is regular in U. We denote by S* the class of functions for which $f(z)$ is univalent and $f(U)$ is starlike with respect to the origin. The condition $\operatorname{Re}[zf'(z)/f(z)] > 0$, $z \in U$, is necessary and sufficient for $f \in S^*$. We denote by C the class of functions for which $f(z)$ is univalent and $f(U)$ is convex. The condition $\operatorname{Re}[zf''(z)/f'(z)+1] > 0$ is necessary and sufficient for $f \in C$.

It is clear that $f \in C$ implies $\mathrm{Re}[zf'(z)/f(z)] > 0$. A.Marx [5] and E.Strohhäcker [9] obtained the stronger conclusion $\mathrm{Re}[zf'(z)/f(z)] > 1/2$. We will prove their result by a simple application of Theorem 5.

THEOREM 9. Let $f(z) = z + a_2z^2 + \dots$ be regular in U.

(i) $\mathrm{Re}[zf''(z)/f'(z) + 1] > 0 \Rightarrow \mathrm{Re}[zf'(z)/f(z)] > 1/2$

(ii) $\mathrm{Re}[zf'(z)/f(z)] > 1/2 \Rightarrow \mathrm{Re}[f(z)/z] > 1/2$

and these bounds are the best possible.

PROOF. (i) Let $p(z) = 2\, zf'(z)/f(z) - 1$. Then $p(z)$ is regular in U, $p(0) = 1$, $zf'(z)/f(z) = (p(z) + 1)/2$ and

$$\frac{zf''(z)}{f(z)} + 1 = \frac{p(z) + 1}{2} + \frac{zp'(z)}{p(z) + 1} = \psi(p(z), zp'(z)), \tag{11}$$

where $\psi(r, s) = (r + 1)/2 + s/(r + 1)$. If we take $n = 1$ and $D = (C - \{-1\}) \times C \times C$ in Definition 1, then $\psi \in \Psi_1$. From (i) and (11) we obtain $\mathrm{Re}\ \psi(p(z), zp'(z)) > 0$ for $z \in U$, and hence by Theorem 5 we must have $\mathrm{Re}\ p(z) > 0$, for $z \in U$. This implies that $\mathrm{Re}\ zf'(z)/f(z) > 1/2$.

(ii) Let $p(z) = 2\, f(z)/z - 1$. Then $p(z)$ is regular in U, $p(0) = 1$ and

$$\frac{zf'(z)}{f(z)} - \frac{1}{2} = \frac{1}{2} + \frac{zp'(z)}{p(z) + 1} = \psi(p(z), zp'(z)), \tag{12}$$

where $\psi(r, s) = 1/2 + s/(r + 1)$. If we take $D = (C - \{-1\}) \times C \times C$, then $\psi \in \Psi_1$. From (ii) and (12) we obtain $\mathrm{Re}\,\psi(p(z), zp'(z)) > 0$, when $z \in U$. Hence by Theorem 5 we obtain $\mathrm{Re}\ p(z) > 0$, for $z \in U$, which proves (ii).

The convex function $f(z) = z/(1 + z)$ shows that the bounds are the best possible.

We now prove a theorem which in its original form was proved by K.Sakaguchi [8]. R.Libera [3] and T.MacGregor [4] have extended it to its present form and are among the many authors who have applied it very successfully. We prove it directly by using Theorem 5.

<u>THEOREM 10</u>. Let $M(z)$ and $N(z)$ be regular in U with $M(0) = N(0) = 0$, and let γ be real. If $N(z)$ maps U onto a (possibly many-sheeted) region which is starlike with respect to the origin then

$$\text{(i) } \operatorname{Re}\frac{M'(z)}{N'(z)} > \gamma,\ z \in U \Rightarrow \operatorname{Re}\frac{M(z)}{N(z)} > \gamma,\ z \in U,$$

and

$$\text{(ii) } \operatorname{Re}\frac{M'(z)}{N'(z)} < \gamma,\ z \in U \Rightarrow \operatorname{Re}\frac{M(z)}{N(z)} < \gamma,\ z \in U.$$

<u>PROOF</u>. (i) If we let $p(z) = M(z)/N(z) - \gamma$, then $\operatorname{Re} p(0) = \operatorname{Re}[M(0)/N(0)] - \gamma = \operatorname{Re}[M'(0)/N'(0)] - \gamma > 0$, and $p(z)$ is regular in U. Setting $1/\alpha(z) = zN'(z)/N(z)$, we have $\operatorname{Re}(1/\alpha(z)) > 0$, $\operatorname{Re}(\alpha(z)) > 0$ and

$$(13) \qquad \frac{M'(z)}{N'(z)} - \gamma = p(z) + \frac{N(z)}{N'(z)}\, p'(z) = p(z) + \alpha\, zp'(z) =$$
$$= \psi(p(z),\ zp'(z)),$$

where $\psi(r, s) = r + \alpha s$. Since $\operatorname{Re}\alpha > 0$ we have $\operatorname{Re}\psi(p(0), 0) = \operatorname{Re} p(0) > 0$, and $\operatorname{Re}\psi(r_2 i, s_1) \leq 0$ when $s_1 \leq 0$. Hence $\psi \in \Psi_1$, and since $\operatorname{Re} M'(z)/N'(z) > \gamma$, from (13) we obtain $\operatorname{Re}\psi(p(z), zp'(z)) > 0$. But by Theorem 5 this implies that $\operatorname{Re} p(z) > 0$ for $z \in U$, that is $\operatorname{Re} M(z)/N(z) > \gamma$ for $z \in U$.

Condition (ii) can be obtained from (i) by replacing $M(z)$ by $-M(z)$.

The next theorem is a result proved by G.M.Golusin [1, Theorem 5] using a very involved series of inequalities. We will prove it very simply by using Theorem 5. This result has many applications in proving distortion properties and coefficient inequalities (see [1]).

<u>THEOREM 11</u>. If $f(z) = z(1 + a_n z^n + a_{n+1} z^{n+1} + \ldots) \in S^*$, with $n \geq 1$, then

$$\mathrm{Re}\left(\left[\frac{f(z)}{z}\right]^{\frac{n}{2}}\right) > \frac{1}{2}$$

<u>PROOF</u>. Let $p(z) = 2(f(z)/z)^{n/2} - 1$. Then

$$p(z) = 2(1 + a_n z^n + a_{n+1} z^{n+1} + \ldots)^{n/2} - 1 = 1 + na_n z^n + \ldots$$

and $p(z)$ is regular in U. A simple calculation yields

$$(14) \qquad \frac{n}{2}\frac{zf'(z)}{f(z)} = \frac{zp'(z)}{p(z)+1} + \frac{n}{2} = \psi(p(z), zp'(z)),$$

where $\psi(r, s) = s/(r+1) + n/2$. Since $\psi(1, 0) = n/2 > 0$, and

$$(15) \qquad \mathrm{Re}\,\psi(r_2 i, s_1) = \frac{s_1}{1 + r_2^2} + \frac{n}{2} \leq \frac{-n(1 + r_2^2)/2}{1 + r_2^2} + \frac{n}{2} \leq 0,$$

when $s_1 \leq -n(1 + r_2^2)/2$, we have $\psi \in \Psi_n$. Since $f(z) \in S^*$, from (14) we have $\mathrm{Re}\,\psi(p(z), zp'(z)) > 0$. Hence by Theorem 5 we must have $\mathrm{Re}\, p(z) > 0$, and this proves the theorem.

Note that (15) in the proof of the theorem requires the stronger form of Definition 1 and Theorem 5 involving Ψ_n instead of Ψ_1.

In what follows we will let $\{f, z\}$ denote the Schwarzian derivative $(f''/f')' - (f''/f')^2/2$. There are several conditions relating the Schwarzian derivative of $f(z)$ to the univalency of $f(z)$ (see [7]). The following theorem relates the Schwarzian derivative of f to the starlikeness (and univalency) of f.

<u>THEOREM 12</u>. Let $u = u_1 + u_2 i$, $v = v_1 + v_2 i$, $w = w_1 + w_2 i$ and let $\theta(u, v, w)$ be a complex-valued function satisfying:

(i) $\theta(u, v, w)$ is continuous in a domain D of $[C - \{0\}] \times C \times C$,

(ii) $(1, 1, 0) \in D$ and $\mathrm{Re}\,\theta(1, 1, 0) > 0$,

(iii) $\mathrm{Re}\,\theta(u_2 i, v_2 i, w_1 + w_2 i) \leq 0$ when $(u_2 i, v_2 i, w_1 + w_2 i) \in D$, $u_2 v_2 \geq (1 + 3u_2^2)/2$ and $u_2 w_2 \geq 0$.

Let $f(z) = z + a_2 z^2 + \dots$ be a function regular in U with $f(z)f'(z)/z \neq 0$ and $(zf'/f,\ zf''/f + 1,\ z^2\{f, z\}) \in D$ when $z \in U$. If

(16) $\quad \operatorname{Re} \Theta(zf'/f,\ zf''/f' + 1,\ z^2\{f, z\}) > 0$ when $z \in U$,

then $\operatorname{Re} zf'/f > 0$, for $z \in U$.

PROOF. If we let $p(z) = zf'/f$ then $p(z)$ is regular in U, $p(0) = 1$ and a simple calculation yields

$$zf''/f' + 1 = p + zp'/p,$$

and $\quad z^2\{f, z\} = (zp' + z^2p'')/p - (zp'/p)^2/2 + (1 - p^2)/2.$

Therefore

(17) $\quad \Theta(zf'/f,\ zf''/f' + 1,\ z^2\{f, z\})$

$$= \Theta(p,\ p + zp'/p,\ (zp' + z^2p'')/p - (zp'/p)^2/2 + (1-p^2)/2)$$

$$= \psi(p,\ zp',\ z^2p''),$$

where

(18) $\quad \psi(r, s, t) = \Theta(r,\ r + s/r,\ (s + t)/r - (s/r)^2/2 + (1-r^2)/2)$

We will now show that ψ satisfies Definition 1. From (i), (ii) and (18) we obtain

(a) $\psi(r, s, t)$ is continuous in a domain

$$D_1 = \{(u, u(v-u), u[w+(v-u)^2/2-(1-u^2)/2-(v-u)]\ |\ (u,v,w) \in D\},$$

and

(b) $(1, 0, 0) \in D_1$ and $\operatorname{Re} \psi(1, 0, 0) = \operatorname{Re} \Theta(1, 1, 0) > 0$.

If $u_2 i = r_2 i$, $v_2 i = (r_2 - s_1/r_2)i$ and

$$w_1 + w_2 i = t_2/r_2 + 3(s_1/r_2)^2/2 + (1 + r_2)^2/2 - (s_1+t_1)i/r_2,$$

then if

(19) $\quad s_1 \leq -(1 + r_2^2)/2$ and $s_1 + t_1 \leq 0$,

then

(20) $\quad u_2 v_2 \geq (1 + 3u_2^2)/2$ and $u_2 w_2 \geq 0$.

From (18), (19), (20) and (iii) we obtain

(c) $\operatorname{Re} \psi(r_2 i,\ s_1,\ t_1 + t_2 i) = \operatorname{Re} \Theta(u_2 i,\ v_2 i,\ w_1 + w_2 i) \leq 0$

wheh $\quad s_1 \leq -(1 + r_2^2)/2$ and $s_1 + t_1 \leq 0$.

Hence from (a), (b) and (c) we see that ψ satisfies Definition 1 and $\psi \in \Psi_1$. From (16) and (17) we obtain $\operatorname{Re} \psi(p, zp', z^2p'') > 0$ for $z \in U$. Therefore by Theorem 5 $\operatorname{Re} p(z) > 0$ for $z \in U$, that is $\operatorname{Re} zf'/f > 0$ and $f \in S^*$.

The following functions satisfy conditions (i), (ii) and (iii)

$$\theta_1(u, v, w) = \alpha u + \beta v + uw, \text{ with } \alpha, \beta \in R, \alpha + \beta > 0,$$
$$\theta_2(u, v, w) = u(v + w).$$

Applying the theorem to θ_1 we obtain

$$\operatorname{Re}\left[\alpha \frac{zf'}{f} + \beta\left(\frac{zf''}{f'} + 1\right) + \frac{zf'}{f} \cdot z^2\{f, z\}\right] > 0 \Rightarrow \operatorname{Re} \frac{zf'}{f} > 0.$$

As a special case, taking $\alpha = 1$ and $\beta = 0$ we obtain

$$\operatorname{Re}\left[\frac{zf'}{f}(1 + z^2\{f, z\})\right] > 0 \Rightarrow \operatorname{Re} \frac{zf'}{f} > 0.$$

Applying the theorem to θ_2 we obtain

$$\operatorname{Re}\left[\frac{zf'}{f}\left(\frac{zf''}{f'} + 1 + z^2\{f, z\}\right)\right] > 0 \Rightarrow \operatorname{Re} \frac{zf'}{f} > 0.$$

Our final result relates the Schwarzian derivative of a function to the convexity (and univalency) of the function. The proof of this theorem is similar to the proof of the previous theorem and will be omitted.

<u>THEOREM 13</u>. Let $u = u_1 + u_2 i$, $v = v_1 + v_2 i$ and let $\rho(u, v)$ be a complex-valued function satisfying:

(i) $\rho(u, v)$ is continuous in a domain $D \subset C^2$,

(ii) $(1, 0) \in D$ and $\operatorname{Re} \rho(1, 0) > 0$,

(iii) $\operatorname{Re} \rho(u_2 i, v_1) \leq 0$ when $v_1 \leq 0$.

Let $f(z) = z + a_2 z^2 + \ldots$ be a function regular in U with $f'(z) \neq 0$ and $(zf''/f' + 1, z^2\{f, z\}) \in D$ when $z \in U$. If

$$\operatorname{Re} \rho(zf''/f' + 1, z^2\{f, z\}) > 0, \text{ for } z \in U$$

then $\operatorname{Re}(zf''/f' + 1) > 0$ for $z \in U$.

The following examples satisfy conditions (i), (ii) and (iii)

$$\rho_1(u, v) = u + \alpha v, \ \mathrm{Re}\,\alpha \geq 0,$$

$$\rho_2(u, v) = u^2 + v,$$

$$\rho_3(u, v) = ue^v.$$

Applying the theorem to these examples we obtain:

$$\mathrm{Re}\left[\left(\frac{zf''}{f'} + 1\right) + \alpha z^2\{f, z\}\right] > 0 \Rightarrow \mathrm{Re}\,\frac{zf''}{f'} + 1 > 0,$$

$$\mathrm{Re}\left[\left(\frac{zf''}{f'} + 1\right)^2 + z^2\{f, z\}\right] > 0 \Rightarrow \mathrm{Re}\,\frac{zf''}{f'} + 1 > 0, \text{ and}$$

$$\mathrm{Re}\left(\frac{zf''}{f'} + 1\right)e^{z^2\{f,z\}} > 0 \Rightarrow \mathrm{Re}\,\frac{zf''}{f'} + 1 > 0.$$

BIBLIOGRAPHY

1. G.M.Golusin, Some estimates for coefficients of univalent functions, Mat. Sb. 3 (45), 2 (1938), 321 - 330.

2. I.S.Jack, Functions starlike and convex of order α, J. London Math. Soc. 3 (1971), 469 - 474.

3. R.J.Libera, Some classes of regular univalent functions, Proc. Amer. Math. Soc., 16 (1965), 755 - 758.

4. T.H.MacGregor, A subordination for convex functions of order α, J. London Math. Soc., (2), 9 (1975), 530 - 536.

5. A.Marx, Untersuchungen über schlichte Abbildungen, Math. Ann., 107 (1932/33), 40 - 67.

6. S.S.Miller, A class of differential inequalities implying boundedness, Ill. J. of Math. (to appear).

7. Ch.Pommerenke, Univalent Functions, Vandenhoeck and Ruprecht, Gottingen, 1975.

8. K.Sakaguchi, On a certain univalent mapping, J. Math. Soc. Japan II (1959), 72 - 75.

9. T.J.Strohhäcker, Beiträge zur Theorie der schlichten Funktionen, Math. Z., 37 (1933), 356 - 380.

Department of Mathematics, State University of New York, Brockport, New York 14420, U.S.A.

Department of Mathematics, Babeş-Bolyai University, Cluj-Napoca, Romania.

AN EXTREMAL PROBLEM FOR THE TRANSFINITE DIAMETER OF A CONTINUUM

by

PETRU T. MOCANU and DUMITRU RIPEANU

In this paper we solve an extremal problem connected with the transfinite diameter of a continuum by using Schiffer's variational method [2] and also the same simple geometric arguments as described in the paper of Reich and Schiffer [1]. As the matter of fact, our problem is quite similar to those solved in [1].

Let

$$(1) \qquad \Phi(c_1,c_2,c_3) = |c_1 - c_2| + |c_2 - c_3| + |c_3 - c_1| ,$$

where c_1,c_2,c_3 are complex numbers. It is obvious that the function (1), which represents the perimeter of the triangle (c_1,c_2,c_3), is invariant under translations and rotations of the plane.

Let E be a continuum in the complex plane, and let c_1,c_2,c_3 be three arbitrary points belonging to E. Our problem is to find

$$(2) \qquad \sup_{c_1,c_2,c_3,E} \frac{\Phi(c_1,c_2,c_3)}{d(E)}$$

where $d(E)$ is the transfinite diameter of E.

The result is the following

THEOREM . <u>If</u> E <u>is a continuum in the plane and</u> c_1, c_2, c_3 <u>belong to</u> E , <u>then</u>

(3) $$|c_1 - c_2| + |c_2 - c_3| + |c_3 - c_1| \leq 3^{3/2} 4^{1/3}\, d(E) .$$

<u>This inequality is sharp, equality being achieved if and only if</u> E <u>is the union of three segments of equal length making angles of</u> $2\pi/3$ <u>with each other, having a common initial point, and</u> c_1, c_2, c_3 <u>as endpoints.</u>

PROOF. It is well-known that if c_1, c_2 belong to E , then $|c_1 - c_2| \leq 4d(E)$, [3]. Hence $\Phi(c_1,c_2,c_3)/\, d(E) \leq 12$, which shows that (2) exist and is assumed. Let E, c_1, c_2, c_3 be extremal for (2), and let D be the complementary domain of E which contains the point at infinity. Consider the conformal mapping

(4) $$w = f(z) = d(E)\left[z + a_0 + \frac{a_1}{z} + \dots \right]$$

of $1 < |z| < \infty$ onto D .

For $w_0 \in D$, we consider the variation

(5) $$w^* = w + \frac{\lambda}{w - w_0}$$

where $|\lambda|$ is sufficently small. Denote by c_1^* , c_2^* , c_3^* , E^* the images of c_1, c_2, c_3, E by (5).

If we set $\Phi = \Phi(c_1,c_2,c_3)$ and $\Phi^* = \Phi(c_1^*,c_2^*,c_3^*)$, then we have the following variational formulas

(6) $$\log \Phi^* = \log \Phi - \mathrm{Re}\left\{ \frac{\lambda}{\Phi} A(c_1,c_2,c_3;w_0) \right\} + o(\lambda) ,$$

where

$$A(c_1,c_2,c_3;w_0) = \frac{|c_1-c_2|}{(c_1-w_0)(c_2-w_0)} + \frac{|c_2-c_3|}{(c_2-w_0)(c_3-w_0)} + \frac{|c_3-c_1|}{(c_3-w_0)(c_1-w_0)}$$

and

(7) $$\log d(E^*) = \log d(E) - \operatorname{Re}\left\{\frac{\lambda}{z_0^2 f'(z_0)^2}\right\} + o(\lambda) ,$$

with $w_0 = f(z_0)$, [2].

Since $\Phi^*/d(E\ ) \leq \Phi/d(E)$, we have

$$\log \Phi^* - \log d(E^*) \leq \log \Phi - \log d(E)$$

and by using (6) and (7) we obtain

$$\operatorname{Re}\left\{\frac{\lambda}{z_0^2 f'(z_0)^2} - \frac{\lambda}{\Phi} A(c_1, c_2, c_3; w_0)\right\} + o(\lambda) \leq 0$$

for all small enough values of $|\lambda|$. From this we conclude that if E, c_1, c_2, c_3 are extremal for (2), then the extremal function (4) satisfies the differential equation

(8) $$\frac{w - \Psi(c_1, c_2, c_3)}{(w-c_1)(w-c_2)(w-c_3)} = \frac{(dz)^2}{z^2 (dw)^2}$$

where

$$\Psi(c_1, c_2, c_3) = \frac{1}{\Phi}\left[c_1 |c_2-c_3| + c_2 |c_3-c_1| + c_3 |c_1-c_2|\right]$$

Since

$$\Psi(c_1+a, c_2+a, c_3+a) = \Psi(c_1, c_2, c_3) + a ,$$

and the extremal points are determined within an additive constant, we can suppose $\Psi(c_1, c_2, c_3) = 0$, i.e.

(9) $$c_1 |c_2-c_3| + c_2 |c_3-c_1| + c_3 |c_1-c_2| = 0 .$$

The differential equation (8) becomes

(10) $$\frac{w(dw)^2}{(w-c_1)(w-c_2)(w-c_3)} = \frac{(dz)^2}{z^2} , \quad |z| > 1 .$$

As in [1] it is easy to show that the extremum continuum E is the set of values omitted by the extremal function f , and the range D of f has no exterior points.

The extremal points c_k are distinct from each other and distinct from 0. Indeed, if $c_1=0$, then from (9) we deduce $c_2|c_3| = -c_3|c_2|$ and we have $\Phi \leq 2\max\{|c_2|,|c_3|\}$, and $\Phi/d(E) \leq 8$. If $c_1=c_2=a$, $c_3=b$, then $\Phi = 2|b-a| \leq 8d(E)$. In each case the value of $\Phi/d(E)$ is not extremal. We remark that the extremal points c_1, c_2, c_3 can not be collinear, since in this case we also have $\Phi = 2|c_1-c_2| \leq 8d(E)$, if we suppose that c_2 lies between c_1 and c_3.

Since Φ is invariant under rotations we can suppose $c_1>0$.

The extremum continuum E consists of the union of three analytic arcs γ_k ,k=1,2,3, having 0 as the only common initial point and c_1, c_2, c_3 as endpoints. The three arcs γ_k meet 0 in equally spaced angles [1]. Using the same topological argument as in [1], we conclude that there exist numbers t_k , $0< t_k < 1$, such that

$$(11) \qquad \operatorname{Im} \frac{c_{k+1} + c_{k+2}}{c_k} t_k = \operatorname{Im} \frac{c_{k+1} c_{k+2}}{c_k^2} \quad , \quad k = 1,2,3 ,$$

where we denote $c_4=c_1$, $c_5=c_2$.

From (9) we obtain

$$\operatorname{Im} \frac{c_{k+2}}{c_k} = - \frac{|c_{k+2} - c_k|}{|c_{k+1} - c_k|} \operatorname{Im} \frac{c_{k+1}}{c_k}$$

and

$$\operatorname{Re} \frac{c_{k+2}}{c_k} = - \frac{|c_{k+2} - c_{k+1}|}{|c_{k+1} - c_k|} - \frac{|c_{k+2} - c_k|}{|c_{k+1} - c_k|} \operatorname{Re} \frac{c_{k+1}}{c_k} .$$

Hence

$$(12) \qquad \operatorname{Im} \frac{c_{k+1} + c_{k+2}}{c_k} = \frac{|c_{k+1} - c_k| + |c_{k+2} - c_k|}{|c_{k+1} - c_k|} \operatorname{Im} \frac{c_{k+1}}{c_k}$$

$$(13) \qquad \operatorname{Im} \frac{c_{k+1} c_{k+2}}{c_k^2} = - \left[\frac{|c_{k+2} - c_{k+1}|}{|c_{k+1} - c_k|} + 2 \frac{|c_{k+2} - c_k|}{|c_{k+1} - c_k|} \operatorname{Re} \frac{c_{k+1}}{c_k} \right] \operatorname{Im} \frac{c_{k+1}}{c_k}$$

Using (9),(11),(12) and (13) we find that the extremal points c_1, c_2, c_3 satisfy the following conditions

$$(14) \quad c_1|c_2-c_1| + c_2|c_3-c_1| + c_3|c_1-c_2| = o$$

$$(15) \quad (|c_3-c_1| - |c_2-c_1|)\, t_1 = |c_3-c_2| + 2|c_3-c_1| \mathrm{Re}\frac{c_2}{c_1}$$

$$(16) \quad (|c_1-c_2| - |c_3-c_2|)\, t_2 = |c_1-c_3| + 2|c_1-c_2| \mathrm{Re}\frac{c_3}{c_2}$$

$$(17) \quad (|c_2-c_3| - |c_1-c_3|)\, t_3 = |c_2-c_1| + 2|c_2-c_3| \mathrm{Re}\frac{c_1}{c_3}$$

where $c_1 > 0$ and $t_k \in (0,1)$.

We shall show that this conditions imply

$$(18) \quad |c_1 - c_2| = |c_2 - c_3| = |c_3 - c_1| .$$

If we let

$$(19) \quad c_2 - c_1 = d = re^{it}, \quad c_3 - c_1 = \delta = \rho e^{i\tau}$$

condition (14) becomes

$$(2o) \quad |c_3 - c_2| = |\delta - d| = -\left[r + \rho + \frac{r\rho}{c_1}(e^{it} + e^{i\tau})\right]$$

From (15) we obtain

$$(21) \quad (r - \rho)(1-t_1) = \frac{r\rho}{c_1}\left[\cos t - \cos\tau - i(\sin t + \sin\tau)\right]$$

If $r\rho = 0$, then from (21) we deduce $r = \rho = 0$ and from (19) we get the trivial solution $c_1 = c_2 = c_3$ which is not possible. Thus $r\rho > 0$ and from (21) we obtain $\sin t + \sin\tau = 0$, which implies $\cos t = \pm \cos\tau$. Suppose $\cos\tau = -\cos t$. Then (2o) becomes $|\delta - d| = -(r + \rho)$, that is, $r = \rho = 0$ which is not possible. Therefore we have only the case $\cos t - \cos\tau = \sin t + \sin\tau = 0$ and from (21) we obtain $r = \rho$. From (19) we deduce

$$(22) \qquad c_2 = c_1 + re^{it} \quad , \quad c_3 = c_1 + re^{-it}.$$

Employing (22) together with (16) and (17), we obtain

$$(23) \qquad ax^2 + bx + c = 0 \quad , \quad a'x^2 + b'x + c' = 0 ,$$

where $x = r/c_1$ and

$$(24) \quad \begin{cases} a = -3 + 4\sin^2 t + (1 - 2|\sin t|)t_2 \\ b = 2\cos t\left[(-3 + (1 - 2|\sin t|)t_2\right] \\ c = -3 + (1 - 2|\sin t|)t_2 \\ a' = 1 + (1 - 2|\sin t|)t_3 \\ b' = 2\cos t\left[1 + 2|\sin t| + (1 - 2|\sin t|)t_3\right] \\ c' = 1 + 4|\sin t| + (1 - 2|\sin t|)t_3 . \end{cases}$$

On the other hand, employing (22) together with (14) or (15) we get

$$(25) \qquad x = -\frac{1 + |\sin t|}{\cos t} .$$

If $\sin t = 0$, then $x = r/c_1 = -1$, which is not possible. If $\sin t > 0$, then from (23), (24) and (25) we obtain

$$(1-t_2)\sin^2 t\,(1 - 2\sin t) = 0$$

and

$$(1-t_3)\sin^2 t\,(1 - 2\sin t) = 0$$

hence $\sin t = 1/2$ and from (22) we get (18). In the case $\sin t < 0$ we obtain

$$(1-t_2)\sin^2 t\,(1 + 2\sin t) = 0$$

$$(1-t_3)\sin^2 t\,(1 + 2\sin t) = 0$$

hence $\sin t = -1/2$ and from (22) we also get (18).

We remark that (18) holds if we only suppose $t_k \neq 1$.

We conclude that the extremal points c_1, c_2, c_3 must satisfy (18), that is, (11) is of the form $0 \cdot t_k = 0$, for $k=1,2,3$. As in [1] this means that the arc γ_k coincides with the segment from 0 to c_k ,for k=1,2,3. We have $c_2 = \omega c_1$, $c_3 = \omega^2 c_1$, where $\omega^3 = 1$ and the differential equation (1o) becomes

$$\frac{z^2 w}{w^3 - c_1^3}\left(\frac{dw}{dz}\right)^2 = 1 .$$

The extremal function will be

$$f(z) = d(E)\ z(1 + z^{-3})^{2/3} ,$$

where $d(E) = 4^{-1/3}c_1$. Moreover the extremal value of Φ is $\Phi = 3\cdot 3^{1/2}c_1 = 3^{3/2}4^{1/3}d(E)$. This completes the proof of our Theorem.

COROLLARY. _If the function_

$$f(z) = z + a_0 + \frac{a_1}{z} + \dots$$

is regular and univalent in $1 < |z| < \infty$ _and_ E _is the complement of its range, then the perimeter of any triangle with vertices in_ E _is less or equal to_ $3^{3/2}4^{1/3}$. _The equality holds if and only if the function_ f _is_

$$f(z) = a_0 + z(1 + e^{it}z^{-3})^{2/3}$$

In this case E _is the union of three segments of equal length_ $L = 4^{1/3}$, _making angles of_ $2\pi/3$ _with each other, having a common intial point. The vertices of the triangle are the endpoints of the three segments._

REFERENCES

1. E. REICH and M. SCHIFFER , Estimates for the transfinite diameter of a continuum , Math.Zeitshr.85(1964),91-1o6.

2. M.SCHIFFER , Hadamard's formula and variation of domain-functions ,Amer.J.Math.68(1946) ,417-448.

3. G.SZEGÖ , Jahresbericht Deutsche Math.Ver.31(1922),problem section,p.42; 32(1923),problem section,p.45.

Faculty of Mathematics
Babeş-Bolyai University
Cluj-Napoca,Romania

ALPHA-CLOSE-TO-CONVEX FUNCTIONS

by N.N.Pascu

1. Let $f(z) = z + a_2 z^2 + \dots$ be a regular function in the unit disc D.

It is said that f(z) is close-to-convex in D, with respect to the convex function h(z), if $\operatorname{Re} \frac{f'(z)}{h'(z)} > 0$ for any $z \in D$. The Property of close-to-convexity, that generalizes that of starlikeness, was introduced by W.Kaplan in [3], where he gave a geometrical interpretation of this property. We denote by K the class of the functions close-to-convex in D.

These functions are univalent and for them Bieberbech's conjecture was proved.

Let α be a number real. The function f(z) is α-starlike-convex in D, if the function $(1-\alpha)f(z)+\alpha zf'(z)$ is starlike in D, and $f(z)f'(z) \neq 0$ for any $z \in D, z \neq 0$. We denote the class of these functions by SC_α and it was introduced in [6]. If S^* respectively C, are the classes of functions univalent starlike, respectively convex in D, then $SC_0 = S^*$, $SC_1 = C$, and for any α, $\alpha \in (0,1)$, $SC_\alpha \subset S^*$ ([6]).

The aim of this note is to put into evidence a generalized property of convexity called α-close-to-convexity. Calling K_α the class of the functions with this property, it is proved that, for any α, $\alpha \in (0,1]$,

$$SC_\alpha \subset K_\alpha \subset K, \text{ and } K_0 = K \tag{1}$$

Finally the results of R.J.Libera ([4]) and S.D.Bernardi ([1]) are generalized.

2. <u>Definition</u>. The function $f(z) = z+a_2z^2+\dots$ is α-close-to-convex in D, if :

a) f(z) is regular in D ;

b) the function (2) $F(z) = (1-\alpha)f(z)+\alpha zf'(z)$ is close-to-convex in D.

The function f(z)=z belongs to the class K_α , for any α , so that, for any α, $K_\alpha \neq \emptyset$.

We know that, if $f(z) \in SC_\alpha$, then the function F(z) defined by (2) is starlike ([6]), so that, for any α , $SC_\alpha \subset K_\alpha$.
For $\alpha = 0$, $K_0 = K$.

<u>THEOREM 1</u>. Let $F(z) = (1-\alpha) f(z) + \alpha zf'(z)$ be close-to-convex with respect to the convex function H(z), and $\alpha \in (0,1]$. Then the function f(z) is close-to-convex with respect to the convex function

$$\varphi(z) = \int_0^z h(t)t^{-1}dt, \text{ where} \tag{3}$$

$$h(z) = 1/(\alpha z^{1/\alpha -1}) \int_0^z t^{1/\alpha -1}H'(t)dt \tag{4}$$

<u>PROOF</u>. If $H(z) \in C$, then $zH'(z) \in S^*$. If F(z) is close-to-convex with respect to the function H(z), then for any $z \in D$ we have

$$\operatorname{Re} \frac{zF'(z)}{zH'(z)} = \operatorname{Re} \frac{F'(z)}{H'(z)} > 0 \tag{5}$$

We know that, if $\alpha \in (0,1]$ and $zH'(z) \in S^*$, then the solution h(z) of the differential equation

$$(1-\alpha) h(z) + \alpha zh'(z) = zH'(z), h(0) = 0 \tag{6}$$

is univalent, starlike in D, and is given by (4). We put $p(z) = zh'(z)/h(z)$ and $r(z) = zf'(z)/h(z)$.

Let's suppose that it exists z_0, $|z_0| = \rho < 1$, so that $\operatorname{Re} r(z) \geq 0$ for $|z| \leq \rho$ and $\operatorname{Re} r(z_0) = 0$. Let u(z) be the meromorphic function in D, defined by

$$(1 + u(z)) / (1 - u(z)) = r(z) \qquad (7)$$

The $u(z)$ carries out the conditions from the hypothesis of Jack's lemma ([2]), therefore the number M, $M = z_0u'(z_0)/u(z_0)$, is real and $M \geq 1$.

A simple calculation yields $:-p(z)r(z)+z^2f''(z)/h(z)+r(z) = = z(zf'(z)/h(z)'=z[(1+u(z))/(1-u(z))]' = (2zu'(z)/u(z))(u(z)/(1-u(z))^2)$.

Because $|u(z_0)| = 1$ and $u(z_0) \neq 1$ (the function f(z) is regular in D), the number N, $N = r(z_0)-p(z_0)r(z_0)+z_0f''(z_0)/h(z_0)$, is real and $N \leq 0$.

We have $z_0f''(z_0)/h(z_0)= N - r(z_0) + p(z_0)\, r(z_0)$, and $\mathrm{Re}[(z_0F'(z_0))/(z_0H'(z_0))]= \mathrm{Re}[(z_0f'(z_0)+\alpha z_0^2f''(z_0))/((1-\alpha)h(z_0)+ +\alpha z_0h'(z_0)) = \mathrm{Re}\,(r(z_0)+\alpha z_0^2f''(z_0)/(h(z_0)\cdot(1-\alpha+\alpha p(z_0))) = = \mathrm{Re}[\alpha N/(1-\alpha+\alpha p(z_0))] + \mathrm{Re}\, r(z_0) \leq 0$. This contradiction proves that $\mathrm{Re}\, r(z) \neq 0$ in D. Because $\mathrm{Re}\, r(0) = 1$, it results that $\mathrm{Re}\, zf'(z)/h(z) = \mathrm{Re}\, f'(z)/(h(z)z^{-1}) = \mathrm{Re}\, r(z) > 0$, for any $z \in D$. If we denote $\varphi(z) = \int_0^z h(t)t^{-1}dt$, we obtain $\mathrm{Re}\, f'(z)/\varphi'(z) > 0$ for any $z \in D$. Since $h(z) \in S^*$, then $\varphi(z) \in C$, and theorem is proved.

<u>COROLLARY 1</u>. For any α, $\alpha \in (0,1]$, $K_\alpha \subset K$.

<u>COROLLARY 2</u>. If $\alpha \in (0,1]$ and $\mathrm{Re}(f'(z)+\alpha zf''(z)) > 0$ in D, then $\mathrm{Re}\, f'(z) > 0$ in D.

The proof results from the Theorem 1, for $H(z) = z$.

For $\alpha = 1$ we obtain :

<u>COROLLARY 3</u>. If the function $zf'(z)$ is close-to-convex with respect to the convex function H(z), then f(z) is close-to-convex, with respect to the convex function $\varphi(z) = \int_0^z H(t)t^{-1}dt$.

<u>THEOREM 2</u>. Let F(z) be close-to-convex with respect to the convex function H(z) and $\alpha \in (0,1]$. Then the solution f(z) of the

differential equation

$$(1-\alpha)\, f(z) + \alpha z f'(z) = F(z),\ f(o) = o \tag{8}$$

is close-to-convex with respect to the convex function (3) and is given by

$$f(z) = 1/(\alpha z^{1/\alpha -1}) \int_0^z t^{1/\alpha -2} F(t)\, dt \tag{9}$$

Proof. If $F(z)=z + b_2 z^2 + \ldots, |z| < 1$, then $g(z)= \int_0^z t^{1/\alpha -2} F(t)dt$ in given by $g(z) = z^{1/\alpha}(1+b_2/(1+\alpha)+\ldots+b_n/(1+\alpha(n-1)))$, and $f(z) = 1/(\alpha z^{1/\alpha -1})\, g(z) = z+b_2/(1+\alpha)z^2+\ldots+b_n/(1+\alpha(n-1))z^n+\ldots$

It results that the function f(z) is well defined and regular in D.

By Theorem 1 results that f(z) is close-to-convex with respect to the convex function $\varphi(z)$, given by (3).

By Theorem 2 we obtain :

COROLLARY 3. If the function F(z) is close-to-convex with respect to the convex function H(z), then, for any $\alpha \in (o,1]$ the function $f(z)=1/(\alpha z^{1/\alpha -1}) \int_0^z t^{1/\alpha -2} F(t)dt$ is close-to-convex with respect to the convex function

$$\varphi(z) = \int_0^z u^{-1}\Big[1/(\alpha u^{1/\alpha -2}) \int_0^u t^{1/\alpha -1} H'(t)dt\Big]\, du$$

Remark. We get from the Corollary 3, for $\alpha=1/2$, the theorem of R.J.Libera ([4]), and for $\alpha=1/(n+1)$, where n is a natural number, we rediscover the theorem of S.D.Bernardi ([1]).

THEOREM 3. If the function F(z) is close-to-convex with respect to the convex function H(z), then, for any $\alpha \in (o,1]$ and for any a >0, the function

$$f_{a,\alpha}(z) = a/\alpha H(z)+1/(\alpha z^{1/\alpha -1}) \int_0^z t^{1/\alpha -2}(F(t)-a\alpha/(1-\alpha)H(t))dt \tag{10}$$

is close-to-convex with respect to the convex function $\varphi(z)$, given by (9)

Proof. From the theorem of Bielecki and Lewandowski ([7]) and from the Corollary 3, it result, the function $1/(\alpha z^{1/\alpha -1}) \int_0^z t^{1/\alpha -2} F(t)dt$ $+a/(\alpha z^{1/\alpha -1}) \int_0^z t^{1/\alpha -1} H'(t)dt=1/(\alpha z^{1/\alpha -1}) \int_0^z t^{1/\alpha -2} F(t)dt+a/\alpha H(z) -$

$- a/((1-\alpha)z^{1/\alpha-1})\int_0^z t^{1/\alpha-2}H(t)dt = f_{a,\alpha}(z)$, is close-to-convex with respect to the convex function $\varphi(z)$ given by (9).

COROLLARY 4. If $f(z)\in K_\alpha$ and $\alpha\in(0,1]$, then there exists a function $h(z)\in SC_\alpha$, so that for any $a>0$, the function $f(z) + ah(z)$ is α-close-to-convex with respect to the function $\varphi(z) = \int_0^z h(t)t^{-1}dt$. The proof results from the Theorem 3, for h(z) given by (4).

The author acknowledges with thanks the valuable suggestions of Professor Petru T. Mocanu, of the University of Cluj-Napoca.

Department of Mathematics, University of Braşov.

Bibliography

1. S.D.BERNARDI, Convex and starlike univalent functions, Trans. Amer.Math.Soc.135 (1969), 428-446.
2. I.S.JACK, Functions starlike and convex of order α, J.London Soc. (2), 3 (1971), 469-474.
3. WILFRED KAPLAN, Close-to-convex Schlicht Functions, The Michingan Math.Journ.vol.1,No.2 July 1952.
4. R.J.LIBERA, Some classes of regular univalent functions,Proc. Math.Soc.L.16 (1965), 755-758.
5. PETRU T.MOCANU and MAXWELL O.READE, On generalized convexity in conformal mappings, Rev.Roum.de Math.Purres et Appl.16 No.1o(1971).
6. N.N.PASCU, Alpha-starlike-convex functions (to appear).
7. A.BIELECKI, Z.LEWANDOWSKI, Sur une théorème concernant les functions univalentes linéairementes accesible de M. Biernacki, Ann.Polon.Math.1962, 12, 61-63.

On a coefficient inequality for schlicht functions

by Albert Pfluger in Zürich.

1. There is the following striking result about the class S of normalized schlicht functions on the unit disc $D = \{z \mid |z| < 1\}$ in view of Bieberbach's conjecture. Let the function f of S have the expansion

$$f(z) = z + a_2 z^2 + \dots + a_k z^k + \dots + a_n z^n + \dots$$

and assume that for some k the coefficients $a_2, \dots, a_k$ are real, then $\mathrm{Re}\, a_n \leqq n$ for $n \leqq 2k+1$ and equality holds only for the Koebe functions $K(z) = \frac{z}{(1-z)^2}$ if n is even, and only for $K(z)$ and $-K(-z)$ if n is odd (conf. [1] and [2]). In case of $n = 2k+1$ there is an inequality due to A. Obrock (conf.[1]) showing how much $\mathrm{Re}\, a_n$ must deviate from its maximum if $\mathrm{Im}\, a_{k+1} \neq 0$:

Theorem 1 . Let $a_2, \dots, a_{k+1}$ be real numbers $(k+1 \geqq 2)$ and denote by $S(a_2, \dots, a_{k+1})$ the class of functions of S having these numbers as coefficients (it is assumed that this class is not empty). Let $n = 2k+1$ and define $A_n = \max \mathrm{Re}\, a_n$ where the maximum is taken over the class $S(a_2, \dots, a_{k+1})$. Then

$$\mathrm{Re}\, a_n \leqq A_n - \frac{k+1}{2}\beta^2 \tag{1}$$

for all functions of S such that

$$f(z) = z + a_2 z^2 + \dots + a_k z^k + (a_{k+1} + i\beta) z^{k+1} + \dots + a_n z^n + \dots$$

for some real β .

Since $A_n \leqq n$ we have $\mathrm{Re}\, a_n \leqq n - \frac{k+1}{2}(\mathrm{Im}\, a_{k+1})^2$ for each function of S having its coefficients $a_2, \dots, a_k$ real, if $n = 2k+1$.

Obrock established inequality (1) by using Jenkins' General Coefficient Theorem (conf. [3]) and a continuity argument. In this paper another proof is presented which replaces the continuity argument by a fundamental property of the n-th coefficient body V_n. It will also be shown that inequality (1) is sharp.

2. To be more precise we first recall two definitions:

1° Let n be an integer > 1. Then the mapping

$$\alpha_n : f \longmapsto (a_2, \ldots, a_n)$$

takes S onto a compact subset V_n of $\mathbb{C}^{n-1}$ which is homeomorphic to a ball. V_n is called the n-th coefficient body of S.

2° A function f of S is called a canonical slit mapping of order n (>1) if $f(D)$ is bounded by piecewise analytic slits and if there is a quadratic differential

$$Q(w)dw^2 = \left(\frac{A_{n-1}}{w^{n+1}} + \cdots + \frac{A_1}{w^3}\right)dw^2 \tag{2}$$

$(A_{n-1}, \ldots, A_1) \neq (0, \ldots, 0)$ such that $Q(w)dw^2 \geqq 0$ along $\partial f(D)$.

There might be several f's having the same Q and several Q's associated to a single f. Obviously a canonical slit mapping of the order n is also of the order $n+1$.

The fundamental property of the body V_n we need later on, is expressed by

Theorem A (conf.[4]). To each boundary point a of V_n there corresponds exactly one function f of S with $\alpha_n(f) = a$; this f is necessarily a canonical slit mapping of order n. Conversely, each canonical slit mapping of order n is taken onto some boundary point of V_n, under the mapping α_n.

In case of a simply connected domain in $\mathbb{C}$, Jenkins' Extended General Coefficient Theorem (cf.[3]) is expressed by

Theorem B. Let k be an integer $\geqq 1$ and $k < n \leqq 2k+1$. Let $\varepsilon = 0$ for $n \leqq 2k$ and $\varepsilon = 1$ for $n = 2k+1$. If f is a canonical slit mapping of order n

then there are numbers F_j, $j=k+1, \dots, n$ such that

$$\mathcal{R}e\left\{\sum_{k+1}^{n} F_j(b_j-a_j) - \varepsilon \frac{k+1}{2} F_n (b_{k+1}-a_{k+1})^2\right\} \geqq 0 \tag{3}$$

for all functions $g(z) = z + b_2 z^2 + \dots + b_n z^n + \dots$ of S such that $b_j = a_j$, $j = 2, \dots, k$. These F_j and the $A_{\nu-1}$ of (2) are related by the equations

$$A_{\nu-1} = \sum_{j=\nu}^{n} a_j^{(\nu)} \bar{F}_j, \quad j = 2, \dots, n \tag{4}$$

, where the $a_j^{(\nu)}$ are defined by

$$f^{\nu}(z) = \sum_{j=\nu}^{\infty} a_j^{(\nu)} z^j .$$

In case of $n \leqq 2k$ and in case of $n = 2k+1$ and $b_{k+1} = a_{k+1}$ equality occurs in (3) only if $g = f$.

3. We now prove two lemma. The first one is about sections of V_n through points $(a_2, \dots, a_k)$ of V_k. Let $(a_2, \dots, a_k)$ be a point of V_k and $n > k$. Then $W_n(a_2, \dots, a_k)$ denotes the set of all points $(a_{k+1}, \dots, a_n)$ of $\mathbb{C}^{n-k}$ such that $(a_2, \dots, a_k, a_{k+1}, \dots, a_n)$ is in V_n. If the point $(a_2, \dots, a_k)$ is on the boundary of V_k, then W_n consists of a single point; otherwise W_n has a non-void interior.

<u>Lemma 1</u> (cf.[5]) If $(a_2, \dots, a_k)$ is not on the boundary of V_k and if $n \leqq 2k$, then $W_n = W_n(a_2, \dots, a_k)$ is a strictly convex body in $\mathbb{C}^{n-k}$.

In fact, if $A = (a_{k+1}, \dots, a_n)$ is on the boundary of W_n, then $(a_2, \dots, a_k, a_{k+1}, \dots, a_n)$ is a boundary point of V_n. Hence Theorem B implies that

$$\mathcal{R}e\left\{\sum_{j=k+1}^{n} F_j(b_j - a_j)\right\} \geqq 0$$

for all points $B = (b_{k+1}, \dots, b_n)$ of W_n, and equality holds only for $B = A$. This shows that W_n lies on one side of some hyperplane through A which contains of W_n only this point A. As this happens for each boundary point of W_n the lemma is proved.

Lemma 2 (conf.[1]). Let $(a_2, \ldots, a_k)$ be a real point of V_k. Let $g(z)$ be a function of S with $\alpha_k(g) = (a_2, \ldots, a_k)$ but having not all its coefficients real. Then there is a canonical slit mapping

$$f(z) = z + a_2 z^2 + \cdots + a_k z^k + a_{k+1} z^{k+1} + \cdots + a_n z^n + \cdots$$

with real coefficients, and a quadratic differential (2) associated to f with real coefficients A_j and $A_{n-1} = -1$ such that $a_j = \mathrm{Re}\, b_j$, $j = k+1, \ldots, n-1$.

Proof of Lemma 2 The point $(a_2, \ldots, a_k)$ is not on the boundary of V_k because if it were, the function g would have all its coefficients real. Hence, according to Lemma 1, $W_{n-1}(a_2, \ldots, a_k)$ is a strictly convex body of $\mathbb{C}^{n-k-1}$ and, the $a_2, \ldots, a_k$ being real, it is symmetric to the real subspace of $\mathbb{C}^{n-k-1}$. Hence, if $B = (b_2, \ldots, b_{n-1})$ is a point of this body W_{n-1}, then so are the points $\bar{B} = (\bar{b}_2, \ldots, \bar{b}_{n-1})$ and $A = \frac{1}{2}(B + \bar{B}) = (a_{k+1}, \ldots, a_{n-1})$.

With the numbers $a_k, \ldots, a_{n-1}$ being defined this way, we consider the set $W_n(a_2, \ldots, a_{n-1})$. According to Lemma 1 it is either a strictly convex region of $\mathbb{C}$ or reduces to a single point. In the latter case the point $(a_2, \ldots, a_{n-1})$ uniquely determines a_n. Moreover, since $(a_2, \ldots, a_{n-1}, a_n)$ is on the boundary of V_n it uniquely determines a function f such that $\alpha_n(f) = (a_2, \ldots, a_n)$ and we conclude that there is only one function f with $\alpha_{n-1}(f) = (a_2, \ldots, a_{n-1})$. This however shows that $(a_2, \ldots, a_{n-1})$ is on the boundary of V_{n-1}.

Let now $g(z) = z + a_2 z^2 + \cdots + a_k z^k + b_{k+1} z^{k+1} + \cdots + b_n z^n + \cdots$ be a function of S having not all its coefficients real. We conclude that the corresponding point

$$A := (\mathrm{Re}\, b_{k+1}, \ldots, \mathrm{Re}\, b_{n-1}) = (a_{k+1}, \ldots, a_{n-1})$$

is not on the boundary of $W_{n-1}(a_2, \ldots, a_k)$. Otherwise the points

$B = (b_{k+1}, \ldots, b_{n-1})$ and $\overline{B} = (\overline{b}_{k+1}, \ldots, \overline{b}_{n-1})$ which are both in W_{n-1} and the point $A = \frac{1}{2}(B + \overline{B})$ would be identical because A would be an extreme point of W_{n-1} if it were on the boundary, and this, of course, would imply that g has all its coefficients real. Hence, by the assumption of Lemma 2, $(a_2, \ldots, a_k, a_{k+1}, \ldots, a_{n-1})$ is an inner point of V_{n-1} and this implies that $W_n(a_2, \ldots, a_{n-1})$ is a closed and strictly convex set of $\mathbb{C}$ which has a non-void interior.

Now, let f_0 maximize $\operatorname{Re} a_n$ among those functions f which satisfy the condition $\alpha_{n-1}(f) = (a_2, \ldots, a_{n-1})$. Then also $\overline{f}_0(z) = \overline{f_0(\overline{z})}$ is extremal and, because by Lemma 1 there is only one extremal function, f_0 and $\overline{f}_0$ are identical and so all coefficients of f are real. Hence, there is a quadratic differential (2) associated to f_0 which has real coefficients $A_{n-1}, \ldots, A_1$. Denoting the n-th coefficient of f_0 by a_n and observing that $F_n = A_{n-1}$ we obtain from Theorem B (with $k = n-1$) that

$$\operatorname{Re}\{A_n(b_n - a_n)\} \geqq 0 \tag{5}$$

for all functions g of S with $\alpha_{n-1}(g) = \alpha_{n-1}(f_0)$. However, there is a point b_n of the region $W_n(a_2, \ldots, a_{n-1})$ such that $b_n < a_n$. Thus inequality (5) excludes the case of $A_{n-1} > 0$, and also the case of $A_{n-1} = 0$ because then we would have equality in (5) and therefore $g = f_0$. Hence $A_{n-1} < 0$. As a positive factor of $Q(w)\,dw^2$ is irrelevant we may assume that $A_{n-1} = -1$ and thus the lemma is proved.

4. Proof of Theorem 1. We may assume that $(a_2, \ldots, a_k)$ is in the interior of V_k because otherwise all functions g with $\alpha_k(g) = (a_2, \ldots, a_k)$ would have real coefficients. Let g be a function of S with coefficients not all real and such that $\alpha_k(g) = (a_2, \ldots, a_k)$. Let f be the corresponding function of Lemma 2. We apply Theorem B.

Since the a_j are all real it follows from (4) that the F_j are real also and that $F_n = A_{n-1} = -1$. Moreover, since $\mathcal{R}e\, b_j = a_j$, $j = k+1, \ldots, n-1$, inequality (3) reduces to

$$\mathcal{R}e\, b_n \leq a_n - \frac{k+1}{2}(\mathcal{I}m\, b_{k+1})^2 \tag{6}$$

for all g of $S(a_2, \ldots, a_k)$ having not all coefficients real. Since (6) is trivial for a g having real coefficients, Theorem 1 is proved.

An almost identical method leads to a result one obtains from Theorem 1 by replacing (1) by

$$\mathcal{R}e\, a_n \geq A_n' - \frac{k+1}{2}\beta^2$$

where $A_n' = \min \mathcal{R}e\, a_n$ and the minimum is taken over the function of $S(a_2, \ldots, a_k)$.

5. We now are going to show that inequality (1) is sharp at least for the point $(a_2, \ldots, a_k) = (0, \ldots, 0)$ and β in some interval. It is well known that for the corresponding class of schlicht functions

$$f(z) = z + a_{k+1} z^{k+1} + \cdots + a_n z^n + \cdots$$

we have $|a_{k+1}| \leq \frac{2}{k}$. We choose a real-valued a_{k+1} from the interval $[-\frac{2}{k}, \frac{2}{k}]$ and maximize $\mathcal{R}e\, a_n$ within the class $S(0, \ldots, 0, a_{k+1})$. Since $n = 2k+1 < 2(k+1)$ it follows from Lemma 1 that there is only one extremal function f. However, for $\varepsilon^k = 1$ also $f_\varepsilon(z) = \varepsilon^{-1} f(\varepsilon z)$ is extremal, hence $\varepsilon f(z) = f(\varepsilon z)$, and this implies that $a_j = 0$ for $j \neq mk+1$, $m = 0, 1, 2, \ldots$. From this we conclude that there is a function

$$F(z) = z + c_2 z^2 + c_3 z^3 + \cdots$$

of S such that $f(z) = (F(z^k))^{1/k}$. Conversely, for each F of S the function $f(z) = (F(z^k))^{1/k}$ also belongs to S we have

$$a_{k+1} = \frac{c_2}{k}, \quad a_n = \frac{1}{k}\left(c_3 - \frac{k-1}{2k} c_2^2\right). \tag{7}$$

This reduces the reasoning to the coefficient body V_3. Since the function

$$f(z) = z + a_{k+1} z^{k+1} + a_n z^n + \cdots$$

maximizing $\operatorname{Re} a_n$ within $S(0, \dots, 0, a_{k+1})$ has all its coefficients real we have to consider only the real points (c_2, c_3) of V_3. Using Schiffer's differential equation (cf [4]) one finds that the set of these points is bounded by a Jordan curve made up by two arcs

$$A_1 : \quad c_3 = c_2^2 - 1, \quad -2 \leq c_2 \leq 2 \qquad \text{and}$$

(8) $$A_2 : \quad \begin{aligned} c_2 &= 2t(1 - \log|t|) \\ c_3 &= 1 + t^2(1 + (1 - 2\log|t|)^2) \end{aligned} \qquad |t| \leq 1 .$$

For our extremal problem only the second arc A_2 is relevant. To the chosen value of a_{k+1} in $[-\frac{2}{k}, \frac{2}{k}]$ there correspond via (7) and (8) unique values for c_2, t and c_3, and we conclude that the quantity A_n of Theorem 1 for our particular case is given by

(9) $$\max a_n = A_n = \frac{1}{k}\left(c_3 - \frac{k-1}{2k} c_2^2\right).$$

Now we use the description of the body V_3 given in chapter XIII of [4]. Let (c_2, c_3) be a point of the arc A_2, let $c_2' = c_2 + i\beta$ for a fixed β such that $|\beta| \leq \sin\varphi - \varphi\cos\varphi$, $\cos\varphi = t$, $0 \leq \varphi \leq \pi/2$. The formula (13.4.4) of [4] immediately implies

$$\max \operatorname{Re} c_3' = c_3 - \beta^2,$$

where the maximum is taken over the class $S(c_2')$. With (c_2', c_3') in place of (c_2, c_3) in (7) it follows from (9) that

$$\max \operatorname{Re} a_n' = A_n - \frac{k+1}{2}\beta^2$$

and this shows that inequality (1) is sharp.

References

[1] A. Obrock, An inequality for certain schlicht functions, Proc.Amer.Math.Soc.vol. 17(1966), 1250-1254.

[2] M. Schiffer, Univalent functions whose n first coefficients are real, J. d'Analyse Math. vol. 18(1967), 329-349.

[3] J.A. Jenkins, An Addendum to the General Coefficient Theorem, Trans.Amer.Math.Soc.vol. 107(1963), 125-128.

[4] A.C. Schaeffer and D.C. Spencer, Coefficient Regions for Schlicht Functions, Amer.Math.Soc. Coll.Publ.vol. 35, 1960.

[5] A. Pfluger, The Convexity of Certain Sections of n-bodies of Coefficients of Univalent Functions, Certain Problems of Mathematics and Mechanics (On the Occasion of the Seventieth Birthday of M.A.Lavrent'ev)(Russian), pp.233-241, Izdat."Nauka",Leningrad 1970.

On holomorphic mappings of annuli into annuli.

H. Renggli

I. Statements.

1. In a paper published in 1951, H. Huber [1] made a comprehensive study of holomorphic mappings of annuli into annuli that incorporated older results and introduced new ones. Subsequently several attempts were made to simplify the presentation. So e.g. E. Reich [2] gave a somewhat more elementary proof for the special case when an annulus is mapped into itself.

2. Here we present a new simple approach that allows to derive all results given in [1] from two basic facts. The first one is a certain principle of "shortest curve":

Proposition 1. Let A denote the annulus $\{z \text{ complex}: 0 < r < |z| < R < \infty\}$. Using a conformal mapping of the universal covering surface of A onto the unit disk one defines in A a non-Euclidean metric (with constant Gaussian curvature -1).

Then for each curve C that is freely homotopic (in A) to a circle $z(t) = \rho e^{it}$, $0 \leq t \leq 2\pi$, $r < \rho < R$, and that has a non-Euclidean length L, the inequality $L \geq d$ holds.

Moreover $d = 2\pi\beta$, where $\beta = \dfrac{\pi}{\log \frac{R}{r}}$, and the equality sign occurs exactly for the circle $z(t) = \sqrt{Rr}\, e^{it}$, $0 \leq t \leq 2\pi$.

Thereafter an application of the Schwarz-Pick Lemma leads to the second fact:

Proposition 2. Let A be an annulus (as in Prop.1), let A^* denote the annulus $\{z \text{ complex}: 0 < r^* < |z| < R^* < \infty\}$, and let f, $f: A \to A^*$, be holomorphic. Let C be an arbitrary curve in A with a non-zero non-Euclidean length L, and let L^* denote the non-Euclidean length in A^* of the image C^* of C under f.

Then $L \geq L^*$, and if the equality sign holds for one such pair C and C^* then it holds for any such pair.

Next such a mapping f is called trivial, if the image of every closed curve in A is homotopic (in A^*) to a point.

3. From the two Propositions one derives quite easily the following Theorem. Here we have normalized the annuli A and A^* in such a way that the positively oriented unit circle C_0 is in both cases the "shortest curve" of lengths d and d^* respectively.

Theorem. Let f, $f: A \to A^*$, be holomorphic, where $A = \{z \text{ complex}: \frac{1}{R} < |z| < R,\ R > 1\}$, $A^* = \{z \text{ complex}: \frac{1}{R^*} < |z| < R^*,\ R^* > 1\}$, and $d = \dfrac{\pi^2}{\log R}$, $d^* = \dfrac{\pi^2}{\log R^*}$.

(i) If $d^* > d$, i.e., $R > R^*$, then f is trivial. It follows especially that two such annuli A and A^* with $R \neq R^*$ are not conformally equivalent.

(ii) If $d^* = d$, i.e., $R = R^*$, then either $L > L^*$ and f is trivial or $L = L^*$. In the latter case f is either a rotation $[f(z) = e^{i\alpha} z$, α real$]$ or an inversion followed by a rotation $[f(z) = e^{i\alpha} \frac{1}{z}]$.

(iii) If $d^* < d$, i.e., $R^* > R$, then either f is trivial or there exists a function g holomorphic in A and an integer m, such that $f = g^m$. Here g maps every curve C freely homotopic (in A) to C_0 onto a curve C* freely homotopic (in A*) to C_0; and the integer m satisfies the inequality $0 < |m| \leq d/d^*$. If especially $d = |m| d^*$, i.e., $R^* = R^{|m|}$, and if $f = g^m$, then g is a rotation $[g(z) = e^{i\alpha} z]$.

This contains albeit in a different form all results mentioned in [1].

II. Proofs.

1. We shall use the notations given in Prop.1. The function $v(z) = \frac{\beta}{2} \log \frac{z}{\sqrt{Rr}}$ maps the universal covering surface of A onto the strip $S = \{v \text{ complex}: -\pi/4 < \Re v < \pi/4\}$. Here the points lying over $\sqrt{Rr}$ will correspond to the points $n\pi\beta i$, n integer. A curve in S that connects a point x_0, x_0 real, in S with $x_0 + \pi\beta i$ can be considered as an image C_v of a curve C in A, and conversely each curve C has an image that can be thus represented.

Next $w(v) = \tan v$ maps S conformally onto the unit disk D. Here the imaginary axis corresponds to the vertical diameter of D, whereas any horizontal segment in S goes onto a circular arc orthogonal to the family of circular arcs in D passing through i and -i. If we denote the point $i \tanh \pi\beta = w(\pi\beta i)$ by P, then $d = 2\pi\beta$ is the non-Euclidean distance between O and P. The image C_w of a curve C_v will connect a point on the horizontal diameter with a point on a circular arc through P. That arc has minimum imaginary part exactly at P. But the set of points that have constant non-Euclidean distance d from the horizontal diameter and lie in the upper half-disk is a Euclidean circular arc through P with endpoints 1 and -1.

That arc has maximum imaginary part exactly at P. Hence each curve C_w, image of some C in A, has a non-Euclidean length that must either be strictly larger than d or can be equal to d only when connecting O and P along the diameter. This proves Prop. 1.

{From the non-Euclidean metric $d\sigma = \dfrac{2|dw|}{1-|w|^2}$

in D one gets in A by an elementary calculation the expression $d\sigma = \beta\left|\frac{dz}{z}\right| \cdot \left[\cos\left(\beta \log \frac{|z|}{\sqrt{Rr}}\right)\right]^{-1}$.

Especially if $|z| = \sqrt{Rr}$, then $d\sigma = \beta \cdot d\Theta$, $\Theta = \arg z$.}

2. Let w(z) be that mapping of the universal covering surface of A onto D given in the proof of Prop. 1. Let $w^*(z^*)$ be similarly defined with respect to A*. We can lift any holomorphic function f, $f: A \rightarrow A^*$, by fixing it at a point and then using analytic continuation, to a function from the universal covering surface of A into that of A*. This induces a function mapping D holomorphically into D. By the Schwarz-Pick Lemma either non-Euclidean distances are always decreased or are always preserved. That validates Prop. 2.

3. We pass to the proof of the Theorem. The image C_o^* of C_o must by Prop. 2 have a non-Euclidean length less than or equal to d. Because C_o is the "shortest curve" in A* and has by hypothesis (i) non-Euclidean length d*, $d^* > d$, it follows that C_o^* is homotopic (in A*) to a point. But homotopic curves have homotopic images under continuous mappings. So f is trivial, and (i) is proved.

The first part of (ii) follows in the same way. If $L^* = L$, then the segment $\overline{OP}$ in D must be mapped by a non-Euclidean motion of D onto some non-Euclidean segment $\overline{O'P'}$, where O' and P' correspond to points lying in A over the same point. By an argument we have used in the proof of Prop. 1 this is only possible when $\overline{O'P'}$ lies on the vertical diameter of D. But to such a motion corresponds in S a translation along the imaginary axis or such a translation followed by a multiplication with -1. Whence the conclusion (ii) follows.

If f is not trivial, then the image C_0^* of C_0 has a winding number m, $m \neq 0$, around the point 0. If $L(C_0^*)$ denotes the non-Euclidean length of C_0^*, $|m|d^* \leq L(C_0^*) \leq d$ by Prop. 1 and 2. Next let $g(z) = [f(z)]^{\frac{1}{m}}$. Since every curve C*, image of some C under f, has the same winding number as C_0^* (around 0), the function g(z) is obviously defined in A and maps each C onto some curve with the same winding number 1. If in addition $R^* = R^{|m|}$, the function g maps A into itself and is by (ii) a rotation. So (iii) is proved.

References.

[1] H. Huber. Ueber analytische Abbildungen von Ringgebieten in Ringgebiete. Compos. Math. 9 (1951), 161-168.

[2] E. Reich. Elementary proof of a theorem on conformal rigidity. Proc. Amer. Math. Soc. 17 (1966), 644-645.

Kent State University
Kent, Ohio 44242, USA

ESTIMATES OF THE RIEMANN MAPPING FUNCTION NEAR A BOUNDARY POINT

by

Burton Rodin[(1)] and S. E. Warschawski[(2)]

Consider the Riemann mapping function $f: R \to S$ of a simply connected region R onto a standard region S. Let ζ be a boundary point or prime end of R, and let $f(\zeta)$ be the boundary point of S which corresponds to ζ. We are interested in finding useful estimates of $f(w)$ when $w \in R$ approaches ζ.

When stated in this general form the problem encompasses many classical investigations. For example, the question of when f has a derivative (respectively, angular derivative) at ζ amounts to obtaining the special estimate $f(w) = f(\zeta) + (w-\zeta)(c + o(1))$ where $o(1) \to 0$ as $w \to \zeta$ in R (respectively, $w \to \zeta$ in every Stoltz angle in R). Recent work has involved estimates which hold in more general situations where f need not possess even an angular derivative.

When viewed in this general way the problem becomes one of estimating conformal quantities in terms of Euclidean ones. Therefore estimates of the conformal quantities in terms of extremal length quantities ought to

---

(1) Research partially supported by the National Science Foundation Grant No. GP38600.

(2) Research partially supported by the National Science Foundation Grant No. GP32156.

be an important intermediate stage. In this paper we announce two general results (Theorems A and B below) which comprise this intermediate stage. The final stage--obtaining sufficiently accurate Euclidean estimates of these extremal length quantities--will be illustrated in a variety of applications.

1. We adopt the normalization of Ahlfors [1] by choosing S to be the parallel strip region $\{z = x + iy: -\infty < x < +\infty,\ 0 < y < 1\}$ and requiring the mapping function $f: R \to S$ to satisfy $\operatorname{Re} f(w) \to +\infty$ as $w \to \zeta$.

Suppose that a family of disjoint cross cuts $\{V_s\}_{s_0 \leqq s < +\infty}$ of R has been chosen so that every sequence from it, $\{V_{s_n}\}$ where $s_n \uparrow +\infty$, is a fundamental sequence converging to the prime end ζ. (The variety of the applications of Theorems A and B result from the various choices that are possible for this family $\{V_s\}$.)

For $s_0 \leq s' < s''$, let $\lambda(s',s'')$ denote the extremal distance from $V_{s'}$ to $V_{s''}$ in the relevant component of $R - V_{s'} - V_{s''}$. We say the cross cuts $\{V_s\}$ are *approximately additive* if to each $\varepsilon > 0$ there is an S_ε such that $S_\varepsilon \leqq s' < s < s''$ implies

$$\lambda(s',s) + \lambda(s,s'') \geqq \lambda(s',s'') - \varepsilon .$$

Theorem A. *The following statements are equivalent:*

(a) *The cross cuts* $\{V_s\}$ *are approximately additive.*

(b) *The oscillation* $\omega(s) = \operatorname{diam} \operatorname{Re} f(V_s)$ *satisfies*

$$\lim_{s \to +\infty} \omega(s) = 0.$$

(c) *There is a constant* C *such that*

$$\lim_{\substack{s \to +\infty \\ w \in V_s}} [\operatorname{Re} f(w) - \lambda(s_0, s)] = C .$$

2. A proof of Theorem A will appear in a forthcoming paper [10]. At present we wish to discuss an application of it to the angular derivative problem. For that purpose we take R to be a strip domain (Ahlfors [1]). Thus R contains an arc $\gamma(t)$ for $-\infty < t < +\infty$ such that $\operatorname{Re} \gamma(t) \to \pm\infty$ as $t \to \pm\infty$, $\operatorname{Re} f(\gamma(t)) \to \pm\infty$ as $t \to \pm\infty$, and $\{V_s\}$ is the family of Ahlfors cross cuts $\{\theta_u\}$ (that is, θ_u is a cross cut of R which is a vertical line segment with real part u). Recall that S is the parallel strip $\{0 < y < 1\}$. Under these normalizations the mapping f is said to have angular derivative C $(-\infty < C < +\infty)$ at $\zeta = +\infty$ if the following two conditions are satisfied:

(1) For each $0 < \delta < \frac{1}{2}$ there is a U_δ such that

$$R_\delta \equiv \{w \mid u > U_\delta,\ \delta < v < 1 - \delta\} \subset R .$$

(2) For each $0 < \delta < \frac{1}{2}$,

$$\lim_{\substack{\operatorname{Re} w \to +\infty \\ w \in R_\delta}} [f(w) - w] = C .$$

If the angular derivative exists for one map $f: R \to S$ then it exists for all such maps; we then say that R has an angular derivative (at $+\infty$).

The *angular derivative problem* (Ahlfors [1, page 32]) asks for geometric properties of R which are necessary and sufficient for R to have an angular derivative.

Suppose R has an angular derivative. Then $\omega(u)$ (the oscillation of $f(\theta_u)$; see (b) of Theorem A) approaches zero as $u \to +\infty$ (Warschawski [16]). Hence $f(\theta_u)$ "approaches" the vertical segment $\{x = U + C, 0 < y < 1\}$ in S, where C is the angular derivative. Therefore

$$(3) \qquad \lambda(u',u'') = u'' - u' + o(1) \quad (o(1) \to 0 \text{ as } u'' > u' \to +\infty).$$

Conversely, suppose (3) holds as well as (1). Then Theorem A can be applied, since (3) guarantees the approximate additivity, and it yields

$$(4) \qquad C = \lim_{\substack{u\to+\infty \\ w\in\theta_u}} [\operatorname{Re} f(w) - \lambda(0,u)] = \lim_{\operatorname{Re} w\to+\infty} [\operatorname{Re} f(w) - \operatorname{Re} w - C_0] .$$

From (4) it follows that $f'(w) \to 1$ in each R_δ, and from this one can infer that $\operatorname{Im} f(w) - \operatorname{Im} w \to 0$ in R_δ. Then f has an angular derivative. We have derived the following result (to appear in Rodin-Warschawski [10]; this result has been obtained indpendently in joint work of J. Jenkins-K. Oikawa [5] to appear):

Theorem 1. *Let* R *be a strip domain. The following properties are necessary and sufficient for the existence of an angular derivative at* ζ*:*

(i) R *contains an* R_δ *as in Equation (1) for each* $0 < \delta < \frac{1}{2}$,

(ii) $\lambda(u',u'') = u'' - u' + o(1)$ *where* $o(1) \to 0$ *as* $u'' > u' \to +\infty$.

3. We do not consider Theorem 1 to be a complete solution to the angular derivative problem because condition (ii), which involves extremal length, is not a purely geometric (Euclidean) property. Nevertheless, the result represents a useful advance on the problem. From it one can obtain,

often in a completely trivial manner, many known results on angular derivatives. In particular, one can obtain all results in Chapter VI of J. Lelong-Ferrand's book [6]. To illustrate this we shall derive a sufficient condition for R to have an angular derivative. The geometric conditions are described by a "*subdivision*." That is, we consider a sequence $0 = u_0 < u_1 < u_2 < \dots$ with $\lim u_n = +\infty$ and the associated geometric data:

$$\delta_n = u_{n+1} - u_n$$

$$v'(u) = \min\{v \mid u + iv \in \theta_u\}$$

$$v''(u) = \max\{v \mid u + iv \in \theta_u\}$$

$$v'_n = \max\{v'(u) \mid u_n \leq u \leq u_{n+1}\}$$

$$v''_n = \min\{v''(u) \mid u_n \leq u \leq u_{n+1}\}$$

$$\Theta_n = v''_n - v'_n$$

$$\theta'_n = v'_n, \; \theta''_n = 1 - v''_n \; .$$

Theorem 2. *If* R *has a subdivision* $\{u_n\}$ *with*

$$\sum \delta_n^2 < \infty \; , \quad \sum (\theta'_n)^2 < \infty \; , \quad \sum (\theta''_n)^2 < \infty \; , \tag{5}$$

then R *has an angular derivative.*

To derive Theorem 2 from Theorem 1 we first note that (5) obviously implies (i) of Theorem 1. To verify (ii) consider a domain $\hat{R} \subset R$ whose associated quantities θ'_n, θ''_n are nonnegative and equal in magnitude to those of R. Thus $\hat{R} \subset S_w = \{w \mid -\infty < u < +\infty,\ 0 < v < 1\}$. Let Δ'_n (respectively Δ''_n) be a disk of radius $2\sqrt{\delta_n^2+|\theta'_n|^2}$ centered at u_n (respectively, of radius $2\sqrt{\delta_n^2+|\theta''_n|^2}$ centered at $u_n + i$). Let $\rho(w)|dw|$ be the linear density $\rho(w) \equiv 2$ in $\bigcup(\Delta'_n \cup \Delta''_n)$ and $\rho(w) \equiv 1$ elsewhere in S_w. For $u' < u''$ sufficiently large, any crosscut γ^* of $\hat{R}$ which separates $\theta_{u'}$ and $\theta_{u''}$ has ρ-length $\geqq 1$. Therefore

$$\lambda_R(u',u'') \leqq \lambda_{\hat{R}}(u',u'') \leqq \frac{A(\rho,\hat{R})}{1} = u'' - u' + o(1) ,$$

where $o(1) \to 0$ as $u'' > u' \to +\infty$ according to (5).

To obtain the opposite inequality consider the rectangle $R_n = \{w \mid u_n \leq u \leq u_{n+1},\ v'_n \leqq v \leqq v''_n\} \subset R$. If v'_n, v''_n are finite there are boundary points P'_n, P''_n of R on the lower and upper edges of R_n. Adjoin to R_n the closed disks of radius δ_n centered at P'_n and P''_n. A closed domain I_n is obtained; the area of I_n is no greater than $\delta_n \Theta_n + \frac{3}{2}\pi\delta_n^2$. Define $\rho(w) \equiv 1$ in I_n and zero elsewhere. The ρ-length of an arc in R joining θ_{u_n} to $\theta_{u_{n+1}}$ is at least δ_n. Therefore $\lambda(u_n,u_{n+1}) \geqq \frac{L^2(\rho)}{A(\rho)} \geqq \dfrac{\delta_n^2}{\delta_n \Theta_n + \frac{3}{2}\pi\delta_n^2} \geqq \frac{\delta_n}{\Theta_n}\left(1 - \frac{3\pi}{2}\frac{\delta_n}{\Theta_n}\right)$. If

$u_{j-1} < u' \leqq u_j < u_{k+1} \leqq u'' < u_{k+2}$ then

$$\lambda_R(u',u'') \geq \sum_{n=j}^{k} (u_n, u_{n+1}) \geq \sum_{n=j}^{k} \frac{\delta_n}{\Theta_n}\left(1 - \frac{3\pi}{2}\frac{\delta_n}{\Theta_n}\right)$$

$$= \sum_{n=j}^{k} \delta_n + \sum_{n=j}^{k} \frac{\delta_n}{\Theta_n}(1-\Theta_n) - \frac{3\pi}{2}\sum_{n=j}^{k} \frac{\delta_n^2}{\Theta_n^2}$$

$$\geqq u'' - u' + o(1) + \sum \delta_n(1-\Theta_n) - \text{const.}\sum \delta_n^2$$

$$= u'' - u' + \sum(\delta_n\theta_n' + \delta_n\theta_n'') + o(1)$$

$$= u'' - u' + o(1) .$$

This completes the verification of (ii) in Theorem 1; we conclude that R has an angular derivative.

Theorem 2 and other angular derivative results will be published in a future paper (Rodin-Warschawski [11]); it will be shown there that the converse of this Theorem 2 is false.

4. Another application of Theorem A is

Theorem 3. *Let* R *be a strip domain. If*

$$\lambda(u',u'') \leq \int_{u'}^{u''} \frac{du}{\theta(u)} + O(1) \tag{6}$$

for all $u_0 \leq u' < u''$ *then there is a real constant* C *such that*

$$\operatorname{Re} f(w) = \int_{u_0}^{u} \frac{du}{\theta(u)} + C + o(1) . \tag{7}$$

To prove this result first note that the function $E(u',u'') \equiv \lambda(u',u'') - \int_{u'}^{u''} du/\theta(u)$ is subadditive,

$$E(u',u) + E(u,u'') \leqq E(u',u'') , \tag{8}$$

positive, and bounded above. It follows that $E(u',u'') \to 0$ as $u'' > u' \to +\infty$. Therefore $O(1)$ in (6) can be replaced by $o(1)$; the resulting inequality together with the obvious one $\int_{u'}^{u''} du/\theta(u) \leqq \lambda(u',u'')$, shows that $\lambda(u',u'')$ is approximately additive. Therefore Theorem A can be applied, and it yields (7).

Theorem 3 yields a short proof of the following ingenious and heretofore quite difficult addition to the Ahlfors Distortion Theorem due to B. G. Eke:

Theorem (Eke [2]). *Let* R *be a strip domain. Then the quantities*

$$\underline{x}(u) - \int_{u_0}^{u} \frac{du}{\theta(u)} \quad \text{and} \quad \overline{x}(u) - \int_{u_0}^{u} \frac{du}{\theta(u)} , \tag{9}$$

(here $\underline{x}(u) = \min_{w \in \theta_u} \operatorname{Re} f(w)$ *and* $\overline{x}(u) = \max_{w \in \theta_u} \operatorname{Re} f(w)$*) tend to a common limit* $\beta\,(-\infty<\beta\leqq+\infty)$ *as* $u \to +\infty$.

For the proof, consider Ahlfors' Distortion Theorem [1]: If $u_0 < u' < u''$ and $\int_{u'}^{u''} du/\theta(u) > 2$ then $\underline{x}(u'') - \overline{x}(u') \geq \int_{u'}^{u''} du/\theta(u) - 4$. An immediate consequence is that the two quantities in (9) are either simultaneously bounded or else tend simultaneously to $+\infty$ as $u \to +\infty$. The second alternative corresponds to $\beta = +\infty$ in the theorem. If the first alternative holds then

$\lambda(u',u'') \leqq \overline{x}(u'') - \underline{x}(u')$

$$= [\overline{x}(u'') - \int_{u_0}^{u''} \frac{du}{\theta(u)}] - [\underline{x}(u') - \int_{u_0}^{u'} \frac{du}{\theta(u)}] + \int_{u'}^{u''} \frac{du}{\theta(u)}$$

$$\leqq O(1) + \int_{u_1}^{u_2} \frac{du}{\theta(u)} .$$

Therefore Theorem 3 applies; it shows that the quantities (9) have the common finite limit C.

5. Before discussing further applications of Theorem A we return to the abstract situation ($\{V_s\}$ is a general family of crosscuts defining the prime end ζ of R) in order to describe our second main result, Theorem B. This theorem provides estimates for the imaginary part of $f\colon R \to S$.

Let $\{H_t\}_{0<t<1}$ be a family of disjoint arcs in R which have their initial points on V_{s_0} and which tend to ζ. For simplicity we require that each H_t intersects each V_s exactly once. Thus the grid $\{V_s\}$ and $\{H_t\}$ in R is topologically the same as that given by the vertical crosscuts and horizontal lines in S. Extend the family $\{H_t\}$ to include the ideal boundary parts H_0 and H_1 which correspond under $f\colon R \to S$ to the lower and upper boundaries of S. If $0 \leqq t' < t'' \leqq 1$, $s_0 \leqq s' < s''$ then the notation $\lambda(V_{s'},V_{s''};H_{t'},H_{t''})$ denotes the extremal length of all arcs in the simply connected region determined by $V_{s'},V_{s''},H_{t'},H_{t''}$ which join $V_{s'}$ to $V_{s''}$; $\lambda(H_{t'},H_{t''};V_{s'},V_{s''})$ denotes the conjugate extremal distance. (Note that $\lambda(V_{s'},V_{s''};H_0,H_1)$ can also be written as $\lambda(s',s'')$.)

We now give a condition on $\{H_t\}$ which is useful for proving that $\operatorname{Im} f(w) \to t$ as $s \to +\infty$ with $w \in V_s \cap H_t$.

Condition B. *Given* $0 < t < 1$ *and* $\varepsilon > 0$, *there is an* $a = a(t,\varepsilon) > 0$, $M = M(t)$, $S_0 = S_0(t,\varepsilon)$ *and a function* $d(s) > s$ *where* $d(s)$ *also depends on* t *and* ε, *such that if* $s > S_0$ *then*

$$\lambda(H_0,H_t;V_s,V_{d(s)}) \geqq (t-\varepsilon)\lambda(H_0,H_1;V_s,V_{d(s)}) , \tag{10}$$

$$\lambda(H_t,H_1;V_s,V_{d(s)}) \geqq (1-t-\varepsilon)\lambda(H_0,H_1;V_s,V_{d(s)}) . \tag{11}$$

$$a(t,\varepsilon) \leqq \lambda(V_s,V_{d(s)};H_0,H_1) \leqq M(t) . \tag{12}$$

Theorem B. *Suppose* $\{V_s\}$ *is approximately additive; suppose* $\{H_t\}$ *satisfies Condition B, and suppose* $\lambda(s_1,s_2)$ *is a continuous function of* s_1 *and* s_2. *Let* $0 < t < 1$. *Then*

$$\lim_{s\to+\infty} [\operatorname{Im} f(w)-t] = 0 \quad (w \in V_s \cap H_t) .$$

6. A proof of Theorem B will be given in [10]. At present we wish to explain a general method for applying Theorems A and B to obtain estimates of $f\colon R \to S$ near ζ.

Suppose an explicit region R and $\zeta \in R$ are given. Let $\underset{\sim}{c} = \underset{\sim}{c}(s,t)$ be a diffeomorphic mapping of the closed half-strip $\{s+it \mid s_0 \leqq s < +\infty, 0 \leqq t \leqq 1\}$ into $R \cup H_0 \cup H_1$. Define V_s by $t \mapsto V_s(t) = \underset{\sim}{c}(s,t)$ for

$0 \leq t \leq 1$, and H_t by $s \to H_t(s) = \underset{\sim}{c}(s,t)$ for $s_0 \leq s < +\infty$. (In practice one tries to choose $\underset{\sim}{c}(s,t)$ so that the corresponding $\{V_s\}$ and $\{H_t\}$ will approximate the level lines Re $f(w)$ = const. and Im $f(w)$ = const.) The extremal distances $(V_{s'},V_{s''};H_{t'},H_{t''})$ are then calculated by known formulas (see Theorem 14 of [9]). In this way the hypotheses of Theorems A and B can be verified and asymptotic estimates of f can be obtained.

Suppose, for example, that R is bounded by the graphs of C^1 functions $\varphi_+ > \varphi_-$,

$$(13) \qquad R = \{w: u + iv \mid -\infty < u < +\infty,\ \varphi_-(u) < v < \varphi_+(u)\} ,$$

and that ζ is the prime end determined by Re $w \to +\infty$, $w \in R$. We may choose

$$(14) \qquad \underset{\sim}{c}(s,t) = (u,v) = (s, t\varphi_+(s) + (1-t)\varphi_-(s)) .$$

An explicit calculation yields ($\theta(s)$ = length of $V_s = \varphi_+(s) - \varphi_-(s)$)

$$(15) \quad \frac{1}{t_2-t_1}\int_{s_1}^{s_2}\frac{ds}{\theta(s)} \leq \lambda(V_{s_1},V_{s_2};H_{t_1},H_{t_2}) \leq \frac{1}{t_2-t_1}\int_{s_1}^{s_2}\frac{ds}{\theta(s)} + \frac{e(s_1,s_2)}{t_2-t_1} ,$$

where

$$(16) \qquad 0 \leq e(s_1,s_2) \leq \int_{s_1}^{s_2} \frac{(\varphi_+'(s))^2 + (\varphi_-'(s))^2}{\theta(s)}\, ds .$$

It can be seen that the hypotheses of Theorems A and B will be satisfied if the integral in (16) approaches zero as $s_2 > s_1 \to +\infty$. In this way we obtain

Theorem 4. *Let* R *be given by (13) and suppose that*

$$\int_{s_0}^{\infty} \frac{(\varphi_+'(s))^2 + (\varphi_-'(s))^2}{\varphi_+(s) - \varphi_-(s)}\, ds < \infty . \tag{17}$$

Let $f\colon R \to S$ *be the conformal mapping with* $\operatorname{Re} f(w) \to +\infty$ *as* $\operatorname{Re} w \to +\infty$. *Then*

$$f(u+iv) = \int_{s_0}^{u} \frac{ds}{\varphi_+(s) - \varphi_-(s)} + C + \frac{v - \varphi_-(u)}{\varphi_+(u) - \varphi_-(u)}\, i + o(1) , \tag{18}$$

where C *is a real constant and* $o(1) \to 0$ *as* $u \to +\infty$.

Theorem 4 is an improvement of previously known results of this type (Warschawski [15] p. 296 and p. 323). In order to obtain (18) there it is necessary to assume, in addition to (17), that R is an L-strip of boundary inclination zero which, in particular, implies that $\varphi_\pm'(s) \to 0$ as $s \to +\infty$. Theorem 4 shows that this additional condition is not needed.

7. A detailed proof of Theorem 4, as well as other asymptotic estimates that can be obtained this way, will appear in [10] and a future paper. Here we shall only sketch the ideas behind some of these applications.

Suppose R, given by (13), is symmetric; i.e., $\varphi_+ = -\varphi_-$. One can then choose the cross cuts V_s to be circular arcs which are normal to the boundary of R; s can be taken as the abscissa of the point where V_s intersects ∂R. The arc H_t can be chosen so that it divides each V_s into the ratio $t: 1 - t$. When Theorems A and B are applied one obtains the result of Gol'dberg-Stročik [3] which they proved with the Teichmüller-Wittich-Belinski Theorem.

Consider again an arbitrary R in the form (13). One can choose V_s to be the straight line cross cut of R which is normal to φ_- at the point $(s,\varphi_-(s))$. H_t can be taken as the arc $t\,\varphi_- + (1-t)\varphi_+$. Under suitable hypotheses (for example, the curvature of $\varphi_\pm$ must be integrable) one obtains asymptotic estimates of the type appearing in Warschawski [15], Theorem IX].

A more sophisticated choice of H_t can be made. Recall that H_t should approximate the level line $\operatorname{Im} f = t$ in R. One has $\operatorname{Im} f = 0$ on φ_-, $\operatorname{Im} f = 1$ on φ_+. We approximate the φ_- and φ_+ curves near the endpoints of V_s by the osculating circles there. Suppose these circles bound an annulus. Then along V_s, $\operatorname{Im} f$ would be expected to behave like the harmonic measure of the annulus. This harmonic measure can be calculated in terms of $\varphi_\pm$; in this way we are led to a formula for H_t which is very different from the previous choice. The resulting asymptotic estimate for f agrees with that in Stročik [13] in the special case when R has constant width.

8. Once again we consider $f: R \to S$ where R is the region in the $w = u + iv$ plane between the graphs of $v = \varphi_-(u)$ and $v = \varphi_+(u)$ (see

(13)). By use of the upper estimate in (15) for $t_0 = 0$, $t_1 = 1$ and an approximation argument we obtain

Theorem 5. *Suppose* $\varphi_+(u)$ *and* $\varphi_-(u)$ *are absolutely continuous for* $a \leqq u \leqq b$ *and* $\theta(u) = \varphi_+(u) - \varphi_-(u) \geqq \ell > 0$ *for* $a < u < b$. *Then*

$$(19) \qquad \lambda(a,b) \leqq \int_a^b \frac{du}{\theta(u)} + \frac{10}{\ell}\left[\int_a^b \frac{(\varphi_+')^2}{1+|\varphi_+'|}\,du + \int_a^b \frac{(\varphi_-')^2}{1+|\varphi_-'|}\,du\right]$$

provided each integral inside the brackets is smaller than $\ell/8$. *In the special case when* $\varphi_+(u) \geqq \ell/2$ *and* $\varphi_-(u) \leqq -\ell/2$ *in* $[a,b]$ *the latter restriction is not needed, and (19) holds with the constant* 10 *replaced by* 5.

We intend to publish the proof of Theorem 5 (and Theorems 6,7 below) in a future paper. From Theorem 5 one can derive an improved version of J. Lelong-Ferrand's extension of the Ahlfors Second Fundamental Inequality [6; p. 201]. The error term will have the form of the bracketed quantities in (19). Note that this error integral is smaller than the variation of $\varphi_\pm$:

$$(20) \qquad \int_a^b \frac{(\varphi')^2}{1+|\varphi'|}\,du \leqq \int_a^b |\varphi'|\,du = V(\varphi;a,b)\ .$$

This indicates that Theorem 5 might also lead to an extension of Ahlfors' Second Fundamental Inequality in the form of Jenkins-Oikawa [4]. Indeed, one can use Theorem 5 to derive the following modified (namely, $\theta_i(u) \leqq L$ is not needed) version of their result:

Theorem 6. *Suppose* $Q = \{-\theta_1(u) < v < \theta_2(u),\ a < u < b\}$ *where* θ_1, θ_2 *are positive functions and have respective finite total variations* V_1, V_2 *on* $[a,b]$; *Suppose* $\theta_i(u) \geqq \theta^{(m)}$ *on* $[a,b]$ $(i=1,2)$. *Let* $\theta(u) = \theta_1(u) + \theta_2(u)$. *Then*

$$\lambda(a,b) \leqq \int_a^b \frac{du}{\theta(u)} + \frac{5}{2\theta^{(m)}}(V_1 + V_2)\ . \tag{21}$$

When these results are combined with Theorems A and B (as in the proof of Theorem 4) the following asymptotic expansion of f is obtained.

Theorem 7. *Let* $R = \{w = u + iv \mid -\infty < u < +\infty,\ \varphi_-(u) < v < \varphi_+(u)\}$ *where* φ_+, φ_- *are absolutely continuous,* $\theta(u) = \varphi_+(u) - \varphi_-(u) \geqq \ell > 0$, *and*

$$\int_{u_0}^{\infty} \frac{(\varphi_+')^2}{1+|\varphi_+'|}\, du < \infty, \quad \int_{u_0}^{\infty} \frac{(\varphi_-')^2}{1+|\varphi_-'|}\, du < \infty\ . \tag{22}$$

Then

$$f(w) = \int_{u_0}^{u} \frac{ds}{\theta(s)} + i\,\frac{v-\varphi_-(u)}{\theta(u)} + C + o(1) \qquad (w = u + iv)\ , \tag{23}$$

where C *is a real constant and* $o(1) \to 0$ *as* $u \to +\infty$.

It should be remarked that in case $\theta(u)$ is bounded away from 0 and $+\infty$, Theorem 7 will have wider applicability than Theorem 4. Theorem 7 is an improved version of Théorème VI.13a of J. Lelong-Ferrand [6] in that it does not require $\theta(u)$ to be bounded.

9. Our final application of Theorems A and B is to obtain a condition on a region R, in the form (13), which will imply that $\operatorname{Im} f$ is approximately linear along each Ahlfors cross cut θ_u.

Theorem 8. *Let* R *be as in (13). Let* H_t $(0 < t < 1)$ *be the graph of* $v = t\,\varphi_-(u) + (1-t)\varphi_+(u)$. *Assume that the three conditions of Theorem A are satisfied, that* $\theta(u) = \varphi_+(u) - \varphi_-(u)$ *is bounded away from zero and infinity, and that* φ_+ *and* φ_- *are uniformly continuous. Then*

$$(24) \qquad \lim_{u\to+\infty} [\operatorname{Im} f(w) - t] = 0 \qquad (w \in H_t \cap \theta(u)) .$$

A proof of this theorem will appear in [10].

REFERENCES

1. Ahlfors, L. *Untersuchungen zur Theorie der konformen Abbildung und der ganzen Funktionen,* Acta Societatis Scientiarum Fennicae, Nova Ser. A, Vol. 1. No. 9: 1-40 (1930).

2. Eke, B. G. *Remarks on Ahlfors' Distortion Theorem,* Journal d'Analyse Math. 19: 97-134 (1967).

3. Gol'dberg, A. and Strŏcik, T. *Conformal mapping of symmetric half-strips and angular domains,* Litovskii Mat. Sbornik 6: 227-239 (1966) (Russian).

4. Jenkins, J. A. and Oikawa, K. *On results of Ahlfors and Hayman,* Illinois Journal of Math. 15: 664-671 (1971).

5. Jenkins, J. A. and Oikawa, K. *Conformality and semi-conformality at the boundary,* (to appear).

6. Lelong-Ferrand, J. *Représentation conforme et transformations a intégrale de Dirichlet bornée, Gauthier-Villars, Paris, 1955.*

7. Obrock, A. E. *On bounded oscillation and asymptotic expansion of conformal strip mappings,* Transactions, Amer. Math. Soc., 173: 183-201 (1972).

8. Oikawa, K. *On angular derivatives of univalent functions,* Kodai Sem. Rep. (to appear).

9. Rodin, B. *The method of extremal length,* Bulletin, Amer. Math. Soc., 80: 587-606 (1974).

10. Rodin, B. and Warschawski, S. E. *Extremal length and the boundary behavior of conformal mappings,* Ann. Acad. Sci. Fenn. (to appear).

11. Rodin, B. and Warschawski, S. E. *Extremal length and univalent functions. I The angular derivative*, (to appear).

12. Stročik, T. V. *On conformal mapping of half-strips of constant width*, Ukrain. Mat. Ž., 21: 60-72 (1969) = Ukrainian Math. J., 21: 48-57 (1969).

13. Stročik, T. V. *On conformal mapping of a class of half-strips*, Sibirskii Mat. Ž., 11: 859-878 (1970) = Siberian Math. J., 11: 647-661 (1970).

14. Teichmüller, O. *Untersuchungen über konforme und quasikonforme Abbildung*, Deutsche Mathematik, 3: 621-678 (1938).

15. Warschawski, S. E. *On conformal mapping of infinite strips*, Transactions, American Math. Soc. 51: 280-355 (1942).

16. Warschawski, S. E. *On the boundary behavior of conformal maps*, Nagoya Math. Journal, 30: 83-101 (1967).

17. Warschawski, S. E. *Remarks on the angular derivative*, Ibid. 41: 19-32 (1971).

University of California, San Diego
La Jolla, California 92093
U.S.A.

PROPERTIES OF STARLIKENESS AND CONVEXITY

PRESERVED BY SOME INTEGRAL OPERATORS

by

Grigore Stefan Sălăgean

1. Introduction. Let $U=\{|z|<1\}$ and S be the class of functions $f(z)=z+a_2z^2+\ldots$ analytic and univalent in U. We say that $f \in S$ is starlike of order α, $0\leq\alpha<1$, if

(1) $\quad \mathrm{Re}(zf'(z)/f(z))>\alpha, \ \forall z \in U,$

and we denote such a class of funktions by $S^*(\alpha)$. We say that $f \in S$ is convex of order α, $0\leq\alpha<1$, if

(2) $\quad \mathrm{Re}(zf''(z)/f'(z)+1)>\alpha, \ \forall z \in U,$

and we denote such a class of functions by $S^c(\alpha)$. For $\alpha=0$ $S^*(0)=S^*$ is the class of starlike functions and $S^c(0)=S^c$ is the class of convex functions.

In 1965, R.J.Libera [2] considered the operator $L:H_o(U)\to H_o(U)$, $H_o(U)=\{f|f(z)=z+a_2z^2+\ldots$ analytic in U$\}$, where for $F \in H_o$, $L(F)=f \in H_o$ and

(3) $$f(z)= \frac{2}{z}\int_0^z F(\zeta)d\zeta \ .$$

He proved that

(4) $\quad L(S^*)\subset S^* \qquad L(S^c)\subset S^c \ .$

S.M. Bernardi [1] obtained a generalization of these results by replacing Libera's operator by the operators

(5) $$L_n:H_o\to H_o, \quad L_n(F)=f, \quad f(z)=nz^{-n+1}\int_0^z \zeta^{n-2}F(\zeta)ds, \ n=1,2,\ldots$$

P.T. Mocanu gave a very simple proof for Libera's results (4) using the following theorem (due to S.S. Miller [3]).

Theorem 1. Let Φ (u,v) be a complex function, $\Phi:D\to\mathbb{C}$, $D\subset\mathbb{C}^2$ ($\mathbb{C}$ is the complex plain), where $u=u_1+iu_2$, $v=v_1+iv_2$. Suppose that Φ verifies the conditions

(i) Φ is continuous in D;

(ii) $(1,0) \in D$ and $\mathrm{Re}\Phi(1,0) > 0$;

(iii) $\mathrm{Re}\,\Phi(u_2 i, v_1) \leq 0$ <u>for all</u> $(u_2 i, v_1)$ <u>in D such that</u> $v_1 \leq -1/2(1+u_2^2)$.

<u>Let $p(z)=1+p_1 z+\ldots$ be analytic in the unit disc U such that $(p(z), zp'(z)) \in D$, for all z in U and furthermore $\mathrm{Re}\,\Phi(p(z), zp'(z)) > 0$, for all z in U. Then $\mathrm{Re}\,p(z) > 0$.</u>

Using this theorem P.T. Mocanu obtained also that

(6) $\quad L(S^*) \subset S^*(\gamma), \quad L(S^c) \subset S^c(\gamma)$,

where γ is the same for both and

(7) $\quad \gamma = (-3+\sqrt{17})/4$.

2. N. Pascu considered the operators L_α, $0 < \alpha \leq 1$,

(8) $$L_\alpha : H_0 \to H_0, \quad L_\alpha(F)=f, \quad f(z) = \frac{1}{\alpha} z^{1-\frac{1}{\alpha}} \int_0^z \zeta^{\frac{1}{\alpha}-2} F(\zeta)\,d\zeta$$

which generalize the operators (3)(for $\alpha=1/2$) and (5)(for $\alpha=1/n$).

It can be shown that Libera's operator transforms largest class than S^*, S^c in S^*, S^c.

In this paper is studied the following problem: Let F be a function in S such that

(9) $\quad \mathrm{Re}(zF'(z)/F(z)) > \beta$

where β is a real number; we consider the function $f=L_\alpha(F)$, $(0 < \alpha \leq 1)$. It must be found those α and β in R (the plane $\alpha O \beta$) such that

(10) $\quad \mathrm{Re}(zf'(z)/f(z)) > \gamma(\alpha,\beta)$

and

(11) $\quad \gamma(\alpha,\beta) \geq 0$,

i.e. $f \in S^*(\gamma)$.

For $\beta < 1$ (more interesting is $\beta \leq 0$) and $\gamma < 1$ we determine the sufficient conditions (a region in the plane $\alpha O \beta$) such that $f \in S^*(\gamma)(\gamma \geq 0)$. The result is the following

Theorem 2. With the above conditions

a) if $0 < \alpha \leq 1/2$ and $\alpha/(2(\alpha-1)) \leq \beta < 1$ then

$$(12) \qquad \gamma=\gamma_A(\alpha,\beta) = \frac{2\alpha\beta+\alpha-2+\sqrt{4\alpha^2\beta^2-12\alpha^2\beta+8\alpha\beta+9\alpha^2-4\alpha+4}}{4\alpha} \text{ and } \gamma_A \geq 0,$$

b) if $1/2 < \alpha \leq 1$ and $(\alpha-1)/(2\alpha) \leq \beta \leq \dfrac{3\alpha - \sqrt{8\alpha}}{2\alpha}$ then

$$(13) \qquad \gamma=\gamma_B(\alpha,\beta) \ \frac{2\alpha\beta+\alpha-\sqrt{4\alpha^2\beta^2-12\alpha^2\beta+9\alpha^2-8\alpha}}{4\alpha} \text{ and } \gamma_B \geq 0;$$

b′) if $1/2 < \alpha \leq 1$ and $(\alpha-1)/(2\alpha) < (3\alpha-\sqrt{8\alpha})/(2\alpha) < \beta < 1$ then $\gamma=\gamma_A$. (See also the diagram).

<u>Proof.</u> First we form the function $\Phi(u,v)$. Set

$$(14) \qquad zf'(z)/f(z)-\gamma=p(z)(1-\gamma), \quad p(z)=1+p_1z+\ldots ,$$

Then

$$(15) \qquad zF'(z)/F(z)-\beta=p(z)(1-\gamma)+\gamma-\beta+\frac{\alpha zp'(z)(1-\gamma)}{1-\alpha+\alpha\gamma+\alpha p(1-\gamma)}$$

We note

$$(16) \qquad u=p, \quad v=zp'$$

and so we form the function

$$(17) \qquad \Phi(u,v)=u(1-\gamma)+\gamma-\beta+\frac{\alpha v(1-\gamma)}{1-\alpha+\alpha\gamma+\alpha u(1-\gamma)} .$$

Now we require that $\Phi(u,v)$ verify the conditions (i), (ii), (iii) of Theorem 1.

For $D=\mathbb{C} \setminus \{-\frac{1-\alpha+\alpha\gamma}{\alpha(1-\gamma)}\} \times \mathbb{C}$ (i) is verified and also (ii).

From the request to verify (iii) and $0 < \alpha \leq 1$, $\beta < 1$ we obtain the statements of this theorem.

Indeed

$$\operatorname{Re}\Phi(u_2 i,v_1) = \gamma-\beta+\operatorname{Re}\frac{\alpha v_1(1-\gamma)}{1-\alpha+\alpha\gamma+\alpha u_2(1-\gamma)i} =$$

$$= \gamma-\beta+(1-\gamma)\frac{\alpha v_1(1-\alpha+\alpha\gamma)}{(1-\alpha+\alpha\gamma)^2+\alpha^2u_2^2(1-\gamma)^2} \leq$$

$$= \gamma-\beta-(1-\gamma)\frac{\alpha(1+u_2^2)(1-\alpha+\alpha\gamma)}{2[(1-\alpha+\alpha\gamma)^2+\alpha^2u_2^2(1-\gamma)^2]} = \frac{A+Bu_2^2}{2C} ,$$

where

$$A = 2(\gamma-\beta)(1-\alpha+\alpha\gamma)^2 - \alpha(1-\gamma)(1-\alpha+\alpha\gamma)$$

$$B = 2\alpha^2(\gamma-\beta)(1-\gamma)^2 - \alpha(1-\gamma)(1-\alpha+\alpha\gamma)$$

$$C = (1-\alpha+\alpha\gamma)^2 + \alpha^2 u_2^2(1-\gamma)^2 > 0$$

$$\operatorname{Re}\Phi(u_2 i, v_1) \leq 0 \iff A \leq 0 \quad \text{and} \quad B \leq 0$$

From $A \leq 0$ we obtain $\gamma \leq \gamma_A(\alpha,\beta)$, where

$$\gamma_A(\alpha,\beta) = \frac{2\alpha\beta+\alpha-2+\sqrt{4\alpha^2\beta^2-12\alpha^2\beta+8\alpha\beta+9\alpha^2-4\alpha+4}}{4\alpha}$$

and $\gamma_A(\alpha,\beta) \geq 0$ if $\beta \geq \frac{\alpha}{2(\alpha-1)}$

Also $B \leq 0$ if $\beta > (3\alpha-\sqrt{8\alpha})/(2\alpha)$ or if

$$\frac{\alpha-1}{2\alpha} \leq \beta \leq \frac{3\alpha-\sqrt{8\alpha}}{2\alpha}, \quad (\text{for } \alpha \geq 1/2)$$

and

$$\gamma \leq \gamma_B(\alpha,\beta) = \frac{2\alpha\beta+\alpha-\sqrt{4\alpha^2\beta^2-12\alpha^2\beta+9\alpha^2-8\alpha}}{4\alpha}$$

$\left(\beta \geq \frac{\alpha-1}{2\alpha} \Rightarrow \gamma_B(\alpha,\beta) \geq 0\right)$.

It can be verified that if

$$\alpha \geq \frac{1}{2} \text{ and } \frac{\alpha-1}{2\alpha} \leq \beta \leq \frac{3\alpha-\sqrt{8\alpha}}{2\alpha} \text{ then } \gamma_B(\alpha,\beta) \leq \gamma_A(\alpha,\beta) .$$

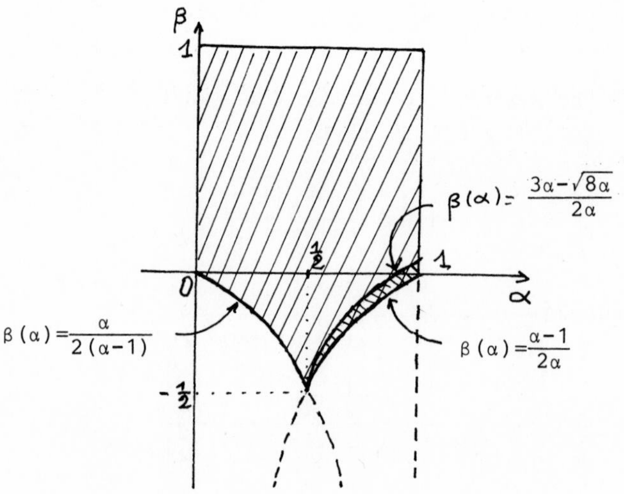

Remarks. (special cases). 1). If $\alpha=1/2$, $\beta=0$ we obtain Libera's result (4):

2). The case $\beta=0$, $0 < \alpha \leq 1$ was studied by Pascu [4].

3). If $\alpha=1$, $\beta=0$ then

$$\mathrm{Re}(zf''(z)/f'(z)+1) > 0 \Rightarrow \mathrm{Re}(zf'(z)/f(z)) > 1/2,$$

a well known result of Marx and Strohhäcker.

4). For $\alpha=1/2$, $\beta=-1/2$ our result is sharp (it can be verified choosing $F(z)=z/(1+z)^3$).

3. An analogue problem as (9) and (10) can be considered for convexity. It is the following

Let F be a function in S such that

(18) $\quad \mathrm{Re}(zF''(z)/F'(z)+1) > \beta,\ \beta < 1;$

we consider the function $f=L_\alpha(F)$, $(0 < \alpha \leq 1)$. It must be found those α and β in R such that

(19) $\quad \mathrm{Re}(zf''(z)/f'(z)+1) > \gamma(\alpha,\beta)$

and

(20) $\quad \gamma(\alpha,\beta) \geq 0,$

i.e. $f \in S^c(\gamma)$.

Using the same method it obtains the same function Φ (u,v) where $u=p$ and $zf''/f'=(p-1)(1-\gamma)$ and the results are the same.

Acknowledgement. The author achnowledges with thanks the valuable suggestions of Professor P.T. Mocanu of the Babes-Bolyai University of Cluj-Napoca.

R E F E R E N C E S

1. S.M. Bernardi, Convex and starlike univalent functions, Trans. A.M.S. 135 (1969), 429-446.

2. R.J. Libera, Some classes of regular univalent functions, Proc. A.M.S. 16 (1965), 755-758

3. S.S. Miller, Differential inequalities and Carathéodory functions, Bull. A.M.S 81 (1975), 79-81

4. N.N.Pascu, Alpha-starlike-convex functions, to appear, (Romanian).

DEPARTMENT OF MATHEMATIKS, BABES-BOLYAI UNIVERSITY,

CLUJ-NAPOCA, ROMANIA

CARLESON-SETS AND FIXED-POINTS OF SCHLICHT FUNCTIONS

Hans Stegbuchner
Institut für Mathematik
Universität Salzburg, Austria

1. Definitions and preliminary results

Let $D = \{z: |z|<1\}$ be the unit disc and $T = \{z: |z|=1\}$ the unit circle in the complex plane $\mathbb{C}$. Let E be a closed subset of T with linear measure zero and

$$T \setminus E = \bigcup_{\nu=1}^{\infty} I_\nu$$

the union of open arcs $I_\nu = \{e^{it}: a_\nu < t < b_\nu\}$. E is called a Carleson-set, if

$$\sum_{\nu=1}^{\infty} l_\nu \log l_\nu > -\infty, \tag{1}$$

where l_ν denotes the length of I_ν: $l_\nu = |b_\nu - a_\nu|$.

There is a close connection between Carleson-sets and sets of uniqueness for functions regular in D and Lipschitz continuous in $\overline{D} = D \cup T$. The classical result of F. and M. Riesz states that $E \subset T$ is a set of uniqueness for the class of bounded analytic functions if and only if E has Lebesgue measure zero. The following theorem is due to A. Beurling and L. Carleson (see [1] and [2]).

Theorem A. *A closed set* $E \subset T$ *of Lebesgue measure zero is a set of uniqueness for the class of functions* $A_\alpha = \{f$*: f analytic in* D *and Lipschitz continuous of order* α *in* $\overline{D}$ $(0 < \alpha \leq 1)\}$ *if and only if* E *is not a Carleson-set.*

In [15] the results of Carleson was generalized as follows:

Theorem B. *A closed set* $E \subset T$ *with Lebesgue measure zero is a set of uniqueness for the class* $A(\omega) = \{f$*: f analytic in* D *and continuous in* $\overline{D}$ *with modulus of continuity* $\omega(\delta)$; $\omega(\delta)=O((\log\delta)^{-\varepsilon}$, $\varepsilon > 0\}$ *if and only if* E *is not a "Carleson-set with modulus of continuity* $\omega(\delta)$*", that is*

$$\sum_{\nu=1}^{\infty} l_\nu \log\omega(l_\nu) = -\infty.$$

It was shown by Ch. Pommerenke [12] and T.A. Metzger [10] recently that there is also a close connection between Carleson-sets and the set of limit points of a Fuchsian group.

We call an infinite sequence of points $\{z_k\}_{k=1}^{\infty}$ with $0<|z_k|<1$ a Blaschke sequence if

$$\sum_{k=1}^{\infty}(1-|z_k|) < \infty. \tag{2}$$

It is known that the Blaschke product B(z)

$$B(z) = \prod_{k=1}^{\infty}\frac{\overline{z}_k}{|z_k|}\cdot\frac{z_k-z}{1-\overline{z}_k z}$$

is convergent if and only if the series (2) is convergent.

Finally let S be the set of normalized schlicht functions $f(z) = z + a_2z^2 + \ldots$ in D.

In the following we need some lemmas:

Lemma 1. (J.D.Nelson [11]) *Let* $\{z_k\} = \{r_k e^{i\theta_k}\}_{k=1}^{\infty}$ *be a Blaschke sequence with limit set* $E_1 \subset T$ *and let* $E_2 = \{e^{i\theta_k}: k=1,2,\ldots\}$ *be the "radial projection" of the sequence* $\{z_k\}$ *on* T. *If* $E = E_1 \cup E_2$

is a Carleson-set, then there is an analytic function f(z) *the zero set of which is exactly the set* $Z = \{z_k\}_{k=1}^{\infty} \cup E_1$, *and its derivatives of all orders are bounded analytic functions.*

Lemma 2. (H.Hornich [9]) *Let* $f(z) = z + a_2 z^2 + ..$ *be a schlicht function. Then there is a bounded analytic function* b(z) *with* $f(z) = z/(1+z\cdot b(z))$.

Lemma 3. (J.D.Nelson [11]) *Let* $\{z_k\}_{k=1}^{\infty}$ *be a Blaschke sequence with limit set* E_1, $Z = \{z_k\}_{k=1}^{\infty} \cup E_1$ *and* $E_2 = \{e^{i\theta_k}: z_k = r_k e^{i\theta_k}\}$ *the radial projection of the sequence* $\{z_k\}$ *on* T. *If*

$$\int_{-\pi}^{\pi} \log \operatorname{dist}(e^{it}, Z)\, dt > -\infty$$

then $E = E_1 \cup E_2$ *is a Carleson-set.*

Lemma 4. (D.J.Caveny [4]) *Let f be a function analytic in* D *with derivative* f' *in the Hardy space* H^1. *Then the boundary zero set* E *of* f *is a Carleson-set.*

2. Fixed-points of schlicht functions

Let us now consider Lemma 2 with regard to the fixed-points of the schlicht function f(z). In the following let f(z) be different from the identical function $f(z) \equiv z$, which has every point as a fixed-point. If ζ is a fixed-point of f, that is $f(\zeta) = \zeta$, we have $b(\zeta) = 0$. Conversely, if ζ is a point with $b(\zeta) = 0$, then ζ is a fixed-point of f. Therefore the set of fixed-points of f(z) is exactly the zero set of b(z). Now we have the factorization theorem for bounded analytic functions: $b(z) = F(z)\cdot B(z)$, where B(z) is the Blaschke product, consisting of all zeros $z_k \in D$ of $b(z)$, and $F(z)$ is a bounded analytic function without zeros in D. This yields together with Lemma 2:

If $\{z_k\}_{k=1}^{\infty}$ $(0 < |z_k| < 1)$ *is the sequence of fixed-points of a normalized schlicht function* $f(z) \not\equiv z$, *then* $\{z_k\}$ *is necessarily a Blaschke sequence.*

The question arises wether a Blaschke sequence $\{z_k\}$ is the set of fixed-points of a schlicht function or not. I am not able to answer this question in general but in some special cases we can find necessary and sufficient conditions. Zinterhof [17] has also considered this problem and he proved the following theorem:

Theorem 1. *Let* $\{z_k\}$ *be a Blaschke sequence with finite limit set* $E = \{y_1, y_2, \ldots, y_L\}$ *so that* $\{z_k\}$ *is the union of* L *subsequences* $\{z_{kl}\}$ $(l = 1,2,..,L)$, *each of which has* y_l *as its limit point on* T. *Further if there are constants* B_l $(l = 1,2,..,L)$ *so that*

$$|z_{kl} - y_l| < B_l(1 - |z_{kl}|), \tag{3}$$

then there is a normalized schlicht function $f(z)$, *of which the set of fixed-points is exactly the sequence* $\{z_k\}$.

Condition (3) means that the convergence of the sequence $\{z_k\}$ happens in a non tangential manner.

Now I will generalize this theorem:

Theorem 2. [13] *Let* $\{z_k\}$ *be a Blaschke sequence with limit set* E_1. *Let* E_2 *be the radial projection of the Blaschke sequence on the unit circle* T *and let* $E = \overline{E_1 \cup E_2}$. *If* E *is a Carleson-set, then there is a normalized schlicht function* $f(z)$ *of which the set of fixed-points is exactly the sequence* $\{z_k\}$. *Further we can construct* $f(z)$ *in such a manner that* f *is convex of order* α *with any* $\alpha \in [0,1)$.

Proof. Since E is a Carleson-set, we can apply Lemma 1 to construct a function $g(z) = B(z)\cdot F(z)$ of which the zero set is exactly the set $Z = \{z_k\}_{k=1}^{\infty} \cup E_1$, so that

$$\sup_{z\in D} |g^{(n)}(z)| \le A_n \quad (n=0,1,\ldots). \tag{4}$$

Now let $f(z) = z + C\cdot z^2\cdot g(z)$ with some positive constant C to be determined later. It is clear that z is a fixed-point of the analytic function f if and only if $z = z_k$ $(k=1,2,..)$. Also we have $f(0) = 0$ and $f'(0) = 1$.

Now every convex function is schlicht; hence we are at the end if we can show that

$$\operatorname{Re}\,\{1 + \frac{z\cdot f''(z)}{f'(z)}\} > \alpha \tag{5}$$

since (5) is necessary and sufficient for f to be convex of order α. Let

$$K(z) = \frac{z\cdot f''(z)}{f'(z)} = \frac{C\cdot z(2g(z) + 4z\cdot g'(z) + z^2 g''(z))}{1 + C(2z\cdot g(z) + z^2 g'(z))}.$$

Fix any $\alpha \in [0,1)$ and choose the constant $C = C(\alpha)$ so that the following inequalities are fulfilled simultaneously:

$$\begin{aligned} &C\cdot(2A_0 + A_1) < \frac{1-\alpha}{2} \\ &C\cdot(2A_0 + 4A_1 + A_2) < \frac{1-\alpha}{2} \end{aligned} \tag{6}$$

Here A_i $i=0,1,2$ are the constants of (4). Hence we have

$$\sup_{z\in D} |K(z)| \le \frac{C\cdot\sup\,|2g(z) + 4z\cdot g'(z) + z^2 g''(z)|}{1 - C\cdot\sup\,|2z\cdot g(z) + z^2 g'(z)|} \le$$

$$\leq \frac{C(2A_0 + 4A_1 + A_2)}{1 - C(2A_0 + A_1)} < \frac{\frac{1-\alpha}{2}}{1 - \frac{1-\alpha}{2}} = \frac{1 - \alpha}{1 + \alpha} \leq 1 - \alpha .$$

Hence we have certainly the inequality

$$\mathrm{Re}\ \{\frac{z \cdot f''(z)}{f'(z)}\} = \mathrm{Re}\ K(z) > \alpha - 1,$$

which is equevalent with

$$\mathrm{Re}\ \{1 + \frac{z \cdot f''(z)}{f'(z)}\} > \alpha,$$

that is f(z) is convex of order α and the theorem is proved completely.

Remark: Let E_1 be the limit set on T of the fixed-points $\{z_k\}$ of the schlicht function f(z). Because of Lemma 2, E_1 is the limit set of the zeros of a bounded analytic function b(z), and hence the Lebesgue measure of E_1 is zero. In our theorem we have assumed that E_1 is a Carleson-set (because every closed subset of a Carleson-set is a Carleson-set). The next theorem shows that there are Carleson-sets of which the Hausdorff measure can be very large.

Theorem 3. [14] *Let* h(t) *be a measure determining function with* $\lim_{t \to 0} h(t)/t = \infty$. *If* E *is a Carleson-set with positive* h*-measure, then*

$$\int_0^1 \frac{dt}{h(t)} < \infty \tag{7}$$

Conversely, to each convex measure function h(t) *with finite integral* (7), *there exists a Carleson-set with positive* h*-measure.*

Particularly there exists Carleson-set with infinite α-dimensional Hausdorff measure for each $\alpha \in (0,1)$.Hence Theorem 2 is much more general then Theorem 1 with respect to the limit set of the fixed-points. But the later theorem is also more general with respect to the convergence of the sequence $\{z_k\}$ to the limit set E_1. One can show that convergence of the series

$$\sum_{k=1}^{\infty} (\operatorname{dist}(e^{i\theta_k}, E_1))^{\beta} \qquad (\beta \geqq 1) \tag{8}$$

is a sufficient condition that the projection E_2 of the Blaschke sequence on T is a Carleson-set. If we state "non tangential convergence" of $\{z_k\}$ to E_1, that is

$$\operatorname{dist}(z_k, E_1) \leq D(1 - |z_k|), \tag{9}$$

then we have the inequalities

$$(\operatorname{dist}(e^{i\theta_k}, E_1))^{\beta} \leq \operatorname{dist}(e^{i\theta_k}, E_1) \leq \operatorname{dist}(z_k, E_1) + (1 - |z_k|) \leq$$

$$\leq (D+1)(1 - |z_k|)$$

and therefore (8) is convergent, that is E_2 is a Carleson-set. But there exists Blaschke sequences $\{z_k\}$ with convergent series (8), where $\{z_k\}$ converges in a "very tangential" manner, e.g.

$$z_k = (1 - \frac{1}{k!}) \cdot \exp(i/k).$$

Obviously there is no inequality for this sequence like (9).

Now we shall consider subclasses of the family of normalized schlicht functions in order to obtain necessary and sufficient conditions for a Blaschke sequence $\{z_k\}$ to be the set of fixed-points of a schlicht function. Let H^p $(p > 0)$ denote the Hardy spaces and S^p the set of schlicht functions with derivative in H^p.

$$S^p = \{f \in S\colon f' \in H^p\}.$$

Then we have the following theorem:

<u>Theorem 4.</u> *A necessary and sufficient condition for a sequence* $\{z_k\}_{k=1}^{\infty}$ *to be the set of fixed-points of a schlicht function in* S^p *with* $p > 1$ *is*

$$\begin{aligned} &(i) \quad \sum_{k=1}^{\infty} (1 - |z_k|) < \infty \\ &(ii) \quad \int_{-\pi}^{\pi} \log \operatorname{dist}(e^{it}, Z)\, dt > -\infty \end{aligned} \tag{10}$$

with $Z = \overline{\{z_k\}_{k=1}^{\infty}}$.

Proof: With Lemma 2 we have for a schlicht function $f(z)$

$$f(z) = z - \frac{z^2 b(z)}{1 + z \cdot b(z)} = z + h(z).$$

Since $f' \in H^p$ $(p > 1)$, we have $h' \in H^p$. With some classical results of Hardy and Littlewood [6] we obtain

$$h'(z) = O((1 - |z|)^{-1/p})$$

and hence $h(z)$ is Lipschitz continuous of order α in the closed disc with $\alpha = 1 - 1/p > 0$. Since $b(z)$ is bounded we obtain that the closure of the set of fixed-points of f is identical with the closure of the zero set of b. The necessity of (i) is evident because of the factorization theorem. With Jensens formula we obtain for $s \in Z$:

$$- \infty < \int_{-\pi}^{\pi} \log |h(t)| \, dt = \int_{-\pi}^{\pi} \log |h(t) - h(s)| \, dt \leq$$

$$\leq \int_{-\pi}^{\pi} \log (\operatorname{dist}(e^{it}, Z))^{\alpha} \leq \text{const.} \int_{-\pi}^{\pi} \log \operatorname{dist}(e^{it}, Z) \, dt .$$

Conversely let us assume that (10) is satisfied. From Lemma 3 we deduce that $E = E_1 \cup E_2$ is a Carleson-set. Therefore we can apply Theorem 2 to construct a schlicht function of which the set of fixed-points is exactely the sequence $\{z_k\}$.

If $p = 1$ we have only the weaker Theorem 5:

Theorem 5. *Let* $\{z_k\}$ *be the fixed-points of a schlicht function in* S^1. *Then we have necessarily*

$$\text{(i)} \quad \sum_{k=1}^{\infty} (1 - |z_k|) < \infty$$

$$\text{(ii)} \quad \int_{-\pi}^{\pi} \log \operatorname{dist}(e^{it}, E) \, dt > -\infty$$

where E *is the limit set of* $\{z_k\}_{k=1}^{\infty}$.

Proof: Because of Lemma 4 we obtain that E is a Carleson-set; it is easy to varify that E is a Carleson-set if and only if the above mentioned integral (ii) is finite.

Conditions (i) and (ii) of Theorem 5 are not sufficient. For a counterexample take the Blaschke sequence

$$z_k = (1 - (k \cdot \log^2 k)^{-1}) \cdot \exp(i/k).$$

Since the limit set is only one point, E is a Carleson-set and so (i) and (ii) are satisfied. But Caughran has shown that there is no function $f \neq 0$ with $f' \in H^1$ so that $f(z_k) = 0$ $(k=1,2,..)$ (see [3]).

3. Fixed-points of convex functions

Now let us consider a familiar class of schlicht functions. We state the following theorem of Eenigenburg and Keogh:

Theorem 6. [5] *Let* f *be a convex function of order* α $(0 \leq \alpha < 1)$ *which is not of the form*

$$f(z) = a + b(1 - z.e^{i\gamma})^{2\alpha-1} \qquad \text{für } \alpha \neq 1/2$$

$$f(z) = a + b\cdot\log(1 - z.e^{i\gamma}) \qquad \text{für } \alpha = 1/2$$

$(a, b \in \mathbb{C}, \gamma \in \mathbb{R})$. *Then there is a* $\delta = \delta(f) > 0$ *such that* $f' \in H^{\beta}$ *with* $\beta = \frac{1}{2(1-\alpha)} + \delta$.

With the aid of this theorem we can prove the next theorem.

Theorem 7. *Conditions* (10) *are necessary and sufficient for a sequence* $\{z_k\}$ *to be the set of fixed-points of a convex function of order* α *with* $1/2 \leq \alpha < 1$. *Further every convex function of order* α *with* $1/2 \leq \alpha < 1$, *which is not of the form* $a + b\cdot\log(1-z\cdot e^{i\gamma})$, *is continuous in the closed disc and satisfies a Lipschitz condition of order* $\lambda = 2\alpha - 1$.

Proof: Since $\alpha \geq 1/2$, we deduce from Theorem 6 that the derivative of f is in H^p with $p = \frac{1}{2(1-\alpha)} + \delta > 1$ if f is not of the mentioned form. Therefore Theorems 4 and 2 can be applied to prove the theorem.

Now let us consider the exceptional functions mentioned above. Since we only consider normalized schlicht functions, these functions reduce to

$$f(z) = -e^{-i\gamma}\log(1 - z\cdot e^{i\gamma}) \quad \text{for } \alpha = 1/2$$

$$g(z) = \frac{e^{-i\gamma}}{2\alpha - 1}(1 - (1 - z\cdot e^{i\gamma})^{2\alpha-1}) \quad \text{for } \alpha > 1/2.$$

Lemma. *The only fixed-point in* D *of the function*

$$f(z) = -e^{-i\gamma}\log(1 - z\cdot e^{i\gamma})$$

is the point $z = 0$.

Proof: $f(z) = z$ is equivalent to $1 - w = e^{-w}$ $(w = z\cdot e^{i\gamma})$. A short computation yields that the fixed-points of $f(z)$ are the points $w = x + iy$ where y satisfies the equation

$$1 + \log(y/\sin y) = y\cdot \mathrm{ctg}\, y \qquad (-1 < y < 1)$$

and x is equal to $-\log(y/\sin y)$. Let $\varphi(y) = 1 + \log(y/\sin y) - y\cdot\mathrm{ctg}\, y$. Since $\varphi(-y) = \varphi(y)$, $\varphi(0) = 0$ and $\varphi(y)$ increases with y $(0 \le y \le 1)$, the lemma is proved.

Now let us consider $g(z)$. Since $g'(z) = 1 - (1 - z\cdot e^{i\gamma})^{2\alpha-2}$ and we have $-1 < 2\alpha-2 < 0$, it follows that $g'(z) = O((1 - |z|)^{2\alpha-2}$, which is equivalent for g to be in Lip $(2\alpha-1)$ $(2\alpha-1 > 0)$. The statement about the Lipschitz continuity is now obvious.

4. Further results concerning fixed-points of schlicht functions

Theorem 8. *If the sequence* $\{z_k\}$ *is the set of fixed-points of a schlicht function* $f(z)$ *which is continuous in the closed disc with modulus of continuity* $\omega(\delta)$, *then we have (i)* $\sum_{k=1}^{\infty} (1 - |z_k|) < \infty$

(ii) $E = E_1 \cup E_2$ *is a Carleson-set with modulus of continuity* $\omega(\delta)$.

If further $\omega(\delta) = O(\delta^{\alpha})$ $(0 < \alpha \le 1)$, *that is* $f(z)$ *is Lipschitz continuous on* $\overline{D}$, *then (i) and (ii) are also sufficient for* $\{z_k\}$ *to be the set of fixed-points of a schlicht function which is Lipschitz continuous in the closed disc.*

Proof: Statement (i) is trivial. Let E_1 be the limit set of the Blaschke sequence on T and $T\backslash E_1 = \bigcup_{\nu=1}^{\infty} I_\nu$, $I_\nu = (e^{ia_\nu}, e^{ib_\nu})$ and $l_\nu = |b_\nu - a_\nu|$. Since $h(z) = f(z) - z$ is continuous with modulus of continuity $\omega(\delta)$, we deduce from Jensens formula for any $s \in E_1$:

$$-\infty < \int_{-\pi}^{\pi} \log |h(t)|\, dt \leqq \sum_{\nu=1}^{\infty} \int_{a_\nu}^{b_\nu} \log |h(t) - h(s)|\, dt \leqq$$

$$C \cdot \sum_{\nu=1}^{\infty} \int_{a_\nu}^{b_\nu} \log \omega(l_\nu)\, dt = C \cdot \sum_{\nu=1}^{\infty} l_\nu \log \omega(l_\nu).$$

Hence E_1 is a Carleson-set with modulus of continuity $\omega(\delta)$. With the same arguments we have

$$\int_{-\pi}^{\pi} \log \omega(\mathrm{dist}(e^{it}, Z))\, dt > -\infty \tag{11}$$

In [16] it has been shown that $E = E_1 \cup E_2$ is a Carleson-set with modulus of continuity $\omega(\delta)$ if $\{z_k\}$ is a Blaschke sequence and the integral (11) is finite. Hence the first part of the theorem is proved.

If $\omega(\delta) = O(\delta^{\alpha})$ $(0 < \alpha \leqq 1)$, Theorem 2 can be applied to construct the desired function.

If we consider the class of bounded schlicht functions, only the trivial necessary conditions on the set of fixed-points $\{z_k\}$ are known:

$$\sum_{k=1}^{\infty} (1 - |z_k|) < \infty \text{ and}$$

$$\lambda(E_1) = 0, \text{ where } \lambda \text{ denotes Lebesgue measure.}$$

It seems not to be easy to obtain some restricted necessary and sufficient conditions, as it is suggested by the following considerations.

A bounded schlicht function is contained in Δ, the class of functions with bounded Dirichlet integral:

$$\int_{-\pi}^{\pi}\int_{0}^{1} |f'(re^{it})|^2 r \, dr \, dt < \infty.$$

Hence $h(z) = f(z) - z$ is in Δ. With a result of Caughran [3] we obtain the following theorem:

Theorem 9. *There exists a Blaschke sequence* $\{z_k\}_{k=1}^{\infty}$ *which converges to* 1 *and of which the radial projection* E_2 *is a Carleson-set with modulus of continuity* $\omega(\delta) = O((\log \delta)^{-1})$, *but there exists no bounded schlicht function with* $f(z_k) = z_k$ $(k=1,2,\ldots)$.

Hence we see that conditions (i) and (ii) of Theorem 8 are not sufficient if

$$\lim_{t \to 0} \omega(t)/t^{\alpha} = \infty \text{ for any } \alpha > 0.$$

But I think that the following is true:

Conjecture. *Let* $\{z_k\}$ *be a sequence in* D *with limit set* E_1 *and radial projection* E_2. *If*

$$\text{(i)} \quad \sum_{k=1}^{\infty} (\omega(1 - |z_k|))^K < \infty \text{ (with some } K \geq 1) \tag{12}$$

(ii) $E = E_1 \cup E_2$ *is a Carleson-set with modulus of continuity* $\omega(\delta)$,

then there exists a schlicht function f, *which is continuous in* $\overline{D}$ *with modulus of continuity* $\omega(\delta)$ *and of which the set of fixed-points is exactely the sequence* $\{z_k\}$.

Remark: If $\omega(\delta)$ is any function with the properties of a modulus of continuity, then we have $t = O(\omega(t))$, and hence every sequence $\{z_k\}$ with convergent series (12) is a Blaschke sequence.

To the end let us consider a Banach space of holomorphic functions introduced by H. Hornich [7]. Let B be the set of analytic and local schlicht functions in D ($f'(z) \neq 0 \ \forall z \in D$) with $f(0) = 0$, $f'(0) = 1$ and $\sup_{z \in D} |\arg f'(z)| < \infty$. With the operations

$$\cdot[f + g] = \int_0^z f'(\zeta) g'(\zeta) \, d\zeta \text{ and}$$

$$[\alpha \times f] = \int_0^z (f'(\zeta))^{\alpha} \, d\zeta \quad (\alpha \in \mathbb{R}) \text{ and } (f'(0))^{\alpha} = 1$$

B is a linear space. With $\|f\| = \sup_{z,w \in D} |\arg f'(z) - \arg f'(w)|$ B is a Banach space. The zero element of B is the constant function $f(z)=z$. Let S_B be the set of schlicht functions in B. One can show that $f \in S_B$ if $\|f\| < \pi$ [7]. Further we have the following estimate [8]: If $f \in B$ and $\|f\| = \gamma$, then $f'(z) = O((1 - |z|)^{2\gamma})$. Therefore the set of functions $f \in B$ with $\|f\| < 1/2$ contains only schlicht functions which are Lipschitz continuous of order $\alpha = 1 - 2\|f\|$ on the closed disc. Hence we can state the following corollary of Theorem 8:

Corollary. *Conditions* (10) *are necessary and sufficient for a sequence* $\{z_k\}$ *to be the set of fixed-points of a schlicht function* f *in the Hornich space* B *with* $\|f'\| < 1/2$.

Analogous to [17] we can construct a sequence $\{f_n\}$ of schlicht functions each of it has exactly $\{z_k\}$ as its set of fixed-points, so that $\{f_n\}$ converges uniformly to the identical function $f(z)=z$ which has every point as a fixed-point. Let us suppose that (10) is satisfied for the sequence $\{z_k\}$. Construct $f_n(z)$ like in Theorem 2: $f_n(z) = z + C_n z^2 g(z)$. Determine the constants C_n so that we have not only (6) but also $C_n A_0 < 1/n$. Each f_n is schlicht with set of fixed-points $\{z_k\}$ and we have the estimate:

$$\sup_{z \in D} |f_n(z) - z| = \|f_n - f\|_\infty \leqq C_n \sup_{z \in D} |g(z)| \leqq C_n A_0 < 1/n.$$

One can also show that $\{f_n\}$ converges in the Hornich space B to the zero element of B:

$$\sup_{z,w \in D} |\arg f_n'(z) - f_n'(w)| \to 0.$$

REFERENCES

[1] Beurling, A.: Ensembles exceptionnels. Acta Math. 72(1940), 1-13.

[2] Carleson, L.: Sets of uniqueness for functions regular in the unit circle. Acta Math. 87(1952), 325-345.

[3] Caughran, J.G.: Two results concerning the zeros of functions with finite Dirichlet integral. Can.J.Math. 21(1969), 312-316.

[4] Caveny, D.J. and W.P.Novinger: Boundary zeros of functions with derivative in H^p. Proc.Am.Math.Soc. 25(1970), 776-780.

[5] Eenigenburg, P.J. and F.R. Keogh: The Hardy class of some univalent functions and their derivatives. Mich.Math.J. 17(1970), 335-346.

[6] Hardy, G.H. and J.E. Littlewood: Some properties of fractional integrals II. Math. Zeitschr. 34(1931), 403-439.

[7] Hornich, H.: Ein Banachraum analytischer Funktionen in Zusammenhang mit den schlichten Funktionen. Monatsh.Math. 73(1969), 36-45.

[8] -- Über einen Banachraum analytischer Funktionen. Man.Math.1 (1969), 79-86.

[9] -- Über die Fixpunkte der schlichten Funktionen. Rend.Ist.di Matem. Univ. Trieste 2(1970), 54-58.

[10] Metzger, T.A.: On vanishing Eichler periods and Carleson-sets. (to appear).

[11] Nelson, J.D.: A characterisation of zero-sets for A^∞. Mich. Math.J. 18(1971), 141-147.

[12] Pommerenke, Ch.: On automorphic forms and Carleson-sets. (to appear).

[13] Stegbuchner, H.: Einige Bemerkungen über die Fixpunkte der schlichten Funktionen. Sb.Österr.Akad.Wiss., math.-nat.Kl., Abt.II (in Druck).

[14] -- Maß, Dimension und Kapazität von Carleson-Mengen.(to appear).

[15] -- Nullstellen analytischer Funktionen und verallgemeinerte Carleson-Mengen I und II. Sb.Österr.Akad.Wiss., math.-nat.Kl., Abt.II,183.Bd.(1974),463-503 und 184.Bd.(1975),83-97.

[16] -- Tangentiale Nullstellenfolge holomorpher Funktionen mit vorgegebenen Stetigkeitsmodul $\omega(\delta)$. Sb.Österr.Akad.Wiss., math.-nat.Kl.,Abt.II (in Druck).

[17] Zinterhof, P.: Konstruktion von schlichten Funktionen mit unendlich vielen Fixpunkten. Rend.Ist.di Matem.Univ.Trieste 3 (1971), 1-10.

Le lieu réduit et le lieu normal d'un morphisme

Constantin Bănică

On utilise les ensembles singuliers d'un faisceau cohérent pour étudier les points réduits et normals d'un morphisme plat d'espaces complexes et d'un morphisme plat de variétés différentiables.

1. Soit $f : X \longrightarrow Y$ un morphisme d'espaces complexes. Pour tout point y de Y on note X_y la fibre de f au-dessus de y. On obtient ainsi une "famille analytique d'espaces complexes", $(X_y)_y$. Considérons une propriété P concernant les espaces complexes. On peut poser la question si l'ensemble

$$\{y \in Y \mid X_y \text{ ait la propriété } P\}$$

est ouvert et si son complémentaire est analytique. C'est une question locale sur Y. De même, si P est une propriété concernant les points d'un espace complexe, alors on peut demander si l'ensemble

$$\{x \in X \mid X_{f(x)} \text{ ait la propriété P au point } x\}$$

est ouvert et si son complémentaire est analytique. C'est une question locale sur X.

Un exemple classique est donné par le théorème de semi-continuité de Remmert [8] :

"Soient $f : X \longrightarrow Y$ un morphisme d'espaces complexes et k un entier. Alors l'ensemble $\{x \in X / \dim_x X_{f(x)} \leq k\}$ est ouvert et son complémentaire est analytique".

Mais en général, si le morphisme f est quelconque, on a peu de chance que des propriétés raisonnable d'une fibre peuvent se transmettre aux fibres voisines. Soit pour cela l'exemple suivant. $Y = \mathbf{C}$, X = le sous-espace de $\mathbf{C}^2$ (de coordonnées z_1, z_2) donné par l'équation $z_1 z_2^2 = 0$ et $f : X \longrightarrow Y$ donné par z_1; la fibre X_o est isomorphe à la droite affine, tandis que les autres fibres X_y sont des points double ($(y,0)$, $\mathbb{C}[z_2]/z_2^2$).

L'hypothèse qu'on doit la faire pour assurer une réponse positive pour beaucoup des propriétés P est la platitude de f pour la deuxième question, respectivement la propriété et la platitude de f pour la première.

On trouve dans ([5], § 9 et § 12) une étude systématique concernant ce sujet dans le cas algébrique.

En particulier on prouve:

<u>Théorème 1</u>. Soit $f : X \longrightarrow Y$ un morphisme plat de type fini, X et Y étant des schémas localement noethériens.Alors les ensembles

$\{x \in X / X_{f(x)}$ soit réduit au point $x\}$, $\{x \in X / X_{f(x)}$ soit normal au point $x\}$

sont ouverts. Si de plus, f est propre, alors les ensembles

$\{y \in Y / X_y$ soit réduit$\}$, $\{y \in Y / X_y$ soit normal$\}$

sont aussi ouverts.

2. Dans le cas analytique on a le même résultat:

<u>Théorème 2</u>. Soit $f : X \longrightarrow Y$ un morphisme plat d'espaces complexes. Alors les ensembles $\{x \in X / X_{f(x)}$ soit réduit au point $x\}$, $\{x \in X / X_{f(x)}$ soit normal au point $x\}$

sont ouverts et leurs complémentaires sont analytiques. Si de plus, f est propre, alors les ensembles

$\{y \in Y / X_y$ soit réduit$\}$, $\{y \in Y / X_y$ soit normal$\}$

sont aussi ouverts et leurs complémentaires analytiques.

Ce théorème est prouvé par Grauert et Kerner [4] quand Y est une surface de Riemann. Dans [6] Kiehl a prouvé l'énoncé général dans le cas non archimédien; comme il est dit dans la préface de cet article, la méthode de démonstration peut être adaptée aussi au cas complexe. Dans [1] on donne une autre demonstration, mais moyenant l'hypothèse supplémentaire que X et Y sont de dimension pure. Esquissons ici une nouvelle démonstration, qui peut être adaptée également au cas différentiel.

Si X est un espace complexe et $F \in \mathrm{Coh}\, X$, alors pour tout entier k on définit l'ensemble singulier $S_k(F) = \{x \in X / \mathrm{prof}\ F_x \leqslant k\}$. $S_k(F)$ est analytique et fermé, de dimension $\leqslant k$. D'après des propriétés générales, l'espace X est normal au point x si et seulement si, pour tout voisinage U de x, l'application $\Gamma(U,O_X) \longrightarrow \Gamma(U \setminus S(X),O_X)$ est bijective (S(X) étant le lieu singulier de X). Ceci et le théorème d'annulation pour la cohomologie locale de Scheja et Trautmann [10] amenent au critère suivant de normalité de Markoe [7] (on trouve dans [9] une première application de la chomologie locale à une question de normalité):

"X est normal au point x si et seulement si $\dim_x(S(X) \cap S_k(O_x)) \leqslant k-2$, pour tout entier k".

De même, X étant réduit au point x si et seulement si les restrictions $\Gamma(U,O_X) \longrightarrow \Gamma(U \setminus S(X),O_X)$ sont injectives,on obtient:

"X est réduit au point x si et seulement si $\dim_x(S(X) \cap S_k(O_X)) \leqslant k-1$, pour tout entier k".

Soient R(f) et N(f) les complémentaires des sous-ensembles de X définis dans l'énoncé du théorème 2. On a donc:

$R(f) = X \setminus \{x \in X / \dim_x(X_{f(x)} \cap S(f) \cap S_k(O_X,f)) \leqslant k-1, \forall k\}$ et

$N(f) = X \setminus \{x \in X / \dim_x(X_{f(x)} \cap S(f) \cap S_k(O_X,f)) \leqslant k-2, \forall k\}$, où

$S(f) = \{x \in X / X_{f(x)}$ soit singulier au point $x\}$,

$S_k(O_X,f) = \{x \in X\ \mathrm{prof}_x\ X_{f(x)} \leqslant k\}$.

Montrons que S(f) et $S_k(O_X,f)$ sont des sous-ensembles analytiques fermés et ce qu'on cherche à prouver concernant R(f) et N(f) résulte du théorème de semi-continuité de Remmert. L'assertion concernant S(f) résulte du fait qu'il est fermé [3] et de l'égalité

$$S(f) = \bigcup_{\ell} (\{x \in X \mid \dim_x X_{f(x)} < \ell\} \cap \{x \mid \dim_{\mathbb{C}}(\Omega_x / m_x \Omega_x) \geq \ell\}),$$

où $\Omega = \Omega^1_{X/Y}$ est le faisceau des différentielles relatives. Concernant $S_k(O_X, f)$ on à une assertion générale :

"Si $f : X \longrightarrow Y$ est un morphisme d'espaces complexes et $F \in \mathrm{Coh}\, X$ est plat sur Y, alors les ensembles $S_k(F, f) = \{x \mid \mathrm{prof}(F_x / m_{f(x)} F_x) \leq k\}$ sont fermés et analytiques".

On prouve ceci en ramenant d'abord le problème au cas où f est une pnojection $Y \times_X U \longrightarrow Y$, U étant un ouvert d'un espace numérique, et puis en exprimant la profondeur en termes de dimension homologique et en utilisant le résultat suivant d'algèbre locale:

"Soient $A \longrightarrow B$ un morphisme local d'anneaux locaux noethériens et M un B-module de type fini. Supposons que M et B sont A-plats. Alors $dh_B M = dh_{B/m_A B}(M/m_A M)$" (on peut trouver des détails dans [2]).

Concernant la deuxième assertion du théorème, les complémentaires des sous-ensembles de Y défini dans l'énoncé sont f(R(f)) et f(N(f)); on conclut en appliquant le théorème de projection de Remmert.

<u>Remarque</u>. On peut trouver dans [12] l'extension en géométrie algébrique des critères de normalité et d'être réduit. En particulier on obtient:

"Soit A un anneau local noethérien, complet et de Cohen-Macaulay. Alors A est réduit (resp.normal) si et seulement si $\dim S(\mathrm{Spec}\, A) \leq \dim A-1$ (resp. $\dim S(\mathrm{Spec}\, A) \leq \dim A-2$), où S est le lieu singulier".

3. Soit $f : X \longrightarrow Y$ un morphisme C^∞ entre deux variétés différentiables. Soit x un point de X. Notons E_x et $E_{f(x)}$ les anneaux de germes de fonctions C^∞ en x, respectivement en f(x). Par F_x et $F_{f(x)}$ notons les complétés par rapport aux idéaux maximals.

Le morphisme f induit un morphisme local $E_{f(x)} \longrightarrow E_x$, donc un morphisme local $F_{f(x)} \longrightarrow F_x$. On dit d'après Tougeron [13] que f est plat en x si le morphisme $F_{f(x)} \longrightarrow F_x$ est plat. On dit que f est plat s'il est plat en tout x de X. Nous disons que f est réduit (resp. normal) au point x si l'anneau $F_x/m_{f(x)} F_x$ est réduit (resp. normal). Si X et Y sont des variétés analytiques (réels ou complexes) et f est une application analytique alors f, regardé comme application C^∞ entre les variétés différentiables associées, est réduit (resp. normal) en x si et seulement si l'espace analytique $X_{f(x)}$ est réduit (resp. normal) en x; en effet, une algèbre analytique est réduite (resp. normale) si et seulement si sa complété est réduite (resp. normale).

Théorème 3. Soit $f : X \longrightarrow Y$ un morphisme plat de variétés différentiables. Alors les ensembles

$$\{x \in X / f \text{ soit réduit en } x\}, \ \{x \in X / f \text{ soit normal en } x\}$$

sont ouverts.

Démonstration. On peut supposer que X et Y sont des ouverts dans des espaces numériques de dimensions n et m, de coordonnées $x_1,\dots,x_n$ et $y_1,\dots,y_m$.

Soit $x^o \in X$. Le morphisme $F_{f(x^o)} \longrightarrow F_{x^o}$ étant plat, d'après ([13],6.1.2 et 6.3.1) $\dim (F_{x^o}/m_{f(x^o)}F_{x^o}) = \dim F_{x^o} - \dim F_{f(x^o)} = n-m$ et $\text{prof}(F_{x^o}/m_{f(x^o)}F_{x^o}) = \text{prof } F_{x^o} - \text{prof } F_{f(x^o)}$. Par conséquent, $F_{x^o}/m_{f(x^o)} F_{x^o}$ est anneau de Cohen-Macaulay, donc il est réduit (resp.normal) si et seulement si son lieu singulier est de dimension $\leq n-m-1$ (resp. $\leq n-m-2$).

Nous avons besoin de équations explicites pour le lieu singulier. Pour cela, utilisons le critère Jacobien de régularité sous la forme de ([13], Ch.II, § 3).

Soient $f_1,\dots,f_m$ les composantes de f. Les éléments $f_1-f_1(x^o),\dots,f_m-f_m(x^o)$ engendrent l'idéal $I_{x^o} = m_{f(x^o)} F_{x^o}$ et de plus, ils forment une suite F_{x^o}-régulière; en effet, c'est l'image par le morphisme plat $F_{f(x^o)} \longrightarrow F_{x^o}$ de la suite réguliere de paramètres $y_1-y_1(f(x^o)),\dots,y_m-y_m(f(x^o))$. On obtient une surjection $F^m_{x^o} \longrightarrow I_{x^o}$, $(\varphi_1,\dots,\varphi_m) \longrightarrow \sum \varphi_i(f_i-f_i(x^o))$, et le noyau est engendré par les relations triviales $\varphi^{ij} = (\varphi^{ij}_k)_{1\leq k\leq m}$, où

$$\varphi^{ij}_k = \begin{cases} 0 & \text{si } k \neq i,j \\ -(f_j-f_j(x^o)) & \text{si } k = i \\ f_i-f_i(x^o) & \text{si } k = j \end{cases}$$

(voir par exemple [13], Ch.I,prop.5.1). Par conséquent on obtient une suite exacte

$$F^p_{x^o} \xrightarrow{\lambda} F^m_{x^o} \longrightarrow I_{x^o} \longrightarrow 0,$$

où l'entier p est indépendant de x^o et la matrice qui donne λ est continue par rapport à x^o. Donc, pour tout entier k, on peut trouver des éléments de F_{x^o} en nombre fini, leur nombre étant indépendant de x^o, qui engendrent l'idéal $\sigma'_k(I_{x^o})$ définit dans ([13],p.30) et, de plus, tels que les éléments varient continûment par rapport à x^o (on regarde sur F_{x^o} la topologie donné par la convergence des coefficients des series). Une assertion analogue vaut pour les idéaux $J_k(I_{x^o})$ de ([13], p.30); en effet, $J_k(I_{x^o})$ est engendré par $f_1-f_1(x^o),\dots,f_m-f_m(x^o)$ et par les jacobiens

$$D(f_{i_1}-f_{i_1}(x^o),\dots,f_{i_k}-f_{i_k}(x^o)) \;/\; D(x_{i_1},\dots,x_{i_k})$$

$1\leq i_1< \dots < i_k\leq n$. En utilisant les deux assertions on peut trouver pour tout entier k des éléments de F_{x^o} en nombre fini, leur nombre étant indépendant de x^o, qui varient continûment par rapport à x^o et dont l'ensemble des zéros dans Spec F_{x^o} coïncide à l'ensemble des zéros de $R_k(I_{x^o}) = \sqrt{J_k(I_{x^o})} \cap \sqrt{\sigma'_k(I_{x^o})}$.

D'après le critère jacobien de régularité, on a(dans Spec F_{x^o})

$$S(F_{x^o} / I_{x^o}) = \bigcap_k V(R_k(I_{x^o})).$$

On peut ainsi trouver des éléments $g_1(x^o),\dots,g_N(x^o)$ de F_{x^o}, leur nombre étant indédendant de x^o, qui varient continûment par rapport à x^o et tels que le lieu singulier $S(F_{x^o}/m_{f(x^o)}F_{x^o})$ s'identifie à l'ensemble des zéros dans $\mathrm{Spec}(F_{x^o}/m_{f(x^o)}F_{x^o})$ du l'idéal $(g_1(x^o),\dots,g_N(x^o))$.

Par conséquent,

$$\dim S(F_{x^o}/m_{f(x^o)}F_{x^o}) = \dim(F_{x^o}/(g_1(x^o),\dots,g_N(x^o))$$

et on finie la démonstration en utilisant ([13], Ch.II, pr.5.3).

Peut être la théorie des espaces différentiables [11] donne un cadre naturel pour généraliser le théorème 3 .

4. On peut généraliser les faits de la section précédente par exemple comme il suit. On va considérer des espaces annelés (X,E_X) qui localement sont de la forme $(\mathrm{Supp}(E_U/I), (E_U/I)|\mathrm{Supp})$, où U est un ouvert d'un espace numérique, E_U le faisceau de germes de fonctions C^∞ sur U et $I\subset E_U$ un faisceau d'idéaux pseudocohérent (cf. SGA 6, I.H.E.S., Bures sur Yvette 1966/1967). Les morphismes $f : (X,E_X) \longrightarrow (Y,E_Y)$ seront les morphismes d'espaces annelés qui localement sont induites par des applications différentiables $f' : U \longrightarrow V$ tels que $f'^*(J)E_U\subset I$.

On dit d'un tel morphisme f qu'il est plat (resp.régulier, réduit, normal) dans le point x si le complété du morphisme $E_{Y,f(x)} \longrightarrow E_{X,x}$ (resp. le complété du l'anneau $E_{X,x}/m_{f(x)}E_{X,x}$) est plat (resp. régulier, réduit, normal). On va supposer par la suite que f est plat dans tous les points de X. De plus, que f est induit par un f'. Soit x_o un point de X. On peut trouver, dans un voisinage de x_o, des entiers n_1 et n_o et des présentations finies

$$E_{U,x}^{n_1} \xrightarrow{\lambda(x)} E_{U,x}^{n_o} \longrightarrow \mathfrak{A}(x) \longrightarrow 0,$$

où $\mathfrak{a}(x)$ est l'idéal de $E_{U,x}$ pour lequel $E_{U,x}/a(x) \simeq E_{X,x}/m_{f(x)}E_{X,x}$, telles que le complété de $\lambda(x)$, $\hat{\lambda}(x) : F_{U,x}^{n_1} \longrightarrow F_{U,x}^{n_o}$, varie continûment par rapport a x quand on regarde sur $F_{U,x}$ (s'identifiant canoniquement à un anneau de séries formelles) la topologie donnée par la convergence des coefficients des séries: on part avec des présentations "continues" convenables pour les idéaux maximaux des anneaux $E_{V,y}$, dans un voisinage de $y_o = f(x_o)$, et on utilise des faits du (Ch.I, § 2; Ch.III, § 1) du SGA 6, en remarquant que la "continuité" peut se préserver. On déduit alors, comme dans 3, l'assertion:

(*) Il existe, dans un voisinage de x_o, un entier p et des éléments $\varphi_1(x),\dots,\varphi_p(x) \in F_{U,x}$ qui varient continûment par rapport à x et tels que dans Spec $(F_{U,x})$ on a:

$$V((\varphi_1(x),\dots,\varphi_p(x)) = S((E_{X,x}/m_{f(x)}E_{X,x})\hat{}\,).$$

On déduit de ([13], Ch.II, pr.5.3):

"Pour tout morphisme plat $f : (X,E_X) \longrightarrow (Y,E_Y)$, la fonction $x \longmapsto \dim S((E_{X,x}/m_{f(x)}E_{X,x})\hat{}\,)$ est semi-continue supérieurement. En particulier, l'ensemble $\{x \in X \mid f$ soit régulier en $x\}$ est ouvert".

Rappelons que pour un anneau A et un entier k on note $S_k^*(A) = \{p \in \operatorname{Spec} A \mid \operatorname{prof} A_p + \dim(A/p) \leq k\}$. Si A est quotient d'un anneau local noethérien régulier, alors $S_k^*(A)$ est fermé (et ainsi prof $A \geq k+1$ si et seulement si $S_k^*(A) = \emptyset$).

Soient $A(= A(x)) = (E_{X,x}/m_{f(x)}E_{X,x})\hat{}$ et $B(= B(x)) = \hat{E}_{U,x} = F_{U,x}$.

Pour un idéal $p \in \operatorname{Spec} A \subset \operatorname{Spec} B$ on a

$$\operatorname{prof} A_p + \dim(A/p) = \dim B - dh_{B_p} A_p.$$

On peut trouver, dans un voisinage de x_o, des résolutions

$$\dots \longrightarrow B(x)^{n_2} \longrightarrow B(x)^{n_1} \longrightarrow B(x)^{n_o} \longrightarrow A(x) \longrightarrow 0$$

telles que les entiers n_q ne dépendent pas de x et telles que les matrices qui définissent les morphismes $B(x)^{n_{q+1}} \longrightarrow B(x)^{n_q}$ varient continûment par rapport à x (on part des résolutions convenables pour les $E_{V,y}$-modules $E_{V,y}/m_y$, dans un voisinage de $y_o = f(x_o)$, et puis on utilise le lemme III.1.1.1 du SGA 6, la platitude de f et de nouveau de lemme).

On obtient l'assertion:

(**) Pour un entier k il existe, dans un voisinage de x_o, un entier q et des éléments $\psi_1(x),\dots,\psi_q(x) \in F_{U,x}$ qui varient continûment par rapport à x et tels que dans Spec $F_{U,x}$ on a

$$V(\psi_1(x),\dots,\psi_q(x)) = S_k^*((E_{X,x}/m_{f(x)}E_{X,x})^\wedge)$$

(ψ_i sont certains mineurs obtenus par les morphismes de la résolution de A(x)).

A l'aide de ceci on prouve:

"Pour tout morphisme plat $f : (X,E_X) \longrightarrow (Y,E_Y)$, la fonction $x \longmapsto \dim S_k^*((E_{X,x}/m_{f(x)}E_{X,x})^\wedge)$ est semi-continue supérieurement. En particulier, l'ensemble $\{x\in X/\text{prof}((E_{X,x}/m_{f(x)}E_{X,x})^\wedge) \geqslant k\}$ est ouvert. De même, l'ensemble $\{x\in X/(E_{X,x}/m_{f(x)}E_{X,x})^\wedge$ soit de Cohen-Macaulay$\}$ est ouvert".

De (*), (**) et en utilisant les critères de normalité et d'être réduit de [12] on conclut enfin:

"Pour tout morphisme plat $f : (X,E_X) \longrightarrow (Y,E_Y)$ les ensembles $\{x\in X/f$ soit réduit en $x\}$, $\{x\in X/f$ soit normal en $x\}$ sont ouverts".

Institut Central de Mathématiques
Bucarest

Bibliographie

1. C.Bănică, Un théorème concernant les familles analytiques d'espaces complexes, Revue Roum. de Math., 10 (1973), 1515-1520.
2. C.Bănică et O.Stănăşilă, Metode algebrice în teoria globală a spaţiilor complexe, Editura Academiei R.S.R., 1974.
3. H.Cartan, Séminaire E.N.S., Paris, 1960-1961.

4. H.Grauert et H.Kerner, Deformationes von Singularitäten komplexer Räume, Math. Annalen 153 (1964), 236-260.

5. A.Grothendieck et J.Dieudonné, Eléments de géométrie algébrique, Ch.IV, Publ. I.H.E.S., No. 20, 24, 28.

6. R.Kiehl, Analitischen Familien Affinoider Algebren, Heidelberg, 1968.

7. A.Markoe, A characterisation of normal analytic spaces by the homological codimension of the structure sheaf, Pacific J. of Math., 52 (1974), 485-489.

8. R.Remmert, Holomorphe und meromorphe Abbildungen Komplexer Räume, Math. Annalen, 133 (1957), 328-370.

9. G.Scheja, Eine Anwendung Riemannscher Hebbarkeitssätze für analytische Cohomologieklassen, Archiv der Math., 12, 341-348. 1961.

10. Y-T-Siu et G.Trautmann, Gap-Sheaves and Extension of Coherent Analytic Subsheaves, Lecture Notes in Math., 172 (1971), Springer Verlag.

11. K.Spallek, Differenzierbare Räume, Math. Ann. 180, 269-296, (1969).

12. M.Stoia, Reduced and normal points of a flat morphism, Revue Roum., de Math., no.9, 1976.

13. J.C.Tougeron, Idéaux de fonctions différentiables, Springer Verlag, 1972.

The Hilbert-Samuel polynomials of a proper morphism

V. Brînzănescu

Let A be a local noetherian ring, $\mathfrak{m}$ its maximal ideal and M an A-module of finite type. Then the A-module $M/\mathfrak{m}^n M$ has finite length and, by a result of Samuel, there exists a unique polynomial P(M) such that $P(M)(n) = \ell(M/\mathfrak{m}^n M)$ for large n; the polynomial P(M) is called the Hilbert-Samuel polynomial (see [4]).

Let $f: X \longrightarrow Y$ be a proper morphism of complex spaces, $\mathcal{F}$ a coherent analytic sheaf on X and y a point of Y. Let $\mathfrak{m}_y$ be the maximal ideal of the local ring $\mathcal{O}_{Y,y}$ and let $\hat{\mathfrak{m}}_y$ be the coherent ideal sheaf on X defined by $\mathfrak{m}_y$. X_y denotes the fiber of f over y, and $\mathcal{F}_y$ denotes the sheaf $\mathcal{F}/\hat{\mathfrak{m}}_y\mathcal{F}$.

From Grauert's coherence theorem, [3] , we have , for all integers $q \geqslant 0$ and $n \geqslant 1$, that the $\mathcal{O}_{Y,y}$-module

$$H^q(X, \mathcal{F}/\hat{\mathfrak{m}}_y^n\mathcal{F}) \simeq H^q(X_y, \mathcal{F}/\hat{\mathfrak{m}}_y^n\mathcal{F}) \simeq R^q f_*(\mathcal{F}/\hat{\mathfrak{m}}_y^n\mathcal{F})_y$$

is of finite type, therefore, of finite length, since it is anihilated by $\mathfrak{m}_y^n$.

Now we want to present some properties of these functions:

$$n \longrightarrow \ell(H^q(X_y, \mathcal{F}/\hat{\mathfrak{m}}_y^n\mathcal{F})) .$$

The first result is the following:

Theorem 1. (Existence theorem) Let $f: X \longrightarrow Y$ be a proper morphism of complex spaces, $\mathcal{F}$ a coherent analytic sheaf on X and y a point of Y. Then:

(a) The function $n \longrightarrow \sum_{q=0}^{\infty} (-1)^q \ell(H^q(X_y, \mathcal{F}/\mathfrak{m}_y^n \mathcal{F}))$ is polynomial (let us denote by $P(\mathcal{F},f,y)$ the associated polynomial).

(b) If, moreover, we assume $\mathcal{F}$ flat over Y, then the function $n \longrightarrow \ell(H^q(X_y, \mathcal{F}/\mathfrak{m}_y^n \mathcal{F}))$ is polynomial (let us denote by $Q^q(\mathcal{F},f,y)$ the associated polynomial).

We recall the semicontinuity and continuity theorems of Grauert [3] :

" Let $f: X \longrightarrow Y$ be a proper morphism of complex spaces, $\mathcal{F}$ a coherent analytic sheaf on X, flat over Y. Then:

(a) (Semicontinuity theorem) For any fixed integer q, the function $y \longrightarrow \dim H^q(X_y, \mathcal{F}_y)$ is upper semicontinuous.

(b) (Continuity theorem) Consider an integer q. If $\mathcal{F}$ is cohomologically flat in dimension q over Y, then the function $y \longrightarrow \dim H^q(X_y, \mathcal{F}_y)$ is locally constant. Conversely, if this function is locally constant and Y is a reduced space, then $\mathcal{F}$ is cohomologically flat in dimension q over Y".

We have the following analogues:

Theorem 2.(Semicontinuity theorem) Let $f: X \longrightarrow Y$ be a proper morphism of complex spaces, $\mathcal{F}$ a coherent analytic sheaf on X , flat over Y. Then:

(a) If $\chi(X_y, \mathcal{F}_y) > 0$ (respective $\chi(X_y, \mathcal{F}_y) < 0$) then the function $y \longrightarrow P(\mathcal{F},f,y)$ is upper semicontinuous (respective lower semicontinuous), where $\chi(X_y, \mathcal{F}_y) = \sum_{q=0}^{\infty} (-1)^q \dim H^q(X_y, \mathcal{F}_y)$ is the Euler-Poincaré characteristic.

(b) If Y is nonsingular, then the function $y \longrightarrow Q^q(\mathcal{F},f,y)$ is upper semicontinuous.

Theorem 3.(Continuity theorem) Let $f: X \longrightarrow Y$ be a proper morphism of complex spaces, $\mathcal{F}$ a coherent analytic sheaf on X, flat over Y. Then:

(a) If Y is nonsingular, then the function $y \longrightarrow P(\mathcal{F},f,y)$ is locally constant. Conversely, if Y is reduced, $\chi(X_y, \mathcal{F}_y) \neq 0$ and $y \longrightarrow P(\mathcal{F},f,y)$ is locally constant, then Y is nonsingular.

(b) Let Y be nonsingular. If, moreover, $\mathcal{F}$ is cohomologically flat in dimension q over Y, then the function $y \longrightarrow Q^q(\mathcal{F},f,y)$ is locally constant. Conversely, if $y \longrightarrow Q^q(\mathcal{F},f,y)$ is locally constant, then $\mathcal{F}$ is cohomologically flat in dimension q over Y.

For details one can consult [1] , [2] .

References

1. C. Bănică, V. Brînzănescu, Sur le polynôme de Hilbert-Samuel d'un morphisme propre, C.R.Acad.Sci. Paris,Ser.A-B,282, 215-217,(1976).
2. C. Bănică, V. Brînzănescu, The Hilbert-Samuel polynomials of a proper morphism, (to appear.Mathematische Zeitschrift 1978 and INCREST Preprint Series in Mathematics 12/1977)
3. H. Grauert, Ein Theorem der analytischen Garbentheorie und die Modulräume komplexer Strukturen, Publ. IHES, No5,(1960)
4. J.P. Serre, Algèbre Locale. Multiplicités, Lecture Notes in Mathematics No 11, Springer-Verlag, Berlin.Heidelberg. New York, (1965).

Polytechnic Institute
Bucharest

Un théorème d'annulation sur les variétés faiblement 1-complètes

Pierre DOLBEAULT
Université de Paris VI

Récemment, S.NAKANO a obtenu des théorèmes d'annulation pour des fibrés vectoriels holomorphes positifs dans des sens divers ([4], [5]) sur une variété faiblement 1-complète X de dimension complexe n. L'un d'eux, pour les fibrés en droites positifs E a le même énoncé que le classique théorème de Kodaira-Akizuki-Nakano pour une variété compacte, i.e. : $H^{p,q}(X,E) = 0$ pour $p+q \geq n+1$ ([2], p.132); comme les variétés de Stein sont aussi des variétés faiblement 1-complètes vérifiant un théorème d'annulation beaucoup plus large (théorème B), un problème naturel est de chercher des théorèmes d'annulation intermédiaires entre le théorème cité et le théorème B des variétés de Stein.

Ceci est une tentative d'adaptation de la méthode de Nakano pour obtenir des conditions suffisantes d'annulation pour la cohomologie de type (0,q).

1. Préliminaires.

1.1. Inégalité de Nakano. Soit X une variété analytique complexe et soit E un fibré holomorphe en droites muni d'une structure hermitienne $\{\ ,\ \}_1$ sur les fibres. Soient D la connexion hermitienne sur E et χ^1 la courbure de la connexion
$\chi^1 = i\,D^2 = i\,(D'\,d'' + d''\,D')$, où D' est de type (1,0).

Supposons maintenant X muni de la métrique kählérienne $d\sigma^2$, de forme fondamentale ω; soient $L = \cdot\wedge\omega$ et $\Lambda = \bar{*}^{1}L*$ où $*$ est défini par $d\sigma^2$; on désigne par $e(\chi^1)$ l'opérateur multiplication extérieure par χ^1 ; alors $(\ ,\ )_1$ étant le produit scalaire global défini par $d\sigma^2$ et $\{\ ,\ \}_1$ sur les formes différentielles sur X à valeurs dans E, pour toute forme C^∞, à support compact, de type (p,q), $\varphi \in \mathcal{D}^{p,q}(X,E)$, on a l'inégalité de Nakano

(1) $A(\varphi) = ((e(\chi^1) \wedge - \Lambda e(\chi^1)) \varphi, \varphi)_1 \leqslant (d''\varphi, d''\varphi)_1 + (\delta''\varphi, \delta''\varphi)_1$.

1.2. $W^{p,q}$-ellipticité. Soit E un fibré en droites, hermitien, sur une variété munie d'une métrique hermitienne ds^2 et soit $\mathcal{L}^{p,q}(X,E)$ la complétion de $\mathcal{D}^{p,q}(X,E)$ par rapport au produit scalaire global (,) . Le fibré E est dit $W^{p,q}$-elliptique ([1], p.89) s'il existe une constante $C > 0$ telle que, pour toute $\varphi \in \mathcal{D}^{(p,q)}(X,E)$, on ait

(2) $(\varphi, \varphi) \leqslant C((d''\varphi, d''\varphi) + (\delta''\varphi, \delta''\varphi))$.

La métrique ds^2 définit une distance sur la variété X compatible avec sa topologie ; ds^2 est dite complète si, pour cette distance , X est un espace métrique complet.
Soit $A^{p,q}(X,E)$ l'espace des formes différentielles C^∞ de type (p,q) sur X, à valeur dans E.

1.3. LEMME (ANDREOTTI-VESENTINI ([1], p.94) . Soit E un fibré en droites $W^{p,q}$-elliptique par rapport à une métrique complète sur X ; alors, si $q \geqslant 1$, pour toute $\varphi \in \mathcal{L}^{p,q}(X,E) \cap A^{p,q}(X,E)$ telle que $d''\varphi = 0$, il existe $\xi \in \mathcal{L}^{p,q-1}(X,E) \cap A^{p,q-1}(X,E)$ telle que $\varphi = d''\xi$.

2. Variétés faiblement 1-complètes ; énoncé.

2.1. Une variété analytique complexe X munie d'une fonction plurisousharmonique Ψ, C^∞ sur X, est dite faiblement 1-complète (par rapport à Ψ) si, pour tout $c \in \mathbb{R}$, les ensembles

$$X_c = \{x \in X \ ; \ \Psi(x) < c\}$$

sont relativement compacts ou vides.

On supposera: $\Psi(x) \geqslant 0$ pour tout $x \in X$.

2.2. THÉORÈME. Si E est un fibré en droites positif sur une variété faiblement 1-complète (X, Ψ) ; si, de plus

(a) la forme de Levi de Ψ possède e valeurs propres strictement positives,

(b) la condition (A_q) ci-dessous (3.6.) est satisfaite ,

alors, pour $1 \leq q \leq e-1$, on a :

$$H^{o,q}(X,E) = H^q(X, \Theta_E) = 0$$

(où Θ_E est le faisceau des sections holomorphes de E).

2.3. LEMME. Dans les hypothèses de 2.2., soit $\varphi \in A^{o,q}(X,E)$, alors il existe une métrique hermitienne $d\sigma^2$ sur X et une structure hermitienne $\{\ ,\ \}_1$ sur E telles que

(a) $d\sigma^2$ soit complète ;

(b) E soit $W^{o,q}$-elliptique par rapport à $d\sigma^2$ et à $\{\ ,\ \}_1$;

(c) $\varphi \in \mathcal{L}_1^{o,q}(X,E)$ (espace $\mathcal{L}^{o,q}$ pour le produit scalaire $(\ ,\ )_1$ défini par $d\sigma^2$ et $\{\ ,\ \}_1$).

Alors, pour toute $\varphi \in A^{o,q}(X,E)$, dans les hypothèses de 2.2., on construit $d\sigma^2$ et $\{\ ,\ \}_1$ telles que (a), (b), (c) de 2.3. soient satisfaites et, d'après 1.3. , il existe $\xi \in A^{p,q-1}(X,E)$ telle que $d''\xi = \varphi$, d'où $H^{o,q}(X,E) = 0$.

3. Condition (A_q) et démonstration de 2.3.

3.1. Soit $\{\ ,\ \}$ une structure hermitienne du fibré E pour laquelle E est positif. Soit (U_j) un recouvrement ouvert de X au-dessus de chaque élément duquel E soit trivial ; alors, sur U_j, il existe une fonction $a_j > 0$, C^∞, telle que, pour toute section locale σ de E, $\{\sigma, \sigma\} = \sigma a_j^{-1} \bar{\sigma}$. La courbure hermitienne de E est donnée par

$$\chi^o = i \sum_{\alpha,\beta} \frac{\partial^2 \log a_j}{\partial z_j^\alpha \, \partial \bar{z}_j^\beta} \, dz_j^\alpha \wedge d\bar{z}_j^\beta$$

(où (z_j^α) est un système de coordonnées complexes locales de X, au-dessus de U_j supposé assez petit).

La positivité de E signifie que, pour tout j, pour tout $x \in U_j$, la matrice $\left(\frac{\partial^2 \log a_j}{\partial z_j^\alpha \, \partial \bar{z}_j^\beta}(x)\right)$ est > 0 ; χ^o est la forme fondamentale d'une métrique kählérienne

$$ds^2 = \sum_{\alpha,\beta} g_{j\alpha\bar\beta} \, dz_j^\alpha \otimes d\bar{z}_j^\beta \quad \text{avec} \quad g_{j\alpha\bar\beta} = \frac{\partial^2 \log a_j}{\partial z_j^\alpha \, \partial \bar{z}_j^\beta}$$

3.2. Soit $\lambda : \mathbb{R} \to \mathbb{R}$ une fonction C^∞. Posons :

$A_j = e^{\lambda(\Psi)} a_j$; $\Gamma_{j\alpha\bar\beta} = \frac{\partial^2 \log A_j}{\partial z_j^\alpha \, \partial \bar{z}_j^\beta}$; $d\sigma^2 = \sum_{\alpha,\beta} \Gamma_{j\alpha\bar\beta} \, dz_j^\alpha \otimes d\bar{z}_j^\beta$; on a

$$\Gamma_{j\alpha\bar\beta} = g_{j\alpha\bar\beta} + \lambda'(\Psi) \frac{\partial^2 \Psi}{\partial z_j^\alpha \, \partial \bar{z}_j^\beta} + \lambda''(\Psi) \frac{\partial \Psi}{\partial z_j^\alpha} \frac{\partial \Psi}{\partial \bar{z}_j^\beta} .$$

Supposons, pour tout $t \in \mathbb{R}$, $\lambda'(t) \geq 0$ et $\lambda''(t) \geq 0$; alors $(\Gamma_{j\alpha\bar\beta}) > 0$; $d\sigma^2$ est une métrique kählérienne.

3.3. LEMME (NAKANO [3]) : Dans les notations ci-dessus, si $\int_0^{+\infty} \sqrt{\lambda''(t)} \, dt = +\infty$, la métrique $d\sigma^2$ est complète.

3.4. Considérons la structure hermitienne modifiée (A_j) sur E, définie en 3.2. et la métrique kählérienne $d\sigma^2$ sur X.

Modifions, de nouveau, la structure hermitienne de E en considérant, sur chaque U_j, la nouvelle fonction $A_j^1 = e^{\eta(\Psi)} A_j$ où $\eta : \mathbb{R} \to \mathbb{R}$ est une fonction C^∞ que l'on choisira ultérieurement ; les (A_j^1) définissent une structure hermitienne $\{\ ,\ \}_1$ sur le fibré E à laquelle est associée une nouvelle courbure

$$\chi^1 = \chi + i \sum_{\alpha,\beta} \left[\eta'(\Psi) \frac{\partial^2 \Psi}{\partial z_j^\alpha \, \partial \bar{z}_j^\beta} + \eta''(\Psi) \frac{\partial \Psi}{\partial z_j^\alpha} \frac{\partial \Psi}{\partial \bar{z}_j^\beta} \right] dz_j^\alpha \wedge d\bar{z}_j^\beta$$

avec $\chi = i \sum_{\alpha,\beta} \Gamma_{j\alpha\bar\beta} \, dz_j^\alpha \wedge d\bar{z}_j^\beta$.

3.5. LEMME. Dans l'hypothèse (a) de 2.2., si $1 \leq q \leq e-1$, il existe une métrique hählérienne $d\sigma^2$ sur X et une structure hermitienne $\{\ ,\ \}_1$ sur E telle que E soit $W^{o,q}$-elliptique.

On va prendre le $d\sigma^2$ ci-dessus sur X et choisir η pour obtenir la conclusion. Désignons par $\langle\ ,\ \rangle$ le produit scalaire ponctuel en $x_o \in X$ pour les formes φ de type (o,q) . Posons

$$A \langle \varphi \rangle = \langle -\Lambda e(\chi^1) \varphi, \varphi \rangle ;$$

on montre que le terme en $\eta''(\Psi)$ dans $A\langle\varphi\rangle$ est ≥ 0 pour $\eta''(\Psi) \leq 0$.

En x_o, il existe une matrice T telle que

$$\left(\Gamma_{j\alpha\bar{\beta}} + \eta'(\Psi)\frac{\partial^2\Psi}{\partial z_j^{\alpha}\,\partial \bar{z}_j^{\beta}}\right) = {}^t T\left(I_n + \eta'(\Psi)\begin{pmatrix} v_1 & & 0 \\ & \ddots & \\ 0 & & v_n\end{pmatrix}\right)\bar{T} \ ;$$

l'ensemble $(v_1(x_o),\ldots,v_n(x_o))$ est indépendant des coordonnées ; e des v_k sont > 0 ; mais les $v_k(x_o)$ dépendent de la fonction λ.

Soit $B = (\ell_1,\ldots,\ell_q)$ un multi-indice ; on pose $\# B = q$;

$M'(B) = [1, \ldots, n] \setminus [\ell_1, \ldots, \ell_q]$; $\ell(x_o) = \inf\limits_{\#B=q} \sum\limits_{\nu\in M'} v_\nu(x_o)$

et $\gamma(c) = \inf\limits_{\Psi(x)=c} \ell(x)$; on a $\gamma(c) > 0$ car $q \leq e-1$.

Un calcul analogue à celui de l'opérateur $L\Lambda - \Lambda L$ en géométrie kählérienne et l'inégalité de Nakano (1) entraîne la $W^{o,q}$-ellipticité de E si

(3) $\quad -\eta'(c) \geq \dfrac{(K+1)(n-q)}{\gamma(c)}$ pour tout $c \in \mathbb{R}^*$; K étant une constante > 0.

3.6. On considère la condition suivante

(A_q) <u>quelle que soit la fonction</u> $\lambda : \mathbb{R} \to \mathbb{R}$ <u>telle que</u> $\lambda' \geq 0$; $\lambda'' \geq 0$, $\int_0^{+\infty} \lambda''(t)\,dt = +\infty$, <u>il existe une fonction</u> $\eta : \mathbb{R} \to \mathbb{R}$ <u>telle que</u> $\eta(0) = 0$; $\eta'' < 0$ <u>sur</u> $\mathbb{R}_+$;

(3) <u>ci-dessus</u> ;

<u>et</u> (4) $\eta^4(c) \leq \frac{1}{2}\lambda'(c)$ <u>pour tout</u> $c \in \mathbb{R}_+$.

3.7. LEMME : <u>Si la condition</u> (A_q) <u>est satisfaite, pour toute</u> $\varphi \in A^{p,q}(X,E)$, <u>il existe des fonctions</u> λ <u>et</u> η <u>pour lesquelles</u> $\varphi \in \mathcal{L}_1^{(o,q)}(X,E)$.

Reprenant le calcul de Nakano [5], on trouve

$$(\varphi,\varphi)_1 \leq \int_X e^{-\eta(\Psi) - \lambda(\Psi)} \frac{\det(\Gamma_{j\alpha\bar{\beta}})}{\det(g_{j\alpha\bar{\beta}})} a_o[\varphi]\,dv_o$$

où $a_o[\varphi]dv_o = \frac{1}{a_j}\varphi_j \wedge *_o \bar{\varphi}_j$, l'opérateur $*_o$ et l'élément de volume dv_o étant calculés pour la métrique ds^2 définie par (a_j).

On choisit, dans le raisonnement de Nakano, $\mu_o(t) = \nu_1(t) + \rho^4(t)$ où $\nu_1(t)$ et $\rho(t)$ sont définis dans [5, § 2] et on forme, comme dans [5] , une fonction λ telle que

$$\lambda'(t) \geqslant 2\mu_o(t)$$

(4) entraîne $$\lambda'(t) \geqslant \mu_o(t) + \eta^4(t)$$

et $-\eta(t) \leqslant \frac{1}{3}\lambda(t)$;

ce qui, compte tenu de l'évaluation de $\frac{\text{dét}(\Gamma_{j\alpha\bar{\beta}})}{\text{dét}(g_{j\alpha\bar{\beta}})} a_o[\varphi] dv_o$ faite par Nakano, entraîne $(\varphi, \varphi)_1 < +\infty$.

4. Remarque : La condition (a) du théorème 2.2. est insuffisante pour assurer l'annulation de $H^{o,q}(X,E)$ comme le montre l'exemple suivant qui m'a été communiqué par J.VAROUCHAS.

Soit Y une surface de Riemann compacte, de genre $g \geqslant 2$; $X = Y \times \mathbb{C}^2$; X est munie de la fonction plurisousharmonique d'exhaustion $\Psi : (y,z) \longmapsto |z|^2$ pour laquelle $e = 2$; soient F le fibré positif défini par le diviseur $1.y_o$ $(y_o \in Y)$; K le fibré canonique sur Y ; $L = K \otimes F^{-1}$; E le fibré positif $\pi^*(L)$ sur X où π est la première projection $Y \times \mathbb{C}^2 \longrightarrow Y$; alors $H^1(X, \mathcal{O}_E) \neq 0$.

Bibliographie

[1] ANDREOTTI (A.) and VESENTINI (E.). - Carleman estimates for the Laplace-Beltrami equation on complex manifolds. Publ. Math., I.H.E.S., 25, (1965), 81-130.

[2] MORROW (J.), KODAIRA (K.). - Complex manifolds, Holt, Rinehart and Winston, 1971.

[3] NAKANO (S.). - On the inverse of a monoidal transformation,Publ. R.I.M.S., Kyoto Univ., 6, (1970-1971), 483-502.

[4] NAKANO (S.). - Vanishing theorems for weakly 1-complete manifolds, Number theory, Alg. Geometry and commutative alg., in honor of Y.Akizuki, Kinokuniya , Tokyo,(1973),169-179.

[5] NAKANO (S.). - Vanishing theorems for weakly 1-complete manifolds II, Publ. R.I.M.S., Kyoto Univ., 10, (1974), 101-110.

REPERES DE FRENET EN GEOMETRIE HERMITIENNE

Simone Dolbeault
Université de Poitiers

INTRODUCTION ET BIBLIOGRAPHIE

Les variétés homogènes G/H, quotient d'un groupe de Lie G par un sous-groupe fermé H interviennent fréquemment en géométrie, ainsi que leurs sous-variétés M qui sont invariantes par G ; E. Cartan a pensé que, dans beaucoup de cas généralisant l'espace euclidien, G peut être identifié à un ensemble de repères sur G/H, ce qui permet d'associer à M, de façon naturelle, un ensemble de repères [5]. D'un autre point de vue, les repères sont les sections du fibré $G \to G/H$ et les repères sur M sont les restrictions à M de ces sections. La donnée d'une connection ω définit la forme de Maurer - Cartan et permet d'appliquer la méthode du repère mobile dont nous allons donner un aperçu général et son application en géométrie hermitienne [3]; puis nous considèrerons les courbes holomorphes d'une variété hermitienne, ce qui nous conduira à introduire les repères de Frenet et à caractériser celles des variétés hermitiennes dont toutes les courbes holomorphes sont munies de repères de Frenet [3]. Ensuite, nous considèrerons les cas particuliers des variétés Kahlériennes et riemanniennes [3], [4] et comme exemple d'application nous nous servirons de repères de Frenet pour retrouver facilement des résultats classiques sur les courbes holomorphes de l'espace projectif complexe [1] [5]. Comme ouvrages de base en géométrie hermitienne, nous citons seulement [2], [6] et [7].

1. E. CALABI. Isometric imbeddings of complex manifolds. Ann. of Math. vol 58, (1953) pp. 1-23.

2. S. S. CHERN. Complex manifolds without potential theory. Van Nostrand Math. Studies Princeton 1957.

3. S. S. CHERN, M. J. COWEN, A. L. VITTER III. FRENET frames along holomorphic curves. Value distribution theory, part A, M. Dekker 25, New-York 1974.

4. S. DOLBEAULT. Moving frames in Hermitian geometry, à paraitre dans Proceedings of Symposia in pure Mathematics (Summer Institute on Several Complex Variables, Williamstown, 1975.

5. S. KOBAYASHI and K. NOMIZU. Foundations of Differential geometry. Interscience, New-York 1963 and 1966.

6. P. GRIFFITHS. On Cartan's methods of Lie groups and moving frames as applied to uniqueness and existence questions in differential geometry. Duke Math. J. vol 41, (1974), pp. 775-814.

7. H. WU. The equidistribution theory of holomorphic curves. Annals of Math. Studies Princeton 1970.

1. Géométrie différentielle hermitienne.

a) Formes de Maurer-Cartan sur un groupe de Lie G. Soit g l'algèbre de Lie sur G des champs de vecteurs sur G, invariants à gauche par l'action de G; g* est le dual de g. Une forme de Maurer-Cartan est une 1-forme sur G, à valeurs dans g, invariante à gauche et satisfaisant à l'équation de Maurer-Cartan :

$$d\,\omega = \tfrac{1}{2}\,[\omega\,,\,\omega]$$

Si $(X_1, \ldots, X_n)$ et $(\omega_1, \ldots, \omega_n)$ sont deux bases duales de g et g* respectivement, alors

$$\omega = \sum_{1 \leq i \leq n} X_1 \otimes \omega_i \;;$$

Si l'on pose

$$d(X_i \otimes \omega_i) = X_i \otimes d\,\omega_i\;, \qquad [X_i \otimes \omega_i,\, X_k \otimes \omega_k] = [X_i,\, X_k] \otimes \omega_i \wedge \omega_k\;,$$

l'équation de Maurer-Cartan se traduit par n conditions exprimant les $d\,\omega_i$ à l'aide des $\omega_j \wedge \omega_k$. En particulier, si ω est la forme canonique (i;e; la forme de Maurer-Cartan déterminée de façon unique par $\omega(A) = A$, $\forall\, A \in \mathbf{g}$) et si les C_{ijk} sont les constantes de structure de E. Cartan (définies par $[X_i,\, X_k] = \sum_{1 \leq j \leq n} C_{ijk}\, X_j$), l'équation de Maurer-Cartan est équivalente au système des équations de structure de E. Cartan

$$d\,\omega_i = \sum_{j<k} C_{jik}\, \omega_j \wedge \omega_k .$$

b) Connections sur un fibré vectoriel complexe E sur une variété différentiable M. Soient TM (resp. T*M) le fibré tangent (resp. cotangent) à M, Γ (E) (resp. Γ(T*M⊗E)) l'espace des sections de E (resp. T*M⊗E) qui satisfait aux conditions suivantes :

$$D(\gamma_1 + \gamma_2) = D\,\gamma_1 + D\,\gamma_2, \qquad \gamma_1,\, \gamma_2 \in \Gamma\,(E),$$

$$D(f\gamma) = df.\,\gamma + f\,D\gamma\;, \qquad \gamma \in \Gamma\,(E),\ f \text{ fonction } C^{\infty} \text{ sur } M,$$

à valeurs complexes.

Localement, la situation est la suivante : soit U un ouvert de M et soit $e = (e_1, \ldots, e_n)$ un champ de repères au-dessus de U ; alors $De = \omega e$, i.e.

$$De_i = \sum_{1 \leq k \leq n} \omega_{ik}\, e_k\;, \qquad 1 \leq i \leq n\;;$$

la matrice $\omega = (\omega_{ik})$ détermine la connection.

La forme de courbure est définie par $\Omega = d\omega - \omega\wedge\omega$, et l'identité de Bianchi est $d\Omega + \Omega\wedge\omega - \omega\wedge\Omega = 0$.

c, <u>Géométrie hermitienne</u>. Une variété hermitienne M est un couple constitué par une variété complexe, notée aussi M, et une métrique hermitienne H dans le fibré tangent. Localement :

$$\langle\cdot,\cdot\rangle = H = \sum_{1\leq k\leq n} \omega_k \bar{\omega}_k ;$$

les n formes $\omega_1, \ldots, \omega_n$ sont de type (1, 0) et déterminées à une transformation unitaire près. Elles constituent une base de T*M ou <u>corepère</u>; la base $(e_1, \ldots, e_n)$ de TM, duale de $(\omega_1, \ldots, \omega_n)$, est un repère. Rappelons aussi que M est Kahlérienne lorsque sa forme de Kahler $\hat{H} = \sum_{1\leq k\leq n} \omega_k\wedge\bar{\omega}_k$ est fermée : $d\hat{H} = 0$. Si l'on considère une connection ω sur le fibré tangent TM, les formes de torsion sont :

$$\tau_i = d\omega_i - \sum_{1\leq k\leq n} \omega_k \wedge \omega_{ki} ;$$

elles sont de type (2,0) si et seulement si les formes ω_{ki}, déterminant ω, sont de type (1,0). On sait qu'il existe sur M une connection unique, appelée <u>connection hermitienne</u> de M, telle que les τ_i soient de type (2,0) et que

$$\omega_{ik} + \bar{\omega}_{ki} = 0, \qquad i,k = 1, 2, \ldots, n.$$

cette dernière condition exprime que la connection conserve la métrique (i;e; dH = 0). C'est cette connection que nous choisirons sur M. Alors la courbure est définie par

$$\Omega_{ik} = d\omega_{ik} - \sum_{1\leq j\leq n} \omega_{ij} \wedge \omega_{jk} ;$$

les formes Ω_{ik} sont de type (1,1) et satisfont à $\Omega_{ik} + \bar{\Omega}_{ki} = 0$.

En chaque point $m \in M$ et à tout couple de vecteurs $X, Y \in T_m M$, on associe à la courbure une transformation linéaire

$$\Omega_m (X, Y) : T_m M \rightarrow T_m M,$$

définie par

$$\langle \Omega_m(X, Y)\, e_i\ ,\ e_k \rangle = \Omega_{ik}\,(X, Y)\ ;$$

cette transformation linéaire est utilisée pour les repères de Frenet.

2. Repères de Frenet le long des courbes holomorphes. Une courbe holomorphe est une application holomorphe $f : Z \to M$ d'une surface de Riemann Z dans M. Soient ζ une coordonnée locale de Z et z^i un système de coordonnées locales de M. Alors f est définie par $z^i = z^i(\zeta)$ et son vecteur tangent à l'origine, $f'(0) \in T_{z(0)} M$, est de la forme $\sum_i \frac{\partial z_i}{\partial \zeta}(0) \frac{\partial}{\partial z^i} = \zeta^\rho V$, où V est un vecteur non nul et ρ un entier non négatif.

Définitions. Un repère unitaire de M le long de f est un ensemble d'applications $e_i : Z \to TM$ tel que $(e_i(\zeta), \ldots, e_n(\zeta))$ soit une base unitaire de $T_{f(\zeta)}M$ pour chaque $\zeta \in Z$. Etant donnée la connection hermitienne ω de M, un repère de Frenet le long de f est un repère unitaire le long de f tel que :

$$e_1(0) = \|V\|^{-1} V, \qquad f^*(\omega_{ik}) = 0 \begin{cases} i = 1, & k > 2, \\ 1 < i < n, & |k-i| > 1. \end{cases}$$

Dans une variété hermitienne de dimension n, il n'existe pas, en général, de repère de Frenet le long d'une courbe holomorphe.

THEOREME : Soit M une variété hermitienne de dimension n; considérons les conditions suivantes :

(A) $\forall\, m \in M,\ \forall\, X \in T_m M,\qquad \Omega_m(X, \bar{X})X = a(X)X$, où $a(X) \in \mathbb{R}$;

(B) $\forall\, m \in M,\ \forall\, X, Y \in T_m M,\ X \perp Y,\ \Omega_m(X, \bar{X})Y = b(X)Y$, où $b(X) \in \mathbb{R}$;

pour que M admette un repère de Frenet le long de toute courbe holomorphe, il faut et il suffit que l'une des trois conditions suivantes soit satisfaite

(i) $n = 2$;

(ii) $n = 3$ et M vérifie (A);

(iii) $n \geq 4$ et M vérifie (A) et (B).

Schéma de la démonstration : Le long de f, on peut prendre $f^*(\omega_1) = h(\zeta)d\zeta$, $f^*(\omega_k) = 0$, $(2 \leq k \leq n)$; le fait que τ est de type (2,0) implique $f^*(\tau_i) = 0$, $(i = 1,\dots,n)$; il en résulte, après un éventuel changement de repère unitaire portant sur $e_2,\dots, e_n$,

$$f^*(\omega_{1k}) = \begin{cases} h_1(\zeta)d\zeta, & k = 2, \\ 0, & k > 2; \end{cases}$$

ceci prouve (i).

Pour n = 3, la différentiation extérieure de $f^*(\omega_{1k})$ montre que $h_1(\zeta)$ satisfait à une équation différentielle dont la condition d'intégrabilité est (A) et il n'y a pas d'autre condition pour n = 3 : ceci prouve (ii). Pour $n \geq 4$, le même procédé, appliqué à $f^*(\omega_{ik})$, conduit à la seule condition (B), de sorte que (iii) est démontré.

Cas particulier : Lorsque la variété hermitienne M est à torsion nulle (i.e. lorsque M est Kählérienne), l'interprétation géométrique est plus simple :

Théorème : Pour qu'une variété Kählérienne de dimension ≥ 3 admette un repère de Frenet le long de toute courbe holomorphe, il faut et il suffit qu'elle soit à courbure sectionnelle holomorphe constante.

Remarque : Le cas où M est une variété riemannienne de dimension n et de classe C^r $(r \geq 4)$ peut être étudié directement.

Courbes holomorphes de l'espace projectif complexe. Soit $\mathbb{P}^n$ l'espace projectif complexe de dimension n, muni de la métrique de Fubini-Study qui en fait une variété Kählérienne de courbure sectionnelle holomorphe constante. Pour une courbe holomorphe $f : Z \to \mathbb{P}^n$, posons $\theta = f^*(\omega_1) = h\, d\zeta$ et $\theta_{ik} = f^*(\omega_{ik})$. Alors f a une métrique conforme $H_o = \theta\bar{\theta}$ et une forme de Kahler $\hat{H}_o = \frac{\sqrt{-1}}{2}\, \theta \wedge \bar{\theta}$; en outre, il existe une connection φ et une seule telle que $d\theta = \theta \wedge \varphi$, $\varphi + \bar{\varphi} = 0$. (Explicitement, on trouve $\varphi = -\partial \log h + \bar{\partial} \log h$) et la forme de Ricci est définie par $\mathrm{Ric}\, H_o = \frac{\sqrt{-1}}{2}\, d\varphi = \sqrt{-1}\, \partial\bar{\partial} \log h$, ce qui équivaut à $\mathrm{Ric}\, H_o = -2K\, \hat{H}_o$, où K est la courbure gaussienne de la métrique. (En fait, θ est défini à une transformation unitaire près : changer θ en $e^{i\psi}\theta$ change φ en $\varphi + id\psi$, mais Ric H_o reste le même). Ecrivons aussi, pour $k = 1, \ldots, n-1$,

$$\hat{H}_k = \frac{\sqrt{-1}}{2}\,\theta_{k\;k+1} \wedge \bar{\theta}_{k\;k+1}, \qquad \varphi_k = \theta_{kk} - \theta_{k+1\;k+1},$$

$$\text{Ric } H_k = \frac{\sqrt{-1}}{2}\, d\,\varphi_k ;$$

ceci permet de donner une expression très simple du second théorème fondamental de la théorie de Nevanlinna :

Théorème : Dans les notations précédentes :

$$\text{Ric } H_o = -2\hat{H}_o + \hat{H}_1 , \quad \text{Ric } \hat{H}_k = \hat{H}_{k-1} - 2\hat{H}_k + \hat{H}_{k+1} \qquad (1 \leq k \leq n-1).$$

Les H_k $(k = 1, \ldots, n-1)$ sont appelées métriques osculatrices de la courbe holomorphe. Comme conséquences immédiates de cette théorie, on retrouve des propositions classiques :

1° Proposition. Pour une courbe holomorphe de $\mathbb{P}^n$, les métriques osculatrices sont déterminées de façon unique par H_o.

2° Théorème de Blaschke. La métrique de Poincaré $(1 - |\zeta|^2)^{-2} d\zeta\, d\bar{\zeta}$ sur le disque unité ne peut pas être obtenue par un plongement isométrique dans $\mathbb{P}^n$.

3° Théorème de Calabi. Une courbe holomorphe non dégénérée est déterminée de **façon unique, à un déplacement** rigide près, par sa première forme fondamentale H_o.

HOLOMORPHIC SPACES

by PAUL FLONDOR

We shall give here some results about holomorphic spaces. The proofs are only sketched.

Let $(X,\mathcal{O}_X)$ be a ringed space and $T\subset X$, $\mathcal{O}_T=\mathcal{O}_{X|T}$. The ringed space $(T,\mathcal{O}_T)$ will be called a restriction subspace of $(X,\mathcal{O}_X)$. Let Q be a Stein compact in $\mathbb{C}^n$. The restriction subspace $(Q,\mathcal{O}_Q)$ of $(\mathbb{C}^n,\mathcal{O}_{\mathbb{C}^n})$ will be called an <u>affine model</u>.(here $\mathcal{O}_{\mathbb{C}^n}$ is the sheaf of germs of holomorphic functions on $\mathbb{C}^n$).A <u>local model</u> is a subspace of finite presentation of an affine model. The structure sheaf of a local model is a coherent sheaf of rings.(in fact of $\mathbb{C}$-algebras).

<u>Definition</u>. A <u>holomorphic space</u> is a $\mathbb{C}$-ringed space $(X,\mathcal{O}_X)$ such that X is separated and for any $x\in X$ there is a compact neighbourhood V of x such that the restriction subspace $(V,\mathcal{O}_V)$ is isomorphic (as a $\mathbb{C}$-ringed space) to a local model.

Holomorphic spaces were first introduced in [1].

The morphisms of holomorphic spaces are the morphisms of $\mathbb{C}$- ringed spaces. In this way we obtain a category, which contains the categories of complex spaces, real analytic spaces and the dual of the category of analytic algebras.

If $(X,\mathcal{O}_X)$ is a holomorphic space then $\mathcal{O}_X$ is a coherent sheaf of $\mathbb{C}$-algebras and the stalks $\mathcal{O}_{X,x}$ are analytic algebras.

It is easily seen, that in the category of holomorphic spaces, finite products always exist.

Let $(X,\mathcal{O}_X)$ be a holomorphic space. A compact set $K\subset X$ will be called a <u>Stein compact</u> if there is a coherent imbedding $(K,\mathcal{O}_K)\longrightarrow(Q,\mathcal{O}_Q)$ where $(Q,\mathcal{O}_Q)$ is an affine model($(K,\mathcal{O}_K)$ is the restriction subspace).

If K_1,K_2 are Stein compacts in X then $K_1\cap K_2$ is a Stein compact.

Observe that if $(K,\mathcal{O}_K)$ is a local model then for every coherent sheaf $\mathcal{F}$ of $\mathcal{O}_K$-modules, $\Gamma(K,\mathcal{F})=\mathcal{F}(K)$ has a natural structure of a DFS space. In this sense one can say that the structure sheaf of a holomorphic space is a sheaf of type DFS on a base of compacts(Stein compacts).

Let $(X,\mathcal{O}_X)$, $(Y,\mathcal{O}_Y)$ be holomorphic spaces. A morphism $f: X\longrightarrow Y$ of holomorphic spaces is called <u>$\mathbb{C}$-analytic</u> if for every $x\in X$

there are open sets $U \ni x, V \supset f(U)$ and a commutative diagram

$$\begin{array}{ccc} U & \xrightarrow{i} & V \times D \\ {\scriptstyle f}\downarrow & \swarrow {\scriptstyle \pi} & \\ V & & \end{array}$$

where D is an open set in $\mathbb{C}^n$, π the canonical projection and i a coherent imbedding.

For example the projection $X \times \mathbb{C}^n \longrightarrow X$ is $\mathbb{C}$-analytic, for every holomorphic space X.

The composition of two $\mathbb{C}$-analytic morphisms is $\mathbb{C}$-analytic. If

$$\begin{array}{ccc} X' & \xrightarrow{g} & X \\ {\scriptstyle f'}\downarrow & & \downarrow{\scriptstyle f} \\ Y' & \xrightarrow{h} & Y \end{array}$$

is a cartesian diagram of holomorphic spaces and f is $\mathbb{C}$-analytic then f' is $\mathbb{C}$-analytic.

The following theorem was proved in [1]

Theorem I. Let $f : X \longrightarrow Y$ be a proper morphism of holomorphic spaces. Assume f to be $\mathbb{C}$-analytic. Then for every coherent sheaf $\mathcal{F}$ of $\mathcal{O}_X$-modules the sheaves $R^n f_*(\mathcal{F})$ are coherent on Y. $(n \geq 0)$

For a more general result see [2]

Theorem I has many important consequences. Among them we note:

Let $(X, \mathcal{O}_X)$ be a holomorphic space. A set $A \subset X$ is called $\mathbb{C}$-analytic if for every $x \in X$ there is an open neighbourhood V of x and sections $f_1, \ldots, f_p \in \Gamma(V, \mathcal{O}_X)$ such that $A \cap V = \{ y \mid f_1(y) = \cdots = f_p(y) = 0 \}$.

Now we have the following: if $f : X \longrightarrow Y$ is a proper $\mathbb{C}$-analytic morphism of holomorphic spaces then $f(X)$ is $\mathbb{C}$-analytic.

For the proof one can apply theorem I observing that $f(X) = \mathrm{Supp}\, f_*(\mathcal{O}_X)$.

Another consequence is:

Let $f : X \rightarrow Y$ be a proper $\mathbb{C}$-analytic morphism of holomorphic spaces and $\mathcal{F}$ a coherent sheaf on X, which is Y-flat. Then for every $q \geq 0$ the function $y \mapsto \dim_{\mathbb{C}} H^q(X_y, \mathcal{F}_y)$ is upper semicontinous on Y. (X_y is the analytic "fiber" of f in y, and $\mathcal{F}_y$ the analytic restriction of $\mathcal{F}$ to X_y).

In order to prove this, one can adapt for holomorphic spaces the proof known for complex spaces. (for example, the proof given in [3]).

One can also obtain a kind of Stein factorisation in the case of holomorphic spaces.

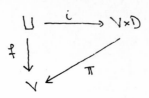

The following graded version of theorem I can be proved:

Theorem 2. Let $f: X \to Y$ be a proper $\mathbb{C}$-analytic morphism of holomorphic spaces. Let $\mathcal{F}$ be a graded coherent sheaf of $\mathcal{O}_X[T]$ -modules($\mathcal{O}_X[T]$ is the graded sheaf of rings associated to $\mathcal{O}_X$, $T=[T_1,..T_m]$). Then the sheaves $R^n f_*(\mathcal{F})$ are $\mathcal{O}_Y[T]$ –coherent for every $n \geq 0$.

The result in the case of complex spaces is in [4] .

The proof of theorem 2 is somewhat similar to the proof of theorem I.

Let $(X,\mathcal{O}_X)$ be a holomorphic space. A complex space $(Y,\mathcal{O}_Y)$ is called a complexification of $(X,\mathcal{O}_X)$ if there is an immersion $X \xrightarrow{i} Y$ such that $(X,\mathcal{O}_X)$ is isomorphic to a restriction subspace of $(Y,\mathcal{O}_Y)$. Observe that, from the definition of holomorphic spaces follows, that locally, complexifications always exist. We have:

Proposition. Let $(X,\mathcal{O}_X)$ be a paracompact holomorphic space. There is a complexification of $(X,\mathcal{O}_X)$(which is separated). Two complexifications Y_1, and Y_2 of $(X,\mathcal{O}_X)$ are isomorphic, in the sense that, there are open neighbourhoods $U_1 \supset X$, $U_2 \supset X$ in $Y_1(Y_2)$ and an isomorphism of complex spaces $U_1 \to U_2$ extending the identity of X.

The proof goes by glueing togbther the local complexifications and applying for example the Bruhat-Whitney procedure in order to obtain separation. [5]

By using the complexification one can obtain the analytic spectrum of a coherent algebra on a holomorphic space. More precisely if $(Y,\mathcal{O}_Y)$ is a holomorphic space and $\mathcal{A}$ a coherent $\mathcal{O}_Y$ -algebra, there is a holomorphic space $S(\mathcal{A}) \xrightarrow{g} Y$ over Y such that g is $\mathbb{C}$ -analytic and the usual universal property holds.

One can also prove a normalization theorem for holomorphic spaces. The canonical morphism $Y \xrightarrow{p} X$ of the normalization is $\mathbb{C}$-analytic.

Bibliography.

I. P.Flondor , M.Jurchescu -Grauert's coherence theorem for holomorphic spaces. Rev.Roum.Math. Pures.et Appl.XVIII, 8, 1973.

2. C.Houzel -Espaces analytiques relatifs et theoremes de finitude, Math.Ann.205, I973.

3.M.Schneider -Halbstetigkeitssatze fur relativ analytische Raume, Inv.Math I6, I972.

4. C.Bănică -Le complété formel d'un espace analytique le long d'un sous-espace:un théoreme de comparaison,Manuscripta math 6, 1972.

5.Bruhat-Whitney -Quelques propriètes fondamentales des ensembles analytiques reël,Comm. Math Helv. 33,1959.

Institute Polytechnic
Bucharest

Holomorphe Transformationsgruppen mit kompakten Bahnen

von Harald Holmann (Freiburg, Schweiz)

0. Einleitung

D.B.A. Epstein zeigt in [2]:

Operiert die additive Gruppe $\mathbb{R}$ *der reellen Zahlen differenzierbar auf der kompakten differenzierbaren Mannigfaltigkeit* X, *so dass alle* $\mathbb{R}$*-Bahnen kompakt* 1*-dimensional* (dh. Kreislinien) *sind, dann gibt es eine differenzierbare Operation der Kreisgruppe* S^1 *auf* X *mit den gleichen Bahnen.*

Das hat zur Folge:

(1) *Alle* $\mathbb{R}$*-Bahnen sind stabil* (dh. jede Umgebung einer Bahn enthält eine invariante Umgebung dieser Bahn).

(2) $X/\mathbb{R}$ *hat eine kanonische Mannigfaltigkeitsstruktur* (dh. unter anderem, dass die kanonische Projektion $\pi: X \longrightarrow X/\mathbb{R}$ differenzierbar ist).

(3) $(X,\pi,X/\mathbb{R})$ *ist ein differenzierbarer Seifertscher Prinzipalfaserraum über* $X/\mathbb{R}$ *mit* S^1 *als Strukturgruppe.*

Dieser Satz von Epstein war die erste positive Antwort auf die folgende Vermutung von A. Häfliger:

Ist X ein kompakte differenzierbare Mannigfaltigkeit mit einer differenzierbaren Blätterung, so dass alle Blätter kompakt sind, dann gelten die folgenden (untereinander äquivalenten) Aussagen:

(1) Die Holonomiegruppen aller Blätter sind endlich.

(2) Alle Blätter sind stabil.

(3) Der zugehörige Blätterraum ist hausdorffsch.

R. Edwards, K. Millet, D. Sullivan (siehe [1]) und E. Vogt (siehe [12]) konnten in Verallgemeinerung des Satzes von Epstein für 2-codimensionale differenzierbare Blätterungen kompakter Mannigfaltigkeiten die Vermutung von Häfliger bestätigen. Analog zum Satz von Epstein ergibt auch hier die Blätterung einen differenzierbaren Seifertschen Faserraum über dem Blätterraum.

Für 3-codimensionale Blätterungen ist die Vermutung von Häfliger noch offen. Für 4-codimensionale Blätterungen wurden jedoch von Sullivan und Thurston (siehe [10], [11]) Gegenbeispiele in Form von 5-dimensionalen kompakten differenzierbaren Mannigfaltigkeiten mit differenzierbaren Blätterungen in Kreislinien, deren Blätter nicht alle stabil sind, angegeben. Diese Gegenbeispiele lassen sich sogar reell-analytisch konstruieren.

Es sei noch bemerkt, dass schon länger 3-dimensionale nicht kompakte differenzierbare und reell-analytische Mannigfaltigkeiten mit differenzierbaren bzw. reell analytischen Blätterungen in Kreislinien bekannt sind, deren Blätter nicht alle stabil sind (siehe z.B. [2], [9]).

Für kompakte komplexe Mannigfaltigkeiten mit holomorphen Blätterungen, so dass alle Blätter kompakt sind, ist Häfligers Vermutung, ob alle Blätter stabil sind, völlig offen. Selbst für nicht kompakte komplexe Mannigfaltigkeiten hat man bisher noch keine Gegenbeispiele gefunden. Die Verhältnisse scheinen hier völlig anders zu liegen. Im komplex-analytischen Kontext ist z.B. der Satz von

Epstein ohne Dimensionseinschränkungen richtig (siehe [7]):

Die additive Gruppe $\mathbb{C}$ operiere holomorph auf dem kompakten zusammenhängenden komplexen Raum X, so dass alle $\mathbb{C}$-Bahnen kompakt, komplex-eindimensional (dh. Tori) *sind. Dann gibt es eine holomorphe Operation einer komplexen Torusgruppe T auf X mit den gleichen Bahnen.*

Hieraus folgt auch wieder, dass alle Bahnen stabil sind und X ein holomorpher Seifertscher Prinzipalfaserraum über dem komplex-analytischen Bahnenraum $X/\mathbb{C}$ mit T als Strukturgruppe ist (siehe [3], [4], [6], [8]).

Für nicht notwendig kompakte komplexe Mannigfaltigkeiten soll hier gezeigt werden:

X sei eine zusammenhängende komplexe Mannigfaltigkeit, G eine zusammenhängende kommutative komplexe Liesche Gruppe, die holomorph auf X operiert. Alle G-Bahnen G(x), $x \in X$, seien kompakt und reell 2-codimensional. Dann sind alle Bahnen stabil und der Bahnenraum X/G ist auf kanonische Weise eine Riemannsche Fläche.

1. Ausgezeichnete Transversalen.

Wir werden in diesem Abschnitt mit Hilfe ausgezeichneter G-Transversalen ein notwendiges und hinreichendes Stabilitätskriterium für kompakte G-Bahnen beweisen, das dann im nächsten Abschnitt beim Beweis des in der Einleitung angekündigten Satzes verwendet wird.

Definition: *X sei ein komplexer Raum, G eine komplexe Liesche Gruppe, die auf X holomorph operiert* ($\psi: G \times X \longrightarrow X$ bezeichne die holomorphe Operationsabbildung).

1) *Eine lokalanalytische Menge S in X heisst eine G-Transversale, wenn gilt:*

   a) $S \cap G(s)$ *ist diskret für alle G-Bahnen* $G(s) := \{g(s);\ g \in G\}$, $s \in S$.

   b) *Es gibt eine offene Umgebung U von S und eine biholomorphe Abbildung* $\varphi: Q \times S \longrightarrow U$, *wobei* $Q \subset \mathbb{C}^m$ *ein Polyzylinder um den Nullpunkt ist, so dass* $\varphi(0,s) = s\ \forall s \in S$.

2) *Eine G-Transversale S heisst ausgezeichnet in* $x_o \in S$, *wenn es auf S eine Umgebungsbasis* $\mathfrak{W}$ *von* x_o *mit folgenden Eigenschaften gibt:*

   a) *Jedes* $V \in \mathfrak{W}$ *ist offen, zusammenhängend und relativ kompakt in S.*

   b) $\partial V \cap \hat{V} = \emptyset$ *für jedes V aus* $\mathfrak{W}$ (dabei ist $\hat{V} := \{g(v); g \in G, v \in V\}$ die invariante Hülle von V).

Hierzu ist zu bemerken: Haben alle G-Bahnen die gleiche Dimension, so gibt es durch jeden Punkt $x_o \in X$ eine G-Transversale S. Die Abbildung $\varphi: Q \times S \longrightarrow U$ in 1)b) der obigen Definition kann dabei wie folgt gewählt werden:

$\varphi(y,s) := \psi(j(y),s)$, wobei $j: Q \longrightarrow G$ eine geeignete holomorphe Einbettung von Q in eine Umgebung des neutralen Elements $e \in G$ ist (siehe [5], S. 102, Theorem 2).

Satz 1: *X sei ein komplexer Raum, G eine zusammenhängende komplexe Liesche Gruppe, die auf X holomorph operiert. Alle G-Bahnen seien kompakt und von gleicher Dimension. Dann gilt:*

Eine G-Bahn $G(x_o)$, $x_o \in X$, ist genau dann stabil, wenn es eine in x_o ausgezeichnete G-Transversale S durch x_o gibt.

Bemerkung: Ein analoger Satz lässt sich für reguläre Blätterungen differenzierbarer Mannigfaltigkeiten bzw. komplexer Räume beweisen. Dabei hat man die Operation von G auf X durch gewisse Holonomieoperationen zu ersetzen.

Beweis: Die Notwendigkeit des Stabilitätskriteriums ist leicht einzusehen. Wir zeigen, dass es auch hinreichend ist. Zu jeder Umgebung W der kompakten G-Bahn $G(x_o)$ ist also eine invariante Umgebung $\tilde{W} \subset W$ von $G(x_o)$ anzugeben. Dabei können wir voraussetzen, dass eine in x_o ausgezeichnete G-Transversale S durch x_o existiert. Wie wir im Anschluss an die obige Definition bemerkt haben, kann man eine Untermannigfaltigkeit A einer Umgebung des neutralen Elements e in G mit $e \in A$ finden, so dass $A \times S$ durch die holomorphe Operationsabbildung $\psi: G \times X \longrightarrow X$ auf eine offene Umgebung U von S abgebildet wird.

Man kann S hinreichend klein wählen, so dass $G(x_o) \cap S = \{x_o\}$. Dann definieren wir wie folgt relativ-kompakte offene Umgebungen $B \subset\subset C \subset\subset D \subset\subset E$ von $e \in G$: Da die Bahn $G(x_o)$ kompakt ist, gibt es eine offene relativ kompakte Umgebung B von e mit $B(x_o) = G(x_o)$; es sei dann $C := B \circ A$, $D := C \circ C$, $E := D^{-1} \circ D$. Wir wählen eine genügend kleine Umgebung $V \in \mathcal{W}$ von x_o, so dass die offene Umgebung $\psi(C \times V)$ von $G(x_o)$ ganz in W liegt.

G_{x_o} bezeichne die (in G abgeschlossene) Isotropiegruppe von x_o. Da $G_{x_o} \cap \bar{E}$ kompakt ist und da $\psi(G_{x_o} \times \{x_o\}) = \{x_o\}$, so kann man bei genügend kleinem V annehmen, dass $\psi((G_{x_o} \cap E) \times V) \subset U$. Wir behaupten, dass dann sogar gilt: $\psi((G_{x_o} \cap E) \times V) \subset \psi(A \times V)$.

Sei $g \in G_{x_o} \cap E$, so ist $g(V) \subset U = \psi(A \times S)$. Bezeichnet $p_2 \colon A \times S \longrightarrow S$ die kanonische Projektion auf die zweite Komponente und $\psi^* \colon U = \psi(A \times S) \longrightarrow A \times S$ die holomorphe Umkehrabbildung zu $\psi | A \times S \longrightarrow U$, so ist $\hat{g} := p_2 \circ \psi^* \circ g \colon V \longrightarrow S$ eine holomorphe Abbildung, wobei s und $\hat{g}(s)$ für $s \in V$ stets auf den gleichen G-Bahnen liegen. Auf Grund der Bedingung 2)b) in der obigen Definition ist folglich $\hat{g}(V) \cap \partial V = \emptyset$. Da V zusammenhängend ist und $\hat{g}(x_o) = x_o$ gilt (wegen $G(x_o) \cap S = \{x_o\}$), so muss $\hat{g}(V)$ in V enthalten sein. Das bedeutet aber gerade $g(V) \subset \psi(A \times V)$.

Wenn wir jetzt $\psi(D \times V) = \psi(C \times V)$ zeigen können, so sind wir fertig, denn hieraus folgt $\psi(C^n \times V) = \psi(C \times V) \subset W$ für alle $n \in \mathbb{N}$. Da $G = \bigcup_{n \in \mathbb{N}} C^n$, so ist $\tilde{W} := \psi(G \times V) = \psi(C \times V) \subset W$ eine offene G-invariante Umgebung von $G(x_o)$.

Zum Nachweis von $\psi(D \times V) = \psi(C \times V)$ geben wir für $(d,v) \in D \times V$ jeweils Elemente $(c,v') \in (C \times V)$ mit $d(v) = c(v')$ an:
Da $G(x_o) = B(x_o)$, so gibt es ein $b \in B$ mit $d(x_o) = b(x_o)$. Da $g := b^{-1} \circ d \in G_{x_o} \cap E$, so existieren $a \in A$ und $v' \in V$ mit $g(v) = a(v')$. Setzen wir $c := b \circ a \in C$, so gilt: $d(v) = b(g(v)) = b(a(v')) = c(v')$.

2. Zur Stabilität kompakter Bahnen.

Ziel dieses Abschnittes ist es, den folgenden schon in der Einleitung aufgeführten Satz zu beweisen:

Satz 2: *X sei eine zusammenhängende komplexe Mannigfaltigkeit* (mit abzählbarer Topologie), *G eine zusammenhängende kommutative komplexe Liesche Gruppe, die holomorph auf X operiert. Alle Bahnen* $G(x)$, $x \in X$, *seien kompakt mit* $\operatorname{codim}_{\mathbb{C}} G(x) = 1$.

Dann sind alle Bahnen stabil und der Bahnenraum X/G ist auf kanonische Weise eine Riemannsche Fläche.

Beweis:

1. Wir behandeln zuerst den Fall, dass alle Isotropiegruppen $G_x := \{g \in G;\ g(x) = x\}$, $x \in X$, diskret sind. Dabei können wir annehmen, dass G effektiv auf X operiert, d.h. $\bigcap_{x \in X} G_x = \{e\}$. Ist G kompakt, so sind wir fertig; X ist in diesem Fall sogar ein holomorpher Seifertscher Prinzipalfaserraum über X/G mit G als Strukturgruppe (siehe [4], Satz 3, S. 148). Ist G nicht kompakt, so sind alle Isotropiegruppen G_x, $x \in X$, von $\{e\}$ verschiedene diskrete Untergruppen von G und G/G_x ist jeweils eine kompakte komplexe Mannigfaltigkeit. Da alle G-Bahnen die gleiche Dimension haben, so gibt es durch jeden Punkt $x_o \in X$ eine G-Transversale S. Auf Grund von Satz 1 genügt es zu zeigen, dass S in x_o ausgezeichnet ist. Da $\operatorname{codim}_{\mathbb{C}} G(x) = 1$ für alle $x \in X$, so ist $\dim_{\mathbb{C}} S = 1$. S ist singularitätenfrei wählbar, da wir X als komplexe Mannigfaltigkeit vorausgesetzt haben.

$M := \{(g,s) \in G \times S;\ g(s) = s,\ g \neq e\}$ ist eine analytische Menge in $(G-\{e\}) \times S$. $p_1: G \times S \longrightarrow G$ und $p_2: G \times S \longrightarrow S$ seien die kanonischen Projektionen. Da $G_s \neq \{e\}\ \forall s \in S$, so ist $p_2(M) = S$. Da alle Isotropiegruppen diskret sind, so muss $\dim M = 1$ sein. M zerfällt in zwei disjunkte analytische Mengen $M = M^1 \cup M^o$, wobei jeweils M^i rein i-dimensional ist. M^o ist eine höchstens abzählbare diskrete Punktmenge. Aus den definierenden Eigenschaften einer G-Transversalen folgt, dass $p_2|M^1 \longrightarrow S$ lokal biholomorph ist. Wir behandeln nun die Fälle (a) $x_o \in p_2(M^1)$, (b) $x_o \notin p_2(M^1)$ getrennt.

(a): Es gibt nach Voraussetzung zu $x_o \in S$ einen Punkt $(g_o,x_o) \in M^1$ und eine offene zusammenhängende Umgebung U_o von (g_o,x_o) auf M^1, die durch p_2 biholomorph auf eine offene Umgebung V_o von x_o auf S abgebildet wird. Da G effektiv auf X operiert, so kann $p_1|U_o \longrightarrow G$ nicht konstant sein, dh. $p_1(U_o)$ ist eine komplex-eindimensionale analytische Menge in einer offenen Umgebung W_o von g_o, wenn wir U_o genügend klein wählen. Wir können dabei annehmen, dass $p_1|U_o \longrightarrow p_1(U_o)$ eine holomorphe eigentliche verzweigte Ueberlagerung darstellt mit $p_1^{-1}(g_o) \cap U_o = (g_o,x_o)$. Folglich ist auch $f := p_1 \circ (p_2|U_o)^{-1}: V_o \longrightarrow p_1(U_o)$ eine holomorphe eigentliche verzweigte Ueberlagerungsabbildung, wobei über g_o nur der Punkt x_o liegt. Hat man U_o klein genug gewählt, so kann man annehmen, dass $U_o = M \cap (W_o \times V_o)$.

Wir können nun auf $p_1(U_o)$ eine Umgebungsbasis $\mathfrak{W}$ von g_o so bestimmen, dass alle $W \in \mathfrak{W}$ offen, zusammehängend und relativkompakt in $p_1(U_o)$ liegen. $\mathfrak{V} := \{f^{-1}(W),\ W \in \mathfrak{W}\}$ ist dann eine Umgebungsbasis von x_o auf S mit den Eigenschaften 2)a) und b) der obigen Definition. Um zu zeigen, dass $\partial V \cap \hat{V} = \emptyset$ für alle

$V \in \mathfrak{W}$, hat man sich wegen der Kommutativität von G nur klar zu machen, dass für Punkte $v \in V$ und $s \in \partial V$ gilt: $G_v \neq G_s$. Das folgt aber sofort aus der Tatsache, dass $f(s) \in G_s$, während $f(s) \notin G_v$.

(b): Wir definieren $K_r := \{w \in \mathbb{C};\ |w| < r\}$ für $r \in \mathbb{R}$, $r > 0$. Da die G-Transversale S durch x_o eine 1-dimensionale komplexe Mannigfaltigkeit ist, gibt es eine biholomorphe Abbildung σ von K_1 auf eine offene Umgebung von x_o auf S mit $\sigma(0) = x_o$. Da $p_2(M^o)$ abzählbar ist, so gibt es eine monoton fallende Folge $(r_\nu)_{\nu\in\mathbb{N}}$ mit $0 < r_\nu < 1$ und $\lim_{\nu\to\infty} r_\nu = 0$, so dass $\sigma(\partial K_{r_\nu}) \cap p_2(M^o) = \emptyset$, dh. $\sigma(\partial K_{r_\nu}) \subset p_2(M^1)$ für alle $\nu \in \mathbb{N}$. Es gibt folglich zu jedem ν eine kompakte Menge C_ν in G, so dass $\sigma(\partial K_{r_\nu}) \subset p_2(M^1 \cap (C_\nu \times S))$.

$M^1_\nu := M^1 \cap (C_\nu \times \sigma(\overline{K}_r))$ ist kompakt in $G \times S$; folglich ist auch $p_2(M^1_\nu)$ kompakt in S. Da $x_o \notin p_2(M^1_\nu)$ und da nach Wahl von C_ν gilt $\sigma(\partial K_{r_\nu}) \subset p_2(M^1_\nu)$, so ist $V_\nu := \sigma(\overline{K}_{r_\nu}) - p_2(M^1_\nu) \subset \sigma(K_{r_\nu})$ eine offene relativ-kompakte Umgebung von x_o auf S. Da $\partial V_\nu \subset p_2(M^1_\nu)$ und $V_\nu \cap p_2(M^1_\nu) = \emptyset$, so folgt $\partial V_\nu \cap \hat{V}_\nu = \emptyset$. Man kann die C_ν so wählen, dass $C_\nu \subset C_{\nu+1}$ für alle $\nu \in \mathbb{N}$. Dann ist auch $V_{\nu+1} \subset V_\nu$ für alle $\nu \in \mathbb{N}$. Die V_ν, $\nu \in \mathbb{N}$, bilden folglich eine Umgebungsbasis $\mathfrak{W}$ von x_o auf S mit den Eigenschaften 2)a) und b) der obigen Definition, dh. die G-Transversale S ist in x_o ausgezeichnet.

2. Wir behandeln nun den Fall, dass $\dim_{\mathbb{C}} G_x = m > 0$ für alle $x \in X$. Eine G-Transversale S durch $x_o \in X$ sei wie unter 1. gewählt. Bezeichnet $\psi: G \times X \longrightarrow X$ die holomorphe Operationsabbildung, so ist $\psi|G \times S \longrightarrow X$ eine reguläre holomorphe Abbildung mit $\mathrm{rk}_{(g,s)}\ \psi = \dim X$ für alle $(g,s) \in G \times S$ (siehe [5], Abschnitt 3). Es gibt folglich eine Umgebung W von e in G, so dass bei passender Wahl von S gilt (1) $g(s) \in S$ für $g \in W$, $s \in S$ impliziert $g(s) = s$.
(2) $M_e := \bigcup_{s \in S} (G_s \cap W) \times \{s\} = \{(g,s) \in W \times S;\ g(s) = s\} = (\psi|W{\times}S)^{-1}(S)$ ist eine m+1 dimensionale komplexe Untermannigfaltigkeit von $W \times S$. Die kanonische Projektion $p_2: W \times S \longrightarrow S$ induziert eine reguläre Abbildung $p_2: M_e \longrightarrow S$, deren m-dimensionale Fasern die Form $(G_s \cap W) \times \{s\}$, $s \in S$, haben. Das induziert eine holomorphe Abbildung $\varphi: S \longrightarrow G(m,n)$, $n = \dim G$, die jedem $s \in S$ den Tangentialraum $T_e(G_s)$ an G_s im neutralen Element e von G zuordnet. Dabei bezeichnet $G(m,n)$ die komplexe Grassmannmannigfaltigkeit der m-dimensionalen komplexen linearen Unterräume des n-dimensionalen Tangentialraumes $T_e(G)$.

Wir können annehmen, dass G effektiv auf X operiert. $\varphi: S \longrightarrow G(m,n)$ ist folglich nicht konstant. Wir können S passend verkleinern, so dass $\varphi(S)$ eine analytische Menge in einer offenen Umgebung V von $\varphi(x_o)$ wird und $\varphi: S \longrightarrow \varphi(S)$ eine verzweigte eigentliche Ueberlagerung mit $\varphi^{-1}(\varphi(x_o)) = x_o$ darstellt. Man kann nun auf $\varphi(S)$ eine Umgebungsbasis $\mathcal{W}$ von $\varphi(x_o)$ finden, so dass alle $W \in \mathcal{W}$ offen, zusammenhängend und relativkompakt auf $\varphi(S)$ sind.
$\mathcal{V} := \{\varphi^{-1}(W);\ W \in \mathcal{W}\}$ ist dann eine Umgebungsbasis von x_o auf S mit den Eigenschaften 2)a) und b) der obigen Definition, dh. S ist eine in x_o ausgezeichnete G-Transversale.

Aus der Stabilität der G-Bahnen von X ergibt sich wie in [3] (vergleiche Satz 15, Seite 350), dass der Quotientenraum X/G auf kanonische Weise die Struktur einer Riemannschen Fläche besitzt.

L i t e r a t u r

[1] Edwards, R., Millet, K., Sullivan, D.: Foliations with all leaves compact. Publ.I.H.E.S., No. 46 (1976)

[2] Epstein, D.B.A.: Periodic flows on three-manifolds. Ann. of Math. 95, 66-82 (1972)

[3] Holmann, H.: Komplexe Räume mit komplexen Transformationsgruppen. Math. Ann. 150, 327-360 (1963)

[4] Holmann, H.: Seifertsche Faserräume. Math. Ann. 157, 138-166 (1964)

[5] Holmann, H.: Local properties of holomorphic mappings. Proceedings Conf. on Complex Analysis, Minneapolis, Springer (1965)

[6] Holmann, H.: Holomorphe Blätterungen komplexer Räume. Comm. Math. Helv. 47, 185-204 (1972)

[7] Holmann, H.: Analytische periodische Strömungen auf kompakten komplexen Räumen (Publikation in Vorbereitung)

[8] Orlik, P.: Seifert Manifolds. Lecture Notes in Math. 291, Springer (1972)

[9] Reeb, G.: Sur certaines propriétés topologiques des variétés feuilletées. Act. Sci. et ind. N° 1183, Hermann, Paris (1952)

[10] Sullivan, D.: A counterexample to the periodic orbit conjecture. Publ. I.H.E.S. (1975)

[11] Sullivan, D.: A new flow. Bull. Am. Math. Soc., 82, 331-332 (1976)

[12] Vogt, E.: Foliations of codimension 2 with all leaves compact. manuscripta math., 18, 187-212 (1976)

VARIÉTÉS MIXTES

de M. Jurchescu

Ce travail, dont l'origine se trouve dans [5] , poursuit essentiellement les deux buts suivants:

Tout d'abord, il s'agit de construire une bonne catégorie (dont les objets seront les variétés mixtes de classe $\mathcal{C}^p$) contenant en tant que sous-catégories pleines celle des variétés différentiables de classe $\mathcal{C}^p$ et celle des variétés analytiques complexes, et telle que sur ses objets fonctionne un calcul différentiel analogue au calcul différentiel réel ou bien complexe. Le point de vue adopté ici est nouveau, en ce sens que l'on renonce à l'idée de considérer explicitement les variétés mixtes comme "familles" sur un espace donné (cf. Douady [4]).

Ensuite, il s'agit de délimiter une sous- catégorie de variétés mixtes de classe $\mathcal{C}^\infty$ et de dimension finie (les variétés de Cartan) contenant comme sous-catégorie pleine celle des variétés de Stein, et qui puisse être caractérisée par des théorèmes du type A et B de Cartan.

Rappelons que le calcul différentiel mixte sur les ouverts de $\mathbb{R}^m \times \mathbb{C}^n$ a été developpé, dans un contexte cohomologique, dans Andreotti-Grauert [1].

1. Espaces de Banach mixtes

Un <u>espace de Banach mixte</u> est un espace de Banach réel E muni d'un sous-espace de Banach E_1 de E et d'une structure complexe sur l'espace de Banach réel E_1 (i.e. d'un endomorphisme continu j de E_1, tel que j^2 = - identité).

L'espace de Banach complexe E_1 est appelé la <u>composante complexe</u> de E, et l'espace de Banach réel $E_2 = E/E_1$ la <u>composante réelle</u> de E. On dit d'un espace de Banach mixte E qu'il est <u>complexe</u> lorsque $E_2 = 0$ et <u>réel</u> lorsque $E_1 = 0$. L'espace euclidien mixte $E = \mathbb{R}^m \times \mathbb{C}^n$ est considéré comme espace de Banach mixte, à composante complexe $\mathbb{C}^n$ et à composante réelle $\mathbb{R}^m$.

Si E et F sont deux espaces de Banach mixtes, un <u>morphisme</u> d'espaces de Banach mixtes de E dans F est une application linéaire continue $u: E \longrightarrow F$ telle que $u(E_1) \subset F_1$ et telle que l'application $u_1: E_1 \longrightarrow F_1$ induite par u soit $\mathbb{C}$-linéaire; par passage au quotient on obtient alors une application $\mathbb{R}$-linéaire continue $u_2: E_2 \longrightarrow F_2$. Lorsque E est réel toute application $\mathbb{R}$-linéaire continue de E dans F est un morphisme d'espaces de Banach mixtes. Il est clair que les espaces de Banach mixtes constituent une catégorie additive, non abélienne, et que les espaces de Banach complexes et les espaces de Banach réels constituent des sous-catégories pleines de celle-ci.

Pour qu'un morphisme $u : E \longrightarrow F$ d'espaces de Banach mixtes soit un <u>isomorphisme</u> il faut que u, u_1 et u_2 soient bijectives, et il suffit que deux de ces trois applications soient bijectives.

Le produit direct de deux espaces de Banach mixtes E et F est l'espace de Banach mixte $E \times F$, à composante complexe $E_1 \times F_1$ et à composante réelle $E_2 \times F_2$.

Si E et F sont deux espaces de Banach mixtes, on dit que F est un <u>sous-espace</u> de Banach mixte de E si F est un sous-espace de Banach réel de E et si en outre la composante complexe de F est $F \cap F_1$ muni de la structure d'espace de Banach complexe induite par celle de E_1.

Lorsque F est un sous-espace de Banach mixte de E, l'espace de Banach quotient E/F admet une structure d'espace de Banach mixte à composante complexe E_1/F_1 et à composante réelle E_2/F_2.

Par exemple, si $u : E \to F$ est un morphisme d'espaces de Banach mixtes, $u^{-1}(0)$ est un sous-espace de Banach mixte de E à composante complexe $u_1^{-1}(0)$, et $\overline{u(E)}$ est un sous-espace de Banach mixte de F à composante complexe $F_1 \cap \overline{u(E)}$.

Il en résulte que dans la catégorie des espaces de Banach mixtes tout morphisme possède un noyau et un conoyau. Un morphisme $u : E \to F$ d'espaces de Banach mixtes est dit <u>monomorphisme direct</u> s'il y a un morphisme $v : G \to F$ d'espaces de Banach mixtes tel que $\binom{u}{v} : E \times G \to F$ soit un isomorphisme d'espaces de Banach mixtes. Dualement, le morphisme u est dit <u>epimorphisme direct</u> s'il existe un morphisme d'espaces de Banach mixtes $v : E \to G$ tel que $(u,v) : E \to F \times G$ soit un isomorphisme d'espaces de Banach mixtes.

Pour que u soit un monomorphisme direct (resp. un epimorphisme direct) il faut et il suffit que u admette un inverse à gauche (resp. à droite) dans la catégoire des espaces de Banach mixtes. Un sous-espace de Banach mixte E de F est dit <u>direct</u> lorsque l'inclusion i : $E \to F$ est un monomorphisme direct.

2. Calcul différentiel mixte

Soit p un entier fixé, $p \geq 1$.

Soient E et F deux espaces de Banach mixtes, D un ouvert de E, D' un ouvert de F, et $f : D \to D'$ une application.

On dit que f est un $\mathcal{C}^p$-morphisme (d'ouverts d'espaces de Banach mixtes) si f est de classe $\mathcal{C}^p$ pour la structure de Banach réelle sous-jacente et si, en outre, pour tout point x de X, la dérivée $f'(x) : E \to F$ est un morphisme d'espaces de Banach mixtes. Lorsque $E = \mathbb{R}^m \times \mathbb{C}^n$ cette dernière condition signifie que $\frac{\partial f}{\partial \bar{z}_k} = 0$ pour $1 \leq k \leq n$, où $z_1, \ldots, z_n$ sont les fonctions coordonnées dans $\mathbb{C}^n$.

Il est clair que les ouverts de Banach mixtes et leurs $\mathcal{C}^p$-morphismes constituent une catégorie $\mathcal{L}$. En localisant $\mathcal{L}$ on obtient la catégorie des variétés mixtes de classe $\mathcal{C}^p$.

Afin de préciser la terminologie, soit X un ensemble. Une carte mixte sur X est une application $\phi : U \to E$, injective et d'image ouverte, avec U un sous-ensemble de X et E un espace de Banach mixte; U sera dit le <u>domaine</u> et $\phi(U)$ <u>l'image</u> de la carte.

Il est alors clair ce qu'on entend par cartes mixtes $\mathcal{L}$-compatibles, atlas mixtes de classe $\mathcal{C}^p$, atlas équivalente, et structures mixtes de classe $\mathcal{C}^p$ sur X.

Une <u>variété mixte</u> de classe $\mathcal{C}^p$ est un ensemble X muni d'une structure de variété mixte de classe $\mathcal{C}^p$ sur X; les cartes appartenant aux divers atlas équivalents constituant la structure de variété mixte de X sont appelées les <u>cartes structurales</u> de X. Si X est une variété mixte de classe $\mathcal{C}^p$, il existe une topologie sur X, unique, telle que pour toute carte structurale $\Phi : U \to E$ de X, U soit ouvert et que l'application $U \to \Phi(U)$ induite soit un homéomorphisme; on dira que c'est la <u>topologie canonique</u> de X.

Par ouvert de X on entend un ouvert par rapport à la topologie canonique; il en est de même lorsqu'on considère la continuité d'une fonction définie ou bien à valeurs dans X.

Si X et Y sont deux variétés mixtes de classe $\mathcal{C}^p$, on appelle <u>morphisme</u> de variétés mixtes de classe $\mathcal{C}^p$ toute application: $f : X \to Y$ ayant la propriété que, pour tout point x de X, il y a une carte structurale $\Phi : U \to E$ de X et une carte structurale $\Psi : V \to F$ de Y avec $x \in U$, $f(U) \subset V$ et telles que $\Psi \circ f \circ \Phi^{-1} : \Phi(U) \to \Psi(V)$ soit un $\mathcal{C}^p$-morphisme d'ouverts d'espaces de Banach mixtes; en particulier f résultera alors continue.

Il est clair que les variétés mixtes de classe $\mathcal{C}^p$ et leurs morphismes constituent une catégorie. Le produit direct de deux objects de cette catégorie se construit de la même manière que dans le cas des variétés différentiables ou bien des variétés analytiques complexes.

Soit X une variété mixte de classe $\mathcal{C}^p$ et X' une partie de X. On dit que X' est une sous-variété mixte de classe $\mathcal{C}^p$ de X si, pour tout $x \in X$, il existe une carte structurale $\phi : U \to E$ de X telle que $x \in U$ et que $\phi(U \cap X')$ soit un ouvert d'un sous-espace de Banach mixte direct de E; en particulier tout ouvert de X est une sous-variété mixte de classe $\mathcal{C}^p$ de X. Si X' est une sous-variété mixte de classe $\mathcal{C}^p$ de X, X' admet une structure évidente de variété mixte de classe $\mathcal{C}^p$, et l'inclusion $i : X' \to X$ est un morphisme de variétés mixtes de classe $\mathcal{C}^p$; en outre, si Z est une variété mixte de classe $\mathcal{C}^p$ et $f : Z \to X'$ une application, alors f est un morphisme de variétés mixtes de classe $\mathcal{C}^p$ si et seulement si $i \circ f$ est un morphisme de variétés mixtes de classe $\mathcal{C}^p$.

Soit X une variété mixte de classe $\mathcal{C}^p$ et soit $x \in X$. On considère des couples (ϕ, u) avec $\phi : U \to E$ une carte structurale sur X dont le domaine U contient le point x et u un vecteur de E. On dit que deux tels couples (ϕ, u) et (ψ, v) sont <u>équivalents</u> si $f'(x)u = v$, où $f = \psi \circ \phi^{-1}$. Soit $T_x X$ l'ensemble des classes d'équivalence de couples (ϕ, u). Alors, pour toute carte structurale $\phi : U \to E$ de X dont le domaine contient le point x on a une bijection $\beta_{\phi,x} : E \to T_x X$ telle que, si $u \in E$, $\beta_{\phi,x}(u)$ soit la classe d'équivalence contenant le couple (ϕ, u). Lorsque $\phi : U \to E$ et $\psi : V \to F$ sont deux cartes structurales sur X dont les domaines U et V contiennent le point x, le diagramme suivant (où $f = \psi \circ \phi^{-1}$) est commutatif.

$$\begin{array}{ccc} & T_xX & \\ \beta_{\Phi,x} \nearrow & & \nwarrow \beta_{\Psi,x} \\ E & \xrightarrow[f'(x)]{} & F \end{array}$$

L'application $f'(x)$ étant un isomorphisme d'espaces de Banach mixtes, il en résulte que T_xX possède une structure d'espace de Banach mixte, unique, telle que pout toute carte structurale Φ sur X dont le domaine contienne le point x, $\beta_{\Phi,x}$ soit un isomorphisme d'espaces de Banach mixtes. Cet espace de Banach mixte T_xX s'appelle <u>espace tangent</u> en x à X et ses éléments vecteurs tangents en x à X; la composante complexe $(T_xX)_1$ de T_xX est <u>l'espace tangent complexe</u> de X en x et la composante réelle $(T_xX)_2$ <u>est l'espace tangent réel</u>.

Lorsque $f : X \to Y$ est un morphisme de variétés mixtes de classe $\mathcal{C}^p$, et x un point de X, on a une application

$$T_xf : T_xX \longrightarrow T_{f(x)}Y$$

telle que, si t est un vecteur tangent au point x à X et si (Φ, u) est un couple représentant t, alors $(\Psi, h'(x)u)$ est un couple représentant $T_xf(t)$ pour toute carte structurale Ψ de Y dont le domaine contient $f(x)$, où $h = \Psi \circ f \circ \Phi^{-1}$: $\Phi(U \cap f^{-1}(V)) \longrightarrow \Psi(V)$.

Cette application T_xf est un morphisme d'espaces de Banach mixte et sera appelée <u>application linéaire tangente</u> à f au point x. On obtient ainsi un foncteur T défini sur la catégorie des couples (X, x), X étant une variété mixte de classe $\mathcal{C}^p$, $x \in X$, ce foncteur étant à valeurs dans la catégorie des espaces de Banach mixtes.

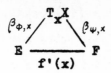

Un morphisme $f : X \to Y$ de variétés mixtes de classe $\mathcal{C}^p$ est <u>étal</u> au point x de X s'il existe un ouvert U dans X contenant le point x tel que f(U) soit un ouvert de Y et que l'application $U \to f(U)$ induite par f soit un isomorphisme de variétés mixtes de classe $\mathcal{C}^p$.

Théorème 2.1. Pour qu'un morphisme $f : X \to Y$ de variétés mixtes de classe $\mathcal{C}^p$ soit étal au point x il faut et il suffit que $T_x f$ soit un isomorphisme d'espaces de Banach mixtes.

On dit qu'un morphisme $f : X \to Y$ de variétés mixtes de classe $\mathcal{C}^p$ est une immersion au point $x \in X$ s'il y a un diagramme commutatif

$$\begin{array}{ccc} U & \xrightarrow{\varphi} & V \times D \\ & {}_{f}\searrow & \downarrow{\scriptstyle pr_1} \\ & & V \end{array}$$

où U est un ouvert de X, avec $x \in U$, V un ouvert de Y avec $f(U) \subset V$, D un ouvert de Banach mixte et φ un isomorphisme de variétés mixtes de classe $\mathcal{C}^p$.

Dualement, on dit que f est une submersion au point $x \in X$ s'il existe un diagramme commutatif

$$\begin{array}{ccc} U \times D & \longrightarrow & V \\ {\scriptstyle i_1}\uparrow & \nearrow_{f} & \\ U & & \end{array}$$

avec U un ouvert de X, $x \in U$, D un ouvert de Banach mixte avec $0 \in D$, i_1 l'application $x \longmapsto (x, 0)$ et φ un isomorphisme de variété mixtes de classe $\mathcal{C}^p$.

Il est évident, vu le théorème 2.1, qu'il suffit dans ces définitions que φ soit étal en x.

On dit que f est une immersion (submersion) si f est une immersion (submersion) en chaque point de X, et un plongement si f est une immersion injective propre.

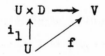

Un point $y \in Y$ est dit valeur régulière pour f si $T_x f$ est un epimorphisme direct pour tout $x \in f^{-1}(y)$, et <u>valeur critique</u> dans le cas contraire.

<u>Corollaire 2.2. Soit</u> $f : X \longrightarrow Y$ <u>un morphisme de variétés mixtes de classe</u> $\mathscr{C}^p$ <u>et soit</u> $x \in X$. <u>Alors</u>:

(1) f <u>est une immersion au point</u> $x \Longleftrightarrow T_x f$ <u>est un monomorphisme direct</u>

(2) f <u>est une submersion au point</u> $x \Longleftrightarrow T_x f$ <u>est un epimorphisme direct</u>.

<u>Corollaire 2.3. Si</u> y <u>est une valeur régulière de</u> f, <u>alors</u> $f^{-1}(y)$ <u>est une sous-variété mixte fermée de</u> X.

Par exemple, dans le cas $f : X \longrightarrow R$, si $T_x f \neq 0$ pour tout $x \in f^{-1}(0)$, alor 0 est une valeur régulière pour f, donc $f^{-1}(0)$ est une sous-variété mixte fermée de X.

La même conclusion vaut lorsque $f : X \longrightarrow \mathbb{C}$, dans l'hypothèse que $(T_x f)_1 \neq 0$ pour tout $x \in f^{-1}(0)$.

3. <u>Faisceaux</u>

Dans cette section on entendra par variété mixte, toute variété mixte de classe C^∞, séparée et localement de dimension finie.

Soit X une variété mixte. On dira que X est de <u>type</u> (m, n) au point x s'il existe une carte structurale $\varphi : U \longrightarrow E$ de x avec $x \in U$ et $E = \mathbb{R}^m \times \mathbb{C}^n$. Il est clair que le type d'une variété mixte est constant sur chaque composante connexe et invariant par rapport aux isomorphismes de variétés mixtes. On dit que la variété X est de type (m, n) si elle est de type (m, n) en chaque point.

Les variétés complexes sont exactement les variétés mixtes du type de la forme (0, n) en tout point, et les variétés différentiables celles dont le type est de la forme (m, 0) en chaque point. Nous dirons, dans le premier cas qu'il s'agit de variétés mixtes <u>purement complexes</u> et dans de second de variétés mixtes <u>purement réelles</u>.

On a deux faisceaux: $\mathcal{O}_X(\mathbb{C})$ et $\mathcal{O}_X(\mathbb{R})$ sur X tels que

$$(U, \mathcal{O}_X(\mathbb{C})) = \{f : U \to \mathbb{C} \mid f \text{ morphisme de variétés mixtes}\}$$

$$(U, \mathcal{O}_X(\mathbb{R})) = \{f : U \to \mathbb{R} \mid f \text{ morphisme de variétés mixtes}\}$$

Vu que l'addition et la multiplication dans $\mathbb{C}$ et dans $\mathbb{R}$ sont des morphismes de variétés mixtes, et que l'inclusion de $\mathbb{R}$ dans $\mathbb{C}$ est un morphisme injectif de variétés mixtes, il résulte que $\mathcal{O}_X(\mathbb{C})$ est un faisceau de $\mathbb{C}$-algèbres sur X, et que $\mathcal{O}_X(\mathbb{R})$ est un sous-faisceau de $\mathbb{R}$-algèbres de $\mathcal{O}_X(\mathbb{C})$. La variété mixte X est purement réelle si et seulement si $\mathcal{O}_X(\mathbb{C}) = \mathcal{O}_X(\mathbb{R}) + i\mathcal{O}_X(\mathbb{R})$ et purement complexe si et seulement si $\mathcal{O}_X(\mathbb{R}) = \mathbb{R}$.

Pour définir les formes différentielles sur X on dispose tant de l'espace tangent mixte T_xX que de l'espace tangent complexe $(T_xX)_1$, $x \in X$. Dans le premier cas on parlera de formes différentielles mixtes ou simplement des formes différentielles sur X; dans le second cas on parlera des formes différentielles complexes. Pour p. q entiers ≥ 0, une forme différentielle complexe ω sur X est dite de type (p,q) si, pour tout $x \in X$, $\omega(x)$ est un tenseur alterné de type (p,q) sur l'espace tangent complexe $(T_xX)_1$.

On notera $\mathcal{E}^{p,q}$ le faisceau des formes différentielles complexes de type (p, q) et de classe $\mathcal{C}^{\infty}$ sur les ouverts de X et $\Omega^{(p)}$ le sous-faisceau de $\mathcal{E}^{p,o}$ des formes à coefficients dans $\mathcal{O}_X(\mathbb{C})$. En particulier $\mathcal{E}^{o} := \mathcal{E}^{o,o}$ est le faisceau des fonctions de classe $\mathcal{C}^{\infty}$ sur les ouverts de X, et $\Omega^{(o)} = \mathcal{O}_X(\mathbb{C})$.

Pour tout entier $p \geqslant 0$ on a une suite exacte de $\mathcal{O}_X(\mathbb{C})$-modules

$$0 \longrightarrow \Omega^{(p)} \longrightarrow \mathcal{E}^{p,o} \longrightarrow \cdots \xrightarrow{\bar{\partial}} \mathcal{E}^{p,q} \longrightarrow \cdots$$

C'est la résolution de Dolbeault de $\Omega^{(p)}$ si X est purement complexe, et la résolution triviale si X est purement réel.

Comme tout $\mathcal{E}^{p,q}$ est un $\mathcal{E}^{o}$-module et comme, pour X paracompact $\mathcal{E}^{o}$ est mou, on a un théorème de Dolbeault dans le cas mixte:

<u>Théorème 3.1</u>. <u>Si</u> X <u>est une variété mixte paracompacte</u>,

$$H^q(X, \Omega^{(p)}) = \mathrm{Ker}\ \bar{\partial}_X^{p,q} / \mathrm{Im}\ \bar{\partial}_X^{p,\, q-1}$$

<u>et</u>, <u>en particulier</u>, <u>si</u> X <u>est de type</u> (m, n) <u>alors</u>

$$H^q(X, \Omega^{(p)}) = 0 \quad \text{pour} \quad q > n.$$

Si X est une variété mixte et d un entier $\geqslant 0$, on dit d'un $\mathcal{O}_X(\mathbb{C})$-module $\mathcal{F}$ qu'il est d-cohérent (pseudo-cohérent, dans la terminologie de Grothendieck) si, pour tout $x \in X$ il existe un ouvert U contenant x et une suite exacte de $\mathcal{O}_U(\mathbb{C})$-modules

$$\mathcal{L}_d \longrightarrow \cdots \longrightarrow \mathcal{L}_o \longrightarrow \mathcal{F}|_U \longrightarrow 0$$

où $\mathcal{L}_i = \mathcal{O}_U(\mathbb{C})^{r_i}$ avec r_i entier $\geqslant 0$; on dit de type "fini" pour 0-cohérent et "de présentation finie" pour 1-cohérent.

$\mathcal{F}$ est <u>cohérent</u> s'il est d-cohérent pour tout entier $d \geqslant 0$. Le lemme suivant est fondamental en ce qui concerne les propriétés élémentaires des faisceaux cohérents (cf. [6]).

Lemme 3.2. Soit X une variété mixte et

$$0 \longrightarrow \mathcal{F}' \longrightarrow \mathcal{F} \longrightarrow \mathcal{F}'' \longrightarrow 0$$

une suite exacte de $\mathcal{O}_X(\mathbb{C})$-modules. Alors:

(1) Si $\mathcal{F}'$ et $\mathcal{F}''$ sont d-cohérents, alors $\mathcal{F}$ est d-cohérent

(2) Si $\mathcal{F}$ est d-cohérent et si $\mathcal{F}''$ est (d+1)-cohérent, alors $\mathcal{F}$ est d-cohérent

(3) Si $\mathcal{F}'$ est d-cohérent et si $\mathcal{F}$ est (d+1)-cohérent, alors $\mathcal{F}'$ est (d+1)-cohérent.

Un compact K de $E = \mathbb{R}^m \times \mathbb{C}^n$ est un polycylindre si $\overset{\circ}{K} \neq \emptyset$ et si en outre $K = K_0 \times K_1 \times \dots \times K_n$, où K_i sont des convexes de $\mathbb{C}$ et K_0 un convexe de $\mathbb{R}^m$.

Lorsque X est une variété mixte, un compact K de X est dit polycylindrable s'il existe une carte structurale $\emptyset : U \longrightarrow E$ de X, où $E = \mathbb{R}^m \times \mathbb{C}^n$, telle que $\emptyset(K)$ soit un polycylindre.

Pour tout $\mathcal{O}_X(\mathbb{C})$-module $\mathcal{F}$, on pose

$$\text{Tor dim}_{\mathcal{O}_X(\mathbb{C})} \mathcal{F} = \sup_{x \in X} \text{Tor dim}_{\mathcal{O}_{X,x}(\mathbb{C})} \mathcal{F}$$

Théorème 3.3. Si X est une variété mixte de type (m,n) et si $\mathcal{F}$ est un $\mathcal{O}_X(\mathbb{C})$-module (m+n+1)-cohérent, alors

$$\text{Tor dim}_{\mathcal{O}_X(\mathbb{C})} \mathcal{F} \leqslant m+n$$

Théorème 3.4. Lorsque X est une variété mixte, d un entier $\geqslant 0$ et $\mathcal{F}$ un $\mathcal{O}_X(\mathbb{C})$-module, les conditions suivantes sont equivalentes:

(i) $\mathcal{F}$ <u>est</u> (d+1)-<u>cohérent et</u> Tor dim $_{\mathcal{O}_X(\mathbb{C})}\mathcal{F} \leq d$

(ii) <u>Pour tout compact polycylindrable</u> $K \subset X$, <u>il existe un ouvert</u> U <u>contenant</u> K <u>et une suite exacte</u>

$$0 \longrightarrow \mathcal{L}_d \longrightarrow \dots \longrightarrow \mathcal{L}_o \longrightarrow \mathcal{F}|_U \longrightarrow 0$$

<u>avec</u> $\mathcal{L}_i = \mathcal{O}_U(\mathbb{C})^{r_i}$ (<u>en particulier</u>, $\mathcal{F}$ <u>est cohérent</u>).

<u>Corollaire</u>. <u>Pour</u> X <u>variété mixte de type</u> (m, n), <u>tout</u> $\mathcal{O}_X(\mathbb{C})$-<u>module</u> (m+n+1)-<u>cohérent est cohérent</u>.

On a ici un théorème de Oka faible sur X. On sait que dans le cas des variétés complexes on a le théorème de Oka fort: "1-cohérent" implique "cohérent".

La question si pour toute variété mixte on a un théorème de Oka fort, i.e. si "m+1-cohérent" implique "cohérent" lorsque X est de type (m, n) reste ouverte.

4. <u>Variétés de Cartan</u>

Dans cette section on supposera en outre que les variétés mixtes possèdent une base dénombrable.

Soit X une variété mixte. On munit le faisceau $\mathcal{E}^o$, des fonctions à valeurs complexes de classe C^∞ sur les ouverts de X, de la topologie de la C^∞-convergence compacte. $\mathcal{E}^o$ devient ainsi un faisceau de Fréchet, et le faisceau structural $\mathcal{O}_X(\mathbb{C})$ est un sous-faisceau de $\mathbb{C}$-algèbres, <u>fermé</u> de $\mathcal{E}^o$; on munit $\mathcal{O}_X(\mathbb{C})$ de la topologie induite par celle de $\mathcal{E}^o$ et $\mathcal{O}_X(\mathbb{C})^r$, r entier ≥ 0 de la topologie produit $\mathcal{O}_X(\mathbb{C})^r$ devient ainsi un faisceau de Fréchet.

Un $\mathcal{O}_X(\mathbb{C})$-module cohérent $\mathcal{F}$ est dit <u>séparé</u> si, pour tout point $x \in X$, il existe un ouvert U contenant le point x et une suite exacte de $\mathcal{O}_U(\mathbb{C})$-modules

$$0 \longrightarrow \mathcal{F}' \longrightarrow \mathcal{O}_U(\mathbb{C})^r \longrightarrow \mathcal{F}|_U \longrightarrow 0$$

avec r entier ≥ 0 et $\mathcal{F}$ un sous-faisceau fermé de $\mathcal{O}_U(\mathbb{C})^r$.

Par exemple, tout $\mathcal{O}_X(\mathbb{C})$-module localement libre est cohérent séparé.

Un morphisme $\varphi : X \to Y$ de variétés mixtes est $\mathbb{C}$-analytique au point $x \in X$ s'il existe un diagramme commutatif:

$$\begin{array}{ccc} U & \xrightarrow{\ i\ } & V \times D \\ & {\scriptstyle\varphi}\searrow \quad \swarrow & \\ & V & \end{array}$$

où U est un ouvert de X contenant le point x, V un ouvert de Y contenant $\varphi(U)$, D un ouvert dans un espace numérique complexe $\mathbb{C}^N$ et i un plongement d'espaces annelés défini par un idéal fermé cohérent $\mathcal{I}$ de $\mathcal{O}_{V\times D}(\mathbb{C})$. On dit que φ est $\mathbb{C}$-analytique si φ est $\mathbb{C}$-analytique en tout point de X; on dit encore dans ce cas que φ définit X comme variété analytique complexe relative au-dessus de Y.

On dit d'un morphisme $\varphi : X \to Y$ de variétés mixtes qu'il est fini en x si φ est $\mathbb{C}$-analytique en x et si x est un point isolé de la fibre $\varphi^{-1}(\varphi(x))$. Par exemple "étal en x" implique "fini en x".

Une variété mixte X est appelée variété de Cartan si:

(C_1) X est $\mathcal{O}_X(\mathbb{C})$-convexe, i.e. pour tout compact K de X l'ensemble

$$\hat{K} = \left\{x \in X \,\middle|\, |f(x)| \leq \sup_K |f|, \text{ pour tout } f \in \Gamma\left(X, \mathcal{O}_X(\mathbb{C})\right)\right\}$$

est encore compact.

(C_2) X est $\mathcal{O}_X(\mathbb{C})$-séparé, i.e. pour tout $x \in X$ il existe un morphisme de variétés mixtes $\varphi : X \to \mathbb{R}^{\mu} \times \mathbb{C}^{\nu}$ fini en x.

Pour X une variété mixte, un ouvert U de X est dit ouvert de Cartan si U est une variété de Cartan pour la structure mixte induite.

Exemples: 1) Les variétés de Stein sont exactement les variétés de Cartan purement complexes;

2) Toute variété différentiable est une variété de Cartan;

3) Le produit de deux variétés de Cartan est une variété de Cartan;

4) Tout sous-variété mixte fermée d'une variété de Cartan est une variété de Cartan;

5) Tout ouvert convexe de $\mathbb{R}^m \times \mathbb{C}^n$ est un ouvert de Cartan.

Pour toute variété mixte X, les ouverts de Cartan constituent une base d'ouverts, stable par rapport aux intersections finies.

Le théorème suivant étend aux variétés de Cartan les théorèmes A et B de Cartan pour les variétés de Stein.

Théorème 4.1. Si X est une variété de Cartan et si $\mathcal{F}$ est un $\mathcal{O}_X(\mathbb{C})$-module cohérent séparé, alors:

A) Pour tout $x \in X$, $\mathcal{F}_x$ est $\mathcal{O}_{X,x}(\mathbb{C})$-engendré par l'image de $\mathcal{F}(X)$ dans $\mathcal{F}_x$.

B) $H^q(X, \mathcal{F}) = 0$ pour $q \geq 1$.

Pour la démonstration on peut suivre l'organisation des démonstrations des théorèmes A et B pour les espaces de Stein dans [6].

Comme $\Omega^{(p)}$ est un $\mathcal{O}_X(\mathbb{C})$-module localement libre, on a, compte tenu du théorème 3.1, le

Corollaire 4.2. Si X est une variété de Cartan, l'équation $\bar{\partial} u = f$ a une solution $u \in \Gamma(X, \mathcal{E}^{p,q})$ pour tout $f \in \Gamma(X, \mathcal{E}^{p,q+1})$ tel que $\bar{\partial} f = 0$.

Pour toute variété mixte X et tout entier $p \geq 0$, on définit le $\mathcal{O}_X(\mathbb{C})$-module $\mathcal{O}_X^{(p)}(\mathbb{C})$ comme suit. Pour tout ouvert W de X, $\Gamma(W, \mathcal{O}_X^{(p)}(\mathbb{C}))$ est le $\Gamma(W, \mathcal{O}_X(\mathbb{C}))$-module des formes différentielles ω de degré p sur W telles que, si $\Phi : U \to E$ de X est une carte structurale de X avec $U \subset W$ et $E = \mathbb{R}^m \times \mathbb{C}^n$, alors

$$(\Phi)^{-1} \omega = \sum_{|I|=p}{}' f_I \, du^I$$

avec $f_I \in \Gamma(\Phi(U), \mathcal{O}_E(\mathbb{C}))$, où $u_1, \ldots, u_m, u_{m+1} = z_1, \ldots, u_{m+n} = z_n$ sont les fonctions coordonnées sur E; en particulier, $\mathcal{O}_X^{(0)}(\mathbb{C}) = \mathcal{O}_X(\mathbb{C})$.

Alors on a une suite exacte

$$0 \to \mathbb{C} \to \mathcal{O}_X(\mathbb{C}) \xrightarrow{d} \mathcal{O}_X^{(1)}(\mathbb{C}) \to \ldots \to \mathcal{O}_X^{(p)}(\mathbb{C}) \to \ldots$$

Comme chaque faisceau $\mathcal{O}_X^{(p)}(\mathbb{C})$ est localement libre, on a, compte tenu du théorème 4.1, le

Corollaire 4.3. Si X est une variété de Cartan, alors

$$H^q(X, \mathbb{C}) = \text{Ker } d_X^q / \text{Im } d_X^{q-1} .$$

En particulier, si X est de type (m, n), alors

$$H^q(X, \mathbb{C}) = 0 \quad \text{pour} \quad q > m+n.$$

Si X' est une sous-variété mixte fermée de X, l'idéal associé est cohérent fermé, donc on a le

Corollaire 4.4. Soit X une variété de Cartan et X' une sous-variété mixte fermée de X; alors pour toute fonction $f' \in \Gamma(X', \mathcal{O}_X(\mathbb{C}))$, il existe une fonction $f \in \Gamma(X, \mathcal{O}_X(\mathbb{C}))$ telle que $f|_{X'} = f'$.

En particulier, les sections globales de $\mathcal{O}_X(\mathbb{C})$ séparent les points de la variété.

En outre, si a est un point fixé de X, p un entier $\geqslant 0$, alors on a un idéal $\mathcal{I}$ de $\mathcal{O}_X(\mathbb{C})$ tel que

$$\mathcal{I}_x = \begin{cases} \mathcal{O}_{X,x}(\mathbb{C}) & \text{pour } x \neq a \\ \underline{m}_x^2(\mathbb{C}) & \text{pour } x = a \end{cases}$$

où $\underline{m}_x(\mathbb{C})$ est l'idéal maximal de $\mathcal{O}_{X,x}(\mathbb{C})$. L'idéal $\mathcal{I}$ étant fermé cohérent, on aura en particulier le

Corollaire 4.5. Si X est une variété de Cartan et a un point de X il existe un morphisme $\varphi : X \to \mathbb{R}^m \times \mathbb{C}^n$ étal en x.

Pour que les deux derniers corollaires soient vrais il suffit que l'on ait le théorème B pour les ideaux fermés cohérents de $\mathcal{O}_X(\mathbb{C})$; en particulier le théorème B caractérise les variétés de Cartan.

Enfin, on a un théorème de plongement des variétés de Cartan

Théorème 4.6. Si X est une variété de Cartan de type (m, n) il existe un plongement $\varphi : X \to \mathbb{R}^{2m+1} \times \mathbb{C}^{m+2n+1}$.

Un théorème de plongement dans le cas X ouvert $\mathcal{O}_X(\mathbb{C})$-convexe de $\mathbb{R}^m \times \mathbb{C}$ a été communiqué à l'auteur par B. Gilligan (Regina University) dans une lettre de 1974. Il a donné également des caractérisations de la $\mathcal{O}_D(\mathbb{C})$-convexité pour un ouvert D de $\mathbb{R}^m \times \mathbb{C}$.

BIBLIOGRAPHIE

[1] Andreotti A., Grauert H. Théorèmes de finitude pour la cohomologie des espaces complexes. Bull. Soc. Math. de France, 90, 193-259 (1962).

[2] Bourbaki N. Variétés différentiables et analytiques. Fascicule des résultats. Hermann 1967 et 1971.

[3] Cartan H. Variétés analytiques complexes et cohomologie. Colloque sur les fonctions de plusieurs variables complexes, Bruxelles 1953.

[4] Douady A. Variétés et espaces mixtes. Séminaire Henri Cartan (Familles d'espaces complexes et fondements de la géométrie analytique), 13e année, 1960/61, no. 2.

[5] Jurchescu M. Espaces annelés transcendants et morphismes analytiques (Les travaux du Séminaire d'espaces analytiques, Bucarest 25 - 30 septembre 1969). Editions de l'Académie de la R.S. de Roumanie, Bucarest 1971.

[6] Jurchescu M. Introduzione agli spazi analitici. Appunti redati da Tancredi. Quaderni dei gruppi di Ricerca Matematica del C.N.R. Instituto di Matematica dell'Università di Perugia, vol. I (1971), vol. 2 (1974).

[7] Malgrange B. Ideals of differentiables functions. Oxford University Press, 1966.

University of Bucharest
Faculty of Mathematics

Projections and liftings of exact holomorphic forms. Applications to the Cauchy problem.

Otto Liess

1. Let P(D) be a constant coefficient partial differential operator of form $P(D) = D_t^m + \sum_{|\alpha|+j\leq m, j<m} a_{\alpha j} D_x^\alpha D_t^j$, $x\in R^n$, $t\in R$, and consider the associated Cauchy problem:

$$(1)\quad \begin{aligned} &P(D)\,u = 0\\ &(\partial/\partial t)^i u|_{t=0} = \varphi_i\ ,\ i=0,\dots,m-1, \end{aligned}$$

where $\varphi_0,\dots,\varphi_{m-1}$ are germs of C^∞ functions at $0\in R^n$, and u is a germ of a C^∞ function at $0\in R^{n+1}$. All equalities from (1), are in the sense of germs.

If P is not hyperbolic with respect to the variable t, then the solution of (1) need not exist. If it exists, it is however unique, in view of Holmgrens unicity theorem (cf. e.g. [1]).

If one wants to study the solvability of (1) for nonhyperbolic operators, then it is natural to introduce the following intermediate problems:

$$(2)_s\quad \begin{aligned} &P(D)\,u = 0\\ &(\partial/\partial t)^i u|_{t=0}=0, i=0,\dots,s-1,\ (\partial/\partial t)^s u|_{t=0} = \psi \end{aligned}$$

for germs u and ψ.

It is clear that, if we can say for which germs ψ the problems $(2)_s$ are solvable, then we will be also able to say, for what $\varphi_0,\dots,\varphi_{m-1}$ we can solve (1).

This leads us to introduce spaces F_s by:

F_s is the space of germs ψ for which there is a solution u for $(2)_s$.

It follows from a theorem of Matsuura-Palamodov (cf. [2]),

that $F_i \subset F_{i-1}$, but in general the inclusion is strict (cf. [3]).

2. The fundamental principle of Ehrenpreis-Palamodov, gives a quite natural guess of what the spaces F_s should be.

To explain this, we introduce two notations.

First, we denote with $\tau_1(\zeta), \dots, \tau_m(\zeta)$, the roots of the equation $P(\zeta,\tau) = 0$, labelled such that $|\mathrm{Im}\,\tau_i(\zeta)| \leqslant |\mathrm{Im}\,\tau_{i+1}(\zeta)|$ for $i=1,\dots,m-1$. Here $P(\zeta,\tau)$ is the polynomial obtained by formally changing $\partial/\partial x_j$ in $P(D)$ with $-\sqrt{-1}\,\zeta_j$ and $\partial/\partial t$ with $-\sqrt{-1}\,\tau$. With these conditions the functions τ_i are not uniquely defined, but the functions $|\mathrm{Im}\,\tau_i|$ already have a precise meaning. It can be shown, that there is $\varepsilon > 0$ and $\delta > 0$, such that
$|\zeta^1 - \zeta^2| \leqslant \varepsilon(1 + |\zeta^1|)^{-\delta} \Rightarrow \big||\mathrm{Im}\,\tau_j(\zeta^1)| - |\mathrm{Im}\,\tau_j(\zeta^2)|\big| \leqslant 1$, for all j.

In the sequel we will always assume that P is without multiple factors.

Next, we use the notation $E^W_{|\mathrm{Im}\,\tau_i(\zeta)|}$ for the space of germs f of C^∞ functions at $0 \in R^n$, for which there is $B > 0, B > 0$ such that for every $b \geqslant 0$ there is a Radon measure μ on C^n, with $\int d|\mu| < \infty$, and such that

$f(x) = \int \exp i\langle x,\zeta\rangle \, d\mu(\zeta)/\exp B|\mathrm{Im}\,\tau_i(\zeta)| + B|\mathrm{Im}\,\zeta| + b\ln(1 + |\zeta|)$,

for $|x| \leqslant B$.

It is now an easy consequence of the fundamental principle that

Proposition 2.1. $E^W_{|\mathrm{Im}\,\tau_1(\zeta)|} = F_0$, $E^W_{|\mathrm{Im}\,\tau_i(\zeta)|} \subset F_{i-1}$, for $i \geqslant 2$.

The equality $E^W_{|\mathrm{Im}\,\tau_i(\zeta)|} = F_{i-1}$, which seems quite natural also for $i \geqslant 2$, is false in general, and in this paper, we try to explain the reason why it is false.

3. At this moment, we need new notations and conventions. For $\varphi : U \to R, U \subset C^n$, φ continuous, we denote

$L^2(U,\varphi) = \{ u \in L^2_{loc}(U);\ u \exp -\varphi \in L^2(U)\}$,

and for $X \subset D'(U)$, we denote $X_{(o,q)}$, the space of (0,q) forms $\sum_{|J|=q} u_J\, d\bar{\zeta}^J, u_J \in X$.

Further, if φ is as before, then we will write $g_q(U,\varphi) = \{0\}$, if for every β', B', there are β, B, K, such that the short sequence

(3) $[L^2(U, \beta'\varphi + B'|\operatorname{Im}\zeta| + b^{+K}\ln(1+|\zeta|)]_{(o,q-1)} \xrightarrow{\bar\partial}$

$\rightarrow [L^2(U, \beta\varphi + B|\operatorname{Im}\zeta| + b\ln(1+|\zeta|)]_{(o,q)} \xrightarrow{\bar\partial} [D'(U)]_{(o,q+1)}$

is exact.

The following result is now quite elementary:

<u>Theorem 3.1.</u> Suppose that $g_1(C^n, |\operatorname{Im}\tau_i(\zeta)|) = \{0\}$, for $i=1,\dots,s$. Then it follows that $F_s = E^W_{|\operatorname{Im}\tau_{s+1}(\zeta)|}$.

The main result upon we report in this paper, is now the following partial converse to theorem 3.1:

<u>Theorem 3.2.</u> Suppose that $g_2(C^n, |\operatorname{Im}\tau_i(\zeta)|) = \{0\}$, for $i=1,\dots, m$, and that $E^W_{|\operatorname{Im}\tau_{k+1}(\zeta)|} = F_k$, for $0 \leqslant k < m$. Then it also follows that $g_1(C^n, |\operatorname{Im}\tau_i(\zeta)|) = \{0\}$, for $i=1,\dots,s$.

4. Before we explain some of the ideas involved in the proof of theorem 3.2, we indicate some results concerning the cohomology of the sequences from (3).

The first is obtained by standard cohomology arguments:

Proposition 4.1. Consider U a domain in C^n and $\varphi_1, \varphi_2 : U \to R$, $\varphi_1 \leqslant \varphi_2$ functions such that $|\zeta^1 - \zeta^2| \leqslant \varepsilon(1+|\zeta^1|)^{-\delta} \Rightarrow |\varphi_i(\zeta^1) - \varphi_i(\zeta^2)| \leqslant 1$, for $i=1,2$. Denote $U_1 = \{\zeta \in U;\ \varphi_1(\zeta) \neq \varphi_2(\zeta)\}$, and suppose that $g_q(U,\varphi_2) = \{0\}$, $g_q(U_1,\varphi_1) = \{0\}$, $g_{q-1}(U_1,\varphi_2) = \{0\}$, for some q. Then it follows that $g_q(U,\varphi_1) = \{0\}$.

Using this and (by now standard) results from [4], we obtain

Proposition 4.2. Suppose $i+j \geqslant m+1$.

Then $g_i(C^n, |\mathrm{Im}\, \tau_j(\zeta)|) = \{0\}$.

We also have the following result

Proposition 4.3. Consider $\varphi : C^n \to R$ such that $|\zeta^1 - \zeta^2| \leq \varepsilon(1 + |\zeta^1|)^{-\delta} \Rightarrow |\varphi(\zeta^1) - \varphi(\zeta^2)| \leq 1$, for some ε, δ. Then there are equivalent:

(i) $g_q(C^n, \varphi) = \{0\}$, for $q=1,\dots,n$,

(ii) For every β', B', there are C, β, B, K with the following property: for every $\zeta^0 \in C^n$, there is $h \in A(C^n)$, such that

$$|h(\zeta)| \leq C \exp \beta'\varphi(\zeta) + B'|\mathrm{Im}\zeta| + (b+K)\ln(1 + |\zeta|)$$

and

$$|h(\zeta^0)| \geq \exp \beta\, \varphi(\zeta^0) + B|\mathrm{Im}\zeta^0| + b\, \ln(1 + |\zeta^0|).$$

We can now give an example of an operator for which $g_2(C^n, |\mathrm{Im}\, \tau_i(\zeta)|) = \{0\}$, for all i, but such that $g_1(C^n, |\mathrm{Im}\, \tau_1(\zeta)|) \neq \{0\}$. The simplest operator of this kind is perhaps associated with $P(\zeta,\tau) = \tau^4 - \zeta_1^2\, \zeta_2^2$. Here $|\mathrm{Im}\, \tau_3| = |\mathrm{Im}\, \tau_4|$ and $|\mathrm{Im}\, \tau_2| = |\mathrm{Im}\, \tau_1|$. The assertion $g_2(C^n, |\mathrm{Im}\, \tau_i(\zeta)|) = \{0\}$ therefore essentially follows from proposition 4.2. On the other hand, it follows from the Phragmén-Lindelöf principle, that property (ii) from proposition 4.3, cannot be satisfied for $|\mathrm{Im}\, \tau_1(\zeta)|$.

5. Let now $u \in D'(C^{n+1})$ be such that the projection from $C^{n+1} \to C^n$ is proper on supp u. For $k \in N$, we consider $\pi_k u \in D'(C^n)$ defined by $(\pi_k u, v) = (u, \tau^k v)$, $v \in C_0^\infty(C^n)$. It is then clear that $\pi_k(\partial/\partial\bar{\zeta}_j u) = (\partial/\partial\bar{\zeta}_j)\pi_k u$ for $1 \leq j \leq n$, and that $\pi_k(\partial/\partial\bar{\tau} u) = 0$.

The next thing, is to find an inverse to π_k, which is adapted for our purpose.

Let then $P(\zeta,\tau)$ be a polynomial as in the above, and choose $f \in C_0^\infty(C^n)\hat{\otimes} C^\infty(C)$ such that $\partial/\partial\bar{\tau}\, f = 0$. We can then find $v_0, \dots, v_{m-1} \in C_0^\infty(C^n)$ such that $P(\zeta,\tau) = 0 \Rightarrow f(\zeta,\tau) = \sum \tau^j v_j(\zeta)$.

Therefore, if $u_o, \ldots, u_{m-1}$ are in $D'(C^n)$, we can define an element $\tilde{u} \in D'(C^n) \hat{\otimes} \Lambda'(C)$ by

(4) $\quad \tilde{u}(f) = \sum_{j=o}^{m-1} u_j(v_j).$

We can now apply Hahn-Banach's theorem, in order to obtain $u \in D'(C^{n+1})$ with $u = \tilde{u}$ in $D'(C^n) \hat{\otimes} \Lambda'(C)$. Since the values of the v_j depend only on those of f at $V = \{(\zeta,z); P(\zeta,z) = 0\}$ (if P has a multiple zero at $(\zeta^o, z^o) \in V$, then v_j depends also on some derivatives of f), it follows that we can choose u such that $P(\zeta,z)u = 0$. It is also clear that $\pi_k\, u = u_k$, for $k=0,\ldots,m-1$.

Let us call η the lifting which associates to $u_o,\ldots,u_{m-1}$ an u such that $Pu = 0$ and $\pi_k(u) = u_k, k=0,\ldots,m-1$. $\eta(u_o,\ldots,u_{m-1})$ is not uniquely defined by these conditions. From results from [5], where maps similar to π_k and η are considered in another context (using another construction for η), it follows that $u=\eta(u_o,\ldots,u_{m-1})$ is unique, modulo elements of form $w=(\partial/\partial\bar{z})v$, $Pv=0$.

6. We now want to extend π_k and η to holomorphic forms. First take $h = \sum_{|J|=q+1} h_J\, d\bar{\zeta}^J + \sum_{|K|=q} h_{(K,1)}\, d\bar{\zeta}^K \wedge d\bar{z}$.

We define $\pi_k h = \sum_{|K|=q} \pi_k(h_{(K,1)})\, d\bar{\zeta}^K$.

Thus the part $(h \wedge d\bar{z})/d\bar{z}$ of h has no contribution in $\pi_k h$. On the other hand, if $g_i = \sum_{|K|=q} g_{i,K}\, d\bar{\zeta}^K$, $i=0,\ldots,m-1$, are forms in $[D'(C^n)]_{(0,q)}$, then we define a $(0,q+1)$ form with coefficients in $D'(C^{n+1})$ by setting

$h = \sum_{|K|=q} \eta(g_{o,K},\ldots,g_{m-1,K})\, d\bar{\zeta}^K \wedge d\bar{z}$.

In this way, we have constructed some h with $\pi_k h = g_k$, $k=0,\ldots,m-1$, and $Ph = 0$.

The nonunicity in this case is still greater then before, and we can use it to obtain the following result:

if $\bar\partial_\zeta g_k = 0$, $k = 0,\ldots,m-1$, then there is $f \in [D'(C^{n+1})]_{(0,q+1)}$ such that $\bar\partial_{\zeta,\tau} f = 0$, $Pf = 0$, and $\pi_k f = g_k$, $k = 0,\ldots,m-1$.
In fact, first we consider $h = \sum_{|K|=q} \eta(g_{0,K},\ldots,g_{m-1,K})\, d\bar\zeta^K \wedge d\bar\tau$, and then we search for $h' = \sum_{|J|=q+1} h'_J\, d\bar\zeta^J$, $Ph'_J = 0$, such that $\bar\partial_{\zeta,\tau}(h+h') = 0$. Now the last system splits to $\bar\partial_\tau h' = -\bar\partial_\zeta h$, $\bar\partial_\zeta h' = 0$, which is easily seen to have a solution in $D'(C^n)\hat\otimes E'(C)$. This solution clearly then also satisfies $Ph' = 0$. This completes our construction, since we may take $f=h+h'$.

7. We can now conclude with some comments concerning the proof of theorem 3.2.

The first step is to reduce, by dualisation, the assertion $g_1(C^n, |\mathrm{Im}\, \tau_i(\zeta)|) = \{0\}$, to the following:
if ν is a Radon measure on C^n of form $\nu = \mu/\exp B'|\mathrm{Im}\, \tau_i(\zeta)| + B'|\mathrm{Im}\zeta| + b'\ln(1+|\zeta|)$, $\int d|\mu| < \infty$, such that $\int \exp i\langle x,\zeta\rangle\, d\nu(\zeta) = 0$, for all $|x| < B'$, then there are Radon measures $\nu_j = \mu_j/\exp B|\mathrm{Im}\, \tau_i(\zeta)| + B|\mathrm{Im}\zeta| + b\ln(1+|\zeta|)$, $\int d|\mu_j| < \infty$, such that $= \sum(\partial/\partial\bar\zeta_j)\, \nu_j$.

This reduction also involves a passage from L^2 estimates to sup-norm estimates.
Next suppose ν is as in the above, and consider ν' such that $\pi_k\nu' = 0$, $k < i-1$, $\pi_{i-1}\nu' = \nu$. If one constructs ν' carefully enough (note that $\pi_i\nu',\ldots,\pi_{m-1}\nu'$ are at our disposal), then we may suppose that ν' satisfies certain decay conditions (in distribution sense), which we don't describe here (cf. [5]). Roughly speaking these conditions say that ν' decays like $\exp - B'|\mathrm{Im}\zeta| - B'|\mathrm{Im}\zeta| - b\ln(1+|(\zeta,\tau)|)$ at infinity. From the hypothesis of theorem 3.2, concerning the F_s, it now follows, that there is $\tilde\nu$, which satisfies similar decay conditions, such that $\pi_k\tilde\nu = 0$, $k \leqslant i-1$, and $\int \exp i\langle x,\zeta\rangle + i\, t\tau\, d(\nu'-\tilde\nu) = 0$, for

small x,t. We conclude, using standard results concerning the $\bar\partial$ operator (in dual form), that $\nu' - \tilde\nu = \sum(\partial/\partial\bar\zeta_j)\varkappa_j + (\partial/\partial\bar z)\varkappa_{n+1}$, with $\varkappa_j$ once again satisfying suitable decay conditions.
We now write $\nu = \pi_{i-1}(\nu' - \tilde\nu) = \sum_j(\partial/\partial\bar\zeta_j)\,\pi_{i-1}(\varkappa_j)$. This is not yet a representation of ν in the desired form, but using $\pi_j(\nu' - \tilde\nu) = 0, j \leq i-2$, and $g_2(C^n, |\operatorname{Im} z_j|) = \{0\}$, we can change $\pi_{i-1}(\varkappa_j)$ with w_j, such that still $\nu = \sum(\partial/\partial\bar\zeta_j)w_j$, with w_j satisfying "in distribution sense", $w_j = \mu_j'/\exp B|\operatorname{Im} z_i(\zeta)| + B|\operatorname{Im}\zeta| + b\ln(1+|\zeta|)$. In the last part of the proof, while changing $\pi_{i-1}(\varkappa_j)$ with w_j, it is convenient to think of ν as an (0,n) form, $\nu', \tilde\nu$ as (0,n+1) forms, etc.

8. The results from this paper have been obtained in [6]. Part of it were obtained, while the author had a grant from the Sweedish I.V.A., at the Mittag-Leffler Institute, in autumn 1974.

References.

1. L. Hörmander, Linear Partial Differential Operators, Springer Verlag, Grundlehren, Bd. 116

2. V.P. Palamodov, Linear Differential Operators with Constant Coefficients (in Russian), Moskow, 1967

3. O. Liess, The microlocal Cauchy problem..., Revue Roumaine de Math. Pure et Appl., tXXI, nr. 9, 1221-1239, 1976

4. L. Hörmander, Existence theorems for the $\bar\partial$ operator by L^2 methods, Acta Math., 113 89-152, 1965

5. O. Liess, The fundamental principle of Ehrenpreis-Palamodov. New proofs, and essential uniqueness of the constructions. Preprint series in Math. No 7/1976, INCREST, INSTITUTUL de MATEMATICA, Bucharest, april 1976

6. O. Liess, teza de doctorat, June 1975

INCREST
Bucharest

Principal holomorphic fiber bundles as applied to topological algebras

Anastasios Mallios

1. The purpose of the following discussion is to give an application of the "Oka's principle" (cf. Theorem 4.1, as well as, Scholium 5.1 below), within the context of topological algebras (not necessarily normed ones; cf., instead [13], [8]) to the extent that one can get therewith conditions, guaranteing the "complementation" of matrices with entries from topological algebras of the type under consideration.

Thus, the subject matter of the present study is certainly well-known, when considered within another concrete context, as it is, for instance, the case of matrices whose entries are real-valued continuous functions on a suitable compact subset of $\mathbb{R}^n$ (: T. Ważewski [29]), and yet some more general cases, concerning the domain of definition of the continuous functions involved, which have been obtained as an application of fiber space theory (:B. Eckmann [6]; C. Ehresmann [7]). On the other hand, there exists a "holomorphic analogon" of the preceding, due to H.Cartan [3], while the latter has also been strengthened in several aspects by K.J. Ramspott [24], as a result of his study on the (homotopic) equivalence between continuous and holomorphic sections of suitable holomorphic fiber bundles (ibid.).

Now, the above result of Ramspott has been recently applied by V.Ya. Lin [13], in considering the problem of complementing matrices with entries from an abstract Banach algebra within his study of a generalized version of the functional calculus for Banach algebras. However, it has already been indicated in [19] that, concerning the Banach algebras theory framework applied by Lin, one can consider instead a much more general class of topological algebras, including several of the important classes of topological algebras occurring in Functional Analysis, so that a similar question of complementing matrices whose entries belong to such more general topological algebras naturally arises, as an application of the above realization, and it is the purpose of the following discussion to give a more detailed presentation of the relevant results, than the one already summarized in (19). In this concern, a similar account to the respective results of Lin has also been recently given in (25), however within the same Banach algebras theory context.

On the other hand, it is to be noticed that the preceding constitutes indeed only a fragmentary aspect of a larger number of results, obtained as an application of a recent amalgamation of techniques from Function Theory of Several Variables, Banach Algebras Theory and Algebraic Topology (cf., for instance, (8), (28), (23) (27)), and it is besides the ultimate goal of this discussion to show that one can actually work within the more general class of topological algebras adopted herewith; A more complete account of the results obtained is planned to be given elsewhere. However, a brief indication thereon has been already reported in (20).

The following constitutes an amplified version of a talk delivered by the author at the "III. Romanian-Finnish Seminar on Complex Analysis" in Bucharest, June 27-July 2, 1976, under title "*Function Theory methods in Topological Algebras*". The author wishes to expess, also at this place, his sincere thanks to the organizer of the Seminar Prof. Cabiria Andreian-Cazacu and to Profs. M. Jurchescu and I Colojoară who, by their kind invitation, offered him the opportunity to give the lecture on which the present material is based.

2. We start with fixing up the terminology, concerning the topological algebras theory context within which we are going to work in the sequel.

Thus, by a *topological algebra* we mean a complex linear, associative algebra, endowed with a Hausdorff vector space topology, with respect to which the ring multiplication of the algebra is in each variable (:separately) continuous.

Now, a crucial object associated with a given topological algebra E (whenever there does exist!) is its *spectrum*, denoted by $\mathfrak{M}(E)$, and being, by definition, the set of all non-zero, continuous, multiplicative, linear forms (:$\mathbb{C}$- valued maps) on E, this set being further equipped with the relative topology induced on it by the weak topological dual E'_s of the given topological algebra E.

Besides, a topological algebra of the kind which we are dealing with below will be, in particular, a topological subalgebra of a suitable projective limit of Banach algebras, i.e. a *locally m-convex algebra* (21).

On the other hand, the appropriate class of topological algebras we apply (cf. § 5 below) is provided by those having the useful property that *the weakly bounded subsets of their spectra are also equicontinuous*. We call such topological algebras (and yet, in general, not necessarily locally convex) *spectrally barrelled* (18), (19). In this respect, every Banach algebra is certainly spectrally barrelled. Even more, every Fréchet or yet every m-barrelled topological algebra, and still more inductive limits in all the preceding classes, are spectrally barrelled algebras (ibid.).

Now, suppose we are given a *commutative topological algebra E with an identity element,* and whose spectrum is $\mathfrak{M}(E)$. Moreover, let

$$a = (a_{ij})$$

be an $m \times n$ matrix whose entries a_{ij}, with $1 \leqq i \leqq m$ and $1 \leqq j \leqq n$, are elements of the given algebra E.

Thus, if $1 \leqq m \leqq n$, we shall say that *the matrix a is complemented in E,* whenever there exists an $n \times n$ matrix $\bar{a}$, the entries of which are elements of E, and in such a way that the following two conditions are satisfied:

1) *The matrix $\bar{a}$ contains the given matrix a,* i.e. the first $m \times n$ "minor matrix" of $\bar{a}$ coincides with a, or what amounts to the same, one has

$$\bar{a}_{ij} = a_{ij}, \text{ with } 1 \leqq i \leqq m, \text{ and } 1 \leqq j \leqq n.$$

2) The *determinant* of $\bar{a}$ (denoted by $\det\bar{a}$), being an element of the algebra E, satisfies the relation

$$f(\det\bar{a}) \neq 0,$$

for every element $f \in \mathfrak{M}(E)$ (i.e., $\det\bar{a}$ E is *spectrally inversible*).

In this regard, the significance of the preceding condition (2.1.2) becomes clearer by the fact that *in case of a commutative, complete, locally m-convex algebra E with and identity element, a given $n \times n$ matrix $a = (a_{ij})$ with $a_{ij} \in E$, $1 \leqq i,j \leqq n$, is a unit* (i.e., an invertible element) *of* $M_n(E)$ (:the algebra of all $n \times n$ matrices with entries from E) *if, and only if,* $\det a \in E$ *is spectrally inversible*, i.e. $f(\det a) \neq 0$, for every $f \in \mathfrak{M}(E)$. (In this concern, cf. also (22; p.11 ff.), and (21; p.22, Theorem 5.4)).

On the other hand, if $a = (a_{ij})$ is an $m \times n$ matrix in E, as in (2.1) above, then one defines the *rank of a in E,* as the greatest $k \in \mathbb{N}$ (:the set of natural numbers), denoted by

$$r(a, E) = k,$$

such that there exists $f \in \mathfrak{M}(E)$, in such a way that the minors of a of order k do not belong all together (i.e., as a subset of E) to $ker(f) \subseteq E$. Thus, we shall say that the given matrix a is of *maximum rank*, if one has:

$$(2.3) \qquad r(a, E) = m$$

Now, the corresponding to the given algebra E, Gel'fand map $g : E \mapsto \mathcal{C}_c(\mathfrak{M}(E))$, defined by the relation:

$$(2.4) \qquad (g(x))(f) \equiv \hat{x}(f) = f(x),$$

for any $x \in E$, and $f \in \mathfrak{M}(E)$, gives rise to the *Gel'fand transform* $\hat{a} = (\hat{a}_{ij})$ *of the given matrix* a, being thus a matrix whose entries are elements of the Gel'fand transform algebra of E, that is the algebra $g(E) \equiv E^{\wedge} \subseteq \mathcal{C}_c(\mathfrak{M}(E))$ (for the notation applied, we refer, for instance, to [19]).

It is immediate by the preceding that if the matrix a as in (2.1), is complemented in the algebra E then $\hat{a}$ is complemented in $\mathcal{C}_c(\mathfrak{M}(E))$ (cf. also [16; p. 478, §4]), so that $\hat{a}$ is of maximum rank, i.e., $r(\hat{a}, \mathcal{C}_c(\mathfrak{M}(E))) = m$. Therefore, we can summarize with the following lemma, the second part of which is also an easy consequence of standard reasoning, regarding locally m-convex algebras. Thus, one gets

Lemma 2.1 *Let E be a commutative topological algebra with an identity element, and whose spectrum is $\mathfrak{M}(E)$. Moreover, let $a = (a_{ij})$ be an $m \times n$ matrix with entries from E and which is complemented in E. Then, the Gel'fand transform matrix $\hat{a} = (\hat{a}_{ij})$ of the given matrix a is of maximum rank, i.e., one has $r(\hat{a}, \mathcal{C}_c(\mathfrak{M}(E))) = m$. In particular, if E is moreover, a complete locally m-convex algebra, then the hypothesis for the given matrix a implies that it is of maximum rank, that is one has $r(a, E) = m$.*

In the special case of numerical (say, complex-valued) matrices, the above lemma actually characterizes the respective situation. Thus, it is the further purpose of the following to provide whithin the present context, conditions guaranteeing a similar conclusion, to the converse direction of that given by Lemma 2.1 (cf. Theorem 5.1 and its Corollary 5.1).

3. We are concerned in this section with some preliminary material needed in the sequel, and which constitutes an extended version in our case of a functional calculus formalism considered by V. Ya. Lin in Ref. [13] within the context of Banach algebras theory.

Thus, suppose that E is a topological algebra whose spectrum is $\mathfrak{M}(E)$, and let X be a complex manifold. Now, a continuous map $\varphi : \mathfrak{M}(E) \mapsto X$ is said to be *spectral* if for every (complex-valued) holomorphic map h on an open neighborhood U of $Im(\varphi) \subseteq X$, one has

(3.1) $$\varphi_* (h) \equiv h \circ \varphi = \hat{x},$$

with $x \in E$, that is, denoting by

$$\mathcal{H}ol\ (Im(\varphi)) = \varinjlim_{U \supseteq Im(\varphi)} \mathcal{H}ol(U)$$

the algebra of holomorphic functions on $Im(\varphi)$, the map φ is said to be spectral in case the following relation holds true:

(3.3) $$\varphi_*(\mathcal{H}ol(Im((\varphi))) \subseteq E^{\wedge} \equiv g(E).$$

On the other hand, the given map φ as above is said to be *weakly spectral*, if one has:

(3.4) $$\varphi_*(\mathcal{H}ol(X)) \subseteq E^{\wedge}.$$

Now, it is evident that *every spectral map is weakly spectral*, while the converse statement is not, in general, true for an arbitrary complex manifold X (13). However, when $X = \mathbb{C}^n$, then it is a consequence of the well-known Šilov-Arens-Calderōn-Waelbroeck theory that *the above two notions are essentially equivalent*: That is, if the map $\varphi: \mathfrak{M}(E) \mapsto \mathbb{C}^n$ is weakly spectral, and $z_i (i = 1, \ldots, n)$ are the coordinate functions on $\mathbb{C}^n$, then $z_i \circ \varphi : \mathfrak{M}(E) \mapsto \mathbb{C}$ $(i = 1, \ldots, n)$ are holomorphic functions on $Im(\varphi) \subseteq \mathbb{C}^n$, so that there exist, by hypothesis, elements $a_i \in E$ $(i = 1, \ldots, n)$, such that one has:

(3.5) $$\hat{a}_i = z_i \circ \varphi : \mathfrak{M}(E) \mapsto \mathbb{C} \ (i = 1, \ldots, n).$$

Thus, the given map φ is determined by the preceding relation (3.5), and hence one obtains:

(3.6) $$Im(\varphi) = S_{p_E}(\mathbf{a} \equiv (a_1, \ldots, a_n)) = \{\hat{\mathbf{a}}(f) : f \in \mathfrak{M}(E)\} \subseteq \mathbb{C}^n$$

(the *joint spectrum* of the elements $a_i \in E$, with $i = 1, \ldots, n$). Therefore, *If E is a commutative, complete, locally m-convex algebra with an identity element and $h \in \mathcal{H}ol(U)$*, where U is an open neighborhood of $Im(\varphi) \subseteq \mathbb{C}^n$, *one obtains the respective relation to (3.3) above* (cf., for instance, (1; p.412, Théorème 3)). Thus, we may summarize the preceding into the form of the following:

Lemma 3.1. *Let E be a commutative, complete, locally m-convex algebra with an identity element, and whose spectrum is $\mathfrak{M}(E)$. Moreover, let $\varphi: \mathfrak{M}(E) \mapsto \mathbb{C}^n$ be a given continuous map. Then, the map φ is a spectral map if, and only if, it is weakly spectral.*

The following result extends the situation described by the preceding Lemma 3.1 to the case of an arbitrary Stein manifold X, and specializes to the analogous result of V. Ya. Lin for the case that E is a (complex) commutative, semisimple, Banach algebra with an identity element (cf. (13; p. 122, Theorem 1)). That is, we have:

Lemma 3.2. *Let E be a commutative, complete, locally m-convex algebra with an identity element whose spectrum is $\mathfrak{M}(E)$, and let X be a Stein manifold. Moreover, consider a continuous map*

$$(3.7)\qquad \varphi : \mathfrak{M}(E) \mapsto X.$$

Then, the map φ is a spectral map if, and only if, it is weakly spectral.

Proof: It is clear that it suffices only to prove the "if part" of the statement. Thus, by Remmert's imbedding theorem (cf., for instance, (10; p.224, Theorem 10)), one may suppose that X is embedded as a closed (complex analytic) subvariety in a suitable complex numerical space $\mathbb{C}^n$. Hence, by applying next the Docquier-Grauert theorem on the normal bundle of the last imbedding (ibid., p. 257, Theorem 8), one concludes that every holomorphic function on an open neighborhood U of $Im(\varphi)$ in X can be extended as such to an open neighborhood, say V of $Im(\varphi) \subseteq \mathbb{C}^n$ with $U \subseteq V \subseteq \mathbb{C}^n$. Therefore, *if the given map (3.7) is spectral relative to $\mathbb{C}^n$, it is also spectral relative to X,* so that, by Lemma 3.1, the proof is reduced to verify that (3.7) is weakly spectral relative to $\mathbb{C}^n$, but this is now immediate by what has been applied above and the hypothesis. ∎

4. The purpose of this section is to provide background material permitting to extend within the present framework the fundamental result of Lin's paper [13; p. 123, Theorem 2], on which our main results in the next section are based.

Thus, suppose we have a topological algebra E whose spectrum is $\mathfrak{M}(E)$, and moreover consider a topological space X and a given continuous map:

$$(4.1)\qquad \varphi : \mathfrak{M}(E) \mapsto X.$$

Now, if $(x_1, \ldots, x_n)$ is a finite sequence of elements of E, one defines an *E-modification of φ corresponding to the element $\mathbf{x} = (x_1, \ldots, x_n) \in E^n := \underbrace{E \times \ldots \times E}_{n\text{-times}}$* by the following map:

$$(4.2)\qquad \varphi' : \mathfrak{M}(E) \mapsto X \times \mathbb{C}^n : f \mapsto \varphi'(f) := (\varphi(f); \hat{x}_1(f), \ldots, \hat{x}_n(f)).$$

Now, in this concern, we first have the following result (cf. also [13; p.125, Lemma 5]):

Lemma 4.1. *Let E be a commutative, complete, locally m-convex algebra with an identity element and whose spectrum is $\mathfrak{M}(E)$. Moreover, let X, Y be Stein manifolds and φ, ψ two given spectral maps of $\mathfrak{M}(E)$ into X, Y respectively. Then, the* *(cartesian product) map:*

(4.3) $$\varphi \times \psi : \mathfrak{M}(E) \longmapsto X \times Y$$

is also a spectral map. In particular, every E-modification of a given spectral map of $\mathfrak{M}(E)$ into a Stein manifold is also spectral.

Proof: Concerning the first part of the statement, it suffices to prove, by Lemma (3.2), that the map (4.3) is a weakly spectral map: Thus, since $X \times Y$ is, by hypothesis, a Stein manifold (cf. also [15; p. 308, Theorem 4.2] and [10; p. 222, Corollary 5]), then it is "regularly imbedable" in a suitable $\mathbb{C}^n$ (Remmert), so that the assertion is now an immediate consequence of the respective "finite cartesian product analogon" of Lemma 3.1, plus H. Cartan's extension theorem for holomorphic functions (when defined, in particular, on Stein submanifolds of $\mathbb{C}^n$; cf., for instance, [10; p. 245, Theorem 18]). On the other hand, if $\varphi : \mathfrak{M}(E) \longmapsto X$ is a given spectral map, with X a Stein manifold, then an E-modification of φ of the form (4.2) is also a spectral map, by directly applying the last argument above, plus Lemma 3.1 for the weakly spectral map $\hat{x} : \mathfrak{M}(E) \longmapsto \mathbb{C}^n : f \longmapsto \hat{x}(f) = S_{p_E}(x) := (\hat{x}_1(f), \ldots, \hat{x}_n(f))$ (cf. also the notation applied in the rel. (4.2) and 3.6 above). ∎

Now, suppose that φ is a given continuous map as in the rel. (4.1) above, where X is, in particular, a complex manifold. Moreover, suppose that (4.2) is a given E-modification of φ, correspodning to a finite sequence $(x_1, \ldots, x_n)$ of elements of E, and *let U be an open neighborhood of $Im(\varphi)$ in X.*

Thus, we shall say that the E-modification φ' of φ (cf. the rel. (4.2) above) is *subordinated to U,* if one has the relation:

(4.4) $$\pi(\widehat{\varphi'(\mathfrak{M}(E))}) \subseteq U,$$

where $\widehat{\varphi'(\mathfrak{M}(E))}$ denotes the holomorphically convex hull of $Im(\varphi')$ in $X \times \mathbb{C}^n$ and $\pi : X \times \mathbb{C}^n \longmapsto X$ the respective canonical projection onto the first factor.

Now, the following result (Lemma 4.2) extends within the present context Lemma 6 in Ref. [13; p. 125] and constitutes besides a similar extension of the well-known "Arens-Calderõn Lemma" (cf., for instance, [9; p. 85. Lemma 5.2]). On the other hand, it is essentially this, which forces upon us to consider the topological algebras we use below, as this is also pointed out, however within another context, by M. Bonnard in Ref. [2; p.29. Lemma 5].

Thus, the topological algebras we consider in the following lemma have compact spectra and their respective Gel'fand maps (cf. the rel. (2.4) above) continuous, so that in case of a commutative, complete, locally m-convex algebra with an identity element, as it will be the case below, such a situation is actually equivalent with that of having the topological algebras involved a "continuous inversion" in the sense of L. Waelbroeck (cf., for instance, (17; p. 108, Corollary 3.3)). In the locally convex case, the later algebras are also called *Waelbroeck algebras* (cf., for instance, (5; p. 147)): They have the crucial property for the functional calculus of being *Arens-Calderõn algebras* in the sense of M. Bonnard (cf. (2; p. 29, Proposition 8, and p. 31, ff.]).

Thus, we now have the following.

Lemma 4.2. *Let E be a commutative, complete, locally m-convex algebra with an identity element and whose spectrum $\mathfrak{M}(E)$ is a compact (Hausdorff) space, having moreover the respective Gel'fand map (cf. (2.4)) continuous. Besides. let $\varphi: \mathfrak{M}(E) \mapsto X$ be a given weakly spectral map, with X being a Stein manifold, and let U be an open neighborhood of Im (φ) in X. Then, there exists an E-modification of the given map φ, which is subordinated to U.*

Proof: Consider that the Stein manifold X is "regularly imbedded" (Remmert) in a suitable $\mathbb{C}^n$, and suppose that the map $\varphi : \mathfrak{M}(E) \mapsto X \subsetneq \mathbb{C}^n$ is given by the elements $x_i \in E$ $(i=1, \ldots, n)$, that is one has the relation:

$$\varphi(f) = (\hat{x}_1(f), \ldots, \hat{x}_n(f)), \tag{4.5}$$

for every $f \in \mathfrak{M}(E)$ (:take into account the hypothesis for φ, and the argumentation before Lemma 3.1 above). Now, by considering X as being indentified with its image in $\mathbb{C}^n$, one concludes that $U = X \cap V$, where V is an open neighborhood of $\varphi(\mathfrak{M}(E))$ in $\mathbb{C}^n$. Therefore, by applying "Arens-Calderõn Lemma" (cf. (2; p.29, Lemme 5)), one gets a finite sequence $(y_1, \ldots, y_m)$ of elements of E, in such a way that the corresponding E-modification of φ, i.e., the map

$$\varphi' : \mathfrak{M}(E) \mapsto X \times \mathbb{C}^m \subsetneq \mathbb{C}^n \times \mathbb{C}^m, \tag{4.6}$$

given by $\varphi'(f) = (\hat{x}_1(f), \ldots, \hat{x}_n(f); \hat{y}_1(f), \ldots, \hat{y}_m(f))$, with $f \in \mathfrak{M}(E)$, satisfies the relation:

$$\pi'(\widehat{Im(\varphi')}) \subseteq V, \tag{4.7}$$

where $\pi' : \mathbb{C}^n \times \mathbb{C}^m \mapsto \mathbb{C}^n$ denotes the respective canonical projection and $\widehat{Im(\varphi')}$ the polynomially convex hull of the compact set $Im(\varphi')$ in $\mathbb{C}^n \times \mathbb{C}^m$. On the other hand, one has the relation:

$$\widehat{Im(\varphi')} = \widetilde{Im(\varphi')}, \tag{4.8}$$

where its second member denotes the holomorphically convex hull of $Im(\varphi')$ in $X \times \mathbb{C}^m$ (cf., for instance, [10; p. 243, Theorem 13,14]), so that, by applying the canonical projection $\pi : X \times \mathbb{C}^m \mapsto X$, one obtains:

(4.9) $$\widetilde{n(Im\,(\varphi'))} \subseteq V \cap X = U,$$

and this proves the assertion. ∎

Now, some more terminology from topological algebras theory will be needed. Thus, the topological algebras we are next going to consider will appropriately be constituted, from finitely-generated ones, having the crucial property that their spectra are (within a homeomorphism) compact polynomially convex subsets of suitable complex numerical spaces. Thus, we first set the following

Definition 4.1. Suppose we are given a topological algebra E whose spectrum is $\mathfrak{M}(E)$. Then, we shall say that E is a *(k)-algebra,* whenever there exists a local basis $(U_\alpha)_{\alpha\in I}$ of E in such a way that the corresponding family $(\mathfrak{M}(E_\alpha))_{\alpha\in I}$ of subsets of $\mathfrak{M}(E)$, with $\mathfrak{M}(E_\alpha) = \mathfrak{M}(E) \cap U_\alpha^0$, $\alpha\in I$ (cf. also [18; p. 156]), defines a *k-covering family* of $\mathfrak{M}(E)$, i.e., the following three conditions are satisfied:

(4.1.1) Each one of the sets $\mathfrak{M}(E_\alpha)$, $\alpha\in I$, is a compact subset of the spectrum $\mathfrak{M}(E)$ (:Alaoglu-Bourbaki).

(4.1.2) The family $(\mathfrak{M}(E_\alpha))_{\alpha\in I}$ defines a covering of $\mathfrak{M}(E)$.

(4.1.3) *For every compact* $K\subseteq \mathfrak{M}(E)$, *there exists* an $\alpha\in I$, *with* $K\subseteq \mathfrak{M}(E_\alpha)$.

In this regard, we also remark that, by hypothesis for $(U_\alpha)_{\alpha\in I}$ we always have the relation:

(4.10) $$\mathfrak{M}(E) = \bigcup_{\alpha\in I} (\mathfrak{M}(E) \cap U_\alpha^\circ) \equiv \varinjlim \mathfrak{M}(E_\alpha),$$

and besides the sets $\mathfrak{M}(E_\alpha)$, $\alpha\in I$, are compact subsets of $\mathfrak{M}(E)$ (:Alaoglu-Bourbaki theorem), that is cond. (4.1.1) and (4.1.2) of the preceding definition are actually redundant, so that the definitive new feature which indeed enters the preceding picture is that *the family* $(\mathfrak{M}(E_\alpha))_{\alpha\in I}$ *satisfies cond. (4.1.3) above:*

Now, a large class of topological algebras, widely applied in several contexts, satisfies the preceding condition, including of course the Banach algebras, the Fréchet algebras and their (topological) inductive limits. Besides, the m-barrelled (and hence also the barrelled) topological algebras, or even the spectrally barrelled ones, previously considered in Ref. (17), (18), share this property, as well. Something more, which actually explains all the preceding instances, one has, as a consequence of (15; p.305, Theorem 3.1) (cf. also (19; p. 460, Lemma 2.1)) that, *every topological algebra with an identity element having the Gel'fand map continuous is a (k)-algebra.*

We now have the following result, decisive for the sequel and also justifing the preceding terminology.

Lemma 4.3. *Let E be a commutative, complete, n-generated, locally m-convex, (k)-algebra, having an identity element and a compact spectrum* $\mathfrak{M}(E)$. *Then* $\mathfrak{M}(E)$ *is homeomorphic to a compact, polynomially convex subset of* $\mathbb{C}^n$.

Proof: If $(x_1, \ldots, x_n)$ is a system of generators of $\boldsymbol{E}$, the asserted homeomorphism is given by the map:

$$\varphi : \mathfrak{M}(E) \longmapsto \mathbb{C}^n : f \longmapsto \varphi(f) = (\hat{x}_1(f), \ldots, \hat{x}_n(f)). \tag{4.11}$$

Now, the proof is based on that of Theorem 4.1 in Ref. [19; p. 469] (cf. also the resulting Corollary 4.2, ibid., p. 470), which in turn depends essentially on the cond. (4.1.3) of the preceding Definition 4.1, the last condition being concluded therein from being the topological algebra involved a spectrally barrelled one (cf., [18; p. 156, Lemma 4.1]).∎

Now, the topological algebras we next apply are suitable "colimits" of those appeared in the above Lemma 4.3, extending thus the familiar situation one has in case of Banach algebras. Thus, we first have

Definition 4.2. Suppose we are given a complete, locally m-convex algebra E, whose spectrum is (E). Moreover, let $(E_\alpha, \varphi_{\beta\alpha})$, $\alpha \in \boldsymbol{I}$, be an inductive system of subalgebras of E, each one being a finitely generated (k)-algebra of the type considered in Lemma 4.3, and in such a way that the following two conditions are satisfied:

(4.2.1) $E = \varinjlim E_\alpha$, *algebraically.*

(4.2.2) $\mathfrak{M}(E) = \varprojlim \mathfrak{M}(E_\alpha)$,

Then, under the preceding circumstances, we shall say that the given topological algebra E is an *(lk)-algebra.*

Thus, it is clear that *every (lk)-algebra is commutative and has an identity element, as well as a compact spectrum.*

On the other hand, suppose that E *is an (lk)-algebra, whose spectrum is* $\mathfrak{M}(E)$. Besides, let

$$h : \mathfrak{M}(E) \longmapsto X \ (\subseteq \mathbb{C}^n) \tag{(4.12)}$$

be *a continuous map of* $\mathfrak{M}(E)$ *into a given topological space X.*

Now, we shall say that *the map h is subordinated to an index* $\alpha \in I$ (cf. Definition 4.2), whenever there exists $\alpha \in I$ in such a way that if the respective algebra E_α is, say n-generated, then the following condition is satisfied:

(4.13) If $(x_1, \ldots, x_n)$ are elements of E which generate the subalgebra E_α of E(E_α has an identity element), the resulting continuous map:

(4.13.1) $$h_\alpha : \mathfrak{M}(E_\alpha) \longmapsto \mathbb{C}^n,$$

with $h_\alpha(f_\alpha) = (\hat{x}_1(f_\alpha), \ldots, \hat{x}_n(f_\alpha))$, and $\hat{x}_i(f_\alpha) = f_\alpha(x_i)$ $(i = 1, \ldots, n)$, for every $f_\alpha \in \mathfrak{M}(E_\alpha)$, satisfies the relation:

(4.13.2) $$h = h_\alpha \circ {}^t\varphi_\alpha$$

where $\psi_\alpha \equiv {}^t\varphi_\alpha : \mathfrak{M}(E) \longmapsto \mathfrak{M}(E_\alpha)$ denotes the transpose of the canonical map $\varphi_\alpha : E_\alpha \longmapsto E$, defined by the cond. (4.2.2).

The preceding technical language will become clearer by the concrete cases, we are going to consider below. Thus, we first have the following result specializing to the main result of V.Ya Lin in Ref. (13; p. 123, Theorem 2), the latter being the initial motivation to the present setting; it also improves a previous version in Ref. (20; Lemma 6.2):

Theorem 4.1. *Let E be an (lk)-algebra, having the respective Gel'fand map continuous. Moreover, let $h : \mathfrak{M}(E) \longmapsto X$ be a weakly spectral map of the spectrum of E into a Stein manifold X, subordinated to a given index $\alpha \in I$, in such a way that the respective algebra E_α is n-generated, with $n \in \mathbb{N}$ being the dimension of the complex numerical space into which X may be "regularly imbedded". Then, the following two assertions are equivalent:*

1) There exists an open neighborhood U of $Im(h)$ in X, and a principal holomorphic fiber bundle $\xi = (V, \pi, U)$ on U, admitting a continuous covering map $\sigma : \mathfrak{M}(E) \longmapsto V$ of h.

2) The bundle ξ, as in 1), admits a spectral covering map $\sigma_* : \mathfrak{M}(E) \longmapsto V$ of the given map h.

That is, in other words, a bundle $\xi = (V, \pi, U)$ as above, admits a spectral covering map of h if, and only if, it admits, for the given map h, a continuous covering map.

Proof : It suffices obviously to prove that 1) $\Rightarrow$ 2): Thus, we first have, by Lemma 3.2, that the given map h is essentially a spectral map. Next, we may assume that X is regularly imbedded in a complex numerical space $\mathbb{C}^n$, so that one may consider the map:

(4.14) $$h : \mathfrak{M}(E) \longmapsto h(\mathfrak{M}(E)) \subseteq U \subseteq X \subsetneqq \mathbb{C}^n,$$

where U is an open neighborhood, with respect to X, of the compact set $Im(h) \subseteq X$.

Now, there exists, by hypothesis, an index $\alpha \in I$, in such a way that the algebra E_α is n-generated, while the rel. (4.13.2) holds true, so that by (4.14) one obtains:

(4.15) $$(h_\alpha \circ \psi_\alpha)(\mathfrak{M}(E)) = h(\mathfrak{M}(E)) \subseteq U \subseteq X.$$

On the other hand, we remark that *the polynomially convex hull of the compact set* $h(\mathfrak{M}(E)) \subseteq h_\alpha(\mathfrak{M}(E_\alpha)) \subseteq \mathbb{C}^n$ (cf. the rel. (4.13.1) and (4.13.2) above) *coincides with the compact polynomially convex set* $h_\alpha(\mathfrak{M}(E_\alpha))$ *which is homeomorphic to* $\mathfrak{M}(E_\alpha)$, the spectrum of the (k)-algebra E_α (cf. Definition 4.2 and Lemma 4.3). Hence, by the preceding relations, one concludes that the range of the map h_α, defined by (4.13.1), is actually contained in $X \subsetneq C^n$, so that *we may also assume that* U *is an open Stein neighborhood of* $Im(h_\alpha)$ *in* X.

Thus, suppose that $\xi = (V, \pi, U)$ is a principal holomorphic fiber bundle on U, admitting a *continuous covering map* $\sigma : \mathfrak{M}(E) \mapsto V$ of the given map h. Hence, by (4.15) and [13; p. 123, Lemma 2], there exists an index $\beta \in I$, with $\beta \geq \alpha$, such that *the map* $h_\beta = h_\alpha \circ \psi_{\alpha\beta}\Big|_{Im(\psi_\beta)} : \psi_\beta(\mathfrak{M}(E)) \mapsto U$ *admits a continuous covering map* $\mu : \Psi_\beta(\mathfrak{M}(E)) \mapsto V$.

Now, if the (k)-algebra E_β is m-generated (Definition 4.2) and $(y_1, \ldots, y_m)$ is a set of generators of E_β, consider the following *modifications* of the preceding maps, defined by the elements $y_i \in E_\beta \subseteq E$ $(i = 1, \ldots, m)$ (cf. the rel. (4.2)). That is, we have:

(4.16) $$\begin{aligned} h' &: \mathfrak{M}(E) \mapsto U \times \mathbb{C}^m, \\ h'_\beta &: \psi_\beta(\mathfrak{M}(E)) \mapsto U \times \mathbb{C}^m, \\ \mu' &: \psi_\beta(\mathfrak{M}(E)) \mapsto V \times \mathbb{C}^m, \end{aligned}$$

as well as, the respective modification of the given bundle ξ:

(4.17) $$\pi' : V \times \mathbb{C}^m \mapsto U \times \mathbb{C}^m$$

(the corresponding fiber for π' being the same with that of π), so that one obtains the following commutative diagram:

(4.18) $$\begin{array}{ccc} \mathfrak{M}(E) & \xrightarrow{\ \psi_\beta\ } & \psi_\beta(\mathfrak{M}(E)) \subseteq \mathfrak{M}(E_\beta) \\ {\scriptstyle h'}\downarrow & \overset{h'_\beta}{\swarrow} & \downarrow{\scriptstyle \mu'} \\ U \times \mathbb{C}^m & \xleftarrow{\ \pi'\ } & V \times \mathbb{C}^m \end{array}$$

That is, one gets the relation:

(4.19) $$h' = h'_\beta \circ \psi_\beta.$$

Now, since $\psi_\beta(\mathfrak{M}(E))$ is a compact set, the map h'_β, being one-to-one, by its definition and the hypothesis for the $y'_i s$, it is actually a homeomorphism onto its range, i.e., by (4.19), onto the set $h'_\beta(\psi_\beta(\mathfrak{M}(E))) = h'(\mathfrak{M}(E))$, so that, if $\rho = (h'_\beta)^{-1}$ is the inverse map of h'_β, that is,

(4.20) $$\rho : h'(\mathfrak{M}(E)) \longmapsto \psi_\beta(\mathfrak{M}(E)),$$

then, by (4.18), one obtains that *the map*

(4.21) $$\mu' \circ \rho : h'(\mathfrak{M}(E)) \longmapsto V \times \mathbb{C}^m$$

is a continuous section of π' (cf. 4.17) over the compact set $h'(\mathfrak{M}(E)) \subseteq U \times \mathbb{C}^m$ (cf. the first of (4.16). Therefore [13; p. 124, Lemma 4], *there exists a continuous extension of* (4.21), still denoted by $\mu' \circ \rho$, *onto an open neighborhood* W of $h'(\mathfrak{M}(E))$ in $U \times \mathbb{C}^m \subseteq X \times \mathbb{C}^m$. Besides, *since* h *is a spectral map, its modification* h', defined by the first of the rel. (4.16), *is also a spectral map* (Lemma 4.1). Hence (Lemma 4.2), *there exists an E-modification of the map:*

(4.22) $$h' : \mathfrak{M}(E) \longmapsto X \times \mathbb{C}^m,$$

say h'', corresponding to a finite sequence $(a_1, \ldots, a_k)$ of elements of E, and *subordinated to the open neighborhood* W of $Im(h')$ in $X \times \mathbb{C}^m$, as above; that is, one has:

(4.23) $$h'' : \mathfrak{M}(E) \longmapsto (X \times \mathbb{C}^m) \times \mathbb{C}^k,$$

in such a way that (cf. (4.4)):

(4.24) $$\mathrm{pr}\,(\widetilde{h''(\mathfrak{M}(E))}) \subseteq W \subseteq X \times \mathbb{C}^m,$$

so that, one finally obtains:

(4.25) $$h''(\mathfrak{M}(E)) \subseteq \widetilde{h''(\mathfrak{M}(E))} \subseteq N \subseteq W \times \mathbb{C}^k,$$

where N denotes an appropriate open Stein neighborhood of the compact holomorphically convex set $\widetilde{h''(\mathfrak{M}(E))}$ in $X \times \mathbb{C}^m \times \mathbb{C}^k$.

On the other hand, by the preceding, there exists a continuous section of π' over W, so that, *the modification of* (4.17), i.e.,

(4.26) $$\pi'' : (V \times \mathbb{C}^m) \times \mathbb{C}^k \longmapsto (U \times \mathbb{C}^m) \times \mathbb{C}^k,$$

admits a continuous section over $W \times \mathbb{C}^k$, and hence a fortiori *over* N, an open Stein subspace of $U \times \mathbb{C}^m \times \mathbb{C}^k$, and consequently, *there* also *exists a holomorphic section*, say s, *of* π'' *over* N (cf. also Scholium 4.1 which follows). Thus, one obtains the following (commutative)diagram:

(4.27)

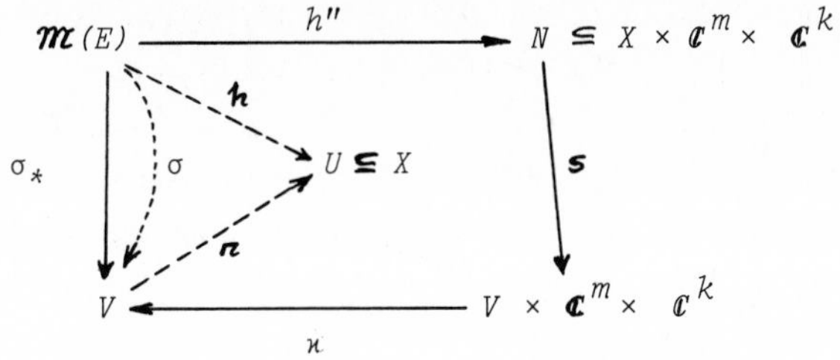

where one defines the map:

(4.28) $$\sigma_* = \varkappa \circ s \circ h'',$$

denoting by $\varkappa$ the respective canonical projection, as indicated in (4.27).

Now, *the map* h'', being a modification of the spectral map h, *is* itself also a *spectral* map (Lemma 4.1). Besides, the maps π and s are holomorphic, by their definition, so that one actually concludes that σ_* *is a spectral map.* Finally, one can easily verify the relation:

(4.29) $$h = \pi \circ \sigma_*$$

that is σ_* *is a* (spectral) *covering map of* h, and this finishes the proof of the theorem. ∎

Scholium 4.1.- The fiber bundles considered in the preceding theorem were *principal holomorphic* ones, defined actually *on Stein manifolds*, this having as a consequence the validity, in this respect, of the "Oka's principle", as the latter has been given, for instance, in the work of K. J. Ramspott (24), on the "equivalence between holomorphic and continuous sections" for such bundles (ibid.; p. 236, Satz. Cf. also (4; p. 102, Théorème 1, and the comments following it). However, this has been so chosen only for simplicity's sake, and in order that the preceding principle to hold true, being besides the case for the bundles, we are going to consider in the next Section, when applying Theorem 4.1 (cf. Theorem 5.1 and its Corollary). Thus, a more general form of the theorem could be given by considering "principal holomorphic fiberings" in the sense of Ref. (13; p. 123, Cf. also, p. 127 and p. 126, Remark 2)).

On the other hand, some remarks concerning the topological algebras considered would be fitted in: Thus, one would have an analogous result to the above theorem, assumming the given map $h : \mathfrak{M}(E) \longmapsto X$ to be subordinated to an appropriate index $\alpha \in I$ (depending on the complex dimension of the manifold X), for which the respective algebra E_α is a *(k)*-algebra of the type applied in Lemma 4.3, this last assumption being not necessarily further essential for the rest of the algebras E_β,

$\beta \in I$, of the inductive system $(E_\alpha, \varphi_{\beta\alpha})$ defining the algebra E (Definition 4.2), the kind of the milder assumptions thus appropriate, being besides easily understood from the context (cf. the proof of the theorem).

In this regard, we also abserve that by using the relevant reasoning applied in Ref. (13; p. 127), one can get, as another application of the above Theorem 4.1 (in conjunction with the preceding Scholium), an extension, within the present setting, of some similar considerations to those of (13), contained in Ref. (12; cf., for instance, p.8, Theorem 1).

5. In this final Section we come to the formulation and proof of the main result, indicating the applications of the preceding discussion, which also motivated the present material.

Thus, we are now in position to state the following, which also generalizes the respective result in Ref. (13; p. 127, Theorem 3). That is, we have:

Theorem 5.1. *Let E be an (lk)-algebra, having the respective Gel'fand map continuous, and whose spectrum is $\mathfrak{M}(E)$. Moreover, let $a = (a_{ij})$ be a given $m \times n$ matrix with entries in E, and in such a way that there exists an index $\alpha \in I$, such that the respective (k)-algebra E_α is (m.n)-generated (cf. Definition 4.2). Then, the following two assertions are equivalent:*

1) The given matrix a is complemented in the algebra E.

2) The Gel'fand transform matrix $\hat{a}$ of a is complemented in the algebra $\mathcal{C}_c(\mathfrak{M}(E))$.

(Terminological) Scholium.- We shall say that *the algebra E*, satisfying the conditions of the preceding theorem concerning the matrix $a = (a_{ij})$, *subordinates the* given *matrix a*. On the other hand, a is said to be *functionally complemented* in E, if cond. 2) of the above statement holds true. Thus, we may express the preceding, in a rather more concise and descriptive manner, by saying that *a matrix $a = (a_{ij})$ with entries from a given algebra E* (Theorem 5.1), *subordinating it, is complemented in the algebra E if, and only if, it is functionally complemented in E.*

Proof of Theorem 5.1.: Since the spectrum of the (locally m-convex) algebra $\mathcal{C}_c(\mathfrak{M}(E))$ is (within a homeomorphism) $\mathfrak{M}(E)$ (cf., for instance,(16; p. 478)), it is clear (cf. also the rel. (2.1.2) in the foregoing), that 1) *implies* 2). On the other hand, suppose that the Gel'fand transform matrix $\hat{a}$ of the given matrix $a = (a_{ij})$ is complemented in the algebra $\mathcal{C}_c(\mathfrak{M}(E))$, and let $V_m(\mathbb{C}^n)$ be the set of "m-frames" in $\mathbb{C}^n$ with $1 \leq m \leq n$, that is, the correspodning (complex) Stiefel manifold (cf., for instance, (26; p. 33)). Then *the map:*

(5.1) $$\pi : GL(n, \mathbb{C}) \longmapsto V_m(\mathbb{C}^n),$$

defined by assigning to each $n \times n$ invertible matrix $\omega = (\omega_{ij}) \in GL(n, \mathbb{C})$, the m-tuple consisting of the first m rows of ω, *gives a principal holomorphic fiver bundle* on the open set $V_m(\mathbb{C}^n)$ *in* $\mathbb{C}^{m.n}$ (ibid., p. 33, § 7.7), such that the respective map:

(5.2) $$\hat{a} : \mathfrak{M}(E) \longmapsto \mathbb{C}^{m.n},$$

defined from the given matrix $a = (a_{ij})$ by the relation

(5.3) $$\hat{a}(f) = (\hat{a}_{ij}(f)),$$

for every $f \in \mathfrak{M}(E)$, *is a continuous map,* and in such a way that by the hypothesis for $\hat{a}$, one obtains:

(5.4) $$\hat{a}(\mathfrak{M}(E)) \subseteq V_m(\mathbb{C}^n) \subseteq \mathbb{C}^{m.n}.$$

On the other hand, *the map* (5.2), given by the rel (5.3), *is a weakly spectral map* (as a matter of fact, it is actually a spectral map; cf., for instance, the comments before Lemma 3.1 above). Now, by the hypothesis for $\hat{a}$, *there exists a continuous map* $\sigma : \mathfrak{M}(E) \longmapsto GL(n, \mathbb{C})$, such that if $\sigma = (\sigma_{ij})$ is the respective matrix representation of σ in $GL(n, \mathbb{C})$, then (σ_{ij}) complements $\hat{a} = (a_{ij})$ (cf. the rel. (5.3) above) in $\mathcal{C}_c(\mathfrak{M}(E))$. In particular, the following diagram:

(5.5)

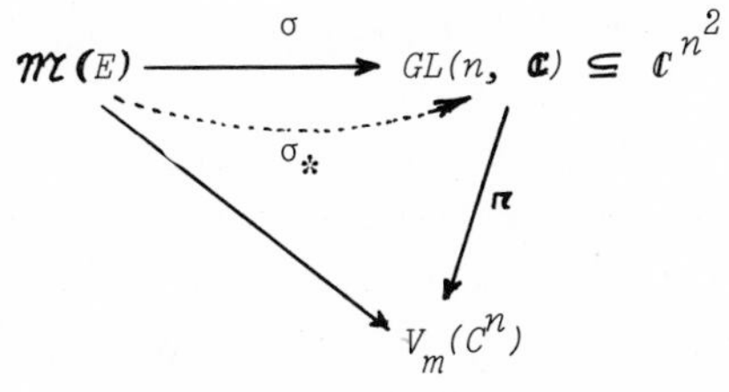

is made commutative, i.e., σ *is a continuous covering map* of $\hat{a}$. Therefore, we may apply by hypotheses for $\hat{a}$ the preceding Theorem 4.1, so that *one obtains a spectral covering map,* say σ_* *of* $\hat{a}$ (cf. the preceding diagram (5.5)); thus, if

(5.6) $$\sigma_* = ((\sigma_*)_{ij})$$

denotes the respective $n \times n$ matrix representation of σ_* in $GL(n, \mathbb{C})$, since σ_* is a spectral map, one concludes, by applying the coordinate functions of the space $\mathbb{C}^{n^2}$, the existence of an (invertible; cf. (2.1.1)) $n \times n$ matrix, say $b = (b_{ij})$, with entries from E, and in such a way that one has, by (5.5), the relation:

(5.7) $$\hat{b}_{ij} = \hat{a}_{ij}$$

with $1 \leq i \leq m$, and $1 \leq j \leq n$ (concerning the respective Gel'fand transforms). Thus, by the rel. (5.7) and (2.1.1), it is now an easy task to check that *the following* $n \times n$ *matrix:*

(5.8) $$\bar{a} = (\bar{a}_{ij}),$$

given by the relations:

(5.9) $$\bar{a}_{ij} = a_{ij}, \text{ with } 1 \leq i \leq m, \text{ and } 1 \leq j \leq n,$$ $$\bar{a}_{ij} = b_{ij}, \text{ with } m+1 \leq i \leq n, \text{ and } 1 \leq j \leq n,$$

defines a complement in E *of the given matrix* a, and this finishes the proof of the theorem. ∎

Scholium 5.1.- It is a consequence of the proof of the preceding theorem, that the complementation of a given $m \times n$ matrix $a = (a_{ij})$ with entries in an *algebra* E, *of the type considered in Theorem 4.1* (and subordinating a; cf. the scholium, preceding the above proof), is reduced to the case of the respective Gel'fand transform algebra $E^{\wedge}$ of E, when the latter is considered as a subalgebra of $\mathcal{C}_c(\mathfrak{M}(E))$ (cf. the rel. (2.4)). Now, this in turn was a consequence of cond. 2) in Theorem 4.1 (: "Oka's principle", as applied to topological algebras), so that, since the latter is a "homotopy invariant" of $\mathfrak{M}(E)$ (cf. also (11; p. 62, and p. 66, Theorem 4.2), as well as (24; p. 236 Satz)), *one obtains the same conclusion for any given space X, which is homotopic to the spectrum of the algebra E.*

Now, the following is immediate by the argumentation in the preceding Scholium (cf. also (13; p. 127, Theorem 3)), recovering the situation one has in the case of (complex) numerical matrices. That is, we obtain:

Corollary 5.1. *Suppose that the conditions of the preceding Theorem 5.1 are satisfied, and moreover assume that the spectrum* $\mathfrak{M}(E)$ *of the given algebra E has the homotopy type of a point (*:contractible space). *Then, the given* $m \times n$ *matrix* $a = (a_{ij})$ *is complemented in E if, and only if, it is of a maximum rank, i.e., one has the relation* $r(\alpha, E) = m$ (cf. the rel. (2.3)).

In this connection, we finally observe that, an application of the reasoning of the above Scholium 5.1 yields the corresponding result within the present context of Theorem 4 in Ref. (13; p. 128), so that we omit the details; in this respect, cf. also the analogous results in Ref. (25).

References

1. M. Bonnard: *Sur le calcul fonctionnel holomorphe multiforme dans les algèbres topologiques*. Ann. scient. Éc. Norm. Sup. (4) 2(1969), 397-422.

2. M. Bonnard: *Sur des applications du Calcul fonctionnel holomorphe*. Bull. Soc. math. France, Mémoire 34(1937), 5-54.

3. H. Cartan: *Sur les matrices holomorphes de n variables complexes*. J. Math. pures appl. 19(1940), 1-26.

4. H. Cartan: *Espaces fibrés analytiques*. Sympos. Intern. Topol. Alg., Mexico, 1958; pp. 97-121.

5. I. Colojoară: *Elemente de Theorie Spectrală*. Bucureşti, 1968.

6. B. Eckmann: *Zur Homotopietheorie gefaserter Räume*. Commen. Hath. Helv. 14(1942), 141-192.

7. C. Ehresmann: *Sur les applications continues d'un espace dans un espace fibré ou dans un revêtement*. Bull. Soc. math. France 72(1944), 27-54.

8. O. Forster: *Funktionentheoretische Hilfsmittel in der Theorie der Kommutativen Banach-Algebren*. Jber. Deutsch. Math. - Verein. 76(1974), 1-17.

9. T. W. Gamelin: *Uniform Algebras*. Prentice-Hall, Inc. Englewood Cliffs, New Jersey, 1969.

10. R.C. Gunning - H. Rossi: *Analytic Functions of Several Complex Variables*. Prentice-Hall, Inc. Englewood Cliffs, New Jersey, 1965.

11. S. T. Hu: *Homotopy Theory*. Academic Press, New York, 1959.

12. Yu. S. Il'yašenko: *Multivalued analytic functions of elements of a commutative normed ring*. Vestnik Mosk. Univ. No 5 (1969), 8-11 (russian).

13. V. Ya. Lin: *Holomorphic fiberings and multivalued functions of elements of a Banach algebra*. Funkt. Anal. Pril. 7(1973), 43-51 (russian) = Funct. Anal. Appl. 7(1973), 122-128 (english transl.).

14. A. Mallios: *Note on the spectrum of topological inductive limit algebras*. Bull. Soc. math. Grèce 8(1967), 127-131.

15. A. Mallios: *On the spectra of topological algebras*. J. Funct. Anal. 3(1969), 301-309.

16. A. Mallios: *On functional representations of topological algebras*. J. Funct. Anal. 3(1969), 301-309.

17. A. Mallios: *On m-barrelled algebras*. Prakt. Akad. Athēnōn 49(1974), 98-112.

18. A. Mallios: *On the barrelledness of a topological algebra relative to its spectrum*. Remarks. Bull. Soc. math. Grèce 15(1974), 152-161.

19. A. Mallios: *On a convenient category of topological algebras*, I. General theory. Prakt. Akad. Athēnōn 50(1975), 454-477.

20. A. Mallios: *On a convenient category of topological algebras*, II. Applications. Prakt. Akad. Athēnōn 51(1976), 245-263.

21. E. A. Michael: *Locally multiplicatively-convex topological algebras*. Mem. Amer. Math. Soc. No 11(1952) (repr. 1968).

22. M. Newman: *Integral Matrices*. Academic Press, New York, 1972.

23. M. E. Novodvorskii: *On certain homotopic invariants of the space of maximal ideals*. Mat. Zametki 1(1967), 487-494 (russian).

24. K. J. Ramspott: *Stetige und holomorphe Schnitte in Bündeln mit homogener* Faser. Math. Zeitschr. 89(1965), 234-246.

25. N. Sibony - J. Wermer: *Generators for $A(\Omega)$*. Trans. Amer. Math. Soc. 194(1974), 103-114.

26. N. Steenrod: *The Topology of Fibre Bundles*. Princeton Univ. Press, Princeton, New Jersey, 1951.

27. J. L. Taylor: *Topological invariants of the maximal ideal space of a Banach algebra*. Adv. Math. 19(1976), 149-206.

28. J. Wagner: *Faisceau structural associé à une algèbre de Banach*. Sém. P. Lelong, 9e ann. (1968/69), 164-167. Springer-Verlag, Lecture Notes in Math. No 116(1970).

29. T. Ważewski: *Sur les matrices dont les elements sont des fonctions continues*. Compos. Math. 2(1935), 63-68.

University of Athens
Mathematical Institute

ON THE DUALIZING COMPLEX

by O. Stănăşilă

§1. In 1955 Serre proved the following duality theorem : Let X be a compact complex manifold of dimension n, Ω the sheaf of germs of holomorphic forms of maximal degree and F an analytic sheaf on X, which is locally free. Then the $\mathbb{C}$-linear spaces $H^p(X,F)$, $H^{n-p}(X,\check{F}\otimes\Omega)$ are in algebraic duality.

If F is an arbitrary coherent sheaf, then $H^{n-p}(X,\check{F}\otimes\Omega)$ is to be substituted by $Ext^{n-p}(X; F,\Omega)$; moreover, if X is not compact, then the invariants must be topologized, H^{n-p}(respectively Ext^{n-p}) is to be substituted by H_c^{n-p} (Ext_c^{n-p} respectively) and the algebraic duality is transformed in a topological one. This idea is due to Serre and Malgrange.

A new problem was the extension of the above dualities to the case of the complex spaces with singularities, where the main difficulties were to find a substitute for Ω and an analogous of the trace map. In this sense J.P.Ramis and G. Ruget built, for any complex space $(X,\mathcal{O})$, a complex of $\mathcal{O}$-modules $K_X^{\bullet}$ (called a dualizing complex of X), which plays the part of Ω and a trace map $H_c^0(X,K_X^{\bullet})\to\mathbb{C}$. In fact A. Grothendieck defined the first dualizing complex in the algebraic geometry and the same general line was used in the analytic case [6]. As a remark we note that from only the existence of a dualizing complex one obtains an extension of some classical duality results.

We now give the precise statements of the Ramis-Ruget results [9] .

<u>THEOREM 1</u> (the existence of the Ramis-Ruget dualizing complex) : Let $(X,\mathcal{O})$ be a complex space. Then there exists a complex $K_X^{\bullet}$ of $\mathcal{O}$-modules such that :

(a) whenever X is a manifold of dimension n, $K^{\bullet}_X$ is a resolution of $\Omega^{\bullet}[n]$ and for any p, the stalks $K^p_{X,x}$ are injective $\mathcal{O}_x$-modules, for any $x \in X$.

(b) whenever $f : X \to Y$ is an immersion of complex spaces, there is an isomorphism $\bar{f} : K^{\bullet}_X \simeq \underline{Hom}_{\mathcal{O}_Y}(f_* \mathcal{O}_X, K^{\bullet}_Y) \mid X$ and the correspondence $f \mapsto \bar{f}$ agrees with the composition of immersions.

(c) the cohomology sheaves $\underline{H}^p(K^{\bullet}_X)$ are coherent for any p and $\underline{H}^p(K^{\bullet}_X) = o$ if $p < -\mathrm{prof}\, \mathcal{O}$.

(d) $K^{\bullet}_X$ is bounded if X is finite dimensional; namely $K^p_X = o$ for $p < -\dim X$ and for $p > o$.

THEOREM 2 (the first theorem of absolute duality on complex spaces):

Let X be a finite dimensional complex space with countable basis and $F \in \mathrm{Coh}(X)$. Then for any p there exist unique structures QFS on $H^p(X,F)$ and QDFS on $\mathrm{Ext}_c^{-p}(X;F,K^{\bullet}_X)$ such that the trace map defines a topological duality between the associated separated spaces, i.e. the $\mathbb{C}$-bilinear map

$$H^p(X,F) \times \mathrm{Ext}_c^{-p}(X;F,K^{\bullet}_X) \xrightarrow{\text{Yoneda}} H^o_c(X,K^{\bullet}_X) \xrightarrow{\text{Trace}} \mathbb{C}$$

induces a perfect coupling between $H^p(X,F)_{sep}$, $\mathrm{Ext}_c^{-p}(X;F,K^{\bullet}_X)_{sep}$.

THEOREM 3 (the second theorem of absolute duality on complex spaces):

Let X be a finite dimensional complex space with countable basis and $F \in \mathrm{Coh}(X)$. Then for any p there exist unique structures QDFS on $H^p_c(X,F)$ and QFS on $\mathrm{Ext}^{-p}(X;F,K^{\bullet}_X)$ such that the trace map induces a topological duality between the associated separated spaces.

In order to build their dualising complex, Ramis and Ruget considered first of all the case of germs of manifolds (V,x), by taking

$$K^{\bullet}_{V,x} = L^{\bullet}_{V,x} \otimes_{\mathcal{O}_{V,x}} \Omega_{V,x}$$

where $L^{\bullet}_{V,x}$ is the Cousin complex of modules over the regular local ring $\mathcal{O}_{V,x}$. Then, for any open subset U of a numerical space $\mathbb{C}^n$ one can define a complex $K^{\bullet}_U$, by glueing together the punctual constructions via Frish's noetherianity theorem. The complex $K^{\bullet}_U$ has "good behaviour" with respect to immersions $U \to U'$ (in sense of theorem 1,(b)) and by suitable immersions, the dualizing complex $K_{\bullet}$

is so obtained.

§2. In the book [3],ch.II a complex of sheaves $\hat{K}^{\bullet}_X$ on any complex space X was built, by making use of some facts of analytic local cohomology. F.Fouché has established in [4] the main result, namely he proved that this complex has all the stalks of its components injective and indeed, $K^{\bullet}_X$ is a dualizing complex.

Here is a sketch of this construction : let $V \subset \mathbb{C}^n$ be an open set and $F \in Coh(V)$; for any open subset $U \subset V$ we put

$\Phi^q(U)$ = the family of all closed analytic subsets of U of codimension $\geq q$.

Consider the complex $\hat{L}^{\bullet}_V$ defined by

$$0 \longrightarrow F \longrightarrow \underline{H}^0_{\Phi^0/\Phi^1}(F) \longrightarrow \underline{H}^1_{\Phi^1/\Phi^2}(F) \longrightarrow \cdots\cdots ,$$

where $\underline{H}^p_{\Phi^q/\Phi^{q+1}}(F)$ is the sheaf on V associated to the presheaf $U \longmapsto H^p_{\Phi^q(U)/\Phi^{q+1}(U)}(U,F)$. Denote $\hat{K}^{\bullet}_V = \hat{L}^{\bullet}_V \otimes_{\mathcal{O}_V} \Omega_V[n]$ and obtain a complex with a "good behaviour" with respect to immersions. For an arbitrary complex space $(X,\mathcal{O})$ one can glue together the complexes $K^{\bullet}_U$, U being an open set of X embedded in an open set V from a numerical space by an immersion $\varphi : U \longrightarrow V$,

$$\hat{K}^{\bullet}_U = \underline{Hom}_{\mathcal{O}_V}(\varphi_* \mathcal{O}_U, \hat{K}^{\bullet}_V) \mid U ;$$

one thus obtains a new dualizing complex $\hat{K}^{\bullet}_X$.

In order to prove that $\hat{K}^{\bullet}_X$ has the property of injectivity of the stalks of all components, F.Fouché used the following criterion : if A is a noetherian local regular ring and M is an A-module such that $Ext^1_A(A/I,M) = 0$ whenever I is an ideal of A generated by a regular sequence, then M is an injective A-module.

§3. Recently V.D.Golovine has given a different dualizing complex in a more direct way [5].

Let $(X,\mathcal{O})$ be a complex space and $q \geq 0$ an integer. In the paper [2] a sheaf D^qF is associated for any sheaf $F \in Coh(X)$, such that whenever $i : U \longrightarrow G$ is a closed immersion ($U \subset X$, $G \subset \mathbb{C}^n$ being open subsets), we have

$$D^qF \,|\, U \simeq \underline{Ext}^{n-q}_{\mathcal{O}_{\mathbb{C}^n}} (i_*F, \Omega)\,|\,U \quad ,$$

particularly $D^oF \,|\, U \simeq \underline{Ext}^{n}_{\mathcal{O}_{\mathbb{C}^n}} (i_*F,\Omega)\,|\,U$. By making use of Ramis-Ruget duality theorems one can easily prove that

$$D^qF \simeq \underline{Ext}^{-q}_{\mathcal{O}} (F,K^\bullet_X) \;, \text{ for any } F \in Coh(X) \quad (\text{see } [3]).$$

$\underline{Ext}_{\mathcal{O}_{\mathbb{C}^n}}(i_*F,\Omega)|U$ can be considered for any F, The groups analytic sheaf (not necessarily coherent), and this family can be glued together and again define a sheaf D^oF for any $\mathcal{O}$-module F.

If $0 \longrightarrow \mathcal{O} \longrightarrow C^\bullet(\mathcal{O})$ is the canonical flabby Godement resolution, then by putting

$$\widetilde{K}^{-p}_X = D^o(C^p(\mathcal{O})) \;,$$

a complex $\widetilde{K}^\bullet_X$ is thus obtained. Golovine proved that this complex is dualizing (in the sense that all its components have their stalks injective, the cohomology sheaves are coherent and much more, the first duality theorem holds).

The complex $\widetilde{K}^\bullet_X$ has the components themselves injective (hence their stalks are also injective) and to prove this another Golovine's result has to be mentioned : if $U \subset \mathbb{C}^n$ is open, then an $\mathcal{O}$-module I is injective if and only if I is flabby and the stalks of the sheaf $\underline{H}^o_S(I)$ are injective, for any closed subset $S \subset U$. By the way, from this criterion it follows a proof of the injectivity of the sheaf of germs of hyperfunctions as module over the sheaf of germs of holomorphic functions.

§4. We now make some remarks.

a) The complex $K^\bullet_X$ from §1 is embedded in $\widehat{K}^\bullet_X$ by a quasi-isomorphism whose arrows are all injective. The complex $K^\bullet_X$ has an algebraic nature and $\widehat{K}^\bullet_X$ has an analytic nature. An useful property of the complex $\widehat{K}^\bullet_X$

is the fact that its components can be endowed by TVS structures [4] while $\tilde{K}^\bullet_X$ has no similar property.

In the work [10] a theorem which exposes a little the dualizing complex $K^\bullet_X$ is proved, namely : there is a quasi-isomorphism $K^\bullet_X \longrightarrow D^{n,\bullet n}_X$ (the currents of bidegree n = dim X) which has injective arrows and is compatible with the immersions. F.Fouché has also proved the existence of a quasi-isomorphism $\hat{K}^\bullet_X \longrightarrow B^\bullet_X$ (the complex $B^\bullet_X$ is the Dolbeault complex with coefficients hyperfunctions on X), which extends the principal-value morphisme in sense of Herrera.

b) Here are some results of analytic geometry involving the dualising complexes.

First of all, in order to apply the duality it is very important to have separation criteria. In the paper [11] two such criteria are proved for $H^\bullet_c(X,F)$, $Ext^\bullet_c(X;F,K^\bullet_X)$, conjectured by Malgrange. A simpler proof of them can be found in [1] . As an application of the analytic duality and the separation one obtains an elegant proof of the following Malgrange-Siu's criterion :"If X is an irreducible non compact complex space of dimension n, then $H^n(X,F) = o$ for any coherent sheaf F on X " [11] .

In the works [10], [11] some relative duality theorems, with respect to a morphism, by making of an essential use of the complex $K^\bullet_X$ are proved; similar things are suggested in [5] for the complex $\tilde{K}^\bullet_X$.
All these facts are useful to obtain the deep significance of the convexity or concavity in complex analytic geometry [8].

The dualizing complex is also used in [1] in order to obtain duality theorems of type X mod Y (Y being a closed part of a complex space).

Other recent parers apply the dualizing complex in problems of Runge approximation or in the study of countable unions of Stein manifolds [7].

c) A natural question is whether do exist other dualizing complexes.

Let $(X, \mathcal{O})$ be a reduced complex space with countable topology and E the sheaf of germs of C^∞-differentiable functions on X with complex values. By Malgrange theorem the stalks E_x are flat $\mathcal{O}_x$-modules, for any $x \in X$. From the Dolbeault resolution

$$0 \longrightarrow \mathcal{O} \longrightarrow E \longrightarrow E^{o,1} \longrightarrow E^{o,2} \longrightarrow \quad \ldots.$$

one then obtains for any $F \in \mathrm{Coh}(X)$, an exact sequence

$$0 \longrightarrow F \longrightarrow F \otimes_{\mathcal{O}} E \longrightarrow F \otimes_{\mathcal{O}} E^{o,1} \longrightarrow F \otimes_{\mathcal{O}} E^{o,2} \longrightarrow \quad \ldots. \quad ,$$

hence a resolution of F by soft Fréchet sheaves. Such Dolbeault resolutions do not have "a good behaviour" with respect to immersions. It would be interesting to put in eviddence natural resolutions of flat soft Fréchet sheaves.

Recall that for any sheaf G on X, $\widetilde{G}$ means the flabby sheaf $\{U \longmapsto \prod_{x \in U} G_x \text{ , } U \subset X \text{ open}\}$; a canonical injection $G \longrightarrow \widetilde{G}$ holds.

Let $i : \mathcal{O} \longrightarrow E \longrightarrow \widetilde{E}$ be the composition of the inclusions and $E_1 = (\widetilde{E}/\mathcal{O})^\sim$. Denote $E_o = \widetilde{E}$, $d_o : \widetilde{E} \longrightarrow \widetilde{E}/\mathcal{O} \longrightarrow E_1$, $E_2 = (E_1/\mathrm{Im}d_o)^\sim$, $d_1 : E_1 \longrightarrow E_1/\mathrm{Im}d_o \longrightarrow E_2$, etc. In this way we obtain an exact sequence

$$(*) \qquad 0 \longrightarrow \mathcal{O} \longrightarrow E_o \xrightarrow{d_o} E_1 \xrightarrow{d_1} E_2 \longrightarrow \ldots.$$

<u>The objects of the resolution $(*)$ of the structural sheaf $\mathcal{O}$ are soft and flat $\mathcal{O}$-modules.</u>

Indeed the E_k's are $\widetilde{E}$- modules, hence E-modules, hence soft. On the other hand, we prove that $\widetilde{E}$ is a flat $\mathcal{O}$-module. Indeed, in [12], ch.I, it is proved the following flatness criterion : "Let A be a noetherian regular local ring and M an arbitrary A-module. M is A-flat if and only if for any A-sequence $a = \{a_1, \ldots. a_k\}$ which generates the ideal $\underline{a}$, it holds $\mathrm{Tor}_1^A(A/\underline{a}, M) = o$". We also recall that $\mathrm{Tor}_1^A(A/\underline{a}, M) \simeq R_M(a)/R_A(a).M$, where $R_M(a)$ means the A-module of the relations between $a_1, \ldots, a_k$ with coefficients in M.

The sheaf E is a flat $\mathcal{O}$-module (by Malgrange's theorem), hence for any $x \in X$ and for any $\mathcal{O}_x$-sequence a we have $R_{E_x}(a) \simeq R_{\mathcal{O}_x}(a).E_x$, whence $R_{\widetilde{E}_x}(a) \simeq R_{\mathcal{O}_x}(a).\widetilde{E}_x$ (it suffices to use trivial relations), that is $\mathrm{Tor}_1^{\mathcal{O}_x}(\mathcal{O}_x/\underline{a},\widetilde{E}_x) = o$ therefore $\widetilde{E}_x$ is $\mathcal{O}_x$-flat module, for any $x \in X$.

Apply for the inclusion $\mathcal{O}_x \to \widetilde{E}_x$ the following result of algebra : „if $A \to B$ is a faithfully flat morphisme of commutative unitary rings and M is an A-module such that there is an exact sequence of A-module $0 \to A \to B \to M \to 0$, then M is A-flat" and so, $\widetilde{E}_x/\mathcal{O}_x$ will be $\mathcal{O}_x$-flat, etc. We can also use the description of the sections of E_q's, regarded as objects of the Godement flabby canonical resolution.

BIBLIOGRAPHY

[1] A.Andreotti, C.Bănică - Relative duality on complex spaces, Revue Roum.de Math.pures et appl. Nr.9,1975; Nr.9,1976.

[2] A.Andreotti, A.Kas - Duality on complex spaces, Annali Sc.Norm. Sup.Pisa,27,1973.

[3] C.Bănică, O.Stănăşilă - Algebraic methods in the global theory of complex spaces - Ed.Academiei 1974 (English edition J.Wiley 1976).

[4] F.Fouché - Un complexe dualisant en géometrie analytique complexe, C.R.Acad.Sc.Paris,t.28o,1975.

[5] V.D.Golovine - The homology of the analytic sheaves (in russian), Dokl.Akad.Nauk SSSR,t.225,Nr.1,1975.

[6] R.Hartshorne - Residues and Duality,Lecture Notes 2o,Springer Verlag,1966.

[7] A.Markoe - Runge families and increasing unions of Stein spaces (will appear) .

[8] J.P.Ramis - Théorèmes de séparation et de finitude pour l'homologie et la cohomologie des espaces (p,q)-convexes-concaves,Annali Sc.Norm.Sup.Pisa,29,1975.

[9] J.P.Ramis,G.Ruget - Complexe dualisant et théorèmes de dualité en géomètrie analytique complexe,Publ.IHES, Nr.38,1971.

[10] J.P.Ramis,G.Ruget - Résidus et dualité,Inventiones math.26,1974.

[11] J.P.Ramis,G.Ruget,J.L.Verdier - Dualité relative en géométrie analytique complexe,Inventiones math.13,1971.

[12] J.C.Tougeron - Idéaux de fonctions différentiables,Springer - Verlag,1972.

Polytechnic Institute
Bucharest

--//--

THE 'RESTRAUM' PROBLEM FOR 1-CONVEX SPACES.

Vo Van Tan.

Univ.degli Studi
Ist.Mat."U.Dini"-Firenze

The restraum problem consists of removing a subvariety Y of pure codimension 1 from a given $\mathbb{C}$-analytic space X satisfying some nice property and asking then whether on the complementary space $Z := X \setminus Y$ that nice property is preserved.

That problem was originated in Algebraic Geometry where the nice property is nothing but to be affine. It was interesting because its solution (resp. the counterexample of it) is closely related to the positive (resp; negative) answer to the so called Hilbert 14th problem. We refer the reader to [1] for more details on this subject. Since the analogue for affine varieties in Analytic Geometry is precisely Stein spaces, the restraum problem is then transposed to the analytic case.

The first positive solution for the restraum problem was given by Simha [2] with the following result

Theorem1 [2]: Let X be a normal 2-dimensional Stein spaces and let Y be a subvariety of pure codim.1 in X. Then $Z := X \setminus Y$ is also a Stein space.

Dimensionwise, that result is sharp since one can look at the following counterexample:

Let $X := \{ z w + z w = o \} \subset \mathbb{C}^4$ where z_1 z_2 w_1 & w_2 are the usual coordinates of $\mathbb{C}^4$. It is clear that X is a normal 3-dimensional Stein space with only one singular point, namely the origin. Let $Y := \{ z_1 = z_2 = o \}$. It is obvious that Y is a (non singular) subvariety of codim.1 in X. However one can see easily that $Z := X \setminus Y$ is not Stein!

Here our purpose is to generalize Simha's result to the class of $\mathbb{C}$-analytic spaces which are 1-convex.

Definition 2: A $\mathbb{C}$-analytic space X is said to be 1-convex if

i) X is holomopphically convex.

ii) X admits a maximal compact analytic subvariety S (i.e. S is a compact analytic subvariety without isolated points and for any compact analytic subvariety without isolated points T in X, then necessarily $T \subset S$)

Example 3: Let X be the blowing up of $\mathbb{C}^n$ at the origin and let S be the proper transform of the origin. ($n \geqslant 2$), then certainly X is a 1-convex space with its maximal compact subvariety $S \cong \mathbb{P}_{n-1}$.

We are now in a position to state our main result:

Theorem 4: Let X be a normal 2-dimensional 1-convex space and let Y be a non compact subvariety of pure codim.1, then $Z := X \setminus Y$ is also 1-convex.

Notice that in Theorem 4 we do not assume Y to be irreducible but it is necessary that Y should be non compact as one can convince oneself by the example 3. In fact there, X is 1-convex and $Y := S = \mathbb{P}_{n-1}$ is a subvariety of codim.1, but $Z := X \setminus Y \cong \mathbb{C}^n \setminus \{o\}$ is not holomorphically convex for $n \geqslant 2$.

Complete proof and more related results to Theorem 4 as well as their analogues in algebraic geometry will appear in [3] .

References.

[1] Nagata, M. : Lectures on the 14th problem of Hilbert. Tata inst. publ. Bombay; no. 31 (1965).

[2] Simha, R. : On the complement of a curve on a Stein space of dimension 2. Math. Zeitsch. 82 (1963)

[3] Vo Van Tan : On the complement of a divisor on a 1-convex space and its analogue in algebraic geometry. To appear.

-.-.-.-.-.-

Ein inverses Problem der Wärmeleitungsgleichung, I.

von

Gottfried Anger in Halle a. d. Saale

Einleitung

In der vorliegenden Arbeit wird die von G. Anger [4] - [9] entwickelte und von B.-W. Schulze [25] - [28] ausgebaute Methode zum Studium inverser Probleme auf die Wärmeleitungsgleichung übertragen. Es handelt sich hierbei um denjenigen Problemkreis, bei dem man aus der Kenntnis gewisser Randwerte die rechte Seite einer Differentialgleichung zu bestimmen versucht. Die Idee besteht darin, am Anfang der Untersuchungen alle positiven Maße ν zu betrachten, die außerhalb eines vorgegebenen Gebietes $\Omega \subset R^n$ das gleiche Potential $\Phi\nu$ erzeugen (Φ Fundamentallösung). Dabei ist meistens eine positive Massenverteilung μ auf dem Rand $\partial\Omega$ vorgegeben. In den Arbeiten [4] - [9] wurde in einigen Beweisen von folgender grundlegenden Eigenschaft der Lösungen elliptischer Differentialgleichungen Gebrauch gemacht: Stimmen zwei auf einem Gebiet Ω_0 erklärte Lösungen u_1 und u_2 auf einer offenen Teilmenge überein, so gilt $u_1 = u_2$ auf Ω_0. Diese Schlußweise ist im Fall der Wärmeleitungsgleichung für das Cauchy-Problem bezüglich des Gebietes $\Omega_{0,T} = \{x \in R^n, 0 < x_n < T\}$ gültig. Daher lassen sich viele früher gewonnenen Ergebnisse auf diesen Spezialfall übertragen, was der Zweck der vorliegenden Arbeit ist. Im Fall eines Zylinders $\Omega_{0,T} = S \times (0,T)$, S Gebiet im R^{n-1}, wird obige Schlußweise durch Betrachtungen über die Eindeutigkeit des Cauchy-Problems [23] ersetzt. Diese Ergebnisse werden im Teil II dieser Arbeit veröffentlicht. Zusammenfassend ist der Inhalt der vorliegenden Arbeit folgender: Im Abschnitt 1 stellen wir die wichtigsten Ergebnisse über Maße, gewisse Räume stetiger Funktionen, Potentiale der Wärmeleitungsgleichung und einen Kapazitätsbegriff zusammen. Im Abschnitt 2 zeigen wir, daß sich jede auf der Ebene $x_n = \tau$ stetige Funktion f, die im unendlichfernen Punkt verschwindet, durch adjungierte Potentiale $\Phi^*\nu$, deren Maße ν auf der Ebene $x_n = T$, $\tau < T$, gelegen sind, gleichmäßig approximieren läßt. Im Abschnitt 3 stellen wir die wichtigsten Ergebnisse über das Cauchy-Problem für die Ebene $x_n = \tau$ zusammen und führen das Balayage-Prinzip als adjungierte Abbildung ein.

Damit wird der Zusammenhang zu unseren früheren Untersuchungen hergestellt. Im Abschnitt 4 folgen Ergebnisse über das bereits erwähnte inverse Problem bezüglich $\Omega_{0,T}$. Es sei μ ein auf der Ebene $x_n = T$ gelegenes positives Maß, $\mathcal{L}(\mu)$ die Menge aller auf $\overline{\Omega_{0,T}}$ gelegenen positiven Maße, deren Potential für $x_n > T$ gleich dem Potential $\Phi\mu$ ist. Nach Satz 4 ist $\mathcal{L}(\mu)$ eine konvexe, schwach kompakte Menge. Jedes Maß $\nu \in \mathcal{L}(\mu)$ ist daher nach dem Satz von Krein-Milman-Choquet als ein gewisses Integral über die extremalen Maße von $\mathcal{L}(\mu)$ darstellbar (Satz 5). Daher spielen die extremalen Maße von $\mathcal{L}(\mu)$, von denen einige Klassen in den Sätzen 6 - 8 charakterisiert werden, eine zentrale Rolle. Nach unserer Auffassung sind die extremalen Maße von Satz 7 und Satz 8 von besonderem Interesse. Die Charakterisierung aller extremalen Maße ist bisher nicht gelungen.

Wegen weiterer Literatur über inverse Probleme, speziell bei der Wärmeleitungsgleichung, vergleiche man [19] - [24], [30]. Allgemeine Betrachtungen über das Dirichlet-Problem und das Cauchy-Problem der Wärmeleitungsgleichung findet man in [10], [15], [18].

1. Bezeichnungen

Wir bezeichnen mit R^n den n-dimensionalen euklidischen Raum, mit $x = (x_1,\dots,x_n)$, $y = (y_1,\dots,y_n)$ Punkte des R^n. In den weiteren Ausführungen betrachten wir Lösungen u der Wärmeleitungsgleichung

$$(1) \qquad Lu = \frac{\partial u}{\partial x_n} - \Delta_{n-1} u = 0$$

und der adjungierten Wärmeleitungsgleichung

$$(2) \qquad L^{*}u = \frac{\partial u}{\partial x_n} + \Delta_{n-1} u = 0 .$$

Hierbei ist

$$\Delta_k u = \sum_{i=1}^{k} \frac{\partial^2 u}{\partial x_i^2}$$

der Laplace-Operator. Es sei $C(R^n)$ der Raum aller im R^n erklärten endlichen, stetigen Funktionen, $C_\omega(R^n)$ der Raum aller $f \in C(R^n)$ mit

$f(x) \to 0$ für $x \to \omega$ (ω unendlichferner Punkt). Weiter sei $C_0(R^n)$ der Raum aller $f \in C(R^n)$ mit kompaktem Träger

$$\text{supp } f = \overline{\{x : f(x) \neq 0\}},$$

$C_0(K,R^n)$ der Raum aller $f \in C_0(R^n)$ mit supp $f \subset K \subset R^n$. Auf $C_\omega(R^n)$ führen wir die Norm

$$(3) \qquad \|f\| = \sup\{|f(x)|, \quad x \in R^n\}$$

ein. Damit wird $C_\omega(R^n)$ ein Banach-Raum. Es gilt

$$(4) \qquad \overline{C_0(R^n)} = C_\omega(R^n).$$

In $C_\omega(R^n)$ existiert eine abzählbare, dichte Menge, die man unter Verwendung der Polynome mit rationalen Koeffizienten und des Lemmas von Urysohn konstruieren kann.

Unter einem Radon-Maß verstehen wir eine Linearform [11], [16]

$$\mu : C_0(R^n) \to R^1,$$

für welche zu jeder kompakten Menge $K \subset R^n$ eine Konstante $M_K \geq 0$ derart existiert, so daß

$$(5) \qquad |\mu(f)| \leq M_K \|f\|$$

gilt. Für $\mu(f)$ schreibt man auch $\int f d\mu$. Das Riemann-Integral ist ein spezielles Radon-Maß. Mit $\mathfrak{M}(R^n)$ bezeichnen wir den Raum aller Radon-Maße, mit $\mathfrak{M}_0(R^n)$ den Raum aller Maße mit kompaktem Träger. Dabei versteht man unter dem Träger supp μ eines Maßes μ das Komplement der größten offenen Menge $\Omega \subset R^n$, so daß $\mu(f) = 0$ für alle f mit supp $f \subset \Omega$ gilt. Es handelt sich bei dem Träger eines Maßes um diejenige Teilmenge des R^n, auf welcher die Masse konzentriert ist. Jedes Maß μ kann als Differenz zweier ausgezeichneter positiver Maße μ^+ und μ^- dargestellt werden

$$\mu = \mu^+ - \mu^-.$$

Setzt man

$$\|\mu\| = \sup\{|\mu(f)|, \ \|f\| \leq 1\},$$

so gilt

$$\|\mu\| = \|\mu^+\| + \|\mu^-\| .$$

Wegen (4) besteht der zu $C_\omega(R^n)$ duale Raum $C_\omega^*(R^n)$ aus allen Maßen μ mit $\|\mu\| < \infty$. Wir betrachten das Dualsystem

$$\langle C_\omega(R^n),\ C_\omega^*(R^n)\rangle$$

bezüglich der Bilinearform

$$(6) \qquad (f,\mu) \to \langle f,\mu\rangle = \int f d\mu , \quad \|\mu\| < \infty ,$$

und die mittels dieser Bilinearform auf $C_\omega^*(R^n)$ erklärte schwache Topologie [12], [16]. Die abgeschlossene Kugel $B_M \subset C_\omega^*(R^n)$ aller Maße μ mit $\|\mu\| \leq M$ ist schwach kompakt [12]. Da in $C_\omega(R^n)$ eine abzählbare, dichte Teilmenge existiert, ist die schwache Topologie auf B_M metrisierbar [12], [16].

Man kann die voranstehenden Räume auch für eine offene Menge $\Omega \subset R^n$ einführen und Ω als lokalkompakten Raum auffassen.

Die Funktion E mit

$$(7) \qquad E(x) = \left(\frac{1}{2\sqrt{\pi x_n}}\right)^{n-1} \exp\left[-\frac{\sum_{i=1}^{n-1} x_i^2}{4x_n}\right], \quad x_n > 0$$

$$E(x) = 0 , \quad x_n \leq 0$$

ist eine Grundlösung von (1), d. h. es gilt, wenn $\Phi(x,y) = E(x - y)$ gesetzt wird,

$$(8) \qquad \int \Phi(x,y) L^* \varphi(x) dx = \varphi(y) \quad \text{für jedes} \quad \varphi \in C_0^\infty(R^n)$$

und jedes $y \in R^n$. Hierbei ist $C_0^\infty(R^n)$ der Raum aller $\varphi \in C_0(R^n)$ mit stetigen partiellen Ableitungen beliebiger Ordnung.

Die Lösung des Cauchy-Problems der Wärmeleitungsgleichung für die Ebene

$$I_n = \{x \in R^n ,\quad x_n = T\}$$

hat die Gestalt [17], [18], [29]

(9) $$u(x) = \int_{I_n} f(y_1,\ldots,y_{n-1})\, \Phi(x,y)\,dy_1\ldots dy_{n-1}\,.$$

Entsprechend lautet die Lösung des Cauchy-Problems für die adjungierte Wärmeleitungsgleichung

(10) $$u(x) = \int_{I_n} f(y_1,\ldots,y_{n-1})\,\Phi(y,x)\,dy_1\ldots dy_{n-1}\,.$$

Für ein positives Maß μ wird das Potential $\Phi\mu$ im Punkt $x \in R^n$ mit Hilfe des Integrals

(11) $$\Phi\mu(x) = \int \Phi(x,y)\,d\mu(y)$$

erklärt. Für ein beliebiges Maß $\mu = \mu^+ - \mu^-$ setzt man

$$\Phi\mu(x) = \Phi\mu^+(x) - \Phi\mu^-(x)\,.$$

Da der Kern $\Phi \geqq 0$ nach unten halbstetig ist, ist für ein positives Maß μ das Potential $\Phi\mu$ ebenfalls nach unten halbstetig [6], [13], [16]. Das adjungierte Potential $\Phi^*\mu$ wird mit Hilfe des Integrals

$$\Phi^*\mu(y) = \int \Phi(x,y)\,d\mu(x) = \int \Phi^*(y,x)\,d\mu(x)$$

eingeführt.

In [2] wurde für den Kern Φ ein Kapazitätsbegriff eingeführt. Eine Borelsche Menge $B \subset R^n$ ist genau dann von der Φ-Kapazität Null, wenn jedes Maß μ mit $\operatorname{supp}\mu \subset B$ ein unstetiges Potential erzeugt. Bezeichnet $F^+(\Phi)$ die Menge aller positiven Maße mit kompaktem Träger und stetigem Potential, so ist B genau dann von der Φ-Kapazität Null, wenn

$$\lambda(B) = \int 1_B(x)\,d\lambda(x) = 0 \quad \text{für jedes } \lambda \in F^+(\Phi)$$

gilt. Hierbei ist 1_B die charakteristische Funktion von B, d. h. $1_B(x) = 0$ für $x \notin B$, $1_B(x) = 1$ für $x \in B$. Es besteht die Beziehung $F^+(\Phi) = F^+(\Phi^*)$. Jede Ebene I_n ist von der Φ-Kapazität Null [2]. Weiter wurde in [2], [18] gezeigt, daß für festes y stets

(12) $$\Phi(x,y) \to 0 \quad \text{für } x \to \omega$$

gilt. Hieraus folgt für $\mu \in \mathfrak{M}_0(R^n)$

(12') $\Phi\mu(x) \to 0$ für $x \to \omega$.

Daher gilt für $\lambda \in F^+(\Phi)$ stets $\Phi\lambda \in C_\omega(R^n)$. Für ein Maß μ mit $\|\mu\| < \infty$ existiert das Integral $\int \Phi\lambda \, d\mu$, und es gilt

$$\text{(13)} \qquad \begin{aligned} \int \Phi^* \mu \, d\lambda &= \int \Phi\lambda \, d\mu = \mu(\Phi\lambda)\,, \\ \int \Phi\mu \, d\lambda &= \int \Phi^*\lambda \, d\mu = \mu(\Phi^*\lambda)\,. \end{aligned}$$

Für ein Maß $\mu \geqq 0$ mit $\|\mu\| < \infty$ ist daher die Menge aller $x \in R^n$ mit $\Phi\mu(x) = \infty$ von der Φ-Kapazität Null.

Unsere späteren Überlegungen beziehen sich auf die Menge

$$\Omega_{0,T} = \{x \in R^n, \quad 0 < x_n < T\}.$$

Wir setzen

$$\begin{aligned} \Omega_{T+} &= \{x \in R^n, \ x_n > T\}, \\ \Omega_{T-} &= \{x \in R^n, \ x_n < T\}, \\ \partial\Omega_T &= \{x \in R^n; \ x_n = T\}. \end{aligned}$$

Es gilt

$$\Omega_{0,T} = \Omega_{0+} \cap \Omega_{T-}\,, \qquad \partial\Omega_T = \partial\Omega_{T+} = \partial\Omega_{T-}\,.$$

Mit $(\operatorname{supp}\mu)_n$ bezeichnen wir die Projektion von $\operatorname{supp}\mu$ auf die x_n-Achse; A_n sei das Supremum von $(\operatorname{supp}\mu)_n$. Weiter betrachten wir folgende Mengen von Maßen bzw. Potentialen

$$\begin{aligned} \mathfrak{M}_o(\Omega_{T-}) &= \{\nu \in \mathfrak{M}_o(R^n), \ A_n < T\}\,, \\ \mathfrak{M}_o(\Omega_{T+}) &= \{\nu \in \mathfrak{M}_o(R^n), \ A_n > T\}\,, \\ H(\Omega_{T-}) &= \{\Phi\nu, \ \nu \in \mathfrak{M}_o(\Omega_{T-})\}\,, \\ H^*(\Omega_{T+}) &= \{\Phi^*\nu, \ \nu \in \mathfrak{M}_o(\Omega_{T+})\}\,, \\ H(\partial\Omega_T) &= \{\Phi\nu, \ \nu \in \mathfrak{M}_o(R^n), \ A_n = T\}\,, \\ H^*(\partial\Omega_T) &= \{\Phi^*\nu, \ \nu \in \mathfrak{M}_o(R^n), \ A_n = T\}. \end{aligned}$$

Betrachten wir in den voranstehenden Räumen nur Maße der Gestalt

$$\nu = \sum_{k=1}^{N} c_k \, \delta_{y^{(k)}} , \quad \delta_{y^{(k)}}(f) = f(y^{(k)}) ,$$

so schreiben wir $H_\delta(\Omega_T-)$, $H^*_\delta(\Omega_T+)$, $H_\delta(\partial\Omega_T)$, $H^*_\delta(\partial\Omega_T)$.
Bezeichnet B eine der Mengen Ω_T+, $\partial\Omega_T$ und A eine der Mengen $\Omega_\tau-$, $\partial\Omega_\tau$, $\tau \leq T$, so sei $H^*(A,B)$ die Menge der Einschränkungen $f|_A$ von Funktionen $f \in H^*(B)$ auf A. Entsprechend sei $H(B,A)$ die Menge der Einschränkungen $f|_B$ von Funktionen $f \in H(A)$.

Weiter sei $C_\omega(\overline{\Omega_T-})$ der Raum aller auf $\overline{\Omega_T-}$ erklärten stetigen Funktionen mit $f(x) \to 0$ für $x \to \omega$, $x \in \overline{\Omega_T-}$, $D_\omega(\overline{\Omega_T-})$ der Raum aller $f \in C_\omega(\overline{\Omega_T-})$, die in Ω_T- der adjungierten Wärmeleitungsgleichung $L^*f = 0$ genügen. Es sei $\mathcal{L}$ eine auf $C_\omega(\overline{\Omega_T-})$ stetige Linearform. Wegen $\|\mathcal{L}\| < \infty$ kann man $\mathcal{L}$ eine auf $C_\omega(R^n)$ stetige Linearform $\tilde{\mathcal{L}}$ mit

$$\tilde{\mathcal{L}}(f) = \mathcal{L}(f|_{\overline{\Omega_T-}})$$

zuordnen. Es ist dann $\tilde{\mathcal{L}}$ bzw. $\mathcal{L}$ mit Hilfe eines auf $\overline{\Omega_T-}$ gelegenen Maßes μ mit $\|\mu\| < \infty$ darstellbar. Also besteht $C^*_\omega(\overline{\Omega_T-})$ aus allen Maßen μ mit $\operatorname{supp}\mu \subset \overline{\Omega_T-}$ und $\|\mu\| < \infty$. Entsprechende Aussagen bestehen für die Räume $C_\omega(\overline{\Omega_T+})$ und $C_\omega(\partial\Omega_T)$. Der Raum $D_\omega(\overline{\Omega_T+})$ besteht aus allen $f \in C_\omega(\overline{\Omega_T+})$, die der Wärmeleitungsgleichung $Lf = 0$ genügen.

2. Ein Approximationssatz

Wir wollen jetzt einen Approximationssatz beweisen. Diese Beweistechnik wurde schon früher für die Laplace-Gleichung verwendet (siehe zum Beispiel [2], [3]). Im Teil II dieser Arbeit wird dieser Satz für einen Zylinder $S \times (0,T)$, $S \subset R^{n-1}$, bewiesen.

Satz 1: Der Raum $H^*(\partial\Omega_\tau;\partial\Omega_T)$, $\tau < T$, ist dicht in $C_\omega(\partial\Omega_\tau)$; der Raum $H(\partial\Omega_T, \partial\Omega_\tau)$ ist dicht in $C_\omega(\partial\Omega_T)$.

Beweis: Die Potentiale $\Phi^*\nu$, $\operatorname{supp}\nu \subset \partial\Omega_T$ kompakt, erzeugen auf $\partial\Omega_\tau$ stetige Funktionen. Angenommen, es sei

$$\overline{H^*(\partial\Omega_\tau;\partial\Omega_T)} \neq C_\omega(\partial\Omega_\tau).$$

Dann existiert nach dem Satz von Hahn-Banach eine auf $C_\omega(\partial\Omega_\tau)$ stetige Linearform $\mathcal{L}$ mit $\mathcal{L}(f) = 0$ für jedes $f \in \overline{H^*(\partial\Omega_\tau;\partial\Omega_T)}$. Nach den früheren Bemerkungen kann $\mathcal{L}$ mit Hilfe eines auf $\partial\Omega_\tau$ gelegenen Maßes $\mu \neq 0$ dargestellt werden, welches von der Gestalt

$$\mu = \mu^+ - \mu^-\,,\quad \mu^+, \mu^- \geqq 0\,,$$

ist. Es gilt also

$$\int f d\mu^+ = \int f d\mu^- \quad \text{für jedes } f \in H^*(\partial\Omega_\tau;\partial\Omega_T).$$

Wegen (12) kann man für f folgende Funktion einsetzen:

$$f(y) = \Phi(x,y)\,,\quad y \in \partial\Omega_\tau\,,\quad x \in \partial\Omega_T\,.$$

Hieraus folgt

$$\Phi\mu^+(x) = \Phi\mu^-(x) \quad \text{für jedes } x \in \partial\Omega_T\,.$$

Für alle x mit $x_n > \tau$ genügen obige Potentiale der Wärmeleitungsgleichung. Da das Cauchy-Problem für diese Potentiale eindeutig lösbar ist, gilt obige Gleichheit für alle x mit $x_n > T$. Aus der Analytizität dieser Potentiale für $x_n > \tau$ und $\Phi(x,y) = 0$ für $x_n \leqq \tau$ folgt

$$\Phi\mu^+(x) = \Phi\mu^-(x) \quad \text{für alle } x \in R^n\,.$$

Wir multiplizieren diese Gleichheit mit $L^*\varphi$ und integrieren die so entstehende Beziehung über den R^n. Hieraus folgt

$$\int \Phi\mu^+(x)L^*\varphi(x)dx = \int \Phi\mu^-(x)L^*\varphi(x)dx\,.$$

Unter Verwendung des Satzes von Fubini und der Beziehung (8) folgt

(14) $$\int \Phi\mu^+(x)L^*\varphi(x)dx = \int\left(\int \Phi(x,y)L^*\varphi(x)dx\right)d\mu^+(y)$$
$$= \int \varphi(y)d\mu^+(y) = \int \Phi\mu^-(x)L^*\varphi(x)dx = \int \varphi(y)d\mu^-(y)\,.$$

Daher gilt

$$\int \varphi(y)d\mu^+(y) = \int \varphi(y)d\mu^-(y) \quad \text{für jedes } \varphi \in C_0^\infty(R^n)\,.$$

Da der Raum $C_0^\infty(R^n)$ in $C_\omega(R^n)$ dicht ist, gilt

$$\mu^+ = \mu^- \quad \text{oder} \quad \mu = \mu^+ - \mu^- = 0\,,$$

was einen Widerspruch zur Annahme darstellt. Also ist

$$\overline{H^*(\partial\Omega_\tau\,;\partial\Omega_T)} = C_\omega(\partial\Omega_\tau)\ .$$

Entsprechend beweist man

$$\overline{H(\partial\Omega_T;\partial\Omega_\tau)} = C_\omega(\partial\Omega_T)\ .$$

Folgerung 1: Es ist $H^*(\partial\Omega_\tau\,;\Omega_{T+})$, $\tau < T$, dicht in $C(\partial\Omega_\tau)$ und entsprechend $H(\partial\Omega_T;\Omega_{\tau-})$ dicht in $C(\partial\Omega_T)$.

Beweis: Für $T_1 > T$ gilt

$$H^*(\partial\Omega_\tau\,;\partial\Omega_{T_1}) \subset H^*(\partial\Omega_\tau;\Omega_{T_1+})\ .$$

Aus Satz 1 folgt für $T = T_1$ die Dichtheit.

Folgerung 2: Satz 1 und Folgerung 1 gelten ebenfalls für die Räume $H^*_\delta(\partial\Omega_\tau\,;\partial\Omega_T)$, $H_\delta(\partial\Omega_T;\partial\Omega_\tau)$, $H^*(\partial\Omega_\tau\,;\Omega_{T+})$, $H_\delta(\partial\Omega_T;\Omega_{\tau-})$.

Beweis: Beim Beweis von Satz 1 wurden nur die speziellen Maße $\nu = \delta_y$, $y \in \partial\Omega_T$, verwendet. Daher ist

$$\overline{H^*_\delta(\partial\Omega_\tau\,;\partial\Omega_T)} = C_\omega(\partial\Omega_\tau)\ .$$

3. Cauchy-Problem und Balayage-Prinzip

Das Cauchy-Problem ist eine korrekt gestellte Aufgabe. Für jedes $g \in C_\omega(\partial\Omega_T)$ ist nach (10) die Lösung v des Cauchy-Problems der adjungierten Wärmeleitungsgleichung ein Element von $D_\omega(\overline{\Omega_{T-}})$, [10], [17], [18]. Es sei

$$I(\partial\Omega_T) = \{f \in C_\omega(\overline{\Omega_{T-}})\ ,\quad f(y) = 0 \ \text{ auf } \partial\Omega_T\}.$$

Setzt man für $f \in C_\omega(\overline{\Omega_{T-}})$

$$Pf = v\ ,$$

wobei v die Lösung des Cauchy-Problems mit den Anfangswerten $g = f|_{\partial\Omega_T}$ ist, so gilt wie im Fall des Dirichlet-Problems der Laplace-Gleichung

$$C_\omega(\overline{\Omega_{T-}}) = D_\omega(\overline{\Omega_{T-}}) \oplus I(\partial\Omega_T)\ .$$

Das Symbol $\oplus$ bedeutet, daß jedes $f \in C_\omega(\overline{\Omega_T})$ in der Form

$$f = v + h\ ,\quad v \in D_\omega(\overline{\Omega_{T-}})\ ,\quad h \in I(\partial\Omega_T)\ ,$$

darstellbar ist und die beiden Räume nur das Nullelement gemeinsam haben. Wegen

$$\sup\{|Pf(x)|,\ x\in\overline{\Omega_{T^-}}\} = \sup\{|v(x)|,\ x\in\overline{\Omega_{T^-}}\}$$
$$\leqq \sup\{|v(y)|,\ y\in\partial\Omega_T\} \leqq \sup\{|v(y)+h(y)|,\ y\in\partial\Omega_T\}$$
$$\leqq \sup\{|v(x)+h(x)|,\ x\in\overline{\Omega_{T^-}}\}$$

ist P eine stetige Projektion auf $C_\omega(\overline{\Omega_{T^-}})$. Bezüglich des Dualsystems

$$\langle C_\omega(\overline{\Omega_{T^-}}),\ C^*_\omega(\overline{\Omega_{T^-}})\rangle$$

mit

$$\langle f,\nu\rangle = \int f d\nu,\quad \operatorname{supp}\nu\subset\overline{\Omega_{T^-}},$$

wird die adjungierte Abbildung Π durch die Beziehung

$$\langle Pf,\nu\rangle = \langle f,\Pi\nu\rangle$$

erklärt. Wegen Pf = 0 für jedes $f\in I(\partial\Omega_T)$ gilt $\operatorname{supp}\Pi\nu\subset\partial\Omega_T$. Setzt man $\mu = \Pi\nu$, so besteht das

<u>Balayage-Prinzip (Integralform):</u>

Zu jedem Maß ν mit $\operatorname{supp}\nu\subset\overline{\Omega_{T^-}}$ existiert ein auf $\partial\Omega_T$ gelegenes Maß $\mu = \Pi\nu$, so daß

(15) $\quad \int f d\nu = \int f d\mu$ für jedes $f\in D_\omega(\overline{\Omega_{T^-}})$

gilt.

Wir formulieren ein Balayage-Prinzip mit Hilfe von Potentialen. In Satz 2 beweisen wir die Äquivalenz beider Prinzipien.

<u>Balayage-Prinzip:</u>

Zu jedem Maß ν mit $\operatorname{supp}\nu\subset\overline{\Omega_{T^-}}$ existiert genau ein auf $\partial\Omega_T$ gelegenes Maß μ, so daß

(16) $\quad \Phi\nu(x) = \Phi\mu(x)$ für jedes x mit $x_n > T$

gilt.

<u>Anmerkung 1:</u> Man kann entsprechend das Balayage-Prinzip für die adjungierte Wärmeleitungsgleichung bezüglich Ω_{T^+} und $D_\omega(\overline{\Omega_{T^+}})$ formulieren und beweisen.

Satz 2: Die Beziehungen (15) und (16) sind äquivalent.

Beweis: a) Es sei $x \notin \overline{\Omega}_{T^-}$. Dann ist f mit $f(y) = \Phi(x,y)$ nach (12) ein Element von $C_\omega(\overline{\Omega}_{T^-})$. Daher folgt aus (15) stets (16).

b) Die Beziehung (16) gelte für jedes $x \notin \overline{\Omega}_{T^-}$, speziell für jedes $x \in \partial\Omega_{T_1}$, $T_1 > T$. Die Menge $H^*_\delta(\partial\Omega_T, \partial\Omega_{T_1})$ ist nach Satz 1 dicht in $C_\omega(\partial\Omega_T)$. Daher läßt sich jedes $f \in C(\partial\Omega_T)$ durch Linearkombinationen der Gestalt

$$v_m(y) = \sum_{k=1}^{m} c_k \Phi(z^{(k)}, y), \quad z^{(k)} \in \partial\Omega_{T_1},$$

gleichmäßig approximieren. Für $y_n < T$ genügen die v_m der adjungierten Wärmeleitungsgleichung. Aus der gleichmäßigen Konvergenz von v_m auf $\partial\Omega_T$ folgt nach dem Maximum-Minimum-Prinzip die gleichmäßige Konvergenz von v_m auf $\overline{\Omega}_{T^-}$ gegen eine Lösung $v \in D_\omega(\overline{\Omega}_{T^-})$ der adjungierten Wärmeleitungsgleichung. Hieraus folgt

$$\int v\,d\nu = \int v\,d\mu \quad \text{für jedes} \quad v \in D_\omega(\overline{\Omega}_{T^-}).$$

Man vergleiche zu den voranstehenden Ausführungen auch [10], [15].

4. Ein inverses Problem für einen Streifen

Es sei $S \subset R^{n-1}$ ein Gebiet,

$$\Omega_{0,T} = S \times (0,T), \quad 0 < T,$$

ein offener Zylinder im R^n, $\partial S \times [0,T]$ der Mantel des Zylinders, $S_0 = S \times \{0\}$, $S_T = S \times \{T\}$, $\partial S_0 = \partial S \times \{0\}$, $\partial S_T = \partial S \times \{T\}$.

In den Anwendungen ist folgendes Problem von Interesse: Gegeben ist eine Funktion $f \in C(\overline{S}_T)$, (oder $L^2(\overline{S}_T)$) mit $f(y) = 0$ auf ∂S_T. Gesucht ist eine Funktion $g \in C(\overline{S}_0)$ mit $g(y) = 0$ auf ∂S_0, so daß die Lösung u des Cauchy-Problems mit den Anfangswerten g und $u(y) = 0$ auf $\partial S \times [0,T]$ die Funktion f auf S_T erzeugt, d. h. $f = u|_{\partial S_T}$. Bekanntlich ist dieses Problem nicht für jedes $f \in C(S_T)$ lösbar. Man kann aber f durch geeignete Lösungen u beliebig genau approximieren [21], [23]. Einen Beweis für diesen Sachverhalt geben wir im Teil II dieser Arbeit. Der Beweis von Satz 1 läßt sich formal auf diesen Sachverhalt übertragen, indem man an Stelle von die Greensche Funktion von $\Omega_{0,T}$ verwendet.

Wir wollen hier ein anderes inverses Problem für einen Streifen studieren [20], [23]. Gesucht sind die in $\Omega_{0,T}$ gelegenen Wärmequellen, die außerhalb $\Omega_{0,T}$ (genauer für $x_n > T$) eine vorgegebene Lösung u der Wärmeleitungsgleichung erzeugen. Wir folgen hier früheren Überlegungen, die wir für die Laplace-Gleichung angestellt haben [4] - [9]. Dabei ist u das Potential eines auf dem Rand $\partial\Omega_T$ gelegenen positiven Maßes μ. Auf diese Weise lassen sich die Ergebnisse der modernen Potentialtheorie anwenden [2], [6].

Inverses Problem (Inverses Balayage-Prinzip):

Es sei μ ein auf $\partial\Omega_T$ gelegenes positives Maß mit $\|\mu\| < \infty$. Es ist die Menge $\mathcal{L}(\mu)$ aller positiven Maße ν mit $\operatorname{supp}\nu \subset \overline{\Omega_{0,T}}$ und

(17) $\Phi\nu(x) = \Phi\mu(x)$ für jedes x mit $x_n > T$

zu studieren.

Anmerkung 2: Nach Satz 2 ist (17) äquivalent

(18) $\int f d\nu = \int f d\mu$ für jedes $f \in D_\omega(\overline{\Omega_T-})$.

Anmerkung 3: Die Funktion $\Phi\mu$ ist für $x_n \leqq T$ Null. Daher gilt

$$\Phi\nu(x) \geqq \Phi\mu(x) \quad \text{für alle } x \in R^n.$$

Satz 3: Es gilt $\|\nu\| = \|\mu\|$ für jedes $\nu \in \mathcal{L}(\mu)$. Hieraus folgt $\mathcal{L}(\mu) \subset C^*_\omega(R^n)$.

Beweis: Auf der Ebene $I_n = \{y \in R^n,\ y_n = T\}$ betrachten wir, wenn $y^T = (0,\dots,0,T)$ gesetzt wird, die Teilmenge

$$K_m(y^T) = \{y \in R^n,\ y_n = T,\ |y - y^T| \leqq m\}$$

und das Maß

$$d\lambda_m(y) = \varphi_m(y)dy_1\dots dy_{n-1}.$$

Hierbei ist $0 \leqq \varphi_m \leqq 1$, stetig auf I_n,

$$\varphi_m(y) = 1 \quad \text{für } y \in K_m$$

$$\varphi_m(y) = 0 \quad \text{für } y \notin K_{m+1}.$$

Für $m \to \infty$ strebt φ_m auf I_n monoton wachsend gegen die Funktion 1. Das Potential $\bar{\Phi}^* \lambda_m$ ist für $x_n < T$ stetig und strebt für $m \to \infty$ monoton wachsend gegen das Potential

$$\bar{\Phi}^* \lambda_o, \quad d\lambda_o = dy_1 \ldots dy_{n-1}, \quad y_n = T .$$

Es gilt

$$\bar{\Phi}^* \lambda_o(x) = 1 \quad \text{für jedes} \quad x_n < T .$$

Wir setzen

$$f_m(x) = \bar{\Phi}^* \lambda_m(x) \quad \text{für} \quad x_n < T$$

$$f_m(x) = \lim_{z \to x} \bar{\Phi}^* \lambda_m(x) = \varphi_m(x), \quad x_n = T .$$

Es ist

$$f_m \in D_\omega(\overline{\Omega_{T^-}}) \subset C_\omega(\overline{\Omega_{T^-}}) .$$

Daher existiert das Integral $\int f_m d\mu$, nach (18) gilt

$$\int f_m d\nu = \int f_m d\mu .$$

Aus $f_m \nearrow f_o = 1$ folgt

$$\begin{aligned} \|\nu\| &= \int d\nu = \int \lim f_m d\nu = \lim \int f_m d\nu \\ &= \lim \int f_m d\mu = \int \lim f_m d\mu = \int d\mu = \|\mu\| . \end{aligned}$$

Daher ist $\nu \in B_M = \{\gamma : \|\gamma\| \leq M\}$, $M = \|\mu\|$. Hieraus folgt $\mathcal{L}(\mu) \subset B_{\|\mu\|} \subset C_\omega^*(R^n)$, was zu beweisen war.

Auf $\mathcal{L}(\mu) \subset B_{\|\mu\|}$ betrachten wir die durch $C_\omega^*(R^n)$ induzierte schwache Topologie. Entsprechend den Überlegungen von Abschnitt 1 ist diese schwache Topologie metrisierbar. Hieraus folgt der

Satz 4: $\mathcal{L}(\mu)$ ist eine konvexe, schwach kompakte Menge.

Beweis: Aus $\nu = t\nu_1 + (1-t)\nu_2$, $\nu_1, \nu_2 \in \mathcal{L}(\mu)$, $0 \leq t \leq 1$ folgt unter Beachtung von (17) oder (18) sofort $\nu \in \mathcal{L}(\mu)$. Da jede abgeschlossene Teilmenge der kompakten Menge B_M selbst kompakt ist, brauchen wir nur die Abgeschlossenheit von $\mathcal{L}(\mu) \subset B_M$ zu zeigen. Aus

$$\int f d\nu_k \to \int f d\mu \quad \text{für jedes} \quad f \in C_\omega(R^n)$$

folgt wegen $\nu_k \in \mathcal{L}(\mu)$

$$\int g d\nu_k = \int g d\mu \quad \text{für jedes} \quad g \in D_\omega(\overline{\Omega_{T^-}})$$

und hieraus

$$\int g d\nu = \int g d\mu \text{ für jedes } g \in D_\omega(\overline{\Omega_{T^-}}) .$$

Weiter ist wegen $\nu_k \geqq 0$ auch $\nu \geqq 0$. Das bedeutet aber $\nu \in \mathcal{L}(\mu)$, womit Satz 4 bewiesen ist.

Wir betrachten in den weiteren Ausführungen den Fall $\mu = \mu_T$, wobei μ_T durch Balayage des im Nullpunkt gelegenen Dirac-Maßes δ entsteht, d. h. $\delta(f) = \mu_T(f)$ oder

$$(19) \qquad f(0) = \int_{I_n} f(y)\Phi(y,0)dy_1 \ldots dy_{n-1} = \int f(y)d\mu_T(y)$$

für jedes $f \in D_\omega(\overline{\Omega_{T^-}})$. Nach Satz 2 ist (19) äquivalent der Beziehung

$$(20) \qquad \int \Phi(x,y)d\mu_T(y) = \int \Phi(x,y)\Phi(y,0)dy_1 \ldots dy_{n-1} = \Phi(x,0), \ x_n > T.$$

Von besonderem Interesse sind die extremalen Elemente von $\mathcal{L}(\mu)$. Ein Maß $\nu \in \mathcal{L}(\mu)$ heißt dabei extremal, wenn es nicht in der Form

$$\nu = t\nu_1 + (1-t)\nu_2 , \quad 0 < t < 1, \quad \nu_1, \nu_2 \in \mathcal{L}(\mu) ,$$

darstellbar ist. Die Menge der extremalen Elemente bezeichnen wir mit ex $\mathcal{L}(\mu)$. Bezeichnet $C(\mathcal{L}(\mu))$ den Raum der auf der kompakten Menge $\mathcal{L}(\mu)$ stetigen Funktionen, $C^*(\mathcal{L}(\mu))$ den Raum der auf $C(\mathcal{L}(\mu))$ stetigen Linearformen (Maße), $\ell \in C(\mathcal{L}(\mu))$ eine affine Funktion, d. h.

$$\ell(t\nu_1 + (1-t)\nu_2) = t\ell(\nu_1) + (1-t)\ell(\nu_2), \quad 0 \leqq t \leqq 1 ,$$

so gilt nach Krein-Milman-Choquet [1], [14] für jedes $\nu_0 \in \mathcal{L}(\mu)$ eine gewisse "Schwerpunktdarstellung". Dieser Schwerpunktbegriff ist eine Verallgemeinerung des klassischen Schwerpunktbegriffes der Mechanik des R^n auf konvexe, kompakte Mengen eines Funktionalraumes.

<u>Satz 5 (G. Choquet):</u> Jedes $\nu_0 \in \mathcal{L}(\mu)$ hat die Gestalt

$$\ell(\nu_0) = \int \ell(\nu) dm(\nu) .$$

Hierbei ist $\ell \in C(\mathcal{L}(\mu))$ eine affine Funktion, $m \in C^*(\mathcal{L}(\mu))$ ein positives Maß mit $\|m\| = 1$ und $m(\mathcal{L}(\mu) \setminus \text{ex } \mathcal{L}(\mu)) = 0$.

Von besonderem Interesse sind die extremalen Elemente von $\mathcal{L}(\mu)$. Wir geben jetzt verschiedene Klassen von extremalen Elementen an. Die Menge aller extremalen Elemente konnte bisher nicht charakterisiert werden. Sie ist nur im Spezialfall der Laplace-Gleichung im R^1 bekannt [7].

Satz 6: Es sei $\mu = \mu_T$. Die in (19) erklärten Maße μ_τ, $0 < \tau \leq T$, sowie das Dirac-Maß δ sind extremale Elemente von $\mathcal{L}(\mu)$.

Beweis: Wir betrachten zuerst die Maße μ_τ, $0 < \tau \leq T$. Angenommen es gelte

(21) $$\mu_\tau = t\nu_1 + (1-t)\nu_2, \quad \nu_1, \nu_2 \in \mathcal{L}(\mu_T), \quad 0 < t < 1 .$$

Da nach Voraussetzung die drei Maße von (21) positiv sind, folgt für $f \geq 0$ aus

$$\int f d\mu_\tau = 0 \quad \text{stets} \quad \int f d\nu_1 = 0 \quad \text{und} \quad \int f d\nu_2 = 0 .$$

Daher ist

$$\operatorname{supp}\nu_1, \ \operatorname{supp}\nu_2 \subset \operatorname{supp}\mu_\tau = \partial\Omega_\tau .$$

Für $x_n > T$ gilt wegen $\nu_1, \nu_2 \in \mathcal{L}(\mu_T)$

(22) $$\Phi\mu_\tau(x) = \Phi\nu_1(x) = \Phi\nu_2(x) .$$

Die voranstehenden drei Potentiale sind für $x_n > \tau$ analytisch. Hieraus folgt das Bestehen der Beziehung (22) für alle x mit $x_n > \tau$. Für $x_n \leq \tau$ sind diese Potentiale identisch Null. Daher gilt (22) für alle $x \in R^n$. Multiplizieren wir diese Beziehung mit $L^*\varphi$, $\varphi \in C_0^\infty(R^n)$, und integrieren sie über den R^n, so folgt wie beim Beweis von Satz 1

$$\int \varphi d\mu_\tau = \int \varphi d\nu_1 = \int \varphi d\nu_2 \quad \text{für jedes } \varphi \in C_0^\infty(R^n) .$$

Das bedeutet aber $\mu_\tau = \nu_1 = \nu_2$, was einen Widerspruch zu $\nu_1 \neq \nu_2$ darstellt. Also sind die μ_τ, $0 < \tau \leq T$, extremale Elemente. Weiter würde aus $\delta = t\nu_1 + (1-t)\nu_2$ sofort $\nu_1 = c_1\delta$, $\nu_2 = c_2\delta$, $c_1, c_2 \geq 0$ folgen. Wegen $\|\delta\| = 1$ müssen $c_1 = c_2 = 1$ sein. Damit ist Satz 6 bewiesen.

Wir geben eine weitere Art von extremalen Elementen an. Wie beim Laplace-Operator [4], [5], [9], [25] kann man vom Rand $\partial\Omega_T$ Masse ins Innere von $\Omega_{0,T}$ ziehen. Im Fall einer Ebene $\partial\Omega_\tau$, $\tau < T$, zieht man die gesamte Masse von $\partial\Omega_T$ auf $\partial\Omega_\tau$. Es sei $K \subset \Omega_{0,T}$ eine abgeschlossene Menge, ν_K die Einschränkung von ν auf K, γ_T die Einschränkung von ν auf $\partial\Omega_T$. Wir suchen extremale Elemente der Gestalt

$$\nu = \nu_K + \gamma_T \in \mathcal{B}(\mu_T).$$

Der Einfachheit wegen beschränken wir uns in den weiteren Betrachtungen auf den R^2. Dabei betrachten wir die Fälle $K = (b,\tau)$ und $K = \{(b,\tau),\ (-b,\tau)\}$. Allgemeinere K wurden bisher nicht untersucht. Jedes Maß $\nu \in \mathcal{B}(\mu_T)$ mit $\operatorname{supp}\nu = \{z\} \cup \partial\Omega_T$ hat die Gestalt

(23) $$\nu = a\delta_z + \gamma_T .$$

Für jedes $f \in D_\omega(\overline{\Omega_{T^-}})$ gilt wegen $\nu \in \mathcal{B}(\mu_T)$ und $\mu = \mu_T$ nach Anmerkung 2

(24) $$\nu(f) = a\delta_z(f) + \gamma_T(f) = \mu_T(f) .$$

Hieraus folgt unter Beachtung von (19), wenn 0 durch (b,τ) ersetzt wird,

(25) $$\gamma_T(f) = \mu_T(f) - a\delta_z(f) = \mu_T(f) - a\mu_{(b,\tau)}(f) .$$

Da die Menge der Einschränkungen $f|_{I_n}$ von Funktionen $f \in D_\omega(\overline{\Omega_{T^-}})$ gleich $C_\omega(\partial\Omega_T)$ ist (Folgerung aus Beziehung (10)), ist das Maß γ_T durch (25) eindeutig bestimmt. Wegen (19) hat das Maß γ_T folgende Gestalt

$$d\gamma_T(y) = d\mu_T(y) - a\,d\mu_{(b,\tau)}(y) = (\Phi(y,0) - a\Phi(y,z))dy_1 \dots dy_{n-1} .$$

Setzen wir $g_a(y) = \Phi(y,0) - a\Phi(y,z)$, so gilt für $z = (b,\tau)$ und $y = (y_1,T)$

$$g_a(y) = \frac{1}{2\sqrt{\pi T}}\exp\left[-\frac{y_1^2}{4T}\right] - \frac{a}{2\sqrt{\pi(T-\tau)}}\exp\left[-\frac{(y_1-b)^2}{4(T-\tau)}\right].$$

Aus $\nu \geqq 0$ folgt notwendig $a \geqq 0$ und $g_a \geqq 0$.

<u>Satz 7:</u> Es sei $z = (b,\tau)$, $y = (y_1,T)$. Jedes Maß $\nu \in \mathcal{B}(\mu_T)$ mit $\operatorname{supp}\nu = \{z\} \cup \partial\Omega_T$ hat die Gestalt $\nu_a = a\delta_z + (\mu_T - a\mu_{(b,\tau)})$, $a \geqq 0$. Im Fall $a_0 = (1-\tau/T)^{1/2}\exp(-b^2/4\tau)$ ist ν_{a_0} ein extremales Element von $\mathcal{B}(\mu_T)$. Es gilt $g_{a_0}(y_1,T) = 0$ für

$y_1^o = \frac{bT}{\tau}$. Hieraus folgt $0 \leqq a \leqq a_o$. Für $a > a_o$ ist ν_a kein positives Maß; es gilt in diesem Fall auch $\Phi\nu_a() = \Phi\mu_T(x)$ für $x_n > T$.

Beweis: Auf Grund der vorangegangenen Überlegungen haben wir nur noch a_o zu berechnen und die Extremalität des entsprechenden Maßes nachzuweisen.

a) Die Funktion g_a ist für $T > 0$ analytisch bezüglich y_1. Aus $g_a(y_1,T) \geqq 0$ und $g_a(y_1^o,T) = 0$ folgt, daß g_a in y_1^o die y_1-Achse berührt. Aus $g_a(y_1^o,T) = 0$ folgt

$$\tau y_1^2 - 2y_1 bT + b^2 T = 4T(T - \tau)\ln a/(1 - \tau/T)^{1/2}\,. \tag{26}$$

Die Gleichung (26) besitzt eine doppelte Nullstelle genau dann, wenn

$$a_o = (1 - \tau/T)^{1/2} \exp(-b^2/4\tau)$$

ist. Es ist $g_{a_o}(y_1,T) = 0$ für $y_1^o = \frac{bT}{\tau}$. Man prüft leicht nach, daß für $o < a < a_o$

$$g_a(y_1,T) \geqq g_{a_o}(y_1,T) \geqq 0$$

und für $a > a_o$ die Gleichung (26) zwei verschiedene Nullstellen y_1', y_1'' besitzt. Dann ist

$$g_a(y_1,T) < 0 \quad \text{für} \quad y_1' < y_1 < y_1''\,.$$

b) Es ist noch zu zeigen, daß

$$\nu_{a_o} = a_o\delta_z + (\mu_T - a_o\mu_{(b,\tau)}) \in \mathrm{ex}\,\mathcal{L}(\mu_T). \tag{27}$$

Angenommen, ν_{a_o} sei nicht extremal. Dann gibt es zwei Maße $\nu_1, \nu_2 \in \mathcal{L}(\mu_T)$ mit

$$\nu_{a_o}(f) = t\nu_1(f) + (1-t)\nu_2(f),\ 0 < t < 1\,, \tag{28}$$

für jedes $f \in C_\omega(R^n)$. Nach (23) und (25) gilt

$$\nu_1(f) = a'\delta_z(f) + (\mu_T(f) - a'\mu_{(b,\tau)}(f))$$
$$\nu_2(f) = a''\delta_z(f) + (\mu_T(f) - a''\mu_{(b,\tau)}(f)).$$

Wegen der Forderung $\nu_i \in \mathcal{L}(\mu_T)$ muß nach a) notwendig $a' \leqq a_o$ und $a'' \leqq a_o$ sein. Weiter folgt aus (28)

$$a_o = ta' + (1-t)a''\,.$$

Wegen $t \neq 0$ muß $a' < a_o < a''$ oder $a'' < a_o < a'$ sein. Im ersten Fall ist $g_{a''}(y_1) < 0$ in einer offenen Menge, im zweiten Fall $g_{a'}(y_1) < 0$ in einer geeigneten offenen Menge. Das stellt einen Widerspruch zu $\nu_{a''}, \nu_{a'} \in \mathcal{L}(\mu_T)$ dar. Also ist $\nu_{a_o} \in \operatorname{ex} \mathcal{L}(\mu_T)$, was zu beweisen war.

Wir betrachten jetzt den Fall zweier Punkte $z = (-b, \tau)$ und $\zeta = (+b, \tau)$, $b > 0$. Jedes Maß $\nu \in \mathcal{L}(\mu_T)$ mit $\operatorname{supp} \nu \subset \{z, \zeta\} \cup \partial \Omega_T$ hat die Gestalt

(29) $$\nu = a_1 \delta_z + a_2 \delta_\zeta + \gamma_T, \quad a_1, a_2 > 0 .$$

Es gilt für jedes $f \in D_\omega(\overline{\Omega_T-})$

$$\nu(f) = a_1 \delta_z(f) + a_2 \delta_\zeta(f) + \gamma_T(f) = \mu_T(f).$$

Wegen

$$\delta_z(f) = \mu_{(-b,\tau)}(f) \quad \text{und} \quad \delta_\zeta(f) = \mu_{(b,\tau)}(f)$$

für jedes $f \in D_\omega(\overline{\Omega_T-})$ ist

(30) $$\gamma_T(f) = \mu_T(f) - a_1 \mu_{(-b,\tau)}(f) - a_2 \mu_{(b,\tau)}(f) .$$

Da die Menge der Einschränkungen $f|\partial\Omega_T$ gleich $C_\omega(\partial\Omega_T)$ ist, ist das Maß γ_T durch (30) eindeutig bestimmt. Es gilt also, wenn wir in Zukunft $\nu = \nu_{a_1,a_2}$ schreiben

(31) $$\nu_{a_1,a_2}(f) = a_1 \delta_z(f) + a_2 \delta_\zeta(f) + (\mu_T(f) - a_1 \mu_{(-b,\tau)}(f) - a_2 \mu_{(b,\tau)}(f)).$$

Das Maß γ_T hat die Gestalt

$$d\gamma_T(y) = g_{a_1,a_2}(y_1, T)\, dy_1$$

mit

(32) $$g_{a_1,a_2}(y_1,T) = \frac{1}{2\sqrt{\pi}} \Big(T^{-1/2} \exp - y_1^2/4T - a_1 (T-\tau)^{-1/2} \exp\left[-(y_1+b)^2/4(T-\tau)\right] - a_2 (T-\tau)^{-1/2} \exp\left[-(y_1-b)^2/4(T-\tau)\right]\Big) .$$

Im folgenden Satz geben wir ein extremales Element der Gestalt ν_{a_1,a_2} an. Eine vollständige Charakterisierung dieser Maße ist noch nicht bekannt.

<u>Satz 8:</u> Es seien $z = (-b, \tau)$, $\zeta = (b, \tau)$, $b < (2\tau(T-\tau)/T)^{1/2}$. Jedes Maß $\nu \in \mathcal{L}(\mu_T)$ mit $\operatorname{supp} \nu = \{z, \zeta\} \cup \partial\Omega_T$ hat die Gestalt

$$\nu_{a_1,a_2} = a_1\delta_z + a_2\delta_\zeta + (\mu_T - a_1\mu_{(-b,\tau)} - a_2\mu_{(b,\tau)}), a_1, a_2 > 0.$$

Für $a_1 = a_2 = \alpha$, $\alpha = \frac{1}{2}(1-\tau/T)^{1/2}\exp(b^2/4(T-\tau))$ ist

$$\nu_{\alpha,\alpha} \in \operatorname{ex} \mathcal{L}(\mu_T).$$

<u>Beweis:</u> a) Die Gestalt von ν_{a_1,a_2} wurde in (31) angegeben. Wir suchen zuerst ein α und ein y_1^o mit

(33) $$g_{\alpha,\alpha}(y_1^o, T) = 0, \quad g_{\alpha,\alpha}(y_1, T) \geqq 0.$$

Für die Ableitung von g_{a_1,a_2} gilt

(34) $$4\pi^{1/2} \frac{dg_{a_1,a_2}(y_1,T)}{dy_1} = -\frac{y_1}{T^{3/2}} \exp\left[-\frac{y_1^2}{4T}\right]$$
$$+ a_1 \frac{y_1 + b}{(T-\tau)^{3/2}} \exp\left[-\frac{(y_1+b)^2}{4(T-\tau)}\right] + a_2 \frac{y_1 - b}{(T-\tau)^{3/2}} \exp\left[-\frac{(y_1-b)^2}{4(T-\tau)}\right].$$

Diese Ableitung ist Null für $a_1 = a_2 = \alpha$ und $y_1 = 0$. Es gilt

$$g_{\alpha,\alpha}(0,T) = 0 \quad \text{für } \alpha = \frac{1}{2}(1-\tau/T)^{1/2}\exp\left[\frac{b^2}{4(T-\tau)}\right].$$

Weiter ist $g_{\alpha,\alpha}(y_1,T) > 0$ für $y_1 \neq 0$. Für ein Minimum von $g_{\alpha,\alpha}$ im Punkt $y_1 = 0$ ist die Positivität der zweiten Ableitung von g, hinreichend. Es gilt

$$4\pi^{1/2} \frac{d^2 g_{\alpha,\alpha}(0,T)}{dy_1^2} = -\frac{1}{T^{3/2}} + \frac{\alpha}{(T-\tau)^{3/2}}\left(2 - \frac{b^2}{(T-\tau)}\right) \exp\left[-\frac{b^2}{4(T-\tau)}\right]$$
$$= \frac{1}{T^{1/2}(T-\tau)}\left(\frac{\tau}{T} - \frac{b^2}{2(T-\tau)}\right) > 0 \text{ für } b < (2\tau(T-\tau)/T)^{1/2}.$$

b) Wir zeigen nun, daß für das angegebene α das Maß $\nu_{\alpha,\alpha}$ extremal ist. Angenommen, $\nu_{\alpha,\alpha}$ sei nicht extremal. Dann gibt es Maße $\nu_1, \nu_2 \in \mathcal{L}(\mu_T)$ mit

(35) $$\nu_{\alpha,\alpha} = t\nu_1 + (1-t)\nu_2, \quad 0 < t < 1.$$

Aus $\nu_1, \nu_2 \geqq 0$ folgt

$$\operatorname{supp}\nu_1, \ \operatorname{supp}\nu_2 \subset \operatorname{supp}\nu_{\alpha,\alpha}.$$

Die ν_i haben nach (31) die Gestalt

$$\nu_1 = c_1\delta_z + d_1\delta_\zeta + (\mu_T - c_1\mu_{(-b,\tau)} - d_1\mu_{(b,\tau)})$$
$$\nu_2 = c_2\delta_z + d_2\delta_\zeta + (\mu_T - c_2\mu_{(-b,\tau)} - d_2\mu_{(b,\tau)}).$$

Aus (35) folgt

$$\alpha = tc_1 + (1 - t)c_2 ,$$
$$\alpha = td_1 + (1 - t)d_2 .$$

Aus der ersteren Gleichung folgt $c_1 = c_2 = \alpha$, oder $c_1 < \alpha < c_2$ oder $c_2 < \alpha < c_1$. Entsprechend folgt aus der zweiten Gleichung $d_1 = d_2 = \alpha$ oder $d_1 < \alpha < d_2$ oder $d_2 < \alpha < d_1$. Im Fall $c_1 = c_2 = \alpha$ muß notwendig $d_1 = d_2 = \alpha$ sein, da für $\alpha < d_1$ oder $\alpha < d_2$ die Bedingung (33) nicht erfüllt ist. Im Fall $c_1 < \alpha < c_2$ muß notwendig $d_2 < \alpha < d_1$ sein, da für $\alpha < c_2$ und $\alpha < d_2$ die Bedingung (33) nicht erfüllt ist. Entsprechendes gilt für den anderen Fall. Aus (35) folgt also

$$(36) \qquad c_1 < \alpha < c_2 , \quad d_2 < \alpha < d_1$$

oder

$$(37) \qquad c_2 < \alpha < c_1 , \quad d_1 < \alpha < d_2 .$$

Weiterhin folgt aus (35)

$$g_{\alpha,\alpha}(y_1^o,T) = tg_{c_1,d_1}(y_1^o,T) + (1 - t)g_{c_2,d_2}(y_1^o,T) .$$

Wegen $g_{c_i,d_i} \geqq 0$ muß notwendig

$$g_{c_1,d_1}(y_1^o,T) = g_{c_2,d_2}(y_1^o,T) = 0$$

gelten. Hieraus folgt wegen $y_1^o = 0$

$$c_i + d_i = (1 - \tau/T)^{1/2} \exp\left[\frac{b^2}{4(T - \tau)}\right] = 2\alpha .$$

Aus (34) folgt

$$4\pi^{1/2} \frac{dg_{c_i,d_i}(0,T)}{dy_1} = (c_i - d_i)b(T-\tau)^{-3/2}\exp\left[-\frac{b^2}{4(T-\tau)}\right] \neq 0$$

für die in (36) und (37) angegebenen Werte. Die Funktion g_{c_i,d_i} besitzt positive und negative Werte, was einen Widerspruch zu (32) darstellt. Also ist $\nu_{\alpha,\alpha}$ extremal. Damit ist Satz 8 bewiesen.

Einige Ergebnisse dieser Arbeit wurden gemeinsam mit Herrn Le Trông Luc (Vietnam) diskutiert und in seiner Diplomarbeit aufgeschrieben.

Literatur

[1] Alfsen E. M., Compact Sets and Boundary Integrals, Berlin-Heidelberg-New York 1971.

[2] Anger G., Funktionalanalytische Betrachtungen bei Differentialgleichungen unter Verwendung von Methoden der Potentialtheorie, I, Berlin 1967

[3] Anger G., Eindeutigkeitssätze und Approximationssätze für Potentiale, II, Math. Nachr. 50, 229 - 244 (1971)

[4] Anger G., Eindeutigkeitssätze und Approximationssätze für Potentiale, III. Sibir.mat.Žurn. 15, 6, 1163 - 1179 (1973) (russisch; englische Übersetzung in Sibir.math.J. 14, 811 - 824 (1973)

[5] Anger G., Direct and Inverse Problems in Potential Theory, published in Nonlinear Evolution Equations and Potential Theory, Praha 1975, 11 - 44

[6] Anger G., Die Rolle der modernen Potentialtheorie in der Theorie der inversen Aufgabenstellungen, Gerlands Beiträge zur Geophysik, 85, 1, 1 - 20 (1976)

[7] Anger G., Das inverse Problem der Laplace-Gleichung im R^1 (erscheint in Sibir. mat. Žurn., russisch)

[8] Anger G., Konvexe Mengen bei inversen Problemen (erscheint in den Proceedings des Petrovskij-Kolloquiums, welches vom 26. - 31. Januar 1976 in Moskau stattfand, russisch)

[9] Anger G. und Schulze B.-W., Einige Bemerkungen über harmonische Maße und inverse Probleme, Beiträge zur Analysis, 8, 13 - 26 (1976).

[10] Bauer H., Zum Cauchyschen und Dirichletschen Problem bei elliptischen und parabolischen Differentialgleichungen, Math. Ann. 164, 142 - 153 (1966)

[11] Bourbaki N., Intégration, Chap. I - IV, Paris 1952

[12] Bourbaki N., Espaces vectoriels topologiques, Chap. I - II, Paris 1953, Chap. III - V, Paris 1955

[13] Brelot M., Eléments de la théorie classiques du potentiel, Paris 1959

[14] Choquet G., Lectures on Analysis, I, II, III, New York-Amsterdam 1969

[15] Constantinescu C. und Cornea A., Potential Theory on Harmonic Spaces, Berlin-Heidelberg-New York 1972

[16] Dieudonné J., Grundzüge der modernen Analysis, II, Berlin 1975 (Übers. a. d. Französischen)

[17] Friedman A., Partial Differential Equations of Parabolic Type, Englewood Cliffs, N. J., 1964

[18] Landis E. M., Gleichungen zweiter Ordnung vom elliptischen und parabolischen Typ, Moskau 1971 (russisch)

[19] Lavrentiev M. M., Some Improperly Posed Problems of Mathematical Physics, Berlin-Heidelberg-New York 1967 (Übers. a. d. Russischen)

[20] Lavrentiev M. M., Romanov O. V. und Vasiliev V. G., Multidimensional Inverse Problems for Differential Equations, Berlin-Heidelberg-New York 1970 (Übers. a. d. Russischen)

[21] Lions J. L. et Magenes E., Problèmes aux limites non homogénes et applications, I, II, III, Paris 1968, 1969

[22] Morozov V. A., Lineare und nichtlineare nichtkorrekte Aufgaben, Itogi nauki i techniki, Matematičeskij analyz 11, Moskva 1973 (russisch)

[23] Payne L. E., Improperly Posed Problems in Partial Differential Equations, Regional Conference Series in Applied Mathematics, vol. 22, Society for Industrial and Applied Mathematics, Philadelphia, 1975

[24] Romanov V. G., Inverse Probleme für Differentialgleichungen, Novosibirsk 1973 (russisch)

[25] Schulze B.-W., Über das inverse Problem beim Laplace-Operator, Math. Nachr. 67, 225 - 235 (1975)

[26] Schulze B.-W., Über die Potentiale für elliptische Gleichungen höherer Ordnung und das inverse Problem (erscheint in Differencial'nye Uravnenija, russisch)

[27] Schulze B.-W., Über das inverse Problem für elliptische Gleichungen höherer Ordnung (ersch. in Differencial'nye Uravnenija, russisch)

[28] Schulze B.-W., Potentialtheoretische Grundlagen eines inversen Problems der Geophysik, Gerlands Beiträge zur Geophysik, 85, 1, 21 - 25 (1976)

[29] Tychonov A. N. und Samarski A. A., Differentialgleichungen der mathematischen Physik, Berlin 1959 (Übers. a. d. Russischen)

[30] Tychonov A. N. und Arsenin V. Ja., Lösungsmethoden nicht-korrekter Aufgaben, Moskau 1974 (russisch).

Pseudo Projections and Balayage in Algebraic Potential Theory

By

Maynard Arsove and Heinz Leutwiler

1. In classical potential theory, M. Brelot [5] has shown that the sweeping-out process, or method of 'balayage', can be characterized in terms of the mapping carrying a superharmonic function u into the regularized reduced function $\hat{R}_u^A$, where A is a subset of the underlying region. The balayage concept carries over naturally to axiomatic potential theory (see e.g. M. Brelot [6] and C. Constantinescu and A. Cornea [7]). More recently, closely related ideas have been used by G. Mokobodzki [8] as the basis for a theory of cones of potentials, and an abstract formulation of 'balayage operators' in H-cones has been introduced by N. Boboc, Gh. Bucur, and A. Cornea (see [3] and [4]).

Our objective here is to define the concept of a balayage operator in the purely algebraic setting of a mixed lattice semigroup and indicate some of the basic properties of such operators. (The theory of mixed lattice semigroups is developed in [2], and certain aspects of the cone case are treated in [1].) Let $\mathcal{U}$ be an abelian semigroup under addition, with identity element 0, having a partial ordering $\leq$ with the following properties: for all u, v, w $\varepsilon\ \mathcal{U}$, (i) $u \geq 0$ and (ii) $u \leq v$ if and only if

$u + w \leq v + w$. Along with the given <u>initial order</u>, essential use is made of the induced <u>specific order</u> $\preccurlyeq$, where $u \preccurlyeq v$ means that $v = u + u'$ for some u' (denoted by $v - u$). In all cases involving order, initial order is to be presumed unless specific order is stated. The ordered triple $(\mathcal{U}, +, \leq)$ (generally abbreviated to $\mathcal{U}$) will be called a <u>mixed lattice semigroup</u> provided it satisfies

<u>Axiom</u> I. For all $u, v \in \mathcal{U}$ there exist the (unsymmetrical) <u>mixed lower</u> and <u>upper envelopes</u>

$$(1.1) \qquad u \curlywedge v = \max\{w \in \mathcal{U} : \ w \preccurlyeq u \ \text{ and } \ w \leq v\}$$

and

$$(1.2) \qquad u \curlyvee v = \min\{w \in \mathcal{U} : \ w \succcurlyeq u \ \text{ and } \ w \geq v\},$$

subject to the identity

$$(1.3) \qquad u \curlywedge v + v \curlyvee u = u + v.$$

It can be shown that the nonnegative superharmonic functions on any harmonic space form a mixed lattice semigroup.

Let $\mathcal{U}$ be fixed as any mixed lattice semigroup. Then arbitrary elements of $\mathcal{U}$ have the following properties:

$$(1.4) \qquad v \preccurlyeq w \implies u \curlywedge v \preccurlyeq u \curlywedge w \quad \text{and} \quad u \curlyvee v \preccurlyeq u \curlyvee w,$$

$$(1.5) \qquad (u+a) \curlywedge (v+a) = u \curlywedge v + a, \quad (u+a) \curlyvee (v+a) = u \curlyvee v + a,$$

$$(1.6) \qquad u \curlywedge (v + w) \leq u \curlywedge v + u \curlywedge w, \quad (u + v) \curlywedge w \preccurlyeq u \curlywedge w + v \curlywedge w,$$

$$(1.7) \qquad u \leq v_1 + v_2 \implies u = u_1 + u_2 \quad \text{with} \quad u_1 \leq v_1, \ u_2 \leq v_2,$$

$$(1.8) \qquad u \preccurlyeq v_1 + v_2 \implies u \preccurlyeq v_1 \curlywedge u + v_2 \curlywedge u.$$

Moreover, mixed lower and upper envelopes can be shown to obey certain associative and distributive inequalities.

In view of the unsymmetrical nature of mixed envelopes, there are two types of orthogonal complements of a set $X \subset \mathcal{U}$. The left orthogonal complement ${}^{\perp}X$ consists of those $u \in \mathcal{U}$ such that $u \curlywedge x = 0$ for all $x \in X$, and the right orthogonal complement $X^{\perp}$ consists of those $u \in \mathcal{U}$ such that $x \curlywedge u = 0$ for all $x \in X$.

A given nonempty set $E \subset \mathcal{U}$ may admit a supremum (sup) or a specific supremum (sp sup). There is also need for a third type of supremum, defined as follows. The set E will be said to have u_0 as a strong supremum (str sup) provided u_0 is a common supremum of E relative to initial and specific orders, having the further property that $\inf (u_0 - E) = 0$. Semigroups in $\mathcal{U}$ solid under initial order are called ideals, and those solid under specific order are called specific ideals.

In this setting, a preharmonic band is defined as any specific ideal $\mathcal{H}$ such that if $E \subset \mathcal{H}$ and $\sup E$ exists in $\mathcal{U}$, then $\sup E$ lies in $\mathcal{H}$. (A preharmonic band $\mathcal{H}$ such that, for all $h \in \mathcal{H}$ and $u \in \mathcal{U}$, $h \leq u$ implies $h \preccurlyeq u$ is called a harmonic band.) Moreover, a potential band is defined as any ideal $\mathcal{P}$ such that if $E \subset \mathcal{P}$ and $\text{str sup } E$ exists in $\mathcal{U}$, then $\text{str sup } E$ lies in $\mathcal{P}$. The nonnegative harmonic functions and potentials on any harmonic space form harmonic and potential bands, respectively.

2. Projections and pseudo projections. Let X now be any nonempty subset of $\mathcal{U}$. We call x_0 the projection of $u \in \mathcal{U}$ on X provided

$$(2.1) \qquad x_0 = \max\{x \in X: \ x \le u\} \quad \text{and} \quad x_0 \preccurlyeq u.$$

Similarly, we call x_0 the <u>specific</u> <u>projection</u> of $u \in \mathcal{U}$ on X provided it is the specific maximum

$$(2.2) \qquad x_0 = \text{sp}\max\{x \in X: \ x \preccurlyeq u\}.$$

A <u>projection</u> <u>band</u> is defined as any preharmonic band $\mathcal{H}$ such that every $u \in \mathcal{U}$ admits a projection on $\mathcal{H}$, and a <u>specific</u> <u>projection</u> <u>band</u> is defined as any potential band $\mathcal{P}$ such that every $u \in \mathcal{U}$ admits a specific projection on $\mathcal{P}$. Projection and specific projection bands arise naturally in connection with Riesz decompositions of $\mathcal{U}$, analogous to the Riesz decompositions of nonnegative superharmonic functions as sums of harmonic functions and potentials.

By a <u>Riesz</u> <u>decomposition</u> of $\mathcal{U}$ we mean any direct sum decomposition of the form

$$(2.3) \qquad \mathcal{U} = \mathcal{H} \oplus \mathcal{P},$$

where $\mathcal{H}$ is a preharmonic band and $\mathcal{P}$ is a potential band. It suffices here merely to assume that $\mathcal{H}$ and $\mathcal{P}$ are sets solid under specific order and initial order, respectively, since the decomposition (2.3) then forces them to be bands of the required types.

<u>Theorem</u> 2.1. <u>If</u> $\mathcal{H}$ <u>is</u> <u>a</u> <u>projection</u> <u>band</u>, <u>then</u> $\mathcal{U}$ <u>admits</u> <u>the</u> <u>Riesz</u> <u>decomposition</u> (2.3) <u>with</u> $\mathcal{P} = \mathcal{H}^{\perp}$; <u>dually</u>, <u>if</u> $\mathcal{P}$ <u>is</u> <u>a</u> <u>specific</u> <u>projection</u> <u>band</u>, <u>then</u> $\mathcal{U}$ <u>admits</u> <u>the</u> <u>Riesz</u> <u>decomposition</u> (2.3) <u>with</u> $\mathcal{H} = {}^{\perp}\mathcal{P}$. <u>Conversely</u>, <u>if</u> $\mathcal{U}$ <u>admits</u> <u>the</u> <u>Riesz</u> <u>decomposition</u> (2.3), <u>then</u> $\mathcal{H}$ <u>is</u> <u>a</u> <u>projection</u> <u>band</u> <u>and</u> $\mathcal{P}$

a *specific projection band*, with $\mathcal{P} = \mathcal{H}^{\perp}$ and $\mathcal{H} = {}^{\perp}\mathcal{P}$

Of course, the requirement $x_0 \preccurlyeq u$ in (2.1) is essential in proving Theorem 2.1. The case in which this condition is discarded also turns out to be of considerable interest. We shall call x_0 the *pseudo projection* of $u \in \mathcal{U}$ on X provided

(2.4) $$x_0 = \max\{x \in X: \ x \leq u\}.$$

An operator T on $\mathcal{U}$ will be called a *projection operator*, *specific projection operator*, or *pseudo projection operator* if, for all $u \in \mathcal{U}$, the element Tu is the projection, specific projection, or pseudo projection, respectively, of $\mathcal{U}$ on the set $X = T\mathcal{U}$. Such operators can be characterized intrinsically. For example, an operator T mapping $\mathcal{U}$ onto a set X is a pseudo projection operator if and only if it is idempotent, contractive, and increasing (i.e. $T^2 = T$, $Tu \leq u$, and $u \leq v$ implies $Tu \leq Tv$). Observe that no additivity requirements are imposed on the operators T.

Let T be a pseudo projection operator mapping $\mathcal{U}$ onto a set X. Then

(2.5) T is superadditive $\Longleftrightarrow$ X is a semigroup,

(2.6) T is subadditive $\Longleftrightarrow$ X is specifically solid.

Hence, T will be additive if and only if X is a specific ideal. More can be said, however, in this connection. Defining a *preharmonic operator* on $\mathcal{U}$ as any pseudo projection operator mapping $\mathcal{U}$ onto a preharmonic band, we have

Theorem 2.2. *The additive pseudo projection operators on* $\mathcal{U}$

<u>are precisely the preharmonic operators on</u> $\mathcal{U}$.

Preharmonic operators are increasing relative to specific order and have the important property of preserving strong suprema.

3. <u>Balayage operators</u>. Henceforth, the mixed lattice semigroup $\mathcal{U}$ will be assumed to be <u>upper complete</u>, in the following sense: every set $E \subset \mathcal{U}$ which is filtering upward and bounded above admits a supremum, and such suprema are translation invariant (i.e. $\sup (E + u) = \sup E + u$ for all $u \in \mathcal{U}$). Let $\hat{X}$ denote the specific ideal generated by X. For each $u \in \mathcal{U}$, the set $\{x \curlywedge u: \ x \in \hat{X}\}$ is filtering upward and bounded above by u, and we define

$$B_X u = \sup_{x \in \hat{X}} (x \curlywedge u). \tag{3.1}$$

The resulting operator B_X will be referred to as the <u>balayage operator</u> determined by X. For example, in the classical setting, X can be taken as the set of all potentials of nonnegative mass distributions supported by a prescribed set E.

The role of balayage operators in the algebraic theory is apparent from the following characterizing property.

<u>Theorem 3.1. The balayage operator</u> B_X <u>is precisely the preharmonic operator mapping</u> $\mathcal{U}$ <u>onto the preharmonic band</u> $\mathcal{H}_X$ <u>generated by</u> X.

Consequently, every preharmonic operator is a balayage operator B_X for some X. Since also

$$B_X u = \max\{h \in \mathcal{H}_X: \ h \leq u\}, \tag{3.2}$$

two balayage operators B_X and B_Y coincide if and only if the sets X and Y generate the same preharmonic band. Another consequence of (3.2) is that the potentials formed relative to $\mathcal{H}_X$, i.e. the elements $p \in \mathcal{U}$ for which the only minorant h of p in $\mathcal{H}_X$ is $h = 0$, are just the elements of $\mathcal{U}$ annihilated by B_X. These potentials comprise the potential band $X^{\perp}$.

An important special case occurs when there is a single generating element x, so that $X = \{x\}$, and the balayage operator will then be denoted simply as B_x. Thus,

$$B_x u = \sup_{n \geq 1} [(nx) \curlywedge u] \tag{3.3}$$

for all $u \in \mathcal{U}$. In addition to the above properties of general balayage operators, the operators B_x satisfy

(3.4) $x \preccurlyeq y \Longrightarrow B_x u \leq B_y u$,

(3.5) $B_{x+y} u \leq B_x u + B_y u$,

(3.6) for any sequence $\{x_n\}$ in $\mathcal{U}$ converging strongly upward to x_0, in the sense that $\{x_n\}$ is increasing relative to specific order and has x_0 as strong supremum, the sequence $\{B_{x_n} u\}$ converges upward to $B_{x_0} u$.

The operators B_x are connected with the theory of infinitesimal generators and quasi-units (see [1]) in the following way. Although a cone structure was assumed in [1], basic aspects of the theory remain valid when $\mathcal{U}$ is merely assumed to be a strongly superharmonic semigroup. This is defined as a mixed lattice semigroup $\mathcal{U}$ which satisfies

Axiom II. Every bounded increasing sequence in $\mathcal{U}$ has a supremum, and the suprema of such sequences are translation invariant.

We suppose that $\mathcal{U}$ is any strongly superharmonic semigroup and fix $x \in \mathcal{U}$. By the generator determined by x we mean the operator A_x such that $A_x u = B_u x$, i.e.

$$(3.7) \qquad A_x u = \sup_{n \geq 1} [(nu) \curlywedge x]$$

for all $u \in \mathcal{U}$. Further, an x-quasi-unit is defined as any $u \in \mathcal{U}$ such that $(2u) \curlywedge x = u$, and we denote the set of all x-quasi-units by $\mathcal{E}_x$.

Theorem 3.2. In any strongly superharmonic semigroup $\mathcal{U}$, the generators A_x map $\mathcal{U}$ onto $\mathcal{E}_x$ and are idempotent, subadditive, and weakly increasing (in the sense that $u \precsim v$ implies $A_x u \leq A_x v$).

Thus, the operators A_x are generators in the sense that they generate x-quasi-units $A_x u$. When $\mathcal{U}$ admits a cone structure, as in [1], the generators A_x turn out to be infinitesimal generators associated with certain one-parameter semigroups of operators on $\mathcal{U}$. Finally, we note that the generators A_x map any preharmonic band $\mathcal{H}$ onto the set of x-quasi-units in $\mathcal{H}$.

REFERENCES

1. M. Arsove and H. Leutwiler, Infinitesimal generators and quasi-units in potential theory, Proc. Nat. Acad. Sci. USA 72 (1975), 2498-2500.

2. ———, Algebraic potential theory, Memoirs of the Amer. Math.

Soc., to appear.

3. N. Boboc, Gh. Bucur, and A. Cornea, Cones of potentials on topological spaces, Rev. Roumaine Math. Pures et Appl. 18 (1973), 815-865.

4. ———, H-cones and potential theory, Ann. Inst. Fourier 25 (1975), 71-108.

5. M. Brelot, Éléments de la théorie classique du potentiel (4e édition), Centre de Documentation Universitaire, Paris, 1969.

6. ———, Axiomatique des fonctions harmoniques, Les Presses de l'Université de Montreal, Montreal, 1966.

7. C. Constantinescu and A. Cornea, Potential theory on harmonic spaces, Springer Verlag, New York, 1972.

8. G. Mokobodzki, Cônes de potentiels et noyaux subordonnés, Potential Theory Stresa Notes, Edizione Cremonese, Rome, 1970.

University of Washington and Universität Erlangen-Nürnberg

HILBERTIAN AND LATTICE THEORETICAL METHODS IN POTENTIAL THEORY.

N.Boboc ,Gh.Bucur, A.Cornea

Introduction. The use of hilbertian methods in modern potential theory was initiated by H.Cartan in 1945 [10].Later, beginning with 1958, A.Beurling and J.Deny [5], using the concept of functional space of N.Aronszajn and K.T.Smith[4]consider the notion of Dirichlet space and develop a good deal of potential theory : the complete maximum principle, the theory of balayage, the theory of capacity,etc A Dirichlet space means a Hilbert space H of classes of real functions on a measurable space X satisfying some relations between the hilbertian norm of H and the space X. In a natural way one may introduce in H an order relation (positive meaning positive almost everywhere) by means of which the potentials are defined : an element p is called potential if $\langle p,h\rangle \geq 0$ for any positive h. The development of a comprehensive potential theory in H necessitates a supplementary condition called " the module contraction operates in H" which means that H is a lattice and $\| |h| \| \leq \| h \|$ for any h.

In this paper we consider the Dirichlet space as a system (H,T) introduced by A.Ancona[1], where H is an ordered Hilbert space and T is a continuous and coercive operator on H such that H is a lattice and such that $\langle Tx,y\rangle \leq 0$ for any $x,y \in H$ for which $x \wedge y = 0$.This last relation stands for the property " the modulus contractions operates in (H,T)".

Among the principal results we mention : the continuity of the lattice operations on a Dirichlet space and a representation theorem of Dirichlet spaces as functional spaces; it is shown that the cone of T-potentials is an H-cone and in its dual, the cone of potentials with respect to the adjoint of T is solid and dense in order from below; the balayage operators on H are characterized ; namely B is a balayage if it is a linear continuous operator on H for which the anihilator Ker B is a solid closed subspace and I-B is the Stampacchia-projection on Ker B. Finally are studied the local Dirichlet spaces i.e. the spaces for which $\langle Tx, y\rangle = 0$ whenever $x \wedge y = 0$. These spaces are characterized by a sheaf property for the associated cone of potentials.

0. Preliminaries

Let $(H, \langle\ \rangle)$ be a real Hilbert space, $H_+ \subset H$ be a closed convex cone such that $H_+ \cap (-H_+) = o$ and let $T: H \longrightarrow H$ be a linear continuous operator such that the bilinear form

$$\langle x,y\rangle \longrightarrow \langle Tx,y\rangle$$

is coercive (i.e. there exists $\alpha > o$ such that $\langle Tx,x\rangle \geq \alpha \|x\|^2, x \in H$)

We shall use the notations $\leq$ for the order relation on H induced by H_+, $\vee, \wedge$ for the union and intersection with respect to this order relation and x_+, x_-, $|x|$ for $x \vee o$, $(-x) \vee o$, $x \vee (-x)$ if they make sense. The adjoint operator of T will be denoted by $T^\vee$ (thus we have $\langle T^\vee x,y\rangle = \langle Ty,x\rangle$ for any $x,y \in H$).

It is known [13] that for any closed convex subset Γ of H and for any $x \in H$ there exists uniquely $x_o \in \Gamma$ such that

$$\langle T(x - x_o), y - x_o\rangle \leq o$$

for any $y \in \Gamma$ We shall denote by π^T_Γ the map $x \rightarrow x_o$. We have

$$\| \pi^T_\Gamma(x) - \pi^T_\Gamma(y) \| \leq \frac{\|T\|}{\alpha} \|x-y\| \quad , x,y \in H.$$

Also if $(\Gamma_\iota)_{\iota\in I}$ is an increasing (resp.decreasing which $\bigcap_{\iota\in I}\Gamma_\iota \neq \phi$) family of closed convex subsets of H then for any $x \in H$ the family $(\pi^T_{\Gamma_\iota}(x))_{\iota\in I}$ converges to $\pi^T_\Gamma(x)$ where $\Gamma = \overline{\bigcup_{\iota\in I}\Gamma_\iota}$ (resp. $\Gamma = \bigcap_{\iota\in I}\Gamma_\iota$)

An element $p \in H$ will be called a T-potential (or simply potential) if $\langle Tp,h \rangle \geq o$ for any $h \in H_+$ and we shall denote by $\mathcal{P}_T = \mathcal{P}$ the set of all potentials. The set $\mathcal{P}$ is a closed convex cone, $\mathcal{P} - \mathcal{P}$ is dense in H and if H_+^o denotes the polar of H_+ (i.e. $H_+^o = \{x \in H \mid \langle x,y \rangle \leq o$ for any $y \in H_+ \}$) we have $\mathcal{P} = T^{-1}(-H_+^o) = -(T^{\vee}(H_+))^o$. Since $p \in \mathcal{P} \Longleftrightarrow p = \pi^T_{p+H_+}(0)$ it follows, using the properties of π^T_Γ that any decreasing (resp. increasing and bounded in norm or in order) family of $\mathcal{P}$ is convergent. For any $x \in H$ we denote $R^T(x) = R(x) =: \pi^T_{x+H_+}(o)$. It is easy to see that $R(x)$ is uniquely determined by the following properties :

a) $R(x) \in \mathcal{P}$

b) $R(x) \geq x$

c) $\langle TR(x), R(x) - x \rangle = o$

Also we have $R(x) = \pi^{T^{\vee}}_{\mathcal{P}}(x)$. Hence

$$\| R(x) - R(y) \| \leq \frac{\| T^{\vee} \|}{\alpha} \| x - y \| \quad , \qquad x,y \in H.$$

We remember now a result of A. Ancona [1].

Theorem. The following assertions are equivalent:

1) $(H, \leq)$ is a vector lattice and $\langle Tx_+, x_- \rangle \leq o$ for any $x \in H$.

2) For any $p,q \in \mathcal{P}$ there exists $p \wedge q$ and $p \wedge q \in \mathcal{P}$.

3) $\mathcal{P} \subset H_+$ and $(\mathcal{P}, \leq)$ is a cone of potentials in the sense of Mokobodzki [12].

Remark .1 Since the assertion 1) holds for T and $T^{\vee}$ simultaneously the same is true for the assertions 2) and 3).

Remark 2. Assume that one of the assertions of the above theorem holds. Then we have

$$R(x) = \wedge \{ p \in \mathcal{P} \mid p \geq x \}$$

and therefore $R(x)$ is the usual reduite of x with respect to the cone $\mathcal{P}$ (see [12]).

Definition . A system $(H,\langle\ \rangle, H_+,T)$ as above is called a Dirichlet space if one of the assertions of the Ancone theorem holds .

§1. Continuity of lattice operations and representations for Dirichlet spaces .

In the sequel $(H,\langle\ \rangle,H_+,T)$ will be a fixed Dirichlet space and $\mathcal{P}$ (resp. $\check{\mathcal{P}}$) will denote the cone of T-potentials (resp. $\check{T}$-potentials).

The following relations are immediate :

$$\|x_+\|^2 + \|x_-\|^2 \leq \frac{\|T\|}{\alpha}\|x\|^2 ,$$

$$\|\,|x|\,\|^2 \leq \frac{\|T\|}{\alpha}\|x\|^2 ,$$

$$\|x \wedge y\| \leq M(\|x\| + \|y\|) \text{ where } M = \frac{1}{2} + \left(\frac{\|T\|}{\alpha}\right)^{1/2},$$

$$p,q \in \mathcal{P},\ p \leq q \implies \|p\| \leq \frac{\|T\|}{\alpha}\|q\| .$$

Theorem 1.1. The lattice operations in H are continuous with respect to the norm.

Proof. If $(x_n)_n \to x$ and $y_n = x - x_n$ then $(y_n)_n \longrightarrow 0$, $(|y_n|)_n \longrightarrow 0$, $((y_n)_+)_n \longrightarrow 0$, $((y_n)_-)_n \longrightarrow 0$.

Also we have

$$(x_n)_+ = x_+ + (y_n)_- - x_+\wedge(y_n)_+ - x_-\wedge(y_n)_-$$

and therefore it is sufficient to show that if $z \in H_+$ and $(z_n)_n \longrightarrow 0$ $z_n \in H_+$ then $(z \wedge z_n)_n \longrightarrow 0$.

Firstly we show that the sequence $(z\wedge z_n)_n$ converges weakly to zero .Indeed it is bounded and we have

$$0 \leq z \wedge z_n \leq z_n,\ n \in \mathbb{N}.$$

Hence for any $p \in \mathcal{P}$ we get

$$0 \leq \langle Tp, z\wedge z_n\rangle \leq \langle Tp, z_n\rangle,\ (\langle Tp, z\wedge z_n\rangle)_n \longrightarrow 0.$$

Since $T(\mathcal{P}-\mathcal{P})$ is dense it follows that the sequence $(z\wedge z_n)_n$ converges weakly to zero.

From

$$\langle Tz,z_n\rangle + \langle T(z\wedge z_n), z\wedge z_n\rangle - \langle Tz, z\wedge z_n\rangle - \langle T(z\wedge z_n), z_n\rangle =$$

$$= \langle T(z - z\wedge z_n), z_n - z\wedge z_n\rangle \leq 0$$

it follows

$$(T(z \wedge z_n), z \wedge z_n)_n \longrightarrow o,$$

and therefore, T being coercive, $(z \wedge z_n)_n \longrightarrow o$.

Corollary: For any $x \in H_+$ there exists a sequence $(x_n)_n$, $x_n \in H_+ \cap (\mathcal{P}-\mathcal{P})$ such that $x = \sum_n x_n$.

It is sufficient to show that for any $\varepsilon > o$ there exists $y \in H_+ \cap (\mathcal{P}-\mathcal{P})$ $y \leq x$ such that $\| x-y \| < \varepsilon$. For this purpose let $(y_n)_n$, $y_n \in \mathcal{P}-\mathcal{P}$ such that $x = \lim_{n\to\infty} y_n$. Then $(y_n - R(y_n - x))_n$ converges to x. Using the above theorem $((y_n - R(y_n - x))_+)_n$ converges also to x and

$$o \leq (y_n - R(y_n - x))_+ \leq x.$$

Remark. The assertion from theorem 1-1 was proved also by A.Ancona for the case of Dirichlet spaces which are represented as Hilbert functional spaces. In fact using theorem 1.1 and the following theorem 1-2 it will be shown that any Dirichlet space my be represented as a Hilbert functional space.

We remember that a positive element u of a vector lattice L is cellae a weak unit if

$$x \in L_+, \quad x \wedge u = o \Rightarrow x = o.$$

It is easy to see that if u is a weak unit of L then for any $x \in L_+$ we have $x = \bigvee_n (x \wedge nu)$ if this union exists.

Theorem 1-2. There exists a family $(H_i)_{i \in I}$ of closed, solid subspaces of H such that :

a) For any $i \in I$, H_i possesses a weak unit element u_i for which the sequence $(h \wedge nu_i)_n$ is convergent for any $h \in H_+$.

b) $i, j \in I$, $i \neq j \Rightarrow H_i \cap H_j = \{o\}$ and $\langle Th_i, h_j \rangle = 0$

for any $h_i \in H_i$, $h_j \in H_j$.

c) <u>If $h \in H_+$ then $h_i =: \pi^T_{H_i}(h) \in H_+$, the family $(h_i)_{i\in I}$ is summable in H and we have</u>

$$h = \sum_{i\in I} h_i$$

Proof.

Let $p \in \mathcal{P}$ and $x \in H_+$. Since

$$x \wedge (np) = x \wedge (R(x) \wedge (np))$$

and since the sequence

$$(R(x) \wedge (np))_n$$

converges in H (being an increasing and bounded sequence in $\mathcal{P}$) we deduce, using theorem 1-1 that the sequence $(x \wedge (np))_n$ is convergent.

For any $x \in H$ we denote $x' =: \lim_{n\to\infty}(x \wedge (np))$. We have $y' \wedge (x-x') = o$ for any $x, y \in H_+$. Indeed

$$p \wedge (x-x') = \lim_{n\to\infty} [(p + x \wedge (np)) \wedge x - (x \wedge (np))] =$$

$$= \lim_{n\to\infty} [((n+1)p) \wedge x - x \wedge (np)] = 0.$$

Hence $o \leq y' \wedge (x-x') = \lim_{n\to\infty} y \wedge (np) \wedge (x-x') \leq \lim_{n\to\infty} (np) \wedge (n(x-x')) = o$

Since in the assertions of the theorem only the order relation on H, the topology and the bilinear form

$$(x,y) \longrightarrow \langle Tx, y\rangle$$

play an essential role, we may assume that the scalar product on H satisfies the relation

$$\langle x,y\rangle = \tfrac{1}{2}(\langle Tx,y\rangle + \langle x,Ty\rangle).$$

In this case the space $(H, \langle\ \rangle, H_+, I)$ is a Dirichlet space i.e. $x, y \in H$, $x \wedge y = o \Rightarrow \langle x,y\rangle \leq 0$.

Let $u \in H_+$ be an I-potential (i.e. $\langle u,h\rangle \leq o$ for any $h \in H_+$) and denote

$$H^u =: \{x \in H \mid |x| = \lim_{n\to\infty} |x| \wedge (nu)\}$$

and by

$$H_u =: \{x \in H \mid |x| \wedge u = o\}.$$

Using theorem 1.1 we see that H^u, H_u are closed solid subspaces of H $H^u \cap H_u = \{o\}$ and u is a weak unit element in H^u.

Further for any $x \in H_+$ we have $x' = \lim_{n\to\infty} x \wedge (nu) \in H^u \cap H_+$ and from the first part of the proof $x-x' \in H_u \cap H_+$. Hence H is the direct ordered sum of H^u and H_u.

We show now that for any $x \in H^u$, $y \in H_u$ we have $\langle x,y\rangle = o$. Since H^u, H_u are solid we may assume $x \geq o$, $y \geq o$. From $x \wedge y = o$ we have $\langle x,y\rangle \leq o$. Hence, for any $n \in N$,

$$0 \geq \langle (nu)\wedge x, y\rangle = -\langle nu - nu\wedge x, y\rangle + \langle nu, y\rangle \geq 0,$$

$$\langle x,y\rangle = \lim_{n\to\infty} \langle (nu)\wedge x, y\rangle = 0.$$

Thus we see that H is also the orthogonal sum of H^u, H_u.

Now if p is a general potential with respect to the identity and $p = p' + p''$, $p' \in H^u$, $p'' \in H_u$, then p', p'' are also potentials with respect to the identity. Indeed let $h \in H$ and $h = h' + h''$, $h' \in H^u$, $h'' \in H_u$. Then we have

$$(p',h) = (p',h') = (p,h') \geq o, \quad (p'',h) = (p'',h'') = (p,h'') \geq o$$

From this last remark we see that for any $x \in H^u$ (respectively $x \in H_u$) the reduite $R(x)$ belongs also to H^u (resp. H_u).

Let now $(u_\iota)_{\iota \in I}$ be a family of potentials with respect to the identity such that for any $i \neq j$, $u_i \wedge u_j = o$ and assume that there exists $x \neq o$, $x \in \bigcap_{\iota\in I} H_{u_\iota}$. Then obviously $|x| \in \bigcap_{\iota\in I} H_{u_\iota}$ and from the preceding remark $u =: R(|x|) \in \bigcap_{\iota\in I} H_{u_\iota}$, $u \wedge u_\iota = o$ for any $i \in I$.

A Zorn argument shows now that for a maximal (with respect to the inclusion relation) family $(u_\iota)_{i\in I}$ as above we have $\bigcap_{\iota\in I} H_{u_\iota} = \{o\}$ or equivalently

$$i \in I, \quad x \in H_+, \quad x \wedge u_\iota = o \Longrightarrow x = o.$$

The required family $(H_\iota)_{\iota\in I}$ from the theorem may be taken of the form $H = H^u$ where $(u_\iota)_{i\in I}$ is a maximal family just considered. Obviously in the modified scalar product H_i, H_j are orthogonal closed solid subspaces for $i \neq j$ and H is the direct orthogonal sum of $(H_\iota)_{\iota\in I}$. If $h \in H_+$ and $h_\iota \in H_\iota$ such that $h = \sum_\iota h_\iota$, we have $\sum_{j\neq i} h_j \in H_{u_i}$, and therefore $h_i \geq 0$. Let $x \in H_i$, $y \in H_j$, $i \neq j$. Then

$$\langle Tx, y\rangle = 0 .$$

Indeed we may assume $x \geq 0$, $y \geq 0$ and the assertion follows immediately from the relations :

$$\langle Tx,y\rangle \leq 0, \quad \langle Ty,x\rangle \leq 0 \quad \text{and}$$

$$0 = \langle x,y\rangle = \frac{1}{2}\left(\langle Tx,y\rangle + \langle Ty,x\rangle\right).$$

The equality $h_\iota = \pi^T_{H_i}(h)$ follows now from $\langle T(h-h_\iota),x\rangle = \sum_{j\neq i}\langle Th_j, x\rangle = 0$ for any $x \in H_i$.

__Corollary__. __With the notations from the above theorem if $p \in \mathcal{P}$ then $\pi^T_{H_i}(p) \in \mathcal{P}$ for any $i \in I$ and if $h \in H_i$ then $R(h) \in H_\iota$. Consequently replacing u_i by $R(u_i)$ we may assume $u_\iota \in \mathcal{P}$.__

Proof. If $h \in H_+$, $h = \sum_{\iota\in I} h_\iota$ we have

$$\langle T(\pi^T_{H_i}(p)), h\rangle = \langle T(\pi^T_{H_i}(p)), h_\iota\rangle = \langle Tp, h_i\rangle \geq 0$$

__Remark 1__ . Using theorem 1-2 and the Kakutani representation one may show that a Dirichlet space $(H, \langle\ \rangle, H_+, T)$ is isomorphic with a Hilbert functional space ([3],[4]) on which the module contraction $x \longrightarrow |x|$ and the unit contraction $x \longrightarrow \inf(x, 1)$ operate ([3]). Indeed denote by X_ι the compact space associated with the vector lattice H_i by the Kakutani representation, where u_i is represented by the function 1 and denote by X the direct topological sum of the family $(X_\iota)_{\iota\in I}$. Any T-potential (resp. $\check{T}$-potential) p yields a Radon measure μ_p on X defined by

$$\mu_p(\tilde{h}) = \langle Tp, h\rangle \quad (\text{resp. } \mu_p(\tilde{h}) = \langle \check{T}p, h\rangle$$

where $\tilde{h}$ is the function on X representing the element $h \in H_+$. We denote by $\mathcal{N}$ the subsets A of X which are null-sets for any measure μ_p. $\mathcal{N}$ will play the role of neglijable sets for the wanted Hilbert functional space.

(One may show that if $(h_n)_n$ is a convergent sequence in H such that $\| h_{n+1} - h_n \| < \frac{1}{2^n}$ then the sequence $(h_n)_n$ is convergent pointwisely with the except of a set form $\mathcal{N}$.

Moreover if $u_\iota \in \mathcal{P}$ for any $\iota \in I$ then the unit contraction operates on this functional space.

(Obviously the module contraction operates on this Hilbert functional space and if $u_i \in \mathcal{P}$ for any $\iota \in I$ then the unit contraction operates too.

§2. H- cones in Dirichlet spaces.

Theorem 2.1 .The cone $\mathcal{P}$ possesses the following properties :

1) $\mathcal{P} \subset H_+$;

2) for any family $(p_i)_{i \in I}$ in $\mathcal{P}$ there exists $\bigwedge_{\iota \in I} p_i$ and belongs to $\mathcal{P}$;

3) for any increasing family $(p_i)_{i \in I}$ dominated in $\mathcal{P}$ there exists $\bigvee_{\iota \in I} p_i$ and belongs to $\mathcal{P}$;

4) for any $p, p_1, p_2 \in \mathcal{P}$, $p \leq p_1 + p_2$ there exist $p', p'' \in \mathcal{P}$ such that

$$p = p' + p'', \ p' \leq p_1, \ p'' \leq p_2$$

(i.e. the Riesz splitting property holds in $\mathcal{P}$)

For the proof see [1]; for 4) see also [12].

Remark 1. If $(p_i)_{i \in I}$ is a decreasing (resp. increasing and dominated) family in $\mathcal{P}$ then it is a convergent family in H and

$$\bigwedge_{\iota \in I} p_i \left(\text{resp. } \bigvee_{\iota \in I} p_i\right) = \lim_{\iota} p_i .$$

Remark 2. The convex cones C of positive elements in an ordered vector space E possessing the properties 2),3),4) from the above theorem were considered in [6],[7] and are called H-cones. Such cones occur frequently in potential theory and constitute a natural framework for the development of a great part of this theory as well as of a natural dual theory.

The essential concept in the dual theory for an H-cone C is the H-integral (i.e. a map

$$\mu : C \longrightarrow \overline{R}_+$$

such that

1) $s,t \in C, \quad \alpha \in R_+ \Rightarrow \mu(s+t) = \mu(s) + \mu(t)$;

2) $s,t \in C, \quad s \leq t \Rightarrow \mu(s) \leq \mu(t)$;

3) $s_i \uparrow s, \quad s_i, s \in C \Rightarrow \mu(s_i) \uparrow \mu(s)$;

4) $s \in C \Rightarrow (\exists) s_i \uparrow s, \quad s_i \in C, \quad \mu(s_i) < +\infty$.

The set C^* of all H-integrals on C endowed with the natural algebraic operations and order relation (as functions) forms also an H-cone called the dual of C. The evaluation map

$$C \xrightarrow{\sim} C^{**}$$

(defined by $\tilde{s}(\mu) = \mu(s)$, $s \in C$, $\mu \in C^*$) is linear and preserves the order and lattice operations.

If C^* separates C then

$$s \leq t \iff \tilde{s} \leq \tilde{t}$$

and C will be identified, through the evaluation map, with its images in C^{**} (See [6],[7]).

In an H-cone C we remember among the most useful concepts; the specific order

$$s \precsim t \iff t-s \in C$$

and the reduit

$$R(s-t) = : \wedge \{u \in C \mid u \geq s-t\}.$$

It is shown [7],[9] that C is a lower complete lattice with respect to the specific order and that for any two elements $\mu_1, \mu_2 \in C^*$ the following calculus formula holds :

$$R(\mu_1-\mu_2)(s) = \sup\{\mu_1(t)-\mu_2(t) \mid t \in C,\ t \leq s,\ \mu_2(t)<+\infty\}.$$

Using the above remark 1 it follows that for any $\check{p} \in \check{\mathcal{P}}$ the numerical function on $\mathcal{P}$

$$q \longrightarrow \check{p}(q) = \langle Tq, \check{p} \rangle$$

is an H-integral on the H-cone $\mathcal{P}$ and from $\mathcal{P} = -(\check{T}(H_+))^{\circ}$ we see that

$$\check{p}_1 \leq \check{p}_2 \iff \check{p}_1(q) \leq \check{p}_2(q) \quad \text{for any } q \in \mathcal{P}$$

Thus we may identify $\check{\mathcal{P}}$ with a subcone of $\mathcal{P}^*$; /obviously/ this identification is order preserving. Since $\check{\mathcal{P}}$ and therefore $\mathcal{P}^*$ separates $\mathcal{P}$, then $\mathcal{P}$ may be imbeded one to one, through the evaluation map, in $\mathcal{P}^{**}$.

Proposition 2.1 Any increasing and additive real function μ on $\mathcal{P}$ is an H-integral on $\mathcal{P}$ which belongs to $\mathcal{P}^{\vee}$.

Proof. If we extend μ linearly on $\mathcal{P}-\mathcal{P}$ then there exists $\beta>0$ such that

$$|\mu(h)| \leq \beta \|h\| , \quad h \in \mathcal{P}-\mathcal{P}.$$

On the contrary case there exists a sequence $(h_n)_n$, $h_n=p_n-q_n$, $p_n, q_n \in \mathcal{P}$ such that

$$\| h_n \| \leq \frac{1}{2^n} , \quad |\mu(h_n)| \geq 1 .$$

If we denote

$$u_n = R(h_n)$$

we have $u_n \in \mathcal{P}$,

$$\| u_n \| \leq \frac{\|T\|}{\alpha} \| h_n \| \leq \frac{\|T\|}{\alpha} \cdot \frac{1}{2^n} ,$$

$$1 \leq \mu(h_n) \leq \mu(u_n) .$$

Hence $\sum_{n=1}^{\infty} u_n \in \mathcal{P}$, which leads to the contradictory relation

$$n \leq \sum_{m=1}^{n} \mu(u_m) \leq \mu\left(\sum_{n=1}^{\infty} u_n\right) < +\infty .$$

From the above considerations we see that μ may be extended continously on H and therefore there exists $h \in H$ such that

$$\mu(q) = \langle q, h \rangle , \quad q \in \mathcal{P}.$$

Taking $p^{\vee} = T^{\vee -1}(h)$ we have

$$\mu(q) = \langle q, T^{\vee}(T^{\vee -1}(h))\rangle = \langle q, T^{\vee} p^{\vee}\rangle = \langle Tq, p^{\vee}\rangle .$$

Since μ is increasing on $\mathcal{P}$, using the corollary of theorem 1-1, it follows $p^{\vee} \in \mathcal{P}^{\vee}$ and therefore $\mu = p^{\vee}$.

Proposition 2.2. Any $\mu \in \mathcal{P}^{*}$ is a lower semicontinuous map from $\mathcal{P}$ into $\overline{R}_{+}$. Conversely any additive, increasing and lower semicontinuous numerical map μ on $\mathcal{P}$ which is finite on a dense set is an H-integral.

Proof. For the first assertion it is sufficient to show that

$$K =: \{ p \in \mathcal{P} \mid \mu(p) \leq 1 \}$$

is a closed subset of $\mathcal{P}$. Let $p \in \overline{K}$ and $(p_n)_n$ be a sequence in K convergent to p. We may assume that

$$\| p_n - p \| < \frac{1}{2^n}, \quad n \in N .$$

Denote

$$q_n =: \bigwedge_{k \geq n} p_k$$

Obviously

$$\sum_{k \geq n} R(p_k - p) \in \mathcal{P},$$

$$q_n \geq p - \sum_{k \geq n} R(p_k - p),$$

$$p_{n+i} \geq q_n, \quad n \in N, \ i \in N.$$

Hence

$$p \geq q_n .$$

Since $(q_n)_n$ is increasing $q_n \in K$ we have

$$p = \lim_n q_n = \bigvee_n q_n,$$

$$\mu(p) = \sup_n \mu(q_n) \leq 1, \quad p \in K .$$

For the second assertion we see that the property 3) for an H-integral follows from the above remark 1.

(For the property 4) let $p \in \mathcal{P}$ and let $(p_n)_n$ be a sequence converging to p such that $\mu(p_n) < +\infty$, $n \in N$. We may assume $\| p_n - p \| < \frac{1}{2^n}$, $n \in N$.

(If we denote

$$q_n =: \bigwedge_{k \geq n} p_k$$

we have, as in the first part of the proof,

$$\bigvee_n q_n = p.$$

Obviously

$$\mu(q_n) \leq \mu(p_n) < +\infty, \quad n \in N.$$

Theorem 2.2 . a) Any $\mu \in \mathcal{P}^*$ finite on $\mathcal{P}$ belongs to $\mathcal{P}^\vee$ and $\mathcal{P}^\vee$ is solid in $\mathcal{P}^*$;

b) for any $\mu \in \mathcal{P}^*$ there exists an increasing family $(\mu_\iota)_{\iota \in I}$ in $\mathcal{P}^\vee$ such that $\mu = \bigvee_{\iota \in I} \mu_\iota$;

c) $\mathcal{P}$ is solid and dense in order from below in $\mathcal{P}^{**}$.

Proof. a) follows immediately from proposition 2.1.

We prove now assertion b). Let

$$K =: \mathcal{P} \cap \{h \in H \mid \|h\| \leq 1\} .$$

Obviously K is a weakly compact convex set .

Let $p_o \in K$, $\eta \in R$, $\eta < \mu(p_o)$ and let $\varepsilon > 0$

Since $\mu/_K$ is an affine lower semicontinuous function (proposition 2-2) it is weakly lower semicontinuous and therefore there exists a finite continuous affine function f on K such that

$$f < \mu/_K \,, \quad -\varepsilon \leq f(0) \,, \quad \eta < f(p_o) .$$

Because the set

$$\{ p^\vee - q^\vee + r \mid p^\vee, q^\vee \in \mathcal{P}^\vee, \; r \in R \}$$

is dense uniformly in the space of all finite continuous affine functions on K we may assume f of the form $p^\vee - q^\vee - \varepsilon$. Denote by $\nu \in \mathcal{P}^*$ the reduite of the element $p^\vee - q^\vee - \mu$. Let now p be arbitrary in K. Since

$$q \in \mathcal{P} \,, \; q \leq \frac{\alpha}{\|T\|} p \;\Rightarrow\; q \in K ,$$

using the calculus formula for the reduite we have

$$\nu\left(\frac{\alpha}{\|T\|} p\right) = \sup \left\{ p^\vee(q) - q^\vee(q) - \mu(q) \mid q \in \mathcal{P}, \, q \leq \frac{\alpha}{\|T\|} p, \, \mu(q) < +\infty \right\} \leq \varepsilon .$$

Hence $\nu(p) \leq \frac{\|T\|}{\alpha} \cdot \varepsilon$ for any $p \in K$ and threfore $\nu \in \mathcal{P}^\vee$ (see proposition 2-1), Let now ν_o be the reduite of $p^\vee - q^\vee - \nu$ Since $\nu_o \leq p^\vee$ we have , from assertion a) , $\nu_o \in \mathcal{P}^\vee$.

(Obviously $\nu_o \leq \mu$ and

$$\nu_0(p_0) \geq (p^\vee - q^\vee - \nu)(p_0) \geq (p^\vee - q^\vee)(p_0) - \varepsilon \ + \varepsilon - \nu(p_0) \geq$$
$$\geq f(p_0) + \varepsilon\left(1 - \frac{\|T\|}{\alpha}\right) > \eta + \varepsilon\left(1 - \frac{\|T\|}{\alpha}\right).$$

Since ε, η and p_0 are arbitrary we have

$$\mu = \vee\{\nu \in \mathcal{P}^\vee \mid \nu \leq \mu\}.$$

Assertion b) follows now from the fact that the set

$$\{\nu \in \mathcal{P}^\vee \mid \nu \leq \mu\}$$

is upper directed.

Assertion c) follows from b) and from the relation

$$\mathcal{P} = (\mathcal{P}^\vee)^\vee.$$

§3. Balayage operators in Dirichlet spaces

Again $(H, \langle\ \rangle, H_+, T)$ is a fixed Dirichlet space.

Let $(C, \leq)$ be an H-cone. We recall [6], [8] that a map

$$B : C \longrightarrow C$$

is called a balayage if it is

a) additive and positively homogeneous ,

b) idempotent ,

c) increasing ,

d) contraction (i.e. $Bs \leq s$, $s \in C$) ,

e) continuous in order from below .

The map

$$B^* : C^* \longrightarrow C^*$$

defined by

$$B^*(\mu)(s) = \mu(Bs)$$

is a balayage on C^*. If C^* separates C then

$$B^{**}|_C = B \ .$$

In this section we shall deal with special aspects of balayages on the H-cone of T-potentials.

Theorem 3-1. *Let $B : \mathcal{P} \longrightarrow \mathcal{P}$ be a map satisfying properties a) $\longrightarrow$ d) from above. Then B is a balayage on the H-cone $\mathcal{P}$ and may be extended to a map $\tilde{B} : H \longrightarrow H$ which satisfies the following properties :*

1) $\tilde{B}$ is linear and continuous

2) $\tilde{B}(H_+) \subset H_+$

3) $\tilde{B}x = o \implies \tilde{B}(|x|) = o$

4) $\langle T(\tilde{B}x), y\rangle = \langle T(\tilde{B}x), \tilde{B}y\rangle$

5) $(\tilde{B})^2 = \tilde{B}$

Proof. For any $p,q \in \mathcal{P}$ we have

$$r =: R\,(Bp + Bq-q) = Bp.$$

Indeed we have

$$r \leq Bp,$$
$$r \geq Bp + Bq-q, \quad r \geq Br \geq Bp .$$

Using the relation

$$\langle TRx, x - Rx\rangle = o$$

we deduce

$$\langle TBp, Bq-q\rangle = o .$$

Hence if $x = p_1 - p_2$, $y = q_1 - q_2$ where , p_i, $q_i \in \mathcal{P}$,

we have

$$\langle TBx, y - By\rangle = o$$

where B is considered extended linearly on $\mathcal{P} - \mathcal{P}$.

Particularly we deduce

$$\alpha \| Bx\|^2 \leq \langle TBx, Bx\rangle = \langle TBx, x\rangle \leq \|T\| \cdot \|Bx\| \cdot \|x\| ,$$

$$\|Bx\| \leq \frac{\|T\|}{\alpha} \cdot \|x\|$$

and therefore B may be extended to a linear continuous operator $\tilde{B}: H \longrightarrow H$. The property e) from the definition of a balayage for B follows from the continuity of $\tilde{B}$ in H and from the fact that any increasing and dominated net in $\mathcal{P}$ is convergent in H. The required properties 2), 4), 5) follow immediately remarking that they hold on $\mathcal{P}-\mathcal{P}$ and using the density of $(\mathcal{P}-\mathcal{P}) \cap H_+$ in H_+. (see the corollary of theorem 1-1).

For the property 3) let $x=p-q$ with $p,q \in \mathcal{P}$ such that $Bx=o$. We have

$$Bp = Bq \leq p \wedge q, \quad Bp=Bq=B(p \wedge q),$$

and therefore

$$\tilde{B}(|x|) = \tilde{B}(p-p\wedge q)+\tilde{B}(q-p\wedge q)=o.$$

For x general, with $\tilde{B}x = o$, let $(x_n)_n$ be a sequence in $\mathcal{P}-\mathcal{P}$ converging to x. Since $(x_n-\tilde{B}x_n)_n$ converges also to x, $x_n - \tilde{B}x_n \in \mathcal{P}-\mathcal{P}$

and

$$\tilde{B}(x_n - \tilde{B}x_n) = o$$

we deduce, from the above consideration and theorem 1-1,

$$\tilde{B}(|x|) = \lim_{n\to\infty} \tilde{B}(|x_n-\tilde{B}x_n|) = o.$$

Definition . A continuous linear operator B on H satisfying properties 1) $\longrightarrow$ 5) from the above theorem will be called a T-balayage on H. Obviously we have

$$\| B \| \leq \frac{\| T \|}{\alpha}$$

In fact theorem 3-1 asserts that any balayage on $\mathcal{P}$ is the restriction to $\mathcal{P}$ of a unique T-balayage on H.

Theorem 3-2. Let B be a linear continuous operator on H and

$$K = \text{Ker } B =: \{x \in H \mid Bx = o\}.$$

Then the following assertions are equivalent:

a) K is solid in H and $B = I - \pi_K^T$;

b) B is a T-balayage on H ;

c) $B(\mathcal{P}) \subset \mathcal{P}$ and $B|_{\mathcal{P}}$ is a balayage on $\mathcal{P}$.

Proof. a) $\Rightarrow$ c) We show first that

$$x \in H_+ \Rightarrow y =: Bx \in H_+ .$$

We have

$$y = x - \pi_K^T x \ , \quad y_- \leq |\pi_K^T x|$$

and therefore $y_- \in K$. Using the definition of π_K^T , we get

$$\langle Ty, y_- \rangle = o$$

From

$$o \leq \langle Ty_-, y_- \rangle = \langle T(-y), y_- \rangle + \langle Ty_+, y_- \rangle \leq o$$

it follows

$$y_- = o, \quad Bx \in H_+.$$

We show now that $B(\mathcal{P}) \subset \mathcal{P}$. Let $B' =: I - \pi_K^{\check{T}}$. Since, obviously $K = \operatorname{Ker} B'$ we see, from the above proof, that $B'(H_+) \subset H_+$. Further for any $x, y \in H$ we have $\langle TBx, \pi_K^{\check{T}}(y) \rangle = o = \langle T(\pi_K^T x), B'y \rangle$ and therefore

$$\langle TBx, y \rangle = \langle TBx, \pi_K^{\check{T}} y + B'y \rangle = \langle TBx, B'y \rangle,$$

$$\langle Tx, B'y \rangle = \langle T(Bx + \pi_K^T x), B'y \rangle = \langle TBx, B'y \rangle$$

Hence for any $p \in \mathcal{P}$ and $h \in H_+$ we have

$$\langle TBp, h \rangle = \langle Tp, B'h \rangle \geqslant o$$

and therefore $Bp \in \mathcal{P}$

Finally for any $p \in \mathcal{P}$ we have $Bp \leqslant p$. Indeed since $(p-Bp)_- \in K$ and since $\langle T(Bp), (p-Bp)_- \rangle = o$ we have

$$\langle T((p-Bp)_-), (p-Bp)_- \rangle = \langle T(Bp-p), (p-Bp)_- \rangle +$$

$$+ \langle T(p-Bp)_+, (p-Bp)_- \rangle \leqslant -\langle T p, (p-Bp)_- \rangle \leqslant o$$

and therefore $(p-Bp)_- = o$, $p \geqslant Bp$. The assertion a) $\Longrightarrow$ c) follows now using the first part of theorem 3-1.

The assertion c) $\Longrightarrow$ b), b) $\Longrightarrow$ a) are immediate from theorem 3-1, from the definition of a T-balayage and from the definition of π_K^T.

Remark 1. Theorems 3-1, 3-2, allow us to identify the set of balayage on $\mathcal{P}$ with the set of T-balayages. We shall identify also the set of all balayages on $\mathcal{P}^{\vee}$ with the set of $\check{T}$-balayages.

Remark 2. Since for any closed subspace K of H the map $I - \pi_K^T$ is a linear continuous operator on H for which K its null set, the above Theorem, gives us a one to one correspondence between the set of all balayages on $\mathcal{P}$ and the set of all closed solid subspaces of H.

Remark 3. If B is a balayage on $\mathcal{P}$ then

$$B_1 = I - \pi_K^{\check{T}}$$

where $K = \operatorname{Ker} B$, is a balayage on $\mathcal{P}^{\vee}$ satisfying the relation

$$\langle T Bx, y \rangle = \langle Tx, B_1 y \rangle .$$

Further from the definition of balayage B^* on $\mathcal{P}^*$ and the imbeding of the H-cone $\mathcal{P}^{\vee}$ into $\mathcal{P}^{*}$ we deduce that

$$B^{*}|_{\mathcal{P}^{\vee}} \doteq B_1 .$$

Thus it is natural to use the notation B for B_1. Obviously we have

$$\text{Ker } B = \text{Ker } B^{*},$$
$$B = BB^{*}, \quad B^{*} = B^{*}B.$$

Remark 4. On the set of all balayages on an H-cone C there exists a natural order relation given by

$$B_1 \leq B_2 \Leftrightarrow (s \in C \Rightarrow B_1 s \leq B_2 s)$$

It was shown [9] that this ordered set is a complete distributive lattice.

If we denote, for balayage B on $\mathcal{P}$,

$$\text{Ker } B =: \{ x \in H \mid Bx = o \}$$

then the following assertions concerning the order relation on the set of all balayages on $\mathcal{P}$, are immediate :

a) $B_1 \leq B_2 \Leftrightarrow B_1 = B_1B_2 = B_2B_1 \Leftrightarrow \text{Ker} B_2 \subset \text{Ker } B_1$;

b) If $(B_i)_{i \in I}$ is a family of balayages on $\mathcal{P}$ then

$$\text{Ker } (\bigvee_{i\in I} B_i) = \bigcap_{i \in I} \text{Ker } B_i$$

and Ker $(\bigwedge_{i\in I} B_i)$ is the closure of the space generated by $\bigcup_{i\in I} \text{Ker } B_i$;

c) for any two balayages B_1, B_2 on $\mathcal{P}$ we have

$$p \in \mathcal{P} \Rightarrow (B_1 \wedge B_2)(p) = \bigwedge_n \left((B_1B_2)^n (p)\right) = \lim_{n\to\infty} (B_1B_2)^n (p) ;$$

d) Any lower (resp. upper) directed family $(B_i)_{i\in I}$ of balayages on $\mathcal{P}$ is convergent pointwise and

$$\bigwedge_{i\in I} B_i \text{ (resp } \bigvee_{i\in I} B_i) = \lim_i B_i .$$

Let B be a balayage on $\mathcal{P}$ and denote $\mathcal{P}_B$ the order subcone in H of all elements of the form p-Bp with $p \in \mathcal{P}$ It was shown [9] that $\mathcal{P}_B$ is an H-cone and for any $p, q \in \mathcal{P}$

$$p \wedge (q - Bq) \in \mathcal{P}_B$$

and if

$$q - Bq \leq p-Bp$$

then

$$q-Bq + Bp \in \mathcal{P}$$

Further denote

$$H_B = :\mathrm{Ker}\, B, \quad T_B =: \pi^{I}_{H_B}(T|_{H_B}), \quad H_B^+ = H_B \cap H_+ .$$

Theorem 3-3. With the above notations the system $(H_B, \langle\ \rangle, H_B^+, T_B)$ is a Dirichlet space and we have

a) $\mathcal{P}_B$ is a solid subcone of $\mathcal{P}_{T_B}$;

b) for any $p \in \mathcal{P}_{T_B}$ there exists a sequence $(p_n)_n$ in $\mathcal{P}_B$ such that $p = \sum_{n=1}^{\infty} p_n$;

c) for any $s \in H_B$ such that $(p \in \mathcal{P}_B \Rightarrow s \wedge p \in \mathcal{P}_B)$ we have

$$s \in \mathcal{P}_{T_B}$$

d) For any $p \in \mathcal{P}$ and any $s \in \mathcal{P}_{T_B}$ we have

$$s \wedge p \in \mathcal{P}_{T_B} .$$

Proof. Let $s \in H_B$. We show that $s \in \mathcal{P}_{T_B}$ if and only if there exists a sequence $(p_n)_n$ in $\mathcal{P}$ such that $s = \lim_{n\to\infty}(p_n - Bp_n)$. Indeed we have

$$s \in \mathcal{P}_{T_B} \iff Ts \in -(\mathrm{Ker}\, B \cap H_+)^{\circ} \iff$$

$$\iff Ts \in \overline{(\mathrm{Ker}\, B)^{\circ} - (H_+)^{\circ}} \iff s \in \overline{T^{-1}((\mathrm{Ker}\, B)^{\circ}) + \mathcal{P}} .$$

Since

$$T^{-1}((\mathrm{Ker}\, B)^{\circ}) = B(H)$$

we have

$$s \in \mathcal{P}_{T_B} \iff s = \lim_{n\to\infty}(p_n + Bh_n)$$

where $(p_n)_n$ is a sequence in $\mathcal{P}$ and $(h_n)_n$ is a sequence in H.

Since $s \in H_B$ we have

$$Bs = o, \quad \lim_{n\to\infty}(Bp_n + Bh_n) = o,$$

$$s \in \mathcal{P}_{T_B} \iff s = \lim_{n\to\infty}(p_n - Bp_n).$$

a) Let $s \in \mathcal{P}_{T_B}$, $p \in \mathcal{P}$ and $(p_n)_n$ be a sequence in $\mathcal{P}$ such that

$$s = \lim_{n\to\infty}(p_n - Bp_n), \qquad s \leq p-Bp.$$

Since $\mathcal{P}_B$ is min-stable in H we may assume, using theorem 1-1, that

$$p_n - Bp_n \leq p - Bp \ , \quad n \in N.$$

We have

$$p_n - Bp_n + Bp \in \mathcal{P} \ , \qquad n \in N.$$

$$s + Bp \in \mathcal{P}$$

and therefore

$$s = (s+Bp) - B(s+Bp) \in \mathcal{P}_B$$

b) Let $s \in \mathcal{P}_{T_B}$. It is sufficient to show that, for any $\varepsilon > 0$, there exists $s_\varepsilon \in \mathcal{P}_{T_B}$ and $p_\varepsilon \in \mathcal{P}_B$ such that

$$s = p_\varepsilon + s_\varepsilon \quad \text{and} \quad \| s_\varepsilon \| < \varepsilon$$

For this purpose let $(p_n)_n$ be a sequence in $\mathcal{P}_B$ such that

$$(p_n)_n \longrightarrow s$$

and denote s_n the reduite in H_B of the element $s - p_n$. Obviously $(\| s_n \|)_n \longrightarrow 0$ and

$$s - s_n \in \mathcal{P}_{T_B} \ , \qquad s - s_n \leq p_n$$

We may choose $s_\varepsilon = s_n$ for a sufficiently large n and $p_\varepsilon = s - s_\varepsilon$.

c) Let s_1 the reduite in H_B of s and let $(p_n)_n$ be a sequence in $\mathcal{P}_B$ such that $(p_n)_n \longrightarrow s_1$. Obviously we have

$$s = s \wedge s_1 = \lim_{n \to \infty} (p_n \wedge s) \in \mathcal{P}_{T_B}.$$

d) follows from c) remarking that

$$p \in \mathcal{P} \ , \ q \in \mathcal{P}_B \Rightarrow p \wedge q \in \mathcal{P}_B.$$

Remark 1. If B is a balayage we have, using the above notations:

$$H_B = H_{B^*} , \qquad (T_B)^\vee = T^\vee_{B^*}$$

and the assertions a)—d) from the above theorem are also true for $\mathcal{P}_{(T_B)^\vee}$ and $\mathcal{P}^\vee_{B^*}$ instead of $\mathcal{P}_{T_B}$ and $\mathcal{P}_B$ respectively.

Remark 2. If $p\in\mathcal{P}$ the family $(s\wedge p)_{s\in\mathcal{P}_B}$ is an increasing family of H-integrals on $\mathcal{P}_{(T_B)}^{\vee}$ such that for any $\varkappa=q-B^*q\in\mathcal{P}_{B^*}^{\vee}$ we have

$$(s\wedge p)(\varkappa)\leq\langle Tp,q\rangle$$

Indeed if $s\wedge p=r-Br$ we have

$$r-Br\leq p,$$

$$(s\wedge p)(\varkappa)=\langle T(r-Br),\ q-B^*q\rangle=\langle T(r-Br),q\rangle\leq\langle Tp,q\rangle .$$

Thus we may define an H-integral μ_p on $\mathcal{P}_{(T_B)}^{\vee}$ by

$$\mu_p(\varkappa)=\sup_{s\in\mathcal{P}_B}(s\wedge p)(\varkappa).$$

We have, obviously

$$\mu_{p_1+p_2}=\mu_{p_1}+\mu_{p_2},\quad p_i\uparrow p\Rightarrow\mu_{p_i}\uparrow\mu_p,\quad \mu_{\widehat{i\in I}p_i}=\widehat{i\in I}\mu_{p_i}.$$

§4. Local Dirichlet spaces.

Among Dirichlet spaces $(H,\langle\ \rangle,H_+,T)$ those for which the contractive inequality relation

$$x\wedge y=0\Rightarrow\langle Tx,y\rangle\leq 0$$

is strengthened to an equality (i.e

$$x\wedge y=0\Rightarrow\langle Tx,y\rangle=0)$$

play an important role having as models the Dirichlet spaces associated with elliptic differential operators of second order. We shall call such a Dirichlet space local Dirichlet space.

It turns out that the local property is closely related with a sheaf property for the associated H-cones of potentials which justify also the term local. Also in the general theory of H-cones this property is equivalent with an axiom of domination which coincides with axiom D of Brelot in axiomatic potential theory.

Let f be a positive element in H. Obviously for any $p\in\mathcal{P}$ the sequence $(R(p\wedge nf))_n$ is increasing and dominated by p and therefore it is convergent to an element of $\mathcal{P}$. We shall denote by $B_f:\mathcal{P}\longrightarrow\mathcal{P}$ the map

$$B_f(p)=:\bigvee_n R(p\wedge nf) .$$

Proposition 4-1. Let $f \in H_+$ and denote

$$K =: \{x \in H \mid |x| \wedge f = o\}.$$

Then we have :

a) B_f is a balayage on $\mathcal{P}$;

b) $K = \operatorname{Ker} B_f$;

c) $B_f(f) \geq f$

Proof .We recall first the following property in general archimedian vector lattices : if $o \leq x \leq y$ and $f \geq o$ then

$$(y-x) \wedge f = o \iff (x \wedge (nf) = y \wedge (nf) \text{ for any } n \in N)$$

To prove that let $u=(y-x) \wedge f$. If $u=o$ then $(y-x) \wedge (nf)=o$ and from $y \wedge nf \leq (y-x) \wedge (nf) + x \wedge (nf)$ we get $y \wedge (nf) = x \wedge (nf)$.

Conversely, assuming that $y \wedge (nf)=x \wedge (nf)$ for any $n \in N$, we have inductively $nu \leq x$. Indeed for $n = 1$ we have

$$u \leq y, \quad u \leq f, \quad u \leq y \wedge f = x \wedge f \leq x$$

From $nu \leq x$ we have

$$(n+1)u \leq x+u \leq y, \ (n+1)u \leq (n+1)f, \ (n+1)u \leq y \wedge (n+1)f = x \wedge (n+1)f \leq x.$$

Thus $u \leq \frac{1}{n} x$ for any $n \in N$ and therefore $u = 0$,

a) From the definition it is immediate that B_f is increasing, idempotent and $B_f(p) \leq p$, for any $p \in \mathcal{P}$

Let $p, q \in \mathcal{P}$. From the inequality $(p+q) \wedge nf \leq p \wedge nf + q \wedge nf$ for any $n \in N$ we deduce $B_f(p+q) \leq B_f(p) + B_f(q)$. Let now $p', q' \in \mathcal{P}$ be such that

$$p' \leq B_f p, \ q' \leq B_f q, \ B_f(p+q) = p'+q'$$

From the obvious relations

$$p'+q' \leq p+q, \ (p'+q') \wedge nf = B_f(p+q) \wedge nf = (p+q) \wedge nf,$$

we deduce, using the above considerations,

$$((p+q)-(p'+q')) \wedge f = o, \ (p-p') \wedge f = o, \ (q-q') \wedge f = o.$$

Hence

$$p \wedge nf = p' \wedge nf, \quad q \wedge nf = q' \wedge nf \quad p' \geq B_f p, \ q' \geq B_f q.$$

and therefore B_f is additive. The assertion a) follows now from theorem 3-1.

b) Let $x = p-q$, $x \geq 0, p,q \in \mathcal{P}$. Using again the remark from the beginning of the proof we have

$$x \in K \Leftrightarrow (p \wedge nf = q \wedge nf \text{ for any } n \in N) \Leftrightarrow B_f\, p = B_f q \Leftrightarrow B_f(x) = 0.$$

Now for a general x the assertion b) follows using the theorem 1-1 and its corollary and theorems 3-1, 3-2.

c) Let $p,q \in \mathcal{P}$ such that $0 \leq p-q \leq f$. Since $(p-B_f\, p) \wedge f = 0$ and

$$p-q \leq (p-B_f p)+(B_f p-B_f q),$$

we get

$$p-q=(p-q) \wedge f \leq (p-B_f p) \wedge f + (B_f p-B_f q) \wedge f \leq B_f(p-q).$$

The assertion c) follows from the corollary of theorem 1-1.

Let B be a balayage on $\mathcal{P}$. Then there exists a balayage B' on $\mathcal{P}$ such that $B \vee B' = I$ and such that B' is the smallest balayage having this property. Indeed if we denote

$$K' =: \{x \in H \mid f \in \operatorname{Ker} B \Rightarrow |x| \wedge |f| = 0\},$$

we see that K' is a closed solid subspace of H and for any balayage B_1 on $\mathcal{P}$ we have

$$B \vee B_1 = I \Leftrightarrow \operatorname{Ker} B_1 \subset K'.$$

The balayage B' for which $\operatorname{Ker} B' = K'$ satisfies the above requirements.

Proposition 4-2. Let B be a balayage on $\mathcal{P}$ and denote

$$\mathcal{G} =: \{g \in H_+ \mid g = (1-\varepsilon)p-Bp,\ p \in \mathcal{P},\ \varepsilon > 0\}.$$

Then we have

a) $g \in \mathcal{G} \Rightarrow B_g \wedge B = 0$;

b) $B' = \vee\{B_g \mid g \in \mathcal{G}\} = \vee\{B_f \mid f \in \operatorname{Ker} B,\ f \geq 0\}$;

c) $B = \wedge\{B'_g \mid g \in \mathcal{G}\}$;

d) $B(f) \geq f$ for any $f \in \operatorname{Ker} B'$, $f \geq 0$;

e) $(B')' \leq B$ and $(B'_f)' = B_f$ for any $f \in H_+$.

Proof. Let $p \in \mathcal{P}$, $\varepsilon > o$ and denote

$$g = ((1-\varepsilon)p - Bp)_+ ,$$
$$h = ((1-\varepsilon)p - Bp)_- ,$$
$$h' = ((1-\tfrac{\varepsilon}{2})p - Bp)_+ .$$

By a straightforward calculus we get

$$\tfrac{\varepsilon}{2} p \leq h+h' , \qquad Bh'=o.$$

Using the proposition 4-1 b) we have

$$B_g(h) = o.$$

Let now $x \in H_+$ and denote

$$x' = \lim_{n\to\infty} x \wedge (np), \quad x''=x-x' .$$

Using the arguments from the beginning of the proof of theorem 1-2 we have

$$x' \wedge x'' = o, \qquad x'' \wedge p = o$$

a) With the above notations we have for any $x \in H_+$,

$$x = x'+ x'', \qquad x' \wedge np \leq \frac{2n}{\varepsilon}(h+h').$$

Hence $x' \wedge np = x_n + x_n'$ where $o \leq x_n \leq \frac{2n}{\varepsilon} h$, $o \leq x_n' \leq \frac{2n}{\varepsilon} h'$.

Since $x_n \in \operatorname{Ker} B_g$, $x'' \in \operatorname{Ker} B_g$, $x_n' \in \operatorname{Ker} B$ it follows that $\operatorname{Ker} B_g + \operatorname{Ker} B$ is dense in H. Hence $B_g \wedge B = 0$.

c). Let $g \in \mathcal{G}$, $g=(1-\varepsilon)p - Bp)_+$. We show first

$$B_g' \geq B .$$

This means

$$\operatorname{Ker} B_g' \subset \operatorname{Ker} B.$$

i.e. we have to show that if $x \in H_+$ is such that $x \wedge y = o$ for any $y \in H_+$ for which $y \wedge g = o$ then $Bx = o$. Using the above considerations we have

$$x = x' + x'' ,$$
$$x'' \wedge g \leq x'' \wedge p = o.$$

Hence

$$x'' = x \wedge x'' = o.$$

From $$x' = \lim_{n\to\infty}(x_n + x'_n)$$

and
$$x_n \leq \frac{2n}{\varepsilon}h,$$

we have
$$g \wedge x_n = o$$

and therefore
$$x_n = x \wedge x_n = o\ .$$

Finally we have
$$B\mathfrak{x} = B\mathfrak{x}' = \lim_{n\to\infty} Bx'_n = o.$$

From
$$(1-\varepsilon)p - Bp = g-h$$

and
$$B'_g(g) = o \text{ (see proposition 4-1.b))}$$

we deduce
$$o \leq B'_g(p) - Bp \leq \varepsilon p.$$

Since the family of balayage $(B'_g)_g$ is lower directed we have
$$o \leq \lim_g B'_g(p) - B(p) \leq \varepsilon p,$$

The assertion follows since ε and p are arbitrary.

b) follows from the definition of B' using proposition 4-1 b), theorem 3-3 b) and the remark 4) to the theorem 3-2.

c) Since $B' \vee B = I$ we have $(B')' \leq B$. For the case $B = B_f$ with $f \in H_+$ we have to show $(B')' \geq B$. From the definition of B' we have $f \in \operatorname{Ker} B'$ and therefore using the assertion b) we get $(B')' \geq B$.

d) Assume first $f = p-q$, p, $q \in \mathcal{P}$. We have
$$p-q \leq (p-Bp) + (Bp - Bq)$$

Since $p-Bp \in \operatorname{Ker} B$ it follows
$$(p-Bp) \wedge f = o$$

and therefore
$$f = p-q \leq (B(p)-B(q)) \wedge f + (p-Bp) \wedge f \leq Bp-Bq = Bf\ .$$

For the general case the assertion follows using the corollary of theorem 1-1.

Proposition 4-3. *Let B_1, B_2 be two balayages on $\mathcal{P}$ such that $B_1 \vee B_2 = I$ and $B_1B_2 = B_2B_1$. Then*

$$I + B_1B_2 = B_1 + B_2, \quad B_1B_2 = B_1 \wedge B_2$$

Proof. Let $x \in H$. We have $x - B_1x \in \operatorname{Ker} B_1$,

$$B_1(B_2(x - B_1x)) = B_2(B_1(x - B_1x)) = 0,$$

$$B_2(x - B_1x) \in \operatorname{Ker} B_1,$$

$$(x - B_1x) - B_2(x - B_1x) \in (\operatorname{Ker} B_1) \cap (\operatorname{Ker} B_2).$$

Since $B_1 \vee B_2 = I$ we have $(\operatorname{Ker} B_1) \cap (\operatorname{Ker} B_2) = \{o\}$ and therefore

$$x - B_1x - B_2(x - B_1x) = o,$$

$$x + B_2B_1x = B_1x + B_2x.$$

The last assertion follows immediately since B_1B_2 is a balayage.

Theorem 4-1. *The following assertions are equivalent :*

a) *The Dirichlet space $\langle H, \langle\ \rangle H_+, T\rangle$ is local ;*

b) *For any two balayages B_1, B_2 on $\mathcal{P}$ such that $B_1 \vee B_2 = I$ we have $B_1B_2 = B_2B_1$;*

c) *For any balayage B on $\mathcal{P}$ we have*

$$BB' = B'B \quad ;$$

d) *For any $f \in H_+$ we have*

$$B_f B_f' = B_f' B_f \quad ;$$

c) *For any $f \in H_+$ we have*

$$B_f(f) = f \quad ;$$

f) *For any balayage B on $\mathcal{P}$ and any $g \in \operatorname{Ker} B'$ we have*

$$Bg = g.$$

Proof. a) $\Rightarrow$ b). From the definition of a local Dirichlet space it follows that

$$x, y \in H, \quad |x| \wedge |y| = o \quad \Rightarrow \langle Tx, y\rangle = o.$$

Using theorem 3-2 we have

$$B_1 = I - \pi^T_{K_1}, \qquad B_2 = I - \pi^T_{K_2}$$

where $K_i = \mathrm{Ker}\, B_i$, $i=1,2$. We shaw first that for $x \in K_1$ we have $\pi^T_{K_2} x = 0$. Indeed from the definition of π^T_F and from the relation $K_1 \cap K_2 = \{0\}$ it follows:

$$\langle T(\pi^T_{K_2} x), \pi^T_{K_2} x \rangle = \langle T x, \pi^T_{K_2} x \rangle = 0$$

Hence

$$\pi^T_{K_2} \pi^T_{K_1} = 0$$

and analogously $\pi^T_{K_1} \pi^T_{K_2} = 0$.

Finally $B_1 B_2 = (I - \pi^T_{K_1})(I - \pi^T_{K_2}) = B_2 B_1$

b) $\Rightarrow$ c) and c) $\Rightarrow$ d) are immediate .

d) $\Rightarrow$ e). From proposition 4-3 we have

$$B_f(I - B_f') = I - B_f'$$

Since $f \in \mathrm{Ker}\, B_f'$ it follows

$$B_f(f) = f$$

e) $\Rightarrow$ a). Let $x, y \in H$ be such that $x \wedge y = 0$. We have

$$\langle Tx, y \rangle = \langle TB_x(x), y \rangle = \langle TB_x(x), B_x(y) \rangle = \langle TB_x(x), 0 \rangle = 0$$

c) $\Rightarrow$ f). From proposition 4-3 we have

$$B(I - B') = I - B$$

and therefore

$$Bg = g$$

and any $g \in \mathrm{Ker}\, B'$.

f) $\Rightarrow$ e) follows using the fact

$$f \in \mathrm{Ker}\, B_f'$$

Theorem 4-2. Assume that $(H, \langle\ \rangle H_+, T)$ is a local Dirichlet space. Then for any balayage B, (using the notations from theorem 3-3), the Dirichlet space $(H_B, \langle\ \rangle, H_B^+, T_B)$ is also local.

The proof is immediate.

Proposition 4-4. Assume that the Dirichlet space $(H, < >, H_+, T>$ is local and let $x \in H_+$ and B be a balayage on $\mathcal{P}$ such that for any $t \in \mathcal{P}_B$ we have $x \wedge t \in \mathcal{P}_B$. Then $x - Bx \in \mathcal{P}_{T_B}$.

Proof. Let $p \in \mathcal{P}$, $\varepsilon > 0$ and denote

$$g =: ((1-\varepsilon)p - Bp)_+ .$$

From proposition 4-2 c), $B'_g \geq B$

Further let q be an element of $\mathcal{P}$ such that

$$x \wedge (p-Bp) = q - Bq$$

and denote

$$f =: ((1-\varepsilon)p - Bp - x)_+ .$$

Obviously $f \leq g$, $B'_f \geq B'_g \geq B$ and

$$(x - (q-Bq)) \wedge f = 0.$$

Hence, using proposition 4-1 b),

$$B_f(x) = B_f(q-Bq).$$

By hypothesis $B'_f B_f = B_f B'_f$ (see theorem 4-2 a) $\Rightarrow$ d)) and therefore, using proposition 4-3,

$$B_f + B'_f = I + B_f B'_f .$$

Hence we have

$$x - B'_f x = B_f x - B_f B'_f B_f x = B_f(q-Bq) - B_f B'_f(B_f(q-Bq)) = B_f(q-Bq-B'_f(q-Bq)) =$$

$$= B_f(q - Bq - B'_f q + B'_f Bq) = B_f(q - B'_f q) ,$$

$$x - B'_f x = q - B'_f q.$$

If we denote

$$f_n = ((1-\varepsilon)p - Bp - \frac{1}{n} x)_+$$

we have

$$f_n \uparrow g , \qquad g - f_n \leq \frac{1}{n} x$$

and therefore

$$B'_{f_n} \downarrow B'_g$$

$$x - B'_g x = q - B'_g q.$$

Let now p_0 be a fixed element of $\mathcal{P}$.

We have, using the last relation,

$$(x - B'_g x + B'_g p_o) \wedge p_o \in \mathcal{P}.$$

Using proposition 4-1 c) we have

$$(x - Bx + Bp_o) \wedge p_o = (x - \lim_g (B'_g x - B'_g p_o)) \wedge p_o =$$

$$= \lim_g \; (x - B'_g x + B'_g p_o) \wedge p_o \in \mathcal{P}.$$

Theorem 4-3. The following assertions are equivalent:

a) The Dirichlet space $(H, \langle\ \rangle, H_+, T)$ is local ;

b) For any $x \in H_+$ and any family $(B_i)_{i \in I}$ of balayages on $\mathcal{P}$ such that $\bigwedge_{i \in I} B_i = 0$ and such that $x \wedge t \in \mathcal{P}_{B_i}$ for any $i \in I$ and any $t \in \mathcal{P}_{B_i}$ we have $x \in \mathcal{P}$.

c) For any balayage B on $\mathcal{P}$ and any $x \in H_+$ such that $x \wedge t \in \mathcal{P}_B$ for any $t \in \mathcal{P}_B$ we have

$$(x + Bq) \wedge p \in \mathcal{P}$$

for any $p, q \in \mathcal{P}$ for which we have

$$(x+q) \wedge p \in \mathcal{P}$$

Proof. a) $\Rightarrow$ b). Let x and $(B_i)_{i \in I}$ satisfying the hypotheses of b). Since $\bigwedge_{i \in I} B_i = 0$ it follows that the space generated by $\bigcup_{i \in I} \operatorname{Ker} B_i$ is dense in H. Using theorem 3-2 c) $\Rightarrow$ a) and theorem 1-1 we deduce that the set

$$A =: \{ h = \sum_{i \in J} h_i \mid J \subset I,\ J \text{ finite},\ h_i \in (\operatorname{Ker} B_i) \cap H_+ \}$$

is dense in H_+. From the preceding proposition it follows that

$$i \in I \Rightarrow x - B_i x \in \mathcal{P}_{T_{B_i}}$$

i.e.

$$\langle T(x - B_i x), h \rangle \geq 0$$

for any $h \in H_+ \cap (\operatorname{Ker} B_i)$. Hence for any $h \in A$, $h = \sum_{i \in J} h_i$ we have

$$\langle Tx, h \rangle = \sum_{i \in J} \langle Tx, h_i \rangle = \sum_{i \in J} \langle T(x - B_i x), h_i \rangle \geq 0$$

and therefore $x \in \mathcal{P}$.

b) $\Rightarrow$ c). Let x and B satisfying the hypotheses of c) and let p, $q \in \mathcal{P}$ such that

$$(x+q)\wedge p \in \mathcal{P}.$$

Further for $u \in \mathcal{P}$ and $\varepsilon > 0$ denote

$$g = ((1-\varepsilon)u - Bu)_{+}.$$

and

$$y = (x + B'_g q) \wedge p.$$

It is sufficient to show that the element y and the family (B, B_g) satisfy the conditions from the assertion b). Indeed, using proposition 4-2 a) we have

$$B_g \wedge B = 0.$$

Let now $t \in \mathcal{P}_{B_g}$. Then we have

$$((x+q)\wedge p - y)\wedge t \le ((x+q)-(x+B'_g q))\wedge t = (q-B'_g q)\wedge t.$$

Since obviously $t \in \operatorname{Ker} B_g$ it follows

$$(q-B'_g q)\wedge t = 0,$$

$$((x+q)\wedge p)\wedge t = y\wedge t,$$

$$y\wedge t \in \mathcal{P}_{B_g}.$$

If $t \in \mathcal{P}_B$ we have

$$y \wedge t = (x + B'_g q)\wedge t \wedge p =$$

$$= (x\wedge t + B'_g (q)\wedge t)\wedge t \wedge p \in \mathcal{P}_B.$$

c) $\Rightarrow$ a) Let B_1, B_2 be two balayages on $\mathcal{P}$ such that $B_1 \vee B_2 = I$. In virtue of theorem 4-1 a) $\Leftrightarrow$ b) we have to show that $B_1B_2 = B_2B_1$.

If we denote

$$A =: B_1B_2 - B_2B_1B_2$$

we get inductively, for any $p \in \mathcal{P}$,

$$p_n = p - n \,.\, Ap \in \mathcal{P}, \quad n \in N.$$

Indeed if $p_n \in \mathcal{P}$ then the pair (B_2, x) where

$$x = p_n - B_1B_2p_n$$

satisfies the conditions from c). Obviously

$$(x + B_1B_2p_n)\wedge p_n = p_n \in \mathcal{P}.$$

Hence

$$(x + B_2B_1B_2p_n) \wedge p_n = p_n - B_1B_2p_n + B_2B_1B_2p_n \in \mathcal{P}.$$

Since $B_1B_2A = o$ we get

$$p_n - B_1B_2p_n + B_2B_1B_2p_n =$$

$$p - n\,Ap - B_1B_2p + B_2B_1B_2\,p = p_{n+1} \in \mathcal{P}.$$

$$Ap \leq \frac{1}{n}\,p$$

e get

$$Ap = o,$$

$$B_1B_2 = B_2B_1B_2,$$

$$(B_1B_2)^2 = B_1B_2,$$

and therefore

$$B_1B_2 = B_1 \wedge B_2.$$

Analogously

$$B_2B_1 = B_1 \wedge B_2$$

and therefore

$$B_1B_2 = B_2B_1.$$

Remark. The assertion b) may be interpreted as a sheaf property for the H-cone $\mathcal{P}$ with respect to a topology for which a basis is given by the system of solid closed subspaces of H.

Bibliography

1. A.Ancona . Contraction module et principe de réduite dans les espaces ordonnées à forme coercive. C.R.Acad.Sc.Paris,t.275(1972) p.701-704

2. A.Ancona.Continuité des contractions dans les espace de Dirichlet C.R.Acad.Sc.Paris,t.282 (1976) p.871 . .

3. A.Ancona.Continuité des contractions dans les espaces de Dirichlet (Séminaire de Théorie du Potentiel) no.563 (1976) p.1-26

4. N.Aronszajn and K.Smith. Functional spaces and functional completion Ann, Inst.Fourier 6 (1956) p.125-185.

5. A.Beurling and J.Deny: Dirichlet spaces.Proc.Nat.Acad.of.Sci.45 (1959), p.259-271.

6. N.Boboc et A.Cornea Cônes convexes ordonnés.H cônes et adjoints de H-cônes C.R.Acad.Sci.Paris 270 (1970) p.598-599.

7. N.Boboc et A.Cornea . Cônes convexes ordonnés. H cônes et biadjoints de H-cônes .C.R.Acad.Sci.Paris 270 (1970) p.1679-1682.

8. N.Boboc et A.Cornea. Cônes convexes ordonnés . Représentations intégrales. C.R.Acad.Sci.Paris 271(1970), p.880-883

9. N.Boboc, Gh.Bucur and A.Cornea, H-cones and potential theory. Ann.Inst.Fourier 25(1975) p.71-108

10. H.Cartan - Théorie du potentiel ... Bull.Soc.Math.de France 73 (1945) p.74-106

11. H.Deny . Méthodes hilbertiennes en théorie du potentiel (Centro Internationale Matematico Estivo, Stresa 1969)

12. G.Mokobodzki . Structures des cônes de Potentiels. Séminaire Bourbaki no.377, 1970.

13. G.Stampacchia . Formes bilinéaires coercives sur les ensembles convexes. C.R.Acad.Sci.Paris 258 (1964) , p.4413-4416

University of Bucharest
Faculty of Mathematics

INCREST
Bucharest

Approximating, majorizing, and extending functions defined on unbounded sets

by

P.M. Gauthier

Let V be defined on a subset F of G , and let u be defined on all of G . From our point of view the notions in the title are related as follows.

Approximation: $u|F - v$ is small.

Majorization: $u|F - v$ is positive.

Extension: $u|F - v$ is harmonic.

The results I shall mention on approximation are joint work with Myron Goldstein and W. Ow. Those on majorization and extension are joint work with Myron Goldstein. Full details will be published elsewhere.

Approximation. G is a domain in $\mathbb{R}^n$ or an open Riemann surface, F the closure in G of an open set in G , ∂F the boundary in G of F , H(F) the harmonic functions on F , $H_c(F)$ the functions continuous on F and harmonic on the interior, and G* the one-point compactification of G .

Theorem 1. H(G) is uniformly dense in H(F) iff $G^*\backslash F$ is connected and locally connected.

Theorem 2. Suppose $G\backslash F$ is not thin at each point of ∂F . Then H(G) is uniformly dense in $H_c(F)$ iff $G^*\backslash F$ is connected and locally connected.

Our theorems cover only those sets F which have an interior. The analogue of Theorem 2 in case $F^{\circ} = \phi$ has been solved by Šaginjan [5]. In case F is compact, both of our theorems are classical and were obtained by Brelot [1] and Deny [2].

Majorization and extension. Let G be a Green space in the sense of Brelot-Choquet and $\hat{G}$ its Martin compactification.

Theorem 3. Let $\hat{G}$ be a Martin space, locally connected at the boundary, and with every boundary point minimal. Then every subharmonic function on G having a local harmonic majorant near each boundary point has a global harmonic majorant.

In proving this theorem, we make use of the equivalence [4] of passing from local harmonic majorants to global harmonic majorants and of extending local potentials to global potentials.

As an application of Theorem 3, we solve a Dirichlet problem. We call a Martin space $\hat{G}$ a regular space if it is regular for the Dirichlet problem. Let f be an extended real valued continuous function on the boundary Δ of $\hat{G}$. We say that the Dirichlet problem for F has a Dirichlet solution u if u is a continuous extended real valued function on G which is harmonic on $\hat{G}$ and agrees with f on Δ.

Theorem 4. Let $\hat{\Omega}$ be a regular Martin space which is locally connected at the boundary Δ and such that each point of Δ is minimal. Let f be an extended real valued continuous function on Δ. Then, the Dirichlet problem for f has a Dirichlet solution if and only if it has a Perron-Weiner-Brelot solution. Moreover, the Dirichlet solution, it it exists, is essentially unique in that its quasibounded component is the P-W-B solution.

For the unit disc, Theorem 3 was obtained in [3].

Theorem 3 and 4 apply, in particular, to bounded Lipschitz domains and to compact bordered Riemann surfaces. However, we show by example that for a domain in $\mathbb{R}^n$, we cannot always replace the Martin boundary by the usual boundary in Theorem 3.

REFERENCES

1. Brelot, M., Sur l'approximation et la convergence dans la theorie des fonctions harmoniques ou holomorphes. Bull. Soc. Math. France 73, 55-70(1945). MR 7-205.

2. Deny, J., Sur l'approximation der fonctions harmoniques. Bull. Soc. Math. France 73, 71-73(1945). MR 7-205.

3. Gauthier, P.M. and Hengartner, W., Local harmonic majorants of functions subharmonic in the unit disc. J. D'Analyse Math., 26(1973), 405-412. MR 48#8816.

4. Hueber, H., Doctoral dissertation, University of Bielefeld(submitted 1975).

5. Šaginjan, A.A., The uniform and tangential harmonic approximation of continuous functions on arbitrary sets. (Russian) Mat. Zametki 9 (1971), 131-142. MR 45#2375.

Université de Montréal
Montréal, Canada

An Almost Everywhere Regular, Metrizable Boundary Supporting the Maximal Representing Measures for Bounded and Quasibounded Harmonic Functions.

Peter A. Loeb

In [12], Hunt and Wheeden show that the Euclidean and minimal Martin boundaries coincide for a Lipshitz domain in Euclidean space, i.e., for a domain D which is determined locally, along with its boundary ∂D, by functions of class Lip. 1. Therefore, not only are points of ∂D regular with respect to the Dirichlet problem, but having chosen $x_0 \in D$, there is for each positive harmonic function h on D a unique Borel measure ν_h (maximal with respect to the Choquet ordering [21]) on ∂D such that ν_h represents h. To be exact, there is a kernel $K(x,y)$ on $D \times \partial D$ with the following properties:

i) For each $x \in D$, $K(x,\cdot)$ is a continuous representative on ∂D of the Radon-Nikodym derivative $\frac{d\mu_x}{d\mu_{x_0}}$ of harmonic measure for x, μ_x, with respect to harmonic measure for x_0. (Here, for each continuous f on ∂D, the harmonic extension h_f at x is $\int_{\partial D} f\, d\mu_x$.)

ii) For each $y \in \partial D$, $K(\cdot,y)$ is a positive harmonic function on D, and the mapping $y \longrightarrow K(\cdot,y)$ from ∂D <u>into</u> the positive harmonic functions h on D with $h(x_0) \leq 1$ is a homeomorphism with respect to the (metrizable) topology of uniform convergence on compact sets, the u.c.c. topology.

iii) For each positive harmonic function h on D, there is a unique Borel measure ν_h supported by points y corresponding to minimal harmonic functions $K(\cdot,y)$ (i.e., if g is harmonic and $0 \leq g \leq K(\cdot,y)$ on D, then g is a multiple of $K(\cdot,y)$) such that for each $x \in D$,

$$h(x) = \int_{\partial D} K(x,y)\, d\nu_h(y).$$

Note that for a Lipshitz domain D, every $y \in \partial D$ is a minimal point, i.e., $K(\cdot,y)$ is a minimal harmonic function. For arbitrary domains, the usual generalization with respect to the above properties of the Euclidean boundary is the Martin boundary (see [19], or [5] or [6]). M. G. Shur ([23], [24]) has shown, however, that there can exist a minimal point z on the Martin boundary such that z is irregular with respect to the Dirichlet problem and yet $\{z\}$ has positive harmonic measure.

In this note, we describe an alternative generalization of the boundary and kernel for a Lipshitz domain. Our kernel satisfies (i) and (ii), and Property (iii) is satisfied not for every positive harmonic function but at least for positive bounded and quasibounded (i.e., limit of an increasing sequence of bounded) harmonic functions on the given domain. Moreover, every minimal point y of our boundary Δ is a regular point, i.e., $\lim_{x \to y} h_f(x) = f(y)$ for each continuous f on Δ. Since harmonic measure for x_0 is ν_1, almost all points of Δ are regular. Complete details of this work will be given in Chapters 5 and 6 of [16]. We will not, however, use nonstandard analysis [22] here, nor is knowledge of nonstandard analysis required for Chapters 5 and the beginning of Chapter 6 in [16].

We first consider a generalization of potential theory itself. Since our results are valid for a number of cases including solutions of rather general elliptic differential equations on Euclidean spaces and solutions of $\Delta u = Pu$, $P \geq 0$, on open Riemann surfaces, we work in the setting of a hyperbolic Brelot harmonic space ([3], [4], [13]) for which the constant function 1 is superharmonic. A reader who is interested in only one of the cases subsumed by Brelot's theory can of course assume that the case of interest is the one under discussion; recall, however, that 1 is here superharmonic, not necessarily harmonic.

Brelot's theory considers a locally compact, but not compact, connected and locally connected Hausdorff space W with a family $\mathcal{H}$ of functions called harmonic functions. An open subset $\Omega \subset W$ is called an inner region if it is connected and has compact closure $\overline{\Omega}$ in W. Such an Ω is called regular if each continuous f on $\partial\Omega$ has a continuous harmonic extension h_f on Ω with $h_f \geq 0$ if $f \geq 0$. For each $x \in \Omega$, μ_x^Ω denotes the corresponding harmonic measure; i.e. $h_f(x) = \int_{\partial\Omega} f \, d\mu_x^\Omega$. Each $h \in \mathcal{H}$ has an open domain in W; letting

$$\mathcal{H}_\Omega = \{h \in \mathcal{H} : \text{domain } h = \Omega\} ,$$

one assumes that $\mathcal{H}_\Omega$ is a real vector space for each open $\Omega \subset W$. Moreover, one assumes that a function h is in $\mathcal{H}$ if and only if h is locally in $\mathcal{H}$, that there is a base for the topology of W consisting of regular inner regions, and that the upper envelope of an increasing sequence of harmonic functions in a region is either identically $+\infty$ or again harmonic.

A lower semicontinuous function v with open domain Ω is called superharmonic, and we write $v \in \overline{\mathcal{H}}_\Omega$, if v is finite at some point in each component of Ω and if for each $x \in \Omega$ and each regular, inner region ω with $x \in \omega \subset \overline{\omega} \subset \Omega$ we have

$$-\infty < \int_{\partial\omega} v \, d\mu_x^\omega \leq v(x).$$

A function $v \in \overline{\mathcal{H}}_W$ for which the greatest harmonic minorant is 0 is called a potential on W. We assume that $1 \in \overline{\mathcal{H}}_W$, that there is a positive potential on W, and that there is a nonzero bounded harmonic function on W. That is, $\mathcal{H}$ is hyperbolic on W in the sense of [13]. These assumptions are weaker than those needed to obtain the usual Martin boundary theory for Brelot harmonic spaces.

Let $\overline{W}$ be the unique (up to homeomorphism) Hausdorff compactification of W such that each bounded $h \in \mathcal{H}_W$ and each bounded continuous potential on W has a continuous extension to $\overline{W}$ and the set of these extensions separates the points of $\overline{W} - W$. (See [6], [14], or [15].) Let Γ be the harmonic part of $\overline{W} - W$, i.e., the set of points at which all positive potentials on W have lim inf 0. The mapping $h \longrightarrow h/\Gamma$ is an isometric isomorphism from the Banach space $\mathcal{H}_W^b$ of bounded harmonic functions on W with the sup. norm <u>onto</u> the space $C(\Gamma)$ of continuous real-valued functions on Γ with the sup. norm. (See Section 2 of [18] or Section 5 of [16].) All positive harmonic functions on W have continuous, extended real-valued extensions to $\overline{W}$. Of course, if $f \in C(\Gamma)$, then f is the restriction to Γ of a unique harmonic function $h_f \in \mathcal{H}_W^b$. For each $x \in W$, let μ_x

denote the Radon measure supported by Γ such that $h_f(x) = \int_\Gamma f\, d\mu_x$ for each $f \in C(\Gamma)$. We call μ_x the harmonic measure for x on $\overline{W} - W$.

Choose $x_0 \in W$. For each point $x \in W$, μ_x is absolutely continuous with respect to μ_{x_0} and the Radon-Nikodym derivative $\frac{d\mu_x}{d\mu_{x_0}}$ is bounded. Moreover, each $L_\infty(\mu_{x_0})$ class on Γ contains a unique continuous representative. Therefore, for each $x \in W$ there is an $h_x \in \mathcal{H}_W^b$ such that $\frac{d\mu_x}{d\mu_{x_0}} = h_x | \Gamma$. For each pair $x,y \in W$, let

$$r(x,y) = h_x(y) = \int_\Gamma \frac{d\mu_x}{d\mu_{x_0}} \frac{d\mu_y}{d\mu_{x_0}}\, d\mu_{x_0}.$$

Clearly $r(x,y) = r(y,x)$.

Let φ denote the unique continuous mapping of W onto the unique quotient $\widetilde{W}$ of $\overline{W}$ such that for each $x \in W$, $\varphi(x) = x$, $r(x,\cdot)$ has a continuous extension to $\widetilde{W}$, and the set $\{r(x,\cdot) : x \in W\}$ of extensions separates the points $\Delta \equiv \widetilde{W} - W$. For each $x \in W$, let $q(x,y)$ denote the extension of $r(x,\cdot)$ at $y \in \widetilde{W}$. Since r is symmetric, we may assume that $q(x,y) = q(y,x)$ for each $x \in W$, $y \in \widetilde{W}$.

The kernel $q(x,y)$, $x \in W$, $y \in \Delta$, satisfies Properties i and ii satisfied by the kernel $K(x,y)$ for a Lipshitz domain. The compactification $\widetilde{W}$ is an example of a general class of compactifications considered by Thomas Armstrong in [1], and Properties i and ii can be established for q as a corollary of results in [1]. Of fundamental importance here, however, is the fact that for $x \in W$, $q(x,y)$ is defined for all $y \in \tilde{W}$. We now sketch our main results.

<u>Theorem 1</u>. Fix $z \in \Delta$ and assume that $q(z,x_0) = 1$. (This is the case if $1 \in \mathcal{H}_W$ or if $z \in \varphi(\Gamma)$.) Then z is a regular point if z is a minimal point.

<u>Proof</u>: For each $x \in W$, let $\tilde{\mu}_x$ denote harmonic measure for x on Δ. If f is continuous on Δ

$$\int_\Delta f\, d\tilde{\mu}_x = \int_\Gamma f \circ \varphi\, d\mu_x = h_{f \circ \varphi}(x).$$

Let x_α be a net converging to z in W and let γ be the weak* limit of an arbitrary weak*-convergent subset $\tilde{\mu}_{x_\beta}$ of $\tilde{\mu}_{x_\alpha}$. For each $w \in W$,

$$\int_\Delta q(s,w)\,d\gamma(s) = \lim_\beta \int_\Delta q(s,w)\,d\tilde{\mu}_{x_\beta}(s) = \lim_\beta q(x_\beta,w) = q(z,w),$$

and $\int_\Delta d\gamma = \int_{\varphi(\Gamma)} d\gamma = q(z,x_0) = 1.$

Thus γ is a probability measure on the set of harmonic functions $\{q(s,\cdot) : s \in \varphi(\Gamma)\}$, and γ represents the minimal harmonic function $q(z,\cdot)$. Therefore $\gamma(\{z\}) = 1$, i.e., z is a regular point.

<u>Theorem 2</u>. For each nonnegative $h \in \mathcal{H}_W$ and each Borel set $A \subset \Delta$, set

$$\nu_h(A) = \int_{\varphi^{-1}[A] \cap \Gamma} h\, d\mu_{x_0}.$$

If $h \geq 0$ is bounded or quasibounded, then ν_h is a (unique) Borel measure on the minimal points z of $\varphi(\Gamma)$ (whence $q(z,x_0) = 1$) such that for each $w \in W$,

$$h(w) = \int_{\varphi(\Gamma)} q(z,w)\, d\nu_h(z).$$

That is, ν_h represents h in the sense of Property iii, and ν_h is maximal with respect to the Choquet ordering [21].

Proof: For each $w \in W$,

$$h(w) = \int_{\Gamma} r(y,w)\, h(y)\, d\mu_{x_0}(y) = \int_{\Delta} q(z,w)\, d\nu_h(z).$$

Since the mapping $h \longrightarrow \nu_h$ is affine, it follows from Corollaries of a result of B. Fuchssteiner [10] that ν_h is maximal with respect to the Choquet ordering and is, therefore, supported by the minimal points of $\varphi(\Gamma)$. (See Section 4 of [16].)

Theorem 3. If h is the greatest harmonic minorant of 1, then ν_h is a multiple of harmonic measure $\tilde{\mu}_{x_0}$. It follows that almost all points (with respect to $\tilde{\mu}_{x_0}$) are minimal points in $\varphi(\Gamma)$ and are therefore regular.

Theorem 4. Let $\{\Omega_n\}$ be a countable exhaustion of W by regular inner regions, i.e., $\overline{\Omega}_n \subset \Omega_{n+1}$ for each n and $\bigcup \Omega_n = W$. (See [9] and Section 4 of [13].) Given a bounded $h \in \mathcal{H}_W$, ν_h is absolutely continuous with respect to $\tilde{\mu}_{x_0}$, and

$$\frac{d\nu_h}{d\tilde{\mu}_{x_0}}(z) = \lim_{n \to \infty} \int_{\partial\Omega_n} h(y)\, q(y,z)\, d\mu_{x_0}^{\Omega_n}(y)$$

for $\tilde{\mu}_{x_0}$-almost all $z \in \Delta$.

Proof: A proof using nonstandard analysis [22] and the usual theory of fine limits ([5] or [11]) is given in [16].

When it exists, the Martin compactification of W may be quite unlike the compactification $\tilde{W}$. For example, given $\mathcal{H}$ and a positive $h \in \mathcal{H}_W$, the class $\mathcal{H}/h = \{f/h : f \in \mathcal{H}\}$ is a harmonic class (see [3], [4], or [13]) with the same Martin boundary for W as $\mathcal{H}$. If h is minimal, however, the only bounded elements of $\mathcal{H}/h$ are multiples of 1. In this case, Γ is a single point and $\tilde{W}$ is the one point compactification of W. For a Lipshitz domain, however, we have the following result.

Theorem 5. Let W be a bounded Lipschitz domain in Euclidean space and $\mathcal{H}$, the class of solutions of Laplace's equation $\Delta u = 0$ on W. Then $\tilde{W}$ is the Euclidean closure $\overline{W}^E$ of W, and q is the kernel K for the Lipshitz domain.

Proof. The functions $K(x,\cdot)$, $x \in W$, are continuous, they separate points of $\partial W = \overline{W}^E - W$, and they have continuous extensions to $\overline{W}^E$. Therefore, $\overline{W}^E$ is the continuous image under the map ψ, with $\psi(x) = x$ for each $x \in W$, of the compactification $\overline{W}$. Since each $h \in \mathcal{H}_W^b$ is the integral (with respect to K) of its fine limits on ∂W, $K \circ \psi = r$ on $\overline{W}$. Therefore, $K = q$ and $\tilde{W} = \overline{W}^E$.

Bibliography

[1] T. E. Armstrong, Poisson Kernels and Compactifications of Brelot Harmonic Spaces, Ph.D. dissertation, Princeton University, 1973.

[2] H. Bauer, Harmonishe Räume und ihre Potential theorie, Springer-Verlag, Berlin, 1966.

[3] M. Brelot, Lectures on Potential Theory, Tata Institute, Bombay, 1961.

[4] ________, Axiomatique des Fonctions Harmoniques, University of Montreal Press, Montreal, 1966.

[5] ________, On Topologies and Boundaries in Potential Theory, Springer-Verlag, Berlin, 1971.

[6] C. Constantinescu and A. Cornea, Ideale Ränder Riemannsher Flächen, Springer-Verlag, Berlin, 1963.

[7] ________________________, Compactifications of harmonic spaces, Nagoya Math. Jour., Vol. 25(1965), pp. 1-57.

[8] ________________________, Potential Theory on Harmonic Spaces, Springer-Verlag, Berlin, 1972.

[9] A. Cornea, Sur la denombrabilite a l'infini d'un espace harmonique de Brelot, C. R. Acad. Sci. Paris, 1967, pp. 190A-191A.

[10] B. Fuchssteiner, Sandwich theorems and lattice semigroups, Journal of Functional Analysis, Vol. 16, No. 1 (1974), pp. 1-14.

[11] K. Gowrisankaran, Fatou-Naim-Doob limit theorems in the axiomatic system of Brelot, Ann. lnst. Fourier, Greenoble, Vol. 16 (1966), pp. 455-467.

[12] R. Hunt and R. Wheeden, Positive harmonic functions on Lipshitz domains, Trans. Amer. Math. Soc., Vol. 147 (1970), pp. 507-526.

[13] P. A. Loeb, An axiomatic treatment of pairs of elliptic, differential equations, Ann. Inst. Fourier, Grenoble, Vol. 16, No. 2 (1966), pp. 167-208.

[14] ________, A minimal compactification for extending continuous functions, Proc. Amer. Math. Soc. Vol. 18, No. 2 (1967), pp. 282-283.

[15] ________, Compactifications of Hausdorff spaces, Proc. Amer. Math. Soc., Vol. 22, No. 3 (1969), pp. 627-634.

[16] P. A. Loeb, Applications of nonstandard analysis to ideal boundaries in potential theory, to appear in The Israel Journal of Mathematics.

[17] P. A. Loeb and B. Walsh, The equivalence of Harnack's principle and Harnack's inequality in the axiomatic system of Brelot, Ann. Inst. Four., Grenoble, Vol. 15 (1965), pp. 597-608.

[18] ______________________, A maximal regular boundary for solutions of elliptic differential equations, Ann. Inst. Fourier, Grenoble, Vol. 18, No. 1 (1968), pp. 283-308.

[19] R. S. Martin, Minimal positive harmonic functions, Trans. Amer. Math. Soc., Vol. 49 (1941), pp. 137-172.

[20] C. Meghea, Compactification des Espaces Harmoniques, Springer-Verlag, Berlin, 1971.

[21] R. Phelps, Lectures on Choquet's Theorem, Van Nostrand, Princeton, 1966.

[22] A. Robinson, Nonstandard Analysis, North-Holland, Amsterdam, 1966.

[23] M. G. Shur, A Martin compact with a non-negligible irregular boundary point, Theory of Probability and its Applications, Vol. 17, No. 2 (1972), pp. 351-355.

[24] __________, An example of a Martin compact with a nonnegligible boundary point, Trudy Moskov. Mat. Obšč 28(1973), 159-179. English translation: Trans. Moscow Math. Soc. 28(1973), 158-178 (1975).

Department of Mathematics
University of Illinois
Urbana, Illinois 61801

WHAT IS THE RIGHT SOLUTION OF THE DIRICHLET PROBLEM ?

Jaroslav LUKEŠ and Ivan NETUKA, Praha

Introduction. The problem of finding a harmonic function (on a bounded open set $U \subset R^n$) taking on preassigned continuous boundary values is known as the Dirichlet problem. It is historically the oldest problem of existence of potential theory and various attacks on this problem brought many important mathematical discoveries. Let us mention at least methods of the calculus of variations connected with the Dirichlet integral, the alternating method of H.A. Schwarz, the method of the arithmetic mean of C. Neumann and "méthode de balayage" of H. Poincaré, or the method of integral equations of I. Fredholm. Each of these methods, however, had its limitations on the shape of the domain or the properties of the boundary values.

At the close of the last century the Dirichlet problem was regarded as always solvable and it was believed that methods of proof powerful enough would be found to confirm this opinion. A new period begins with the recognition that this opinion is not justified. In 1903, M. Bôcher proved that an isolated singularity was removable for bounded harmonic functions. As remarked by S. Zaremba in 1911, this meant the existence of open sets for which the classical Dirichlet problem was not solvable for all continuous boundary data. More specifically, Zaremba pointed out that domains with isolated boundary points are not regular for the Dirichlet problem. An even more striking example of a non-regular domain was given for three-dimensional space by H. Lebesgue in 1913. In these examples, some boundary points turn out to be exceptional for the Dirichlet problem. It was recognized that it is connected with the behaviour of the Green function of the set in question. A point of the boundary at which the Green function (with a fixed pole) approaches 0 is said to be regular. Regular sets are exactly those open sets having only regular boundary points.

It was H. Lebesgue who explicitely proposed to separate the investigation of the Dirichlet problem into two parts: At first produce a harmonic function depending in a way on the given boundary condition and then investigate the boundary behaviour of the resulting candidate for a solution. Some old methods for a construction of such a harmonic function were mentioned above, but none of them applied to the case of general domains. On the other hand, in the twenties of this century, two completely different new methods without any limitation on the region were proposed.

O. Perron 1923 [14] and R. Remak [15] considered an arbitrary bounded function f on the boundary of a general bounded open set U and defined the upper class for f as the set of all continuous superharmonic functions in U whose lower limit dominates f at each boundary point of U. The infimum $\overline{H}_f^U$ of such a class was shown to be a harmonic function called the upper solution. The lower solution $\underline{H}_f^U$ was defined similarly and the relation $\underline{H}_f^U \leqq \overline{H}_f^U$ was established. The equality of the upper and lower solutions for a continuous f was proved by N. Wiener 1925 [18].

One year before, N. Wiener defined in [17] another type of solution, the idea of which was based on the following observations: Given a continuous function f on ∂U, one can always extend f to a continuous function F on $\bar{U}$. Further, it is possible to exhaust U by a sequence $\{V_n\}$ of regular sets. To each V_n then there corresponds a classical solution h_n of the Dirichlet problem taking on the values of F on ∂V_n. It was shown by Wiener that the sequence $\{h_n\}$ converges to a harmonic function h on U which is independent of the particular choice of F and $\{V_n\}$. Wiener proved that his generalized solution h tends to the prescribed boundary value at any regular point and established in [18] the equality of his solution with that obtained by Perron. (The question whether such a generalized solution is the only bounded harmonic function having preassigned boundary values at each regular point was settled later.)

A more general treatment of Perron's method was given by M. Brelot 1940 [3]. He considered for an arbitrary numerical function on ∂U the Perron upper class formed by lower semicontinuous superharmonic functions and characterized functions for which upper and lower solutions coincide.

A.F. Monna noticed in 1938, 1939 (see [13] where relevant references and interesting comments on the subject may be found) that the methods of Perron and Wiener are special constructions only and investigated the unicity of the Dirichlet problem from the functional analysis point of view. He asked whether an operator of the Dirichlet problem (submitted to certain natural conditions) was uniquely determined. The similar question was posed and solved by M.V. Keldych 1941 [8], [9] by proving that there is exactly one positive linear operator sending continuous functions on ∂U into harmonic functions on U such that its value is the classical solution, if it exists. (An operator possessing these properties is termed a Keldych operator.) For the Laplace equation there is therefore the only "reasonable" solution of the Dirichlet problem and Perron's or Wiener's methods appear thus as its special constructions. (The same unicity result holds for a wide class of partial differential equations of elliptic type as shown by M. Brelot 1960 [4].)

The Perron's method was extended by W. Sternberg 1929 [16] to the heat equation and later by many authors to abstract theories of harmonic spaces (cf. [1], [5], [6]). In this general situation, uniqueness questions turn out to be more delicate. In contrast to the Laplace case, there is e.g. for some open sets $U \subset R^{n+1}$ more than one reasonable solution of the Dirichlet problem for the heat equation, so that the Keldych theorem fails. Consequently, one can ask what is in fact "the right solution" of the Dirichlet problem.

We shall investigate questions of unicity and the exceptional role of Perron's generalized solution in the set of all Keldych operators. Moreover, we shall introduce another special Keldych

The Wiener type solution has not been studied in the context of abstract potential theory. The reason lies in the fact that a direct application of Wiener's method is not possible because, in general, there are open sets G not containing sufficiently large regular sets. It means that one cannot insert a regular set V between G and an arbitrarily chosen compact $K \subset G$ in the sense that $K \subset V \subset \bar{V} \subset G$. This was observed by H. Bauer 1966 [1] (see p. 147). (To give an example for the case of the heat equation in R^2, take $G = \{[x,y];\ 0 < x^2+y^2 < 2\}$ and $K = \{[x,y];\ x^2+y^2 = 1\}$.) On the other hand, in any elliptic space with the positive potential, the existence of an exhaustion formed by regular sets is guaranteed by a result of R. -M. Hervé 1962 [7] (see Proposition 7.1; cf. [6], Exercise 3.1.14) and in fact Wiener's method may be repeated without changes in the frame of elliptic spaces.

An attempt to modify Wiener's procedure for the heat equation and more general parabolic equations is due to E.M. Landis 1969 [10] and 1971 [11]. His method may be considered as a special case of the following scheme. Given a relatively compact open set U (in a harmonic space) and a continuous function f on ∂U, choose (as in the classical case) a continuous extension F of f to $\bar{U}$. Take any exhaustion $\{V_n\}$ of U consisted of arbitrary open sets with $\bar{V}_n \subset U$. Since V_n's need not be regular and there is no preferable solution on V_n, consider an arbitrary Keldych operator A_n on V_n and investigate the sequence of harmonic functions $h_n = A_n(F/\partial V_n)$. It can be shown that the sequence $\{h_n\}$ converges to a harmonic function W_f^U on U. The function W_f^U is independent of the special choice of F, $\{V_n\}$ and, which is perhaps surprising, even of $\{A_n\}$. Moreover, the equality $W_f^U = H_f^U$ holds.

Keldych sets. In what follows, x will denote a $\mathcal{P}$-harmonic space with countable base in the sense of the axiomatics of C. Constantinescu and A. Cornea. The corresponding harmonic sheaf is denoted by $\mathcal{H}$, $U \subset X$ will be a fixed relatively compact open set such that the system $H(\bar{U})$ of all functions continuous on $\bar{U}$ and harmonic on U contains a strictly positive function and separates points of $\bar{U}$

Recall that a positive linear mapping $A : C(\partial U) \to \mathcal{H}(U)$

such that $A(h \nearrow \partial U) = h$ on U for any $h \in H(\bar{U})$ is termed a <u>Keldych operator</u>. Thus, A gives the classical solution of the Dirichlet problem, provided such a solution exists. If A is a Keldych operator on U and $x \in U$, then the mapping

$$f \longmapsto Af(x) \ , \qquad f \in C(\partial U) \ ,$$

is a positive Radon measure on ∂U which will be called the <u>Keldych measure</u> and will be denoted by α_x.

Clearly, $f \longmapsto H_f^U$ is a Keldych operator. It follows, that the corresponding measure ε_x^{CU} (which is obtained by balayage of the Dirac measure ε_x on the complement of U) is a Keldych measure. It should be mentioned that ε_x^{CU} is the <u>maximal</u> Keldych measure with respect to the partial ordering induced by the smallest min-stable wedge containing $H(\bar{U})$. This shows a special role of the Perron solution among all Keldych operators.

It was proved by J. Bliedtner and W. Hansen [2] that the system $H(\bar{U})$ is simplicial. Thus, for every $x \in U$ there is a unique <u>minimal</u> measure δ_x^U. If we define $D_f^U(x) = \delta_x^U(f)$ for any $x \in U$ and any $f \in C(\partial U)$, then $f \longmapsto D_f^U$ is also a Keldych operator. Another type of a generalized solution was introduced in [12]. The so-called <u>principal solution</u> L_f^U (which is determined on the cone of continuous potentials as the greatest idempotent operator smaller than the usual balayage) is again a Keldych operator.

We shall say that U is a <u>Keldych set</u> if there is exactly one Keldych operator on U. Of course, in this case, the generalized Perron solution H_f^U is the only Keldych operator on U. The following theorem characterizes completely Keldych sets and represents a generalization of the classical theorem of Keldych.

<u>Theorem.</u> The following assertions are equivalent:

(i) U is a Keldych set;

(ii) the set of all irregular points of U is of harmonic measure zero for each $x \in U$

(iii) $D_f^U = H_f^U$ on U for any $f \in C(\partial U)$;

(iv) $L_f^U = H_f^U$ on U for any $f \in C(\partial U)$;

(v) $(\varepsilon_x^{CU})^{CU} = \varepsilon_x^{CU}$ for each $x \in U$.

Denote by $\mathfrak{M}$ the collection of all Keldych measures. Thus, $\alpha \in \mathfrak{M}$ if and only if there is a Keldych operator A on U and $x \in U$ such that $\alpha = \alpha_x$. Further, C_R will be the set of all restrictions of $H(\bar{U})$ to ∂U.

Suppose now that there is a topology τ on $C(\partial U)$ such that the following two conditions hold:

(1) C_R is τ-dense in $C(\partial U)$

(2) for each $\alpha \in \mathfrak{M}$, the functional defined on $C(\partial U)$ by

$$f \longmapsto \alpha(f)$$

is τ-continuous.

Then, obviously, there is exactly one Keldych operator on U. The functional character of the unicity result is explained by the following assertion.

Theorem. The following conditions are equivalent:

(i) U is a Keldych set ;

(vi) there exists a topology τ on $C(\partial U)$ satisfying (1) and (2).

The Wiener type solution. Now we shall propose a modified method of Wiener's type solution for the Dirichlet problem. A sequence $\{V_n\}$ of subsets of U is said to be an exhaustion of U, if each V_n is a relatively compact open set, $\bar{V}_n \subset U$ and if to any compact set $K \subset U$ there corresponds a positive integer $n(K)$ such that $K \subset V_n$ for each $n \geqq n(K)$. Given $f \in C(\partial U)$, let F be an arbitrary continuous extension of f to $\bar{U}$. Taking an arbitrary exhaustion $\{V_n\}$ of U and considering an arbitrary Keldych operator A_n on V_n, it can be proved that

$$W_f^U : x \longmapsto \lim A_n(F/\partial V_n)(x)$$

is a harmonic function on U. The following important lemma enables to prove that W_f^U is independent of the particular continuous extension of f, of the exhaustion $\{V_n\}$ and also of the choice of Keldych operators A_n used in the construction, and combined with some reasonings concerning simpliciality leads to the main result stated below.

Lemma. If $K \subset U$ is an arbitrary compact set, then there is a Keldych set Y with $K \subset Y \subset \bar{Y} \subset U$.

Theorem. The equality $W_f^U = H_f^U$ holds for any $f \in C(\partial U)$.

In other words, the described Wiener type construction and the Perron type construction lead to the same result.

Using our theorem it is easily seen that the Landis´solution defined for parabolic equations in [10], [11] concides with the Perron one. Thus, Landis´regularity conditions of boundary points refer to the notion of regularity usually adopted in potential theory.

In the following Corollary it is, in a certain sense, included an answer to the title of our lecture: The Perron type solution has a priority position among all Keldych operators provided a stability condition is required.

Corollary. Let A be a Keldych operator on U. Then the statements (a) and (b) are equivalent:

(a) A satisfies the following "interior stability condition":

If f is continuous on $\bar{U}$, $\{V_n\}$ is an exhaustion of U and A_n is a Keldych operator on V_n, then

$$\lim A_n(f/\partial V_n)(x) = A(f/\partial U)(x), \quad x \in U.$$

(b) A is the Perron type solution.

All details of this brief exposition are contained in papers:

J. Lukeš, Functional approach to the Brelot-Keldych theorem, to appear in Czechoslovak Math. J.

J. Lukeš and I. Netuka, The Wiener type solution of the Dirichlet problem in potential theory, to appear

REFERENCES

[1] H. Bauer: Harmonische Räume und ihre Potentialtheorie, Lecture Notes in Math. 22, Springer-Verlag, Berlin, 1966

[2] J. Bliedtner and W. Hansen: Simplicial cones in potential theory, Invent. Math. 29(1975), 83-110.

[3] M. Brelot: Famille de Perron et problème de Dirichlet, Acta Sci. Math. (Szeged) 9(1938-40), 133-153.

[4] M. Brelot: Sur un théorème du prolongement fonctionnel de Keldych concernant le problème de Dirichlet, J. Analyse Math. 8(1961), 273-288.

[5] M. Brelot: Axiomatique des fonctions harmoniques, Les Presses de l'Université de Montréal, Montréal, 1966.

[6] C. Constantinescu and A. Cornea: Potential theory on harmonic spaces, Springer-Verlag, Berlin, 1972.

[7] R. -M. Hervé: Recherches axiomatiques sur la théorie des fonctions surharmoniques et du potentiel, Ann. Inst. Fourier 12(1962), 415-571.

[8] M. V. Keldych: On the resolutivity and the stability of Dirichlet problem (Russian), Uspechi Mat. Nauk 8(1941) 172-231.

[9] M. V. Keldych: On the Dirichlet problem (Russian), Dokl. Akad. Nauk SSSR 32(1941), 308-309.

[10] E. M. Landis: Necessary and sufficient conditions for the regularity of a boundary point for the Dirichlet problem for the heat equation (Russian), Dokl. Akad. Nauk SSSR 185(1969), 517-520.

[11] E. M. Landis: Equations of the second order of elliptic and parabolic types (Russian), Nauka, Moscow, 1971.

[12] J. Lukeš: Théorèm de Keldych dans la théorie axiomatique de Bauer des fonctions harmoniques, Czechoslovak Math. J. 24(99)(1974), 114-125.

[13] A.F. Monna: Note sur le problème de Dirichlet, Nieuw Arch. Wiskunde 19(1971), 58-64.

[14] O. Perron: Eine neue Behandlung der ersten Randwertaufgabe für $\Delta u = 0$, Math. Z. 18(1923), 42-54.

[15] R. Remak: Über potentialkonvexe Funktionen, Math. Z. 20(1924), 126-130.

[16] W. Sternberg: Über die Gleichung der Wärmeleitung, Math. Ann. 101(1929), 394-398.

[17] N. Wiener: Certain notions in potential theory, J. Math. Massachussetts 3(1924), 24-51.

[18] N. Wiener: Note on a paper of O. Perron, J. Math. Massachussetts 4(1925), 21-32.

Jaroslav Lukeš and Ivan Natuka

Katedra matematické analýzy a jejich aplikaci

Odděleni teorie funkci a potanciálu
Matematicko-fyzikálni fakulta KU

Sokolovská 83
186 00 Praha 8
Czechoslovakia

Banach-Stone-type theorems for harmonic spaces

U. Schirmeier, Erlangen, F.R. of Germany

In this lecture we are going to formulate and prove a theorem of Banach-Stone type in the theory of harmonic spaces.

By this well-known theorem two compact spaces X and $\tilde{X}$ are homeomorphic iff their rings $\mathcal{C}(X)$ and $\mathcal{C}(\tilde{X})$ of continuous functions are isomorphic. Analogous statements in the theory of complex functions are known : A theorem of Bers says, that two domains in $\mathbb{C}$ are conformally equivalent (up to conjugation) iff the corresponding rings of holomorphic functions are isomorphic. An isomorphism φ between the algebras $\mathcal{K}(X)$ and $\mathcal{K}(\tilde{X})$ of all meromorphic functions on two compact Riemannian surfaces X and $\tilde{X}$ with the property $\varphi(1) = 1$ induces a biholomorphic map $\psi : \tilde{X} \longrightarrow X$ such that $\varphi(h) = h \circ \psi$ holds for all $h \in \mathcal{K}(X)$.

When trying to get a Banach-Stone-type theorem for harmonic spaces the first question which arises is the following :

Which are the "right" function spaces to consider ?
It is not very useful to consider the space of all harmonic functions, because there exist very different harmonic spaces, whose sets of all globally defined harmonic functions contain only the constants. The following example shows, that there cannot be a positive answer to the problem stated above, when we consider certain subcones of the cone of all globally defined hyperharmonic functions :

Example 1. Let X be the half-open interval $[0,1[$ and $\tilde{X}$ the open interval $]0,1[$.
For every interval V open in X we define the number

$$m_V := \begin{cases} \sup\left\{m \in \mathbb{N} : \frac{1}{m} \in V\right\} \subset \mathbb{N} \cup \{\infty\} & \text{if } V \cap \left\{\frac{1}{m} : m \in \mathbb{N}\right\} \neq \emptyset \\ 1 & \text{otherwise.} \end{cases}$$

The harmonic structures $\mathcal{H}^*$ and $\tilde{\mathcal{H}}^*$ on X and $\tilde{X}$ are defined as follows :
A lower semi-continuous, lower finite function u defined on such an interval V belongs to $\mathcal{H}^*(V)$ ($= \tilde{\mathcal{H}}^*(V)$, if $0 \notin V$) iff u is decreasing on

$V \cap [\frac{1}{m_V}, 1[$ and concave on the intervals $V \cap]\frac{1}{n+1}, \frac{1}{n}[$, $n \in \mathbb{N}$.

Especially a continuous function $h : V \longrightarrow \mathbb{R}$ is harmonic iff h is constant on $V \cap [\frac{1}{m_V}, 1[$ and its restriction to the interval $V \cap [0, \frac{1}{m_V}[$ is an affine function. With respect to the sheaves $\mathcal{H}^*$ and $\tilde{\mathcal{H}}^*$ of harmonic functions described in this way, the intervals $]a,b[$ $(0 < a < b < 1)$ and $[0,a[$ (for $\mathcal{H}$) form a base of regular sets.

In this example the restriction map

$$\varphi : p \longmapsto p|_{]0,1[}$$

defines an isomorphism of the cones

$\mathcal{H}^*(X)$ and $\tilde{\mathcal{H}}^*(X)$, respectively

$\mathcal{S}(X)$ and $\tilde{\mathcal{S}}(\tilde{X})$, respectively

$\mathcal{P}(X)$ and $\tilde{\mathcal{P}}(\tilde{X})$,

although the spaces X and $\tilde{X}$ are not homeomorphic.

Since φ does not define an isomorphism between the cones $\mathcal{P}^c$ and $\tilde{\mathcal{P}}^c$ of all continuous real-valued potentials nor between the cones $\mathcal{P}_{\mathbb{R}}$ and $\tilde{\mathcal{P}}_{\mathbb{R}}$ of <u>all</u> real-valued potentials (the potential $\tilde{p} \in \tilde{\mathcal{P}}^c \subset \tilde{\mathcal{P}}_{\mathbb{R}}$, defined by

$\tilde{p}(x) := 2n-n.(n+1)x$ for $\frac{1}{n+1} \leq x < \frac{1}{n}$, $n \in \mathbb{N}$,

has no inverse φ-image in $\mathcal{P}_{\mathbb{R}}$!), we may hope to find a Banach-Stone theory with one of these cones (if these cones are "big" enough, that is, if the harmonic spaces in consideration are $\mathfrak{p}$-harmonic spaces).

In the following we denote by $(X, \mathcal{H}^*)$ and $(\tilde{X}, \tilde{\mathcal{H}}^*)$ two harmonic spaces. The terminology used is essentially that of the book [3]. In addition we need the following

<u>Definition 1</u>. Let P and $\tilde{P}$ be two convex, inf-stable subcones of the cones $\mathcal{P}_{\mathbb{R}}$ and $\tilde{\mathcal{P}}_{\mathbb{R}}$ of all real-valued potentials on X and $\tilde{X}$. An <u>epimorphism</u> (<u>of cones</u>) is an additive, positive-homogeneous, surjective map $\varphi : P \longrightarrow \tilde{P}$ such that $\varphi(\inf(p,q)) = \inf(\varphi(p), \varphi(q))$ holds for all $p, q \in P$.

If in addition φ is injective, then φ is called an <u>isomorphism</u> (<u>of cones</u>).

2. A continuous map $\psi : \tilde{X} \longrightarrow X$ is a <u>harmonic map</u> iff for every open subset U of X the inclusion
$$\mathcal{H}^*(U) \circ \psi := \{h \circ \psi_{|\psi^{-1}(U)} : h \in \mathcal{H}^*(U)\} \subset \tilde{\mathcal{H}}^*(\psi^{-1}(U))$$
holds.
If ψ is bijective and ψ as well as ψ^{-1} are harmonic, then ψ is called a <u>biharmonic map</u>.

<u>Remark</u>. Let f be a continuous, strictly positive function on X. If we denote by $\mathcal{H}^*_f$ the sheaf $\mathcal{H}^*$ divided by f, then $(X, \mathcal{H}^*_f)$ again is a harmonic space. Its cone $\mathcal{P}^c_f$ of all continuous real-valued potentials is isomorphic to $\mathcal{P}^c$ in spite of the fact, that there exists no biharmonic map between the spaces $(X, \mathcal{H}^*)$ and $(X, \mathcal{H}^*_f)$ in general.
This is the reason why in the following theorem such a dividing function turns up.

<u>Theorem 1</u>. Let $(X, \mathcal{H}^*)$ and $(\tilde{X}, \tilde{\mathcal{H}}^*)$ be two σ-compact, $\mathfrak{P}$-harmonic spaces and let

$$\varphi : \mathcal{P}^c \longrightarrow \tilde{\mathcal{P}}^c$$

be an epimorphism of the cones $\mathcal{P}^c$ and $\tilde{\mathcal{P}}^c$ of all continuous real-valued potentials on X and $\tilde{X}$.

Then there exists one and only one continuous injective map

$$\psi : \tilde{X} \longrightarrow X$$

and a continuous, strictly positive function $\tilde{f}$ on $\tilde{X}$ such that $\varphi(p) = \tilde{f} . (p \circ \psi)$ holds for all $p \in \mathcal{P}^c$.

Additionally the following properties are true :

(1) ψ is a homeomorphism from $\tilde{X}$ onto $\psi(\tilde{X})$.

(2) $\psi(\tilde{X})$ is a closed subset of X.

(3) If X' denotes the interior (which may possibly be empty) of $\psi(\tilde{X})$, then

$$\psi_{|\psi^{-1}(X')} : \psi^{-1}(X') \longrightarrow X'$$

is a biharmonic map between the two spaces
$(\psi^{-1}(X'), \tilde{\mathcal{H}}^*_{\tilde{f}}|_{\psi^{-1}(X')})$ and $(X', \mathcal{H}^*|_{X'})$.

If φ is injective we have the required Banach-Stone-type theorem :

Corollary : Every isomorphism $\varphi: \mathcal{P}^c \longrightarrow \tilde{\mathcal{P}}^c$ between the cones of all continuous real-valued potentials of two σ-compact, $\mathfrak{P}$-harmonic spaces $(X, \mathcal{H}^*)$ and $(\tilde{X}, \tilde{\mathcal{H}}^*)$ induces a biharmonic map $\psi: (\tilde{X}, \tilde{\mathcal{H}}^*_{\tilde{f}}) \longrightarrow (X, \mathcal{H}^*)$, where $\tilde{f}$ is a continuous, strictly positive function on $\tilde{X}$.

Remarks. 1. In general the map ψ is not harmonic, as the following example shows :

Example 2. We consider the spaces $X :=]-\infty, 1[$ and $\tilde{X} := [0,1[$. Let $(X, \mathcal{H}^*)$ be the $\mathfrak{P}$-harmonic space of the solutions of the Laplace equation. For every open connected subset U of X with $0 \in U$ resp. $0 \notin U$ we denote by $\tilde{\mathcal{H}}^*(U)$ the set of all decreasing concave functions $f : U \longrightarrow \mathbb{R} \cup \{\infty\}$ (i.e. $\tilde{\mathcal{H}}(U)$ contains only the constants) resp. $\tilde{\mathcal{H}}^*(U) = \mathcal{H}^*(U)$. It is easy to see that the map

$$\varphi : \mathcal{P}^c \longrightarrow \tilde{\mathcal{P}}^c$$
$$p \longmapsto p|_{\tilde{X}}$$

defines an epimorphism of cones. (Every potential $\tilde{p} \in \tilde{\mathcal{P}}^c$ is the restriction of the potential $p \in \mathcal{P}^c$, defined by

$$p = \begin{cases} \tilde{p} & \text{on } \tilde{X} \\ \tilde{p}(0) & \text{on } X \setminus \tilde{X} \end{cases}).$$

The corresponding continuous injection is the canonical embedding

$$\psi : \tilde{X} \longrightarrow X$$

which is not a harmonic map.

2. If the topology of X has a countable base, then the following two properties can be shown to be equivalent :

(a) ψ is a harmonic map.

(b) $\psi(\tilde{X})$ is an absorbing set of X.

3. Again let X have a countable base. Then each point $x \in \partial(\Psi(\tilde{X}))$ is a regular boundary point of the open set $X \setminus \Psi(\tilde{X})$.

4. Let $(X, \mathcal{H}^*)$ and $(\tilde{X}, \tilde{\mathcal{H}}^*)$ be two harmonic spaces with a base of regular sets. If X and $\tilde{X}$ are open subsets of $\mathbb{R}^n$ and all harmonic functions are of class $\mathcal{C}^k$ $(k \geq 2)$, then every biharmonic map $\Psi : \tilde{X} \longrightarrow X$ is k-times continuously differentiable on a dense open subset $\tilde{X}_o$ of $\tilde{X}$. The proof of this fact depends on the results of BONY [2].

5. It may happen that the set $\Psi(\tilde{X})$ has no interior points, as the following example shows :

<u>Example 3.</u> Let $X =]0,1[\times \mathbb{R}$ and $\tilde{X} =]0,1[$. For every open subset U of X we denote by $\mathcal{H}^*(U)$ the set of all lower semi-continuous functions $f : U \rightarrow \mathbb{R} \cup \{\infty\}$, whose restrictions to $U \setminus (\tilde{X} \times \{0\})$ are hyperharmonic with respect to the classical harmonic structure and whose restrictions to every connecting component of $U \cap (\tilde{X} \times \{0\})$ are concave.

It can be shown, that $(X, \mathcal{H}^*)$ is a $\mathfrak{P}$-harmonic space. In this space the harmonic functions are exactly those continuous functions, which are solutions of the Laplace equation outside $\tilde{X} \times \{0\}$ and which are locally affine on $\tilde{X} \times \{0\}$. The set of all rectangles $]a,b[\times]c,d[$ $(0 < a < b < 1, \quad c < d)$ is a strong base of regular sets.

If we endow $\tilde{X}$ with the harmonic structure $\tilde{\mathcal{H}}^*$ defined by the solutions of the Laplace equation on $\mathbb{R}$, then $\tilde{X}$ can be embedded into X by the harmonic map

$$\Psi : \tilde{X} \longrightarrow X$$
$$x \mapsto (x,0)$$

such that the restriction

$$\varphi : \mathfrak{P}^c \longrightarrow \tilde{\mathfrak{P}}^c$$
$$p. \mapsto p \circ \Psi(.) = p(.,0)$$

is an epimorphism of cones. In fact, $\tilde{X} \times \{0\}$ is an absorbing set of X.

This shows, that we are here in the situation of theorem 1, but the embedded set $\psi(\tilde{X})$ has no interior points.

6. Statements analogous to those in theorem 1 for injective, non-surjective morphism of cones are false in general.

   Let us consider the harmonic space $(X,\mathcal{H}^*)$ of example 2. We shall now define a second harmonic structure $\bar{\mathcal{H}}^*$ on X as follows : For every open connected set $U \subset X$ we denote by $\bar{\mathcal{H}}^*(U)$ the set of all lower semi--continuous functions $u : U \longrightarrow \mathbb{R} \cup \{\infty\}$ which are concave on $U \cap]0,1[$ and on $U \cap]-\infty,0[$, and which are decreasing on $U \cap [0,1[$ if $0 \in U$.

   The cone $\bar{\mathcal{P}}^c$ of all continuous real-valued potentials corresponding to $(X,\bar{\mathcal{H}}^*)$ contains $\mathcal{P}^c$, but neither the inclusion $\mathcal{H}(X) \subset \bar{\mathcal{H}}(X)$ nor the inclusion $\bar{\mathcal{H}}(X) \subset \subset \mathcal{H}(X)$ hold.

The proof of theorem 1 is carried out in the following steps :

1. $E := \mathcal{P}^c - \mathcal{P}^c$ is an adapted space. This property implies the existence and uniqueness of a continuous injective map $\psi : \tilde{X} \longrightarrow X$, which is a homeomorphism from $\tilde{X}$ onto $\psi(\tilde{X})$, and the existence of a continuous, strictly positive function $\tilde{f}$ on $\tilde{X}$ such that $\varphi(p) = \tilde{f}.(p \circ \psi)$ holds for all $p \in \mathcal{P}^c$.

2. If $p \in \mathcal{P}^c$ is strictly positive, then $\varphi(p)$ is strictly positive. This follows immediately from 1. By dividing $\mathcal{H}^*$ by p and $\tilde{\mathcal{H}}^*$ by $\varphi(p)$ we may therefore always assume, that $1 \in \mathcal{P}^c$ and $\varphi(1) = 1$.

3. If p and $q \in \mathcal{P}^c$ coincide outside a compact set, the same holds for $\varphi(p)$ and $\varphi(q)$. This property implies the closedness of $\psi(\tilde{X})$ as a subset of X.

4. The last (and rather technical) step in the proof is to show the biharmonicity of the map

$$\psi|_{\psi^{-1}(X')} : \psi^{-1}(X') \longrightarrow X'.$$

As so far we have considered the cones $\mathcal{P}^c$ and $\tilde{\mathcal{P}}^c$. The question

"What happens, if we replace $\mathcal{P}^c$ and $\tilde{\mathcal{P}}^c$ by the cones $\mathcal{P}_{\mathbb{R}}$ and $\tilde{\mathcal{P}}_{\mathbb{R}}$, as suggested by example 1 ?"

is (partially) answered by the following

Theorem 2. Let $(X,\mathcal{H}^*)$ and $(\tilde{X},\tilde{\mathcal{H}}^*)$ be two σ-compact, $\mathfrak{P}$-harmonic spaces and let P and $\tilde{P}$ be two inf-stable convex subcones of $\mathcal{P}_{\mathbb{R}}$ and $\tilde{\mathcal{P}}_{\mathbb{R}}$ such that $\mathcal{P}^c \subset P$, $\tilde{\mathcal{P}}^c \subset \tilde{P}$ and $1 \in P$.
Then to every isomorphism of cones

$$\varphi : P \longrightarrow \tilde{P}$$

with $\varphi(1) = 1$ corresponds one and only one bi-harmonic map

$$\psi : \tilde{X} \longrightarrow X$$

such that

$$\varphi(p) = p \circ \psi$$

holds for all $p \in P$.

Remarks. 1. The proof of this theorem is more complicated than that of theorem 1 because of the following fact: After establishing the existence of a bijection $\psi : \tilde{X} \longrightarrow X$ such that $\varphi(p) = p \circ \psi$ for all $p \in P$, all information on the topologies of X and $\tilde{X}$ must be obtained from the cones P and $\tilde{P}$, in order to prove the continuity of ψ. Once this is carried out, theorem 1 can be used to complete the proof.

2. The assumption $\varphi(1) = 1$ is rather unnatural. For a general isomorphism $\varphi : P \longrightarrow \tilde{P}$ the property $\varphi(1) = 1$ cannot be achieved by a "dividing procedure" unless there exists a strictly positive potential $p \in \mathcal{P}^c$ such that $\varphi(p)$ is continuous too.
In a special case the existence of such a potential can be proved :

Theorem 3. Let φ be an isomorphism of the cones of all real-valued fine potentials of two $\mathfrak{P}$-harmonic spaces $(X,\mathcal{H}^*)$ and $(\tilde{X},\tilde{\mathcal{H}}^*)$ with a countable base and satisfying the axiom of domination.

Then there exists a strictly positive continuous function $\tilde{f}$ on $\tilde{X}$ and a biharmonic map Ψ between the harmonic spaces $(\tilde{X},\tilde{\mathcal{H}}^*_f)$ and $(X,\mathcal{H}^*)$ such that $\varphi(p) = \tilde{f}.(p \circ \Psi)$ holds for every real-valued fine potential p.

Bibliography.

[1] H. BAUER : Harmonische Räume und ihre Potentialtheorie; LN 22, Springer, Berlin (1966).

[2] J.-M. BONY : Détermination des axiomatiques de théorie du potentiel dont les fonctions harmoniques sont différentiables ; Ann. Inst. Fourier, Grenoble, 17, 1 (1967), pp. 353 - - 382.

[3] C. CONSTANTINESCU, A. CORNEA : Potential Theory on Harmonic Spaces ; Die Grundlehren der math. Wiss. in Einzeldarstellungen, Bd. 158, Springer, Berlin (1972).

[4] U. SCHIRMEIER : Isomorphie harmonischer Räume ; Math. Ann. (to appear).

U. Schirmeier

Mathematisches Institut
der Universität Erlangen-
-Nürnberg

Bismarckstraße 1 1/2

D8520 Erlangen

Multiplicative Properties of Elliptic Boundary Value Problems

by Bert-Wolfgang Schulze

We consider elliptic boundary value problems of the type

$$\mathcal{A} = \begin{pmatrix} r^+A+r'B & K \\ r'T & Q \end{pmatrix} : \begin{matrix} \Gamma(X,E) \\ \oplus \\ \Gamma(Y,J) \end{matrix} \longrightarrow \begin{matrix} \Gamma(X,F) \\ \oplus \\ \Gamma(Y,G) \end{matrix} \qquad (1)$$

Operators of the form (1) are studied in [3],[10],[11] and many other papers. X is a smooth compact manifold with boundary Y. E,F are complex vector bundles over X; J,G complex vector bundles over Y. $\Gamma(X,E),\ldots$ denote the spaces of smooth sections in the corresponding bundles. r^+A is an elliptic pseudodifferential operator, defined roughly speaking as an operator in a neighbourhood of X acting on an extension of $u \in \Gamma(X,E)$ by zero and restriction r^+ of the result to X. $r'B$ is a so called Green's operator, $r'T$ a trace operator (r' is the "restriction" to Y), K is a potential operator and Q is a pseudodifferential operator on Y. All these operators are defined by symbols like ordinary pseudodifferential operators (see for instance [3],[7]). The homogeneous principal symbols shall be denoted by

$$\sigma_A(x,\xi)\ ,\ \sigma_B(x',\xi,\tau)\ ,\ \sigma_T(x',\xi)\ ,\ \sigma_K(x',\xi)\ ,\ \sigma_Q(x',\xi').$$

Here (x,ξ) are local coordinates in T^*X (the cotangent bundle to X) and (x',ξ') local coordinates in T^*Y (the cotangent bundle to Y). τ is an additional variable in $\mathbb{R}$. The symbols σ_B, σ_T, σ_K, σ_Q can be thought to be given in a neighbourhood of Y.

We suppose

$$\operatorname{ord}\sigma_A(x,\xi) = \alpha \in \mathbb{Z}\ ,\ \operatorname{ord}\sigma_B(x',\xi,\tau) = \alpha - 1,$$
$$\operatorname{ord}\sigma_T(x',\xi) = \gamma \in \mathbb{Z}\ ,\ \operatorname{ord}\sigma_K(x',\xi) = \lambda \in \mathbb{Z},$$
$$\operatorname{ord}\sigma_Q(x',\xi') = 1-\alpha+\lambda+\gamma\ .$$

For abbreviation we write

$$\operatorname{ord}\mathcal{A} = \begin{pmatrix} \alpha & \lambda + 1/2 \\ \gamma + 1/2 & 1-\alpha+\gamma+\lambda \end{pmatrix} \qquad (2)$$

The transmission property explained in [3] is supposed. We use here the symbol classes which are considered in [7]. All operators are considered modulo negligible operators in the sense of [3] . Note that $\mathcal{A}$ induces continuous operators in Sobolev spaces

$$\mathcal{A} : \begin{array}{c} H^s(X,E) \\ \oplus \\ H^{t+\lambda+1/2}(Y,J) \end{array} \longrightarrow \begin{array}{c} H^t(X,F) \\ \oplus \\ H^{s-\gamma-1/2}(Y,G) \end{array} \qquad (3)$$

where $t = s-\alpha$ and s sufficiently large.
With an operator (1) is connected a so called boundary symbol

$$\sigma_Y(\mathcal{A}) = \begin{pmatrix} \Pi^+\sigma_A+\Pi'\sigma_B & \sigma_K \\ \Pi'\sigma_T & \sigma_Q \end{pmatrix} : \begin{array}{c} p^*E'\otimes H^+ \\ \oplus \\ p^*J \end{array} \longrightarrow \begin{array}{c} p^*F'\otimes H^+ \\ \oplus \\ p^*G \end{array} \qquad (4)$$

Here $p: S^*(Y) \rightarrow Y$ is the canonical projection of the unit cosphere bundle in T^*Y with respect to a Riemannian metric, $E'=E|_Y$, $F'=F|_Y$, H^+ is the space of the Fourier transforms of functions on the x_n-axis which are C^∞ for $x_n \geqq 0$, zero for $x_n<0$ and rapidly decreasing for $x_n \rightarrow \infty$. Π^+ and Π' are the Fourier transforms in x_n-direction of the operators r^+ and r' respectively (cf.[7]).
The ellipticity of a boundary value problem of the type (1) is per def. ellipticity of σ_A in the usual sense and further that $\sigma_Y(\mathcal{A})$ is an isomorphism (this is a generalization of the Lopatinski condition).If $\mathcal{A}$ is elliptic,the index is finite,i.e.

$$\text{ind}\ \mathcal{A} < \infty \qquad (5)$$

In this paper we study a multiplication between certain equivalence classes of elliptic operators of the form (1) and analogous equivalence classes of of elliptic pseudodifferential operators on another closed compact C^∞- manifold M.The multiplication is defined on operator level in the sense of the well known external multiplication of the K-theory

$$K(X,Y) \times K(M) \longrightarrow K(X\times M, Y\times M) \quad .$$

Analogous to the multiplication of elliptic operators on closed compact manifolds we obtain also complexes of operators of the form (1) for which an ellipticity holds in the sense of exactness of the interior symbol sequence and of the induced boundary symbol sequence.
In [3] is constructed an elliptic operator

$$r^+(\Lambda_E^-)^{\varrho} \; : \; \Gamma(X,E) \longrightarrow \Gamma(X,E)$$

($\varrho \in \mathbb{Z}$) with index zero. Let $\varrho' \in \mathbb{Z}$ and $\Delta_J^{\varrho'/2}$ be an elliptic pseudodifferential operator on Y with homogeneous principal symbol $|\xi'|^{\varrho'}$ times identity in $\pi_Y^* J$ (π_Y: $T^*Y \to Y$ the projection). Then

$$\mathcal{L}_{E,J}^{\varrho,\varrho'} = \begin{pmatrix} r^+(\Lambda_E^-)^{\varrho} & 0 \\ 0 & \Delta_J^{\varrho'/2} \end{pmatrix} : \begin{matrix} \Gamma(X,E) \\ \oplus \\ \Gamma(Y,J) \end{matrix} \longrightarrow \begin{matrix} \Gamma(X,E) \\ \oplus \\ \Gamma(Y,J) \end{matrix} \qquad (6)$$

defines an elliptic boundary value problem and

$$\operatorname{ind} \mathcal{L}_{E,J}^{\varrho,\varrho'} = 0 \quad . \qquad (7)$$

In order to define equivalence classes of operators of the form (1) we call two operators $\mathcal{A}_0, \mathcal{A}_1$ homotopic ($\mathcal{A}_0 \simeq \mathcal{A}_1$), if they have the same orders and if there exists a continuous family of elliptic operators $(\mathcal{A}_t)_{0 \leq t \leq 1}$ of the type (1) with $\operatorname{ord} \mathcal{A}_0 = \operatorname{ord} \mathcal{A}_t$ $(0 \leq t \leq 1)$ connecting $\mathcal{A}_0$ and $\mathcal{A}_1$.

Let τ, τ' be integers. Then

$$\operatorname{ord} \mathcal{L}_{F,G}^{\tau,\tau'} \mathcal{A} \mathcal{L}_{E,J}^{\varrho,\varrho'} = \begin{pmatrix} \alpha+\tau+\varrho & \lambda+\tau+\varrho'+1/2 \\ \gamma+\tau'+\varrho+1/2 & 1-\alpha+\gamma+\lambda+\tau'+\varrho' \end{pmatrix} \qquad (8)$$

For operators $\mathcal{A}_1, \mathcal{A}_2$ of the type (1) there is defined in a canonical way the direct sum $\mathcal{A}_1 \oplus \mathcal{A}_2$. Let L be a complex vector bundle over X and V a complex vector bundle over Y. Put

$$\mathcal{A}(\varrho,\varrho',\tau,\tau',L,V) = \mathcal{L}_{F,G}^{\tau,\tau'} \mathcal{A} \mathcal{L}_{E,J}^{\varrho,\varrho'} \oplus \mathcal{L}_{L,V}^{\varkappa,\varkappa'}$$

with $\varkappa = \alpha+\tau+\varrho$, $\varkappa' = 1-\alpha+\gamma+\lambda+\tau'+\varrho'$. Two elliptic boundary value problems

$$\mathcal{A}_i : \begin{matrix} \Gamma(X,E_i) \\ \oplus \\ \Gamma(Y,J_i) \end{matrix} \longrightarrow \begin{matrix} \Gamma(X,F_i) \\ \oplus \\ \Gamma(Y,G_i) \end{matrix} \qquad (i=1,2)$$

with $\alpha_i, \gamma_i, \lambda_i \in \mathbb{Z}$ are called equivalent, if there exist vector bundles L_i over X, V_i over Y and isomorphisms

$$a : E_1 \oplus L_1 \longrightarrow E_2 \oplus L_2 \;, \quad a' : J_1 \oplus V_1 \longrightarrow J_2 \oplus V_2 \;,$$
$$b : F_1 \oplus L_1 \longrightarrow F_2 \oplus L_2 \;, \quad b' : G_1 \oplus V_1 \longrightarrow G_2 \oplus V_2$$

such that for all $\rho_i, \rho_i', \tau_i, \tau_i' \in \mathbb{Z}$ (i=1,2) with

$$\alpha_1+\tau_1+\rho_1 = \alpha_2+\tau_2+\rho_2 ,$$
$$1-\alpha_1+\gamma_1+\lambda_1+\tau_1'+\rho_1' = 1-\alpha_2+\gamma_2+\lambda_2+\tau_2'+\rho_2'$$

there exists a homotopy

$$\mathcal{A}_1(\rho_1,\rho_1',\tau_1,\tau_1',L_1,V_1) \simeq \begin{pmatrix} b & 0 \\ 0 & b' \end{pmatrix}_*^{-1} \mathcal{A}_2(\rho_2,\rho_2',\tau_2,\tau_2',L_2,V_2) \begin{pmatrix} a & 0 \\ 0 & a' \end{pmatrix}_*$$

$(\dots)_*$ denote the isomorphisms of spaces of sections induced by the corresponding bundle morphisms.
By Ell(X,Y) we denote the set of all equivalence classes. The class represented by $\mathcal{A}$ is denoted by $[\mathcal{A}]$. Analogous considerations as in the K-theory show the following

Theorem 1. The set Ell(X,Y) is an Abelean group with respect to $\oplus$. If $\mathcal{A}^{-1}$ is a parametrix of $\mathcal{A}$, we have $-[\mathcal{A}] = [\mathcal{A}^{-1}]$.

It is easily seen that any $\mathcal{A}$ can be represented by some $\mathcal{A}_+$, for which the corresponding orders $\alpha_+, \lambda_+, \gamma_+$ satisfy the conditions

$$\alpha_+ - 1 > 0 , \quad \lambda_+ > 0 , \quad \gamma_+ > 0 , \quad 1 - \alpha_+ + \gamma_+ + \lambda_+ > 0 .$$

Moreover we have

$$\operatorname{ord} \mathcal{A}^{-1} = \begin{pmatrix} -\alpha & -\gamma-1/2 \\ -\lambda-1/2 & \alpha-\lambda-\gamma-1 \end{pmatrix} . \tag{9}$$

If we put $\tilde{\mathcal{A}}^{-1} = \mathcal{L}^{\tau,\tau'}_{F,G} \mathcal{A}^{-1} \mathcal{L}^{\rho,\rho'}_{E,J}$ for $\tau=\rho=\alpha$, $\rho'=\tau'=-\alpha+\lambda+\gamma+1$, we obtain

$$\operatorname{ord} \tilde{\mathcal{A}}^{-1} = \operatorname{ord} \mathcal{A} . \tag{10}$$

Let M be a closed compact C^∞-manifold. Then analogous to the above constructions we can define the Abelean group Ell(M) of equivalence classes of elliptic pseudodifferential operators over M (which is isomorphic to $K(T^*M)$).
In order to define the external multiplication of operators we recall an analogous procedure of the K-theory (cf.[1]). Let (A,A'),(B,B') pairs of compact spaces. Then any $a \in K(A,A')$ can be represented by a complex $0 \rightarrow E \xrightarrow{\alpha} F \rightarrow 0$ which is exact over A' (E,F vector bundles over A). If $b \in K(B,B')$ is represented by

$0 \to V \xrightarrow{\beta} W \to 0$, exact over B' (V,W vector bundles over B), then the complex

$$0 \longrightarrow \begin{matrix} E\otimes V \\ \oplus \\ 0 \end{matrix} \xrightarrow{\begin{pmatrix} \alpha\otimes 1 & 0 \\ 1\otimes\beta & 0 \end{pmatrix}} \begin{matrix} F\otimes V \\ \oplus \\ E\otimes W \end{matrix} \xrightarrow{\begin{pmatrix} 0 & 0 \\ -1\otimes\beta & \alpha\otimes 1 \end{pmatrix}} \begin{matrix} 0 \\ \oplus \\ F\otimes W \end{matrix} \longrightarrow 0$$

is exact over $(A\times B')\cup(A'\times B)$ (the $\otimes$ mean the external tensor products, cf. [2]) and defines an element

$$a\otimes b \in K(A\times B, A\times B'\cup A'\times B) \quad .$$

$a\otimes b$ can be represented also by the following short complex

$$0 \longrightarrow \begin{matrix} E\otimes V \\ \oplus \\ F\otimes W \end{matrix} \xrightarrow{\begin{pmatrix} \alpha\otimes 1 & -1\otimes\beta^{-1} \\ 1\otimes\beta & \alpha^{-1}\otimes 1 \end{pmatrix}} \begin{matrix} F\otimes V \\ \oplus \\ E\otimes W \end{matrix} \longrightarrow 0 ,$$

where $\alpha^{-1}: F \to E$ ($\beta^{-1}: W \to V$) are bundle morphisms which are inverse to α over A' (β over B'). Over $A\times B'\cup A'\times B$ is

$$\frac{1}{2}\begin{pmatrix} 0 & -1\otimes\beta^{-1} \\ 0 & \alpha^{-1}\otimes 1 \end{pmatrix} \text{ right inverse to } \begin{pmatrix} 0 & 0 \\ -1\otimes\beta & \alpha\otimes 1 \end{pmatrix} ,$$

and we have

$$\begin{pmatrix} \alpha\otimes 1 & -1\otimes\beta^{-1} \\ 1\otimes\beta & \alpha^{-1}\otimes 1 \end{pmatrix}^{-1} = \frac{1}{2}\begin{pmatrix} \alpha^{-1}\otimes 1 & 1\otimes\beta^{-1} \\ -1\otimes\beta & \alpha\otimes 1 \end{pmatrix} .$$

Now we define the multiplication

$$\mathrm{Ell}(X,Y)\times \mathrm{Ell}(M) \longrightarrow \mathrm{Ell}(X\times M, Y\times M) \quad , \tag{11}$$

Let $\mathcal{A}$ be elliptic and $\alpha-1>0$, $\lambda>0$, $\gamma>0$, $1-\alpha+\lambda+\gamma>0$. Moreover let

$$S : \Gamma(M,V) \longrightarrow \Gamma(M,W)$$

be an elliptic pseudodifferential operator on M with ord $S = \alpha$. If Δ_W is an operator on M with homogeneous principal symbol $|\mu|^2$ times identity in π_M^*W ($(m,\mu) \in T^*M$, $\pi_M : T^*M \to M$ the projection) we set

$$\tilde{S}^{-1} = S^{-1}\Delta_W^{\alpha}$$

(S^{-1} is a parametrix of S) and $S_\delta = \Delta_W^{\frac{1}{2}(\delta-\alpha)} S$, $\delta \in \mathbb{Z}$. We set

$$\mathcal{A}\otimes 1_V = \begin{pmatrix} (r^+A+r'B)\otimes 1_V & K\otimes 1_V \\ (r'T)\otimes 1_V & Q\otimes 1_V \end{pmatrix} : \begin{matrix}\Gamma(X\times M, E\otimes V)\\ \oplus \\ \Gamma(Y\times M, J\otimes V)\end{matrix} \longrightarrow \begin{matrix}\Gamma(X\times M, F\otimes V)\\ \oplus \\ \Gamma(Y\times M, G\otimes V)\end{matrix}$$

and

$$\mathcal{A}\otimes S = \left(\begin{array}{cc|cc} \multicolumn{2}{c|}{\mathcal{A}\otimes 1_V} & -1\otimes\tilde{S}^{-1} & 0 \\ & & 0 & -1\otimes\tilde{S}_\delta^{-1} \\ \hline 1\otimes S & 0 & \multicolumn{2}{c}{\tilde{\mathcal{A}}^{-1}\otimes 1_W} \\ 0 & 1\otimes S_\delta & & \end{array}\right) : \begin{matrix}\Gamma(X\times M, E\otimes V)\\ \oplus \\ \Gamma(Y\times M, J\otimes V)\\ \oplus \\ \Gamma(X\times M, F\otimes W)\\ \oplus \\ \Gamma(Y\times M, G\otimes W)\end{matrix} \longrightarrow \begin{matrix}\Gamma(X\times M, F\otimes V)\\ \oplus \\ \Gamma(Y\times M, G\otimes V)\\ \oplus \\ \Gamma(X\times M, E\otimes W)\\ \oplus \\ \Gamma(Y\times M, J\otimes W)\end{matrix}$$

where δ = ord Q. Then we have

<u>Theorem 2</u>. $\mathcal{A}\otimes S$ is an elliptic boundary value problem on $X\times M$ with respect to $Y\times M$ and

$$\operatorname{ind}\mathcal{A}\otimes S = \operatorname{ind}\mathcal{A}\cdot\operatorname{ind} S \quad . \tag{12}$$

The ellipticity of $\mathcal{A}\otimes S$ follows immediately from the exactness of the interior and boundary symbol sequence of the complex

$$0 \longrightarrow \begin{matrix}\Gamma(X\times M, E\otimes V)\\ \oplus \\ \Gamma(Y\times M, J\otimes V)\\ \oplus \\ 0\end{matrix} \xrightarrow{\begin{pmatrix}\mathcal{A}\otimes 1 & 0 \\ 1\otimes S_0 & 0\end{pmatrix}} \begin{matrix}\Gamma(X\times M, F\otimes V)\\ \oplus \\ \Gamma(Y\times M, G\otimes V)\\ \oplus \\ \Gamma(X\times M, E\otimes W)\\ \oplus \\ \Gamma(Y\times M, J\otimes W)\end{matrix} \xrightarrow{\begin{pmatrix} 0 & 0 \\ -1\otimes S_0 & \mathcal{A}\otimes 1\end{pmatrix}} \begin{matrix}0\\ \oplus \\ \Gamma(X\times M, F\otimes W)\\ \oplus \\ \Gamma(Y\times M, G\otimes W)\end{matrix} \longrightarrow \cdots$$

$\longrightarrow 0$ (the meaning of the abbreviation $1\otimes S_0$ is obvious). The index of the complex is defined as Euler characteristic of this Fredholm complex. The assertion about the index follows from analogous considerations as in the case without boundary (cf. [6]).

For elliptic operators on M we have a well known difference construction

$$\gamma_M : \operatorname{Ell}(M) \longrightarrow K(B(M), S(M)) \; .$$

An analogous construction for boundary value problems is given in [3]

$$\gamma_{(X,Y)} : \operatorname{Ell}(X,Y) \longrightarrow K(B(X), \partial B(X)) \; .$$

An immediate calculation shows that the following diagram commutes

$$\begin{array}{ccc} Ell(X,Y)\times Ell(M) & \longrightarrow & Ell(X\times M, Y\times M) \\ \Big\downarrow \gamma_{(X,Y)}\times \gamma_M & & \Big\downarrow \gamma_{(X\times M, Y\times M)} \\ K(B(X),\partial B(X))\times K(B(M),S(M)) & \longrightarrow & K(B(X\times M),\partial B(X\times M)) \end{array} \qquad (13)$$

The multiplication (11) admits to consider an action of Ell(Y) on Ell(X,Y).Consider in (11) the special case $M = 2X$ (the double of X).By $l(x) = (x,x)$ we define an embedding

$$l : (X,Y) \longrightarrow (X\times 2X, Y\times 2X) .$$

We obtain an induced homomorphism

$$l^* : Ell(X\times 2X, Y\times 2X) \longrightarrow Ell(X,Y) ,$$

generated roughly speaking by restriction of interior and boundary symbols to (X,Y). If we consider also

$$l^* : K(B(X\times 2X,\partial B(X\times 2X)) \longrightarrow K(B(X),\partial B(X)) ,$$

the following diagram commutes

$$\begin{array}{ccc} Ell(X\times 2X, Y\times 2X) & \longrightarrow & Ell(X,Y) \\ \Big\downarrow \gamma_{(X\times 2X, Y\times 2X)} & & \Big\downarrow \gamma_{(X,Y)} \\ K(B(X\times 2X),\partial B(X\times 2X)) & \longrightarrow & K(B(X),\partial B(X)) \end{array}$$

By the inclusion $i : Y \longrightarrow 2X$ follows the homomorphism

$$i_! : K(B(Y),S(Y)) \longrightarrow K(B(2X),S(2X)) ,$$

defined in [2] .A suitable definition of $i_!$ for operators gives a commutative diagram

$$\begin{array}{ccc} Ell(Y) & \xrightarrow{i_!} & Ell(2X) \\ \Big\downarrow \gamma_Y & & \Big\downarrow \gamma_{2X} \\ K(B(Y),S(Y)) & \xrightarrow{i_!} & K(B(2X)) \end{array}$$

From (13) we obtain a commutative diagram

$$\begin{array}{ccc} \mathrm{Ell}(X,Y)\times \mathrm{Ell}(Y) & \longrightarrow & \mathrm{Ell}(X,Y) \\ \Big\downarrow \gamma_{(X,Y)}\times\gamma_Y & & \Big\downarrow \gamma_{(X,Y)} \\ K(B(X),\partial B(X))\times K(B(Y),S(Y)) & \longrightarrow & K(B(X),\partial B(X)) \end{array}$$

i.e. an action of Ell(Y) on Ell(X,Y) commuting with the difference construction. The author don't know, if the index of the product is equal to the product of the indices.

Literature

[1] M.F.Atiyah: Лекции по K-теории (перевод с англ.), изд-во „Мир", Москва 1967

[2] M.F.Atiyah, I.M.Singer: The index of elliptic operators.I. Ann.of Math. 87, 484 - 530 (1968)

[3] L.Boutet de Monvel: Boundary problems for pseudodifferential operators. Acta Math. 126, 11 - 51 (1971)

[4] А.С. Дынин: Эллиптические краевые задачи для псевдодифференциальных комплексов. Функц. Анал. и Прилож. 6, 1, 75-76 (1972)

[5] Г.И. Эскин: Краевые задачи для эллиптических псевдодифференциальных уравнений Изд.-во „Наука" Москва 1973

[6] R.S.Palais: Семинар по теореме Атьи-Зингера об индексе (перевод с англ.) „Мир", Москва 1970

[7] B.-W.Schulze: Elliptic operators on manifolds with boundary. Contr. to a School on "Global Analysis" Ludwigsfelde. Berlin 1977

[8] B.-W.Schulze: On the set of all elliptic boundary value problems for an elliptic pseudodifferential operator on a manifold. Math.Nachr. 75, 271 - 282 (1976)

[9] Б.В. Шульце: Эллиптические краевые задачи для операторов специального вида. Дифф. Уравн. (to appear)

[10] М.И. Вишик, Г.И. Эскин: Уравнения в свертках в ограниченной области. УМН 20, 3, 89 - 152 (1965)

[11] М.И. Вишик, Г.И. Эскин: Эллиптические уравнения в свертках в ограниченной области и их приложения. УМН 22, 1, 15-76 (1967)

Akad.der Wiss.der DDR ZI
Math-Mech-Berlin

On the hyperharmonic functions associated with a degenerated elliptic operator

by L. Stoica

Introduction

For an elliptic not totally degenerated second order differential operator, the family of all open sets for which the Poisson-Dirichlet problem is resoluble forms a basis of the topology (see [4], [5], [9]). Such an operator may have no property of Harnack type.

On an open set U , for which the Poisson - Dirichlet problem is resoluble, it is possible to define the harmonic measure ρ_x^U and a rezolvent $(G_\lambda)_{\lambda \geq 0}$; the analogous of a hyperharmonic function on U is a Borel function $s : U \longrightarrow \overline{R}_+$ such that $s(x) = \sup \int s \, d\rho_x^W$, where the supremum is taken over the family of all sets $W, \overline{W} \subset U$.

The purpose here is to show that the above functions and the $(G_\lambda)_{\lambda \geq 0}$ excessive functions on U are identical.

In section 0 there are presented some preliminary results concerning differential operators. In fact the rest of the paper may be developed on a locally compact space, in an axiomatic aproach for which the main axioms are exactly the properties presented in 0 .

The sections 1. and 2. aim to show how the well known techniques from potential theory may be used without a Harnack property. Section 3. contains a criterion that may characterise excessive functions, and that is parallel with those established in [3] , ch 2 , sec. 5 and [6] , ch 12 . As a consequence it is obtained the main result of this paper (corollary 3.2).

Finally I wish to thank Professor N. Boboc, who pointed out to me the nice properties of elliptic degenerated differential operators, and Professor A. Cornea for his helpful remarks and suggestions during the preparation of this paper.

0. Let D be an open set in R^d, and

$$L = \sum_{i,j=1}^{d} a_{ij} \, \partial^2 / \partial x_i \partial x_j + \sum_{k=1}^{d} b_k \partial / \partial x_k + c$$

a linear second order differential operator with real coefficients such that $c \leq 0$.

We suppose that the family of all open sets U, that satisfy the property (R) (stated below) forms a basis of the topology.

(R) U is open, $\overline{U}$ is compact and contained in D and for every $\lambda \geq 0$, every function f in $C(\overline{U})$ and every g in $C(\partial U)$ there exists a function u in $C(\overline{U})$ such that $(L-\lambda)u = -f$ in the sense of distributions on U and $u = g$ on ∂U. The function u is uniquely determined by the above conditions and if $f \geq 0$ and $g \geq 0$ then $u \geq 0$. Besides the family of all u in $C(\overline{U})$ such that $-Lu$ is in $C_+(\overline{U})$ separates the points of U.

The above hypothesis is fulfilled for a large class of operators. For example the proofs of theorems 1.6.1. and 1.8.1. in [9] apply to open balls with sufficiently small diameters. Then it results that already the following conditions are sufficient to ensure the validity of our hypothesis :

- $(a_{ij}(x))$ $i, j = 1,\dots,d$ differes from the 0 matrix, and is nonnegative definite at every point x.
- all the coeficients a_{ij}, b_k, c are smooth enough.
- supp $c \leq M < 0$, where M depends only on a_{ij} and b_k.

Another property that we shall use is the following one :

(C) For every x in D and every $\varepsilon > 0$ there exists a neighbourhood V of x such that for every W with the property (R), $\overline{W} \subset V$ we have $1 - \varepsilon \leq u \leq 1$, where u is the continuous function such that $u=1$ on ∂W and $Lu = 0$ in W.

(To check this property it is sufficient to consider the function $f_t(y) = 1 + t(\langle y-x, \xi \rangle)^2$, where $\xi \in R^d$ satisfies the condition $\sum_{i,j} a_{ij}(x) \xi^i \xi^j > 0$, and t is chosen such that $t > 0$ and $Lf_t(x) > 0$. Then V is chosen such that $Lf_t > 0$ and $f_t \leq 1 + \varepsilon$ on V. Clearly $1-\varepsilon \leq \frac{1}{1+\varepsilon} f_t \leq u \leq 1$.)

1. Throughout this section U will be an open set that satisfies (R). For $\lambda \geq 0$ we define a linear operator G_λ on $C(\overline{U})$ in the following way : if $f \in C(\overline{U})$ then $G_\lambda f$ satisfies the conditions $G_\lambda f = 0$ on ∂U, $(L - \lambda) G_\lambda f = -f$ in U. Clearly $G_\lambda f \geq 0$ if $f \geq 0$ and $G_\lambda - G_\beta = (\beta - \lambda) G_\lambda G_\beta$ because $(L - \lambda)(G_\beta f + (\beta - \lambda) G_\lambda G_\beta f) = -f$. We remark also that $C_k^2(U)$ is contained in the range of G_λ.

For a point x in U the harmonic measure ρ_x^U is defined on ∂U in the following way : $\int g \, d\rho_x^U = u(x)$, where $g \in C(\partial U)$ and $u \in C(\overline{U})$, $u = g$ on ∂U and $Lu = 0$ in U. ρ_x^U is a positive measure and $\rho_x^U(1) \leq 1$.

Definition 1.1. Let V be an open set in D and $f : V \longrightarrow \overline{R}_+$ a Borel function. We say that f is hyperharmonic if $f(x) \geq \int f \, d\rho_x^W$ for every W, which satisfies the property (R), $x \in W$, $\overline{W} \subset V$, and $f(x) = \sup \int f \, d\rho_x^W$, where the sup is taken over the family of all these W.

If f is in $C_+(\overline{U})$ then $G_0 f$ is hyperharmonic on U. Indeed for W with the property (R), $\overline{W} \subset U$, the function $g(x) = G_0 f(x) - \int G_0 f \, d\rho_x^W$ is in $C(\overline{W})$, $g = 0$ on ∂W and $Lg = -f$ on W, hence $g \geq 0$. The condition $G_0 f(x) = \sup \int G_0 f \, d\rho_x^W$ there results from (C).

Every function u in $C_+(\overline{U})$ such that $Lu = f \leq 0$, f in $C(\overline{U})$, is hyperharmonic, in particular the constant function 1 is hyperharmonic.

The relation $(L - \lambda)(1 - \lambda G_\lambda 1) = c \leq 0$ leads to $\lambda G_\lambda 1 \leq 1$, hence the resolvent $(G_\lambda)_{\lambda \geq 0}$ is submarkovian.

Every operator G_λ may be extended as a kernel on $\overline{U}$. In order to show that all the kernels G_λ are supported by U we need the following minimum principle :

Proposition 1.2. Let V be an open set, f a function in $C(\overline{V})$, $f \geq 0$ on ∂V and such that for every x in V, there exists W with property

(R), $x \in W \subset \overline{W} \subset V$ and $f(x) \geq \int f \, d\rho_x^W$. If the family $\{v \in C(\overline{V}) \mid v$ hyperharmonic on $V\}$ separates the points in V, then $f \geq 0$ on V.

Proof. The main idea go back to H. Bauer ([1] , 1.3). We only sketch the main steps.

If $\min f = -\alpha$, $\alpha > 0$, then we define $K = \{x \in V \mid f(x) = -\alpha\}$; K is a compact subset of V. The cone

$$\mathcal{C} = \{\beta f + v / \beta \in R_+, v \in C(\overline{V}) \text{ is hyperharmonic on } V\}$$

contains the constant positive functions and separates the points in V. The Choquet boundary associated to $\mathcal{C}$ on K is nonempty : there exists a point x_0 in K such that every positive measure μ on V, with the property $\int s \, d\mu \leq s(x_0)$ for any s in $\mathcal{C}$, is of the form $\mu = r\,\varepsilon_{x_0}$. But the hypothesis gives a measure of the form $\mu = \rho_{x_0}^W \neq \varepsilon_{x_0}$ and we observe that $\rho_{x_0}^W$ must be supported by K.

Proposition 1.3. $G_\lambda(\partial U) = 0$ for every $\lambda \geq 0$.

Proof. Let ϕ_n, $n \in N$, be a sequence of continous functions on $\overline{U}$ such that $\partial U = \bigcap_n \operatorname{supp} \phi_n$, $0 \leq \phi_n \leq 1$, $\phi_n = 1$ on ∂U.

For a given $\varepsilon > 0$ we choose m such that $G_0 1 \leq \varepsilon$ on $\operatorname{supp} \phi_m$, hence $G_0 \phi_m \leq \varepsilon$ on $\operatorname{supp} \phi_m$. We put $V = U \setminus \operatorname{supp} \phi_m$ and $f = \varepsilon - G_0 \phi_m$. The above proposition gives us $G_0 \phi_m \leq \varepsilon$ on V. Letting m to tend to infinity we get $G_0(\partial U) = 0$. The resolvent equation gives $G_\lambda(\partial U) = 0$ for every $\lambda > 0$.

Theorem 1.4. Let u be an excessive function on U, with respect to $(G_\lambda)_{\lambda \geq 0}$, then u is hyperharmonic.

Proof. It is sufficient to show that $G_0 \phi$ is hyperharmonic for every bounded Borel function ϕ, because any excessive function is the limit of an increassing sequence of such functions (after a theorem of Hunt (see [3] p - 73)). We consider a sequence $(\emptyset_n)$ $n \in N$ of Borel functions such that $G_o \emptyset_n$ and $G_o(1-\emptyset_n)$ are hyperharmonic, $0 \leqslant \emptyset_n \leqslant \emptyset_{n+1} \leqslant 1$. Let $\emptyset = \sup_n \emptyset_n$, then $G_o \emptyset$ and $G_o(1-\emptyset)$ are also hyperharmonic. Now a monotone class argument shows that $G_o \emptyset$ is hyperharmonic for every bounded Borel function.

2. From now on we make use of Markov processes procedures and results, and in this area we refer, also for notations, to [3] . As in the preceding section let U be an open set that fulfils (R). We consider the usual Hunt process $(\Omega, \mathcal{M}, \mathcal{M}_t, X_t, \theta_t, P^x)$ on U , that is associated with $(G_\lambda)_{\lambda \geqslant 0}$ ([3] , p. 46).

<u>Lemma 2.1.</u> Let U' be an open set that fulfils (R) such that $\overline{U'} \subset U$. Further let $u \in \mathcal{C}(\overline{U})$, be a hyperharmonic function and define u' in the following way :

$$u'(x) = \begin{cases} u(x) & \text{if } x \in U \setminus U' \\ \int u \, d\rho_x^{U'} & \text{if } x \in U' \end{cases}$$

Then u' is an excessive function.

<u>Proof.</u> (see [6] , 12.b and [8] , p. 199). The inequality $\lambda G_\lambda u' \leqslant u'$ is equivalent with $f = u' + \lambda G_o((u' - \lambda G_\lambda u')^-) - \lambda G_o((u' - \lambda G_\lambda u')^+) \geqslant 0$.

If we put $V = U \setminus \{x \in U / u'(x) \geqslant \lambda G_\lambda u(x)\}$, Proposition 1.2 gives us $\lambda G_\lambda u' \leqslant u'$. On the other hand for x in U we write u' in the from $u' = \emptyset + \Psi$, where $\emptyset, \Psi \in C_+(U)$, $\emptyset = 0$ on ∂U and $\emptyset \leqslant u'$, $\emptyset(x) = u'(x)$.

Then $\lambda G_\lambda \emptyset(x) \longrightarrow \emptyset(x)$ for $\lambda \longrightarrow \infty$ and $\lambda G_\lambda \emptyset(x) \leqslant \lambda G_\lambda u'(x) \leqslant u'(x)$.

Corollary 2.2. If U fulfils (R) and $\overline{U'} \subset U$ then $\rho_x^U = P^x_{U \setminus U'}$. The process on U', obtained from X by courtailment of its lifetime up to $T_{U \setminus U'}$, is associated with the rezolvent of U' (given via property (R)).

Proof. If u is a hyperharmonic function in C $(\overline{U})$ the preceding lemma shows that u (the function constructed there) is excessive. From 1.4 we know that every excessive function v is greater than u' if $v \geq u'$ on $U \setminus U'$.

Hence $u = \inf \{ v/v$ is excessive and $v \geq u$ on $U \setminus U' \}$ and Hunt's theorem ([3], p. 141) shows that $u' = P_{U \setminus U'} u$.

The Stone - Weierstrass theorem shows now that $\rho_x^{U'} = P^x_{U \setminus U'}$. The rest of proof is now straightforward.

3. In this section U is a fixed set that fulfils (R), and $(\Omega, \mathcal{M}, \mathcal{M}_t, X_t, \theta_t, P^x)$ its associated Hunt process.

Theorem 3.1. Let $s : U \longrightarrow \overline{R}_+$ be a Borel function and $\mathcal{F}$ a family of open sets in U that is a topological basis.

We suppose also that every U' in $\mathcal{F}$, has the property (R), that $\overline{U'} \subset U$ and $s(x) \geq \int s \, d\rho_x^{U'}$ for every x in U'.

If $s(x) = \sup \{ \int s \, d\rho_x^{U'} / \; x \in U', \; U' \in \mathcal{F} \}$, then s is excessive.

Proof. Let V be an open set. For a fixed $n \in N$, we choose $U_1, \ldots, U_p$ in $\mathcal{F}$ such that $\overline{V} \subset \bigcup_{i=1}^{p} U_i$, and, such that the diameter of every U_i is smaller than 1/n.

On $\bigcup_{i=1}^{p} U_i$ we define the function

$$s_n(x) = \inf \{ s \, d\rho_x^{U_i} / \; i \leq p, \; x \in U_i \},$$

and we note

$$\Gamma = \overline{V}, \quad T = T_{U \setminus V}, \quad T_i = T_{U \setminus U_i}.$$

One may construct a Borel function $\eta : \Gamma \to \{1, \dots p\}$ such that $\delta = \inf_{x \in \Gamma} d(x, U \setminus U_{\eta(x)}) > 0$. Further we define a sequence of stopping times (R_k), $k = 1, \dots,$ in the following way :

$$R_o = 0, \quad R_{k+1} = R_k \qquad \text{if } T \leqslant R_k$$

and

$$R_{k+1} = R_k + T_{\eta(X_{R_k})} \circ \theta_{R_k} \qquad \text{if } T > R_k.$$

It is clear that $X_{T_i} \in \partial U_i$ and that $X_{R_k} \in \Gamma$ if $T > R_k$, hence $d(X_{R_k}, X_{R_{k+1}}) \geqslant \delta$ on $T > R_k$.

We shall prove inductively that for every $k = 0, 1, \dots$ the next inequality holds for every x_o in V and $t > 0$:

$$(*) \qquad s(x_o) \geqslant E^{x_o}[\, s_n(X_t);\ t \leqslant R_k;\ t < T] \ +$$

$$+ E^{x_o}[s(X_{R_k});\ t > R_k;\ T > R_k].$$

For $k = 0$ the inequality is trivial. To pass from the k - step to the $(k+1)$ - step we shall use the strong Markov property in the form

$$E^x[G(\cdot, \theta_\tau(\cdot)) \mid \mathcal{M}_\tau](\omega) = E^{X_\tau(\omega)}[G(\omega, \cdot)]$$

where τ is a stopping time and $G(\omega, \omega')$ is a $\mathcal{M}_\tau \otimes \mathcal{M}$ measurable positive function and $x \in U$.

For x in Γ and $r \in R_+$ we have $X_{T_{\eta(x)}} = X_{T_{\eta(x)}} \circ \theta_{t-r}$ if $T_{\eta(x)} > t - r$, hence

$$E^x [\, s(X_{T_{\eta(x)}});\ T_{\eta(x)} > t-r \,] = E^x [\, E^{X_{t-r}} [\, s(X_{T_{\eta(x)}}) \,];$$

$$T_{\eta(x)} > t-r \,] \geq E^x [\, s_n(X_{t-r});\ T_{\eta(x)} > t-r \,]$$

because $E^y [s(X_{T_{\eta(x)}})] \geq s_n(y)$ for y in $U_{\eta(x)}$.

If in the above inequality we take $r = R_k(\omega)$

$x = X_{R_k}(\omega)$ and we integrate with respect to $dP^{x_o}(\omega)$

on the set $\{t \wedge T(\omega) > R_k(\omega)\}$, we obtain :

$$\int \chi_{\{t \wedge T(\omega) > R_k(\omega)\}} \int s(X_{T_{\eta(X_{R_k}(\omega))}}(\omega')) \cdot \chi_{\{T_{\eta(X_{R_k}(\omega))}(\omega') > t - R_k(\omega)\}} \cdot$$

$$\cdot\, dP^{X_{R_k}(\omega)}(\omega')\, dP^{x_o}(\omega) \geq$$

$$\int \chi_{\{t \wedge T(\omega) > R_k(\omega)\}} \int s_n(X_{t-R_k(\omega)}(\omega')) \cdot \chi_{\{T_{\eta(X_{R_k}(\omega))}(\omega') > t - R_k(\omega)\}} \cdot$$

$$\cdot\, dP^{X_{R_k}(\omega)}(\omega')\, dP^{x_o}(\omega) .$$

Using the mentioned form of the strong Markov property the last term is

$$= E^{x_o} [\, s_n(X_t)\ ;\ T_{\eta(X_{R_k})} \circ \theta_{R_k} > t - R_k;\ t \wedge T > R_k \,]$$

Since $s(x) \geq E^x [\, s(X_{T_{\eta(x)}})\,]$ for x in Γ , by the use of the definition of R_{k+1}, and of the preceding inequality, we get

$$E^{x_o} [\, s(X_{R_k})\ ;\ t \wedge T > R_k \,] \geq \int \chi_{\{t \wedge T(\omega) > R_k(\omega)\}} \cdot$$

$$\cdot\, s(X_{T_{\eta(X_{R_k}(\omega))}}(\omega'))\, dP^{X_{R_k}(\omega)}(\omega')\, dP^{x_o}(\omega) \geq$$

$$E^{x_o} [\, s_n(X_t)\ ;\ R_{k+1} > t\ ;\ t \wedge T > R_k \,] +$$

$$+ \int \chi_{\{t \wedge T(\omega) > R_k(\omega)\}} \int \chi_{\{T_{\eta}(X_{R_k}(\omega))(\omega') \leq t - R_k(\omega)\}} \cdot$$

$$\cdot s(X_{T_{\eta}(X_{R_k}(\omega))}(\omega')) \, . \, dP^{X_{R_k}(\omega)}(\omega') \, dP^{x_o}(\omega)$$

Using here again the strong Markov property, this last term is transformed in :

$$E^{x_o}[s(X_{T_{\eta}(X_{R_k})} \circ \theta_{R_k}) \; ; \; T_{\eta}(X_{R_k}) \circ \theta_{R_k} \leq t - R_k \; ;$$

$$t \wedge T > R_k] = E^{x_o}[s(X_t) \; ; \; R_{k+1} = t, \; t \wedge T > R_k] +$$

$$+ E^{x_o}[s(X_{R_{k+1}}) \; ; \; R_{k+1} < t \; ; \; T > R_k]$$

Now we can write

$$E^{x_o}[s(X_{R_k}) \; ; \; t \wedge T > R_k] \geq E^{x_o}[s_n(X_t); \; R_{k+1} \geq$$

$$\geq t > R_k; \; T > t] + E^{x_o}[s(X_{R_{k+1}}) \; ; \; R_{k+1} < t \; ; \; T > R_{k+1}] ,$$

that leads to ($*$) written for (k+1).

The quasileft-continuity of the process implies that $\sup_k R_k \geq T$, hence $s(x_o) \geq E^{x_o}[s_n(X_t) \; ; \; t < T]$.

Since $s = \lim s_n$ on V we get $s(x_o) \geq \liminf_{n \to \infty} E^{x_o}[s_n(X_t);$ $t < T] \geq E^{x_o}[s(X_t) \; ; \; t < T]$ and letting V to increase to U we get $s(x_o) \geq$ $\geq E^{x_o}[s(X_t)]$.

Let now W be in $\mathfrak{F}$ and define s' on W by $s'(x) = \int s \, d\rho_n^W$. s' is excesive on W (see [3], p. 73) and we have

$$s'(x_o) = \lim_{t\to 0} E^{x_o}[\, s'(X_t) \; ; \; t < T_{U\setminus W}\,] \; ;$$

then $s(x_o) \geq \limsup_{t\to 0} E^{x_o}[\, s(X_t)] \geq \liminf_{t\to 0} E^{x_o}[\, s(X_t)] \geq$

$$\geq s'(x_o) = \int s \, d\rho^W_{x_o}$$; W being arbitrary we get

$$s(x_o) = \lim_{t\to 0} E^{x_o}[\, s(X_t)] \, .$$

Corollary 3.2. The excesive functions and the hyperharmonic functions on U coincide.

L. Stoica
INCREST
Bd. Păcii nr. 220
București
ROMANIA

REFERENCES :

[1] H. Bauer, Harmonishe Räume und ihre Potentialtheorie. Lecture Notes in Math. 22 (1966)

[2] H. Bauer, Harmonic spaces and associated Markov pre precesses, C.I.M.E. Stresa (1969).

[3] R.M. Blumenthal, R.K. Getoor, Markov Processes and Potential Theory, Academic Press, New York - London (1969)

[4] J.M. Bony, Sur la régularité des solutions du problème de Dirichlet ... C.R.A.S. 267 (1968)

[5] J.M. Bony, Operateurs elliptiques dégénérés associés aux axiomatiques de la théorie du potentiel. C.I.M.E. Stresa (1969).

[6] E.B. Dynkin, Markov Processes, Springer-Verlag, Berlin-Götingen-Heidelberg (1965)

[7] S. Itô, On definitions of superharmonic functions. Ann Inst. Fourier nr. 3-4 (1975).

[8] P.A. Meyer, Probabylity and Potentials, Blaisdell Publ. Comp. Waltham-Toronto-London (1966).

[9] O.A. Oleinik, E.V. Radkevich, Second order equations with non negative characteristics. (in Russian). Itogy Nauky, seria matematika, Moskva (1971).

Boundary value problems for composite type systems of first order partial differential equations

HEINRICH BEGEHR

Dedicated to Professor Dr. Ludwig Bieberbach on the occasion of his ninetieth birthday

1. Introduction. A system of partial differential equations of first order of the form

$$(1) \qquad \sum_{\nu=1}^{n} [a^{\mu}_{\nu} u^{\nu}_{x} + b^{\mu}_{\nu} u^{\nu}_{y}] = f^{\mu}(x,y,u^{1},\ldots,u^{n}) \quad (1 \le \mu \le n)$$

is called elliptic if

$$(2) \qquad \det [a^{\mu}_{\nu} \lambda + b^{\mu}_{\nu}] = 0 \qquad (\det a^{\mu}_{\nu} \neq 0)$$

has no real solution λ (so n must be even) and it is called hyperbolic if (2) has n real solutions different from each other. The system (1) is called a system of mixed kind, if (2) has as well real as non-real solutions. Here we will consider systems for which every $\lambda = \lambda(x,y)$ is real respective non-real within the concerned domain. The simplest case occurs for $n = 3$ when (2) has the solutions $\lambda_1+i\lambda_2, \lambda_1-i\lambda_2, \lambda_3$ ($\lambda_k (k = 1,2,3)$ real). This system has been considered by Vidic [6] in the linear case

$$f^{\mu} = \sum_{\nu=1}^{n} c^{\mu}_{\nu} u^{\nu} + d^{\mu} .$$

In view of studying such systems he mentioned that each elliptic equation of second order

$$\phi_{xx} + \phi_{yy} + a\phi_{x} + b\phi_{y} + c\phi = f$$

which in general connot be reduced to an elliptic system of first order, can be reduced by the transformation

$$u = \phi_{y} \ , \ v = \phi_{x}$$

to the first order system

$$u_x - v_y = 0,\ u_y + v_x + bu + av + c\phi = f,\ \phi_x - v = 0$$

of mixed kind.
Let in the following G be a simply connected bounded domain of the complex plane $\mathbb{C}$ with continuously differentiable boundary ∂G, G_o a domain with

$$\hat{G} := G \cup \partial G \subset G_o$$

and

$$a_\nu^\mu\ ,\ b_\nu^\mu \in C(G_o),\ f^\mu \in C(\hat{G} \times \mathbb{R}^3) \qquad (1 \le \mu, \nu \le 3).$$

The system of real curves given by the solutions of the ordinary differential equation

$$(3) \qquad \frac{dy}{dx} + \lambda_3(x,y) = 0 \qquad (x+iy \in G_o)$$

are the real characteristic curves of (1). We suppose that all those characteristics which meet $\hat{G}$ intersect ∂G in exactly two points with the exception of two curves which only touch ∂G in one of two different points. These two points divide ∂G in two arcs $\Gamma_k (k=1,2)$. Becaurse of the uniqueness of the initial value problem for (3) (observe $\lambda_3 \in C(G_o)$) for every point $z \in G$ there exists one and only one characteristic passing through z. This characteristic will be denoted by $\gamma(z)$.
By a differentiable transformation of the independent variables with nonvanishing Jacobian and linear combinations of the unknown functions the system (1) can be transformed in such a way that the elliptic part appears with Cauchy-Riemannian main part. To see what will happen to the third equation in (1) for n = 3 one has to observe that (1) is equivalent to

$$U_x^\mu - \tilde{\lambda}^\mu U_y^\mu = \sum_{\nu=1}^{3} C_\nu^\mu f^\nu \quad (1 \le \mu \le 3; \tilde{\lambda}^1 = \lambda_1 + i\lambda_2, \tilde{\lambda}^2 = \overline{\tilde{\lambda}^1}, \tilde{\lambda}^3 = \lambda_3)$$

where

$$U^\mu := \sum_{\nu=1}^{3} A_\nu^\mu u^\nu\ ,\ A_\nu^\mu := \sum_{\varkappa=1}^{3} C_\varkappa^\mu a_\nu^\varkappa \qquad (1 \le \mu, \nu \le 3)$$

and C^μ_ν with det $C^\mu_\nu \neq 0$ are solutions of

$$\sum_{\varkappa=1}^{3} C^\mu_\varkappa (\tilde{\lambda}^\mu a^\varkappa_\nu + b^\varkappa_\nu) = 0 \qquad (1 \le \mu, \nu \le 3).$$

The third equation here is real

$$U^3_x - \lambda_3 U^3_y = \sum_{\nu=1}^{3} C^3_\nu f^\nu \qquad (C^1_\nu = \overline{C^2_\nu} (1 \le \nu \le 3))$$

so that the left hand side may be written as a derivative in the direction

$$\alpha = \frac{1 - i\lambda_3}{\sqrt{1+\lambda_3^2}},$$

which is the direction of the real characteristics,

$$U^3_\alpha := D^\alpha U^3 := \frac{1}{\sqrt{1+\lambda_3^2}} (U^3_x - \lambda_3 U^3_y).$$

By the mentioned transformation of the variables (see Vekua[5],p. 104 ff.) the canonical form of the system (1) appears:

(4) $\quad u_x - v_y = f_1(z,w,\omega), u_y + v_x = f_2(z,w,\omega), \ \omega_\lambda = f_3(z,w,\omega).$

Here we use the variables

$$z = x+iy, \ \bar{z} = x-iy, \ w = u+iv, \ \omega = \bar{\omega}$$

and the direction

$$\lambda = \lambda(z) = \lambda_1(x,y) + i\lambda_2(x,y) \qquad (\lambda_1^2(x,y)+\lambda_2^2(x,y)=1)$$

where λ_1, λ_2 are the coefficients of the direction of λ. Let again G be the (transformed) domain in which (4) is considered and G_o is a domain containing $\hat{G}$. G has the same properties as the domain has had before being transformed. In G_o the function λ defines a vector field. If γ is a curve in this field it is a real characteristic of (4). Let s be the parameter of arc length of γ, then we have

$$\frac{dx}{ds} = \lambda_1(x,y) \ , \quad \frac{dy}{ds} = \lambda_2(x,y)$$

or in complex form

$$\frac{dz}{ds} = \lambda(z) \qquad (|\lambda(z)|=1).$$

As λ is a continuous function in G_o this differential equation has exactly one soluton $\zeta(s)$ with initial value $\zeta(s_1) = z_1$, so that two different curves of the vector field don't have common points. Moreover the solution depends continuously on the initial values. As usual we write

$$\zeta(s) = \zeta(s;s_1,z_1)$$

for the solution with $\zeta(s_1) = z_1$. With this notations the part of that real characteristic of (4) through the point ζ_o of Γ_1, where as before Γ_1 is one of the two parts of ∂G defined by the system of real characteristics in G_o, belonging to $\hat{G}$ and having (finite) arc leugth $L(\zeta_o)$ can be described by

$$\gamma(\zeta_o) := \{\zeta : \zeta=\zeta(s;0,\zeta_o),\ 0 \le s \le L(\zeta_o),\ \frac{d}{ds}\,\zeta(s;0,\zeta_o) = \lambda(\zeta(s;0,\zeta_o))\}.$$

For $z \in \gamma(\zeta_o)$ we denote by γ_z that part from $\gamma(\zeta_o)$ between ζ_o and z

$$\gamma_z := \{\zeta : \zeta=\zeta(t;0,\zeta_o),\ 0 \le t \le s\} \quad (z=\zeta(s;0,\zeta_o)).$$

As ∂G is continuously differentiable and $\lambda \in C(\hat{G})$ $L(\zeta_o)$ is a continuous function of $\zeta_o \in \Gamma_1$. If φ is a continuous function on $\hat{G}$ then

$$f(z) := \int_{\gamma_z} \varphi(\zeta)\,|d\zeta| = \int_0^s \varphi(\zeta(t;s,z))\,dt$$

is a continuous function in $\hat{G}$, differentiable in the direction of λ with

$$f_\lambda(z) = \varphi(z) \quad (z \in G).$$

Continuity of f in $\hat{G}$ except the two end points of Γ_1 follows directly by continuity of $\partial G, \gamma(\zeta_o)$ and φ. If ζ_1 is one of the end points of Γ_1 then

$$\lim_{\zeta_o \to \zeta_1} L(\zeta_o) = 0 \qquad (\zeta_o \in \Gamma_1).$$

This can be concluded from the continuity of $\zeta(t;0,\zeta_o)$ with respect to $\zeta_o \in \Gamma_1$. If namely K denotes a sufficiently small circle around ζ_1 and $l(\zeta_o)$ respectively $l(\zeta_1)$ denotes the length of the part of the characteristic through ζ_o respectively ζ_1 in K then $l(\zeta_o) - l(\zeta_1)$ is small if $|\zeta_o - \zeta_1|$ is small. A simple calculation shows that $\gamma(\zeta_o)$ lies in K for sufficiently small $|\zeta_o - \zeta_1|$ so that $L(\zeta_o) \le l(\zeta_o)$. As K shrinks to ζ_1 $l(\zeta_1)$ tends to zero so that $L(\zeta_o)$ tends to zero when ζ_o tends to ζ_1. From this it follows

$$\lim_{z \to \zeta_1} f(z) = 0$$

so that f is also continuous in the end points of Γ_1. The differentiability in λ-direction follows by direct computation.

2. Equivalent system of integral equations. System (4) can be written in complex form

$$(5) \qquad w_{\bar{z}} = f(z,w,\omega) \quad , \quad \omega_\lambda = \varphi(z,w,\omega).$$

Here f is a complex, φ a real function, w is an unknown complex, ω an unknown real function. Starting from the Green-Pompeiu formula

$$w(z) = \frac{1}{2\pi i} \int_{\partial E} w(\zeta) \frac{d\zeta}{\zeta - z} - \frac{1}{\pi} \int_E w_{\bar{\zeta}}(\zeta) \frac{d\xi d\eta}{\zeta - z} (z \in E, \zeta = \xi + i\eta)$$

for a function $w \in C(\hat{E})$ with $w_{\bar{z}} \in L_p(\hat{E})$ $(2<p)$ where $\hat{E}$ is the closure of the unit disc E and $w_{\bar{z}}$ means the generalized derivative in the Sobolev sense (see Vekua [5], I, § 6), and adding the complex conjugate of

$$0 = \frac{1}{2\pi i} \int_{\partial E} w(\zeta) \frac{\bar{z}}{1-\bar{z}\zeta} d\zeta - \frac{1}{\pi} \int_E w_{\bar{\zeta}}(\zeta) \frac{\bar{z}}{1-\bar{z}\zeta} d\xi d\eta \quad (z \in E)$$

One has the representation formula $(z \in E)$

$$(6) \quad \begin{aligned} w(z) &= \frac{1}{\pi i} \int_{\partial E} \operatorname{Re} w(\zeta) \frac{d\zeta}{\zeta - z} - \frac{1}{2\pi i} \int_{\partial E} \overline{w(\zeta)} \frac{d\zeta}{\zeta} - \\ &- \frac{1}{\pi} \int_E [w_{\bar{\zeta}}(\zeta) \frac{1}{\zeta - z} + \overline{w_{\bar{\zeta}}(\zeta)} \frac{z}{1 - z\bar{\zeta}}] d\xi d\eta \,. \end{aligned}$$

If ϕ denotes the conformal mapping of the domain G (G bounded, simply connected with smooth boundary) onto E and $G^I(\zeta,z)$ and $G^{II}(\zeta,z)$ the first and second Green's function of G, which have the forms

$$G^I(\zeta,z) = -\frac{1}{2\pi} \log \left| \frac{\phi(\zeta)-\phi(z)}{1-\overline{\phi(\zeta)}\phi(z)} \right| \qquad (\zeta, z \in G),$$

$$G^{II}(\zeta,z) = -\frac{1}{2\pi} \log |(\phi(\zeta)-\phi(z))(1-\overline{\phi(\zeta)}\phi(z))| \quad (\zeta, z \in G)$$

(see Haack-Wendland [3]), transformation of (6) leads to the representation formula

$$(7) \quad \begin{aligned} w(z) &= -\int_{\partial G} \operatorname{Re} w(\zeta)(d_n G^I - i dG^{II})(\zeta,z) - i \int_{\partial G} \operatorname{Im} w(\zeta) d_n G^{II}(\zeta,z) + \\ &+ 2 \int_G [w_{\bar{\zeta}}(\zeta)(G^I_\zeta + G^{II}_\zeta)(\zeta,z) + \overline{w_{\bar{\zeta}}(\zeta)}(G^I_{\bar{\zeta}} - G^{II}_{\bar{\zeta}})(\zeta,z)] d\xi d\eta \quad (z \in G) \end{aligned}$$

where

$$d := \frac{\partial}{\partial \zeta} d\zeta + \frac{\partial}{\partial \bar{\zeta}} d\bar{\zeta} \,, \quad d_n := -i\left(\frac{\partial}{\partial \zeta} d\zeta - \frac{\partial}{\partial \bar{\zeta}} d\bar{\zeta}\right)$$

are real operators. This formula gives us an integral equation of Fredholm type for solutions of differential equations of the form

$$w_{\bar{z}} = f(z,w)$$

which were studied in [1],[3],[4],[5],[7].

By integrating the derivative of the real function ω in the direction of λ along the curves γ_z from ζ_o to z with respect to the arc length one gets

$$(8)\qquad \omega(z) = \omega(\zeta_o) + \int_{\gamma_z} \omega_\lambda(\zeta)\,|d\zeta| \qquad (|d\zeta| = \lambda_1 d\xi + \lambda_2 d\eta).$$

Thus a solution of a differential equation of the form

$$\omega_\lambda = \varphi(z,\omega)$$

with siutable function φ will fulfil the Volterra integral equation

$$\omega(z) = \omega(\zeta_o) + \int_{\gamma_z} \varphi\,(\zeta,\omega(\zeta))\,|d\zeta| \quad (z\in G, \zeta_o = \Gamma_1 \cap \gamma(z)).$$

With formulas (7) and (8) it is possible to find a system of integral equations equivalent to system (5) when for the solutions of (5) the integrability of f and φ are guaranteed. Under siutable asumptions given in the next paragraph we have

$$(9)\qquad \begin{aligned} w(z) &= -\,\Theta_1(z) + (\underset{\sim}{P}_1(w,\omega))(z) \\ \omega(z) &= \Theta_2(z) + (\underset{\sim}{P}_2(w,\omega))(z) \end{aligned} \qquad (z\in G)$$

with

$$\Theta_1(z) := \int_{\partial G} \mathrm{Re}\, w(\zeta)(d_n G^{I} - i dG^{II})(\zeta,z) + i \int_{\partial G} \mathrm{Im}\, w(\zeta) d_n G^{II}(\zeta,z) \quad (z\in G),$$

$$\Theta_2(z) := \omega(\zeta_o) \qquad (z\in\hat{G},\ \zeta_o = \Gamma_1 \cap \gamma(z)),$$

$$(\underset{\sim}{P}_1(w,\omega))(z) := 2\int_G [f(\zeta,w(\zeta),\omega(\zeta))(G^{I}_{\zeta} + G^{II}_{\zeta})(\zeta,z) + \\ + \overline{f(\zeta,w(\zeta),\omega(\zeta))}\,(G^{I}_{\bar\zeta} - G^{II}_{\bar\zeta})(\zeta,z)]\,d\xi d\eta \quad (z\in G),$$

$$(\underset{\sim}{P}_2(w,\omega))(z) := \int_{\gamma_z} \varphi(\zeta,w(\zeta),\omega(\zeta))\,|d\zeta| \qquad (z\in G).$$

The form of this functions and operators suggests the following boundary value problem.

3. First boundary value problem. Let f be a complex valued function defined on $\hat{G} \times \mathbb{C} \times \mathbb{R}$ such that

$$(10) \qquad f(\cdot,w(\cdot),\omega(\cdot)) \in L_p(\hat{G}) \quad (2<p) \text{ for all } w,\omega \in C(\hat{G})$$

where w is a complex and ω a real function and

$$(11) \quad |f(z),w,\omega)-f(z,\tilde{w},\tilde{\omega})| \le \frac{1}{c}\, g(z,w-\tilde{w},\omega-\tilde{\omega})\; (z\in\hat{G};w,\tilde{w}\in\mathbb{C};\omega,\tilde{\omega}\in\mathbb{R}).$$

Here g is a siutable non negative function with properties to be specified below and c is a constant greater than or equal to $\frac{2}{\pi}$ such that

$$\left|\frac{\phi'(\zeta)(\zeta-z)}{\phi(\zeta)-\phi(z)}\right| \le \frac{\pi}{2}\, c \qquad (z,\zeta\in\hat{G}).$$

Analogically we ask

$$(12) \qquad \varphi(\cdot,w(\cdot),\omega(\cdot)) \in C(\hat{G}) \text{ for all } w,\omega \in C(\hat{G}),$$

$$(13) \quad |\varphi(z,w,\omega)-\varphi(z,\tilde{w},\tilde{\omega})| \le h(z,w-\tilde{w},\omega-\tilde{\omega}) \quad (z\in\hat{G};w,\tilde{w}\in\mathbb{C};\omega,\tilde{\omega}\in\mathbb{R})$$

with non negative sufficiently good function h. Under special asumptions on g and h the boundary value norm problem

$$(14) \qquad w_{\bar{z}} = f(z,w,\omega)\;, \quad \omega_\lambda = \varphi(z,w,\omega),$$

$$(15) \qquad \mathrm{Re}\, w|_{\partial G} = \chi\,, \quad \omega|_{\Gamma_1} = \psi\;, \quad \int_{\partial G} \mathrm{Im}\, w(\zeta)\, d_n G^{II}(\zeta,z) = C$$

for

$$\chi \in C(\partial G)\;, \quad \psi \in C(\Gamma_1)\;, \quad C \in \mathbb{R}$$

which is equivalent to the system of integral equations (9) where now by the boundary conditions

$$(16) \qquad \Theta_1(z) := \int_{\partial G} \chi(\zeta)(d_n G^I - i dG^{II})(\zeta,z) + iC \quad (z\in G)$$

$$(17) \qquad \Theta_2(z) := \psi(\zeta_o) \qquad (z\in G, \zeta_o = \Gamma_1 \cap \gamma(z))$$

are known functions. Θ_1 is an analytic function in G, continuous in $\hat{G}$ and satisfies the boundary conditions

$$\mathrm{Re}\ \Theta_1|_{\partial G} = \chi\ ,\ \int_{\partial G} \mathrm{Im}\ \Theta_1(\zeta) d_n G^{II}(\zeta,z) = C$$

(see Haack-Wendland [3], 10.5). As the solution

$$\zeta(t;0,\zeta_o) = \zeta(t;s,z) \quad (\zeta_o = \Gamma_1 \cap \gamma(z))$$

of the ordinary differential equation

$$\frac{d}{dt}\zeta(t) = \lambda(\zeta(t))$$

depends continuously on the initial values $\zeta_o \in \Gamma_1$ respectively $z \in \hat{G}$ and ∂G itself is continuously differentiable Θ_2 is a continuous function of z in $\hat{G}$.

4. Existence of solutions. Let L be the upper limit of the arc length of the curves γ_z

$$L := \sup_{z \in G} \int_{\gamma_z} |d\zeta| < +\infty$$

and let g and h be two continuous functions of (w,ω) in the point (0,0) of $\mathbb{C} \times \mathbb{R}$ for all $z \in \hat{G}$ uniformly in $z \in \hat{G}$, which vanish at $(w,\omega)=(0,0)$ identically in $z \in \hat{G}$ and have the properties

$$(18) \qquad g(z,w,\omega) \le g(z,K_1,K_2)\ ,\ h(z,w,\omega) \le h(z,K_1,K_2) \qquad (z \in \hat{G}, |w| \le K_1, |\omega| \le K_2)$$

for fixed positive constants K_1 and K_2 and

$$|\Theta_1(z)| + \int_G \{c|f(\zeta,0,0)| + g(\zeta,K_1,K_2)\} \frac{d\xi d\eta}{|\zeta - z|} \le K_1 \qquad (z \in G)$$

(19)

$$|\Theta_2(z)| + \int_{\gamma_z} \{|\varphi(\zeta,0,0)| + h(\zeta,K_1,K_2)\} |d\zeta| \le K_2 \qquad (z \in G).$$

Theorem 1. If f and φ have the properties (10),(11) respectively (12),(13) and (18),(19) then there exists a solution of the system (9) together with (16),(17).

Proof. Consider the compact and convex subset

$$A := \{(w,\omega) : (w,\omega) \in B;\ ||w|| \le K_1, ||\omega|| \le K_2;$$
$$\operatorname{Re} w|_{\partial G} = \chi, \omega|_{\Gamma_1} = \psi, \frac{1}{2\pi} \int_{\partial G} \operatorname{Im} w(\zeta)\, |d\phi(\zeta)| = -C\}$$

of the Banach space $B = C_c(\hat{G}) \times C_r(\hat{G})$ with maximum norm

$$||(w,\omega)|| := ||w|| + ||\omega||.$$

Here $C_c(\hat{G})$ denotes the class of complex valued continuous function in $\hat{G}$ and $C_r(\hat{G})$ the real valued functions of $C(\hat{G})$. Becaurse $(-\Theta_1, \Theta_2) \in A$ this subset is non empty. The non linear operator $\underset{\sim}{P}$ on B given by

$$\underset{\sim}{P}(w,\omega) := (-\Theta_1 + \underset{\sim}{P}_1(w,\omega), \Theta_2 + \underset{\sim}{P}_2(w,\omega))$$

is continuous as follows from the continuity of g and h in $(w,\omega)=(0,0)$ and the inequalities

$$|(\underset{\sim}{P}_1(w,\omega) - \underset{\sim}{P}_1(\tilde{w},\tilde{\omega}))(z)| \le \int_G g(\zeta, (w-\tilde{w})(\zeta), (\omega-\tilde{\omega})(\zeta)) \frac{d\xi d\eta}{|\zeta - z|},$$

$$|(\underset{\sim}{P}_2(w,\omega) - \underset{\sim}{P}_2(\tilde{w},\tilde{\omega}))(z)| \le \int_{\gamma_z} h(\zeta, (w-\tilde{w})(\zeta), (\omega-\tilde{\omega})(\zeta))\, |d\zeta|$$

and maps A into itself as can be seen by

$$|\underset{\sim}{P}_1(w,\omega)| \le \int_G [c|f(\zeta,0,0)| + g(\zeta, w(\zeta), \omega(\zeta))] \frac{d\xi d\eta}{|\zeta - z|},$$

$$|\underset{\sim}{P}_2(w,\omega)| \le \int_{\gamma_z} [|\varphi(\zeta,0,0)| + h(\zeta, w(\zeta), \omega(\zeta))]\, |d\zeta|$$

and inequalities (18),(19). The fix point theorem of Schauder guarantees the existence of a solution of

$$(w,\omega) = \underset{\sim}{P}(w,\omega)$$

which is the system (9). This solution belongs to A.

5. Uniqueness of solution. Under special asumptions on g and h the solution of (9),(16),(17) is unique. To show this we state the following lemma.

Lemma. Let g and h defined on $\hat{G} \times \mathbb{C} \times \mathbb{R}$ have the properties

1. $$g(z,0,\omega) = 0 \ , \ g(z,w,\omega) \le g(z,\tilde{w},\tilde{\omega})$$
$$(z\in\hat{G}, |w| \le |\tilde{w}|, |\omega| \le |\tilde{\omega}|)$$
$$h(z,w,0) = 0 \ , \ h(z,w,\omega) \le h(z,\tilde{w},\tilde{\omega})$$

2. $g(\cdot,w,\omega)$ is mesurable in G for all $w \in \mathbb{C}$, $\omega \in \mathbb{R}$,
   $g(z,\cdot,\cdot)$ is continuous in $\mathbb{C} \times \mathbb{R}$ for almost all $z\in\hat{G}$,
   h is continuous on $\hat{G} \times \mathbb{C} \times \mathbb{R}$
3. There exist positive constants K_1, K_2 such that
$$\int_G g(\zeta,K_1,K_2) \frac{d\xi d\eta}{|\zeta-z|} \le K_1 \ , \quad \int_{\gamma_z} h(\zeta,K_1,K_2)\,|d\zeta| \le K_2 \qquad (z\in\hat{G})$$
4. For all pais of functions $(x,y)\in C_{1,2}(\hat{G})-\{(0,0)\}$,
$$C_{1,2}(\hat{G}) := \{(x,y) : (x,y)\in B, |x(z)| \le K_1, |y(z)| \le K_2 (z\in\hat{G})\},$$
   $\underset{\sim}{I}(x,y)$ is different from $(|x|,|y|)$ where the operator $\underset{\sim}{I}$ is defined by
$$\underset{\sim}{I}(x,y) := (\underset{\sim}{I}_1(x,y), \underset{\sim}{I}_2(x,y)),$$
$$(\underset{\sim}{I}_1(x,y))(z) := \int_G g(\zeta,x(\zeta),y(\zeta))\frac{d\xi d\eta}{|\zeta-z|} \qquad (z\in\hat{G}).$$
$$(\underset{\sim}{I}_2(x,y))(z) := \int_{\gamma_z} h(\zeta, x(\zeta),y(\zeta))\,|d\zeta|$$

Then the pair of integral inequalities
$$(20) \qquad |\delta_k(z)| \le (\underset{\sim}{I}_k(\delta_1,\delta_2))(z) \qquad (z\in\hat{G}; k=1,2)$$
in the space $C_{1,2}(\hat{G})$ only has the trivial solution $(\delta_1,\delta_2) = (0,0)$.

Proof. Let $(x_0,y_0) \in C_{1,2}(\hat{G})$ be a solution of (20). Define iteratively by
$$(x_n,y_n) := \underset{\sim}{I}(x_{n-1},y_{n-1}) \quad (n\in \mathbb{N})$$
a sequence $((x_n,y_n))$ of pairs of functions. As from

$$(x_{n-1},y_{n-1}) \in C_{1,2}(\hat{G}) \quad (n\in \mathbb{N})$$

and properties 1. and 3. it follows

$$0 \le \underset{\sim}{I}_k(x_{n-1},y_{n-1}) \le K_k \quad (k=1,2;n\in \mathbb{N})$$

so that with respect to the continuity of $\underset{\sim}{I}_k(x,y)$ in $\hat{G}$ for k=1,2

$$(x_n,y_n)\in C_{1,2}(\hat{G}) \quad (n\in \mathbb{N}).$$

(x_n) and (y_n) are increasing sequences of non negative continuous functions as can be seen by induction. Thus the functions

$$x:= \lim_{n\to+\infty} x_n \quad , \quad y:= \lim_{n\to+\infty} y_n$$

exist in $\hat{G}$ and

$$0 \le x \le K_1 \ , \ 0 \le y \le K_2 \ .$$

From the monotoneity of the sequences and the continuity of g and h in (w,ω) (condition 2.) one can applicate the theorem of monotone convergence to get

$$x = \underset{\sim}{I}_1(x,y) \quad , \quad y = \underset{\sim}{I}_2(x,y),$$

so that $(x,y) \in C_{1,2}(\hat{G})$. From here and

$$0 \le |x_o| \le x \ , \ 0 \le |y_o| \le y$$

and 4. $x_o = y_o = 0$ can be concluded.

As an example we mention the functions

$$g(z,w,\omega):= \frac{K_1^{1-\alpha}}{(K_1+K_2)^{\alpha}}\, g_o(z)\,|w|^{\alpha}(|w|+|\omega|)^{\alpha}$$

$$(\alpha= \tfrac{1}{2};\alpha\ge 1)$$

$$h(z,w,\omega):= \frac{K_2^{1-\alpha}}{(K_1+K_2)^{\alpha}}\, h_o(z)\,|\omega|^{\alpha}(|w|+|\omega|)^{\alpha}$$

where g_o and h_o are non negative functions with

$$g_o \in L_p(\hat{G})\,(2<p) \quad , \int_G g_o(\zeta)\,\frac{d\xi d\eta}{|\zeta-z|} < 1 \quad (z\in G),$$

$$h_o \in C(\hat{G}) \quad , \int_{\gamma_z} h_o(\zeta)\,|d\zeta| < 1 \;(z\in G).$$

Here property 4. and the conclusion of the lemma can be verified. Under similar asumptions the functions

$g(z,w,\omega) := g_o(z)(|w|+|\omega|)$, $h(z,w,\omega) := h_o(z)(|w|+|\omega|)$

also fulfil property 4. and the conclusion of the lemma. With this functions the conditions (11) and (13) have the form of a Lipschitz condition.

<u>Theorem 2</u>. If the functions g and h from (11) and (13) have the properties of the lemma and moreover

$$(21) \quad \int_G g(\zeta,2K_1,2K_2)\,\frac{d\xi d\eta}{|\zeta-z|} \le K_1, \int_{\gamma_z} h(\zeta,2K_1,2K_2)\,|d\zeta| \le K_2$$

then the system (9),(16),(17) is uniquely solvable in A.

<u>Proof</u>. Let $(w,\omega),(\tilde{w},\tilde{\omega})\in A$ be two solutions.

Then from

$$w-\tilde{w} = \underset{\sim}{P}_1(w,\omega) - \underset{\sim}{P}_1(\tilde{w},\tilde{\omega}), \omega - \tilde{\omega} = \underset{\sim}{P}_2(w,\omega) - \underset{\sim}{P}_2(\tilde{w},\tilde{\omega}),$$

(11),(13), and (21) it follows that

$$(w-\tilde{w},\omega-\tilde{\omega})\in C_{1,2}(\hat{G})$$

and that $(w-\tilde{w},\omega-\tilde{\omega})$ is a solution of (20). So by the lemma $(w,\omega)=(\tilde{w},\tilde{\omega})$.

The asumptions (19) and (21) are restrictions to the dimension of the domain G. For sufficiently small G the regarded integrals are small. In the special case of Lipschitz condition one can take any positive constants for K_1 and K_2, so that the theorems are fulfilled not only in a bounded subset A but in the whole space B.

Here I want to express my thanks to Professor R.P. Gilbert who drew my attention to the paper [6] of Vidic and to I. Begehr, H. Giebel, and V.E. Krüger of the "Zentralblatt für Mathematik" who often inform me about new papers in my fields of interest.

References

[1] Begehr, H.-Gilbert, R.P. Über das Randwert-Normproblem für ein nichtlineares elliptisches System. Erscheint demnächst.

[2] Begehr, H.-Gilbert, R.P. Randwertaufgaben ganzzahliger Charakteristik für verallgemeinerte hyperanalytische Funktionen. Erscheint demnächst.

[3] Haack, W.-Wendland, W. Vorlesungen über partielle und Pfaffsche Differentialgleichungen. Birkhäuser, Basel-Stuttgart, 1969, 555 S.

[4] Tutschke, W. Über Fixpunktmethoden in der Theorie partieller komplexer Differentialgleichungen. Beiträge komplex. Analysis Anw. Diff.-geom. Schriftenr. Zentralinst. Math. Mech. Akad. Wiss. DDR 18, 31-41 (1974).

[5] Vekua, I.N. Verallgemeinerte analytische Funktionen. Akademie Verlag, Berlin. 1963, 538 S.

[6] Vidic, Ch. Über zusammengesetzte Systeme partieller linearer Differentialgleichungen erster Ordnung. Dissertation D 83, Technische Universität Berlin, 1968, 46 S.

[7] Wendland, W. An integral equation method for generalized analytic functions. In: Constructive and computational methods for differential and integral equations. E d i t e d by D.L. Colton and R.P. Gilbert. Lecture Notes in Mathematics 430, Springer, Berlin-Heidelberg-New York. 1974, 414-542.

Freie Univ.Berlin
Inst.Math.I

Sur le mouvement en milieux poreux avec charge variable sur le contour.

Şt.I. Gheorghiţă

1. La théorie du mouvement stationnaire linéaire plan d'un fluide à travers un milieux poreux conduit à un problème aux limites mixte de la théorie des fonctions analytiques. En général, on admet que l'écoulement en milieu poreux a un débit très faible, de sorte que sur les surfaces d'alimentation on peut prendre la partie réelle du potential complexe $f(z) = \varphi(x,y) + i\psi(x,y)$ ($z = x + iy$, φ-le potentiel des vitesses, ψ - la fonction de courant, l'axe Oy du système de réference cartésian orthogonal Oxy étant prise suivant la verticale ascendante) comme étant constante (par exemple, [1], [2]). Mais, d'un part, du point de vue théorique, il serait intéressant de considérer que φ est variable même sur les contours d'alimentation, et, d'autre part, dans certains cas, importants du point de vue pratique, le potentiel des vitesses n'est pas constant sur de tel contours. Ici on va traiter quelques cas de mouvements en milieux poreux, dans les conditions spécifiés au début, quand φ est variable sur les contours d'alimentation, en supposant l'absence des contours libres et des contours de suintement.

2. Si on désigne par p la pression, par γ le poids specifique du fluide, par w la vitesse complexe ($= u - iv$), et par k le coefficient de filtration, constant pour l'instant, alors on a les relations bien connues

$$\varphi = -k(\gamma^{-1} p + y), \qquad (1)$$

$$w(z) = df(z)/dz . \qquad (2)$$

En dehors des contours d'alimentation on aura des contours impermables, et sur ces contours Γ

$$\Psi\Big|_{\Gamma} = \text{const.} \qquad (3)$$

Pour fixer les idées, supposons que le mouvement a lieu entre deux contours d'alimentation, C_1 et C_2 , et deux contours imperméables, Γ_1 et Γ_2 . Sans aucune restriction, on peut prendre ψ nul sur Γ_1 et ψ egal à $\pm Q$, Q étant le débit du fluide qui s'écoule entre C_1 et C_2, sur Γ_2. Alors on est conduit a un problème aux limites du type suivant: déterminer une fonction analytique $f(z) = \varphi + i\psi$, quand

$$\left.\begin{aligned} \varphi(x,y)\Big|_{C_1} &= F_1(x,y)\ , & \varphi(x,y)\Big|_{C_2} &= F_2(x,y), \\ \psi(x,y)\Big|_{\Gamma_1} &= 0 \ , & \psi(x,y)\Big|_{\Gamma_2} &= Q\ . \end{aligned}\right\} \quad (4)$$

La methode qui peut être suivie pour resoudre ce problème est de représenter conformément le domaine du mouvement sur le demi-plan supérieur de la variable Z , de telle sorte que un point de Γ_1 correspond au point à l'infini. Alors on aura un problème aux limites de la forme suivante: detérminer une fonction analytique F, définie partout dans le demi-plan supérieur, où elle est regulière, y compris le point à l'infini, quand on connaît sur la frontière du demi-plan que

$$\left.\begin{aligned} \operatorname{Re}\{F\} &= \begin{cases} \Phi_1(X) & \text{pour } a < X < b\ , \\ \Phi_2(X) & \text{pour } c < X < d\ , \end{cases} \\ \operatorname{Im}\{F\} &= \begin{cases} Q & \text{pour } b < X < c\ , \\ 0 & \text{pour } -\infty < X < a \text{ et } d < X < \infty . \end{cases} \end{aligned}\right\} \quad (5)$$

Alors, en introduisant le polynome $P(Z) = (Z - a)(Z-b)(Z-c)(Z-d)$, la solution du problème sera fournie par la formule de A. Signorini

$$F(Z) = \frac{\sqrt{P(Z)}}{\pi} \left\{ \int_a^b \frac{\Phi_1(S)dS}{\sqrt{|P(S)|}\,(S-Z)} - Q \int_b^c \frac{dS}{\sqrt{P(S)}\,(S-Z)} - \right.$$

$$- \int_c^d \frac{\bar{\Phi}_2(s)ds}{\sqrt{|F(s)|}\,(s-z)} \Bigg\} , \qquad (6)$$

ou la détermination du radical est positive pour Z réel et plus petit que a ([3], p. 112). Un problème particulier, conduisant à des intégrales elliptiques, a été considéré par Sergiu Vasilache [4], et puis quelques cas, envisagés pour vérifier une condition aux limites en hydrodynamique, ont été étudiés par St.I. Gheorghtza ([5], [6]).

3. Le premier exemple sera relatif à un milieu qui occupe, dans le plan du mouvement, un demi-disque de rayon R, de sorte que le segment $-R < x < R$ est une frontière d'alimentation, mais l'arc $|z| = R$, $\theta \in (0, \pi)$ représente une frontière imperméable. Si sur la première frontière on a une variation linéaire de φ,

$$\varphi(x, 0) = Kx \ , \qquad -R < x < R \ ,$$

et si on tient compte que sur la frontière imperméable on peut prendre $\psi = 0$, on est conduit à la problème aux limites suivante: déterminer une fonction analytique dans le demi-plan $y > 0$, où elle est regulière, quand sur l'axe Ox on a la condition

$$\mathrm{Re}\,\{f(z)\} = \begin{cases} Kx & \text{si} \quad -R < x < R \\ KR^2/x & \text{si} \quad x < -R \ \text{et} \ x > R . \end{cases} \qquad (7)$$

La solution de ce problème est

$$f(z) = \frac{K}{\pi i}\left(\frac{R^2}{z}\ \ln\frac{R+z}{R-z} + z\ln\frac{z-R}{z+R}\right) , \qquad (8)$$

et on peut vérifier aisément que cette expression satisfait la condition (7).

Le débit du fluide qui passe à travers le milieu poreux s'obtient de (8),

$$Q = 2KR/\pi . \qquad (9)$$

4. Soit, maintenant, le milieu poreux homogène qui occupe le

domaine $0 < \arg z < \pi/n$ (n impar), les rayons $y = 0$, $x > R$ et $\arg z = \pi/n$, $r > R$ $(r = \sqrt{x^2 + y^2})$, étant des frontières imperméables sur lesquelles on a $\psi = 0$, mais sur les segments restants de la frontière le potentiel des vitesses varie, en valeur absolue, linéairement avec le module de la distance à l'origine à la puissance n, > 0 sur l'axe Ox et < 0 sur l'autre segment d'alimentation. Après une représentation conforme du domaine du mouvement sur le demi-plan $Y > 0$, on est conduit au problème suivant: déterminer une fonction analytique $F(Z)$, définie dans le demi-plan supérieur, où elle est régulière, y compris le point à l'infini, quand sur la frontière du demi-plan on connaît que

$$\left.\begin{aligned} \operatorname{Re}\left\{F(Z)\right\} &= KX \quad \text{şi} \quad -R^n < X < R^n, \\ \operatorname{Im}\left\{F(Z)\right\} &= 0 \quad \text{şi} \quad X < -R^n \quad \text{et} \quad X > R^n. \end{aligned}\right\} \tag{10}$$

La solution de ce problème est

$$f(z) = K\left[z^n - (z^{2n} - R^{2n})^{1/2}\right], \tag{11}$$

et en ce cas le débit du fluide qui passe à travers le milieu poreux est

$$Q = KR^n ; \tag{12}$$

quand $n = 1$, on obtient un résultat connu [6].

On peut obtenir des solutions sous forme finie quand le domaine occupé par le mulieu poreux est $0 < \arg z < (2n-1)\pi/n$ (n impair), les rayons $y = 0$, $x > R$ et $\arg z = (2n-1)\pi/n$, $r > R$ sont des frontières imperméables sur lesquelles $\psi = 0$, et sur les frontières d'alimentation φ varie en valeur absolue avec la puissance n de la distance à l'origine, < 0 sur le segment de l'axe Ox et > 0 sur l'autre segment. Alors on doit resoudre le problème suivant: déterminer une fonction analytique $F(Z)$ dans le demi-plan supérieur, régulière, y compris le point à l'infini, quand sur la frontière du demi-plan on sait que

$$\left.\begin{array}{l} \mathrm{Re}\left\{F(Z)\right\} = KX^{2n-1} \quad \text{si} \quad -R^{n/(2n-1)} < X < R^{n/(2n-1)} \\ \mathrm{Im}\left\{F(Z)\right\} = 0 \;, \quad \text{si} \quad X < -R^{n/(2n-1)} \quad \text{et} \quad X > R^{n/(2n-1)} \end{array}\right\} \quad (13)$$

La solution de ce problème peut être obtenue immediatement en utilisant un résultat dû a Caius Jacob [7]. On aura dans le plan Z

$$\left.\begin{array}{l} F(Z) = K \left\{ Z^{2n-1} - (Z^{2(n-1)} + R^{2n/(2n-1)} Z^{2(n-2)} + \ldots \right. \\ \left. \quad \ldots + R^{2(n-1)n/(2n-1)}) \left[Z^2 - R^{2n/(2n-1)} \right]^{1/2} \right\}, \end{array}\right\} \quad (14)$$

et dans le plan physique

$$f(z) = K \left\{ z^n - (z^{2(n-1)n/2n-1)} + R^{2n/(2n-1)} z^{2(n-2)n/(2n-1)} + \right.$$

$$\left. + \ldots + R^{2(n-1)n/(2n-1)}) \left[z^{2n/(2n-1)} - R^{2n/(2n-1)} \right]^{1/2} \right. . \quad (15)$$

Pour le cas particulier $n = 1$ on obtient le résultat mentioné de [6], tandis que pour $n = 3$ et $n = 5$ on aura, respectivement,

$$f_3(z) = K \left[z^3 - (z^{12/5} + R^{6/5} z^{6/5} + R^{12/5})(z^{6/5} - R^{6/5})^{1/2} \right],$$

$$f_5(z) = K \left[z^5 - (z^{40/9} + R^{10/9} z^{30/9} + R^{20/9} z^{20/9} + R^{30/9} z^{10/9} + \right.$$

$$\left. + R^{40/9})(z^{10/9} - R^{10/9})^{1/2} \right].$$

Le débit du fluide qui passe à travers le milieu poreux sera, en général,

$$Q_n = KR^n . \quad (16)$$

Il paraît qu'entre (12) et (16) il existe une contradiction. Mais quand on a le domaine $0 < \arg z < \pi/n$, la vitesse est nulle dans l'origine, et dans le cas du second domaine la vitesse est infinie au même point.

Bibliographie

1. P.Ia Poloubarinova - Kotchina : La théorie du mouvement des eaux souterraines, Moscou, 1952

2. St.I. Gheorghitza : Introduction mathématique dans l'hydrogasodynamique souterraine, Bucarest, 1966

3. Caius Jacob : Introduction mathématique à la mécanique des fluides, Bucarest - Paris, 1959

4. Sergiu Vasilache : Sur certains problèmes de la théorie des infiltrations, Bul. Şt. Secţ. Şt. Mat. şi Fiz., tome VII, nr. 2, 1955, 365-385 (en roumain)

5. St.I. Gheorghitza : Some plane motions in porous media with variable head on boundaries, J. of Math. and Phys. Sci., vol. 4, nr.3, 1970, 227-236

6. St.I. Gheorghitza : Sur une condition aux limites en hydrodynamique, Zbornik Radova X Jugoslovenskog Kongresa za Racionalnu i Primenjenu Mehaniku, 1970, 153-161.

University of Bucharest
Faculty of Mathematics

ON VARIATIONAL METHODS FOR HAMMERSTEIN EQUATIONS

by

DAN PASCALI

Institute of Mathematics

University of Bucharest

An integral equation of Hammerstein type has the form

$$(1) \qquad u(x) + \int_{\Omega} k(x,y)f(y,u(y))\, dy = w(x) ,$$

where Ω is a domain in $\mathbb{R}^N$ of σ - finite measure. Here the kernel $k: \Omega \times \Omega \longmapsto \mathbb{R}$ and the given function $f: \Omega \times \mathbb{R} \longmapsto \mathbb{R}$ are measurable. We assume both the inhomogeneous term w and the unknown function u belong to dual space X^* of a real Banach space X.

As a rule, every elliptic boundary value problem such that the linear part processes a Green's function can be put into the form of a Hammerstein equation. Among these we mention the problem of the forced oscillation of finite amplitude of a pendulum.

If we introduce the liniar integral operator

$$Av = \int_{\Omega} k(.,y)\ v(y)\ dy$$

and the nonlinear superposition (Niemitskyi operator)

$$Fu = f(.,u(.)) ,$$

in operator terms, the Hammerstein equation (1) can be formally rewritten as

$$(2) \qquad (I + AF)u = w ,$$

where I is the identity operator in X^*.

Since its inception in the work of A.Hammerstein [8] the study of integral equations (1) has profited of various tools, from which the calculus of variation and the fixed point theory for compact maps must be specially mentioned. All these methods were inserted in the first survey paper [6].

The breaking up of the operator $AF: X^* \longmapsto X^*$ into two constituent parts $F: X^* \longmapsto X$ and $A: X \longmapsto X^*$ has lead to a continuous interdependence between the profress of the theory of nonlinear equations (1) and the investigation of monotone mappings. Since 1968, the monotonicity methods has found significant applications in this field. An unified treatment of Hammerstein equations by means of operators of monotone type is accomplished in the second survey paper [4]. In all these works, the Banach spaces have been assumed reflexive, because in their weak topology dispense with nets (generalized sequences) and consider only ordinary sequences [3 ,pp.81].

The wider case of non-reflexive Banach spaces has been taken into account in a recent paper of H.Brézis - F.E.Browder [2]. A survey of the latest investigations in this field is given in [9].

The variational method in the treatment of Hammerstein equations involving unbounded linear operators has been used by Russian mathematicians [10]. The extension to general Banach spaces of this approach due to C.P.Gupta [7] is based on a splitting of liniar monotone operator given by F.E.Browder - C.P.Gupta [5]. Our purpose is a version of C.P.Gupta's result.

Let X be a real Banach space, X^* its dual space and (u,x) the duality pairing between $u \in X^*$ and $x \in X$. We shall denote by $\|\cdot\|$ the norm both in X^* and in X. Strong and weak convergence are expressed by " $\longrightarrow$ " and " $\rightharpoonup$ " respectively.

We consider multivalued mappings T from X into X^*, i.e., subsets or graphs $G(T)$ in $X \times X^*$. For a multivalued mapping $T: X \mapsto 2^{X^*}$ we have the <u>effective domain</u> $D(T) = \{x \in X \mid Tx \neq \emptyset\}$ and the <u>range</u> $R(T) = \{Tx \mid x \in D(T)\}$. The mapping T is said to <u>monotone</u> if $(u_1 - u_2, x_1 - x_2) \geq 0$ for all $[x_1, u_1], [x_2, u_2]$ in $G(T)$ and T is said to be <u>maximal monotone</u> if $G(T)$ is maximal (in the sense of inclusion) among all monotone graphs in $X \times X^*$.

Let X^{**} be the bidual space of X. A sequence $\{u_n\}$ of X^* is called <u>weakly* convergent</u> to u, as is written $u_n \rightharpoonup u$, if $(u_n, x) \to (u, x)$ for every $x \in X$. A functional $f: X^* \mapsto \mathbb{R}$ is said to be <u>weakly* lower-semicontinuous</u> if

$$f(u) \leq \liminf_{n \to \infty} f(u_n)$$

for any $\{u_n\} \subset X^*$ with $u_n \rightharpoonup u$.
We shall use further the generalized Weierstrass theorem, [10]: Let K be a bounded weakly* closed subset of the dual space X^* of a separable or reflexive Banach space X. The any functional f defined and weakly* lower semicontinuous on K is bounded from below and assumes its infimum there.

For a propre (convex) function $f: X^* \mapsto \mathbb{R}$ we set

$$\partial f(u) = \{x \in X \mid f(v) - f(u) \geq (x, v - u), \text{ for any } v \in X^*\}.$$

Elements x which fulfil this inequality are called <u>subgradients*</u> of f at u. The set of all subgradients* of f at u define a monotone mapping (in general multivalued) $\partial f: X^* \mapsto 2^X$ called <u>subdifferential*</u> of f. For the case where X is reflexive, subgradients* and subgradients are the same. In non-reflexive Banach spaces, the subgradients are elements of X^{**}, hence, in general, a subgradient is not a subgradient*.

Variational methods for Hammerstein equations in Banach spaces are based on

PROPOSITION ([1], [5]) Let $A: X \mapsto X^*$ be a linear monotone symmetric operator, densely defined in X. Then there exist a Hilbert space H and a linear map $S:X \mapsto H$ such that $A = S^*S$, where $S^*:H \mapsto X^*$ is the adjoint map of S. Moreover, S^* is one-to-one and $R(S) \subseteq D(S^*)$.

In the following result we suppose that the unit ball in the dual space X^* is weak*- sequentially compact. This condition will be realized when X is either separable or reflexive Banach space.

THEOREM. Let $A:X \mapsto X^*$ be a liniar monotone symmetric densely defined operator and $\partial f:X^* \mapsto 2^X$ be the subdifferential* of a (convex) weakly* lower semicontinuous function $f:X^* \mapsto \mathbb{R}$. Suppose that f is coercive, that is, there is a function $c:\mathbb{R}_+ \mapsto \mathbb{R}$ bounded from below, with $c(r) \to \infty$ as $r \to \infty$, such that

$$f(v) \geq c(\|v\|), \quad \text{for all } v \in X^*.$$

Then there exist at least a solution $u \in X^*$ of Hammerstein equation

$$(I + A\,\partial f)u \ni w,$$

for every $w \in X^*$.

Proof. We may assume without loss of generality that $w = 0$, by replacing $f(u)$ by $f(u+w)$. Let $\langle .,. \rangle$ be the inner product and $|\cdot|$ be the norm of the Hilbert from the previous proposition. Define

$$\Phi(u) = \frac{1}{2}|u|^2 + f(S^*u),$$

where $D(\Phi) = D(S^*)$. If we denote $M = \inf\{c(r) \mid r > 0\}$, one gets

$$\Phi(u) \geq \frac{1}{2}|u|^2 + c(\|S^*u\|) \geq \frac{1}{2}|u|^2 + M, \quad \text{for all } u \in D(\Phi).$$

Let $d = \inf\{\Phi(u) \mid u \in D(\Phi)\}$ and $\{u_n\}$ be a minimizing sequence of Φ. For n sufficiently large, we have

$$d + 1 \geq \Phi(u_n) \geq \frac{1}{2}|u_n|^2 + c(\|S^*u_n\|) \geq M.$$

Consequently, the sequences $\{u_n\}$ and $\{S^*u_n\}$ are bounded. By our assumption on the weak*-sequentially compactness of the closed balls in X^*, passing to subsequences if it is necessary, $u_n \to u_o$ in H and $S^*u_n \to g$ in X^*. We claim that $u_o \in D(S^*)$ by proving that the linear functional $\ell:D(A) \mapsto \mathbb{R}$ defined by $\ell(x) = \langle u_o, Sx \rangle$ is bounded. Really, $\langle u_o, Sx \rangle = \lim_{n\to\infty} \langle u_n, Sx \rangle = \lim_{n\to\infty} (S^*u_n, x) = (g,x)$,

$\langle u_o, Sx \rangle \leq \|g\| \|x\|$ and therefore $S^*u_o = g$.

Next we show that $\Phi(u)$ realizes its infimum at u_o. In fact, the norm in H and the function f being weakly* lower semi-continuous, we obtain

$$\frac{1}{2}|u_o|^2 + f(S^*u_o) \leq \liminf_{n\to\infty}\left\{\frac{1}{2}|u_n|^2 + f(S^*u_n)\right\} = \lim_{n\to\infty}\Phi(u_n) = d,$$

that is, $\Phi(u_o) = d$.

Finally, in order to have a minim of Φ at u_o it is necessary that $0 \in \partial\Phi(u_o)$. As $\partial\Phi(u_o) = u_o + S\partial f(S^*u_o)$ we obtain

$$S^*u_o + A\partial f(S^*u_o) \ni 0.$$

Therefore, S^*u_o is a solution of the equation $(I + A\partial f)u \ni 0$.

Unlike Gupta's approach, in this theorem no hypotheses of maximality was necessary.

REFERENCES

[1] AMANN H., Ein Existenz-und-Eindeutigkeitsatz für die Hammersteinsche Gleichung in Banachräumen, Math.Zeit.,111 (1969),175-190.

[2] BRÉZIS,H. and BROWDER,F.E., Nonlinear integral equations and systems of Hammerstein type, Adv.in Math.,18 (1975) 115-147.

[3] BROWDER,F.E., Nonlinear operators and nonlinear equations of evolution in Banach spaces, Proc.Sympos.Pure Math., Chicago 1968,vol.XVIII,part 2,Amer.Math.Soc.,Providence, (1976).

[4] BROWDER,F.E., Nonlinear functional analysis and nonlinear integral equations of Hammerstein and Urysohn type, in "Contributions to Nonlinear Functional Analysis",Edited by E.Zarantonello,Academic Press,1971,425-500.

[5] BROWDER,F.E.and GUPTA,C.P., Nonlinear monotone operators and integral equations of Hammerstein type, Bull.Amer. Math.Soc.,75 (1969),1347-1353.

[6] DOLPH,G.L. and MINTY,G.J., On nonlinear integral equations of the Hammerstein type,in "Nonlinear integral equations", Univ.of Wisconsin Press,Madison 1964,99-154.

[7] GUPTA,C.P., On nonlinear integral equations of Hammerstein type with unbounded linear mappings, Lect.Notes Math.,384, Springer-Verlag,1974,184-238.

[8] HAMMERSTEIN,A.,Nichtlineare Integralgleichungen nebst Anwendungen,Acta Math.,54 (1930),117-176.

[9] PASCALI,D., Hammerstein equations in general Banach spaces, Seminari di Analisi,1974/75,Istituto Matematico,Roma

[10] VAINBERG,M.M., Variational method and method of monotone operators in the theory of nonlinear equations, John Wiley & Sons,Inc.,1973.

DIE PERMANENZEIGENSCHAFTEN DER TENSORPRODUKTE VON BANACHVERBÄNDEN

Nicolae Popa

Der Zweck dieser Arbeit ist die Untersuchung einigen Permanenzeigenschaften des m-Tensorprodukts zwischen einem Banachraum und einem Banachverband.

Das wichtigste Ergebnis in dieser Richtung ist der bemerkenswerte Satz von U. Schlotterbeck [9] :

Seien E und F die reflexive Banachverbände. Dann ist $E \hat{\otimes}_m F$ auch ein reflexiver Banachverband.

Dafür benutzt man ein Darstellungssatz des Duals des Raumes $E \hat{\otimes}_m F$. Dieser Darstellungssatz kann man verallgemeinert wird; man erhält dann, neben einigen weiteren Anwendungen, ein Kriterium für die schwache Konvergenz einer Folge aus $c_0(E)$ oder $L^p_E(\mu)$, $1<p<\infty$, wenn E' ein Banachraum mit der Radon-Nikodymsche Eigenschaft ist.

Benutzend eine Adaptation des Faktorisierungssatzes von Davi Figiel und Johnson [1] , erhalten wir dass $L^p_E(\mu)$, $1\leq p<\infty$, ein schwach kompakterzeugter Banachraum ist, wenn E auch diese Eigenschaft hat.

So verallgemeinert es sich ein Ergebnis von J. Diestel [2].

Wir erhalten auch einige neue Permanenzergebnisse für Banachverbände.

Im folgende benutzen wir die Terminologie aus [8].

§1 - Die Permanenzeigenschaften des Tensorprodukts von Banachverbänden

Seien E und F zwei Banachverbände. Mit $E\hat{\otimes}_m F$ wird die Komplettierung in der Norm $\|z\|_m = \inf\left\{ \left\| \sum_{i=1}^{n} \|x_i\| \cdot |y_i| \right\| : z = \sum_{i=1}^{n} x_i \otimes y_i \right\}$ des Tensorprodukts von E und F bezeichnet.

Dann wird $E\hat{\otimes}_m F$ ein Banachverband [8] .

Falls E nur ein Banachraum sei, ist $E\hat{\otimes}_m F$ auch ein Banachraum.

Eine Norm auf einem Dedekind-vollständigen Vektorverband heisst ordnungsstetig, wenn für jedes Netz $(x_\alpha)_{\alpha \in A}$ mit $x_\alpha \downarrow 0$, auch $\lim_\alpha \|x_\alpha\| = 0$ gilt.

Sei (X, μ) ein endlicher Massraum.

Sei F ein Ideal in $L^1(X, \mu)$,der $L^\infty(X, \mu)$ enthält und die Norm auf F sei ordnungsstetig.

Zum Beispiel sei F ein Banachverband mit ordnungsstetiger Norm und schwacher Ordnungseinheit u. (Ein Element $u \in F$ heisst schwache Ordnungseinheit, wenn u positiv ist und aus $u \wedge x = 0$ und $x \geq 0$, $x = 0$ gilt).

Tatsächlich, unter diesen Voraussetzungen besitzt F' eine schwache Ordnungseinheit u'. (Vgl. [8] -Satz 6.6 -Kap.II).

Weil die von u' erzeugte Band mit F' zusammenfält, ist u' eine strikt positive Linearform auf F. Dann ist (F, u')- die Komplettierung in der Norm $|||x||| = \langle u', |x| \rangle$ von F - ein AL- Raum mit schwacher Ordnungseinheit u.

Weil die Norm auf F ordnungsstetig ist, ist $F_u = \{ z \in F \mid |z| \leq \lambda u, \ \lambda > 0 \}$ ein dichtes Ideal in F, und auch in (F, u'). Dann gibt es ein endlicher Massraum (X, μ) sodass (F, u') mit $L^1(X, \mu)$, und F_u mit $L^\infty(X, \mu)$ zusammenfällen.

Offenbar ist die kanonische Abbildung $F \longrightarrow (F, u')$ eine Einbettung. F ist ein Ideal in $L^1(X,\mu)$, weil die Ordnungsintervalle in F schwach kompakt sind. (Vgl. [8]- Satz 5.10 - Kap. II).

Unter diesen Voraussetzungen auf F, E seiend ein Banachraum bezeichnen wi mit $L_E^F(X,\mu)$ der Raum derjenigen μ-messbaren Funktionen $g:X \longrightarrow E$ (modulo μ-Nullfunktionen), für die, die Abbildung $x \longrightarrow \|g(x)\|$ μ-fast überall mit einer Funktion aus F übereinstimmt.

Auf $L_E^F(X,\mu)$ führen wir die Norm $g \longrightarrow \|u_g\|$ ein, wo u_g die Funktion $x \longrightarrow \|g(x)\|$ aus F ist.

Wenn E ein Banachverband ist, so ist $L_E^F(X,\mu)$ unter der natürlichen Ordnung ein normierter Vektorverband.

Damn hat U. Schlotterbeck [9] (unter die Voraussetzung dass E ein Banachverband sei) bewiesen:

Satz 1.1 Sei E ein Banachraum (ein Banachverband) und F wie früher. Dann ist $E \hat{\otimes}_m F$ kanonisch norm - (bzw. verband-) isomorph zu $L_E^F(X,\mu)$.

Aus diesem Satz folgt:

Satz 1.2 Seien E und F zwei Banachverbände mit ordnungsstetiger Norm. Dann ist $E \hat{\otimes}_m F$ auch ein Banachverband mit ordnungsstetiger Norm.

Beweis Zunächst beweisen wir die Behauptung unter die Voraussetzung dass F eine schwache Ordnungseinheit besitzt.

Tatsächlich, aus dem Satz 5.10 - Kap. II - [8] sind E und F Dedekind- vollständig.

Dann ist $E \hat{\otimes}_m F = \bigcup_{y \in F_+} E \hat{\otimes}_\varepsilon F_y$ auch ein Dedekind - vollständiger Banachverband.

Wegen des Satzes 1.1 ist $E \hat{\otimes}_m F$ dann verbandsisomorph zu $L_E^F(X,\mu)$.

Sei jetzt $(x_n)_1^\infty$ eine positive monoton wachsende ordnungsbeschränkte Folge in $L_E^F(X,\mu)$, und sei y sodass

$x_n \leq y \in L_E^F(X,\mu)$.

Dann gibt es $x(t) = \sup_n x_n(t) \leq y(t)$ μ-fast überall und $t \longrightarrow x(t)$ ist eine μ-messbare Funktion.

Weil $(\|x_n(t)\|)_1^\infty$ eine monoton wachsende majorisierende Folge aus $F \subset L^1(X,\mu)$ ist, die μ-fast überall gegen $\|x(t)\|$ konvergiert, ist, wegen des Satzes von Lebesgue, $t \longrightarrow \|x(t)\|$ eine Funktion aus $L^1(X,\mu)$ und $(\|x_n(t)\|)_1^\infty$ ordnungskonvergiert gegen $\|x(t)\|$ in $L^1(X,\mu)$.

Aber F ist ein Ideal in $L^1(X,\mu)$, also $\|x(t)\|$ ist eine Funktion aus F und $\|x_n(t)\|$ ordnungskonvergiert gegen $\|x(t)\|$ in F.

Dann ist $\|x(t)\| = \sup_n \|x_n(t)\|$ in $L_E^F(X,\mu)$.

Nach dem Satz 8 - [5] ist es genug dass wir zeigen können, dass jede Folge $(g_n)_1^\infty \in L_E^F(X,\mu)_+$ paarweise disjunkter Vektoren mit $g_n \leq g$, für ein festes $g \in L_E^F(X,\mu)$ und alle n, gegen Null konvergiert.

Dann gelten, μ-fast überall in X, die Beziehungen:

$0 \leq g_n(t) \leq g(t)$ und $g_n(t) \wedge g_m(t) = 0$ für $m \neq n$.

Also, wegen des Satzes 8 - [5], $\|g_n(t)\| \underset{n\to\infty}{\longrightarrow} 0$ und

$\|g_n(t)\| \leq \|g(t)\|$ μ-fast überall in X.

Dann ordnungskonvergiert gegen Null in $L^1(X,\mu)$ die Folge $u_{g_n}(t) = \|g_n(t)\|$, $n \in N$. Weil F_o ein Ideal in $L^1(X,\mu)$ ist, ordnungskonvergiert die Folge $(u_{g_n})_1^\infty$ gegen Null auch in F_o, also $\|u_{g_n}\| \underset{n\to\infty}{\longrightarrow} 0$. Damit wird der Satz in diesem speziellen

Fall bewiesen.

Im allgemeine seien $0 \leq T_n \leq U$ in $E \hat{\otimes}_m F = \bigcup_{z \in F_+} E \hat{\otimes}_\varepsilon F_z$,

Also gibt es $z \in F_+$ sodass $U \in E \hat{\otimes}_\varepsilon F_z$, und sei F_o das abgeschlossene von z in F erzeugte Ideal.

Dann ist F_o ein Banachverband mit ordnungsstetiger Norm und schwacher Ordnungseinheit und auch, eine Projektionsband in F. Weil $0 \leq T_n \leq U$ in $E \hat{\otimes}_m F_o$, gibt es $T = \sup_n T_n$ in $E \hat{\otimes}_m F_o$.

Wenn $T_n \leq V \in E \hat{\otimes}_m F$ für alle n, dann gilt

$$V \geq (I \otimes P)V \geq (I \otimes P)T_n = T_n \text{ , wo P die Projektion auf } F_o$$

ist. So ist T das Supremum von $(T_n)_1^\infty$ auch in $E \hat{\otimes}_m F$.

Wieder benutzen wir den Satz 8 -[5] und seien $0 \leq T_n \leq U \in E \hat{\otimes}_m F$, eine Folge paarweise disjunkter Vektoren.

Wenn F_o ist wie früher, gilt $0 \leq T_n \leq U \in E \hat{\otimes}_m F_o$, und $\lim_n T_n = 0$. Aber $E \hat{\otimes}_m F_o$ ist ein Unterbanachraum in $E \hat{\otimes}_m F$, und damit werden alle gezeigt.

Korollar 1.3 Seien E und F zwei Dedekind-vollständige Banachverbände, die keinen zu l^∞ -verbandsisomorphen Untervektorverband enthälten.

Dann enthält auch $E \hat{\otimes}_m F$ keinen zu l^∞ -verbandsisomorphen Untervektorverband.

Die Aussage dieses Korollars folgt aus dem Korollar 10 - [5].

Bemerkung 1.4 Falls die Norm $\| \ \|_m$ durch

$$\|u\|_p = \inf\left\{ \sum_{i=1}^n \|x_i\| \cdot \|y_i\| : |u| \leq \sum_{i=1}^n x_i \otimes y_i,\ x_i \in E_+,\ y_i \in F_+,\ u \in L(E', F) \right\}$$

ersetzt wird, verliert der Satz 1.2 die Gültigkeit.

Zum Beispiel, wenn $E = F = L^2[0,1]$, dann ist $\| \ \|_p$

keine ordnungsstetige Norm auf $E \hat{\otimes}_p F$. (Vgl. [3]).

<u>Satz 1.5 Seien E und F zwei schwach folgenvollständige Banachverbände. Dann ist</u> $E \hat{\otimes}_m F$ <u>auch schwach folgenvollständig.</u>

<u>Beweis</u> Nach dem Korollar 1.8 - [6] ist ein Banachverband E schwach folgenvollständig dann und nur dann wenn jede monotone und normbeschränkte Folge konvergent ist.

Sei $(T_n)_1^\infty$ eine monoton wachsende Folge aus $E \hat{\otimes}_m F$ mit $\|T_n\| \leq M$ für alle $n \in N$.

Dann gilt es $T_n \in E \hat{\otimes}_\varepsilon F_{z_n}$ mit $\|z_n\| \leq M$ für alle n.

Es genügt also die Behauptung, falls F ein quasi-innerer Punkt besitzt, zu beweisen.

Weil jeder schwach folgenvollständiger Banachverband eine ordnungsstetige Norm besitzt (vgl. Satz 5.10 - Kap. II - [8] und der Korollar 1.8 - [6]), dann ist, wegen des Satzes 1.1 $E \hat{\otimes}_m F_o$ verbandsisomorph zu $L_E^{F_o}(X,\mu)$.

Sei jetzt eine monoton wachsende positive Folge $(g_n)_1^\infty$ mit $\|g_n\| \leq M$ aus $L_E^{F_o}(X,\mu)$.

Dann ist $(g_n(s))_1^\infty$ eine monoton wachsende Folge μ-fast überall auf X, und $(u_{g_n})_1^\infty$ ist eine normbeschränkte Folge in F_o, also in $L^1(X,\mu)$.

Daraus folgt dass $\int_X \|g_n(s)\| \, d\mu \leq M$ für alle $n \in N$.

Der vollbekanntliche Lebesguesche Satz impliziert dass $s \longrightarrow \sup_n \|g_n(s)\|$ eine Funktion aus $L^1(X,\mu)$ ist.

Dann ist $(g_n(s))_1^\infty$, für $s \in X \setminus N$, N seiend eine μ - Nullmenge, eine normbeschränkte Folge in E.

Wegen des Voraussetzungen auf E folgt es dass die Folge $g_n(s)$ gegen $g(s) \in E$, für $s \in X \setminus N$, konvergiert.

Die Funktion $g:X \longrightarrow E$ ist also μ-messbar.

Dazu ist $\|g_n(s)\|$ eine monoton wachsende , gegen $\|g(s)\|$ konvergierende Folge , μ-fast überall auf X .

Weil F_o ein schwach folgenvollständiger Banachverband ist und weil $(u_{g_n})_1^\infty$ eine monoton wachsende Folge mit $\|u_{g_n}\| \leq M$ ist, konvergiert u_{g_n} gegen f in F_o.

Dann gilt $\|g_n(s)\| \xrightarrow[n\to\infty]{} f(s)$ μ-fast überall auf X, also stimmen die Funktionen f(s) und $\|g(s)\|$ (ausserhalb einer gewissen μ-Nullmengen) überein.

Damit gilt $g_n \longrightarrow g$ in $L_E^{F_o}(X,\mu)$, mit $g \in L_E^{F_o}(X,\mu)$.

Korollar 1.6 Seien E und F zwei Banachverbänden die c_o nicht verbandsisomorph enthalten.

Dann enthält $E \hat{\otimes}_m F$ keinen zu c_o verbandsisomorphen Untervektorverband.

Das Korollar folgt wegen des Korollars 1.8 [6] und wegen des Satzes 1.5 .

§2 - Die Kriterien für die schwache Kompaktheit in einigen Funktionenräume .

Sei F ein Banachverband und E ein Banachraum.

Dann bezeichnen wir mit $S_+(F,E)$ der Raum der Operatoren $T:F \longrightarrow E$ sodass, für jede positive summierbare Folge (x_n) aus F, die Folge (Tx_n) absolut-summierbar ist.

Der Raum $S_+(F,E)$, versehen mit der Norm

$$\|T\|_1 = \sup\left\{ \sum_{i=1}^{n} \|Tx_i\| \; ; \; \sum_{i=1}^{n} x_i = x \, , \, x_i \geq 0 \, , \, \|x\| \leq 1 \right\}$$

wird ein Banachraum.

Man kann zeigen dass $S_+(F,E')$ der Dualraum des $E \hat{\otimes}_m F$ ist.

Endlich nennt man einen Operator $S:E' \longrightarrow F'$ <u>majorisierend</u> wenn für jede Nullfolge $(x_n')_1^\infty$ aus E', $(Sx_n')_1^\infty$ eine majorisierte Nullfolge in F' ist.

Mit $M(E',F')$ wird der Raum der majorisierenden Operatoren von E' in F', versehen mit der Norm

$$\|S\|_m = \|\sup_{\|x'\|\leq 1} S(x')\|$$

bezeichnet.

Dann wird $M(E', F')$ ein Banachraum, und $T \in S_+(F,E)$ dann und nur dann wenn $T' \in M(E', F')$.

Im folgende verallgemeinern wir einen Zwischensatz von Nagel [7].

<u>Lemma 2.1</u> <u>Sei F ein Banachverband mit ordnungsstetiger Norm und schwacher Ordnungseinheit.</u>

<u>Es existiert ein endlicher Massraum</u> (X,μ) <u>sodass F und F' als Vektorverbände isomorph sind zu je einem Ideal in</u> $L^1(X,\mu)$ <u>welches</u> $L^\infty(X,\mu)$ <u>umfasst, und dass nach Identifikation von E und E' mit dem entsprechenden Ideal in</u> $L^1(X,\mu)$ <u>das Produkt xx' eines Elements</u> $x \in E$ <u>mit einem</u> $x' \in E'$ <u>jeweils enthalten ist in</u> $L^1(X,\mu)$ <u>mit</u> $\langle x, x'\rangle = \int xx'\,d\mu$.

<u>Beweis</u> Wir haben schon bemerkt in Sektion 1. dass F ein Ideal in $L^1(X,\mu)$ das $L^\infty(X,\mu)$ umfasst, ist.

Dazu ist μ eine schwache Ordnungseinheit in F', also gilt für jedes $x' \in F'$ die Gleichung $x' = \sup_n x' \wedge n\mu$.

Sei u die schwache Ordnungseinheit in F, d.h. die Einsfunktion auf X. Dann, weil die Norm auf F ordnungsstetig ist, ist u eine strikt positive Linearform auf F'.

Also ist die kanonische Abbildung $F' \longrightarrow (F', u)$ eine Einbettung.

Nach dem Korollar 2 -Satz 1.4 -[9] ist F' ein Ideal in (F', u).

Andererseits ist, wegen des Satzes 5.10 - Kap. II - [8],

$(F', u)' = (F'')_u = F_u = L^{\infty}(X, \mu)$.

Dann ist (F', u) isomorph zu $L^1(X, \mu)$, und

$(F', u)_{\mu} = F'_{\mu} = L^{\infty}(X, \mu)$.

Daraus folgt der erste Teil der Behauptung sowie die Identität von $\langle x, x' \rangle$ und $\int xx' d\mu$ für alle $x \in F$ und $x' \in F'_{\mu}$.

Wenn $x' \in F'$ belibig ist, so existiert eine Folge (x'_n) aus $L^{\infty}(X, \mu)$ mit $|x'_n| \leq |x'|$, die in F' ordnungs- und μ-fast überall gegen x' konvergiert. Dann konvergiert x'_n gegen x' in der Topologie $\sigma(F', F)$ und $(x'_n x)$ ist eine Folge in $L^1(X, \mu)$, die μ-fast überall gegen $x'x \in L^1(X, \mu)$ konvergiert.

Dazu

$$\langle x, x' \rangle = \lim_n \langle x'_n, x \rangle = \lim_n \int x'_n x \, d\mu = \int x'x \, d\mu .$$

Wir können jetzt den Darstellungssatz von Schlotterbeck verallgemeinern.

Wenn für jeden Kompaktraum X, jeder Integraloperator im Sinne von Grothendieck [4] von $C(X)$ auf E nuklear ist, sagt man dass E die Radon-Nikodymsche Eigenschaft hat.

<u>Satz 2.2</u> <u>Seien E ein Banachraum (bzw. ein Banachverband) mit Radon-Nikodymsche Eigenschaft und F ein Banachverband mit ordnungsstetiger Norm und schwacher Ordnungseinheit.</u>

<u>Dann ist</u> $S_+(F,E)$ <u>isometrisch (verbandsisometrisch) zu</u> $L_E^{F'}(X, \mu)$, <u>zu jeder Abbildung</u> $T \in S_+(F, E)$ <u>existiert ein Element</u> $g \in L_E^{F'}(X, \mu)$ <u>sodass:</u>

$$\langle Ty, x' \rangle = \int y(s) \langle g(s), x' \rangle \, d\mu(s), \quad y \in F, \ x' \in E'.$$

Beweis Wegen Lemma 2.1 existiert ein endlicher Massraum (X,μ) sodass

$$L^{\infty}(X,\mu) \subset F \subset L^{1}(X,\mu)$$

F seiend ein Ideal in $L^{1}(X,\mu)$, versehen mit ordnungsstetiger Norm.

Dazu F' ist ein Ideal in $L^{1}(X,\mu)$ das $L^{\infty}(X,\mu)$ umfasst.

Die Einschränkung T_o von $T \in S_{+}(F, E)$ auf $L^{\infty}(X,\mu)$ lässt sich zerlegen:

$$\begin{array}{ccc} L^{\infty}(X,\mu) & \xrightarrow{T_o} & E \\ & {\scriptstyle T_1}\searrow \quad \nearrow{\scriptstyle T_2} & \\ & L & \end{array}$$

mit T_1 ein positiver Operator und L ein AL - Raum.

Ein Ergebnis von Grothendieck [4] impliziert dass T_o ein Integraloperator ist.

Weil E die Radon-Nikodymsche Eigenschaft hat, ist T_o sogar nuklear. Dann ist T_o durch ein Element aus $L^{\infty}(X,\mu)'\hat{\otimes}_{\pi}E$ gegeben.

Sei $P: (L^{\infty})' \longrightarrow L^{1}$ die Projektion durch den Satz von Radon-Nikodym gegeben. Dann ist P' die kanonische Einbettung von L^{∞} in $(L^{\infty})''$, also gilt $T_o = (P_o T_o')'$.

Endlich, weil $P \circ T_o'$ kompakt ist, umfasst $L^{1}(X,\mu)$ der Raum $T_o'(E')$, und T_o ist durch ein Element $g \in L^{1}(X,\mu)\hat{\otimes}_{\pi}E = L^{1}_{E}(X,\mu)$ gegeben.

Dann gilt:

$$T_o y = \int_X g(s)y(s)d\mu(s) \quad \text{für} \quad y \in L^{\infty}(X,\mu) \text{, nämlich}$$

$$(*) \quad \langle Ty\, ,\, x'\rangle = \int y(s)\langle g(s),\, x'\rangle d\mu(s) \quad \text{für alle} \quad y \in L^{\infty}(X,\mu)$$

und $x' \in E'$.

Wenn $y \in F$, dann gibt eine monoton wachsende Folge (y_n) aus $L^{\infty}(X,\mu)$ mit $y_n \longrightarrow y$ in $L^1(X,\mu)$.

Also die Beziehung (*) gibt uns:

$$\langle Ty, x'\rangle = \lim_n \langle Ty_n, x'\rangle = \lim_n \int_X y_n(s)\langle g(s), x'\rangle d\mu(s) =$$

$$= \int_X y(s)\langle g(s), x'\rangle d\mu(s) \quad \text{für alle } x' \in E' \text{ und } y \in F.$$

Offenbar ist die Abbildung $T \longrightarrow g$ positiv, falls E ein Banachverband ist.

Es bleibt noch zu zeigen dass die Abbildung $s \longrightarrow \|g(s)\|$ μ-fast überall mit einer Funktion aus F' übereinstimmt.

Sie ist aus $L^1(X,\mu)$. Also mus man zeigen dass diese Funktion bei einer Funktion aus F' μ-fast überall majorisiert wird.

Man kann voraussetzen dass $g(X)$ in einem separablen Teilraum E_o von E enthalten ist. Tatsächlich $E_o = T_o(L^{\infty})$ ist ein separabler Teilraum von E, der die Radon-Nikodymsche Eigenschaft hat. Andererseits T_1 seiend ein Intègraloperator ist $T_o : L^{\infty} \longrightarrow E_o$ auch ein Integraloperator, also nuklear.

Es existiert die Folge $x'_n \in E'$ aus der Einheitskugel sodass

$$(**) \quad \|x\| = \sup_n \langle x, x'_n\rangle \quad (\forall)\ x \in E_o.$$

Dann gilt es, wegen Lemma 2.1 :

$$\int T'(x')(s)\, y(s)\, d\mu(s) = \langle T'x', y\rangle = \langle Ty, x'\rangle =$$

$$= \int y(s)\langle g(s), x'\rangle d\mu(s) \quad \text{für alle } y \in F \text{ und } x' \in E'.$$

Also $(T'x')(s) = \langle g(s), x'\rangle$ ausserdem einer Nullmenge, der von y' abhängt.

Wegen der Gleichung (**) existiert eine Nullmenge N, für die es gilt:

$$\|g(s)\| = \sup_n T'x'_n(s) \qquad \text{für } s \in X \setminus N.$$

Andererseits $\sup_n T'x_n' \leq z' \in F'$, also $\|g(s)\| \leq z'(s)$ μ-fast überall. Die Gleichung $\|z'\| = \|T\|_1$ impliziert $\|g\| \leq \|T\|_1$ Gegenseitig, wenn $g \in L_E^{F'}(X,\mu)$ ist, dann existiert

$$\int_X y(s) \cdot g(s)\, d\mu(s)$$

für alle $y \in L^{\infty}(X,\mu)$.

Für $0 \leq y \in F$, existiert eine Folge $y_n \in L^{\infty}(X,\mu)$ mit $y_n(s) \longrightarrow y(s)$ und $|y_n(s)| \leq |y(s)|$ μ-fast überall.

Dann gilt es

$$y_n(s)\, g(s) \xrightarrow[n]{} y(s) \cdot g(s) \quad \text{und}$$

$$\|y_n(s) g(s)\| \leq y(s) \|g(s)\| \quad \mu\text{-fast überall.}$$

Wegen Lemma 2.1 und des Satzes von Lebesgue existiert $Ty = \int y(s) g(s)\, d\mu(s) = \lim_n \int y_n(s) \cdot g(s)\, d\mu(s) \in E$, für alle $y \in F$.

Für $0 \leq y_i \in F$, $i \leq n$, gilt es , wegen Lemma 2.1:

$$\sum_{i=1}^{n} \|Ty_i\| = \sum_{i=1}^{n} \sup_{\|x'\| \leq 1} \left| \int_X y_i(s) \langle g(s), x' \rangle \cdot d\mu(s) \right| \leq$$

$$\leq \sum_{i=1}^{n} \int_X y_i(s) \|g(s)\| d\mu(s) = \langle u_g , \sum_{i=1}^{n} y_i \rangle \leq \|g\| \cdot \|\sum_{i=1}^{n} y_i\| .$$

Also ist T in $S_+(F,E)$ und $\|T\|_1 \leq \|g\|$, und der Satz ist bewiesen.

Im folgenden nennen wir eine Teilmenge A eines Banachraumes E schwach folgenpräkompakt, wenn aus jeder Folge in A eine Teilfolge ausgewählt werden kann, die eine schwache Cauchyfolge ist.

Korollar 2.3 1) Sei E ein Banachraum (ein Banachverband) sodass E' die Radon-Nikodymsche Eigenschaft hat und F wie im

Satz 2.2 . Dann ist der Dualraum von $L_E^F(X,\mu)$ isometrisch (verbandsisometrisch) zum Raum $L_{E'}^{F'}(X,\mu)$.

2) Wenn dazu die Einheitskugel von F schwach folgenpräkompakt ist, fallen die Räume $(E \hat{\otimes}_m F)'$, $L_{E'}^{F'}(X,\mu)$ und $E' \hat{\otimes}_m F'$ alle zusammen.

Beweis 1) ist klar von dem Satz 2.2 .

2) Wegen des Korollars 1.4 -[6] ist die Norm auf F' ordnungsstetig, also folgt die Aussage unmittelbar aus Sätze 1.1 und 2.2 .

Korollar 2.4 (Schlotterbeck) Wenn E ein reflexiver Banachraum und F ein reflexiver Banachverband sind, ist $E \hat{\otimes}_m F$ ein reflexiver Banachraum.

Beweis Wegen des Satzes von Eberlein kann man F mit schwacher Ordnungseinheit versehen annehmen.

Dann folgt die Aussage unmittelbar aus dem Korollar 2.3 - 2).

Korollar 2.5 1) Sei E ein Banachverband mit ordnungsstetiger Norm sodass E' die Radon-Nikodymsche Eigenschaft hat, und F wie im Satz 2.2 .

Eine beschränkte Teilmenge $A \subset L_E^F(X,\mu)$ ist schwach folgenpräkompakt dann und nur dann wenn für jede ordnungsbeschränkte Folge (g_n) paarweiser disjunkter Funktionen aus $L_{E'}^{F'}(X,\mu)$, gilt:

$$\sup_{f \in A} \int_X \langle |g_n(s)|, |f(s)| \rangle \, d\mu(s) \xrightarrow[n \to \infty]{} 0$$

2) Wenn E ein reflexiver Banachverband und F ein schwach folgenvollständiger Banachverband mit schwacher Ordnungseinheit sind, ist eine beschränkte Teilmenge $A \subset L_E^F(X,\mu)$ schwach relativ kompakt dann und nur dann für jede ordnungsbeschränkte

Folge (g_n) paarweiser disjunkter Funktionen aus $L_{E'}^{F'}(X,\mu)$ gilt es:

$$\sup_{f\in A} \int_X \langle |g_n(s)|, |f(s)|\rangle \, d\mu(s) \xrightarrow[n\to\infty]{} 0$$

3) Sei E ein reflexiver Banachverband. Eine beschränkte Teilmenge aus $L_E^1(X,\mu)$ ist schwach relativ kompakt dann und nur dann wenn für jede ordnungsbeschränkte Folge (g_n) paarweiser disjunkter Funktionen aus $L_{E'}^{\infty}(X,\mu)$ gilt es:

$$\sup_{f\in A} \int_X \langle |g_n(s)|, |f(s)|\rangle \, d\mu(s) \xrightarrow[n\to\infty]{} 0 .$$

Beweis 1) Wegen des Satzes 1.2 ist die Norm auf $L_E^F(X,\mu)$ ordnungsstetig. Das Korollar 2.3 zeigt uns dass $[L_E^F(X,\mu)]' = L_{E'}^{F'}(X,\mu)$ und die Aussage folgt aus dem Satz 1.3 -[6].

2) Wegen des Satzes 1.3 ist $L_E^F(X,\mu)$ ein schwach folgenvollständiger Banachverband und nach 1) ist alles gezeigt.

3) folgt unmittelbar aus 2).

Korollar 2.6 1) Sei E ein Banachraum sodass E' die Radon-Nikodymsche Eigenschaft hat und F ein Banchverband mit schwacher Ordnungseinheit sodass die Normen auf F und F' ordnungsstetig sind.

Dann konvergiert eine Folge (f_n) aus $L_E^F(X,\mu)$ schwach gegen Null dann und nur dann wenn (f_n) beschränkt ist und $\langle f_n, x'\rangle$ konvergiert schwach gegen Null in F für alle $x'\in E'$.

2) Sei E ein Banachraum sodass E' die Radon-Nikodymsche Eigenschaft hat. Dann konvergiert eine beschränkte Folge (g_n) aus $c_0(E)$ schwach gegen Null dann und nur dann wenn (g_n^k) konvergiert schwach gegen Null in E für alle $k\in N$.

3) Unter denselben Voraussetzungen über E konvergiert eine Folge (f_n) aus $L_E^p(X,\mu)$ $(1<p<\infty)$ schwach gegen Null dann

und nur dann wenn $\langle f_n , x' \rangle$ *konvergiert schwach gegen Null in* $L^p(X,\mu)$ *für alle* $x' \in E'$.

Beweis 1) Wegen des Korollars 2.3 - 2) ist $(E \hat{\otimes}_m F)' = E' \hat{\otimes}_m F'$, so genügt es für die schwache Konvergenz der Folge (f_n) dass $\langle f_n , x' \rangle$ gegen Null für $x' \in E'$ schwach konvergiert.

Weil c_o und $L^p(X,\mu)$ für $1 < p < \infty$ die Voraussetzungen des Punktes 1) erfüllen, die Punkte 2) und 3) folgen aus 1)

§ 3 - Die Permanenzeigenschaften des Tensorprodukts von einem Banachverband und einem Banachraum

Benutzend das Korollar 2.4 wollen wir beweisen dass das Tensorprodukt einem schwach kompakt erzeugter schwach folgenvollständiger Banachverband und einem schwach kompakt erzeugter (abkürzlich SKE) Banachraum noch ein SKE Banachraum ist.

Dafür beweisen wir einen Faktorisierungssatz der schwach kompakten Operatoren mit den Werten in einem schwach folgenvollständigen Banachverband.

Dieser Faktorisierungssatz ist mit dem Faktorisierungssatz von Davis, Figiel und Johnson [1] vergleichbar.

Satz 3.1 (Davis , Figiel , Johnson) *Seien* E *und* F *zwei Banachräume und* $T:F \longrightarrow E$ *ein schwach kompakter Operator. Dann gibt es ein reflexiver Banachraum* G *und die Operatoren* S , U *sodass das folgende Diagramm kommutativ sei:*

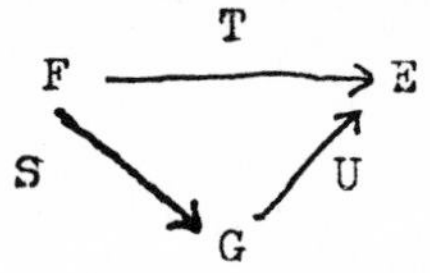

Sei E ein schwach folgenvollständiger Banachverband, B(E) seine Einheitskugel und W eine absolutkonvexe, beschränkte

Teilmenge von E .

Wenn $[W]$ die solide Hülle von W ist, bezeichnen wir mit U_n die Menge $2^n [W] + 2^{-n} B(E)$ für alle $n \in N$.

Dann ist $\| \|_n$, das Minkowski Funktional von U_n, eine solide Norm auf E .

Für $x \in E$ setzen wir $|||x||| = \left(\sum_{n=1}^{\infty} \|x\|_n^2 \right)^{1/2}$,
$F = \{ x \in E \mid |||x||| < \infty \}$, A die Einheitskugel von F und $j : F \longrightarrow E$ die kanonusche Einbettung.

Dann erhalten wir den wichtigen Hilfssatz:

Lemma 3.2 Unter den oberen Voraussetzungen gilt es:

1) $[W] \subset A$.

2) $(F, ||| \; |||)$ ist ein Banachverband, der ein Ideal in E ist, und j ist ein injektiver und stetiger Verbandshomomorphismus

3) j'' ist ein injektiver und stetiger Verbandshomomorphismus und $(j'')^{-1}(E) = F$.

4) Wenn W eine schwach kompakte Teilmenge von E ist, ist $(F, ||| \; |||)$ reflexiv.

Beweis 1) Für $w \in [W]$ gilt es $\|w\|_n \leq 2^{-n}$ für alle $n \in N$, sodass $|||w||| \leq 1$.

2) Weil $[W]$ solid ist, ist $\| \|_n$ eine solide Norm auf E, und weil $B(E) \subseteq 2^n U_n \subset a_n B(E)$, ist $E_n = (E, \| \|_n)$ ein Banachraum verbandsisomorph zu E .

Sei $G = \left(\sum_{n=1}^{\infty} E_n \right)_{l^2} = \left\{ (x_n)_1^{\infty} \mid x_n \in E_n , \sum_{n=1}^{\infty} \|x_n\|_n^2 < \infty \right\}$.

Dann ist G ein Banachverband. Setzen wir $\varphi : F \longrightarrow G$ durch $\varphi(y) = (j(y), j(y), \ldots)$.

Offenbar ist φ eine Verbandsisometrie auf dem abgeschlossemem Untevektorverband von G

$$\varphi(F) = \left\{ z \in G \mid z = (x_n)_1^{\infty} \text{ mit } x_n = x_1 \ (\forall)\ n \in N \right\} .$$

Also ist F ein Banachverband der ein Ideal in E ist.

Es ist klar dass j ein injektiver Verbandshomomorphismus ist. Deswegen $j = P \circ \varphi$, wo $P:G \longrightarrow E$ die Projektion auf der erste Komponente ist, folgt es die Stetigkheit von j.

3) Weil φ'' das Element $y'' \in F''$ auf $(j''(y''), j''(y''), ..)$ $\in (\sum_{n=1}^{\infty} E_n'')_{l^2}$ abbildet und weil φ'' eine Verbandsisometrie ist, wird die erste Behauptung bewiesen.

Andererseits ist $(\varphi'')^{-1}(G) = F$ und $(P'')^{-1}(E) = G$.

Daraus folgt $(j'')^{-1}(E) = F$.

4) Bemerken wir zunächst dass das Abschlissen von $j(A)$ für die Topologie $\sigma(E'', E')$ mit $j''(A'')$ zusammenfällt.

Wegen des Korollars 1.5 -[6] ist $[W]$ schwach relativ kompakt und die Teilmenge

$$2^n\,[\overline{W}] + 2^{-n}\{x'' \in E'' \mid \|x''\| \leq 1\}$$

enthalten alle A und sind abgeschlossen für die Topologie $\sigma(E'', E')$.

Dann gilt es

$$j''(A'') \subset \bigcap_n (2^n\,[\overline{W}] + 2^{-n}\{x'' \in E'' \mid \|x''\| \leq 1\}) \subseteq E.$$

Aus 3) folgt es dass $F'' \subset (j'')^{-1}(E) = F$.

<u>Satz 3.3</u> <u>Sei</u> H <u>ein Banachraum</u>, E <u>ein schwach folgenvollständiger Banachverband und</u> $T:H \longrightarrow E$ <u>ein schwach kompakter Operator</u>.

<u>Dann existieren ein reflexiver Banachverband F, der ein Ideal in</u> E <u>ist, ein Operator</u> $S:H \longrightarrow F$ <u>und ein injektiver und stetiger Verbandshomomorphismus</u> $U:F \longrightarrow E$ <u>sodass</u> $T = U \circ S$.

<u>Beweis</u> Sei C die Einheitskugel von H. Setzen wir in Lemma 3.2, $W = T(C)$. Lemma 3.2 gibt uns den reflexiven Banachverband F und den Verbandshomomorphismus j.

Setzen wir dann $S = j^{-1} \circ T$ und $U = j$.

<u>Satz 3.4</u> <u>Sei K eine schwach kompakte Teilmenge von</u>

einem schwach folgenvollständigen Banachverband E .

Dann ist K affin und verbandshomeomorph (für die schwache Topologie) zu einer schwach kompakte Teilmenge eines reflexiven Banachverbandes.

Beweis Setzen wir $W = \mathrm{co}(K \cup -K)$ und anwenden Lemma 3.2 für die schwach kompakte Teilmenge W .

Wegen 1) Lemma 3.2 ist $K' = j^{-1}(K)$ eine beschränkte, also schwach relativ kompakte Teilmenge von F . Dazu ist K' schwach abgeschlossen und damit wird der Satz bewiesen.

Korollar 3.5 Seien E ein Banachraum, F ein schwach folgenvollständiger Banachverband , K_1 und K_2 die schwach kompakte Teilmenge von E und F .

Dann ist $\overline{\Gamma}(K_1 \otimes K_2)$ auch eine schwach kompakte Teilmenge in $E \hat{\otimes}_m F$.

Beweis Seien X_1 ein reflexiver Banachraum, $j_1 : X_1 \longrightarrow E$ ein affiner und injektiver Homomorphismus und $K_1' = j^{-1}(K_1)$ durch den Satz 2 - §4 - Kap. V - [1] gegeben.

Setzen wir auch X_2 ein reflexiver Banachverband , $j_2 : X_2 \longrightarrow F$ ein injektiver Verbandshomomorphismus und $K_2' = j_2^{-1}(K_2)$ gegeben durch den Satz 3.4 .

Offenbar ist $j_1 \otimes_m j_2 : X_1 \hat{\otimes}_m X_2 \longrightarrow E \hat{\otimes}_m F$ eine stetige Injektion.

Es ist klar dass $\overline{\Gamma}(K_1 \otimes K_2) \supset (j_1 \otimes j_2)(\overline{\Gamma}(K_1' \otimes K_2'))$.

Andererseits weil $K_1' \otimes K_2'$ eine beschränkte Teilmenge des reflexiven (wegen des Korollars 2.4) Banachraumes $X_1 \hat{\otimes}_m X_2$ ist, ist $\overline{\Gamma}(K_1' \otimes K_2')$ sogar eine schwach kompakte Teilmenge.

Daraus folgt dass $(j_1 \otimes j_2)(\overline{\Gamma}(K_1' \otimes K_2'))$ schwach abgeschlossen ist und es gilt $\overline{\Gamma}(K_1 \otimes K_2) = (j_1 \otimes j_2)(\overline{\Gamma}(K_1' \otimes K_2'))$.

Korollar 3.6 Seien E_1, E_2, F_1 Banachräume, F_2 ein schwach folgenvollständiger Banachverband und $T_i : E_i \longrightarrow F_i$ schwach kompakten Operatoren $(i = 1\ ,\ 2)$.

Dann ist $T_1 \otimes T_2 : E_1 \hat{\otimes}_\pi E_2 \longrightarrow F_1 \hat{\otimes}_m F_2$ ein schwach kompakter Operator.

Beweis Wenn U_i die Einheitskugel von E_i sind, bildet $T_1 \otimes T_2$ die Teilmenge $\overline{\Gamma}(T_1 \otimes T_2)$ auf $\overline{\Gamma}(T_1(U_1) \otimes T_2(U_2))$ ab. Wegen des Korollars 3.5 wird der Beweis beendet.

Korollar 3.7 Seien E_1, E_2, F_1 Banachräume und $T_1 : E_1 \longrightarrow F_1$, $T_2 : E_2 \longrightarrow L^1(X, \mu)$ schwach kompakten Operatoren. Dann ist

$$T_1 \otimes_\pi T_2 : E_1 \hat{\otimes}_\pi E_2 \longrightarrow F_1 \hat{\otimes}_\pi L^1 (X, \mu)$$

ein schwach kompakter Operator.

Der folgende Satz gibt uns eine Charakterisierung eines SKE Banachverbandes.

Satz 3.8 Ein schwach folgenvollständiger Banachverband E ist SKE dann und nur dann wenn ein reflexiver Banachverband F und ein injektiver und stetiger Verbandshomomorphismus $T : F \longrightarrow E$, sodass $\overline{T(F)} = E$, existieren.

Beweis Seien E ein SKE Banachverband und K dis schwach kompakte Teilmenge sodass $\overline{Sp}(K) = E$.

Bezeichnen wir mit W $co(K \cup -K)$ und anwenden Lemma 3.2.

Andererseits, wenn die Voraussetzungen des Satzes erfüllt werden, setzen wir $T(U) = K$, wo U die Einheitskugel von F ist.

Dann ist K schwach kompakt und $\overline{Sp}\ (K) = \overline{Im\ \ T} = E$.

Das folgende Korollar verallgemeinert ein Ergebnis von J.

Diestel.

Korollar 3.9 1) Seien E_1 ein SKE Banachraum und E_2 ein schwach folgenvollständiger SKE Banachverband.

Dann ist $E_1 \widehat{\otimes}_m E_2$ ein SKE Banachraum.

2) Sei E ein SKE Banachraum. Dann ist $L^p_E(X,\mu)$, $1 \leq p < \infty$ ein SKE Banachraum.

Beweis 1) Wegen des Satzes 3.8 und des Satzes 3 - §4 - Kap. V -[1] existieren ein reflexiver Banachverband F_2 , ein reflexiver Banachraum F_1 , ein Verbandshomomorphismus $T_2 : F_2 \longrightarrow E_2$ und ein stetiger Operator $T_1 : F_1 \longrightarrow E_1$ sodass $\overline{\text{Im } T_i} = E_i$, $i=1$, 2 .

Dann ist $T_1 \otimes T_2 : F_1 \widehat{\otimes}_m F_2 \longrightarrow E_1 \widehat{\otimes}_m E_2$ eine stetige Injektion und ist $F_1 \widehat{\otimes}_m F_2$ reflexiv.

Weil $\| \ \|_m$ eine Krossnorm ist, gilt es offenbar

$$\overline{\text{Im } (T_1 \otimes T_2)} = E_1 \widehat{\otimes}_m E_2.$$

Man bleibt nur den Satz 3 - §4 - Kap. IV -[1] anzuwenden.

2) folgt unmittelbar aus 1) .

LITERATUR

[1] - Diestel J. - Geometry of Banach spaces - Selected Topics - Lecture Notes in Mathematics - No. 485 - Springer Verlag -Berlin 1975.

[2] - Diestel J. - L^1_X is weakly compact generated if X is P. A.M. S. (to appear).

[3] - Fremlin H. - Tensor products of Banach lattices - Math. Ann. 212, 3, p. 87-106 (1974).

[4] - Grothendieck A. - Produits tensoriels topologiques et espaces nucléaires -Mem. A.M.S. -16, 1955.

[5] - Meyer-Nieberg P. - Charakterisierung einige topologische und ordnungstheoretischer Eigenschaften von Banachverbänden mit Hilfe disjunkter Folgen -Arch. d. Math., 24, p. 640-647 (1973).

[6] - Meyer-Nieberg P. - Zur schwachen Kompaktheit in Banachverbänden -Math. Z., 134, p.303-315, (1973).

[7] - Nagel R. J. - Beiträge zur Theorie der Banachverbänden Habilitationsschrift -Tübingen, 1972.

[8] - Schaefer H. H. - Banach lattices and positive operators Springer Verlag - Berlin -1974.

[9] - Schlotterbeck U. - Tensorprodukte von Banachverbänden und positive Operatoren - Habilitationsschrift -Tübingen, 1974.

CONSTRAINT STRONGLY MONOTONE OPERATORS

by

S. F. SBURLAN
Centre of Mathematical Statistics
National Institute of Metrology - Bucharest

The ideea of a constraint strongly monotone operator appears in connection with boundary displacement problem of elastic equilibrium from finite theory of elasticity, [4].

The deformation of a body $\mathcal{B}$ in Euclidean space $\mathbb{R}^3$ can be described by a mapping

(1) $$x = f(X),$$

where $X = (X^1, X^2, X^3)$ gives the position of a particle $P \in \mathcal{B}$ in reference configuration and $x = (x^1, x^2, x^3)$ is the position of the same particle in the deformed state induced by the action of forces.

Suppose that the reference configuration is a bounded domain D in $\mathbb{R}^3$ with smooth enough boundary ∂D, (for our purpose it is sufficient to have the cone property). The boundary displacement problem leads to study the following system:

(2) $$\begin{cases} -\sum_{\alpha} \dfrac{\partial}{\partial X^{\alpha}} h^{\alpha}(X, B(X)) + b(X, f(X)) = 0 & \text{in } D, \\ f(X) = f_0(X) & \text{on } \partial D, \end{cases}$$

where $h = (h_i^{\alpha})_{3x3}$ - the Piola-Kirchhoff stress tensor, $b = (b_1, b_2, b_3)$ -the density of body forces, $f_0 = (f_0^1, f_0^2, f_0^3)$ - the shape of ∂D, are known functions and $B = (B^{km})_{3x3}$ is the Cauchy-Green strain tensor:

$$(3) \qquad B^{km} = \sum_{\alpha} \frac{\partial f^k}{\partial X^\alpha} \frac{\partial f^m}{\partial X^\alpha} = x^k_{,\alpha} x^m_{,\alpha} ,$$

(the last term is an abreviate explicit form of B^{km}). Therefore in finite elastic behaviour two types of nonlinearities occur, a physical one - the dependence of Ω on B - and a geometrical one - the dependence of B on grad f.

In the case of physical nonlinear matherials existence results for problem (2) has been obtained by many authors. This problem in a more general form is largerly treated in [4] where one can also find a wide bibliography. For finite deformation case we mention that the associated operator is not coercive. In this case we note the uniqueness result from [1] proved for small strains and the path-dependent uniqueness or "weak uniqueness" from [3]. Concerning the existence of variational solution we mention as earlier results, those from [8], [9] and [10]. In [6] by means of locally monotone operators one proves the existence of a path-dependent solution of finite plane deformations provided that the deformation path of the boundary as well as the path of body forces are known.

Using a related notion - the constraint strongly monotone operator- (the two notions are equivalent for differentiable operators) one can proves the existence of a variational solution for problem (2) in the case of "sufficiently small strains", [8]-[9], and which can be extended to a path-dependent solution [10]. Our aim is to point out these results in a general form.

First we mention some abstract results. Let $\mathfrak{X}$ be a real reflexive Banach space, $\mathfrak{X}^*$ its dual space and (.,.) the pairing between $\mathfrak{X}$ and $\mathfrak{X}^*$. The map $T:\mathfrak{X} \longmapsto \mathfrak{X}^*$ is said to be constraint strongly monotone if there exists a closed ball $\bar{B}(o,R) \subset \mathfrak{X}$ such that

$$(4) \qquad (Tx - Ty, x - y) \geq c \| x - y \| , \qquad (\forall)\ x,y \in \bar{B}(o,R),$$

where $c > o$ depends on R.

PROPOSITION ([9]). Let $\mathfrak{X}$ be a reflexive separable Banach space and $T: \mathfrak{X} \longmapsto \mathfrak{X}^*$ a bounded demicontinuous constraint strongly monotone operator. If $\|T(o)\| \leq c$ (the constant of constraint monotonicity), then there exists uniquely x_o in $\bar{B}(o,R)$ such that $Tx_o = o$.

To avoid the "coercivity" condition $\|T(o)\| \leq c$ we use the homotopy argument from [2].

THEOREM. Let $\mathfrak{X}$ be a reflexive separable Banach space and $A_t x = A(x,t): \mathfrak{X} \times [o,1] \longmapsto \mathfrak{X}^*$ a mapping with the following properties:

(i) For any fixed $t \in [o,1]$, the operator $A_t: \mathfrak{X} \longmapsto \mathfrak{X}^*$ is bounded demicontinuous and constraint monotone with respect to the same ball.

(ii) $A(x,t)$ is continuous in t uniformly with $x \in \bar{B}(o,R)$.

(iii) $(A_1 x, x) \geq o$, (or $A_1(-x) = -A_1 x$), for all $x \in \partial B(o,R)$.

If $A_t x \neq o$ for all $x \in \partial B(o,R)$ and all $t \in [o,1]$, then there exists uniquely $x_o \in B(o,R)$ such that $A_o x_o = o$.

Proof. The result follows directly from the cited work. Indeed, it suffices to show that A_t satisfies the condition $(S)_+$ for each t:

$$x_n \rightharpoonup x_o \text{ and } \overline{\lim}(A_t x_n - A_t x_o, x_n - x_o) \leq o \text{ imply } x_n \longrightarrow x_o ,$$

but this is obvious for all sequences $\{x_n\} \subset \bar{B}(o,R)$ because of the constraint monotonicity condition with respect to the same ball $B(o,R)$.

Recall that an operator $S: \mathfrak{X} \longmapsto \mathfrak{X}^*$ is quasimonotone provided

$$x_n \rightharpoonup x_o \quad \text{implies} \quad \limsup\,(Sx_n - Sx_o, x_n - x_o) \geq o.$$

A necessary and sufficient condition for quasimonotonicity of S is that the map $A_o = S + \varepsilon J$ be bounded demicontinuous and of type $(S)_+$.

COROLLARY. Suppose that S is a quasimonotone operator and

$$(5) \qquad (Sx, x) \geq o , \quad (\forall)\, x \in \partial B(o,R).$$

If $S(\bar{B}(o,R))$ is a closed set in $\mathfrak{X}^*$, then there exists $x_o \in \bar{B}(o,R)$ so that $Sx_o = o$.

Proof. For any $\varepsilon > o$ denote $A_t = (1-t)S + \varepsilon J$, where J is the normalized duality map $Jx = \{ f \in \mathfrak{X}^* \mid (f,x) = \|x\|^2, \|f\| = \|x\| \}$. It is obvious that A_t fullfils the conditions of the theorem, so that there exists x_ε the unique solution of $A_o x_\varepsilon = (S+\varepsilon J)x_\varepsilon = o$. Now, for $\varepsilon \longrightarrow o$ one obtains $Sx \longrightarrow o$ and thus there exists $x_o \in \bar{B}(o,R)$ so that $Sx_o = o$ because of closedness of $S(\bar{B}(o,R))$.

Let $w(X)$ be a continuous differentiable function which satisfies the boundary data, i.e., $w(X) = f_o(X)$ on ∂D. Suppose that h fullfils those conditions which allow us to define a bounded demicontinuous map F_w from the Sobolev space $W_o^{1,p}(D)$ into its dual space $W^{-1,p'}(D)$, where p is an integer , $1 < p < \infty$, as usually:

$$(6) \qquad (F_w(z), v) = \sum_{i,\alpha} \int_D h_\alpha^i(X, B(w+z))\, v^i_{,\alpha}(X)\, dX, \qquad (\forall) v \in W_o^{1,p}$$

A sufficient condition for this is that h be a polynomial in the components of grad f. Similarly, if $b(X,f)$ is continuous in its arguments and it satisfies the growning condition

$$|b| = m(1+|f|^{q-1}), \quad m > o, \quad 1 < q \leq p$$

one can define the map $K: W^{1,q}(D) \longmapsto (W^{1,q}(D))^*$ by setting

$$(7) \qquad (Kf, v) = \sum_i \int_D b_i(X,f)\, v^i(X)\, dX, \qquad (\forall)\ v \in W^{1,q}(D).$$

By Green's formula it results that any solution of problem (2) is a variational solution, i.e.,

$$(8) \qquad (F_w + K_w)z = o,$$

where $K_w(z) = K(w+z)$.

Supposing that for certain boundary data, (those for which grad w is enough close to identical matrix I), there exists a closed ball $\bar{B}(o,R)$ in $W_o^{1,p}(D)$ such that for all $z \in \bar{B}(o,R)$

$$\sum_{i,j,\alpha,\beta} \int_D \left[\frac{\partial h_i^\alpha}{\partial x^j_{,\beta}}(w+z)\, z^i_{,\alpha} z^j_{,\beta} + \frac{\partial b_i}{\partial f_j}(w+z)\, z^i\, z^j\right] dX \geq c \sum_{k,\alpha} \int_D |z^k_{,\alpha}|^2 dX$$

with $c > o$ depending on R , one can prove that $S_w = F_w + K_w$ is constraint strongly monotone. Such a condition is not a formal one because in the case of hyperelastic materials and conservative body forces, i.e.,

$$h_i^\alpha(X,B) = \frac{\partial \mathcal{W}}{\partial x^i_{,\alpha}} \quad \text{and} \quad b_i(X,f) = \frac{\partial \vartheta}{\partial f^i}$$

the nonnegativity of first member in above inequality is a necessary condition for stability of equilibrium configuration,([11]). As an exemple we mention that from [9] ,i.e.,

$$\vartheta(B) = a(I_B - 3) + b(II_B - 3) + c(III_B - 1),$$

where $a > o$, $b \geq o$, c are real numbers and I_B, II_B, III_B are the principal invariants of the matrix B.

Now, if $\|S_w(o)\| \leq c$, then by the proposition there exists an unique variational solution. If it is not the case we apply the corollary. In fact, as a property of Niemitskyi operator one deduces that K_w is a completely continuous operator. Since the operator F_w still remains constraint strongly monotone,the sum $S_w = F_w + K_w$ is quasimonotone and we may choose the body forces such that

$$(9) \qquad \| F_w(o) + K_w(z)\| \leq cR \quad (\forall)\ z \in \partial B(o,R).$$

In this case, which provides a criteria for the admissibility of the body forces, we obtain

$$(S_w(z), z) = (F_w(z), z) + (K_w(z),z) = (F_w(z) - F_w(o), z) +$$
$$+ (F_w(o) + K_w(z), z) \geq (c\|z\| - \| F_w(o) + K_w(z)\|)\|z\|$$

for all $z \in \partial B(o,R)$, that is, the condition (5) is fulfilled. Therefore if $S_w(\bar{B}(o,R))$ is a closed set, then there exists $z_o \in \bar{B}(o,R)$ such that $S_w(z_o) = (F_w + K_w)\ z_o = o$.

Since K_w is a compact mapping it results that $K_w(\bar{B}(o,R))$ is closed. Hence $S_w(\bar{B}(o,R))$ is closed if $F_w(\bar{B}(o,R))$ will be so. One way to prove the closedness of $F_w(\bar{B}(o,R))$ is to extend F_w at a manotone mapping on whole space. If the new mapping, $\tilde{F}_w$, still remains demicontinuous then $\tilde{F}_w$ is maximal monotone and thus $\tilde{F}_w(\bar{B}(o,R))$ is a closed set. From mechanical point of view this extension corresponds to physical nonlinear effects in finite deformations. This observation allows us to reformulate the problem (2) and to extend the existence theory to path dependent solutions.

Let $z_o \in \bar{B}(o,R)$ be the variational solution of problem (2) and let $z_* \in C_o^\infty$ be an ε- approximant of z_o, i. e.,

$$\|z_o - z_*\| < \varepsilon .$$

Consider the transformation

(1o) $$x = X + w(X) + z_*(X), \quad (\forall)\ X \in \bar{D},$$

and denote $Z = \sup\{ |z_*(X)| , |\mathrm{grad}\ z_*(X)| \mid X \in \bar{D} \}$.

LEMMA ([8]). Suppose that $w \in C^1(\bar{D})$ and that there exists a constat $r > Z$ such that

$$\sum_{i,\alpha} \frac{\partial w^i(X)}{\partial X^\alpha} \xi^i \xi^\alpha \geq r \sum_i |\xi^i|^2 , \quad X \in D, \ \xi \in \mathbb{R}^3,$$

then the transformation (1o) is one-to-one. If, in addition, D is a convex set, then (1o) is a global transformation.

As a consequence, if $w \in C^2(D) \cap C^1(\bar{D})$ and S_w is continuous then there exists $\theta_\varepsilon \in L^p(D)$ with $\theta_\varepsilon \longrightarrow o$ in L^p as $\varepsilon \longrightarrow o$, such that

$$\sum_{i,\alpha} \int_D \Big[- \frac{\partial}{\partial X^\alpha} h_i^\alpha(X,B) + b_i(X,f) - \theta_\varepsilon^i(X) \Big] v^i(X)\, dX = o.$$

Let Ω be a new reference configuration of $\mathcal{B}$. With respect to this configuration the equations of boundary displacement problem, (2), remain the same. Consider two deformations of $\mathcal{B}$ satisfying the same boundary data, i.e.,

$$x = f_1(X), \quad x = f_2(X), \quad f_1(X) = f_2(X) = f_o(X) \text{ for } X \in \partial\mathcal{B},$$

both being taken with respect to new configuration. Let w(X) be a continuous differentiable function which satisfies the boundary data. We shall say that the body $\mathcal{B}$ is monotone elastic in vecinity of reference configuration if there exist a ball $B(o,R) \subset W_o^{1,p}(\Omega)$ and a constant $c > o$, depending on R, such that

$$\sum_{i,\alpha} \int_\Omega [h_i^\alpha(X,B(f_1)) - h_i^\alpha(X,B(f_2))](f^i_{1,\alpha} - f^i_{2,\alpha})dX \geq c \sum_{i,\alpha} \int_\Omega (f^i_{1,\alpha} - f^i_{2,\alpha})^2 dX$$

for all $z_k = f_k - w, (k=1,2)$, in $\bar{B}(o,R)$.

Suppose now, that we know the deformation path of the boundary of body $\mathcal{B}$ from initial reference configuration to equilibrium configuration, i.e., the configuration Ω correspond to $t = t_1$ in the map:

$$x(t) = f_o(X,t), \quad t \in [o,T].$$

Consider the following reformulation of problem (2): <u>find a map</u> $x = f(X,t)$, <u>satisfying an initial condition</u> $x_o = f(X,o)$, <u>such that</u>

$$\begin{cases} -\sum_{\alpha} \frac{\partial}{\partial X^{\alpha}} h^{\alpha}(X,B(X,t)) + b(X,f(X,t)) = o, & X \in D, \\ f(X,t) = f_o(X,t), & X \in \partial D, \end{cases}$$

<u>for all</u> $t \in [o,T]$.

If the body $\mathcal{B}$ is monotone elastic in vecinity of any reference configuration, the parameters c and R remaining the same in each configuration, and if the body forces are admissible in the sense given by (9), then in physical hypotheses which allow the existence of variational solution, there exists a solution of elastic equilibrium problem. This solution is obtained step by step taking into account at each step only physical nonlinear effects and applying the above approximation procedure. We mention also, that the above condition on c and R is too strong, the result remaining valid if one asks that the product cR does not depend on reference configuration.

REFERENCES

[1] JOHN, F., Uniqueness of non-linear elastic equilibrium for prescribed boundary displacements and sufficiently small st strains, Comm.Pure Appl.Math.,25 (1972), 617-634.

[2] HESS, P., A homotopy argument for mappings of monotone type in Banach spaces, Math.Ann.,2o7 (1974), 63-65.

[3] MAZILU, P., Uniqueness theorems for the first boundary value problem in non-linear elasticity, (to appear).

[4] MAZILU, P. and SBURLAN, S. F., Metode funcţionale în rezolvarea ecuaţiilor teoriei elasticitaţii, Ed.Acad.RSR, 1973.

[5] MARINESCU, G., Tratat de analiză funcţională, vol.II, Ed.Acad. RSR, 1972.

[6] NEČAS, J., Theory of locally monotone operators modeled on the finite displacement theory for hyperelasticity, Beiträge zur Analysis, 8 (1976), 1o3-114.

[7] PASCALI, D., Operatori neliniari, Ed.Acad.RSR, 1974.

[8] SBURLAN, S. F., The Dirichlet problem of elastic equilibrium, Rev.Roum.Sci.Techn.-Méc.Appl., 19 (1974), 833-847.

[9] SBURLAN, S. F., Some remarks on existence theorems for functional equations with odd operators, Rev.Roum.Math.Pures Appl., 21 (1976),

[10] SBURLAN, S. F., Metode aproximative în studiul echilibrului elastic, (to appear), Stud.Cercet.Mat.

[11] TRUESDELL, C. A. and NOLL, W., The nonlinear field theories of mechanics, Handbuch der Physik, III/3, Springer-Verlag,1965.

FATOU AND SZEGŐ THEOREMS FOR OPERATOR VALUED FUNCTIONS

by

Ion Suciu and Ilie Valuşescu

1. Introduction

The celebrated Fatou and Szegö theorems play an important role in the study of non-normal operators on Hilbert spaces. Fatou theorem was the principal tool from the analytic function theory used by B. Sz.-Nagy and C. Foiaş [12] in construction on theirs functional calculus with functions in H^∞. In theirs functional model for contractions they used also, in decisive way, the variants of this theorem for vector or operator valued analytic functions. Szegö theorem and their implications in factorizations are also very intimately related with basic problems in operator theory, like structure of invariant subspaces, Jordan models, cyclicity, etc. The applications of the operatorial methods in prediction, cross also through ideas contained in this very important theorem.

Therefore it is not surprising that several efforts were made in order to obtain clear variants of these theorems for the operator valued functions (see for instance [12], [13], [3]).

In this paper, following the treatement given in [12] for the bounded (operator valued) analytic functions, we intend to point out and some how to overcame in a new way the difficulties which appear in the non bounded case.

After some necessary preliminaries given in Section 2, we prove in section 3 an analogous, for the non bounded case, of B. Sz.-Nagy and C. Foiaş Lemma on Fourier representation of operators which intertwine unilateral shifts (Lemma Q). Section 4 contains the results from [9] about factorization of semi- spectral measures by means of L^2-bounded analytic functions. We prove also that any L^2-contractive analytic function can be factorized into a contractive analytic function and an evaluation function [10]. These theorems are used in section 5 to obtain variants for Fatou and Szegö theorems for operator valued functions.

During the preparation of this paper we benefited by helpfull discutions with Ghe. Bucur, A. Cornea and C. Foiaş.

2. Preliminaries

Let us recall the classical Fatou and Szegö theorems, in a particular case which will be convenient in understanding the variants which we propose for such type of theorems in operator valued case.

Denote by $\mathbb{T}$ the one-dimensional torus $\{z \in \mathbb{C};\ |z| = 1\}$ in the complex plane and by $\mathbb{D}$ the open unit disc $\{z \in \mathbb{C};\ |z| < 1\}$. By L^2 we denote the usual Hilbert Space of measurable complex valued functions v on $\mathbb{T}$ which are square integrable in modulus, with the norm

$$(2.1) \qquad \|v\|^2_{L^2} = \frac{1}{2\pi} \int_0^{2\pi} |v(e^{it})|^2\, dt$$

when dt is the one-dimensional Lebesque measure. By L^2_+ we denote the closed subspace of L^2 consisting from all function in L^2 whose negative Fourier coefficients are zero. Denote by H^2 the Hilbert space of all complex valued functions f on $\mathbb{D}$ which are analytic in $\mathbb{D}$ and verify

$$(2.2) \qquad \|f\|^2_{H^2} = \sup_{0 \le r < 1} \frac{1}{2\pi} \int_0^{2\pi} |f(re^{it})|^2\, dt < \infty .$$

The map

$$f(z) = \sum_{n=0}^{\infty} a_n z^n \longrightarrow f_+(e^{it}) = \sum_{n=0}^{\infty} a_n e^{int}$$

is an isometric isomorphism between H^2 and L^2_+ and we have

$$(2.3) \qquad \|f\|^2_{H^2} = \sum_{n=0}^{\infty} |a_n|^2 = \|f_+\|^2_{L^2_+} .$$

For a function $f \in H^2$ let $\tilde{f}(\lambda) = \int_0^{\lambda} f(z)\, dz$ be its primitive. Then f is an Lipschitzian function on $\mathbb{D}$, thus it can be extended to an absolutely continuous function on $\mathbb{D}$. The restriction of this function to $\mathbb{T}$ gives rise to a complex valued finite Borel measure on $\mathbb{T}$ denoted by μ_f which is absolutely continuous with respect to Lebesque measure.

The variant of Fatou theorem to keep in mind is the following :

THEOREM F. Let $f \in H^2$, f_+ be its correspondent in L^2_+ and μ_f be its primitive measure. Then

(1) $\quad d\mu_f = f_+ dt$

(2) $\quad f(re^{it}) = \frac{1}{2\pi} \int_0^{2\pi} P_r(t-s)\, d\mu_f(s) = \frac{1}{2\pi} \int_0^{2\pi} P_r(t-s)\, f_+(s)\, ds$

where $P_r(t)$ is the Poisson kernell

$$P_r(t) = \frac{1 - r^2}{1 - 2r\cos t + r^2}$$

(3) $f(z)$ tends to $f_+(e^{it})$ as z tends to e^{it} non-tangentially with respect to the unit circle at every point t such that

$$\frac{1}{2s}\int_{t-s}^{t+s} d\mu_f(\tau) = \frac{1}{2s}\int_{t-s}^{t+s} f_+(\tau)\,d\tau \longrightarrow f_+(t)$$

thus a.e.

If we consider instead of the spaces of the scalar valued functions L^2 and H^2 the similar spaces $L^2(\mathcal{E})$ and $H^2(\mathcal{E})$ of $\mathcal{E}$-valued functions, where $\mathcal{E}$ is a locally convex vector space (with suitable definition for the measurability, analyticity and square integrability), then we can look for the existence of measure μ_f and eventually for its derivative f_+ as in the Fatou theorem. In case $\mathcal{E}$ is a separable Hilbert space, we can transpose Theorem F with the same proof as in the scalar case, the isometric isomorphism between the Hilbert space $H^2(\mathcal{E})$ and $L^2_+(\mathcal{E})$ being also preserved. We are not interested in the generalisation of Fatou theorem along this line, for a larger class of locally convex vector spaces, because of two reasons : firstly, the conditions we must impose to $\mathcal{E}$ in order to obtain consistent Fatou theorems are of such type that permit the same proof as in the scalar case; secondly, the space (of the maximal interest for us) of linear bounded operators, both in the norm or strong topology, do not satisfies such a type of conditions.

These are the reasons why we shall study variants of Fatou theorem for operator valued functions with pure operator methods.

It is not surprising that these methods work better in the case of another famous theorem of classical function theory, namely the Szegö theorem. Let us recall Szegö theorem in a variant which contains Kolmogorov-Krein generalisations (cf. [5]).

THEOREM Sz. Let μ be a positive measure on $\mathbb{T}$ such that $\mu(\mathbb{T}) = 1$ and let $d\mu = \frac{1}{2\pi}\cdot h\,dt + d\mu_s$ be the Lebesque decomposition of μ with respect to Lebesque measure. Then

(1) $$\Delta = \inf_p \int_0^{2\pi} |1-p|^2 d\mu = \inf_p \frac{1}{2\pi}\int_0^{2\pi} |1-p|^2\, h\,dt = \exp\left[\frac{1}{2\pi}\int_0^{2\pi} \log h\,dt\right]$$

where the infimum is taken over all analytic polynomial p which wanish in origin.

(2) In order to exists a function $f \in H^2$ such that $|f_+|^2 = h$ it is necessary and suf-

ficient that log $h \in L^1$ or equivalently $\Delta > 0$. In this case, there exist an outer function f in H^2 such that $|f_+|^2 = h$ and $\Delta = |f(0)|^2$.

We shall recognise parts of Theorem F and Theorem Sz. in the results we shall give in operator valued case. But the Fatou - Szegö problematic in general case is far to be elucidate, the nature of the obstructions being variate and mysterious.

3. Operator valued analytic functions

Let $\mathcal{E}$ and $\mathcal{F}$ be two separable Hilbert spaces. A function defined on $\mathbb{D}$ where values are bounded operators $\Theta(\lambda)$ from $\mathcal{E}$ to $\mathcal{F}$ will be called analytic provided it has a power series expansion

(3.1) $$\Theta(\lambda) = \sum_{n=0}^{\infty} \lambda^n \Theta_n \qquad \lambda \in \mathbb{D}$$

where Θ_n are bounded operators from $\mathcal{E}$ to $\mathcal{F}$. The series is supposed to be convergent weakly, strongly or in norm which amounts to the same for the power series. As in [12] we shall denote such a function by the triplet $\{\mathcal{E}, \mathcal{F}, \Theta(\lambda)\}$.

We shall introduce the following three types of boundedness for operator valued analytic functions.

The analytic function $\{\mathcal{E}, \mathcal{F}, \Theta(\lambda)\}$ will be called bounded provided

(3.2) $$\|\Theta(\lambda)\| \leq M \qquad \lambda \in \mathbb{D}.$$

If $\{\mathcal{E}, \mathcal{F}, \Theta(\lambda)\}$ verifies

(3.3) $$\sup_{0 \leq r < 1} \frac{1}{2\pi} \int_0^{2\pi} \|\Theta(re^{it})\|^2 \, dt \leq M^2$$

then it will be called L^2-norm bounded analytic function.

If $\{\mathcal{E}, \mathcal{F}, \Theta(\lambda)\}$ verifies

(3.4) $$\sup_{0 \leq r < 1} \frac{1}{2\pi} \int_0^{2\pi} \|\Theta(re^{it}) a\|^2 \, dt \leq M^2 \|a\|^2_{\mathcal{E}}$$

for any $a \in \mathcal{E}$, then it will be called L^2-strongly bounded or shortly L^2-bounded analytic function.

It is easy to verify that (3.3) and (3.4) are respectively equivalent to

(3.3)' $$\sum_0^{\infty} \|\Theta_n\|^2 \leq M^2$$

and

(3.4)' $$\sum_0^{\infty} \|\Theta_n a\|^2_{\mathcal{F}} \leq M^2 \|a\|^2_{\mathcal{E}}$$

Let us remark that (3.4) and (3.4)' may be stated with M = M(a) depending of a (which corresponds to the term "strongly bounded"). Indeed if we consider the convergent series

$$S(a) = \left(\sum_0^\infty \|\Theta_n a\|^2\right)^{1/2} \qquad (a \in \mathcal{E})$$

then $S_N(a) = \left(\sum_0^N \|\Theta_n a\|^2\right)^{1/2}$, is a continuous function on $\mathcal{E}$ and consequently $S(a) = \sup_N S_N(a)$ is a lower semi-continuous semi-norm on $\mathcal{E}$. It is known then that S(a) is bounded i.e.

$$S(a) \leq M \|a\| \qquad (a \in \mathcal{E})$$

with M independent of a.

Clearly (3.2) $\Longrightarrow$ (3.3) $\Longrightarrow$ (3.4). If we consider the function $\{\mathcal{E}, \mathcal{E}, \Theta(\lambda)\}$ defined as

$$\Theta(\lambda) a = \delta(\lambda) a \qquad (\lambda \in \mathbb{D},\ a \in \mathcal{E})$$

where $\delta(\lambda)$ is a scalar valued function from H^2 which is not bounded, then clearly $\{\mathcal{E}, \mathcal{E}, \Theta(\lambda)\}$ verifies (3.3) but not (3.2).

Let now $\mathcal{E} = H^2$, $\mathcal{F} = \mathbb{C}$ and $\{\mathcal{E}, \mathcal{F}, \Theta(\lambda)\}$ the analytic function defined as :

$$\Theta(\lambda)\, h = h(\lambda) \qquad (\lambda \in \mathbb{D},\ h \in H^2).$$

For a fixed λ in $\mathbb{D}$ and any $h \in H^2$ of the form $h(z) = \sum_0^\infty c_k z^k$ we have :

$$\|\Theta(\lambda) h\|_{\mathcal{F}} = |h(\lambda)| \leq \sum_0^\infty |\lambda|^k |c_k| \leq \left(\sum_0^\infty |\lambda|^{2k}\right)^{1/2} \left(\sum_0^\infty |c_k|^2\right)^{1/2} \leq \frac{1}{\sqrt{1-|\lambda|^2}} \|h\|_{H^2}.$$

Thus $\Theta(\lambda)$ is a bounded operator from $\mathcal{E}$ into $\mathcal{F}$ and

$$\|\Theta(\lambda)\| \leq \frac{1}{\sqrt{1-|\lambda|^2}} .$$

If we put $\Theta_k h = c_k$ then clearly Θ_k is a bounded operator from $\mathcal{E}$ into $\mathcal{F}$ and

$$\Theta(\lambda) = \sum_0^\infty \lambda^k \Theta_k$$

For any h in H^2 we have :

$$\sum_0^\infty \|\Theta_k h\|^2 = \sum_0^\infty |c_k|^2 = \|h\|^2_{H^2}$$

thus $\{\mathcal{E}, \mathcal{F}, \Theta(\lambda)\}$ verifies (3.4)'. By the other way if we take $h(z) = \sum_1^\infty \frac{1}{k} z^k$, then clearly $h \in H^2$ and for n sufficiently large such that $\frac{1}{\sqrt{n}} < \left(\sum_1^\infty \frac{1}{k^2}\right)^{1/2}$ we have

$$\|\Theta_n\| \geq \frac{\|\Theta_n h\|}{\|h\|} = \frac{1/n}{\left(\sum_1^\infty 1/k^2\right)^{1/2}} \geq \frac{1}{\sqrt{n}} .$$

It results that $\sum_0^\infty \|\Theta_n\|^2$ is divergent i.e. $\{\mathcal{E}, \mathcal{F}, \Theta(\lambda)\}$ does not vierifies (3.3).

Let now $\{\mathcal{E}, \mathcal{F}, \Theta(\lambda)\}$ be an L^2-bounded analytic function. We can define the operator V_Θ from $\mathcal{E}$ into $H^2(\mathcal{F})$ by

(3.5) $$(V_{\Theta} a)(\lambda) = \Theta(\lambda)\, a \qquad (a \in \mathcal{E}).$$

We have

$$\|V_{\Theta} a\|^2_{H^2(\mathcal{F})} = \sup_{0 \le r < 1} \frac{1}{2\pi} \int_0^{2\pi} \|(V_{\Theta} a)(re^{it})\|^2_{\mathcal{F}}\, dt = \sup_{0 \le r < 1} \frac{1}{2\pi} \int_0^{2\pi} \|\Theta(re^{it})a\|^2\, dt \le$$
$$\le M^2 \|a\|^2 .$$

Thus V_{Θ} is bounded. Conversely, if V is a bounded operator from $\mathcal{E}$ into $H^2(\mathcal{F})$, then setting

$$\Theta(\lambda) a = (Va)(\lambda)$$

we obtain an L^2-bounded analytic function $\{\mathcal{E}, \mathcal{F}, \Theta(\lambda)\}$ such that $V_{\Theta} = V$.

Thus (3.5) establish on one-to-one correspondence between L^2-bounded analytic functions $\{\mathcal{E}, \mathcal{F}, \Theta(\lambda)\}$ and the bounded operators from $\mathcal{E}$ into $H^2(\mathcal{F})$.

Let us remark that we can consider V_{Θ} as the multiplication by the operator valued function $\Theta(\lambda)$ on the constant functions <u>a</u> from $H^2(\mathcal{E})$. Our next intention is to analyse the maximal multiplication operator on $H^2(\mathcal{E})$ generated by $\{\mathcal{E}, \mathcal{F}, \Theta(\lambda)\}$ and to give an intrinsic characterization of such operators. We shall obtain in Lemma Q the corresponding result for non bounded case of the B. Sz.-Nagy and C. Foiaş lemma of Fourier representation of the operators which intertwine unilateral shifts (cf.[12] pp. 195-198).

Let us call <u>evaluation operator</u> e_{λ} on $H^2(\mathcal{E})$ the operator defined for a fixed $\lambda \in \mathbb{D}$ by

$$e_{\lambda} h = h(\lambda).$$

We have already seen that e_{λ} is a bounded operator from $H^2(\mathcal{E})$ into $\mathcal{E}$ and

$$\|e_{\lambda} h\| \le \frac{1}{\sqrt{1-|\lambda|^2}} \|h\|_{H^2(\mathcal{E})}.$$

If no confusion will arises we shall denote with the same symbol e_{λ} the corresponding evaluation operator for different spaces $H^2(\mathcal{E})$.

Let Q be a linear operator defined on the subspace D(Q) of $H^2(\mathcal{E})$ with values in $H^2(\mathcal{F})$. We say that Q <u>intertwines the evaluations</u> on $H^2(\mathcal{E})$ and $H^2(\mathcal{F})$ if the following conditions hold :

(i) $\mathcal{E} \subset D(Q)$ and $Q|_{\mathcal{E}}$ is a bounded operator from $\mathcal{E}$ into $H^2(\mathcal{F})$.

(ii) D(Q) contains any function $h \in H^2(\mathcal{E})$ for which the function h_Q defined as

$$h_Q(\lambda) = e_{\lambda} Q\, e_{\lambda} h \qquad (\lambda \in \mathbb{D})$$

belongs to $H^2(\mathcal{F})$.

(iii) For any $\lambda \in \mathbb{D}$ and $h \in D(Q)$ we have

$$e_{\lambda} Qh = e_{\lambda} Q e_{\lambda} h .$$

Any operator Q which intertwines the evaluations on $H^2(\mathcal{E})$ and $H^2(\mathcal{F})$ is closed. Indeed, if $h_n \in D(Q)$ such that $h_n \to h$ in $H^2(\mathcal{E})$ and $Qh_n \to g$ in $H^2(\mathcal{F})$ then $e_\lambda Q e_\lambda h_n \to e_\lambda Q e_\lambda h$ for any $\lambda \in \mathbb{D}$, because of (i) it results that $e_\lambda Q e_\lambda$ is bounded. By the other way from (iii) it results that $e_\lambda Q e_\lambda h_n = e_\lambda Q h_n \longrightarrow e_\lambda g$ in $H^2(\mathcal{F})$. Thus for any $\lambda \in \mathbb{D}$ we have $e_\lambda Q e_\lambda h = e_\lambda g$, i.e. $h_Q(\lambda) = e_\lambda Q e_\lambda h = e_\lambda g = g(\lambda)$.

From (ii) it results $h_Q \in D(Q)$ and $Qh = g$.

If Q intertwines the evaluations on $H^2(\mathcal{E})$ and $H^2(\mathcal{F})$ then $D(Q)$ contains any analytic ($\mathcal{E}$-valued) polynomial $p(z) = \sum_0^m z^k a$. Indeed, for such a p we have

$$p_Q(\lambda) = e_\lambda Q e_\lambda p = e_\lambda Q \sum_0^m \lambda^k a_k = e_\lambda \sum_0^m \lambda^k Q a_k = \sum_0^m \lambda^k e_\lambda (Qa_k) = \Big(\sum_0^m z^k Q a_k\Big)(\lambda).$$

Since clearly $\sum_0^m z^k Qa_k \in H^2(\mathcal{F})$ from (ii) it results $p \in D(Q)$.

Thus any operator Q which intertwines evaluations on $H^2(\mathcal{E})$ and $H^2(\mathcal{F})$, is a closed operator with dense domain. It results that Q is bounded on its domain $D(Q)$ if and only if $D(Q) = H^2(\mathcal{E})$. In this case (iii) is equivalent to the fact that Q intertwines the shift operators on $H^2(\mathcal{E})$ and $H^2(\mathcal{F})$ i.e. with :

$$zQh = Qzh \qquad h \in H^2(\mathcal{E}) \ . \tag{3.6}$$

Indeed, if $h = \sum_{k=0}^\infty z^k a_k$, then using (3.6) we have :

$$e_\lambda Qh = (Qh)(\lambda) = \Big(\sum_0^\infty Q(z^k a_k)\Big)(\lambda) = \Big(\sum_0^\infty z^k Qa_k\Big)(\lambda) = \sum_0^\infty \lambda^k e_\lambda (Qa_k) =$$

$$= e_\lambda Q\Big(\sum_0^\infty \lambda^k a_k\Big) = e_\lambda Q e_\lambda h.$$

Thus (3.6) $\Longrightarrow$ (iii). Conversely, from (iii) it results

$$(zQh)(\lambda) = \lambda (Qh)(\lambda) = \lambda e_\lambda Qh = e_\lambda Q(\lambda h) = e_\lambda Q e_\lambda(\lambda h) = e_\lambda Q e_\lambda (zh) =$$

$$= e_\lambda Q(zh) = [Q(zh)](\lambda) \ .$$

LEMMA Q. There exists an one-to-one correspondence between L^2-bounded analytic functions $\{\mathcal{E}, \mathcal{F}, \Theta(\lambda)\}$ and the operators Q which intertwines the evaluations on $H^2(\mathcal{E})$ and $H^2(\mathcal{F})$ given by

$$(Qh)(\lambda) = \Theta(\lambda)\, h(\lambda), \qquad \lambda \in \mathbb{D},\ h \in D(Q). \tag{3.7}$$

The L^2-bounded analytic function $\{\mathcal{E}, \mathcal{F}, \Theta(\lambda)\}$ is bounded if and only if the corresponding operator Q is bounded.

Proof. Let $\{\mathcal{E}, \mathcal{F}, \Theta(\lambda)\}$ be an L^2-bounded analytic function. Denote by $D(\Theta)$ the subspace of $H^2(\mathcal{E})$ consisting from all function h in $H^2(\mathcal{E})$ for which the function $\lambda \to \Theta(\lambda)h(\lambda)$ is in $H^2(\mathcal{F})$, and Q be the operator defined on $D(\Theta)$ by (3.7). Since $\Theta(\lambda)\, a = (V_\Theta a)(\lambda)$, we have $\mathcal{E} \subset D(\Theta)$ and $Qa = V_\Theta a$. Thus Q verifies (i). For any $h \in H^2(\mathcal{E})$ we have :

$$h_Q(\lambda) = e_\lambda Q e_\lambda h = e_\lambda (V_\Theta e_\lambda h) = \Theta(\lambda)\, h(\lambda).$$

Thus, if $h_Q \in H^2(\mathcal{F})$, we have $h \in D(\Theta) = D(Q)$ i.e. Q verifies (ii). For any $h \in D(\Theta)$ we have :

$$e_\lambda Q h = \Theta(\lambda) h(\lambda) = \Theta(\lambda) e_\lambda h = [Q(e_\lambda h)](\lambda) = e_\lambda Q e_\lambda h.$$

Thus Q verifies (iii). Hence Q intertwines the evaluations on $H^2(\mathcal{E})$ and $H^2(\mathcal{F})$.

Conversely, if Q verifies (i)–(iii), then if we put $V_\Theta = Q\,|\,\mathcal{E}$ we obtain a bounded operator from $\mathcal{E}$ into $H^2(\mathcal{F})$ and we already seen that $\Theta(\lambda)a = (V_\Theta a)(\lambda)$ defines an L^2-bounded analytic function $\{\mathcal{E}, \mathcal{F}, \Theta(\lambda)\}$. Since for any $h \in H^2(\mathcal{E})$ we have

$$h_Q(\lambda) = e_\lambda Q e_\lambda h = e_\lambda (V_\Theta h) = (V_\Theta h)(\lambda) = \Theta(\lambda)\, h(\lambda),$$

it is clear that $D(\Theta) \subset D(Q)$. From (iii) it results that for any $h \in D(Q)$ we have

$$e_\lambda Q h = e_\lambda Q e_\lambda h = e_\lambda (V_\Theta h) = \Theta(\lambda) h(\lambda).$$

Thus $D(Q) = D(\Theta)$ and for any $h \in D(Q)$

$$(Qh)(\lambda) = \Theta(\lambda) h(\lambda) \qquad (\lambda \in \mathbf{D}).$$

Suppose now that $\{\mathcal{E}, \mathcal{F}, \Theta(\lambda)\}$ is bounded. Then for any $h \in D(Q)$ we have

$$\|Qh\|^2_{H^2(\mathcal{F})} = \sup_{0 \le r < 1} \frac{1}{2\pi} \int_0^{2\pi} \|\Theta(re^{it})\, h(re^{it})\|^2_{\mathcal{F}}\, dt \le \sup_{0 \le r < 1} \frac{1}{2\pi} \int_0^{2\pi} M^2 \|h(re^{it})\|^2_{\mathcal{E}}\, dt =$$

$$= M^2 \|h\|^2_{H^2(\mathcal{E})} .$$

Thus Q is a bounded operator.

Suppose now that Q is bounded. Then for all analytic scalar valued polynomial $\underline{p}$ and $\underline{a} \in \mathcal{E}$ we have

$$\frac{1}{2\pi} \int_0^{2\pi} |p(e^{it})|^2 \|(V_\Theta a)(e^{it})\|^2 dt = \|p V_\Theta a\|^2_{H^2(\mathcal{F})} = \|pQa\|^2 = \|Qpa\|^2 \le$$

$$\le \|Q\|^2 \|pa\|^2_{H^2(\mathcal{E})} = \|Q\|^2 \frac{1}{2\pi} \int_0^{2\pi} |p(e^{it})|^2 \|a\|^2 dt.$$

It results that for any trigonometric polynomial $\underline{p}$ we have

$$\int_0^{2\pi} |p(e^{it})|^2 \|(V_\Theta a)(e^{it})\|^2 dt \le \int_0^{2\pi} |p(e^{it})|^2 \|a\|^2_{\mathcal{E}} \|Q\|^2\, dt$$

which implies

$$\|(V_\Theta a)(e^{it})\|_{\mathcal{F}} \le \|Q\|\, \|a\| \qquad \text{a.e.}$$

Using known properties of Poisson kernel we obtain

$$\|\Theta(\lambda)a\|_{\mathcal{F}} = \left\|\frac{1}{2\pi} \int_0^{2\pi} P_r(t-s)\, (V_\Theta a)(s) ds\right\| \le \frac{1}{2\pi} \int_0^{2\pi} P_r(t-s) \|(V_\Theta a)(s)\|\, ds \le$$

$$\leq \frac{1}{2\pi}\int_0^{2\pi} P_r(t-s)\,\|Q\|\,\|a\|\;ds = \|Q\|\,\|a\|_{\mathcal{E}}\;,$$

i.e. $\{\mathcal{E}, \mathcal{F}, \Theta(\lambda)\}$ is bounded.

The proof of the Lemma Q is complete.

We shall denote by Θ_+ the operator which intertwines the evaluation on $H^2(\mathcal{E})$ and $H^2(\mathcal{F})$ uniquely associated to the L^2-bounded analytic function $\{\mathcal{E}, \mathcal{F}, \Theta(\lambda)\}$ as in Lemma Q.

Using the natural isomorphism between $H^2(\mathcal{E})$ and $L^2_+(\mathcal{E})$ we consider Θ_+ as an operator from $L^2_+(\mathcal{E})$ into $L^2_+(\mathcal{F})$. The operator Θ_+ is closed and its domain $D(\Theta_+)$ contains any analytic polynomial in $L^2_+(\mathcal{E})$.

Let now p be a trigonometric polynomial in $L^2(\mathcal{E})$. There exists an integer $n \geq 0$ such that $e^{int}p$ is an analytic polynomial. Let us define

$$\Theta p = e^{-int}\,\Theta_+(e^{int}p) \tag{3.8}$$

If $e^{int}p$ and $e^{imt}p$, $(n, m \geq 0)$, belong to $L^2_+(\mathcal{E})$ then if $n \geq m$ we have

$$\Theta_+(e^{int}p) = \Theta_+(e^{i(n-m)t}e^{imt}p) = e^{i(n-m)t}\,\Theta_+(e^{int}p)$$

because Θ_+ is an multiplication operator on $H^2(\mathcal{E})$. Then clearly (3.8) defines a linear operator Θ from the subspace of trigonometric polynomials in $L^2(\mathcal{E})$ into $L^2(\mathcal{F})$.

In case Θ_+ is bounded then clearly Θ is a bounded operator from $L^2(\mathcal{E})$ into $L^2(\mathcal{F})$ and $\Theta_+ = \Theta \,|\, L^2_+(\mathcal{E})$. If Θ_+ is non bounded then Θ is not in general closable. We shall see later that this problem is related to the existence of the boundary limit for the L^2-bounded analytic function $\{\mathcal{E}, \mathcal{F}, \Theta(\lambda)\}$.

A simple example which shows that Θ is not, in general, closable is the following: let $\mathcal{E} = H^2$, $\mathcal{F} = \mathbb{C}$ and $\{\mathcal{E}, \mathcal{F}, \Theta(\lambda)\}$ defined by

$$\Theta(\lambda)\,h = h(\lambda) \qquad\qquad h \in H^2.$$

Consider in $L^2(\mathcal{E})$ the sequence of polynomials

$$p_n = \frac{1}{\log n}\sum_{k=1}^{n} e^{-ikt}a_k$$

were a_k is the function $\frac{1}{k}e^{ikt}$ over $\mathcal{E} = H^2$. Then $\|p_n\|^2 = \frac{1}{(\log n)^2}\sum_{k=1}^{n}\frac{1}{k^2} \longrightarrow 0$

Thus $p_n \longrightarrow 0$ in $L^2(\mathcal{E})$. But

$$\Theta p_n = e^{-int}\frac{1}{\log n}\sum_{k=1}^{n}(e^{i(n-k)t}a_k) = \frac{1}{\log n}\sum_{k=1}^{n}\frac{1}{k}\;.$$

Thus $\Theta p_n \longrightarrow 1$, i.e. Θ is not closable.

The L^2-bounded analytic function $\{\mathcal{E}, \mathcal{F}, \Theta(\lambda)\}$ is called inner if the attached operator $\underset{+}{\Theta}$ is an isometry. Such a function is thus necessary bounded. The L^2-bounded analytic function $\{\mathcal{E}, \mathcal{F}, \Theta(\lambda)\}$ is called L^2-bounded outer function provided

(3.9) $$\bigvee_0^\infty z^n V_\Theta \mathcal{E} = H^2(\mathcal{F}).$$

If $\{\mathcal{E}, \mathcal{F}, \Theta(\lambda)\}$ is an L^2-bounded outer function then for any $\lambda \in \mathbb{D}$ we have $\overline{\Theta(\lambda)\mathcal{E}} = \mathcal{F}$. If $\{\mathcal{E}, \mathcal{F}, \Theta(\lambda)\}$ is simultaneously inner and outer, then it is a unitary constant function.

4. Attached semi-spectral measures and factorizations

Recall that an $\mathcal{L}(\mathcal{E})$- valued semi-spectral measure is a map $\omega \to F(\omega)$ from the family $B(\mathbb{T})$ of all Borel subsets of $\mathbb{T}$ into $\mathcal{L}(\mathcal{E})$ such that for any $a \in \mathcal{E}$ the map $\omega \to (F(\omega)a, a)$ is a positive Borel measure on $\mathbb{T}$. If for any two Borel sets ω_1, ω_2, we have $F(\omega_1 \cap \omega_2) = F(\omega_1) F(\omega_2)$ then we say that the semi-spectral measure is spectral. We shall denote usualy by F a semi-spectral measure and by **E** a spectral measure.

If $\mathcal{K}$ is a Hilbert space, E an $\mathcal{L}(\mathcal{E})$- valued spectral measure on $\mathbb{T}$, and V a bounded operator from $\mathcal{E}$ into $\mathcal{K}$, then if we put for any $\omega \in B(\mathbb{T})$, $F(\omega) = V^* E(\omega) V$, then clearly we obtain an $\mathcal{L}(\mathcal{E})$-valued semi-spectral measure on $\mathbb{T}$. Conversely, using the celebrated Naimark dilation theorem [7], if F is an $\mathcal{L}(\mathcal{E})$-valued semi-spectral measure on $\mathbb{T}$, then there exist a Hilbert space $\mathcal{K}$, a bounded operator V from $\mathcal{E}$ into $\mathcal{K}$ and an $\mathcal{L}(K)$-valued spectral measure E on $\mathcal{K}$ such that

(4.1) $$F(\omega) = V^* E(\omega) V \qquad (\omega \in B(\mathbb{T})).$$

Let us remark that any semi-spectral measure is completely-positive in the following sense : for any finite system $\varphi_1, \ldots, \varphi_m$ in $C(\mathbb{T})$ and any $a_1, \ldots, a_n$ in $\mathcal{E}$ we have

(4.2) $$\sum_{i,j} \int_{\mathbb{T}} \varphi_j \overline{\varphi_i} \, d(F(t) a_j, a_i) \geq 0 .$$

Indeed we have :

$$\sum_{i,j} \int_{\mathbb{T}} \varphi_j \overline{\varphi_i} \, d(F(t)a_j, a_i)_{\mathcal{E}} = \sum_{i,j} \int \varphi_j \overline{\varphi_i} \, d\,(E(t)Va_j, Va_i)_{\mathcal{K}} =$$
$$= \| \sum_i \int \varphi_i \, d\,E(t) a_i \|^2 \geq 0 .$$

The triplet $[\mathcal{K}, V, E]$ is called spectral dilation of F. In the supplementary condition of minimality $\mathcal{K} = \bigvee_{\omega \in B(\mathbb{T})} E(\omega) V\mathcal{E}$, the spectral dilation of F is unique up to a unitarity which conserves the operator V.

Let now $\{\mathcal{E}, \mathcal{F}, \Theta(\lambda)\}$ be an L^2-bounded analytic function and V_Θ the bounded operator from $\mathcal{E}$ into $H^2(\mathcal{F})$, associated to $\{\mathcal{E}, \mathcal{F}, \Theta(\lambda)\}$ as in section 3. We shall consider V_Θ as an operator from $\mathcal{E}$ into $L^2(\mathcal{F})$ (via the geometric isomorphism between $H^2(\mathcal{F})$ and $L^2_+(\mathcal{F})$. Let $E^\times_{\mathcal{F}}$ be the spectral measure attached to the shift operator (multiplication by e^{it}) in $L^2(\mathcal{F})$.

We shall denote by F_{Θ} the $\mathcal{L}(\mathcal{E})$- valued semi-spectral measure on $\mathbb{T}$ defined as

(4.3) $$F_{\Theta}(\omega) = V_{\Theta}^{*} E_{\mathcal{F}}^{\times}(\omega) V_{\Theta} \qquad (\omega \in B(\mathbb{T})) .$$

We shall call F_{Θ} the semi-spectral measure attached to $\{\mathcal{E}, \mathcal{F}, \Theta(\lambda)\}$.

If F is an $\mathcal{L}(\mathcal{E})$ - valued semi-spectral measure which admits as a spectral dilation a triplet $[L^2(\mathcal{F}), V, E_{\mathcal{F}}^{\times}]$ such that $V\mathcal{E} \subset L_{+}^{2}(\mathcal{F})$ then, if we construct as in section 3 the L^2-bounded analytic function $\{\mathcal{E}, \mathcal{F}, \Theta(\lambda)\}$ which verifies $V_{\Theta} = V$, then clearly $F_{\Theta} = F$.

THEOREM 1. Let $\{\mathcal{E}, \mathcal{F}, \Theta(\lambda)\}$, $\{\mathcal{E}, \mathcal{F}_1, \Theta_1(\lambda)\}$ be two L^2-bounded analytic functions, the second one being outer, F_{Θ}, F_{Θ_1} be theirs semi-spectral measures. Suppose

(4.4) $$F_{\Theta} \leq F_{\Theta_1} .$$

Then there exists a contractive analytic function $\{\mathcal{F}_1, \mathcal{F}, \Theta_2(\lambda)\}$ such that

(4.5) $$\Theta(\lambda) = \Theta_2(\lambda)\, \Theta_1(\lambda) \qquad (\lambda \in \mathbb{D}) .$$

If in (4.4) the equality holds, then $\{\mathcal{F}_1, \mathcal{F}, \Theta_2(\lambda)\}$ is inner. If moreover $\{\mathcal{E}, \mathcal{F}, \Theta(\lambda)\}$ is outer then $\{\mathcal{F}_1, \mathcal{F}, \Theta_2(\lambda)\}$ is unitary constant.

Proof. For any function $h \in H^2(\mathcal{F}_1)$ of the form $h = \sum_{0}^{n} e^{ikt} V_{\Theta_1} a_k$, let us put

(4.6) $$Qh = \sum e^{ikt} V_{\Theta} a_k$$

we have

$$\| Qh \|^2_{H^2(\mathcal{F})} = \| \sum_0^n e^{ikt} V_{\Theta} a_k \|^2 = \sum_{k,j} \int_0^{2\pi} e^{i(k-j)t} d(F_{\Theta} a_k, a_j) \leq$$
$$\leq \sum_{k,j} \int_0^{2\pi} e^{i(k-j)t} d(F_{\Theta_1} a_k, a_j) = \| h \|^2_{H^2(\mathcal{F}_1)} .$$

We have used here (4.4) in the completely-positivity form (4.2). Since $\{\mathcal{E}, \mathcal{F}_1, \Theta_1(\lambda)\}$, is outer it results that (4.6) gives rise to a contraction Q from $H^2(\mathcal{F}_1)$ into $H^2(\mathcal{F})$. Clearly $Qe^{it} = e^{it}Q$. From Lemma Q it results that there exists a contractive analytic function $\{\mathcal{F}_1, \mathcal{F}, \Theta_2(\lambda)\}$ such that

$$(Qh)(\lambda) = \Theta_2(\lambda) h(\lambda) \qquad (h \in H^2(\mathcal{F}_1)).$$

If we take $h(\lambda) = \Theta_1(\lambda) a$ we obtain

$$\Theta(\lambda) a = (V_{\Theta} a)(\lambda) = (QV_{\Theta_1} a)(\lambda) = \Theta_2(\lambda)\, \Theta_1(\lambda)\, a.$$

If in (4.4) equality holds then clearly $Q = \Theta_{2+}$ is an isometry, thus $\{\mathcal{F}_1, \mathcal{F}, \Theta_2(\lambda)\}$ is inner.

If moreover $\{\mathcal{E}, \mathcal{F}, \Theta(\lambda)\}$ is outer, then clearly $\{\mathcal{F}_1, \mathcal{F}, \Theta_2(\lambda)\}$ is outer, thus it is unitary constant.

THEOREM 2. Let F be an $\mathcal{L}(\mathcal{E})$ - valued semi-spectral measure on $\mathbb{T}$ and $[\mathcal{K}, V, E]$ its minimal spectral dilation. There exists a unique L^2-bounded outer function $\{\mathcal{E}, \mathcal{F}_1, \Theta_1(\lambda)\}$ with the properties :

(1) $$F_{\Theta_1} \leq F .$$

(2) For any L^2-bounded analytic function $\{\mathcal{E}, \mathfrak{F}, \Theta(\lambda)\}$ such that $F_\Theta \leq F$, we have also $F_\Theta \leq F_{\Theta_1}$.

The equality holds in (1) if and only if

(4.7) $$\bigcap_{n\geq 0} U^n \mathcal{K}_+ = \{0\}$$

where U is the unitary operator which corresponds to the spectral measure E and $\mathcal{K}_+ = \bigvee_0^\infty U^n V\mathcal{E}$.

Proof. Aplying Wold decomposition (cf [12]) to the isometry $U_+ = U|\mathcal{K}_+$ we obtain

$$\mathcal{K}_+ = \bigoplus_0^\infty U^n \mathfrak{F}_1 \oplus \mathcal{R} ,$$

where $\mathfrak{F}_1 = \mathcal{K}_+ \ominus U\mathcal{K}_+$ and $\mathcal{R} = \bigcap_{n\geq 0} U^n \mathcal{K}_+$.

Clearly

$$\mathcal{K} = \bigoplus_{-\infty}^{+\infty} U^n \mathfrak{F}_1 \oplus \mathcal{R} .$$

Let P be the orthogonal projection of $\mathcal{K}$ onto the subspace $\bigoplus_{-\infty}^{+\infty} U^n \mathfrak{F}_1$. Then we have

(4.8) $$PU = UP.$$

Denote by $X_{\mathfrak{F}_1}$ the canonical isomorphism between $\bigoplus_{-\infty}^{+\infty} U^n \mathfrak{F}_1$ and $L^2(\mathfrak{F}_1)$ (Fourier representation), and define $V_{\Theta_1} : \mathcal{E} \to L^2(\mathfrak{F}_1)$ by

$$V_{\Theta_1} a = X_{\mathfrak{F}_1} PVa \qquad (a \in \mathcal{E}).$$

Clearly then $V_{\Theta_1}\mathcal{E} \subset L^2_+(\mathfrak{F}_1)$ and

$$\bigvee_0^\infty e^{int} V_{\Theta_1}\mathcal{E} = \bigvee_0^\infty e^{int} X_{\mathfrak{F}_1} PV\mathcal{E} = X_{\mathfrak{F}_1} \bigvee_0^\infty U^n PV\mathcal{E} = X_{\mathfrak{F}_1} P \bigvee_0^\infty U^n V\mathcal{E} =$$
$$= X_{\mathfrak{F}_1} P\mathcal{K}_+ = X_{\mathfrak{F}_1} \bigoplus_0^\infty U^n \mathfrak{F}_1 = L^2_+(\mathfrak{F}_1) .$$

We obtain that the L^2-bounded analytic function $\{\mathcal{E}, \mathfrak{F}_1, \Theta_1(\lambda)\}$ corresponding to V_{Θ_1} as in section 3, is outer. For any analytic polynomial p we have

$$\int_0^{2\pi} |p|^2 d(F_{\Theta_1} a, a)_{\mathcal{E}} = \|pV_{\Theta_1} a\|^2_{L^2(\mathfrak{F}_1)} = \|pX_{\mathfrak{F}_1} PVa\|^2_{L^2(\mathfrak{F}_1)} = \|p(U)PVa\|^2_{\mathcal{K}} =$$
$$= \|Pp(U)Va\|^2_{\mathcal{K}} \leq \|p(U)Va\|^2 = \int_0^{2\pi} |p|^2 d(Fa, a)_{\mathcal{E}} .$$

Thus

$$F_{\Theta_1} \leq F.$$

We have equality iff $X_{\mathfrak{F}_1} PV = X_{\mathfrak{F}_1} V$, i.e. iff $PV = V$, i.e. iff $\mathcal{R} = \{0\}$, i.e. iff $\bigcap_{n\geq 0} U^n \mathcal{K}_+ = \{0\}$.

Let now $\{\mathcal{E}, \mathfrak{F}, \Theta(\lambda)\}$ be another L^2-bounded analytic function such that $F_\Theta \leq F$. Let us put for an element $k \in \mathcal{K}_+$ of the form $k = \sum_0^n U^k V a_k$

(4.9) $$Xk = \sum e^{ikt} V_\Theta a_k .$$

We have

$$\|Xk\|^2 = \sum_{\ell, j} \int_0^{2\pi} e^{i(\ell - j)t} d(F_\Theta(t) a_\ell, a_j) \leq$$

$$\leq \sum_{\ell,j} \int e^{i(\ell-j)t} d(F(t)a_\ell, a_j) = \| \sum_j U^j V a_j \|^2 = \|k\|^2.$$

Thus (4.9) gives rise to a contraction X from $\mathcal{K}_+$ into $L^2_+(\mathfrak{F})$ such that

$$XU = e^{it}X.$$

We have

$$X\mathcal{R} = X \bigcap_n U^n \mathcal{K}_+ \subset \bigcap_n X U^n \mathcal{K}_+ = \bigcap_n e^{int} X \mathcal{K}_+ \subseteq \bigcap_n e^{int} L^2_+(\mathfrak{F}) = \{0\}.$$

Thus XP = X. We have then for any analytic polynomial $\underline{p}$

$$\int |p|^2 d(F_\Theta a, a) = \|pV_\Theta a\|^2 = \|X p(U) V a\|^2 = \|XPp(U) V a\|^2 \leq \|Pp(U) V a\|^2 =$$

$$= \|X_{\mathfrak{F}_1} Pp(U) V a\|^2 = \|p(e^{it}) X_{\mathfrak{F}_1} P V a\|^2 = \|p V_{\Theta_1} a\|^2 = \int |p(e^{it})|^2 d(F_{\Theta_1} a, a)$$

i.e. $\qquad F_\Theta \leq F_{\Theta_1}$.

Clearly any L^2-bounded outer functions $\{\mathcal{E}, \mathfrak{F}_1', \Theta_1'(\lambda)\}$ which verifies (1) and (2), verifies also $F_{\Theta_1'} = F_{\Theta_1}$ and from Theorem 1 it results that they differ by a unitary constant factor.

COROLLARY. <u>Any</u> L^2-<u>bounded analytic function</u> $\{\mathcal{E}, \mathfrak{F}, \Theta(\lambda)\}$ <u>has a unique factorization of the form</u>

$$\Theta(\lambda) = \Theta_i(\lambda)\Theta_e(\lambda) \qquad\qquad (\lambda \in \mathbb{D})$$

<u>in the inner and the outer parts.</u>

An L^2-bounded outer function $\{\mathcal{E}, \mathfrak{F}, \Delta(\lambda)\}$ will be called <u>evaluation function</u> of $\mathcal{E}$ in $\mathfrak{F}$ if V_Θ is an isometry from $\mathcal{E}$ into $H^2(\mathfrak{F})$.

PROPOSITION 1. <u>An</u> L^2-<u>bounded analytic function</u> $\{\mathcal{E}, \mathfrak{F}, \Delta(\lambda)\}$ <u>is an evaluation function if and only if</u> $\mathcal{E}$ <u>can be isometricaly embeded in</u> $H^2(\mathfrak{F})$ <u>as a cyclic subspace for the shift operator in</u> $H^2(\mathfrak{F})$ <u>such that</u>

$$\Theta(\lambda)a = a(\lambda) \qquad\qquad (\lambda \in \mathbb{D}) .$$

In this case we have necessary $\dim \mathfrak{F} \leq \dim \mathcal{E}$.

<u>Proof.</u> If $\mathcal{E}$ is a cyclis subspace of $H^2(\mathfrak{F})$ then clearly (4.8) defines an evaluation function $\{\mathcal{E}, \mathfrak{F}, \Theta(\lambda)\}$. Conversely, if $\{\mathcal{E}, \mathfrak{F}, \Theta(\lambda)\}$ is an evaluation function then V_Θ is an isometricaly embeding of $\mathcal{E}$ in $H^2(\mathfrak{F})$. Since $\{\mathcal{E}, \mathfrak{F}, \Theta(\lambda)\}$ is by definition outer, then $V_\Theta\mathcal{E}$ is a cyclic subspace for the shift operator on $H^2(\mathfrak{F})$.

Let $\mathcal{E}$ be a cyclic subspace in $H^2(\mathfrak{F})$ and denote by P the orthogonal projection of $H^2(\mathfrak{F})$ on $\mathfrak{F}$. If $f \in \mathfrak{F}$ and for any $a \in \mathcal{E}$ we have

$$(f, Pa) = 0$$

then clearly $(f, \lambda^n a) = 0$ for any $n \geq 0$ and $a \in \mathcal{E}$. From the cyclicity of $\mathcal{E}$ it results f=0. It results that $\overline{P\mathcal{E}} = \mathfrak{F}$, i.e. $\dim \mathfrak{F} \leq \dim \mathcal{E}$.

THEOREM 3. Any L^2-contractive analytic function can be factorized in the form

$$\Theta(\lambda) = M(\lambda)\Delta(\lambda) \qquad \lambda \in \mathbb{D}, \tag{4.10}$$

where $\{\mathcal{E}, \mathcal{E}_1, \Delta(\lambda)\}$ is an evaluation function and $\{\mathcal{E}_1, \mathcal{F}, M(\lambda)\}$ is a contractive analytic function.

Proof. Denote by $D_\Theta = [I - V_\Theta^* V_\Theta]^{1/2}$ and put

$$dF = dF_\Theta + D_\Theta^2\, dt$$

Since $F_\Theta(\mathbb{T}) = V_\Theta^* V_\Theta$ we have $F(\mathbb{T}) = I$. Moreover $F = F_\Omega$ where $\{\mathcal{E}, \mathcal{F}_1, \Omega(\lambda)\}$ is the L^2-bounded analytic function defined as

$$\Omega(\lambda) = \Theta(\lambda) \oplus D_\Theta$$

and $\mathcal{F}_1 = \mathcal{F} \oplus \overline{D_\Theta \mathcal{E}}$. Let $\{\mathcal{E}, \mathcal{E}_1, \Delta(\lambda)\}$ be the outer part of $\{\mathcal{E}, \mathcal{F}_1, \Omega(\lambda)\}$. Then clearly $F_\Omega = F_\Delta$ and consequently $V_\Delta^* V_\Delta = F_\Delta(\mathbb{T}) = I$ and $F_\Theta \leq F_\Delta$. Applying Theorem 1, we obtain the desired factorization.

5. Boundary limits

In section 4, we attached to any L^2-bounded analytic function the bounded operator V_Θ from $\mathcal{E}$ into $L^2_+(\mathcal{F})$ and the semi-spectral measure F_Θ. If F is an $\mathcal{L}(\mathcal{E})$-valued semi-spectral measure, we attached to F its maximal outer function $\{\mathcal{E}, \mathcal{F}, \Theta_1(\lambda)\}$ such that $F_{\Theta_1} \leq F$. In this way we obtained elements for both theorems Fatou and Szegö. We need "only" a desintegration for the operator V_Θ of the following type : there exists a strongly (or in norm) measurable function $t \longrightarrow \Theta(e^{it})$ defined on $\mathbb{T}$ with values in $\mathcal{L}(\mathcal{E}, \mathcal{F})$ such that for any $a \in \mathcal{E}$ we have :

$$(V_\Theta a)(e^{it}) = \Theta(e^{it})a \qquad \text{a.e.} \tag{5.1}$$

In this case, clearly we have $\Theta(\lambda) \longrightarrow \Theta(e^{it})$ strongly a.e. when $\lambda \longrightarrow e^{it}$ nontangentially,

$$dF_\Theta = \Theta(e^{it})^* \Theta(e^{it})\, dt \tag{5.2}$$

and

$$\Theta(re^{it}) = \frac{1}{2\pi} \int_0^{2\pi} P_r(t-s)\, \Theta(e^{is})\, ds \tag{5.3}$$

in strong sense.

If in Theorem 2 we have $F_{\Theta_1} = F$ then

$$dF = \Theta_1(e^{it})^* \Theta_1(e^{it})\, dt$$

which corresponds to Szegö factorization theorem.

In case $\{\mathcal{E}, \mathcal{F}, \Theta(\lambda)\}$ is a bounded function, we have a such desintegration for V_Θ in strong sens (cf.[12]). Indeed, let $\{a_n\}$ be a dense set in $\mathcal{E}$ and ω a total set in $\mathbb{T}$, such that

$\Theta(\lambda)a_n \longrightarrow (V_\Theta a_n)(e^{it})$ for any $t\in\omega$ and a_n, as λ tends to e^{it} non-tangentially. Since $\|\Theta(\lambda)\| \le M$ then $\Theta(\lambda)$ tends strongly to a bounded operator $\Theta(e^{it})$ when λ tends nontangentially to e^{it}. Then the function $t \longrightarrow \Theta(e^{it})$ is clearly strongly measurable and $(V_\Theta a)(t) = \Theta(e^{it})a$ a.e. Even in this case $t \longrightarrow \Theta(e^{it})$ is not necessary norm-measurable, thus $\Theta(e^{it})$ is not a boundary function in norm sense.

For a general L^2-bounded analytic function $\Theta(\lambda)$ does not exist a desintegration for the operator V_Θ. If, for example, $\{\mathcal{E}, \mathcal{F}, \Theta(\lambda)\}$ is the evaluation function, $\mathcal{E} = H^2$, $\mathcal{F} = \mathbb{C}$ and $\Theta(\lambda) = a(\lambda)$, and $t \longrightarrow \Theta(e^{it})$ is the boundary function (in strong sense) for $\{\mathcal{E}, \mathcal{F}, \Theta(\lambda)\}$, then it results that there exists a total set ω in $\mathbb{T}$ such that any function a in H^2 has a radial limit when $\lambda \longrightarrow e^{it}$, $t\in\omega$, which is clearly impossible.

Theorem 3 permits us to reduce the difficulties in construction of boundary limit to such a type of obstructions.

THEOREM 4. Let $\{\mathcal{E}, \mathcal{F}, \Theta(\lambda)\}$ be an L^2-contractive analytic function. Then there exist a Hilbert space $\mathcal{E}_1$ and a bounded analytic function $\{\mathcal{E}_1, \mathcal{F}, M(\lambda)\}$ such that $\mathcal{E}$ can be isometrically embedded in $H^2(\mathcal{E}_1)$,

$$\Theta(\lambda)a = M(\lambda)a(\lambda) \qquad a\in\mathcal{E}\subset H^2(\mathcal{E}_1)$$

and

$$(V_\Theta a)(e^{it}) = M(e^{it})a(e^{it}) \qquad \text{a.e.}$$

where $t \longrightarrow M(e^{it})$ is the boundary function of $\{\mathcal{E}_1, \mathcal{F}, M(\lambda)\}$.

We can interpret $t \longrightarrow M(e^{it})$ as a boundary function of $\{\mathcal{E}, \mathcal{F}, \Theta(\lambda)\}$ (modulo the evaluation of $\mathcal{E}$ into $H^2(\mathcal{E}_1)$). If $\mathcal{E}_1\subset\mathcal{E}$, then $M(\lambda) = \Theta(\lambda)|\mathcal{E}_1$, thus $\Theta(\lambda)$ has strong boundary limit on the subspace of constant functions in $\mathcal{E}$. In general we obtained a desintegration for V_Θ by composing the simultan desintegration of the elements of $\mathcal{E}$ as analytic functions on $\mathcal{E}_1$ and of V_M as a multiplication by $\mathcal{L}(\mathcal{E}_1, \mathcal{F})$-valued strongly measurable and bounded function $t \longrightarrow M(e^{it})$ on $L^2_+(\ )$.

If an L^2-bounded analytic function $\{\mathcal{E}, \mathcal{F}, \Theta(\lambda)\}$ has boundary limit in the Fatou (strong) sense, then clearly the operator $\Theta : D(\Theta) \subset L^2(\mathcal{E}) \longmapsto L^2(\mathcal{F})$ constructed as in the last part of section 3 is closable. The converse assertion seems to be an interesting probleme whose solution we don't know up to now.

As we already remarked, Szegö Theorem for an $\mathcal{L}(\mathcal{E})$- valued semi-spectral measure F consists- modulo the above discussions-actually characterises in terms of F the fact that in the factorization theorem (Theorem 2) one has $F_{\Theta_1} = F$ (or at least $\Theta_1 \neq 0$).

Let F be an $\mathcal{L}(\mathcal{E})$ - valued semi-spectral measure on $\mathbb{T}$ and $[\mathcal{K}, V, U]$ its unitary dilation. Let us put

(5.4) $$(\Delta[F]a,a) = \inf \sum_{k,j=0}^{n} \int_0^{2\pi} e^{i(j-k)t} d(F(t)a_j, a_k)$$

where the infimum is taken over all finite system $a_0, a_1, \ldots, a_n$ in $\mathcal{E}$ such that $a_0 = a$.

We have

$$(\Delta[F]a,a) = \inf_{a_0=a, a_1, \ldots, a_n \in \mathcal{E}} \sum_{k,j} \int_0^{2\pi} e^{i(k-j)t} d(F(t)a_k, a_j)_{\mathcal{E}} =$$

$$= \inf_{a_0=a, a_1, \ldots, a_n \in \mathcal{E}} \sum_{k,j} \int_0^{2\pi} e^{i(k-j)t} d(E(t)Va_k, Va_j)_{\mathcal{K}} = \inf_{a_0=a, a_1, \ldots, a_n \in \mathcal{E}} \sum (U^{k-j}Va_k, Va_j)_{\mathcal{K}} =$$

$$= \inf_{a_1, \ldots, a_n \in \mathcal{E}} \| Va - \sum_{k=1}^{n} U^k Va_k \|^2 = \| (I-P_1) Va \|^2 = (V^*(I-P_1)Va, a)$$

where P_1 is the orthogonal projection of $\mathcal{K}$ on $\bigvee_1^\infty U^k V\mathcal{E}$.

It results that $\Delta[F]$ is a positive operator on $\mathcal{E}$, and we call it the Szegö (or prediction - error) operator of F. The name is justified because if $F = \mu$ is scalar valued, then

$$\Delta[\mu] = \inf_{p_0} \int_0^{2\pi} |1 - p_0|^2 d\mu$$

where the infimum is taken over all analytic polynomial p_0 which vanish in origin.

Now we can state the following generalization of Szegö-Komogorov-Krein theorem:

THEOREM 5. Let F be an $\mathcal{L}(\mathcal{E})$ - valued semi-spectral measure on $\mathbb{T}$ and $\Delta[F]$ its prediction-error operator. Then we have :

(i) $\Delta[F] = 0$ if and only if does not exist an L^2-bounded analytic function $\{\mathcal{E}, \mathcal{F}, \Theta(\lambda)\}$ $\Theta(\lambda) \not\equiv 0$ such that $F_\Theta \leq F$.

(ii) If $\Delta[F] \neq 0$ there exists an unique maximal (in the sense of Theorem 2) outer L^2-bounded analytic function $\{\mathcal{E}, \mathcal{F}_1, \Theta_1(\lambda)\}$ such that $F_{\Theta_1} \leq F$, $\dim \mathcal{F}_1 = \dim (\Delta\mathcal{E})$, and

$$\Delta[F] = \Delta[F_\Theta] = \Theta(0)^* \Theta(0) .$$

Proof. If $\{\mathcal{E}, \mathcal{F}, \Theta(\lambda)\}$ is an L^2-bounded analytic function and $\{\mathcal{E}, \mathcal{F}_1, \Theta_1(\lambda)\}$ its outer part then $F_\Theta = F_{\Theta_1}$ and

$$(\Delta[F_\Theta]a,a) = (\Delta[F_{\Theta_1}]a,a) = \inf_{a_1,\ldots,a_n \in \mathcal{E}} \| V_{\Theta_1} a - \sum_{k=1}^{n} e^{ikt} V_{\Theta_1} a_k \|^2 =$$

$$= \inf_{h \in H^2(\mathcal{F}_1),\, h(0)=0} \| V_{\Theta_1} a - h \|^2 = \| (V_{\Theta_1} a)(0) \|^2 = \| \Theta_1(0) a \|^2$$

Hence if $F_\Theta \leq F$ we have

$$\Theta_1(0)^* \Theta_1(0) = \Delta[F_\Theta] \leq \Delta[F] .$$

Let now $\{\mathcal{E}, \mathcal{F}_1, \Theta_1(\lambda)\}$ be the maximal outer function of F given by Theorem 2. Then if P is the orthogonal projection of $\mathcal{K}$ onto $\bigoplus_{-\infty}^{+\infty} U^n \mathcal{F}_1$ we have

$$(I - P_1)P = P(I - P_1) = I - P_1 .$$

Therefore

$$(\Delta[F]a, a) = \|(I - P_1)Va\|^2 = \|(I - P_1)PVa\|^2 =$$

$$= \|(P - P_1P)PVa\|^2 = (\Delta[F_{\Theta_1}]a, a)$$

$$\Delta[F] = \Delta[F_{\Theta_1}] = \Theta_1(0)^* \Theta_1(0).$$

REFERENCES

[1] Fatou, P., Séries trigonométriques et séries de Taylor, Acta Math., vol. 30, 1906.

[2] Gohberg, I.C. and Krein, M.G., Teoria volterovîh operatorov v ghilbertovom prostranstve i ee prilojenia, Mockva, 1967.

[3] Helson, H., Lectures on invariant subspaces, New–York, London, 1964.

[4] Helson, H. and Lowdenslager, D.B., Prediction theory and Fourier series in several variables I and II, Acta Math., 99 (1958), 165-202; 106(1961), 175-213.

[5] Hoffman, K., Banach spaces of analytic functions, Engelwood Cliffs, N.J. 1962.

[6] Lowdenslager, D.B., On factoring matrix valued functions, Ann. of Math., 78 (1963), 450-454.

[7] Naimark, M.A., Positive definite operator functions on a commutative group, Bull. (Izvestia) Acad. Sci. URSS (ser. math.), 7 (1943), 237-244.

[8] Rozanov, Yu.A., Stationary random processes, Holden - Day, Inc., San Francisco, Calif. 1967.

[9] Suciu, I. and Valuşescu, I., Factorization of semi-spectral measures, Rev. Roum. Math. Pures et Appl. XXI, No. 6, 773-793 (1976).

[10] Suciu, I. and Valuşescu, I., Essential parameters in prediction, Rev. Roum. Math. Pures et Appl. (to appear).

[11] Szegö, G., Uber die Randwerte analytischer Functionen, Math. Ann., 84 (1921), 232-244.

[12] Sz.-Nagy, B. and Foiaş, C., Harmonic Analysis of Operators on Hilbert Space Acad. Kiadó Budapest - North Holland Company-Amsterdam-London, 1970.

[13] Wiener, N. and Masani P., The prediction theory of multivariate stochastic processes I and II, Acta Math. 98 (1957), 111-150 99 (1958), 93-139.

[14] Wold, H., A study in the analysis of stationary time series, Stocholm 1938, 2-nd ed., 1954.

INCREST
Bucharest

ON A THEOREM OF P.LÉVY

by

Silviu Teleman

The following theorem of P.Lévy plays an important role in probability theory :

Let $(F_n)_{n\geq 0}$ be a sequence of distribution functions and $(\varphi_n)_{n\geq 0}$ the corresponding sequence of their characteristic functions. If $\lim_{n\to\infty} \varphi_n(s) = \varphi(s)$, for any $s \in \mathbb{R}$, where φ is the characteristic function of the distribution function F , then

$$\lim_{n\to\infty} \int_{\mathbb{R}} f\, dF_n = \int_{\mathbb{R}} f\, dF$$

for any bounded, continuous function $f : \mathbb{R} \to \mathbb{C}$ (see [4], Ch.IV, § 12.2 for a proof).

The preceding theorem obviously belongs to Harmonic Analysis, since the characteristic functions φ_n are the Fourier-Stieltjes transforms of the Radon measures μ_n which correspond to the distribution functions F_n, $n \in \mathbb{N}$.

It is therefore only natural to extend the theorem to Abstract Harmonic Analysis. It is P.Martin-Löf who first tried such an extension in 1965 (see [5] ; and also

[3] , p.153) but since in [5] Tomita's disintegration theory, as exposed in [6] , was used, theory which proved not to hold in general (see [7]), the extension obtained in [5] remained under doubt.

In what follows we intend to give a corrected proof to the generalization of P.Lévy's theorem, stated in [5] , by using the disintegration theory from [8] . The difficulties which arise in the proof are due to the fact that one has to distinguish between Borel and Baire measurable sets.

1. Let G be any topological, locally compact group, and $U : G \to \mathcal{L}(H)$ a unitary, continuous representation of G , where H is an arbitrary, complex, Hilbert space.

For any $\xi, \eta \in H$ the complex function

$$g \mapsto (U_g \xi \mid \eta) , \quad g \in G ,$$

is continuous and bounded on G .

Let $\mathcal{M}^1(G)$ be the complex vector space of all finite, complex Radon measures on G . For any $\mu \in \mathcal{M}^1(G)$ the integral

$$\int_G (U_g \xi \mid \eta) \, d\mu(g) , \quad \xi, \eta \in H,$$

is defined and it is easy to see that it exists, and it is unique a continuous linear operator $\hat{\mu}(U) \in \mathcal{L}(H)$, such that

$$(\hat{\mu}(U) \xi \mid \eta) = \int_G (U_g \xi \mid \eta) \, d\mu(g),$$

for any $\xi, \eta \in H$. The operator $\hat{\mu}(U)$ is the Fourier-Stieltjes transform of the measure μ by the representation U (see [2], §13.3 ; [3], p.147), whereas the mapping $\mu \mapsto \hat{\mu}(U)$ is a $*$- representation of the involutive algebra $\mathcal{M}^1(G)$ into $\mathcal{L}(H)$ (see [2], proposition 13.3. 1).

It is well known that on $\mathcal{M}^1(G)$ the following topologies can be considered.

a) The vague topology : since $\mathcal{M}^1(G)$ is the conjugate space to the normed space $\mathcal{K}(G)$ of all continuous, complex functions, defined on G and having compact supports, endowed with the norm

$$f \mapsto \|f\|_\infty = \sup_{s \in G} |f(s)|, \quad f \in \mathcal{K}(G),$$

the vague topology is the topology

$$\sigma(\mathcal{M}^1(G); \mathcal{K}(G)).$$

b) The $*$-weak topology : $\mathcal{M}^1(G)$ is, at the same time, the conjugate space to the Banach space $C_0(G)$ of all continuous, complex functions, defined on G and vanishing at infinity, endowed with the same norm as before. The $*$-weak topology is the topology $\sigma(\mathcal{M}^1(G); C_0(G))$, and it is obviously stronger than the vague topology. The two topologies coincide on the norm-bounded subsets of $\mathcal{M}^1(G)$.

c) The narrow topology : any measure from $\mathcal{M}^1(G)$

can be extended, by use of regularity, as a Radon measure on the space $C^b(G) = C(\beta G)$ of all bounded continuous, complex functions, defined on G, space which can be canonically identified with the space of all continuous complex functions which are defined on the Stone-Čech compactification βG of G. The mentioned extension provides an isometric embedding of $\mathcal{M}^1(G)$ into $C(\beta G)^*$, and the narrow topology is the topology induced by $\sigma(C(\beta G)^*; C(\beta G))$. Obviously, the narrow topology is stronger than the $*$-weak topology (see [1], ch.IX §5.3).

d) The Fourier topology. Let $\mathcal{F}(G) \subset C^b(G)$ be the complex vector space of all functions of the form

$$g \mapsto \sum_{i=1}^{n} (U_g^{(i)} \xi_i \mid \eta_i)$$

where $U^{(i)} : G \to \mathcal{L}(H_i)$ are irreducible, continuous, unitary representations, whereas $\xi_i, \eta_i \in H_i$, $i = 1, 2, \ldots, n$; $n \in \mathbb{N}$. The Fourier topology on $\mathcal{M}^1(G)$ is the topology $\sigma(\mathcal{M}^1(G); \mathcal{F}(G))$. Obviously, the Fourier topology is weaker than the narrow topology (see [3], p.152), but, nevertheless, it is a Hausdorff topology.

2. In what follows we shall use the notations and the results from [8], §3.

Let $\mathcal{C}$ be a C^*-algebra (possibly without unit element) and $\pi : \mathcal{C} \to \mathcal{L}(H)$ a cyclic representation of $\mathcal{C}$ on the Hilbert space H. Let $x_0 \in H$ be the considered cyclic vector, such that $\|x_0\| = 1$.

Let $\mathcal{Z} \subset (\pi(\mathcal{C}))'$ be a maximal Abelian, von Neumann subalgebra and $\mathcal{B}$ the C^*-algebra generated by $\pi(\mathcal{C})$ and by $\mathcal{Z}$. Then $\mathcal{B}$ is the norm closure of the set of all sums of the form

$$z_0 + \sum_{i=1}^{n} z_i \pi(c_i),$$

where $z_i \in \mathcal{Z}$, $i = 0, 1, 2, .., n$ and $c_i \in \mathcal{C}$, $i = 1, 2, .., n$. We obviously have $\mathcal{B}' = \mathcal{Z}$ and $\mathcal{Z} \subset \mathcal{B} \subset \mathcal{Z}'$; in [8] it was proved that for any $p \in P(\mathcal{B})$ one has the equality

(1) $$\pi_p(\mathcal{B}) = \pi_p(\pi(\mathcal{C})) + \mathbb{C}\mathbb{1}_{H_p}$$

where π_p is the irreducible representation of $\mathcal{B}$ associated to the pure state p, H_p is the space of the representation π_p, whereas $\mathbb{1}_{H_p}$ is the identical operator on H_p.

From formula (1) it immediately follows that the representation $\pi_p \circ \pi$ is irreducible, maybe degenerate (v. [8], lemma 3.1).

<u>Lemma 1.</u> <u>For any</u> $p \in P(\mathcal{B})$ <u>we have</u> $p \circ \pi \in \in P(\mathcal{C}) \cup \{0\}$.

<u>Proof</u>. Let $\xi_p^0 \in H_p$ be the cyclic vector associated to the pure state p and f_p the pure state defined on $\pi_p(\mathcal{B})$ by

$$f_p(\pi_p(b)) = (\pi_p(b)\xi_p^0 \mid \xi_p^0) = p(b), \quad b \in \mathcal{B}.$$

Since $\pi_p(\pi(\mathcal{C}))$ is a norm closed two-sided ideal of $\pi_p(\mathcal{B})$, from proposition 2.11.7 from [2] it follows that there exists a decomposition of the form

(2) $$f_p = f_p' + f_p'',$$

where f_p', f_p'' are positive linear functionals on $\pi(\mathcal{B})$, such that

$$\| f_p' \| = \| f_p' \mid \pi_p(\pi(\mathcal{C})) \|$$

and

$$f_p'' \mid \pi_p(\pi(\mathcal{C})) = 0.$$

Since f_p is pure, from (2) we infer that there exists a number $\lambda \in [0,1]$, such that

$$f_p' = \lambda f_p, \quad f_p'' = (1-\lambda) f_p.$$

It follows that

$$\lambda = \lambda \| f_p \| = \lambda \| f_p \mid \pi_p(\pi(\mathcal{C})) \|$$

and

$$(1-\lambda) \| f_p \mid \pi_p(\pi(\mathcal{C})) \| = 0.$$

Consequently, $\lambda \neq 0$ implies $\lambda = 1$; i.e., $\lambda = 0$ implies that $p \circ \pi = 0$ and $\lambda \neq 0$ implies that

$\|p \circ \pi\| = 1$. Q.E.D.

Let $E_1(\mathcal{B}) = \{ f \in E(\mathcal{B}) ; \| f \backslash \pi(\mathcal{C}) \| = 1 \}$. Obviously $E_1(\mathcal{B})$ is a convex subset of $E(\mathcal{B})$. We shall denote $P_1(\mathcal{B}) = P(\mathcal{B}) \cap E_1(\mathcal{B})$ and

$P_0(\mathcal{B}) = \{ p \in P(\mathcal{B}); \ p \circ \pi = 0 \}$.

Let α be the central measure corresponding to the state $f_0 \in E(\mathcal{B})$, given by

$$f_0(b) = (b x_0 \backslash x_0), \quad b \in \mathcal{B}.$$

We obviously have $f_0 \in E_1(\mathcal{B})$.

<u>Lemma 2.</u> <u>There exists a convex subset</u> $Q_1 \subset E_1(\mathcal{B})$, <u>which is Baire measurable and such that</u> $f_0 \in Q_1$ <u>and</u> $\alpha(Q_1) = 1$.

<u>Proof.</u> Let $\{u_i\}_{i \in I}$ be an approximate unit in $\mathcal{C}$. Then (see [2], proposition 2.1.5) we have

$$\lim_i f_0(\pi(u_i)) = 1 ;$$

hence for any $n \in \mathbb{N}^*$ there exists an $i_n \in I$, such that

$$(3) \qquad 1 - \frac{1}{n} < f_0(\pi(u_{i_n})).$$

By induction we choose a sequence $(j_n)_{n \in \mathbb{N}^*}$ of indices from I, such that $j_1 = i_1$ and

$$i_n \leq j_n \leq j_{n+1}, \qquad n \in \mathbb{N}^*.$$

It follows that $\pi(u_{j_n}) \leq \pi(u_{j_{n+1}})$ and therefore, we have

$$0 \leq \lambda(\pi(u_{i_n})) \leq \lambda(\pi(u_{j_n})) \leq \lambda(\pi(u_{j_{n+1}})) \leq 1.$$

Let $\varphi = \lim_{n\to\infty} \lambda(\pi(u_{j_n}))$. Then φ is an affine, Baire measurable function $\varphi : E(\mathcal{B}) \to [0,1]$ and we have

(4) $$\int_{E(\mathcal{B})} \varphi \, d\alpha = 1.$$

Let $Q_1 = \{ f \in E(\mathcal{B}) \; ; \; \varphi(f) = 1 \}$. From (4) we infer that we have $\alpha(Q_1) = 1$. On the other hand, for any $f \in Q_1$ we have

$$\lim_i f(\pi(u_i)) = 1 ,$$

and this implies that $\| f \circ \pi \| = 1$; consequently , we have $f \in E_1(\mathcal{B})$. Obviously, Q_1 is a convex , Baire measurable set and from (3) we infer that $f_0 \in Q_1$. The lemma is proved.

Corollary 1. $\beta(P_0(\mathcal{B})) = 0$.

Proof. With the notations from lemma 2 we have

(5) $$1 = \alpha(Q_1) = \beta(Q_1 \cap P(\mathcal{B}))$$

and

(6) $$Q_1 \cap P(\mathcal{B}) \subset P_1(\mathcal{B}).$$

Since the σ-algebra $\mathcal{A}$ is complete, from (5) and (6) we infer that $P_1(\mathcal{B}) \in \mathcal{A}$ and $\beta(P_1(\mathcal{B})) = 1$. Consequently, we have $P_0(\mathcal{B}) \in \mathcal{A}$ and $\beta(P_0(\mathcal{B})) = 0$. Q.E.D.

<u>Corollary 2.</u> $E_1(\mathcal{B})$ is α -<u>measurable and</u> $\alpha(E_1(\mathcal{B}))=1$.

<u>Proof.</u> It is an immediate consequence of lemma 2.

<u>Corollary 3. Any bounded, continuous function</u> $t: E_1(\mathcal{B}) \to \mathbb{C}$ is α -<u>measurable; the function</u> $t|P_1(\mathcal{B})$ <u>is</u> β -<u>measurable and the equality</u>

$$\int_{E_1(\mathcal{B})} t \, d\alpha = \int_{P_1(\mathcal{B})} t \, d\beta$$

<u>holds.</u>

<u>Proof.</u> Since the measure α is regular, there exists an increasing sequence $(K_n)_{n\in\mathbb{N}}$ of compact sets $K_n \subset E_1(\mathcal{B})$, such that $\alpha(K_n)\uparrow 1$.

Let $K=\bigcup_{n\geq 0} K_n$ and $t: E_1(\mathcal{B})\to\mathbb{C}$ be defined by

$$t_n(f) = \begin{cases} t(f), & f\in K_n \\ 0, & f\in E_1(\mathcal{B})\setminus K_n . \end{cases}$$

The functions t_n are Borel measurable and we have

$$\lim_{n\to\infty} t_n = t\chi_K .$$

It follows that t is α-measurable, because

$$\alpha(E_1(\mathcal{B})\setminus K)=0 .$$

From lemma 2 we now immediately infer that

$$\int_{E_1(\mathcal{B})} t\, d\alpha = \int_{Q_1} t\, d\alpha .$$

Since the set Q_1 is Baire measurable, it follows that there exists an increasing sequence of compact sets $L_n \subset Q_1$, $n \in \mathbb{N}$, which are Baire measurable and such that $\alpha(L_n) \uparrow 1$. Since $t \mid L_n$, $n \in \mathbb{N}$, are continuous functions, the functions $s_n : E(\mathcal{B}) \to \mathbb{C}$, $n \in \mathbb{N}$, given by

$$s_n(f) = \begin{cases} t(f), & f \in L_n \\ 0, & f \in E(\mathcal{B}) \setminus L_n \end{cases}$$

are Baire measurable and we have

$$\int_{E(\mathcal{B})} s_n\, d\alpha = \int_{P(\mathcal{B})} s_n\, d\beta = \int_{P_1(\mathcal{B})} s_n\, d\beta .$$

By denoting $L = \bigcup_{n \geq 0} L_n$, we have

$$\lim_{n \to \infty} s_n = t\chi_L ,$$

and $Q_1 \setminus L$ is a Baire measurable set for which $\alpha(Q_1 \setminus L) = 0$, and this implies that

$$\beta((Q_1 \setminus L) \cap P_1(\mathcal{B})) = 0.$$

It follows that

$$\int_{E_1(\mathcal{B})} t\, d\alpha = \lim_{n \to \infty} \int_{E_1(\mathcal{B})} s_n\, d\alpha = \lim_{n \to \infty} \int_{P_1(\mathcal{B})} s_n\, d\beta =$$

$$= \int_{P_1(\mathcal{B})} t\chi_L\, d\beta = \int_{P_1(\mathcal{B})} t\, d\beta ,$$

and the corollary is proved.

Lemma 3. For any affine, bounded, continuous function $t : E_1(\mathcal{B}) \to \mathbb{C}$ we have

$$t(f_0) = \int_{E_1(\mathcal{B})} t \, d\alpha .$$

Proof. Since t is continuous at f_0, for any $\varepsilon > 0$ there exists a finite subset $\{b_1, b_2, \ldots, b_n\} \subset \mathcal{B}$, such that

$$f \in E_1(\mathcal{B}) \quad \text{and} \quad |f(b_i) - f_0(b_i)| < 1, \; i = 1, 2, \ldots, n,$$

implies

$$|t(f) - t(f_0)| < \varepsilon .$$

Let $\{A_1, \ldots, A_m\}$ be a finite partition of $E(\mathcal{B})$, consisting of measurable subsets $A_i \subset E(\mathcal{B})$, such that

$$f', f'' \in A_i \Rightarrow |f'(b_j) - f''(b_j)| < 1, \; j = 1, 2, \ldots, n,$$

for any $i = 1, 2, \ldots, m$. If necessary, we can refine the partition in such a manner that by an arbitrary selection of the points $f_i \in A_i \cap Q_1$, $i = 1, 2, \ldots, m$, we have

$$\left| \int_{E_1(\mathcal{B})} t \, d\alpha - \sum_{i=1}^{m} \alpha(A_i) t(f_i) \right| < \varepsilon .$$

Let γ be the Radon measure on $E(\mathcal{B})$ given by

$$\gamma = \sum_{i=1}^{m} \alpha(A_i) \, \varepsilon_{f_i} .$$

It is obvious that the barycenter $b(\gamma)$ of γ is the point $b(\gamma) = \sum_{i=1}^{m} \alpha(A_i) f_i$ and we have $b(\gamma) \in Q_1$. It follows that

$$|b(\gamma)(b_j) - f_0(b_j)| = |\sum_{i=1}^{m} \alpha(A_i) f_i(b_j) - \int_{E(\mathcal{B})} f(b_j)\, d\alpha(f)| =$$

$$= |\sum_{i=1}^{m} \alpha(A_i) f_i(b_j) - \sum_{i=1}^{m} \int_{A_i} f(b_j)\, d\alpha(f)| =$$

$$= |\sum_{i=1}^{m} \int_{A_i} (f_i(b_j) - f(b_j))\, d\alpha(f)| \leqslant$$

$$\leqslant \sum_{i=1}^{m} \int_{A_i} |f_i(b_j) - f(b_j)|\, d\alpha(f) < 1,$$

for any $j = 1, 2, \ldots, n$, and, therefore, we have

$$|t(b(\gamma)) - t(f_0)| < \varepsilon.$$

Since the function t is affine, we have

$$t(b(\gamma)) = \sum_{i=1}^{m} \alpha(A_i) t(f_i)$$

and this implies that

$$|\int_{E_1(\mathcal{B})} t\, d\alpha - t(f_0)| < 2\varepsilon.$$

The lemma is proved.

<u>Corollary</u>. <u>For any affine, bounded function $t : E(\mathcal{B}) \to \mathbb{C}$, whose restriction to $E_1(\mathcal{B})$ is continuous, we have</u>

$$t(f_0) = \int_{P(\mathcal{B})} t\, d\beta.$$

<u>Proof</u>. It is an immediate consequence of lemmas 2 and 3, and of the fact that $\beta(P_0(\mathcal{B}))=0$.

3. Let now G be an arbitrary locally compact group and $\mathcal{P}(G)$ the set of all continuous functions of positive type, defined on G . Let

$P_1(G)=\{\varphi\in\mathcal{P}(G);\ \varphi(e)=1\}$. It is well known that there exists an affine bijection

$\tau : \mathcal{P}(G)\to(C^*(G))^*_+$ between $\mathcal{P}(G)$ and the set $(C^*(G))^*_+$ of all positive, continuous linear functionals defined on the C^*-algebra $C^*(G)$ of the group G ; it is given by the formula

$$\tau(\varphi)(q(x))=\int_G \varphi x\, d\mu ,$$

where μ is a left-invariant Haar measure on G , $x\in L^1(G)$, $\varphi\in\mathcal{P}(G)$ and $q : L^1(G)\to C^*(G)$ is the canonical injection (see [2] , §2.7 , 13.4 and 13.9). Moreover, we have $\|\tau(\varphi)\|=\varphi(e)=\|\varphi\|_\infty$, for any $\varphi\in\mathcal{P}(G)$. (see [2], §§2.7 . 5 and 13.4.3).

Let $\pi : G\to\mathcal{L}(H)$ be a unitary, continuous, cyclic representation of the group G and ρ : $C^*(G)\to\mathcal{L}(H)$ the corresponding cyclic representation of the C^*-algebra $C^*(G)$(see [2] , § 13.9).

Let $\mathfrak{Z}\subset(\rho(C^*(G)))'$ be a maximal Abelian , von Neumann subalgebra and $\mathcal{B}$ the C^*- algebra generated by $\mathfrak{Z}$ and $\rho(C^*(G))$. Let $x_0\in H$ be the considered cyclic vector, such that $\|x_0\| = 1$; let $f_0 \in E(\mathcal{B})$ be the state of $\mathcal{B}$, given by the formula $f_0(b) = (bx_0 \mid x_0)$, $b \in \mathcal{B}$.

For any measure $\nu \in \mathcal{M}^1(G)$ we shall define the mapping $\tilde{\nu} : E(\mathcal{B}) \to \mathbb{C}$ by the formula

$$\tilde{\nu}(f) = \nu(\tau^{-1}(f \circ \rho)), \quad f \in E(\mathcal{B});$$

it is obvious that $\tilde{\nu}$ is a bounded affine function.

<u>Lemma 4</u>. <u>The function</u> $\tilde{\nu} \mid E_1(\mathcal{B})$ <u>is continuous</u>.

<u>Proof</u>. Let $(f_\alpha)_{\alpha \in A}$ be a generalized sequence, whose members belong to $E_1(\mathcal{B})$, such that

$$\lim_{\alpha \in A} f_\alpha = g \in E_1(\mathcal{B}),$$

in the topology $\sigma(\mathcal{B}^* ; \mathcal{B})$. Then the sequence $(f_\alpha \circ \rho)_\alpha$ converges to $g \circ \rho$ in the topology $\sigma((C^*(G))^*; C^*(G))$, and this implies that the sequence $(\tau^{-1}(f_\alpha \circ \rho))_{\alpha \in A}$ converges to $\tau^{-1}(g \circ \rho)$ in the topology $\sigma(\mathcal{P}(G); L^1(G))$. Since we have

$$\|\tau^{-1}(f_\alpha \circ \rho)\|_\infty = \|\tau^{-1}(g \circ \rho)\|_\infty = 1 , \quad \alpha \in A,$$

with D.A.Raĭkov's theorem (see [2] , theorem 13.5.2) we infer that

(7) $$\lim_{\alpha \in A} \tau^{-1}(f_\alpha \circ \rho) = \tau^{-1}(g \circ \rho)$$

in the topology of compact convergence on G .

Since the measure ν is regular, from (7) we

immediately infer that

$$\tilde{\nu}(g) = \nu(\tau^{-1}(g \circ \rho)) = \lim_{\alpha \in A} (\tau^{-1}(f_\alpha \circ \rho)) = \lim_{\alpha \in A} \tilde{\nu}(f_\alpha),$$

and the lemma is proved.

Corollary. For any $\nu \in \mathcal{M}^1(G)$ we have

$$\tilde{\nu}(f_o) = \int_{P(\mathcal{B})} \tilde{\nu}(p)\, d\beta(p).$$

Proof. It is an immediate consequence of the preceding lemma and of the corollary to lemma 3.

Proposition 1. Let $(\nu_n)_{n \in \mathbb{N}}$ be a sequence of positive measures $\nu_n \in \mathcal{M}^1_+(G)$, $n \in \mathbb{N}$, converging to $\nu \in \mathcal{M}^1_+(G)$ in the Fourier topology. Then we have

$$\lim_{n \to \infty} \nu_n(\varphi) = \nu(\varphi),$$

for any $\varphi \in \mathcal{P}(G)$.

Proof. Let us first remark, that the function $\varphi_o : G \to \{1\}$ is continuous, of positive type and belongs to $\mathcal{P}_1(G)$. Moreover, φ_o corresponds to the trivial representation of G, and, therefore, $\varphi_o \in \mathcal{F}(G)$. Consequently, we have

$$\|\nu\| = \nu(\varphi_o) = \lim_{n \to \infty} \nu_n(\varphi_o) = \lim_{n \to \infty} \|\nu_n\|,$$

and this shows that the sequence $(\|\nu_n\|)_{n \in \mathbb{N}}$ is bounded.

Let now $\varphi \in \mathcal{P}(G)$; without any loss of generality, we can assume that $\varphi \in \mathcal{P}_1(G)$; let $f_o = \tau(\varphi)$

and $\pi : G \to \mathcal{L}(H)$ be the cyclic, continuous unitary representation which corresponds to f_o. By hypothesis, we have

$$\lim_{n\to\infty} \tilde{\nu}_n(p) = \tilde{\nu}(p), \quad p \in P(\mathcal{B}),$$

and, in virtue of the preceding result, the sequence $(\|\tilde{\nu}_n\|)_{n\in\mathbb{N}}$ is bounded. With the dominated convergence theorem of Lebesgue we infer that

$$\tilde{\nu}(f_o) = \int_{P(\mathcal{B})} \tilde{\nu}(p)\, d\beta(p) = \lim_{n\to\infty} \int_{P(\mathcal{B})} \tilde{\nu}_n(p)\, d\beta(p) = \lim_{n\to\infty} \tilde{\nu}_n(f_o),$$

i.e.,

$$\nu(\varphi) = \lim_{n\to\infty} \nu_n(\varphi),$$

and the proposition is proved.

We can now prove the generalization of P.Lévy's theorem.

Theorem. Let G be an arbitrary locally compact group, and $(\nu_n)_{n\in\mathbb{N}}$ a sequence of finite, positive Radon measures on G, such that for any pure, continuous function $\varphi : G \to \mathbb{C}$, of positive type, we have

$$(*) \qquad \lim_{n\to\infty} \nu_n(\varphi) = \nu(\varphi),$$

where ν is a finite, positive Radon measure. Then the relation $(*)$ holds for any bounded, continuous function $\varphi : G \to \mathbb{C}$.

Proof. a) If $\varphi, \psi \in \mathcal{K}(G)$, then $\varphi * \psi$ is a linear combination of four continuous functions of positive type (see [2], corollary 13.6.5). From the preceding proposition we infer that

$$\lim_{n\to\infty} \nu_n(\varphi * \psi) = \nu(\varphi * \psi).$$

b) Let now $\varphi \in \mathcal{K}(G)$ and $(u_i)_{i\in I}$ be an approximate unit in $\mathcal{K}(G)$ (with respect to convolution). We then have

$$\lim_i u_i * \varphi = \varphi,$$

uniformly on G, and, therefore,

$$\lim_{n\to\infty} \nu_n(\varphi) = \nu(\varphi),$$

for any $\varphi \in \mathcal{K}(G)$.

c) Let $\varphi \in C^b(G)$ and $\varepsilon > 0$ be an arbitrary positive number. Then there exists a compact subset $K \subset G$ such that

$$\nu(K) \geq \|\nu\| - \varepsilon .$$

Let $\varphi_0 \in \mathcal{K}_+(G)$ be such that

$$\chi_K \leq \varphi_0 \leq 1.$$

Then we have

$$\|\nu\| - \varepsilon < \nu(\varphi_0) \leq \|\nu\|$$

and this implies that

$$\nu(1 - \varphi_0) < \varepsilon .$$

Since $\lim_{n\to\infty} \|\nu_n\| = \|\nu\|$, from what we have already proved it follows that there exists a number $n_0 \in \mathbb{N}$, such that

$$\|\nu\| - \varepsilon < \|\nu_n\| < \|\nu\| + \varepsilon$$

and

$$\nu(\varphi_0) - \varepsilon < \nu_n(\varphi_0) < \nu(\varphi_0) + \varepsilon,$$

for any $n \geq n_0$. We immediately infer that

$$|\nu_n(\varphi) - \nu(\varphi)| \leq 3\|\varphi\|_\infty + |\nu_n(\varphi\varphi_0) - \nu(\varphi\varphi_0)|,$$

for any $n \geq n_0$. The theorem now immediately follows from the remark that $\varphi\varphi_0 \in \mathcal{K}(G)$. Q.E.D.

Acknowledgments. We acknowledge useful discussions with H.Heyer about Lévy's theorem and its generalizations.

Bibliography

1. N.Bourbaki, Intégration Ch.IX.Élém. de mathématique, Fasc. XXXV, Hermann, Paris, 1969.

2. J.Dixmier, Les C^*-algèbres et leurs représentations, Gauthier-Villars, Paris, 1969.

3. H.Heyer, L'analyse de Fourier non-commutative et applications à la théorie des probabilités, Ann.Inst. Henri Poincaré, vol.IV, no.2, 1968, p.143-164.

4. M.Loève, Probability theory, Van Nostrand, Princeton, Toronto, New York, London, 1960.

5. P.Martin-Löf, The continuity theorem on a locally compact group, Theory of Prob. and its Applications, X, 1965, p.338-341.

6. M.A.Naimark, Normierte Algebren, VEB Deutscher Verlag der Wissenschaften, Berlin, 1969.

7. J.L.Taylor , The Tomita decomposition of rings of operators, Trans. Amer. Math. Soc., vol.113, no.1, 1967, 30-39.

8. S.Teleman, On reduction theory, Rev.Roum, Math.pures et appl., t.XXI, no.4, 1976, p.465-486.

INCREST
Bucharest

ANALYTIC FUNCTIONAL CALCULUS AND

MARTINELLI'S FORMULA

by

F.-H. VASILESCU

1. INTRODUCTION. Let U be an arbitrary open set in $\mathbb{C}^n$ $(n \geq 1)$ and $\Delta \subset U$ an open, relatively compact subset of U, such that the boundary $\partial\Delta$ of Δ is contained in U and is a finite union of smooth surfaces. If $A(U)$ denotes the set of all analytic complex-valued functions defined in U then for any $f \in A(U)$ we have the representation

$$(1.1) \qquad f(w) = \frac{(n-1)!}{(2\pi i)^n} \int_{\partial\Delta} f(z) \sum_{j=1}^{n} (-1)^{j-1} \frac{\bar{z}_j - \bar{w}_j}{\|z-w\|^{2n}} \bigwedge_{\substack{1 \leq k \leq n \\ k \neq j}} d\bar{z}_k \wedge d\underset{\sim}{z} ,$$

where $w = (w_1, \ldots, w_n)$ is an arbitrary point of Δ, $d\underset{\sim}{z} = dz_1 \wedge \ldots \wedge dz_n$ and $\|z-w\|^2 = |z_1 - w_1|^2 + \ldots + |z_n - w_n|^2$ (The symbol "$\wedge$" denotes, as usually, the exterior product.)

The formula (1.1) is known as Martinelli's formula [2]. For $n = 1$ it obviously reduces to the Cauchy formula.

We shall deal in the sequel with the possibility of extending the formula (1.1) in order to obtain a suitable variant which will produce a functional calculus with analytic functions for linear bounded operators.

As it is known, there is an immediate extension of the Cauchy formula which provides the functional calculus with analytic functions for one operator (see, for instance, [1]). The problem of the existence of a functional calculus for several commuting operators, acting on a Banach space, has been completely solved by J.L.Taylor by means of what he called the Cauchy-Weil integral, whose integral kernel is a cohomology class of differential forms [4] .

However, we shall present in what follows a Martinelli type formula for the functional calculus with analytic functions associated with a commuting system of linear continuous operators on a Hilbert space. We mention that this work is not intended to be complete; we shall only outline the proofs. A complete version of this work will be published elsewhere.

Let us recall some definitions and notations. If X is a complex linear space and $s=(s_1,\ldots,s_n)$ a system of indeterminates then we denote by $\Lambda[s,X]$ the space of all exterior forms generated by $s_1,\ldots,s_n$, with coefficients in X, and by $\Lambda^p[s,X]$ the subspace of all homogeneous forms of degree p, $p=0,1,\ldots,n$. Let $a=(a_1,\ldots,a_n)$ be a system of endomorphisms of X. Then we can define the map $\delta_a : \Lambda[s,X] \longrightarrow \Lambda[s,X]$ by the formula

$$\delta_a \xi = (a_1 s_1 + \ldots + a_n s_n)\wedge \xi \quad , \quad (\xi \in \Lambda[s,X]) .$$

Since $a=(a_1,\ldots,a_n)$ is always supposed to be commuting, it follows easily that $\delta_a^2=0$, hence $R(\delta_a)\subset K(\delta_a)$, where R and K denote the range and the kernel of the corresponding endomorphism, respectively.

The system $a=(a_1,\ldots,a_n)$ is said to be <u>nonsingular</u> on X if $R(\delta_a)=K(\delta_a)$ [3] (otherwise a is said to be <u>singular</u> [3]).

From now on we suppose that X is a Hilbert space and $a=(a_1,\ldots,a_n)$ is a commuting system in $L(X)$ (which denotes the algebra of all linear continuous operators on X).

The (joint) spectrum of a on X, denoted by $\sigma(a,X)$, is the set of all $z\in\mathbb{C}^n$ such that $z-a$ is singular on X [3]. In this case it is clear also that the space $\Lambda[s,X]$ has a natural Hilbert structure (as a matter of fact, the space $\Lambda[s,X]$ can be viewed as a direct sum of 2^n copies of X) so that we can consider the adjoint δ_a^* of the operator δ_a and let us define

$$(1.2) \qquad \alpha(a) = \delta_a + \delta_a^* \qquad .$$

As we have noticed in [5], the system a is nonsingular on X if and only if the operator $\alpha(a)$ is invertible on $\Lambda[s,X]$. This result will be essential for our purpose.

2. A MARTINELLI TYPE FORMULA. Let $a=(a_1,\dots,a_n)$ be a fixed system of commuting operators on X. Consider an open set $U\subset\mathbb{C}^n$ such that $U\cap\sigma(a,X)=\emptyset$. If $z=(z_1,\dots,z_n)\in U$ is an arbitrary point, $d\bar z=(d\bar z_1,\dots,d\bar z_n)$ and

$$\bar\partial = \frac{\partial}{\partial\bar z_1}d\bar z_1 + \cdots + \frac{\partial}{\partial\bar z_n}d\bar z_n \quad ,$$

then we may define the operator

$$M_a(z) \;:\; \Lambda[(s,d\bar z),C^\infty(U,X)] \longrightarrow \Lambda[(s,d\bar z),C^\infty(U,X)]$$

by the formula

$$(2.1) \qquad M_a(z)\xi = \alpha(z-a)^{-1}(\bar\partial\alpha(z-a)^{-1})^{n-1}S\xi \quad ,$$

where $\xi\in\Lambda[(s,d\bar z),C^\infty(U,X)]$ is arbitrary and the operator S is given by $S\xi=\xi\wedge s_1\wedge\dots\wedge s_n$. The definition (2.1) is correct since $z\longrightarrow\alpha(z-a)^{-1}$ is a function in $C^\infty(U,L(\Lambda[s,X]))$. Moreover, if $x\in X$ then $M_a(z)x\in\Lambda^{n-1}[d\bar z,C^\infty(U,X)]$. We can prove the following

2.1. THEOREM. *Let* $a=(a_1,\ldots,a_n)\in L(X)$ *be a commuting system and* U *an open neighbourhood of* $\sigma(a,X)$. *Then the formula*

$$(2.2)\qquad f(a)x=\frac{1}{(2\pi i)^n}\int_{\partial\Delta} f(z)M_a(z)x\wedge dz_1\wedge\ldots\wedge dz_n \quad (x\in X)$$

defines a continuous homomorphism from the unital algebra $A(U)$ *into the unital algebra* $L(X)$ *with the property* $Z_j(a)=a_j$, *where* $Z_j(z)=z_j$ $(j=1,\ldots,n)$.

Moreover, the integral (2.2) *does not depend on the set* Δ, *where* $\Delta\supset\sigma(a,X)$ *is an open subset of* U, *whose boundary is a finite union of smooth surfaces contained in* U.

In other words, the formula (2.2) provides a functional calculus with analytic functions for the system $a=(a_1,\ldots,a_n)$. Furthermore, the kernel (2.1) has a canonical form. The proof of Theorem 2.1 can be done after having proved the following

2.2. PROPOSITION. *For any open* $U\subset\mathbb{C}^n$, $U\cap\sigma(a,X)=\emptyset$, $M_a(z)x\in\Lambda^{n-1}[d\bar z,C^\infty(U,X)]$ *and* $\bar\partial M_a(z)x=0$, *where* $x\in X$ *is arbitrary*.

Outline of the proof. Let us define

$$\eta_j=(-1)^j\alpha(z-a)^{-1}(\bar\partial\alpha(z-a)^{-1})^j\,Sx \quad (j=0,1,\ldots,n-1),$$

each η_j being a form of degree j in $d\bar z_1,\ldots,d\bar z_n$ and of degree $n-j-1$ in $s_1,\ldots,s_n$. Then we can show, by induction, that $\eta_j\in K(\delta^*_{z-a})$ and $\bar\partial\eta_j\in K(\delta_{z-a})$ for any j and z. Since $\bar\partial\eta_{n-1}\in K(\delta_{z-a})$ and η_{n-1} is of degree zero in $s_1,\ldots,s_n$ we must have $\bar\partial\eta_{n-1}=0$, hence $\bar\partial M_a(z)x=0$.

Notice also that

$$(2.3)\qquad (\delta_{z-a}+\bar\partial)(\eta_0+\eta_1+\cdots+\eta_{n-1})=Sx\quad .$$

We can sketch now the proof of Theorem 2.1. Suppose that $\psi \in C^{\infty}(U)$ is zero in a neighbourhood of $\sigma(a,X)$ and $\psi = 1$ outside another neighbourhood of $\sigma(a,X)$ which contains $\partial\Delta$. Then the support of the form $\xi = Sx - (\delta_{z-a} + \bar{\partial})(\psi\eta)$ is compact and ξ is $(\delta_{z-a} + \bar{\partial})$-equivalent to Sx, where $\eta = \eta_0 + \eta_1 \ldots + \eta_{n-1}$. If

$$P : \Lambda[(s,d\bar{z}), C^{\infty}(U,X)] \longrightarrow \Lambda[(s,d\bar{z}), C^{\infty}(U,X)]$$

is the map assigning to each form its part containing only the terms in $d\bar{z}_1, \ldots, d\bar{z}_n$ then by the construction of Taylor's [4] we have

$$f(a)x = \frac{1}{(2\pi i)^n} \int_U f(z)(-1)^n P\xi \wedge dz_1 \wedge \ldots \wedge dz_n =$$

$$\frac{1}{(2\pi i)^n} \int_U f(z)\, \bar{\partial}(\psi M_a(z)x) \wedge dz_1 \wedge \ldots \wedge dz_n =$$

$$\frac{1}{(2\pi i)^n} \int_{\partial\Delta} f(z)\, M_a(z)x \wedge dz_1 \wedge \ldots \wedge dz_n \quad .$$

Finally, let us note that the integral (2.2) converges actually in $L(X)$.

In particular, for $n = 2$ a direct calculation gives the formula

$$M_a(z) = -\gamma(z)(z_2-a_2)^* \gamma_*(z) d\bar{z}_1 + \gamma(z)(z_1-a_1)^* \gamma_*(z) d\bar{z}_2 \quad ,$$

where

$$\gamma(z) = ((z_1-a_1)^*(z_1-a_1) + (z_2-a_2)^*(z_2-a_2))^{-1}$$

and

$$\gamma_*(z) = ((z_1-a_1)(z_1-a_1)^* + (z_2-a_2)(z_2-a_2)^*)^{-1} \quad .$$

2.3.PROPOSITION. <u>Suppose that</u> $X=\mathbb{C}$ <u>and let</u> $w\in\mathbb{C}^n$ <u>be arbitrary. Then for any</u> $z\neq w$ <u>we have</u>

$$(2.4)\quad M_w(z)=(n-1)!\sum_{j=1}^{n}(-1)^{j-1}\frac{\bar z_j-\bar w_j}{\|z-w\|^{2n}}\bigwedge_{\substack{1\leq k\leq n\\ k\neq j}}d\bar z_k\ .$$

The equality (2.4) shows that the formula (2.2) reduces to the Martinelli formula in the scalar case.

<u>Outline of the proof</u>. With no loss of generality one may suppose that $w=0$ .It is known that $\alpha(z)^{-1}=\alpha(z)\|z\|^{-2}$,for any $z\in\mathbb{C}^n$, $z\neq 0$ [5] .By the proof of Proposition 2.2 it follows that

$$\alpha(z)^{-1}(\bar\partial\alpha(z)^{-1})^{n-1}\ S(1)=\frac{\delta_z^*}{\|z\|^2}(\bar\partial\frac{\delta_z^*}{\|z\|^2})^{n-1}\ S(1)\ .$$

If we denote $R_j\xi=\bar\partial(\bar z_j\|z\|^{-2}\xi)$ and $S_k\xi=s_k\wedge\xi$ $(j,k=1,\dots,n)$, since $R_jS_k^*=-S_k^*R_j$,we obtain

$$(\bar\partial\frac{\delta_z^*}{\|z\|^2})^{n-1}=$$

$$(-1)^{\frac{(n-1)(n-2)}{2}}\sum_{j=1}^{n}(\sum_{K\in\Pi(j)}(-1)^{\sigma K}R_K)\prod_{k\neq j}S_k^*\ ,$$

where $\Pi(j)$ is the set of all permutations of the index $(1,\dots,j-1,j+1,\dots,n)$, $K=(k_1,\dots,k_{n-1})$ is such a permutation, σK is the parity of K and $R_K=R_{k_1}\dots R_{k_{n-1}}$ .Since

$$R_{k_1}R_{k_2}\dots R_{k_{n-1}}(1)=$$

$$\bar\partial(\bar z_{k_1}\|z\|^{-2})\wedge\bar\partial(\bar z_{k_2}\|z\|^{-2})\wedge\dots\wedge\bar\partial(\bar z_{k_{n-1}}\|z\|^{-2})=$$

$$\|z\|^{-2(n-1)} d\bar{z}_{k_1} \wedge \ldots \wedge d\bar{z}_{k_{n-1}} \quad +$$

$$\|z\|^{-2(n-2)} \sum_{p=1}^{n-1} (-1)^{p-1} \bar{z}_{k_p} \bar{\partial} \|z\|^{-2} \wedge \bigwedge_{k_q \neq k_p} d\bar{z}_{k_q} \quad ,$$

we obtain (2.4) by a direct calculation.

3. SOME COMMENTS AND REMARKS. The Martinelli kernel (2.4) has another interesting property. Indeed, if we consider the formal adjoint of the operator $\bar{\partial}$, namely

$$\bar{\partial}^* = - \sum_{j=1}^{n} \frac{\partial}{\partial z_j} (d\bar{z}_j)^* \quad ,$$

where $(d\bar{z}_j)^*$ acts by the formula $(d\bar{z}_j)^*(\xi_j' + d\bar{z}_j \wedge \xi_j'') = \xi_j''$ for every $\xi = \xi_j' + d\bar{z}_j \wedge \xi_j''$, ξ_j' and ξ_j'' not containing $d\bar{z}_j$, we have also $\bar{\partial}^* M_w(z) = 0$ for any $z \neq w$ (we have already noticed that $\bar{\partial} M_w(z) = 0$). Indeed, the assertion is based upon the identity

$$\frac{\partial}{\partial z_k} \left(\frac{\bar{z}_j}{\|z\|^{2n}} \right) - \frac{\partial}{\partial z_j} \left(\frac{\bar{z}_k}{\|z\|^{2n}} \right) = 0 \quad ,$$

for any k and j.

Unfortunately, the property $\bar{\partial}^* M_a(z) = 0$ is not true for an arbitrary system of commuting operators a acting on a Hilbert space. Indeed, for $n = 2$ the property $\bar{\partial}^* M_a(z) = 0$ is equivalent to the equality

$$(3.1) \qquad \begin{aligned} &(z_1-a_1)^* \gamma_*(z)(z_2-a_2)^* + (z_2-a_2)^* \gamma(z)(z_1-a_1)^* = \\ &(z_1-a_1)^* \gamma(z)(z_2-a_2)^* + (z_2-a_2)^* \gamma_*(z)(z_1-a_1)^* \quad . \end{aligned}$$

Let us show that the equality (3.1) does not always hold.

3.1. EXAMPLE. Let D be a closed disc in the complex plane, $D \not\ni 0$, and denote by X the space $L^2(D) \oplus L^2(D)$, where $L^2(D)$ is the usual Hilbert space of all square integrable classes of measurable functions with respect to the Lebesgue measure in D.

Consider on X the following system of operators :

$$\begin{aligned}(a_1 f\oplus g)(w) &= wf(w)\oplus wg(w)\\ (a_2 f\oplus g)(w) &= g(w)\oplus 0\end{aligned} \qquad (w\in D\ ;\ f,g\in L^2(D)) .$$

It is clear then that

$$\begin{aligned}(a_1^* f\oplus g)(w) &= \bar{w}f(w)\oplus \bar{w}g(w)\\ (a_2^* f\oplus g)(w) &= 0\oplus f(w)\end{aligned} \qquad (w\in D\ ;\ f,g\in L^2(D)) .$$

Since $0\notin D$, the operator a_1 is invertible on X, hence the system (a_1,a_2) is nonsingular. Notice also that

$$\gamma(0)^{-1} f\oplus g = (a_1^*a_1 + a_2^*a_2) f\oplus g = |w|^2 f\oplus |w|^2 g + 0\oplus g ,$$

therefore

$$\gamma(0) f\oplus g = \frac{1}{|w|^2} f\oplus \frac{1}{|w|^2+1} g .$$

Analogously,

$$\gamma_*(0) f\oplus g = \frac{1}{|w|^2+1} f\oplus \frac{1}{|w|^2} g .$$

We have then

$$(a_1^*\gamma_*(0)a_2^* + a_2^*\gamma(0)a_1^*) f\oplus g = 2(0\oplus \frac{\bar{w}}{|w|^2} f) ,$$

while

$$(a_1^*\gamma(0)a_2^* + a_2^*\gamma_*(0)a_1^*) f\oplus g = 2(0\oplus \frac{\bar{w}}{|w|^2+1} f) ,$$

hence (3.1) is not satisfied in this case.

REFERENCES

1. N.Dunford & J.Schwartz,Linear operators,Part I ,Interscience Publishers,New York,London,1958.

2. E.Martinelli,Alcuni teoremi integrali per le funzioni analitiche di più variabili complesse,Memor.Accad. Ital.,9,1938,269-283.

3. J.L.Taylor,A joint spectrum for several commuting operators, J.Funct.Analysis,6,1970,172-191.

4. - ,The analytic functional calculus for several commuting operators,Acta Math.,125,1970,1-38.

5. F.-H.Vasilescu,A characterization of the joint spectrum in Hilbert spaces,Rev.Roum.Math.Pures Appl.,22,1977, 1003-1009.

INCREST
Bucharest

SOME RESULTS ON ABSTRACT DIFFERENTIAL EQUATIONS AND INEQUALITIES.

by

S. Zaidman, Department of Mathematics, Université de Montréal

This is an expository talk; a few results of mine where some complex analysis is involved will be here explained.

1. To start with I shall state an older result [2] where existence theorems for abstract differential equations $u'(t) = Au(t) + f(t)$, A a linear operator in Hilbert space, are derived from the knowledge of the behaviour of the resolvent operator $(\lambda - A)^{-1} = R(\lambda)$, as analytic function of λ. The following definition is borrowed from Agmon-Nirenberg's paper [1] :

Def. 1. Let be F a family of lines in the complex-plane, j a positive integer, s a positive real, and let B a linear operator with domain $D(B)$ in the Hilbert space H.

We say that $R(\lambda;B) = (\lambda-B)^{-1}$ is (j,s) bounded by the number $L > 0$ on F if, for any line of F, $R(\lambda;B) \in L(H,H)$ for λ outside j-intervals of length s, and for these lines, $\|R(\lambda;B)\|_{L(H,H)} \leq L$

We can now state the following

<u>Local existence theorem.</u> Let be A a linear closed operator of dense domain $D(A) \subset H$ and A^* its adjoint operator. Assume $R(\lambda;A^*)$ to be (j,s) bounded by L on a (single) line $\mathrm{Re}\lambda = \sigma$. Let be $-\infty < a < b < +\infty$ a given, finite real interval. Then, for any $f \in L^2(a,b;H)$, there exists at least one function $u(t) \in L^2(a,b;H)$ in such a way that the equality $\int_a^b (u(t), \varphi'(t) + (A^*\varphi)(t))_H dt = -\int_a^b (f(t), \varphi(t))_H dt$ is verified for any $\varphi \in C_0^\infty(a,b;D(A^*))$ where $D(A^*)$ is the linear space equipped with the graph norm: $\|h\|_{D(A^*)} = \|h\|_H + \|A^*h\|_H$, $\forall\, h \in D(A^*)$.

<u>Global existence theorem.</u> Let be A a linear closed operator with dense domain $D(A) \subset H$, such that $R(\lambda;A^*)$ is (j,s) bounded by L on a sequence of lines $\mathrm{Re}\lambda = \sigma_n$, $\sigma_n \to +\infty$ as $n \to \infty$, $\sigma_n \to -\infty$ as $n \to -\infty$

Then, $\forall\, f \in L^2_{loc}(-\infty,\infty;H)$, there exists at least a function $u(t) \in L^2_{loc}(-\infty,\infty;H)$ is such a way that the equality

$$\int_{-\infty}^{\infty}(u(t),\varphi'(t) + (A^*\varphi)(t))_H dt = -\int_{\infty}^{\infty}(f(t),\varphi(t))_H dt \quad \text{is verified,}$$

$\forall\, \varphi \in C_0^\infty(a,b;D(A^*))$.

The method of the proof goes back to B. Malgrange (Lecture Notes at C.I.M.E., Saltino di Valombrosa, Italy, Sept. 1961).

2. A more recent paper of mine (to appear in Rend. Semin. Mat. Universita Padova) discusses an asymptotic result for weak differential inequa-

lities: this is a somewhat "weakened" form of a theorem given by Agmon-Nirenberg in their already quoted paper.

As previously we consider a linear closed operator A in the Hilbert space H, of dense domain, and denote by A^* its adjoint operator. We proved the following.

Theorem. Let us assume that $u(t)$ and $f(t)$ are strongly continuous functions, $0 \le t < \infty \to H$, related through the integral identity

$$\int_0^\infty (u(t), \varphi'(t) + (A^*\varphi)(t))_H dt = -\int_0^\infty (f(t), \varphi(t))_H dt \quad \text{for any } \varphi \in K_{A^*}(o,\infty).^{(1)}$$

Assume also that on a sequence of vertical lines in the complex plane: $\mathrm{Re}\lambda = \sigma_n$, and $\sigma_n \to -\infty$ as $n \to +\infty$, the resolvent operator $(\sigma_n + i\tau - A)^{-1}$ belongs to $L(H;H)$ for $n = 1,2,\dots$, $-\infty < \tau < \infty$, and verifies an estimate $\|R(\lambda,A)\| \le M$, $\mathrm{Re}\lambda = \sigma_n$, $n = 1,2,\dots$. In these conditions, if inequality $\|f(t)\| \le \lambda(t)\ \|u(t)\|_H$, $0 \le t < \infty$ is verified, where $\lambda(t) \le c < \frac{1}{M}$, $0 \le t < \infty$, and if $\sup\limits_{t \ge 0} e^{-at}\ \|u(t)\| < \infty$ for any real number a, then $u(t) = 0\ \forall\ t \ge 0$.

3. Our last result to be explained here concerns bounded solutions of some abstract differential equations.

In the Hilbert space H we consider a resolution of the identity $\{E_\theta\}_{-\infty}^{\infty}$; to it a self-adjoint operator $A = \int_{-\infty}^{\infty} \theta\, d\, E_\theta$ is associated; then,

(1) $K_{A^*}(o,\infty)$ is a "natural" class of test-functions, vector-valued.

given a complex-valued continuous function $Z(\theta) = x(\theta) + iy(\theta)$, $-\infty < \theta < \infty$, we consider the operator $B = Z(A) = \int_{-\infty}^{\infty} Z(\theta)dE_\theta$, which is in general unbounded but linear.

Our aim is to consider strong solutions $u(t)$, $-\infty < t < \infty \to D(B)$ of the differential equation $u'(t) = Bu(t)$, which are bounded over the real line, i.e. $\sup_{-\infty<t<\infty} \|u(t)\| < \infty$.

In fact we prove that if $x(\theta)$, the real part of $Z(\theta)$ has a finite number of zeros: $\theta_1, \theta_2, \ldots \theta_N$, then there exist N orthogonal vectors in H, $h_1, h_2, \ldots h_N$, such that $u(t) = \sum_{j=1}^{N} \exp(iy(\theta_j)t)hj$.

Similar previous results are in my paper [3] and in Agmon-Nirenberg [1] - Ch II. The paper should appear in the Annali dell'Università di Ferrara.

REFERENCES

[1] S. AGMON, L.NIRENBERG: "Properties of solutions of ordinary differential equations in Banach spaces" Comm. Pure Appl. Math., v. 16, May 1963.

[2] S.ZAIDMAN: Equations différentielles abstraites, Les Presses de l'Université de Montréal, 1966.

[3] S.ZAIDMAN: "Uniqueness of bounded solutions for some abstract differential equations" , Annali dell'Università di Ferrara, v. 14, no 10, 1969.

A REMARK ON THE WEAK BACKWARD CAUCHY PROBLEM

S. Zaidman (*)

Department of Mathematics, Université de Montréal

Introduction: In this Note we shall give a certain "weak" extension to theorem 1.1 in Agmon-Nirenberg's paper [1].

A similar idea was already used by us in a previous paper ([4] - th. 4.2), but here we give a straightforward proof.

Let us remark also that in [1] the authors wrote about their result:

"It holds also for weak solutions, ... but we shall consider only regular solutions" (Actually, no further indication in [1] or later was given about the class of weak solution were they thinking about, or of any method of proof, but in writing [4] and the present paper the author had to use a simple but fundamental device communicated to him orally in Pisa, Italy, by professor S. Agmon).

1. Let us remember Agmon-Nirenberg's result, as given in [1]

"Let X be a Banach space, and $u(t) \in X$ be a solution of $Lu = (\frac{1}{i}\frac{d}{dt} - A)\,u = 0$ for $0 \le t \le T$, with $u(T) = 0$. Assume that there is a simple Jordan arc Γ going to ∞, lying in a closed angle in the open upper half λ-plane, on which $R(\lambda)$ is defined and satisfies $|R(\lambda)| = O(e^{\alpha Im\lambda})$ for some constant $\alpha \ge 0$. If $T > \alpha$, then $u(t) = 0$ for $t \ge \alpha$".

(*) This research is supported through a grant of the National Research Council, Canada.

Let us explain now our "weak" extension:

We shall assume that A is a linear closed operator with dense domain $D(A) \subset X$, mapping $D(A)$ into X. We assume also that A^*, the dual operator to A is densely defined in $D(A^*) \subset X^*$ - the dual space to X. Consider now the class $K_{A^*}(0,T)$ of "test-functions" consisting of functions $\phi^*(t)$, $0 \leq t \leq T \to X^*$, continuously differentiable, vanishing near $t = 0$ and $t = T$; also $\phi^*(t) \in D(A^*)$ $\forall\, t \in [0,T]$ and $(A^*\phi^*)(t)$ is X^* - continuous.

We shall prove the following.(*)

<u>Theorem</u>: <u>Let</u> $u(t)$ <u>be a continuous function,</u> $0 \leq t \leq T \to X$, <u>such that</u> $u(T) = 0$ <u>and the integral identity</u>

1.1) $$\int_0^T < \frac{1}{i}\frac{d\phi^*}{dt} + A^*\phi^*, u > dt = 0$$

$\forall\, \phi^* \in K_{A^*}(0,T)$ is satisfied.

<u>Assume that there is a simple Jordan arc</u> Γ <u>going to</u> ∞ <u>lying in the open upper half</u> λ<u>-plane, on which</u> $R(\lambda) = (\lambda - A)^{-1} \in L(X;X)$ <u>and satisfies an estimate</u> $\|R(\lambda)\|_{L(X;X)} = 0(e^{\alpha \mathrm{Im}\lambda})$ <u>for some constant</u> $\alpha \geq 0$. <u>If</u> $T > \alpha$ <u>then</u> $u(t) = 0$ <u>for</u> $t \geq \alpha$.

Remark that in 1.1) the $<\ ,\ >$ means duality between X and X^* and the integral is in Riemann's sense.

We need firstly a result similar to Lemma 5.1 in [4]. Define the

(*) Compare also with theorem 5.1 in [4].

class $K_{A^*}(0,T)$ of functions $\phi^*(t)$, $0 \le t \le T \to D(A^*)$, which are X^* - continuously differentiable, with $A^*\phi^*$ X^*-continuous, vanishing near $t = 0$ (but not necessarily near $t = T$) so that $K_{A^*}(0,T) \subset K_{A^*}(0,T]$. We have then

Lemma 1.1. The integral identity 1.1) is verified for any $\phi^* \in K_{A^*}(0,T]$.

Let be, $\forall \epsilon > 0$, a scalar-valued function $\nu_\epsilon(t)$, continuously differentiable on $[0,T]$, $= 0$ for $T - \epsilon \le t \le T$, $= 1$ for $0 \le t \le T - 2\epsilon$ verifying tne estimate $|\nu_\epsilon'(t)| \le \frac{C}{\epsilon}$ on $[0,T]$, and $|\nu_\epsilon(t)| \le C$ on $[0,T]$.

Then, given any function $\phi^*(t) \in K_{A^*}(0,T]$ we see that $\nu_\epsilon(t)\,\phi^*(t)$ belongs to $K_{A^*}(0,T)$. Consequently we can write

1.2) $$\int_0^T < \frac{1}{i}\frac{d}{dt}(\nu_\epsilon\phi^*) + A^*(\nu_\epsilon\phi^*) , u > dt = 0 \quad \forall \epsilon > 0$$ and also

1.3) $$\int_0^T < \frac{1}{i}(\nu_\epsilon'\phi^* + \nu_\epsilon\frac{d}{dt}\phi^*) + \nu_\epsilon A^*\phi^* , u > dt = 0 .$$ This reduces to

1.4) $$\int_{T-2\epsilon}^{T-\epsilon} \nu_\epsilon' < \frac{1}{i}\phi^* , u > dt + \int_0^{T-2\epsilon} < \frac{1}{i}\frac{d}{dt}\phi^* , u > dt$$
$$+ \int_{T-2\epsilon}^{T-\epsilon} \nu_\epsilon \frac{1}{i}\frac{d}{dt}\phi^* , u > dt + \int_0^{T-2\epsilon} < A^*\phi^* , u > dt$$
$$+ \int_{T-2\epsilon}^{T-\epsilon} \nu_\epsilon < A^*\phi^* , u > dt = 0$$

When let $\epsilon \to 0$ and use properties of ν_ϵ and relation $u(T) = 0$, we get

1.5) $$\int_0^T < \frac{1}{i} \frac{d\phi^*}{dt} , u > dt + \int_0^T < A^*\phi^* , u > dt = 0$$

where ϕ^* was arbitrarily choosen in $K_{A^*}(0,T]$.

At this stage, using the same idea as in our paper [4] we shall consider the extended function $\tilde{u}(t)$ which equals $u(t)$ on $[0,T]$ and is null for $t \geq T$ (as a function $0 \leq t < \infty \to X$) . We have then

Lemma 1.2. The integral identity

1.6) $$\int_0^\infty < \frac{1}{i} \frac{d\phi^*}{dt} + A^*\phi^* , \tilde{u} > dt = 0$$ is verified for any $\phi^* \in K_{A^*}(0,\infty)$

(the class $K_{A^*}(0,\infty)$ has a similar definition to the previously defined class $K_{A^*}(0,T]$).

In fact 1.6) is same as

1.7) $$\int_0^T < \frac{1}{i} \frac{d}{dt} \phi^* + A^*\phi^* , u > dt = 0$$

which is true $\forall \phi^* \in K_{A^*}(0,T]$. But the restriction to $[0,T]$ of any function in $K_{A^*}(0,\infty)$ belongs obviously to $K_{A^*}(0,T]$. This proves Lemma.

Let us remark here that 1.6) becomes, after multiplication by $\sqrt{-1}$

1.8) $$\int_0^\infty < \frac{d\phi^*}{dt} + (iA)^* \phi^* , \tilde{u} > dt = 0$$

because $(\lambda A)^* = \lambda A^*$ $\forall \lambda \in C$ when A^* is the Banach-dual to A. (so that $< A^*x^* , x > = < x^* , Ax > \forall x \in D(A)$ and $\forall x^* \in D(A^*)$).

We shall apply from now on a "regularization" technique which was used by us previously (see for ex. [4] - Lemma 2.2 and [5] - Th. 2. and Final remark).

Let be $\alpha_\epsilon(t)$ a continuously differentiable scalar-valued function on $-\infty < t < \infty$, which vanishes for $|t| \geq \epsilon$, and consider then the convolution

1.9) $$(\tilde{u}* \alpha_\epsilon)(t) = \int_{t-\epsilon}^{t+\epsilon} \tilde{u}(\tau)\alpha_\epsilon(t-\tau)\, d\tau$$

which is well-defined as a function $\epsilon \leq t < \infty \to X$ and is continuously differentiable there.

Then, if J is the canonical map of X into its bidual space X^{**} , it is proved that $J(\tilde{u}*\alpha_\epsilon)(t) \in D(A^{**})$ and the equality

1.10) $\frac{1}{i}\frac{d}{dt} J(\tilde{u}*\alpha_\epsilon) = A^{**}J(\tilde{u}*\alpha_\epsilon)$ holds in X^{**} , $\forall\, t \in (\epsilon,\infty)$ (remember that we assume here that A^* is densely defined in X^* so that A^{**} is well-defined in X^{**} (on a "total" set $D(A^{**})$ but this fact is not used here).

Remark also that $(\tilde{u}*\alpha_\epsilon)(T+\epsilon) = 0$ in X as $\tilde{u} = 0$ for $t > T$; hence $J(\tilde{u}*\alpha_\epsilon)(T+\epsilon) = 0$ in X^{**}.

Let us effectuate now in the equation 1.10) the substitution $t = \sigma + \epsilon$ and $J(\tilde{u}*\alpha_\epsilon)(t) = V_\epsilon(\sigma)$. We see that $\epsilon \leq t \leq T + \epsilon \Leftrightarrow 0 \leq \sigma \leq T$ and also that $V_\epsilon(T) = J(\tilde{u}*\alpha_\epsilon)(T+\epsilon) = \theta$ in X^{**}.

It is also

1.11) $$\frac{1}{i}\frac{d}{d\sigma} V_\epsilon = \frac{1}{i}\frac{d}{dt} J(\tilde{u}*\alpha_\epsilon) \cdot \frac{dt}{d\sigma} = \frac{1}{i}\frac{d}{dt} J(\tilde{u}*\alpha_\epsilon) = A^{**} J(\tilde{u}*\alpha_\epsilon) = A^{**} V_\epsilon$$

so that

1.12) $\frac{1}{i}\frac{d}{d\sigma} V_\epsilon = A^{**} V_\epsilon$, $V_\epsilon(T) = 0$

is true in X^{**} in the strong sense.

Let us use now Theorem 2.16.5 in Hille-Phillips [2] ; from our hypothesis and that theorem we can deduce that $R(\lambda;A^*) = (\lambda-A^*)^{-1}$ is well-defined on the Jordan arc Γ as an operator in $L(X^*,X^*)$ (because A^* is closed). Furthermore $R(\lambda;A^*) = (R(\lambda;A))^*$; consequently we can derive estimate

1.13) $\|R(\lambda;A^*)\|_{L(X^*;X^*)} = \|(R(\lambda;A))^*\|_{L(X^*;X^*)} = \|R(\lambda;A)\|_{L(X;X)} = 0(e^{\alpha \mathcal{I}m\lambda})$

In the same way, using density of $D(A^*)$ in X^*, we obtain existence of $R(\lambda;A^{**}) = (R(\lambda;A^*))^*$ for any $\lambda \in \Gamma$ and the estimate

1.14) $\|R(\lambda;A^{**})\|_{L(X^{**};X^{**})} = 0(e^{\alpha \mathcal{I}m\lambda})$, $\lambda \in \Gamma$.

We can thus apply to 1.12) the uniqueness result of Agmon-Nirenberg (see beginning of 1.) ; remark that they do not ask for density of their $D(A)$ in X , and we do not assume density of $D(A^{**})$ in X^{**}. Consequently we get, if $T > \alpha$, that $V_\epsilon(\sigma) = 0$ for $\sigma \geq \alpha$. This gives relation $J(\tilde{u}*\alpha_\epsilon)(t) = \theta$ for $t \geq \alpha + \epsilon$ and from isometry of J we get too that $(\tilde{u}*\alpha_\epsilon)(t) = 0$ for $t \geq \alpha$ in the given space X.

We are now ready to prove completely our theorem: take a sequence $\{\alpha_n(t)\}_{n=1}^{\infty}$ of non-negative C^1-functions, which $= 0$ for $|t| \geq \frac{1}{n}$ and having integral from $-\frac{1}{n}$ to $+\frac{1}{n}$ equal one. Then we see that
$(\tilde{u}*\alpha_n)(t) = \theta$ for $t \geq \alpha + \frac{1}{n}$, $\forall\, n = 1,2,\ldots$

Let us take now any t in the open interval (α,T). We have obviously the relations

$$u(t) = \tilde{u}(t) = \int_{t-\frac{1}{n}}^{t+\frac{1}{n}} \tilde{u}(t)\ \alpha_n(t-\tau)d\tau \ ;$$

$$\tilde{u}(t) - (\tilde{u}*\alpha_n)(t) = \int_{t-\frac{1}{n}}^{t+\frac{1}{n}} (\tilde{u}(t) - \tilde{u}(\tau))\ \alpha_n(t-\tau)d\tau$$

We derive estimate

1.15) $\|\tilde{u}(t) - (\tilde{u}*\alpha_n)(t)\|_X \leq \sup_{|t-\tau|\leq\frac{1}{n}} \|\tilde{u}(t) - \tilde{u}(\tau)\|_X$ so that

1.16) $\|\tilde{u}(t)\|_X \leq \sup_{|t-\tau|\leq\frac{1}{n}} \|\tilde{u}(t) - \tilde{u}(\tau)\|_X + \|(\tilde{u}*\alpha_n)(t)\|_X$.

Actually, when n is sufficiently large, we get for $t \in (\alpha,T)$ that $t > \alpha + \frac{1}{n}$ and also that $(t - \frac{1}{n}, t + \frac{1}{n}) \subset (\alpha,T)$. Then $(\tilde{u}*\alpha_n)(t) = 0$ and $\tilde{u}(t) = u(t)$ is X-continuous. Hence $\|\tilde{u}(t)\|$ is arbitrarily small, so $u(t) = \theta$ in (α,T) hence in $[\alpha,T]$ too by continuity.

REFERENCES

[1] S. Agmon-L. Nirenberg: Properties of solutions of ordinary differential equations in Banach Spaces; Comm. Pure Appl. Math., Vol. XVI, no 2, May, 1963.

[2] E. Hille-R.S. Phillips: Functional Analysis and Semi-Groups, Amer. Math. Soc. Colloq. Publ., no 31, 1957.

[3] S.G. Krein: Linear differential equations in Banach spaces, NAUKA, Moscou 1967.

[4] S. Zaidman: The weak Cauchy problem for abstract differential equations, to appear Rend. Sem. Mat. Univ. Padova.

[5] S. Zaidman: Remarks on weak solutions of differential equations in Banach spaces Boll. U.M.I (4) 9(1974), 638-643.

Vol. 580: C. Castaing and M. Valadier, Convex Analysis and Measurable Multifunctions. VIII, 278 pages. 1977.

Vol. 581: Séminaire de Probabilités XI, Université de Strasbourg. Proceedings 1975/1976. Edité par C. Dellacherie, P. A. Meyer et M. Weil. VI, 574 pages. 1977.

Vol. 582: J. M. G. Fell, Induced Representations and Banach *-Algebraic Bundles. IV, 349 pages. 1977.

Vol. 583: W. Hirsch, C. C. Pugh and M. Shub, Invariant Manifolds. IV, 149 pages. 1977.

Vol. 584: C. Brezinski, Accélération de la Convergence en Analyse Numérique. IV, 313 pages. 1977.

Vol. 585: T. A. Springer, Invariant Theory. VI, 112 pages. 1977.

Vol. 586: Séminaire d'Algèbre Paul Dubreil, Paris 1975–1976 (29ème Année). Edited by M. P. Malliavin. VI, 188 pages. 1977.

Vol. 587: Non-Commutative Harmonic Analysis. Proceedings 1976. Edited by J. Carmona and M. Vergne. IV, 240 pages. 1977.

Vol. 588: P. Molino, Théorie des G-Structures: Le Problème d'Equivalence. VI, 163 pages. 1977.

Vol. 589: Cohomologie l-adique et Fonctions L. Séminaire de Géométrie Algébrique du Bois-Marie 1965–66, SGA 5. Edité par L. Illusie. XII, 484 pages. 1977.

Vol. 590: H. Matsumoto, Analyse Harmonique dans les Systèmes de Tits Bornologiques de Type Affine. IV, 219 pages. 1977.

Vol. 591: G. A. Anderson, Surgery with Coefficients. VIII, 157 pages. 1977.

Vol. 592: D. Voigt, Induzierte Darstellungen in der Theorie der endlichen, algebraischen Gruppen. V, 413 Seiten. 1977.

Vol. 593: K. Barbey and H. König, Abstract Analytic Function Theory and Hardy Algebras. VIII, 260 pages. 1977.

Vol. 594: Singular Perturbations and Boundary Layer Theory, Lyon 1976. Edited by C. M. Brauner, B. Gay, and J. Mathieu. VIII, 539 pages. 1977.

Vol. 595: W. Hazod, Stetige Faltungshalbgruppen von Wahrscheinlichkeitsmaßen und erzeugende Distributionen. XIII, 157 Seiten. 1977.

Vol. 596: K. Deimling, Ordinary Differential Equations in Banach Spaces. VI, 137 pages. 1977.

Vol. 597: Geometry and Topology, Rio de Janeiro, July 1976. Proceedings. Edited by J. Palis and M. do Carmo. VI, 866 pages. 1977.

Vol. 598: J. Hoffmann-Jørgensen, T. M. Liggett et J. Neveu, Ecole d'Eté de Probabilités de Saint-Flour VI – 1976. Edité par P.-L. Hennequin. XII, 447 pages. 1977.

Vol. 599: Complex Analysis, Kentucky 1976. Proceedings. Edited by J. D. Buckholtz and T. J. Suffridge. X, 159 pages. 1977.

Vol. 600: W. Stoll, Value Distribution on Parabolic Spaces. VIII, 216 pages. 1977.

Vol. 601: Modular Functions of one Variable V, Bonn 1976. Proceedings. Edited by J.-P. Serre and D. B. Zagier. VI, 294 pages. 1977.

Vol. 602: J. P. Brezin, Harmonic Analysis on Compact Solvmanifolds. VIII, 179 pages. 1977.

Vol. 603: B. Moishezon, Complex Surfaces and Connected Sums of Complex Projective Planes. IV, 234 pages. 1977.

Vol. 604: Banach Spaces of Analytic Functions, Kent, Ohio 1976. Proceedings. Edited by J. Baker, C. Cleaver and Joseph Diestel. VI, 141 pages. 1977.

Vol. 605: Sario et al., Classification Theory of Riemannian Manifolds. XX, 498 pages. 1977.

Vol. 606: Mathematical Aspects of Finite Element Methods. Proceedings 1975. Edited by I. Galligani and E. Magenes. VI, 362 pages. 1977.

Vol. 607: M. Métivier, Reelle und Vektorwertige Quasimartingale und die Theorie der Stochastischen Integration. X, 310 Seiten. 1977.

Vol. 608: Bigard et al., Groupes et Anneaux Réticulés. XIV, 334 pages. 1977.

Vol. 609: General Topology and Its Relations to Modern Analysis and Algebra IV. Proceedings 1976. Edited by J. Novák. XVIII, 225 pages. 1977.

Vol. 610: G. Jensen, Higher Order Contact of Submanifolds of Homogeneous Spaces. XII, 154 pages. 1977.

Vol. 611: M. Makkai and G. E. Reyes, First Order Categorical Logic. VIII, 301 pages. 1977.

Vol. 612: E. M. Kleinberg, Infinitary Combinatorics and the Axiom of Determinateness. VIII, 150 pages. 1977.

Vol. 613: E. Behrends et al., L^p-Structure in Real Banach Spaces. X, 108 pages. 1977.

Vol. 614: H. Yanagihara, Theory of Hopf Algebras Attached to Group Schemes. VIII, 308 pages. 1977.

Vol. 615: Turbulence Seminar, Proceedings 1976/77. Edited by P. Bernard and T. Ratiu. VI, 155 pages. 1977.

Vol. 616: Abelian Group Theory, 2nd New Mexico State University Conference, 1976. Proceedings. Edited by D. Arnold, R. Hunter and E. Walker. X, 423 pages. 1977.

Vol. 617: K. J. Devlin, The Axiom of Constructibility: A Guide for the Mathematician. VIII, 96 pages. 1977.

Vol. 618: I. I. Hirschman, Jr. and D. E. Hughes, Extreme Eigen Values of Toeplitz Operators. VI, 145 pages. 1977.

Vol. 619: Set Theory and Hierarchy Theory V, Bierutowice 1976. Edited by A. Lachlan, M. Srebrny, and A. Zarach. VIII, 358 pages. 1977.

Vol. 620: H. Popp, Moduli Theory and Classification Theory of Algebraic Varieties. VIII, 189 pages. 1977.

Vol. 621: Kauffman et al., The Deficiency Index Problem. VI, 112 pages. 1977.

Vol. 622: Combinatorial Mathematics V, Melbourne 1976. Proceedings. Edited by C. Little. VIII, 213 pages. 1977.

Vol. 623: I. Erdelyi and R. Lange, Spectral Decompositions on Banach Spaces. VIII, 122 pages. 1977.

Vol. 624: Y. Guivarc'h et al., Marches Aléatoires sur les Groupes de Lie. VIII, 292 pages. 1977.

Vol. 625: J. P. Alexander et al., Odd Order Group Actions and Witt Classification of Innerproducts. IV, 202 pages. 1977.

Vol. 626: Number Theory Day, New York 1976. Proceedings. Edited by M. B. Nathanson. VI, 241 pages. 1977.

Vol. 627: Modular Functions of One Variable VI, Bonn 1976. Proceedings. Edited by J.-P. Serre and D. B. Zagier. VI, 339 pages. 1977.

Vol. 628: H. J. Baues, Obstruction Theory on the Homotopy Classification of Maps. XII, 387 pages. 1977.

Vol. 629: W. A. Coppel, Dichotomies in Stability Theory. VI, 98 pages. 1978.

Vol. 630: Numerical Analysis, Proceedings, Biennial Conference, Dundee 1977. Edited by G. A. Watson. XII, 199 pages. 1978.

Vol. 631: Numerical Treatment of Differential Equations. Proceedings 1976. Edited by R. Bulirsch, R. D. Grigorieff, and J. Schröder. X, 219 pages. 1978.

Vol. 632: J.-F. Boutot, Schéma de Picard Local. X, 165 pages. 1978.

Vol. 633: N. R. Coleff and M. E. Herrera, Les Courants Résiduels Associés à une Forme Méromorphe. X, 211 pages. 1978.

Vol. 634: H. Kurke et al., Die Approximationseigenschaft lokaler Ringe. IV, 204 Seiten. 1978.

Vol. 635: T. Y. Lam, Serre's Conjecture. XVI, 227 pages. 1978.

Vol. 636: Journées de Statistique des Processus Stochastiques, Grenoble 1977, Proceedings. Edité par Didier Dacunha-Castelle et Bernard Van Cutsem. VII, 202 pages. 1978.

Vol. 637: W. B. Jurkat, Meromorphe Differentialgleichungen. VII, 194 Seiten. 1978.

Vol. 638: P. Shanahan, The Atiyah-Singer Index Theorem, An Introduction. V, 224 pages. 1978.

Vol. 639: N. Adasch et al., Topological Vector Spaces. V, 125 pages. 1978.

Vol. 640: J. L. Dupont, Curvature and Characteristic Classes. X, 175 pages. 1978.

Vol. 641: Séminaire d'Algèbre Paul Dubreil, Proceedings Paris 1976–1977. Edité par M. P. Malliavin. IV, 367 pages. 1978.

Vol. 642: Theory and Applications of Graphs, Proceedings, Michigan 1976. Edited by Y. Alavi and D. R. Lick. XIV, 635 pages. 1978.

Vol. 643: M. Davis, Multiaxial Actions on Manifolds. VI, 141 pages. 1978.

Vol. 644: Vector Space Measures and Applications I, Proceedings 1977. Edited by R. M. Aron and S. Dineen. VIII, 451 pages. 1978.

Vol. 645: Vector Space Measures and Applications II, Proceedings 1977. Edited by R. M. Aron and S. Dineen. VIII, 218 pages. 1978.

Vol. 646: O. Tammi, Extremum Problems for Bounded Univalent Functions. VIII, 313 pages. 1978.

Vol. 647: L. J. Ratliff, Jr., Chain Conjectures in Ring Theory. VIII, 133 pages. 1978.

Vol. 648: Nonlinear Partial Differential Equations and Applications, Proceedings, Indiana 1976–1977. Edited by J. M. Chadam. VI, 206 pages. 1978.

Vol. 649: Séminaire de Probabilités XII, Proceedings, Strasbourg, 1976–1977. Edité par C. Dellacherie, P. A. Meyer et M. Weil. VIII, 805 pages. 1978.

Vol. 650: C*-Algebras and Applications to Physics. Proceedings 1977. Edited by H. Araki and R. V. Kadison. V, 192 pages. 1978.

Vol. 651: P. W. Michor, Functors and Categories of Banach Spaces. VI, 99 pages. 1978.

Vol. 652: Differential Topology, Foliations and Gelfand-Fuks-Cohomology, Proceedings 1976. Edited by P. A. Schweitzer. XIV, 252 pages. 1978.

Vol. 653: Locally Interacting Systems and Their Application in Biology. Proceedings, 1976. Edited by R. L. Dobrushin, V. I. Kryukov and A. L. Toom. XI, 202 pages. 1978.

Vol. 654: J. P. Buhler, Icosahedral Golois Representations. III, 143 pages. 1978.

Vol. 655: R. Baeza, Quadratic Forms Over Semilocal Rings. VI, 199 pages. 1978.

Vol. 656: Probability Theory on Vector Spaces. Proceedings, 1977. Edited by A. Weron. VIII, 274 pages. 1978.

Vol. 657: Geometric Applications of Homotopy Theory I, Proceedings 1977. Edited by M. G. Barratt and M. E. Mahowald. VIII, 459 pages. 1978.

Vol. 658: Geometric Applications of Homotopy Theory II, Proceedings 1977. Edited by M. G. Barratt and M. E. Mahowald. VIII, 487 pages. 1978.

Vol. 659: Bruckner, Differentiation of Real Functions. X, 247 pages. 1978.

Vol. 660: Equations aux Dérivée Partielles. Proceedings, 1977. Edité par Pham The Lai. VI, 216 pages. 1978.

Vol. 661: P. T. Johnstone, R. Paré, R. D. Rosebrugh, D. Schumacher, R. J. Wood, and G. C. Wraith, Indexed Categories and Their Applications. VII, 260 pages. 1978.

Vol. 662: Akin, The Metric Theory of Banach Manifolds. XIX, 306 pages. 1978.

Vol. 663: J. F. Berglund, H. D. Junghenn, P. Milnes, Compact Right Topological Semigroups and Generalizations of Almost Periodicity. X, 243 pages. 1978.

Vol. 664: Algebraic and Geometric Topology, Proceedings, 1977. Edited by K. C. Millett. XI, 240 pages. 1978.

Vol. 665: Journées d'Analyse Non Linéaire. Proceedings, 1977. Edité par P. Bénilan et J. Robert. VIII, 256 pages. 1978.

Vol. 666: B. Beauzamy, Espaces d'Interpolation Réels: Topologie et Géometrie. X, 104 pages. 1978.

Vol. 667: J. Gilewicz, Approximants de Padé. XIV, 511 pages. 1978.

Vol. 668: The Structure of Attractors in Dynamical Systems. Proceedings, 1977. Edited by J. C. Martin, N. G. Markley and W. Perrizo. VI, 264 pages. 1978.

Vol. 669: Higher Set Theory. Proceedings, 1977. Edited by G. H. Müller and D. S. Scott. XII, 476 pages. 1978.

Vol. 670: Fonctions de Plusieurs Variables Complexes III, Proceedings, 1977. Edité par F. Norguet. XII, 394 pages. 1978.

Vol. 671: R. T. Smythe and J. C. Wierman, First-Passage Perculation on the Square Lattice. VIII, 196 pages. 1978.

Vol. 672: R. L. Taylor, Stochastic Convergence of Weighted Sums of Random Elements in Linear Spaces. VII, 216 pages. 1978.

Vol. 673: Algebraic Topology, Proceedings 1977. Edited by P. Hoffman, R. Piccinini and D. Sjerve. VI, 278 pages. 1978.

Vol. 674: Z. Fiedorowicz and S. Priddy, Homology of Classical Groups Over Finite Fields and Their Associated Infinite Loop Spaces. VI, 434 pages. 1978.

Vol. 675: J. Galambos and S. Kotz, Characterizations of Probability Distributions. VIII, 169 pages. 1978.

Vol. 676: Differential Geometrical Methods in Mathematical Physics II, Proceedings, 1977. Edited by K. Bleuler, H. R. Petry and A. Reetz. VI, 626 pages. 1978.

Vol. 677: Séminaire Bourbaki, vol. 1976/77, Exposés 489–506. IV, 264 pages. 1978.

Vol. 678: D. Dacunha-Castelle, H. Heyer et B. Roynette. Ecole d'Eté de Probabilités de Saint-Flour. VII-1977. Edité par P. L. Hennequin. IX, 379 pages. 1978.

Vol. 679: Numerical Treatment of Differential Equations in Applications, Proceedings, 1977. Edited by R. Ansorge and W. Törnig. IX, 163 pages. 1978.

Vol. 680: Mathematical Control Theory, Proceedings, 1977. Edited by W. A. Coppel. IX, 257 pages. 1978.

Vol. 681: Séminaire de Théorie du Potentiel Paris, No. 3, Directeurs: M. Brelot, G. Choquet et J. Deny. Rédacteurs: F. Hirsch et G. Mokobodzki. VII, 294 pages. 1978.

Vol. 682: G. D. James, The Representation Theory of the Symmetric Groups. V, 156 pages. 1978.

Vol. 683: Variétés Analytiques Compactes, Proceedings, 1977. Edité par Y. Hervier et A. Hirschowitz. V, 248 pages. 1978.

Vol. 684: E. E. Rosinger, Distributions and Nonlinear Partial Differential Equations. XI, 146 pages. 1978.

Vol. 685: Knot Theory, Proceedings, 1977. Edited by J. C. Hausmann. VII, 311 pages. 1978.

Vol. 686: Combinatorial Mathematics, Proceedings, 1977. Edited by D. A. Holton and J. Seberry. IX, 353 pages. 1978.

Vol. 687: Algebraic Geometry, Proceedings, 1977. Edited by L. D. Olson. V, 244 pages. 1978.

Vol. 688: J. Dydak and J. Segal, Shape Theory. VI, 150 pages. 1978.

Vol. 689: Cabal Seminar 76–77, Proceedings, 1976–77. Edited by A.S. Kechris and Y. N. Moschovakis. V, 282 pages. 1978.

Vol. 690: W. J. J. Rey, Robust Statistical Methods. VI, 128 pages. 1978.

Vol. 691: G. Viennot, Algèbres de Lie Libres et Monoïdes Libres. III, 124 pages. 1978.

Vol. 692: T. Husain and S. M. Khaleelulla, Barrelledness in Topological and Ordered Vector Spaces. IX, 258 pages. 1978.

Vol. 693: Hilbert Space Operators, Proceedings, 1977. Edited by J. M. Bachar Jr. and D. W. Hadwin. VIII, 184 pages. 1978.

Vol. 694: Séminaire Pierre Lelong – Henri Skoda (Analyse) Année 1976/77. VII, 334 pages. 1978.

Vol. 695: Measure Theory Applications to Stochastic Analysis, Proceedings, 1977. Edited by G. Kallianpur and D. Kölzow. XII, 261 pages. 1978.

Vol. 696: P. J. Feinsilver, Special Functions, Probability Semigroups, and Hamiltonian Flows. VI, 112 pages. 1978.

Vol. 697: Topics in Algebra, Proceedings, 1978. Edited by M. F. Newman. XI, 229 pages. 1978.

Vol. 698: E. Grosswald, Bessel Polynomials. XIV, 182 pages. 1978.

Vol. 699: R. E. Greene and H.-H. Wu, Function Theory on Manifolds Which Possess a Pole. III, 215 pages. 1979.

Vol. 700: Module Theory, Proceedings, 1977. Edited by C. Faith and S. Wiegand. X, 239 pages. 1979.

Vol. 701: Functional Analysis Methods in Numerical Analysis, Proceedings, 1977. Edited by M. Zuhair Nashed. VII, 333 pages. 1979.

Vol. 702: Yuri N. Bibikov, Local Theory of Nonlinear Analytic Ordinary Differential Equations. IX, 147 pages. 1979.

Vol. 703: Equadiff IV, Proceedings, 1977. Edited by J. Fábera. XIX, 441 pages. 1979.

Vol. 704: Computing Methods in Applied Sciences and Engineering, 1977, I. Proceedings, 1977. Edited by R. Glowinski and J. L. Lions. VI, 391 pages. 1979.

Vol. 705: O. Forster und K. Knorr, Konstruktion verseller Familien kompakter komplexer Räume. VII, 141 Seiten. 1979.

Vol. 706: Probability Measures on Groups, Proceedings, 1978. Edited by H. Heyer. XIII, 348 pages. 1979.

Vol. 707: R. Zielke, Discontinuous Čebyšev Systems. VI, 111 pages. 1979.

Vol. 708: J. P. Jouanolou, Equations de Pfaff algébriques. V, 255 pages. 1979.

Vol. 709: Probability in Banach Spaces II. Proceedings, 1978. Edited by A. Beck. V, 205 pages. 1979.

Vol. 710: Séminaire Bourbaki vol. 1977/78, Exposés 507-524. IV, 328 pages. 1979.

Vol. 711: Asymptotic Analysis. Edited by F. Verhulst. V, 240 pages. 1979.

Vol. 712: Equations Différentielles et Systèmes de Pfaff dans le Champ Complexe. Edité par R. Gérard et J.-P. Ramis. V, 364 pages. 1979.

Vol. 713: Séminaire de Théorie du Potentiel, Paris No. 4. Edité par F. Hirsch et G. Mokobodzki. VII, 281 pages. 1979.

Vol. 714: J. Jacod, Calcul Stochastique et Problèmes de Martingales. X, 539 pages. 1979.

Vol. 715: Inder Bir S. Passi, Group Rings and Their Augmentation Ideals. VI, 137 pages. 1979.

Vol. 716: M. A. Scheunert, The Theory of Lie Superalgebras. X, 271 pages. 1979.

Vol. 717: Grosser, Bidualräume und Vervollständigungen von Banachmoduln. III, 209 pages. 1979.

Vol. 718: J. Ferrante and C. W. Rackoff, The Computational Complexity of Logical Theories. X, 243 pages. 1979.

Vol. 719: Categorial Topology, Proceedings, 1978. Edited by H. Herrlich and G. Preuß. XII, 420 pages. 1979.

Vol. 720: E. Dubinsky, The Structure of Nuclear Fréchet Spaces. V, 187 pages. 1979.

Vol. 721: Séminaire de Probabilités XIII. Proceedings, Strasbourg, 1977/78. Edité par C. Dellacherie, P. A. Meyer et M. Weil. VII, 647 pages. 1979.

Vol. 722: Topology of Low-Dimensional Manifolds. Proceedings, 1977. Edited by R. Fenn. VI, 154 pages. 1979.

Vol. 723: W. Brandal, Commutative Rings whose Finitely Generated Modules Decompose. II, 116 pages. 1979.

Vol. 724: D. Griffeath, Additive and Cancellative Interacting Particle Systems. V, 108 pages. 1979.

Vol. 725: Algèbres d'Opérateurs. Proceedings, 1978. Edité par P. de la Harpe. VII, 309 pages. 1979.

Vol. 726: Y.-C. Wong, Schwartz Spaces, Nuclear Spaces and Tensor Products. VI, 418 pages. 1979.

Vol. 727: Y. Saito, Spectral Representations for Schrödinger Operators With Long-Range Potentials. V, 149 pages. 1979.

Vol. 728: Non-Commutative Harmonic Analysis. Proceedings, 1978. Edited by J. Carmona and M. Vergne. V, 244 pages. 1979.

Vol. 729: Ergodic Theory. Proceedings, 1978. Edited by M. Denker and K. Jacobs. XII, 209 pages. 1979.

Vol. 730: Functional Differential Equations and Approximation of Fixed Points. Proceedings, 1978. Edited by H.-O. Peitgen and H.-O. Walther. XV, 503 pages. 1979.

Vol. 731: Y. Nakagami and M. Takesaki, Duality for Crossed Products of von Neumann Algebras. IX, 139 pages. 1979.

Vol. 732: Algebraic Geometry. Proceedings, 1978. Edited by K. Lønsted. IV, 658 pages. 1979.

Vol. 733: F. Bloom, Modern Differential Geometric Techniques in the Theory of Continuous Distributions of Dislocations. XII, 206 pages. 1979.

Vol. 734: Ring Theory, Waterloo, 1978. Proceedings, 1978. Edited by D. Handelman and J. Lawrence. XI, 352 pages. 1979.

Vol. 735: B. Aupetit, Propriétés Spectrales des Algèbres de Banach. XII, 192 pages. 1979.

Vol. 736: E. Behrends, M-Structure and the Banach-Stone Theorem. X, 217 pages. 1979.

Vol. 737: Volterra Equations. Proceedings 1978. Edited by S.-O. Londen and O. J. Staffans. VIII, 314 pages. 1979.

Vol. 738: P. E. Conner, Differentiable Periodic Maps. 2nd edition, IV, 181 pages. 1979.

Vol. 739: Analyse Harmonique sur les Groupes de Lie II. Proceedings, 1976-78. Edited by P. Eymard et al. VI, 646 pages. 1979.

Vol. 740: Séminaire d'Algèbre Paul Dubreil. Proceedings, 1977-78. Edited by M.-P. Malliavin. V, 456 pages. 1979.

Vol. 741: Algebraic Topology, Waterloo 1978. Proceedings. Edited by P. Hoffman and V. Snaith. XI, 655 pages. 1979.

Vol. 742: K. Clancey, Seminormal Operators. VII, 125 pages. 1979.

Vol. 743: Romanian-Finnish Seminar on Complex Analysis. Proceedings, 1976. Edited by C. Andreian Cazacu et al. XVI, 713 pages. 1979.

This series reports new developments in mathematical research and tea
ing – quickly, informally and at a high level. The type of material conside
for publication includes:

1. Preliminary drafts of original papers and monographs
2. Lectures on a new field or presentations of a new angle in a classical f
3. Seminar work-outs
4. Reports of meetings, provided they are
   a) of exceptional interest and
   b) devoted to a single topic.

Texts which are out of print but still in demand may also be conside
if they fall within these categories.

The timeliness of a manuscript is more important than its form, wh
may be unfinished or tentative. Thus, in some instances, proofs may
merely outlined and results presented which have been or will later
published elsewhere. If possible, a subject index should be incluc
Publication of Lecture Notes is intended as a service to the internatic
mathematical community, in that a commercial publisher, Springer-
lag, can offer a wide distribution of documents which would otherv
have a restricted readership. Once published and copyrighted, they
be documented in the scientific literature.

Manuscripts

Manuscripts should be no less than 100 and preferably no more than 500 pages in length.
They are reproduced by a photographic process and therefore must be typed with extreme care. Sy
not on the typewriter should be inserted by hand in indelible black ink. Corrections to the type
should be made by pasting in the new text or painting out errors with white correction fluid. Authors re
75 free copies and are free to use the material in other publications. The typescript is reduced slig
size during reproduction; best results will not be obtained unless the text on any one page is kept
the overall limit of 18 x 26.5 cm (7 x 10½ inches). On request, the publisher will supply special pape
the typing area outlined.

Manuscripts should be sent to Prof. A. Dold, Mathematisches Institut der Universität Heidelberg, Im N
heimer Feld 288, 6900 Heidelberg/Germany, Prof. B. Eckmann, Eidgenössische Technische Hochs
CH-8092 Zürich/Switzerland, or directly to Springer-Verlag Heidelberg.

Springer-Verlag, Heidelberger Platz 3, D-1000 Berlin 33
Springer-Verlag, Neuenheimer Landstraße 28–30, D-6900 Heidelberg 1
Springer-Verlag, 175 Fifth Avenue, New York, NY 10010/USA

ISBN 3-540-**09550**-0
ISBN 0-387-**09550**-0